Handbook of Web Based
Energy Information and Control Systems

Handbook of Web Based Energy Information and Control Systems

Editors

Barney L. Capehart, PhD, CEM

Timothy Middelkoop, PhD, CEM

Associate Editors

Paul J. Allen, MSISE

David C. Green, MA

Routledge
Taylor & Francis Group

LONDON AND NEW YORK

Published 2020 by River Publishers
River Publishers
Alsbjergvej 10, 9260 Gistrup, Denmark
www.riverpublishers.com

Distributed exclusively by Routledge
4 Park Square, Milton Park, Abingdon, Oxon OX14 4RN
605 Third Avenue, New York, NY 10017, USA

First issued in paperback 2023

Library of Congress Cataloging-in-Publication Data

Handbook of web based energy information and control systems / editors Barney L. Capehart, Timothy Middelkoop ; associate editors, Paul J. Allen, David C. Green.

   p. cm.

Includes bibliographical references and index.

ISBN-10: 0-88173-669-4 (alk. paper)

ISBN-13: 978-8-7702-2298-3 (electronic)

ISBN-13: 978-1-4398-7684-8 (Taylor & Franics distribution : alk. paper)

   1. Energy conservation--Automation. 2. Energy conservation--Data processing. 3. Remote control. 4. Management information systems. I. Capehart, B. L. (Barney L.) II. Middelkoop, Timothy.

TJ163.3.H3565 2011

658.2'6--dc23

2011012773

Handbook of web based energy information and control systems, by Barney L. Capehart and Timothy Middlekoop; associate editors, Paul Allen, David Green
First published by Fairmont Press in 2011.

©2011 River Publishers. All rights reserved. No part of this publication may be reproduced, stored in a retrieval systems, or transmitted in any form or by any means, mechanical, photocopying, recording or otherwise, without prior written permission of the publishers.

Routledge is an imprint of the Taylor & Francis Group, an informa business

Publisher's Note
The publisher has gone to great lengths to ensure the quality of this reprint but points out that some imperfections in the original copies may be apparent.

ISBN-13: 978-87-7022-912-8 (pbk)
ISBN-13: 978-1-4398-7684-8 (hbk)
ISBN-13: 978-8-7702-2298-3 (online)
ISBN-13: 978-1-0031-5167-8 (ebook master)

While every effort is made to provide dependable information, the publisher, authors, and editors cannot be held responsible for any errors or omissions.

Contents

Foreword

We live in exciting and complex times, and in an age when we are using information technology in new and astounding ways. Information technology has penetrated many sectors, changing the way we work, communicate, and see the world. Information technology is now demonstrating that it can produce great value for optimizing and minimizing energy use in buildings and industrial facilities. New "energy aware" control systems ensure that facilities provide intended end-use services using only as much energy as is needed for the job at hand. The evidence is growing that new information technology, with faster, better, and cheaper platforms and applications, are critical to meet national energy goals and spur international business value.

This new book—the *Handbook of Web Based Energy Information and Control Systems*—is the final volume in a four-part series that shows how quickly these new technologies are developing. One of the most important concepts in this field is the great need to move beyond simple use of monthly utility bills as our primary source of information to manage energy use. We need to know when and why equipment is operating, and how much energy it is using. Is it controlled properly? Are the sensors properly calibrated? Consider also the trends that lead to greater complexity in energy markets. New smart grid systems, more sub-hourly energy data, demand response, dynamic rates and tariffs, and more renewable and on-site generation will continue to challenge our scope of applications. Greater standardization in interoperability of smart grid communications, wireless mesh networks, and greater penetration of internet technologies create a continuously evolving set of needs and capabilities for energy management. The methods for optimal, information-based energy management have moved well beyond theory and into best practices. The business imperatives are becoming clear. Organizations and institutions that fail to embrace these practices are wasting money. Our nation and our world cannot afford to waste resources, whether they are electricity, natural gas, oil, water, or others. The economic and environmental opportunities to better use existing resources are clearer than ever.

Today many of these systems are oriented toward motivating a person to act on the information the systems provide. In the future we are moving toward fully automated responses. But today, the majority of these systems require a person to interpret the information—a metric, a graph, a signal. As these technologies improve in their ability to turn raw data into actionable information, we continue to ask the question—who needs what information? Our exploration of this question has broadened from owners and operators to tenants, energy managers, portfolio mangers, third-party service providers, maintenance engineers, real estate developers, account managers and a growing list of participants. In summary—the opportunity is great and the challenge is impressive. We need to understand both the needs of the people and the performance of the engineered systems. This book provides a handbook to understand and accelerate the use of best practices at a time when it is greatly needed.

Mary Ann Piette
Lawrence Berkeley Laboratory
Berkeley, California
2011

Foreword

Our clouded future includes new virtual connections to our buildings from the communities they are part of with both physical and social interactions. Buildings connected with open protocols to the powerful internet cloud and its web services are redefining our industry. The reach and the visibility of our industry have never been greater nor has change been so rapid. Our clouded future includes new virtual connections to our buildings from the communities they are part of with both physical and social interactions. An example is digitally displayed energy/environmental dashboards to inform all occupants of the building's impact in real time energy use, plus the percentage generated from renewal sources. Connections to the smart grid make our buildings a physical part of their supply energy infrastructure. Our ability to operate buildings efficiently via the internet cloud from anywhere allows the complexity of our industry to be better managed and appear greatly simplified. Web services or software as a service (SaaS), as it is sometimes called, coupled with powerful browser presentation is changing how we appear and interact with our clients. Building information model (BIM) software allows the power of visual relational databases to improve decisions throughout the building design. New visualization and simulation tools reveal the effects of decisions made prior to the commitment of funds. In similar fashion, cloud computing provides a collaborative process that leverages web-based BIM capabilities and traditional document management to improve coordination.

The data cloud for our industry has become real. As we see applications and services moved 'off-site' you can imagine the opportunities for managing real estate, reducing energy and providing value-added applications for buildings. We have a long, long way to go to move out of the deeply rooted vendor centric unconnected world of traditional marketing of our products and services. We need to define the services and recognize the transaction and the performance we can deliver. We need to "Give our customers the ability to do something new, that they couldn't do before, but would have wanted to do; if only they knew they had the ability to do it." Imagine every single possible bit of information at your fingertips, perfectly contextualized, and perfectly personalized. This may never be realized, but the journey to find perfection will be the change we seek.

Our community of change agents needs to continue to educate customers and especially those who have never been our customers, with all types of events and in all forms of media. We need to explore new initiatives like Open Source, which is not vendor or hardware dependent, so we can become the enablers. The open protocol work is mostly done but we need to build on the flat line created by the data pushers and find an elegant entrance to the browser and the IT world. We need virtual space in which will occur the creativity of the change we seek. We are the people who will give the kids a place to build applications quickly and easily. We are the people who will create Building 2.0, the version that still has not yet been defined. Interoperability becomes the driver when it is economical. The overwhelming operational requirement to have all buildings connected to the smart grid is a great example. The technology is here, we just need to create ways for our industry to be the one that figures out how to use technology to sell solutions.

This book presents many diverse industry views of the complex changes that are occurring in the energy information and control industry. It should help you understand the future of web based energy information and control systems and the great changes occurring in our industry. I am a great fan of these books because, as much as I believe in the speed of the electronic media, the information must be organized and available in a logical form. We need links and strong connections to the traditional methods of learning to help kick start those that have lost connection with rapidly involving web-based control. The format of this book allows a physical separation from the virtual side of web-based ways to what is reality. With this book and the web you should have ready access to some of the most critical information on this new wave of web based systems technology.

Ken Sinclair Publisher/Owner
www.AutomatedBuildings.com
sinclair@automatedbuildings.com
January 2011
Vancouver Island, Canada
Published in Part in ASHRAE Show News AHR 2010 Expo
http://www.automatedbuildings.com/news/feb10/articles/ksin/100128050606ksin.htm

z

Introduction

In the last ten years, building and facility energy costs have risen dramatically and our energy supply chain has become quite tenuous. Only the relentless march of advancing technology has continued as a positive factor in our economy. But it is the utilization of new technologies that helps us to produce more efficient equipment, more efficient processes, and much more complex and effective control systems. It is this latter advancement that is addressed in this handbook.

The goal of this *Handbook of Web Based Energy Information and Control Systems* is to continue promoting the huge benefits to energy and facility managers from the emergence of one of the most powerful technological waves of our time. This technological wave has been sweeping through the development and application of energy information and control systems, and has brought with it the internet, the world wide web, TCP/IP, web browsers, relational databases, and a host of highly capable, timesaving software tools to help energy and facility managers operate buildings and facilities much more efficiently, much more cost effectively, and much more successfully. This wave of information technology (IT) and web based energy information and control systems (web based EIS/ECS) continues to roll on with increasing speed and intensity. A large number of new web based EIS/ECS system supplier companies have come on the scene and many new, exciting applications and adoptions of web-based technology have taken place.

What started as basic web based energy information systems has expanded into web based energy information and control systems, and finally to enterprise energy and facility management systems and building automation systems (BAS). Technological progress in information technology and use of the internet and world wide web will continue to be made at a rapid rate. Applying these advancements to computerized facility and energy management systems requires the innovative skills of many people in both the IT and the energy and facility management fields. If history in this area is a good indicator of what will happen in the future, we are in for a fantastic ride on this new and powerful technological wave.

This *Handbook of Web Based Energy Information and Control Systems* is both a presentation of technological advancements in this field over the last several years, as well as a compilation of what the editors believe is the best of the information from the previous three books in this area.

The combined thrust of this new information plus relevant contents of the three previous volumes is that the highest-level functions of the building and facility automation system are delivered by a web based EIS/ECS system that provides energy management, facility management, over all facility operational management, and ties in with the enterprise resource management system for the entire facility or the group of facilities being managed. This is where we have been headed with our use of IT, IP networks, XML, and web based systems to help us operate our buildings and facilities better—where better means higher energy efficiency and lower operating costs, improved occupant satisfaction, and higher productivity of our buildings and facilities through better indoor environmental quality, and a more direct tie to the business functions of our operations.

Improving overall energy and facility management begins with data collection. All three previous volumes in this series discuss and emphasize the need for metering, monitoring, and measuring devices to obtain data that are then sent to a centralized database for processing and storage. IP and XML are critical data transmission protocols and structures to obtain and transmit the huge amount of data that are necessary to provide the information needed to operate our facilities better. Data is the starting place. It is only when we can turn data into information to help us make decisions which will result in our buildings and facilities operating better that we're making real progress toward accomplishing our goals. A description and discussion of the equipment and devices to collect these data and to transmit them to a centralized database is still a basic and very important purpose of this handbook.

Advanced metering requirements for most federal buildings over 5000 square feet come from EPACT 2005, EISA 2007, and EO 13514 October 2009. These requirements specify that these federal buildings must install advanced metering for electricity, gas, steam, hot water, and chilled water by dates varying from 2012 to 2016. The required metering systems must 1) measure and record interval data, 2) be able to send data to a managing agency, 3) have time-stamped data with data storage for two years, and 4) have automated backups. Fortunately, many of the energy managers at federal facilities have already recognized the value and are putting in multifunction, smart, and network meters that will collect a wide range of data automatically, and send them to a computer system for processing and storage. Even using a basic EIS/ECS or enterprise energy management system can have tremendous savings potential, as well as facility operating improvements from utilizing the data to make data-driven operational decisions.

Once we have these large amounts of data, what is needed in our facilities is a highly capable, highly connected system to take these massive amounts of data, process them and change them into a few selective pieces of information. Compare the old operator workstation of the past, which was a screen full of numbers, to a modern GUI which shows a visual schematic of the system of interest. And under this schematic is a message area that says "Change the filter in AH3B," or "Valve V5 in chilled water line CW2 is stuck open." With today's information technology providing high-speed and high-capacity computers and communication networks, we have the ability to take thousands of data points as inputs, and process and store them in huge relational databases in our web based EIS/ECS and BAS systems. Using IP networking and the internet we can easily collect the data from these hundreds or thousands of data points, and our EIS/ECS systems can process these data to create information that can help us operate our facilities more efficiently and much more productively.

With the information produced by these web based EIS/ECS and BAS systems, we now have the hardware and the software technology to help us accomplish energy management, maintenance management, and overall building and facility management. Now, the building and facility management staff has enough information to operate the facility efficiently and cost effectively. The technology for overall building and facility management is here today. It is the wave of the future. The move to use web based systems and modern information technology is a wave of even greater magnitude than the DDC wave of 20 to 30 years ago. Previously, those who did not get on the wave of DDC were rolled over by it. It is now time for building and facility energy managers to get on the wave of IT and web based systems in order to capture the benefits of reduced energy and operational costs and improved building and facility productivity. This is now not only enterprise energy and facility management, it is enterprise resource management. A description and discussion of the data processing, data storage in centralized databases, and display functions is the second basic and very important purpose of this handbook.

The capability and use of information technology and the internet in the form of web based energy information and control systems continues to grow at a very rapid rate. New equipment and new suppliers have appeared rapidly, and existing suppliers of older equipment are offering new web based systems. Building and facility managers, maintenance managers, and energy managers are all interested in knowing what problems and what successes are coming from the use of these web based systems, and need to be prepared for current and future installations of web based EIS and ECS technologies in their buildings and facilities. Knowing what is being implemented in other buildings and facilities and knowing what is actually being accomplished is important information for the energy and facility managers if they are going to successfully purchase, install, and operate complex web based energy information and control systems.

The third major purpose of this handbook is to document the operational experience with web based systems in actual facilities and varied applications. Web based EIS and ECS systems have allowed the development of many new opportunities for energy and facility managers to quickly and effectively control and manage their operations. The case studies and applications described in this handbook should greatly assist all energy managers, facility managers, and maintenance managers as well as consultants and control systems and development engineers. These case studies and applications presented have shown conclusively that web based energy information and control systems are feasible, cost effective, and can create significant improvement in the energy and operational related performance of a facility. Documented benefits include reduced energy costs, reduced maintenance costs, improved operational performance, and reduced capital investments for these energy, maintenance, and operational savings. It is also clear that early adopters of these web based systems are seeing that is giving them a competitive advantage over their non-adopting business and organizational peers.

Where the previous series of books had to convince the users of the value of web based systems and the power of IP networks, it is now an accepted reality. This is in part due to the ubiquity of the internet and more recently the smart phone revolution. This has been driven by the dramatic cost reduction and increased capability of embedded hardware where even the simplest of devices now have the capability to become a part of the building's IP network yielding a wealth of information (embedded processors now contain everything from accelerometers to GPS receivers—all on the same chip). Almost everything has an Ethernet port now. The ability to collect, store and process this data has also dramatically increased since the release of the first book, with a simultaneous increase in capacity and a reduction in cost. To put this in perspective, today it is possible to store the data from 100,000 points sampled every minute, for ten years, for about the cost of the average textbook. The hardware to collect, store, and analyze data is no longer a barrier.

This is also happening for sensor networks (WiFi, IEEE 802.15.4, ZigBee, and Bluetooth) but has not yet hit critical mass. This may change when the smart grid (smart meters) hits consumers, creating a demand for sensors to monitor in-house energy usage, which will commoditize the technology bringing down the cost for building applications. We are already starting to see the beginnings of this with products such as Google Power and other home energy management systems. We can only now guess what the impact of smart phones running "apps" tied to the building will have. Unless there are systems in place to manage and use this data, they will only add cost without deriving any value, only to sit idle in the building. The principles discussed in this *Handbook for Web Based Energy Information and Controls* will also be applicable (if not more important) for buildings with large-sensor network capabilities.

Just as hardware has improved in the past decade so has software. Operating systems, browsers, web application platforms, data servers, and even software development environments have all matured and stabilized. Gone are the days where applications and machines would crash for no apparent reason. This does not mean that innovation has ended, just that there are products and standards that are mature. The Apache web server and PHP are such examples for software, built upon the mature data standards HTML and XML, and data protocols HTTP, SOAP, and AJAX. The open source software movement has been a part of this maturation with stable projects bringing down costs and preventing vendor locking by providing irrevocable access to the source code. Long-term product support and access is especially important for building systems which can have lifetimes of decades, compared to the 6- to 18-month product life cycles of modern hardware and software. It is infeasible to replace sensor and control systems every few years in buildings.

The realization that proprietary systems come and go (and of course firsthand experience) is one factor that is driving the use of open standards, such as Ethernet and BACnet, in the building controls industry. Truly open standards are not standards that hide important parts in propitiatory implementations behind so-call open interfaces. Truly open standards allow customers to choose alternative implementations or develop their own without resorting to reverse engineering or being hindered by license or patent restrictions. This not only applies to controls but also to the energy management systems that monitor and operate the building. Companies must make an investment in the energy management platform, whether in purchasing a system or developing their own. If this platform goes away, what is the impact on the user? For proprietary systems, this can mean being forced to find a new solution or forced to use the technology until it breaks. Using open standards for controls and data collection insulates the user from having to recommission the building in the event that the energy management system needs to be replaced. Using open source infrastructure insulates the user even further. This is best illustrated by the recent acquisition of Sun by Oracle. Oracle decided not to continue the support and development of a number of open source technologies that it acquired from Sun. Since these tools had active communities and the source could not be taken away, alternate stewards and support companies sprang up almost over night.

The development of all the standards, infrastructure, and software has taken time. We believe we are now at a point where we will see community-developed open source energy management systems on top of open protocols and open source infrastructure (Linux, Apache, PHP, Python, Java, Tomcat, MySQL, and PostgreSQL are all examples of enterprise quality open source software). This handbook is ideal for those starting their own efforts (open or not) or joining others. The time is right; the standards, software and hardware are all in place, and the economic and energy climate is the motivation (along with the knowledge that saving energy is also good for the environment). Intelligent energy systems, such as those we describe here, are what will be needed if our society is going to meet the environmental and energy challenges of the future.

How will we accomplish this? What will these systems look like? Many of the articles in this handbook provide this vision. We believe that these systems will, and must, take a systems perspective. This is driven in part by the increased use of renewable energy, which is often variable in nature, and storage technologies. The smart grid will also drive this by allowing consumers to take advantage of (or be forced to use) real time pricing. All of this requires the intelligent scheduling of energy consumption and storage. By taking a systems perspective, one can take advantage of the interactions between equipment and processes. This is important since the local optimization of equipment may not always mean that the entire system will run optimally. A systems perspective takes all equipment and operations into account, simultaneously scheduling usage appropriately. With these technologies in place, energy producers will be able to interact with consumers to optimize energy usage across the entire system—a very interesting optimization problem to solve!

Integrating building energy information and control systems requires sensing and control to have a single uniform representation across the entire building in real time. This may take the form of a single centralized database or a distributed system. In either case a united system perspective for information is necessary for both users (single seat for operation) and algorithms (entire building control and optimization) to control the building effectively. We the editors, strongly believe that this is a necessary condition for the success of any web based energy information and control system. We also believe that many of the other ideas expressed in this handbook will help guide you in building effective, easy-to-use and maintain, systems that stand the test of time. Building systems on open standards and open source technologies is a sure way to make this happen and keeps with the spirit of web based applications.

Finally, all four of us on the editorial team for this handbook hope that it contributes to the successful development, implementation and application of new web based energy information and control systems in many of your buildings and facilities. The editors are extremely pleased to have received a chapter from Mr. Jim Lewis of Obvius, formerly with Veris Industries, about the development of the Enercept electric power meter that is so prevalent today and is used in most building and facility energy information and control systems. Because of the significant impact this electric meter has had on allowing operational people to easily install and use these very accurate and flexible measuring devices, this history chapter from Jim is of interest to all of us in the web based EIS/ECS field. This story is also a good example of the innovative process that energy and facility managers should strive for in their effort to save energy. We appreciate Jim's willingness to spend the time and effort to go back 15 years and put together the story on the "Evolution of the Enercept Meter." With the historical nature of this story, we have placed it as Chapter 2 in this *Handbook of Web Based Energy Information and Control Systems*.

It has been our pleasure to work with our two associate editors on this important contribution to the IT and web based systems education and training of working energy and facility managers. Mr. Paul J. Allen and Mr. David C. Green have both played a major role in getting this handbook prepared and completed. Our most sincere thanks go to each of them for making our job much easier then it could have been. Thanks also to our two foreword writers, Ms. Mary Ann Piette, and Mr. Ken Sinclair, who have made major contributions of their own to this dynamic field. Also we want to thank each of the 55 individual authors who has written material that appears in this handbook. Without their kind and generous help in writing these detailed chapters, this book would not have been possible. Each of these authors is identified in the alphabetic list of authors at the end of this book.

Barney L Capehart, Ph.D., C.E.M.
Timothy Middelkoop, Ph.D., C.E.M.
University of Florida, Gainesville, Florida
Spring 2011

Section I

Introduction to Web Based Energy Information and Control Systems

Chapter 1

Introduction to the Handbook of Energy Information and Control Systems

Barney L Capehart
Timothy Middelkoop

In the last ten years building and facility energy costs have risen dramatically and our energy supply chain has become quite tenuous. Only the relentless march of advancing technology has continued as a positive factor in our economy. But it is the utilization of new technologies that helps us to produce more efficient equipment, more efficient processes, and much more complex and effective control systems. It is this latter advancement that is addressed in this handbook.

The goal of the *Handbook of Web Based Energy Information and Control Systems* is to continue promoting the huge benefits to energy and facility managers from the emergence of one of the most powerful technological waves of our time. This technological wave has been sweeping through the development and application of energy information and control systems, and has brought with it the internet, the world wide web, TCP/IP, web browsers, relational databases, and a host of highly capable, timesaving software tools to help energy and facility managers operate buildings and facilities much more efficiently, much more cost effectively, and much more successfully. This wave of information technology (IT) and web-based energy information and control systems (web based EIS/ECS) continues to roll on with increasing speed and intensity. A large number of new web based EIS/ECS system supplier companies have come on the scene and many new, exciting applications and adoptions of web-based technology have taken place.

What started as basic web based energy information systems has expanded into web based energy information and control systems, and finally to enterprise energy and facility management systems and building automation systems (BAS). Technological progress in information technology and use of the internet and world wide web will continue to be made at a rapid rate. Applying these advancements to computerized facility and energy management systems requires the innovative skills of many people in both the IT and the energy and facility management fields. If history in this area is a good indicator of what will happen in the future, we are in for a fantastic ride on this new and powerful technological wave.

The *Handbook of Web Based Energy Information and Control Systems* is both a presentation of technological advancements in this field over the last several years, as well as a compilation of what the editors believe is the best of the information from the previous three books in this area. The combined thrust of this new information plus relevant contents of the three previous volumes is that the highest level functions of the building and facility automation system are delivered by a web based EIS/ECS system that provides energy management, facility management, over all facility operational management, and ties in with the enterprise resource management system for the entire facility or the group of facilities being managed. This is where we have been headed with our use of IT, IP networks, XML, and web based systems to help us operate our buildings and facilities better; where better means higher energy efficiency and lower operating costs, improved occupant satisfaction, and higher productivity of our buildings and facilities through better indoor environmental quality, and a more direct tie to the business functions of our operations.

Improving overall energy and facility management begins with data collection. All three previous volumes in this series discuss and emphasize the need for metering, monitoring, and measuring devices to obtain data that are then sent to a centralized database for processing and storage. IP and XML are critical data transmission protocols and structures to obtain and transmit the huge amount of data that are necessary to provide the information needed to operate our facilities better. Data is the starting place. It is only when we can turn data into information to help us make decisions which will result in our buildings and facilities operating better that we're making real progress toward accomplishing our goals. A description and discussion of the equipment and devices to collect these data and to transmit them to a centralized database is still a basic and very important purpose of this handbook.

Advanced metering requirements for most federal buildings over 5000 square feet come from EPACT 2005, EISA 2007, and EO 13514 October 2009. These requirements specify that these federal buildings must install advanced metering for electricity, gas, steam, hot water, and chilled water by varying dates from 2012 to 2016. The required metering systems must 1) measure and record interval data, 2) be able to send data to a managing agency, 3) have time stamped data with data storage for two years, and 4) have automated backups. Fortunately, many of the energy managers at federal facilities have already recognized the value and putting in multifunction, smart, and network meters that will collect a wide range of data automatically, and send it to a computer system for processing and storage. Even using a basic EIS/ECS or enterprise energy management system can have tremendous savings potential, as well as facility operating improvements from utilizing the data to make data driven operational decisions.

Once we have these large amounts of data what is needed in our facilities is a highly capable, highly connected system to take these massive amounts of data, process it and change it into a few selective pieces of information. Compare the old operator workstation of the past which was a screen full of numbers, to a modern GUI which shows a visual schematic of the system of interest. And under this schematic is a message area that says "Change the filter in AH3B," or "Valve V5 in chilled water line CW2 is stuck open." With today's information technology providing high speed and high capacity computers and communication networks, we have the ability to take thousands of data points as inputs, and process and store them in huge relational databases in our web based EIS/ECS and BAS Systems. Using IP networking and the internet lets us easily collect the data from these hundreds or thousands of data points, and EIS/ECS systems can process these data to create information that can help us operate our facilities more efficiently and much more productively.

With the information produced by these web based EIS/ECS and BAS systems, we now have the hardware and the software technology to help us accomplish energy management, maintenance management, and overall building and facility management. Now, the building and facility management staff has enough information to operate the facility efficiently and cost effectively. The technology for overall building and facility management is here today. It is the wave of the future. The move to use web based systems and modern information technology is a wave of even greater magnitude than the DDC wave of 20 to 30 years ago. Previously, those who did not get on the wave of DDC were rolled over by it. It is now time for

building and facility energy managers to get on the wave of IT and web based systems in order to capture the benefits of reduced energy and operational costs and improved building and facility productivity. This is now not only enterprise energy and facility management, it is enterprise resource management. A description and discussion of the data processing, data storage in centralized databases, and display functions is the second basic and very important purpose of this handbook.

The capability and use of information technology and the internet in the form of web based energy information and control systems continues to grow a very rapid rate. New equipment and new suppliers have appeared rapidly, and existing suppliers of older equipment are offering new web based systems. Building and facility managers, maintenance managers, and energy managers are all interested in knowing what problems and what successes are coming from the use of these web based systems, and need to be prepared for current and future installations of web based EIS and ECS technologies in their buildings and facilities. Knowing what is being implemented in other buildings and facilities and knowing what is actually being accomplished is important information for the energy and facility managers if they are going to successfully purchase, install, and operate complex web based energy information and control systems.

The third major purpose of this handbook is to document the operational experience with web based systems in actual facilities and varied applications. Web based EIS and ECS systems have allowed the development of many new opportunities for energy and facility managers to quickly and effectively control and manage their operations. The case studies and applications described in this handbook should greatly assist all energy managers, facility managers, and maintenance managers as well as consultants and control systems and development engineers. These case studies and applications presented have shown conclusively that web based energy information and control systems are feasible, cost effective, and can create significant improvement in the energy and operational related performance of a facility. Documented benefits include reduced energy costs, reduced maintenance costs, improved operational performance, and reduced capital investments for these energy, maintenance, and operational savings. It is also clear that early adopters of these web based systems are seeing that is giving them a competitive advantage over their non-adopting business and organizational peers.

Where the previous series of books had to convince the users of the value of web based systems and the power of IP networks, it is now an accepted reality. This is in part due to the ubiquity of the internet and more recently the smart phone revolution. This has been driven by the dramatic cost reduction and increased capability of embedded hardware where even the simplest of devices now have the capability to become a part of the building's IP network yielding a wealth of information (embedded processors now contain everything from accelerometers to GPS receivers—all on the same chip). Almost everything has an Ethernet port now. The ability to collect, store and process these data has also dramatically increased since the release of the first book with a simultaneous increase in capacity and a reduction in cost. To put this in perspective, today it is possible to store the data from 100,000 points sampled every minute, for ten years, for about the cost of the average textbook. The hardware to collect, store, and analyze data is no longer a barrier

This is also happening for sensor networks (WiFi, IEEE 802.15.4, ZigBee, and Bluetooth) but has not yet hit critical mass. This may change when the smart grid (smart meters) hit consumers creating a demand for sensors to monitor in-house energy usage, which will commoditize the technology bringing down the cost for building applications. We are already starting to see the beginnings of this with products such as Google Power and other home energy management systems. We can only now guess what the impact of smart phones running "apps" tied to the building will have. Unless there are systems in place to manage and use this data (like those in this handbook) they will only add cost without deriving any value, only to sit idle in the building. The principles discussed in this handbook for web based energy information and controls will also be applicable (if not more important) for buildings with large sensor network capabilities.

Just as hardware has improved in the past decade so has software. Operating systems, browsers, web application platforms, data servers, and even software development environments all have matured and stabilized. Gone are the days where applications and machines would crash for no apparent reason. This does not mean that innovation has ended, just that there are products and standards that are mature. The Apache web server and PHP are such examples for software, built upon mature data standards HTML and XML, and data protocols HTTP, SOAP, and AJAX. The open source software movement has been a part of this maturation with stable projects bringing down costs and preventing vendor locking by providing irrevocable access to the source code. Long term product support and access is especially important for building systems which can have lifetimes of decades, compared to the 6 to 18 month product life cycles of modern hardware and software. It is infeasible to replace sensor and control systems every few years in buildings.

The realization that proprietary systems come and go (and of course firsthand experience) is one factor that is driving the use of open standards, such as Ethernet and BACnet, in the building controls industry. Truly open standards are not standards that hide important parts in propitiatory implementations behind so-called open interfaces. Truly open standards allow customers to choose alternative implementations or develop their own without resorting to reverse engineering or being hindered by license or patent restrictions. This does not only apply to controls but also to the energy management systems that monitor and operate the building. Companies must make an investment in the energy management platform, whether in purchasing a system or developing their own. If this platform goes away what is the impact on the user? For proprietary systems this can mean being forced to find a new solution or forced to use the technology until it breaks. Using open standards for controls and data collection insulates the user from having to recommission the building in the event that the energy management system needs to be replaced. Using open source infrastructure insulates the user even further. This is best illustrated by the recent acquisition of Sun by Oracle. Oracle decided not to continue the support and development of a number of open source technologies that it acquired from Sun. Since these tools had active communities and the source could not be taken away, alternate stewards and support companies sprang up almost over night

The development of all the standards, infrastructure, and software has taken time. We believe we are now at a point where we will see community developed open source energy management systems on top of open protocols and open source infrastructure (Linux, Apache, PHP, Python, Java, Tomcat, MySQL, and PostgreSQL are all examples of enterprise quality open source software). This handbook is ideal for those starting their own efforts (open or not) or joining others. The time is right; the standards, software and hardware are all in place and the current economic and energy climate is the motivation together with the knowledge that saving energy is also good for the environment. Intelligent energy systems, such as those we describe here, are what will be needed if our society is going to meet the environmental and energy challenges of the future.

How will we accomplish this? What will these systems look like? Many of the articles in this handbook provide this vision. We believe that these systems will, and must, take a systems perspective. This is driven in part by the increased use of renewable energy, which is often variable in nature, and storage technologies. The smart grid will also drive this by allowing consumers to take advantage of (or be forced to use) real time pricing. All this requires the intelligent scheduling of energy consumption and storage. By taking a systems perspective one can take advantage of the interactions between equipment and processes. This is important since the local of optimization equipment may not always mean the entire system will run optimally. A systems perspective takes all equipment and operations into account; simultaneously scheduling usage appropriately. With these technologies in place energy producers will be able to interact with consumers to optimize energy usage across the entire system; a very interesting optimization problem to solve!

Integrating building energy information and control systems requires sensing and control to have a single uniform representation across the entire building in real time. This may take the form of a single centralized database or a distributed system. In either case a united system perspective for information is necessary for both users (single seat for operation) and algorithms (entire building control and optimization) to control the building effectively. We the editors, strongly believe that this is a necessary condition for the success of any web based energy information and control system. We also believe that many of the other ideas expressed in this handbook will help guide you in building effective, easy to use and maintain, systems that stand the test of time. Building systems on open standards and open source technologies is a sure way to make this happen and keeps with the spirit of web based applications.

Finally, all four of us on the Editorial Team for this handbook hope that it has helped contribute to the successful development, implementation and application of new web based energy information and control systems in many of your buildings and facilities. It has been our pleasure to work with our two associate editors on this important contribution to the IT and web based systems education and training of working energy and facility managers. Mr. Paul J. Allen and Mr. David C. Green have both played a major role in getting this handbook prepared and completed. Our most sincere thanks go to each of them for making our job much easier then it could have been. Thanks also to our two foreword writers, Ms. Mary Ann Piette, and Mr. Ken Sinclair, who have made major contributions of their own to this dynamic field. Also we want to thank each of the 55 individual authors who have written material that appears in this handbook. Without their kind and generous help in writing these detailed chapters, this book would not have been possible. Each of these authors is identified in the alphabetic list of authors at the end of this book.

Barney L Capehart, Ph.D., C.E.M.
Timothy Middelkoop, Ph.D., C.E.M.
University of Florida
Gainesville, Florida
Spring 2011

Chapter 2

Evolution of the Enercept™ Meter

Jim Lewis

Author's note: the origin of this article was a request from Barney Capehart, the editor of this book to provide a history of the development and production of the Enercept™ electrical submeter. The meter was first designed in 1994 and released to the market in 1996 by Veris Industries of Portland, OR (Veris is now a division of Schneider Electric, which purchased Veris in 1999 and owns all rights to the Enercept™ meter). The author was the President and co-founder of Veris and was a co-holder of the Enercept™ patent with Paul Stoffregen prior to the sale of the company to Schneider Electric.

Writing a history of the development of the Enercept at first seemed like a fairly straightforward task, but has proved more challenging than one would expect for several reasons:

- Time—the events described occurred over 15 years ago and trying to reconstruct an accurate record is quite challenging

- Process—as will be described in more detail later, the development of any new product is rarely the result of an "ah hah!" moment where the whole of the product is revealed in a miraculous burst of insight. The process is more often a series of revelations and small insights that ultimately result in a final product—a process which extends over considerable time

- Ego—Over the years, at least four different people outside of Veris and many people employed at the company at the time have claimed to be the inspiration for the Enercept™ meter concept in whole or in part. It is virtually certain that someone else writing this history would produce a greatly different version of the events than the author simply due to differing memories. As the old saying goes, "Success has many fathers, but failure is an orphan."

The bottom line is that the story told here is only the view of one individual and will no doubt exaggerate the role of the author while understating the contributions of many other people to the ultimate success of the meter. With this disclaimer and apologies to those whose efforts are overlooked or downplayed, we can start.

BACKGROUND

Veris Industries, Inc. was started in 1992 by Jim Lewis and Kent Holce in Portland, Oregon. The company's initial product was a current sensor designed to provide proof of flow in air or water systems by detecting the amount of current flowing through a conductor supplying a fan or pump motor. The design of the current sensor provided a contact closure output in the event of a significant drop in amperage due to a broken belt, failed motor or other failure. This output was then connected to a building automation system to send an alarm to a building operator in the event the current level fell below the threshold. Typical customers for

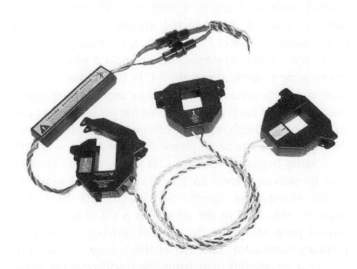

The Enercept

7

these products included manufacturers of building automation systems (BAS) such as Honeywell, Johnson Controls, Siemens, Andover, Alerton and many others.

This initial product led to the development of a line of current sensing devices for a range of applications, including simple go/no go sensors which indicated the presence of current flow above a minimum level and analog output current sensors designed to provide an indication of the amount of current flowing through the conductor. These sensors proved very successful in the BAS market as replacements for a variety of other mechanical devices used to provide proof of flow and as a means to measure the electrical load on various parts of the system.

At the core of all of these current sensors is a current transformer (CT) which measures the amount of current flowing in a conductor without having physical contact with the conductor. This is accomplished by measuring the electromagnetic field generated by the flow of electricity in a conductor using a steel or composite core and a series of windings mounted on the core perpendicular to the direction of the flow in the conductor. While a detailed description of how this is done is beyond the scope of this paper, the key to the application of these products is that a CT produces a low current or voltage output proportional to the current produced in the primary conductor. This output signal can then be used by an electronic circuit mounted in the CT housing to produce a contact closure or analog output for connection to the BAS.

In addition to being used for standalone devices such as those described above, the CT can be used as part of a larger system for monitoring power consumption (kW) if the CT provides an industry standard output such as 5 amps, 1 volt or .333 volts. In this capacity one CT is connected around each conductor in an electrical panel (1 or 2 CTs for single phase and 3 CTs for 3 phase systems). In a typical 3 phase power monitoring system, there are 3 CTs to measure current flow on each phase and 3 voltage inputs to measure the voltage on each phase. Power for each phase is calculated by multiplying the voltage on the phase times the current for the same phase (Volts x Amps = Watts).

Veris began producing CTs designed for connection to power meters in the same physical platform as the standalone current sensors in late 1993. The industry standard at the time was a CT with a 5 amp output proportional to the current flowing through the primary conductor. The use of the 5 amp CT was primarily a legacy left over from the traditional mechanical power meter mounted on an electrical enclosure

which required a significant amount of power from the CT. The 5 amp CT had several disadvantages in the submetering market, in particular:

1. Shock hazard—the output of the CT (although much lower than the primary conductor) provided a very significant hazard unless the circuit was de-energized prior to contact

2. Poor accuracy at low current levels—the nature of the CT design provides high accuracy at high currents, but the resistance in the secondary winding produces poor accuracy at low current levels

3. Cost—because of the shock hazard outlined above, the installation of the CTs required a disconnect switch to de-energize the output of the CT to eliminate the danger, adding significantly to the total installation cost

The company's sensors provided a significant advantage over industry 5 amp CTs by providing a millivolt (.333 volt) maximum output signal. The circuitry in this type of CT incorporates a high accuracy resistor to produce an accurate voltage signal proportional to the current in the primary conductor. In addition to providing a highly accurate output, the use of this resistor makes the secondary output from the CT inherently safe as no high potential power is produced. The nature of the windings on the CT (multiple turns of very small gauge wire) also provides for a much better dynamic range for the CT with high accuracy at as little as 3% of the primary current.

The story of the development of the Enercept meter begins with these CTs and the company's decision to produce power meters targeted at sub-metering applications in commercial environments.

THE NEW PRODUCT DEVELOPMENT PROCESS

Before delving into the details of the Enercept™ development, it is probably useful to spend some time on the general product development process from the author's perspective. As the creator of a number of successful (and unsuccessful) products at two different companies, the author has considerable experience in how a WOW product comes to life and the processes involved in bringing a product from initial conception to production.

Many people who have not been involved in a

new product development/design process have misconceptions about how the process evolves, particularly in a small company. The biggest advantage to a small company is the ability to be nimble, responding to needs in the marketplace much quick=er than larger competitors and taking advantage of niche opportunities that might be too small for the bigger companies. Some of the most common views of how product ideas evolve include the following:

Focus groups—usually a gathering of customers and other stakeholders, focus groups are intended to solicit customers' thoughts on what new products and features are needed in the marketplace. These groups are generally useful for incremental "tweaks" to existing products and rarely produce the kind of breakthrough products that are so valuable to the industry. Changes suggested are usually incremental improvements and rarely result in game-changing products.

Management retreats—the internal equivalent of focus groups, the idea here is to send a team of employees to a resort or cabin in the woods to brainstorm new ideas. Liberally sprinkled with catch phrases like "thinking outside the box," retreats usually include team-building exercises along with making smores and singing Kum-Ba-Ya along with product thoughts. This process can be a critical part of product development if managed correctly, but in the author's experience has rarely produced major big product ideas.

The "ah-ha" moment—in this model, the inventor is suddenly struck with a fully formed vision of what the new product looks like and the only remaining work is dreaming up the marketing literature and pounding out a couple of prototypes. This probably does happen in certain instances, particularly when the product is relatively simple and lacks the complexity of most industrial products (think Snuggies™).

In the author's experience, new product breakthroughs are not instantaneous, nor can they be driven by a timeline ("we need some new stuff, let's head for the mountains for a couple of days"), but are rather the result of a combination of free association and incremental improvements. One of the critical keys to successful WOW products is the ability to step back from some of the traditional methods outlined above and instead to start with a tabula rasa (clean slate) in approaching the problem. Rather than starting with the solutions available today and trying to make them better, cheaper, faster, it's much more useful to approach the problem from a whole new perspective. This process produces a whole bunch of little "ah-ha" moments that eventually can add up to something very new and valuable.

To illustrate this approach, it is useful to provide some more background on the metering world as it existed in the early 90's and build an example from that.

SUBMETERING IN THE EARLY 90s

As mentioned above, Veris Industries in the early 1990's was making a number of products based on the use of current transformers to indicate loads on motors or other electrical equipment. The logical progression for the company was to add the electronics and voltage sensors to the CTs and produce a meter that would measure not just current, but power. This was hardly a brea-kthrough concept and the company began with a mix of products that was very similar to many of the devices available from a number of other manufacturers. Basically the products had the following components:

- Current transformers (CTs)—mounted in the electrical panel around the primary conductor the CTs measured the current flow through one or more phases (generally three phases in a commercial environment

- Voltage taps—wires are connected to sense the voltage of each phase being monitored so that power can be calculated (Volts x Amps = Watts)

- Electronics—the leads from the CTs and voltage taps are brought into a box mounted near the electrical panel and a processor is used to calculate power on each phase at a very high rate of speed, the total power for all phases is then calculated

- Output/communications—once the power is calculated, it is can then be displayed or communicated to other master devices such as BAS, PLCs or other communications interfaces. This output can be in the form of a simple pulse (e.g. one pulse = 1 kW) or can be more sophisticated using serial communications and either an open protocol such as Modbus or a proprietary protocol to report to a proprietary master device.

As this market was perceived to have high growth potential, the challenge for the product team at Veris was twofold:

1. How to grow the market in general

2. How to differentiate the Veris offering from one of many available in the market from a variety of manufacturers

To attack these challenges the product team (consisting of a product manager, sales executives and senior management) first approached existing customers (an informal focus group) to determine why they weren't buying more meters. Was it lack of demand from customers, wrong feature set, cost? Almost without exception, the response was: "The customers see the need for this information, but your meters are too expensive and we can't justify the cost."

Armed with this feedback, the team embarked on a several-months long quest to identify ways to reduce the meter cost, which was typically in the $800 to $1000 range. Options for cost reduction included different housings, simplified electronics, new means of voltage sensing, cheaper CTs, outsourcing manufacturing, eliminating local displays and a variety of other potential cost-saving ideas. At the end of the day, it was determined that costs could be reduced by 5 to 10 % which would translate to a selling price savings of $40 to $100 per meter.

Follow up conversations with customers and the general response was tepid at best. Most indicated that while they would be happy with any cost reduction, the likelihood that the 5 to 10% would generate a significant upturn in business was very low. Given the level of capital investment required to achieve the 5 to 10% savings, it was very difficult for the company's management to justify the expense in light of the limited potential for return on the investment. At this point, it appeared that there was little opportunity to get the kind of breakthrough result needed to accelerate growth in the market and grab market share and a new approach was needed.

WE ASKED THE RIGHT QUESTION, BUT...

The team began to focus on whether we were missing something in our dialogue with the focus group customers. Had we asked the wrong question? Did they not understand the question? It was during this re-examination that the first ah-ha moment occurred.

It seems simple, but the bottom line was we had asked the right question, but the context meant that the customers were answering the question they thought we had asked and we heard the answer we thought we

were looking for. The reality was that while customers said they needed to have a cheaper meter, what they actually meant was that the total cost of buying and installing the meter was prohibitive. By the time a customer finished buying, installing and commissioning the meter the total cost was not $1000, but closer to $2,000. Installation required several hours of an electrician's time at $50 to $100 per hour and then also required someone to program the BAS or other system that would read the data.

This ah-ha moment shifted the focus of the development effort from lowering the $1000 product cost to reducing the total cost of installation. This was what customers were asking for, we just weren't listening. The math is pretty simple, if you could reduce the total cost by 10%, you'd be saving $200+ rather than just the $50 achieved by reducing the product cost.

THE HARD WORK BEGINS

The first response to this insight was a basic slap to the forehead, how did we miss this kind of moment followed by a revised plan to gather information. While the team was knowledgeable about the hardware required to sub-meter, the simple fact was that none of us had spent much time in the field to understand what else was involved in completing the installation. With this in mind, the author and others arranged with customers to observe typical meter installations from purchase to commissioning with a goal of determining where costs could be reduced.

In theory, the process of installing a meter is fairly straightforward. In a three-phase panel installation the electrician:

1. Turns off the power

2. Installs the CTs around the primary conductors,

3. Connects the voltage leads to the primary conductors,

4. Pulls all the wires out of the panel (2 per CT plus one per voltage tap and an optional neutral),

5. Mounts the meter box on the wall,

6. Terminates the conductors to the appropriate terminals in the meter box

7. Turns the power back on

8. Verifies the reading on the meter

9. Connects the communications wiring

After reviewing several installations, there were some things that were very consistent and applied to virtually all of the projects. In looking at where the time and money were spent from an installation perspective, the team determined that for all installations:

- Installation of CTs and voltage taps required roughly 30% of the installation time

- Getting the wires out of the panel including conduit accounted for 40% of the installation time

- The remaining 30% of the time was spent verifying the accuracy of the readings and cleaning up the installation

What was most interesting (and somewhat unexpected) was that in more than half of the installations we observed the data being reported by the meter was inaccurate due to wiring errors. The most common error was mixing the phases, i.e. getting the wrong voltage lead with the CT. If, for example Phase A voltage is connected to Phase B current, the power calculation may be off significantly and the electrician has to go back and revisit the wiring to correct the problem. The time required to troubleshoot the problem was typically as much or more than the time required to do the original installation. Other common errors included reversing the CT (many CTs are directionally sensitive) which results in a negative current measurement and improper voltage connections.

In follow up conversations with installers, it was discovered that the price quotes for meter installations that they were giving their customers included not only the direct installation time, but also a fudge factor to cover the cost of troubleshooting and fixing installation problems. The amount of the fudge factor varied, but was typically in the range of 50% of the installation labor time.

In reviewing this data, the team surmised the following:

- $300 of the installation budget was spent on installing the CTs and voltage taps

- $400 of the installation budget was spent on wiring between the meter and the box

- $300 of the installation budget was allocated for troubleshooting and repairing errors in the installation process

The assumption as this point was that the dollars spent on CTs and voltage taps would be difficult to reduce or eliminate since these connections were required. The major opportunity for savings was in reducing the time spent on wiring and eliminating the requirement for troubleshooting and repairs. If the wiring cost could be cut in half and the fudge factor eliminated, the total cost of installation would drop from $2,000 to $1,500 or less (a 25% savings).

IN A PERFECT WORLD

As mentioned earlier, the most common way to come up with product ideas is to examine the way people are solving the problem now and try to add or remove features to make it more attractive and cost effective (better, cheaper, faster). This traditional approach can result in solid incremental improvements, but will rarely produce WOW products.

The alternative is to start from scratch and build a model that is ideal from the customer perspective by adding features and costs as needed to meet the customer's requirement. In this case the perfect meter would have the following features:

- The hardware would be free
- The meter would install itself
- The meter would require no wiring
- The installation would be idiot-proof

In examining the feature set, several things are immediately apparent. First, even if the meter could be built and sold for free, it would be difficult for the manufacturer to build a business model based on a product that produced no revenue so this was off the table. Second, it is clear that the meter can't install itself, but improvements to the feature set could result in reduced installation time, complexity and costs. Third, while wireless meters are around today the technology in 1994 was not available so the next best option would be a meter that had less wiring requirements. Finally, every time a manufacturer tries to produce an idiot-proof product, the world produces better quality idiots so the best one can do is to produce something that is easy to understand and less prone to error.

In sum, most of the installation time and cost was not connected to the physical installation, but was associated with getting the right wires in the right order to the right terminals on the box located outside the panel.

This produced the second ah-ha idea: what if you moved the box inside the electrical panel? This would make it much easier for the installer to associate the correct phases and greatly reduce the likelihood of errors in pulling multiple sets of wires out of the panel as well as making troubleshooting much easier. There were at least two immediate problems with this brilliant solution: 1) there wasn't room in the typical electrical panel for the typical 6" x 8" x 2" box and 2) electrical inspectors weren't likely to be too thrilled to find large quantities of low voltage wires terminated in an open box in the panel. A number of potential solutions were explored, but it seemed that there was no practical means of getting around these issues.

After struggling with potential solutions, along came ah-ha number 3. Probably the most important breakthrough feature of the Enercept came from a simple question: why can't you put the electronics from the box in the CTs? While there were immediate concerns about whether there would be enough room and whether it would meet code requirements, the benefits were obvious: you could mount the meter with no additional hardware in the panel, there would be much shorter wiring runs, it would be easy to make sure that the phases weren't mixed up, and there would only be one pair of wires running out of the panel.

This single feature resolved almost all of the challenges associated with installing the meters by greatly reducing the costs of wiring, minimizing the likelihood of phase mixing and eliminating the need for installing an external box.

There were several other smaller ah-ha's as the product moved through development, including:

- Color coded voltage taps and CTs to match phases

- Intelligence to automatically adapt for CTs in the wrong direction

- Choosing a Modbus serial output to eliminate the need for pulse scaling

- Warning LEDs to flash if phases were swapped

DEVELOPMENT

One of the biggest challenges for non-engineers involved in design and development of industrial/electronic products is the time required for the actual building and programming of the product once the feature set is determined. In the case of the Enercept, the time from the decision to move forward with prod-

uct development to final product release was roughly two years, largely due to the fact that the project was not a single development utilizing one resource but rather a series of smaller projects requiring a variety of resources. Major tasks involved in developing the Enercept included:

- Hardware (circuit board) design

- Software

- Tooling to modify the existing injection molded CTs to accommodate the additional electronics and wiring

- Certification by UL

- Test equipment and fixtures to build the meters

- Marketing collateral materials

- Training for production and technical support personnel

A couple of key points to remember are that very few of these tasks can be run concurrently and that some of the tasks require the bulk of the time. In the case of the Enercept, the most crucial steps were the software development and the board layout. As an example, no work could be done to modify the injection molding tool until the circuit board layout was completed since the amount of real estate required would be dictated by the parts and spacing required for the boards.

Even the steps outlined above are greatly simplified in that they imply a fairly straightforward progression (i.e. lay out the board, then write the software, then modify the tool). In fact, the process involves multiple iterations—you can't finalize the hardware until you know exactly how the software is going to work and vice versa. Some of the functions that you plan to accomplish in software will dictate that certain hardware is included and the limitations of the hardware (memory, speed, etc.) will drive some of the software functions so it is almost impossible to complete one task and move sequentially to the rest.

Some of the critical questions facing the design team were:

1. Could the circuit board be designed small enough to fit inside the CT?

2. Was there enough horsepower in the microprocessor to meet the overall need for accuracy (ANSI C 12.1)?

3. Could the meter be designed and built to meet UL requirements?

4. Typical CT/meter combinations required disconnect switches to protect the installer or service personnel. How could you provide protection from dangerous voltages and currents and protect the circuitry from surges in voltage?

5. The design called for the meter to be powered from the circuit being monitored. Was it possible to design a power supply that could meet a wide range of input power (240-480 VAC)?

6. Since the meter would be installed completely within an electrical panel, it would not have any sort of LCD display to assist the installer. How could the meter be designed to let the installer know if everything was put in correctly?

7. What sort of test and calibration system would be needed to produce the meters in large volumes in a timely manner with acceptable accuracy?

To help in understanding the challenges, we will briefly look at these major questions and how they came to be resolved in the eventual design.

1. Could the circuit board be designed small enough to fit inside the CT?

This was probably the biggest challenge of the whole project as the team was basically taking electronic circuitry that was housed in an external box of approximately 80 cubic inches and putting it into a space of roughly 12 to 15 cubic inches. This space limitation was ultimately resolved by using a very small 8 bit PIC microprocessor with limited memory, but requirement for using this microprocessor was a significant additional burden on the software development. A lot of functions that could be accomplished in a larger box using additional hardware had to be met by the use of highly creative software design. One of the development breakthroughs was to ignore this issue by building prototypes in a separate box without the real estate constraints of mounting in the CT. This allowed the designer to test the functionality of all the components in a board that had much more real estate and proof of concept could be demonstrated without the additional challenges inherent in the CT mounting.

2. Was there enough horsepower in the microprocessor to meet the overall need for accuracy (ANSI C 12.1)?

In addition to the size limitations mentioned in the previous question, issues regarding power requirements and cost also contributed to the selection of the 8 bit microprocessor. This presented a couple of different challenges—1) the microprocessor had to have a high enough sample frequency on the six inputs to meet accuracy needs while also doing the math to calculate power and handling the communication on the serial port; and 2) having the processing power and memory to calculate power on a sinusoidal AC waveform. The first issue was resolved by having the processor be able to interrupt communications transmissions to insure that sample frequencies were high enough to meet accuracy needs—no simple process and a task that added considerable time to the development. The second challenge was more fundamental and challenging. AC power is a sinusoidal waveform which requires calculating the RMS (root mean square) value of a series of samples from the currents and voltages. Without going into the details of the math, calculating RMS means taking a series of samples, squaring the values, taking the mean of the squares and calculating the square root. The first two pieces (squaring and calculating the mean) are relatively straightforward and easy to accomplish using any microprocessor. Square roots, on the other hand, are very complicated and require significant amounts of processor time and memory—too much for the PIC microprocessor the engineer had chosen. The solution was to bypass the RMS calculation and use a look-up table that required a fraction of the resources, but still met the accuracy requirements.

3. Could the meter be designed and built to meet UL requirements?

This issue probably best illustrates the challenges and tradeoffs inherent in development more than any other. The challenge is this: UL requires a certain amount of spacing between circuit traces on a circuit board, the amount of spacing increases as the voltage increases. Given the limited real estate available on the board, bringing 480 V onto the board meant that a large amount of real estate would not be available to the board layout designer as no circuits could be routed too close to the high voltage. Further complicating the design issue is that certain electronic components needed to be placed close to others to meet the sampling frequency requirements. Changes to the location of the components also impacted the firmware as timing changed, so the need for 480 V to power the meter drove changes to the location and selection of components as well as the software.

4. *Typical CT/meter combinations required disconnect switches to protect the installer or service personnel. How could you provide protection from dangerous voltages and currents and protect the circuitry from surges in voltage?*

Obviously, having 480 V present on a hot wire introduces a significant hazard risk to installers and service personnel. In addition, since the primary voltage is used to power the meter it was important to protect the meter from voltage surges that could travel down the voltage taps. The ah-ha moment here was coming up with a solution that would meet the needs for safety and protection of the circuits without taking up additional space in the already crowded CT housing. The answer was to provide an in-line fuse on each voltage tap between the primary and the meter with a fuse pack that allowed the installer to be protected from the voltage. When the two part fuse pack was unscrewed, the female end of the fuse pack was no longer connected to the primary voltage so the installer was insulated from the dangerous power, and when the fuse pack was connected with an in-line fuse, the electronic circuit was protected from dangerous voltage surges.

5. *The design called for the meter to be powered from the circuit being monitored. Was it possible to design a power supply that could meet a wide range of input power (240—480 VAC)?*

This was one of the more significant hardware challenges in designing the Enercept as the problem is actually more complicated than it appears. Although the nominal operating voltage was 240—480 V, the design actually had to function accurately at +/- 10% of nominal so the effective operating voltage range was 216 to 528 V. Stepping this voltage down to the 5 volts required, without generating additional noise (and inaccuracy) into the electronics, proved to be a very complicated task. Details of the solution are beyond the scope of this article, but suffice it to say the answer resulted from a very ingenious power supply design.

6. *Since the meter would be installed completely within an electrical panel, it would not have any sort of LCD display to assist the installer. How could the meter be designed to let the installer know if everything was put in correctly?*

This was actually a two-part issue: 1) finding a means to make the installation as fool-proof as possible in the first place and 2) providing status indication to the installer/operator with a minimum of additional hardware. From a design perspective, there were several improvements that were developed along the way to make installation and communication easier:

*   Color coded voltage wires and CTs to match the individual phases of three phase installations making it easier for installers to get the wiring right

*   Automatic correction of CT orientation—this innovation meant that the installer did not have to be concerned about whether the CT was installed "right" side towards the load

*   Detection of out of phase wiring (wrong voltage phase to current phase)

*   Dip switches for setting the Modbus address of the meter without the need for a laptop or other computer or software

*   A single, dual color LED to indicate status of the installation: slow green blink for OK, slow red blink for wiring issues or poor power factor, and fast red blink for overcurrent operation

7. *What sort of test and calibration system would be needed to produce the meters in large volumes in a timely manner with acceptable accuracy?*

This issue proved to be very challenging for several reasons. First, the industry standard was to use a calibrated meter to measure the input power, but these meters were typically only 1-2% accuracy and thus the total error could be as much as 3-4% for the Enercepts being calibrated. This issue was ultimately resolved by designing an automated test and calibration system using highly accurate laboratory power supplies to produce voltage and current signals. An additional complication was that each CT had slightly different characteristics in measuring currents due to issues such as winding orientation, differences in cores, etc. and this made it difficult to calibrate across the whole amperage range of the meter. This was solved by designing the meter in calibration mode to "learn" the characteristics of each CT and have a lookup table of calibration to correct the readings from the CT through a specified amperage levels. The meter automatically applies this calibration factor for each of the samples taken from the CT.

In short, the development process continued the string of small ah-ha's from the earliest conceptual design to the final production process, very typical for a significant design project. All stages of the process had input from a wide variety of stakeholders, from engineering to sales to production and purchasing that made the project a major success.

CONCLUSION

One of the most interesting parts of this development was the final cost of the product. The reader may recall that the initial customer response to how to improve meters was to "make the meter cheaper." In fact, the Enercept was more expensive than its predecessors and represented an approximately 10% increase in hardware cost. Because the total installed cost for the meter was reduced by roughly 40%, other meter products were discontinued as customers were happy to pay the 10% hardware premium for the additional benefit and total cost reduction, so Veris ultimately delivered exactly the opposite of what customers said they wanted.

The effort required to get from concept to product took over a year and a half and required the remarkable talents of an exceptional team dedicated to delivering the product. Although there were many, many people involved in the development and release of the Enercept, the key individual without whom the Enercept would never have been brought to market was Paul Stoffregen, design engineer. Paul was able to take a remarkably small amount of real estate and processing power and produce a reliable and accurate product. The difficulty of the task can be summarized by a conversation the author had with Paul late in the process when he told me that he probably couldn't add more features as he only had 3 bits left in memory (not a misprint, it was bits not bytes!).

There have been many improvements to the Enercept over the last 15 years, but the basic product as designed continues to be sold today and is one of the most popular sub meters ever produced.

Chapter 3

Web Based Energy Information and Control Systems for Energy and Facility Management

Barney L Capehart, PhD, CEM

INTRODUCTION

Energy and facility management are extremely important tasks and opportunities for every building and facility to pursue. Particularly with the expanded interface areas in recent years from the related topics of sustainability, green buildings and facilities, energy efficiency and energy conservation, renewable energy, greenhouse gas reduction, carbon reduction, energy cost reduction, and energy productivity improvement. Energy and facility management play critical roles in all of these areas, and need all the assistance they can get from modern technology—especially modern information technology.

New technology in the form of web based energy information and control systems is available, cost effective and highly functional to accomplish these tasks and to provide additional opportunities to operate buildings and facilities in an optimum manner. Use of advanced web based energy information and control systems provides the opportunity to reduce building and facility operating costs by 10-15% or more without requiring new lights, air conditioners, boilers, or similar equipment.

Using new web based energy information and control systems offers increased profits for many buildings and facilities in the private sector. As an example, if your company is getting a 15% return on sales, saving $15,000 in energy and operating costs is like increasing your sales $100,000 a year. Also, the EPA/DOE Energy Star Program tells us that reducing the energy and operating cost $50,000 a year for a building with a capital asset factor of ten would increase the capital asset value of that building $500,000.

The purchase, installation and use of this new web based energy information and control technology developed over the last ten years can provide significant operating cost reductions and improved operational performance for 99% of all of our buildings and facilities. As the old cartoon character Pogo said many years ago, "We are surrounded by insurmountable opportunities."

REDUCING BUILDING ENERGY AND OPERATING COSTS AND MAKING MORE PROFIT WITH WEB BASED ENERGY INFORMATION AND CONTROL SYSTEMS

The starting place for energy and facility management includes a list of activities that must be accomplished prior to setting up the web based energy information and control systems to best produce cost effective results to control energy costs and to increase profits. These activities include:

- Develop an energy master plan
- Set energy use reduction goals
- Meter and monitor energy use
- Move energy cost out of overhead and down to individual cost centers
- Perform energy audits and assessments
- Incorporate energy use into lean Manufacturing

Energy Master Planning

One of the best resources available for helping set up an Energy Master Plan is the US EPA/DOE Energy Star Program. Their website has a wealth of resources for helping buildings and facilities get started with energy and facility management. For those facilities looking for world class energy and facility management, the soon to be approved ISO Standard 50001 should be approved and out in mid to late 2011. Until then, the ANSI Standard for MSE 2008 can provide help.

Set Energy Reduction Goals

Setting energy reduction and cost reduction goals is critical for focusing the task on controlling these building and facility energy and operating costs. For energy use reduction, a minimum of 2-3% per year reduction is easily accomplished. Federal facilities have been meeting 3% per year goal for several years, and private companies like 3M have recently increased their goals to 4% per year savings.

Even 5% is a reasonable, but aggressive goal for annual energy use and operating cost reductions. Use of Energy Star Energy Performance Benchmark numbers works very well for many buildings or facilities to use to evaluate their energy use reductions.

Meter and Monitor Energy Use and Cost

You can't manage energy use and cost if you don't measure it. You must get the use and cost of energy moved down to individual cost centers. Then submeter all major processes in that cost center. Get it out of overhead. Even if you know what you are using and spending on energy, you can't do the management job that is needed if the cost of energy is in overhead. Here is where you need to implement a building or facility wide information and control system.

Energy Audits and Assessments

This is where we find out how much energy is being used by a building or facility; and where that energy is being used. We need detailed metering and monitoring equipment and systems, and then we can see how energy use and cost can be reduced. An energy audit examines the way energy is currently used in a facility and identifies some alternatives for reducing energy costs. The goals of an energy audit are to clearly identify the types and costs of energy use, to understand how that energy is being used—and possibly wasted, to identify and analyze alternatives such as improved operational techniques and/or new equipment that could substantially reduce energy costs, and to perform an economic analysis on those alternatives and determine which ones are cost effective for the business or industry involved.

Substantial basic data and metering data are required to conduct an initial energy audit, and continuing collection data is required to understand the operation of the facility and to determine that the facility is operating correctly. An existing web based EIS/ECS or BAS can tremendously increase the speed and value of an energy audit of an operating facility.

Incorporate Energy Use into Lean Manufacturing

Lean manufacturing is defined as the systematic elimination of waste from all aspects of an organization's operations. Waste is viewed as any use or loss of resources that does not lead directly to creating the product or service a customer wants when they want it. This definition clearly includes energy. Lean manufacturing—also known as lean, agile manufacturing, or *just-in-time* production— was originally developed by the Toyota Motor Company in Japan based on concepts pioneered by Henry Ford.

Substantial energy savings typically ride the coat-tails of lean. By eliminating manufacturing wastes, such as unnecessary processing and transportation, facilities also reduce the energy needed to power equipment, and provide lighting, heating, and cooling. Unnecessary processing could occur as a result of poor quality product being made which requires rework, or is scrap that goes back to the start of the manufacturing process—such as ground up plastic pipe or aluminum to be recycled. And all of these require excess energy or energy waste.

A key step in effective lean manufacturing and energy waste identification efforts is learning where to target energy-reduction activities. This is what we do as energy or facility managers. This is the energy audit or a plant assessment. Methods for assessing energy use and identifying opportunities to save energy in the context of lean include energy treasure hunts, value stream mapping, Six Sigma, and Kaizen events. Data collection is one of the most important tasks in the application of lean to energy waste in manufacturing.

THE BUSINESS CASE FOR ENERGY AND FACILITY MANAGEMENT WITH WEB BASED ENERGY INFORMATION AND CONTROL SYSTEMS

Demands on buildings and facilities continually increase in many areas, including the need to:
*   Reduce operating costs of:
        Energy—in spite of increasing costs
        Maintenance and training
        All other facility resources
*   Improve indoor environmental quality
        Improved comfort
        Improve air quality
        Improve productivity
*   Greater tie-in to the business function of the organization

The Need for Building and Facility Operational Data

Our need for huge amounts of operational data about our facility's operation comes from the formal or informal use of the Deming Continuous Improvement (CI) Cycle, lean, or the Motorola Six Sigma Method, and one of their fundamental principles: Make only data driven decisions. Measurements are required. We have all heard the statement that "You can't manage what you don't measure." A more positive version is that "If you measure it, you can manage it." Data driven decision making provides a structure—a set of guidelines for knowing what decisions to make.

Because of the large amount of data that can and needs

to be collected, there is a need to use remote, electronic, methods to automatically collect this data and send it to a data storage and processing system so that useful information can be sent to decision makers at the buildings or facility. Effective use of web based energy information and control systems allows the collection of large amounts of data, which is then processed into information, and results in the capacity to share data and information leading to a deeper understanding of facility operational decisions for success.

The Need for Information

What is Information? Information comes from processed data. Information is what helps us make decisions to operate our facility better. Better relates to higher energy efficiency and lower operating costs, improved satisfaction and productivity of our facilities through better indoor environmental quality, and a more direct tie in to the business functions of our facilities.

Facility Improvement Process

It is important to set measurable goals and targets toward key performance indicators and benchmarks, such as normalized Btu/square foot/year, or Btu/pound of product. Many other KPIs can be created and use to effectively manage the energy use and operational costs of a facility, as well as other important facility operation parameters, and other opportunities for improvements. Some other key performance indicator examples could be:

- Consumption based
  - Btu/month, kWh/month
  - Btu/sq ft/yr, kWh/sq foot/yr
  - Btu/sq ft/yr weather-adjusted
  - Gal of water/month
  - Gal of waste water/month
- Production based
  - Btu/lb of product, kWh/lb of product
  - Btu/person in an office, kWh/person
  - Btu or kWh per item produced
  - Btu or kWh per student per year
  - Gal or water and waste water/month per person
- Cost based
  - $/hr at equipment, AHU, building levels
  - $/ton-hr at equipment, chiller, plant levels
  - kW/ton of cooling, CFM/HP, kWh/gal pumped
  - $/mo of water and sewer cost/person
- Comfort/IAQ based
  - Comfortable hours/month (ASHRAE temperature & humidity)
  - Ventilation ratio (delivered vs. ASHRAE 62.1)

Service based—number of hot & cold calls/month

The starting place for quickly getting the large quantity of data we need to work with is to use a web based energy and facility information system. This is our first task.

WEB BASED ENERGY INFORMATION AND CONTROL SYSTEMS FOR ENERGY AND FACILITY MANAGEMENT

What is Really Needed?

There are two critical features that are needed for a highly successful facility information and control system:
1. Single seat operation
2. A common database for all operational data

This means that an operator for the EIS/ECS system can go to any PC in the building or facility and operate all energy and operational equipment and systems from that one computer; and that all data and information goes into and out of the common database for those systems. Thus, in particular, there is not one energy management computer, and not one maintenance management computer. These two functions are controlled from one computer with one common data base.

What about manufacturing plants? Manufacturing plants usually have good to excellent process control systems, and manufacturing management control systems, but not good facility information systems! Why? What is the problem? No single seat operation. No common database for all facility data. All facility operational data should be housed in a common data base that is accessible from a single seat.

How do we do that? Bring all facility data into one facility information system from all of the various meters, monitors, sensors, process control systems, manufacturing control systems, HVAC and lighting control systems, maintenance management systems, SCADA systems, etc. Then put them into a common database. This idea of operation of the system with a single seat and a common data base is not obtained with a collection of different systems that are simply web enabled as shown in Figure 1.

GETTING THE DATA IN TO THE WEB BASED ENERGY INFORMATION AND CONTROL SYSTEM

The data collection process must automatically and remotely bring in data from all of the meters, monitors, transducers, sensors, and other systems in the building or

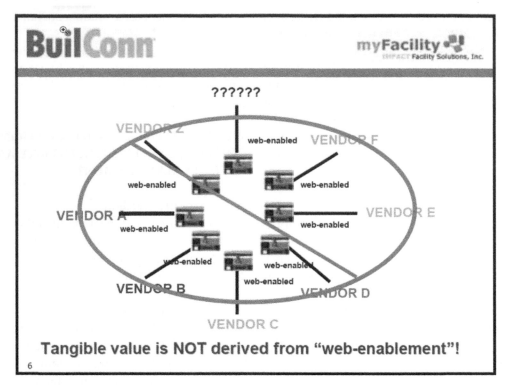

Figure 3-1. Illustration of a Web Enabled multivendor information and control system

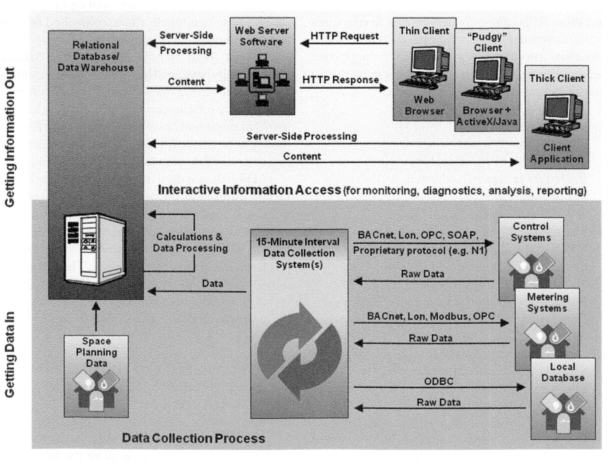

Figure 3-2. A common database energy information and control system

facility to the web based energy information and control system. One source of data for collection is from the basic metering devices in the building or facility.

Data Collection—Metering
- Direct Analog
    4-20 ma, 0-10 v from sensors
    Consumption is measured by averaging rate over time
- Pulse Output
    Pulse for every pre-defined amount of consumption
    Example: 1 kWh/pulse
- Digital, Network-based
    Meter calculates and stores the totaled consumption
    Data are retrieved via network

Other Data Collection Devices
- Maintenance/Energy Data Needed
    Differential pressure across filters
    Vibration sensors on all motors over 10 HP
    Real time monitors on steam traps

Data collection from other building or facility systems

There may be many different systems that have data available for collection over the web from all of the various process control systems, manufacturing control systems, HVAC and lighting control systems, maintenance management systems, SCADA systems, and other special system controls for the building or facility. All of this data should be collected remotely and automatically over the internet or intranet.

Using the Data and Creating Information to Make Operational Decisions

Collecting the raw data is in important part of the ability to operate a building or facility and an optimum level. However, the next step is also critical, and that is taking the raw data and making information out of it that can provide displays and tools to determine what decisions need to be made to improve the operation of the building or facility.

Example 1

A facility the size of a supermarket is a nice example to show that web based energy and operational information and control systems can be used quite effectively in smaller facilities. In this example, a major supermarket operator in central Virginia wanted to implement a monitoring and management system in their stores that would provide a standardized user interface from one store to another, provide local and remote monitoring, and easily integrate data from all facilities infrastructure in the

Electric Meters

Chilled/Hot Water Meters

Gas Meters

Water Meters

© Barney L. Capehart 2008

Figure 3-3. Common examples of building or facility direct metering devices

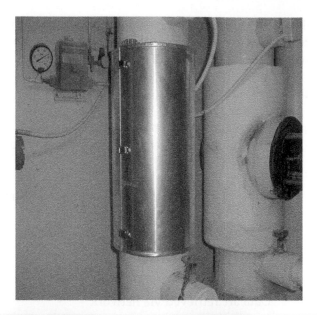

Figure 3-4. Electric three phase meters for large equipment loads such as chillers

store. The new system needed to be intuitive, utilize open systems, and easily integrate with other facilities systems without dependence on a single vendor. The types of systems monitored in each store include cold storage systems in the stock room, generators, HVAC units, refrigerator and freezer cabinets, lighting, and facility temperatures. With the rich graphical user interface, any store employee with basic training can view alarms, react to them quickly to take corrective action, and notify maintenance. By reducing reaction times to problems, spoilage has been reduced, increasing the profitability of the stores. More information on this application can be found in reference [4].

Example 2—Walt Disney World Utility Reporting System Chiller Plant Operation

Paul Allen, the energy manager at Walt Disney World (WDW) in Orlando, Florida, has initiated a wide spread

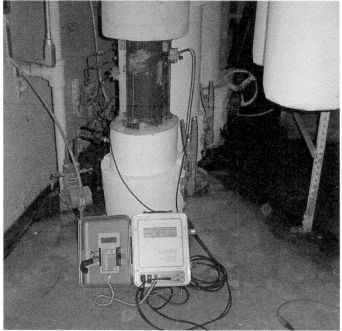

Figure 3-5. Portable Flow/Btu meters for chilled water

program for saving energy at WDW by using a web based energy information and control system. As part of the overall energy management system process improvement activity he performs a BTU (building tune-up) activity—essentially the same as retrocommissioning. See Section VII. One Btu project he conducted looked at the HVAC system in the Coronado Springs Exhibit Hall and involved a systematic review of the HVAC time and schedule setpoints and the repair of defective energy management system controls. The use of information from the web based system provided the ability to substantially improve the operation of the HVAC system for that facility.

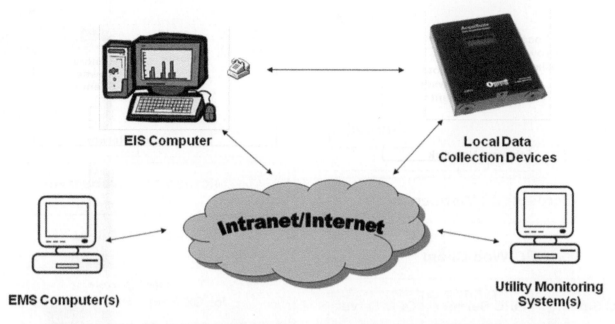

Figure 3-6. Other data sources provide a wealth of data for collection from the internet or intranet

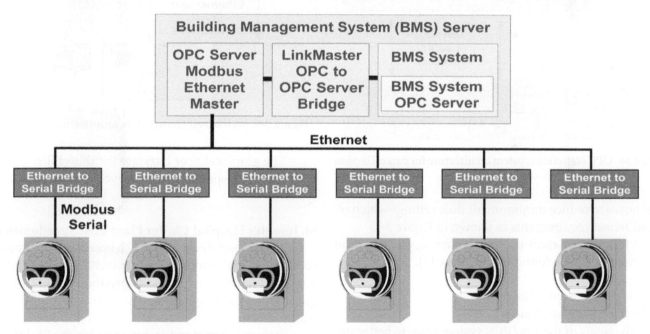

Figure 3-7. Metering data from a Building Management System server made available to a Local Area Network using an OPC server

The graphical user interface for the Walt Disney World chiller plant operation is shown in Figure 3-11.

From this graphical user interface for the EIS/ECS system, data, information, and analysis of the HVAC system resulted in the following optimization measures:

- Changed single zone HVAC systems to fixed supply temperature, VAV.
- Raised CHWS from 40°F to 44°F
- Lowered minimum chiller flow from 600 gpm to 400 gpm.

- Turned CHW plant off from midnight to 6 a.m.
- Lowered Minimum HVAC speed from 75% to 20%.
- Turned CHW plant off when OSA<60°F, used air handler economizers

The results were a 34% energy reduction was achieved.

A second project on improving the hot water system performance also produced some significant savings for HVAC system reheat. Data and information from the web based system resulted in actions being taken to re-calibrate

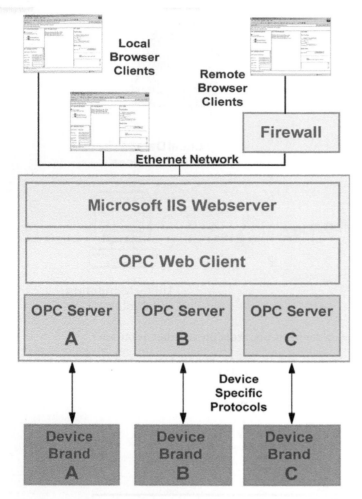

Figure 3-8. OPC web client system architecture for data collection

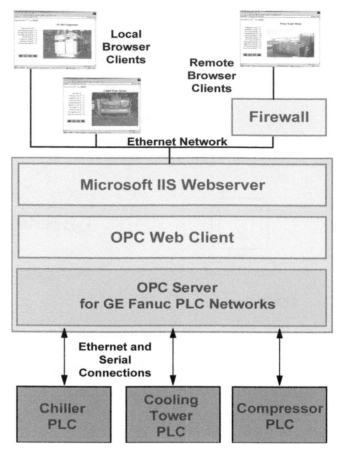

Figure 3-9. Web based facility data monitoring application

VAV boxes to reduce minimum air flow settings which reduced reheat requirements as shown in Figure 3-13.

More information on this system application and performance can be found in Reference [5].

Example 3—St. Joseph's Hospital
Chiller Plant Cost Speedometer

The goals of the EIS/ECS system were to better understand and measure chiller plant performance/costs; optimize operations, verify or refute need for proposed construction.

- St. Joseph's Hospital Environment
    5,700 tons chilling capacity
    1 gas chiller (800 tons)
    3 electric chillers (1,500 tons each)
    1 absorber (400 tons)
    4 secondary pumps (75 hp each)
    2 million gsf
    On/off peak electric rates

The graphical user interface for this chiller plant at St. Joseph's Hospital shows the overall schematic of the system.

St. Joseph's Hospital Chiller Plant Cost Speedometer

This project developed a web based information and control system that took the data from this chiller plant and displayed in the form of a cost speedometer that showed quickly and easily what the cost of running this system was in real time.

Data collection and technology for the St. Joseph's chiller speedometer:

- 15-minute interval data collection
    800 points within plant
    18,000 total points across hospital
    164 calculations in speedometer

- Technology
    Johnson Controls, Metasys
    Johnson Controls, OPC server
    Matrikon, OPC client
    Microsoft, SQL server
    Interval data systems, energy witness

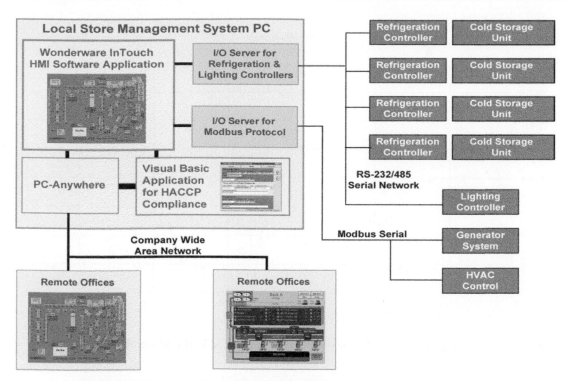

Figure 3-10. Supermarket facility web based information and control system

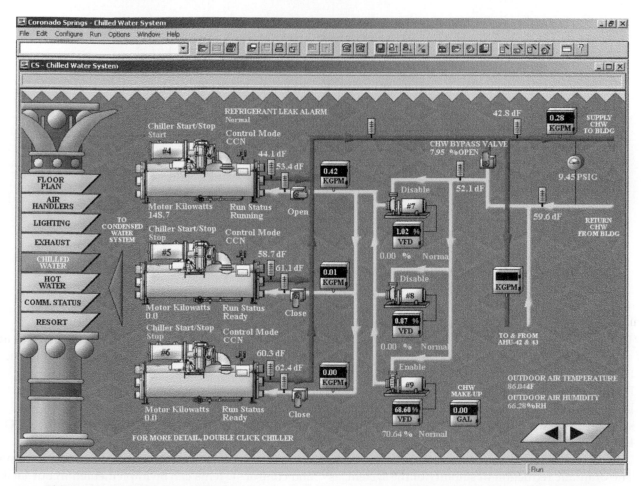

Figure 3-11. Graphical user interface for the chilled water plant for the Coronado Springs Exhibit Hall

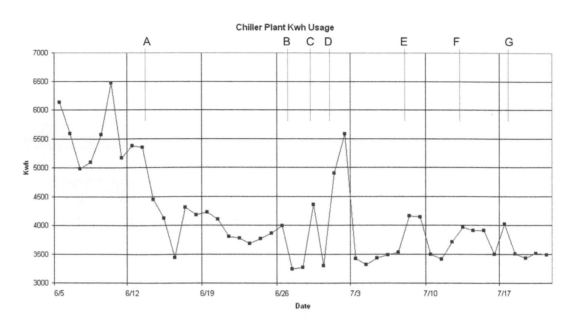

Figure 3-12. Chilled water plant kWh reduced in the Btu project

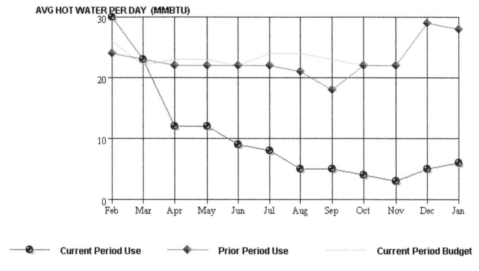

Figure 3-13. Hot water system cost reduction from reducing HVAC system reheat

The results of using the St. Joseph's Hospital chiller plant cost speedometer have been:

- Eliminated system instability
- Reduced cooling tower electric use by $40K/yr
- Identified $220K/yr in energy waste and comfort issues, resulting in construction project approval to fix
- Expanded system use throughout buildings

The dramatic reduction in stability of the system, as well as improved operation of the chillers and the cooling towers is shown in Figure 3-16. More information about this system and its operation can be found in Reference [6].

WEB BASED ENERGY INFORMATION AND CONTROL SYSTEM IMPLEMENTATION OPTIONS

In terms of how these web based energy information and control systems can be implemented in our facilities, here are three EIS implementation options: build it, out-

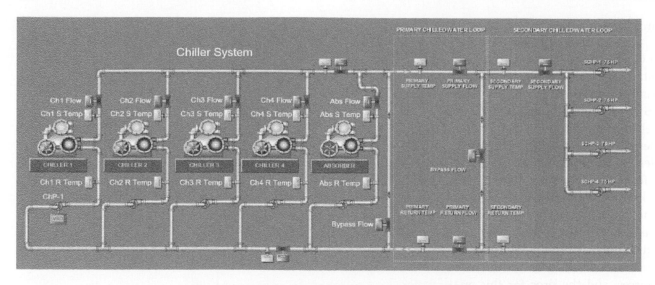

Figure 3-14. Chiller system at St. Joseph's Hospital

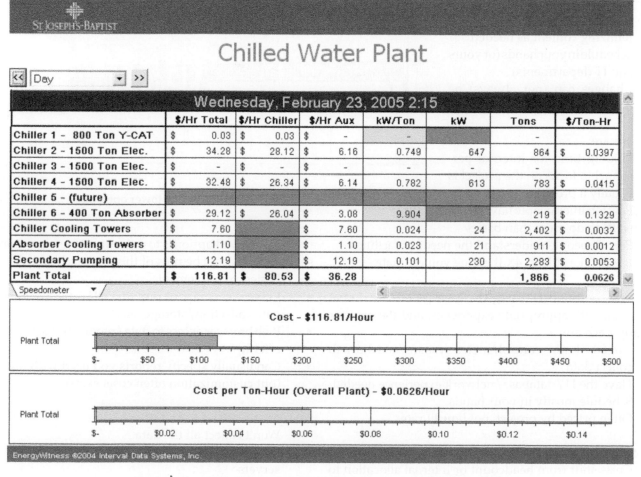

Chilled Water Plant

Wednesday, February 23, 2005 2:15	$/Hr Total	$/Hr Chiller	$/Hr Aux	kW/Ton	kW	Tons	$/Ton-Hr
Chiller 1 – 800 Ton Y-CAT	$ 0.03	$ 0.03	$ –	–		–	
Chiller 2 – 1500 Ton Elec.	$ 34.28	$ 28.12	$ 6.16	0.749	647	864	$ 0.0397
Chiller 3 – 1500 Ton Elec.	$ –	$ –	$ –	–	–	–	
Chiller 4 – 1500 Ton Elec.	$ 32.48	$ 26.34	$ 6.14	0.782	613	783	$ 0.0415
Chiller 5 – (future)							
Chiller 6 – 400 Ton Absorber	$ 29.12	$ 26.04	$ 3.08	9.904		219	$ 0.1329
Chiller Cooling Towers	$ 7.60		$ 7.60	0.024	24	2,402	$ 0.0032
Absorber Cooling Towers	$ 1.10		$ 1.10	0.023	21	911	$ 0.0012
Secondary Pumping	$ 12.19		$ 12.19	0.101	230	2,283	$ 0.0053
Plant Total	$ 116.81	$ 80.53	$ 36.28			1,866	$ 0.0626

Speedometer ▼

Cost – $116.81/Hour

Plant Total

$- $50 $100 $150 $200 $250 $300 $350 $400 $450 $500

Cost per Ton-Hour (Overall Plant) – $0.0626/Hour

Plant Total

$- $0.02 $0.04 $0.06 $0.08 $0.10 $0.12 $0.14

EnergyWitness ©2004 Interval Data Systems, Inc.

Figure 3-15. Chiller Plant Cost Speedometer for St. Joseph's Hospital

source it, or buy it.

The simplest EIS system is one that you can build, and once data are available on the network, the easiest EIS is to just transfer the data from the DAS to your own PC, and input it to Excel. Then you can graph them, make tables, slice them and dice them (pivot tables), and produce many more kinds of information depending on how well you know Excel. But, you miss most of the value of having a web based EIS and ECS system unless you put the data into a common database.

You can build IT with internal resources, and the pros for this approach are:

- Driven by you—your needs, priorities, business value gets addressed
- Control over capabilities & customizing content as needed
- Schedule in your hands (or yours and IT departments)
- Changes can (not always) be done quickly & cost effectively

The cons are:

- Finding time for internal personnel to focus on the project is usually a problem
- Few facilities people have the IT skills to do more than basics
- Few IT people understand the needs of facilities
- It will take at least 4X the time you estimate

You can build IT using the services of an external consultant with appropriate experience, and the pros for this approach are:

- Already know control/meter systems and how to collect data
- Have the IT/database/network experience needed
- Schedule mostly in your hands
- Often priced by project, not hourly rates

The cons are:

- Costs shift from headcount or internal allocation to IT department, to expenses
- Few facilities-savvy IT consultants exist
- They will need help understanding the data
- Will take 2X the estimated time

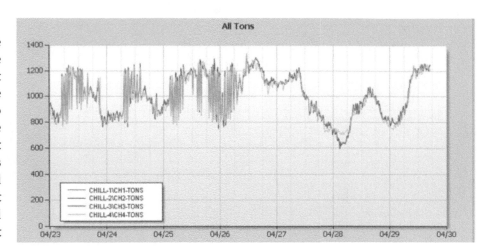

Figure 3-16a. Improved chiller system stability and operation

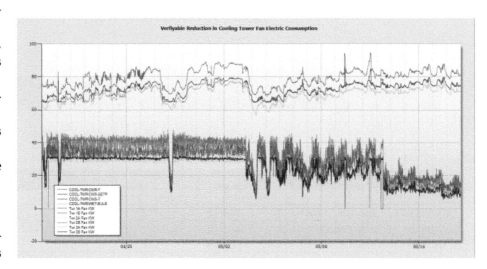

Figure 3-16b. Improved cooling tower operation

The second option is to outsource IT using a monthly service purchase. The pros of this approach are:

- Easiest to implement
- EIS service company provides all IT-related functions, data collection/storage, etc.
- Web browser access to data/reports (for services that offer it)
- Can usually specify reports you need with provider (but customization often costs extra)

The cons are:

- Won't collect all the data
- You lose ownership of data as it goes back to their servers
- Not all give access to data
- Outsources knowledge and expertise of facilities
- Monthly report model focuses on problems, not positives

You could also outsource it with system providers leasing you the system. The third option is to buy it, and purchase a commercial EIS package. The pros of this approach are:

- Choice of products that meet your needs & budget
- Collect data from and work with most control/meter systems
- Data kept/managed on site
- Capture/build knowledge throughout facilities org.
- High-end systems go way beyond engineering, operations, or even energy

The cons of this approach are:

- Often perceived as expensive, possibly requiring capital cost

- Require real work to get value/benefits/payback

- Configuration/customization takes time, effort, budget

- Some built by people who don't really understand facilities

- Customizations/changes in vendors hand, not yours

ENERGY AND FACILITIES MANAGEMENT IS A BUSINESS

Operating a Facility is a Business

Most facilities' business issues are driven by increasing demands such as increasing business pressures, more space, higher energy costs, limited headcount, higher labor rates, comfort complaints, under-funded maintenance, new initiatives (i.e. LEED), and changing customer needs. Upper management is demanding better information and better results. The facilities industry has focused its technology advancements on hardware efficiency and control technology, but still has not seriously embraced information technology (IT) or web based energy information and control systems. Facilities management is not just about how the HVAC system performs, but how the business performs. *The activity of energy and facilities management should be considered to be a business providing products and services, such as chilled water and thermal comfort. As a business, these products and services should be provided with cost of service data and information.* Facilities as an organization are delivering customer service, and decisions about the operation of the energy management and facilities management should be based on data, not just an opinion.

Management by fact (using data) is a management concept for preventing management by opinion. The analysis of relevant data allows informed decisions to be made and significantly reduces the risk of decisions made only on someone's opinion, and not based on existing facts. Facts are unknown until they are established through the collection of measurement data that shows verifiable results. A critical principle from the formal or informal use of the Deming Continuous Improvement (CI) Cycle or the Motorola Six Sigma Method stated earlier in this chapter is to—**make only data driven decisions**. Most energy and facility managers realize the need to make data driven decisions even if they have never heard of Deming or the Continuous Improvement Cycle, or Six Sigma. But Deming and others have formalized this process for us, and have helped convince us that this is what we should be doing in our buildings and facilities.

Smaller buildings and facilities are more likely to run without adequate data than bigger buildings and facilities. But, even very large facilities like universities and airports often run with only a few percent of the data that they need to do the job well. Data driven decision making provides a structure—a set of guidelines for knowing what decisions to make. Effective use of IT and web based EIS/ECS systems allows the collection of large amounts of data from sensors, meters, monitors and other operational systems. This results in the capacity to share data and information resulting in a more complete and useful understanding of building and facility operations, and therefore more correct and optimal decisions to operate buildings and facilities better.

The power of data and information from an energy and facility business information system about the operation of our buildings and facilities should allow energy and facility managers enhanced abilities to:

- Assess current and future needs of their building or facility and its operation
- Identify root causes of problems
- Decide what to change
- Determine if goals are being met
- Engage in continuous building and facility improvement

Energy and facilities business information systems that are required to meet these five operational features needs to be the same web based EIS/ECS that provide the basic data and information to operate the buildings or the facilities. It is more than just energy; and it is really about running the facilities business better.

From a business systems perspective the structure of the overall energy and facilities management system should look similar to Figure 3-17.

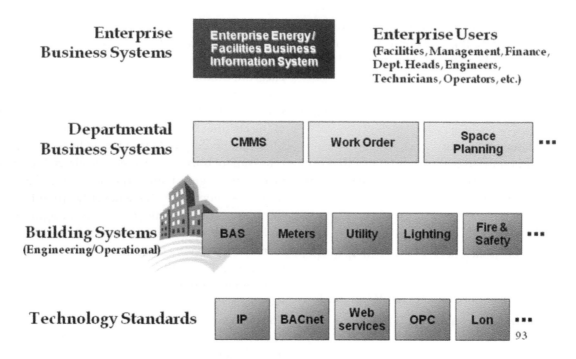

Figure 3-17. General structural considerations for an overall energy and facility information and control system

The Value of Highly Accessible Information from EIS/ECS and BAS Systems

Facilities business systems should provide highly accessible information that is highly interactive, well organized, actionable information, and delivered quickly. Figure 3-18 shows how this is accomplished for the chiller speedometer project for St. Joseph's Hospital discussed earlier. Basic data from the chiller plant is collected automatically and processed to provide information that can be used to make decisions about what to change to operate the chillers more efficiently and more effectively. The highly interactive graphical user interface with this information is especially designed to show the data and information quickly and easily to derive the maximum benefit for making any changes needed to improve the operation of the chiller plant.

The utilization of highly accessible information from advanced web based energy and facility information and control systems produces many benefits associated with energy efficiency and energy cost reduction, but they also provide a number of benefits related to simpler and more mundane activities that are everyday events going on at buildings and facilities. Some of these activities and events are shown in Figure 3-19, together with the estimates of time required to perform these activities and events with and without the highly accessible information from an advanced web based energy and facility information and control system.

CONCLUSION

Energy and facility management are extremely important tasks and opportunities for every building and facility. New technology in the form of web based energy information and control systems is available, cost effective and highly functional to accomplish these tasks and to provide additional opportunities to operate buildings and facilities in an optimum manner. Use of advanced web based energy information and control systems provides the opportunity to reduce building and facility operating costs by 10-15% or more without requiring new equipment. The purchase, installation and use of this new web based energy information and control technology developed over the last ten years can provide significant operating cost reductions and improved operational performance for 99% of all of our buildings and facilities. As the old cartoon character Pogo said many years ago, "We are surrounded by insurmountable opportunities."

Acknowledgments

Much of the information in this chapter has come from other chapters that have appeared in the previous three volumes on web based energy information and control systems. In addition, three particular people have contributed the most to this chapter—Mr. Paul Allen, Energy Manager at Walt Disney World in Orlando, Florida;

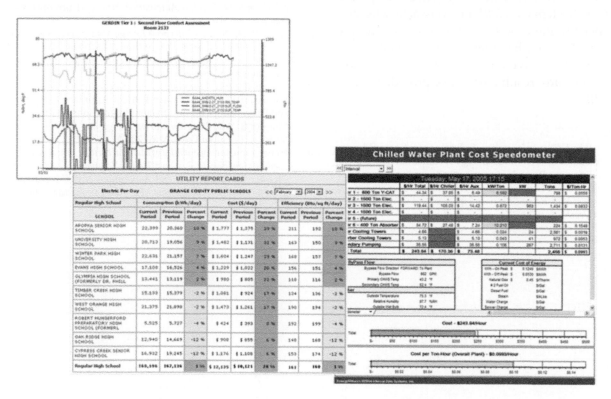

Figure 3-18. Example of highly interactive display of information for the Chiller Water system at St. Joseph's Hospital

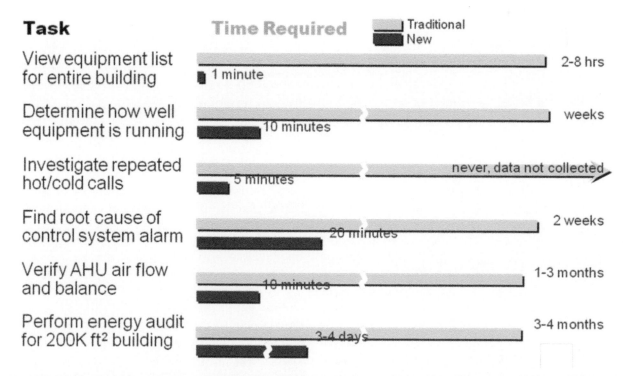

Figure 3-19. Productivity gains from highly accessible information. Traditional is without a web based EIS/ECS, and new is with the use of an advanced web based EIS/ECS.

Mr. Kevin Fuller, Vice President for System Development at Interval Data Systems in Waltham, Massachusetts; and Mr. John Weber, President, Software Toolbox, in Matthews, North Carolina. I owe each of them a debt of gratitude for all the help they have given me, and all the great information and case study results they have provided to me. Thank you all!

References

[1]   Capehart, B.L., Editor, Information Technology for Energy Managers: Understanding Web-Based Energy Information and Control Systems, (Fairmont Press, November, 2003).

[2]   Capehart, B.L., and Lynne C. Capehart, Editors, *Case Studies and Applications of Web-Based Energy Information and Control Systems,* (Fairmont Press, May, 2005).

[3]   Capehart, B.L., and Lynne C. Capehart, Editors, *Web Based Enterprise Energy and Building Automation Systems,* (Fairmont Press, May, 2007).

[4]   Allen, Paul, *Disney's Enterprise Energy Management Systems,* Chapter 28; *Using the Web for Energy Data Acquisition (with David Green and Jim Lewis),* Chapter 27, *Web Based Enterprise Energy and Building Automation Systems,* (Fairmont Press, May, 2007).

[5]   Gnerre, Bill and Kevin Fuller, *How Can a Building be Intelligent if it Has Nothing to Say?,* Chapter 4, An IT Approach to Optimization and Diagnosing Complex System Interactions, (with Greg Cmar) Chapter 8, *Web Based Enterprise Energy and Building Automation Systems,* (Fairmont Press, May, 2007).

[6]   Weber, John, *Interoperability of Manufacturing Control and Web Based Facility Management Systems: Trends, Technologies and Case Studies,* Chapter 20, *Case Studies and Applications of Web-Based Energy Information and Control Systems,* (Fairmont Press, May, 2005).

Section II

The Case for Web Based Energy Information and Control Systems

Chapter 4

Defining the Next Generation Enterprise Energy Management System

Bill Gnerre
Gregory Cmar

ABSTRACT

This chapter highlights the key functional requirements of an Enterprise Energy Management System (EEMS) and describes how this functionality can be used to reduce costs, increase efficiency, and improve energy planning and cost allocation, all while improving or maintaining building comfort. Appendices to this chapter contain case studies showing the value of using interval data from the building automation system (BAS) as a tool to diagnose and monitor facility operations.

INTRODUCTION

The challenge is simple: how do facility organizations find new and innovative ways to ensure maximum operational efficiency, reduce deferred maintenance budgets by extending the life of systems and equipment, be good stewards of the building assets, forecast energy needs more accurately, and achieve the lowest energy purchase?

The answer is an Enterprise Energy Management System.

Today, large campuses and facilities typically have one or more building automation and control systems, campus metering systems (automated and/or manual read), a lighting management system, and some form of space management system. In addition, they deal with several utility companies, each of which has changing rates and rate structures. These systems generate an enormous amount of valuable operational information—which is nearly all thrown away without even being looked at because it is difficult to capture and access data.

While historically organizations have attempted to control energy and building maintenance costs by managing each individual energy source and energy consumer (e.g. building automation system), without a comprehensive 360° view of the facility's current energy consumption true energy optimization cannot be achieved.

An EEMS provides actionable insight through the consolidation of data from all of the institution's disparate energy and building management systems and the interactive access to that data, providing the facility's operations and engineering departments with an accurate picture of operations. With facts in hand, they can steward their assets, lower total energy consumption and operational costs quickly and effectively, and have the ability to verify and measure results.

EEMS DEFINED

At the most simplified level, an Enterprise Energy Management System consolidates *all* energy related data (sources, costs, control and monitoring points) into a data warehouse and provides tools to access and truly *interact* with the data. Conceptually straightforward, but today's energy management systems just do not do it.

It is worth noting what an EEMS is not. It is not a control system and should not be confused with building automation systems (BAS). An EEMS is much broader in scope than control systems, reaching well beyond the BAS. It provides data collection, data access, diagnostic and monitoring capabilities, a historical data warehouse, and a lot more as detailed throughout this paper. Similarly, an EEMS should not be confused with utility billing systems. It encompasses billing and meter data, but extends far beyond and connects billing information directly to the related operational data.

The EEMS makes data available so that the end

user is able to perform in-depth diagnostics, analysis, and monitoring in a small fraction of the time it took with earlier methods. This, in turn, provides facilities' staffs with actionable information; i.e. information that enables them to make informed decisions to reduce energy consumption, accurately identify energy costs by cost center, or forecast energy costs in the future.

A true Enterprise Energy Management System is based upon five simple, but crucial, principles:

1. All energy related data must be consolidated into a centralized data warehouse.

2. The collected data must be 'normalized' and 'structured' to be usable.

3. Access to data must be 'interactive' and the information presented must be 'actionable.'

4. The system must measure and verify results.

5. The system must provide a platform that embraces industry standards for data collection, management, analysis, and publication.

EEMS DESIGN PRINCIPLE 1: CONSOLIDATE ALL ENERGY RELATED DATA INTO A DATA WAREHOUSE

Energy data come from purchased utilities, generated utilities, building automation systems (BAS), metering systems (both advanced and manually read), weather, and space planning systems. (There are also calculated data, but that will be covered later.) Additionally the EEMS must manage rate and billing data, users, and organizational information.

In order to be able to utilize energy data, the first step is to identify and collect the right data into a data warehouse so that accurate and actionable information can be available. The EEMS needs to collect *all* data, as one cannot optimize the whole by optimizing each component.

This section identifies the different data sources and attributes that define an EEMS and populate its data warehouse.

Purchased Utilities

For each utility within a campus or hospital, consumption data and billing information is generated.

Consumption information is typically time-based, regardless of the type of utility. For example, electric

utilities use 15-minute intervals, natural gas utilities use a daily time interval, oil uses time between tank fill-ups, and water uses monthly or quarterly intervals. Other utilities such as steam or chilled water often use 15-minute to hourly interval data. Eventually these different time series need to be "normalized" (an issue that is discussed later in this white paper) so that information can be presented in consistent intervals.

Billing information is equally complex: for large campuses, there are often multiple vendors for each utility type, each with differing rates, billing cycles, and pricing structures.

An EEMS must manage both consumption and billing information, and present this data in an intelligible, clear, and actionable format. The diagram below is a simplified view of the issues related to collecting utility data.

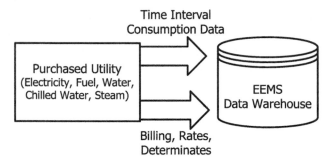

Figure 4-1. Billing & consumption data must be collected in tandem.

Time Interval Consumption Data

Many utility companies do not offer a way to track interval consumption data as it happens. The interval data are made available at the end of the billing cycle through reports or spreadsheets, which leaves the EEMS without current consumption data for a month at a time (with water being much worse). Other companies provide interval data through Web-based reports, which although more current, are far from ideal for populating the EEMS data warehouse.

The best option for collecting consumption interval data is for the meter to provide the data directly at regular intervals, or to attach a reading device that can provide the consumption data to the EEMS.

Bills, Rates, Determinants

The EEMS must understand and track the hierarchy of meter data that comes from a purchased utility. The data hierarchy goes from utility type, to supplier, to account, to meter and rate. A single rate is typically used for multiple meters spanning multiple accounts.

A large facility will have a significant amount of billing data to collect and manage. For example a university may have 1,000-3,000 utility bills per year. Today this is often captured in spreadsheets—limiting the accessibility and usability of that information. The EEMS should consolidate all the billing information, including the bill's underlying determinants.

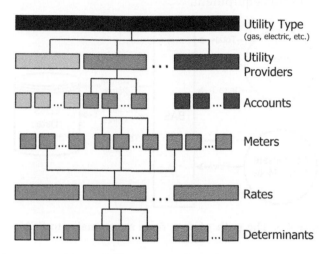

Figure 4-2. The EEMS must understand the hierarchy of utility meter data—from utility type, down to meter, and then to rates and determinants.

Billing Determinants				
Determinant	Data Type	Rate 1	Rate 2	Rate 3
Start Date	Date	■	■	■
End Date	Date	■	■	■
Total kWh	Integer	■	■	■
Billed kWh	Integer	■		
Total kW	Number	■	■	■
Billed kW	Number	■		
Rate Billing	Money	■	■	■
Customer Charge	Money		■	
Fuel Charge	Money	■	■	■
Sales Tax	Money	■	■	■
Municipal Franchise Adj.	Money	■	■	
Total Current Bill	Money	■	■	■

Figure 4-3. The EEMS must have the flexibility to handle multiple rates with varying determinants.

The bill is made up of a varying number of determinants. It is common for a large campus to have a dozen or more different rates from a single utility company. As an example, Figure 4-3 shows the billing determinants for three different rates in use at a university from a single electric company.

Electricity

Today most electric utilities quantify consumption by averaging the demand over a 15-minute period (standard interval). The majority of electric utilities make the interval data available electronically to the customers, although again, not always in convenient ways to collect it.

Both 15-minute average demand and month-to-date consumption are required for the EEMS. Similarly, electric bills with determinants must be stored in the database too (for reasons discussed later) and, because billing rates change over time, it is important that the EEMS can accommodate this dynamic data and propagate these adjustments.

Natural Gas

Due to the fact that natural gas utility companies rarely bill based upon readily obtainable standard time intervals (a fact which has led many institutions to install their own gas meters to validate billing), it is important that the natural gas meters installed throughout the site are connected to the automated metering or building automation system for data collection. It is also critical that the meter configuration and BAS point configuration collect running totals of consumption flow, etc. as well as instantaneous readings. Running total data is required to make it possible to reconstruct the inevitable gaps and missed readings.

Chilled Water

While many organizations generate their own chilled water for air conditioning, etc., some chilled water is purchased from third-party utility companies. Similar problems exist concerning metering, again leading some organizations to purchase their own meters to validate bills. Because chilled water generation is tied to electric consumption, suppliers are increasingly moving towards more accurate time-series billing.

Steam

Like chilled water, steam is often produced by an organization itself, but, when purchased, is typically billed based upon time intervals ranging from 15 minutes to one hour.

Generated Utilities

Many large campuses/facilities have their own power plants that generate chilled water or steam that is subsequently distributed to the buildings. The EEMS must collect data from the control system(s) of the power plant as well as from the distribution network and end users of the energy.

Billing information must also be collected, much the same as it is for purchased utilities. Facilities that generate their own chilled water, steam, or even electricity will have their own rate structure and determinants and bill internally based on consumption.

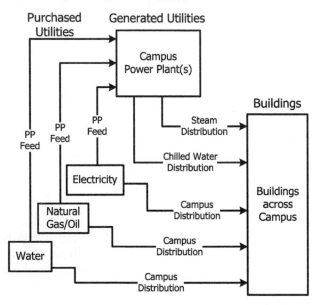

Figure 4-4. Institutions typically deliver purchased and generated utilities to the campus.

Chilled Water

Chilled water is distributed throughout the campus to different buildings. Typically, a building automation system will control the chiller plant (chillers, cooling towers, primary distribution loops, etc.).

The EEMS system must collect data from both the generating plant and the distribution network (buildings). In some cases the chiller plant will be run by a different BAS than the building and an EEMS must be able to display chiller plant efficiency as well as allow the user to determine how the chilled water is being used in the distribution system. Operating the chiller plant at maximum efficiency does not necessarily mean that the distribution system can benefit from a high delta T chiller operation, for example.

Steam

Like chilled water, steam is distributed throughout the site to different buildings. In order to determine

the steam usage of a building, both a steam flow meter and condensate return meter are required: where the steam meter measures the energy delivered to the building while the condensate meter measures that which is returned. An EEMS then accesses the steam and condensate return meters via the building automation system that is normally used to monitor and control this equipment.

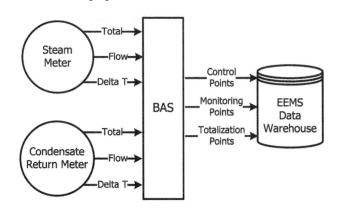

Figure 4-5. The path of data from steam meter to EEMS.

Building Automation Systems

First, an important point of distinction between a Building Automation System (BAS) and an Enterprise Energy Management System (EEMS) should be noted—BAS controls the HVAC equipment, lighting, security systems, etc. whereas the EEMS provides management information derived from the BAS, as well as all other energy systems across the site. An EEMS is not a control system.

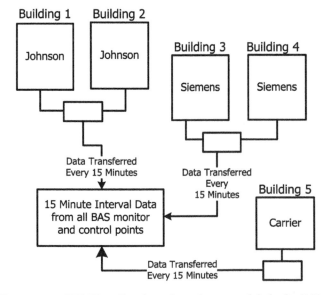

Figure 4-6. EEMS collecting data from multiple building automation systems.

Many institutions operate more than one BAS and so an EEMS provides a complete, holistic view by collecting data from all systems and overcoming the limitations of relying on the BAS for data. It is this comprehensive 360° view that enables organizations to more quickly and effectively diagnose energy usage.

As important as data collection is, it is equally important for the EEMS to provide fast access to the data, and to present a holistic view in a comprehensible form. This is covered in greater detail in "EEMS Design Principle 3."

Extracting Data from the BAS

Virtually all BASs (particularly modern systems) allow external applications to collect data without negatively impacting performance. To provide a comprehensive picture, an EEMS requires data at 15-minute intervals from control and monitoring points.

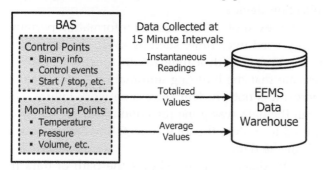

Figure 4-7. The EEMS collects various kinds of data from control and monitoring points every 15 minutes.

It is important to recognize that gathering data only from monitoring points means that, while you may identify something to improve, it is unlikely that you will gather sufficient information to know with certainty how to improve it. (This is what happens when advanced metering systems are installed instead of an EEMS.) Control points must also be gathered to be able to monitor any changes and understand how control changes affect behavior throughout the systems.

For operations at a fairly large facility or campus, it is not unusual to have 20,000 points or more. With such a large number of data collection points, each generating one record every 15 minutes for 365 days a year, the total data set comprises over 700 million records per year. From a user's data-handling perspective, 700 million records are overwhelming, and without an EEMS, much, if not all, of this information was thrown away.

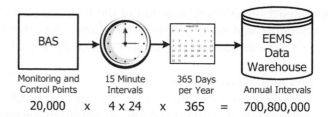

Figure 4-8. The ability of an EEMS to store, manipulate, and display very large volumes of data in an efficient manner is mission critical.

The Importance of Trend Data

Trend data are the foundation of diagnostics, monitoring, measurement and verification, and building a historical record of facility operations. While BASs typically contain a function called trend logs, they are a poor mechanism to collect trend data for the EEMS to operate against.

Trend logs provide an excellent medium for viewing real-time BAS data, particularly where a known problem exists. The operator simply identifies the points required for the trend log and initiates data collection. However, there are several issues that severely limit their ability to serve as a useful data collection device or as a diagnostic or monitoring tool:

- All data are not collected. At any point in time, trend logs are typically only active for a few hundred points out of the thousands within the BAS.

- Trend logs cannot collect all data. They were not designed to be active for all data points, and attempting to do so severely impacts the performance of the BAS server, affecting its ability to properly execute control functions.

- Data are limited to the BAS. Trend logs have no ability to combine data from multiple BASs or other energy data sources critical to facilities operations.

- There is no meaningful historical data. With no constant data collection, there is no ability to look back in time at points of interest. Logs are turned on and then facilities staff must wait until enough data are collected. Earlier data are lost forever.

Fortunately there are approaches to collecting trend data without relying on trend logs. The EEMS needs to access the BAS monitoring and control points directly or through an independent server that does not impact the primary function of the BAS—to control the building and energy system operations. How this is done will be discussed later, under "Design Principle 5."

Meters and Metering Systems

Meters and metering systems are typically located at buildings and throughout the distribution system to measure usage of electricity, steam, chilled water, fuel, etc. Different energy types will require different approaches to move the data from the meter into the EEMS warehouse.

Most electric meter manufacturers have software applications that consolidate the data and display consumption demand and power quality data. In effect, metering systems are designed to simply collect and display data needed to understand usage. The EEMS requires this data to also be present, necessitating a connection between the metering systems' databases and the EEMS data warehouse.

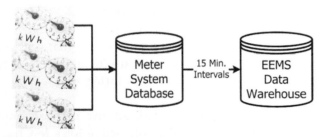

Figure 4-9. A connection between the metering databases and EEMS must exist.

For electricity, access to metering data via the metering database is a relatively straightforward process, but for those utilities that do not leverage advanced metering systems, data can be obtained and transferred through the building automation system; i.e., the BAS collects the data from the meter and the EEMS acquires the data from the BAS system. In reality, however, this connectivity is not a simple task. Meters must be reconfigured to collect and total the data, and the BAS points must be configured to access and deliver the required data to the EEMS.

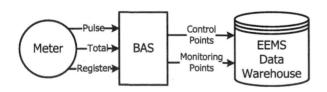

Figure 4-10. Data can be obtained & transferred through a building automation system.

In addition to the automated data collection described above, there is also the information from manually read meters that must be added to the EEMS warehouse in order to provide a complete picture for error diagnosis and identification. Many organizations already do this at some level, but the data are typically put into spreadsheets where there is limited access and no connection to other operational data. For manually read meters, like other meters, a configuration is required; i.e., the data being collected by the meter reader must be the right data for the EEMS.

Considerations for Meter Data Collection

Whenever meters are used for EEMS data collection there are a number of important considerations to make certain the right data are collected, even in the event of network failure. A decision must be made concerning the type of data needed. This can then be used as the basis for selecting the most appropriate collection device.

For example, if the desired information pertains to energy used over time, such as kWh, the device must be configured with a totalization function. While it is possible that the building automation system can total energy, it is not prudent since a network failure would mean the data was gone for good. With the meter configured to contain the total energy, one can go back and "fill in the data."

The diagram above shows the path of data from the device to the EEMS database. It is important to recognize that there are a number of potential points of failure—at each hardware device and each network link. One must plan for a loss of data and ensure that the data collection process incorporates the appropriate safeguards to minimize data loss.

As the diagram shows, data can be collected at

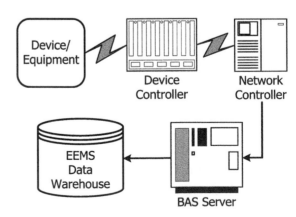

Figure 4-11. The communication path of data from device to EEMS.

the device, device controller, network controller (sometimes referred to as the panel board), or the BAS server (via OPC). In selecting and specifying data collection points it is crucial to consider the goals and objectives for the project. For example, if you need to know the total amount of energy consumed you must configure the meter for register data *and* pulse data.

One other consideration when collecting meter data is deduct meters. This occurs when there is a main meter and then a series of sub-meters that cover part, but not all, of the energy consumption through the main meter. In this case the EEMS will want to collect interval data directly from the sub-meters, but then must deduct their consumption from the main meter before apportioning energy usage to other spaces.

Weather Data

To enable accurate energy forecasting, it is important to understand the context within which energy usage occurs. Weather is possibly the single biggest factor in this equation. For the most accurate results, weather data from a local airport (official METAR—Meteorological Terminal Aviation Routine Weather Report) should be used in the EEMS. While BASs typically have temperature and humidity sensors, invariably they do not provide quality reference information necessary for a number of reasons: temperature sensors may be located too close to the building, in the sun, on a roof, on the south side of the building; humidity sensors may be broken, etc.

An EEMS therefore should be configured to receive a weather feed from a local airport so that actual weather data can be used in energy forecasting applications and in assessing building performance characteristics. At a minimum, weather feeds should include the following data:

* Hourly temperature and dew point
* Wind speed and direction
* Barometric pressure
* Sky conditions (clear to cloudy)

Space Planning Data

Space planning data are required by the EEMS in order to identify energy costs at the space level. Space planning systems (SPS) contain information concerning the use and allocation of all areas within a campus or facility. They map the hierarchy of the campus by site, zone, building, floor and room.

Space planning systems also understand the relationships between space and cost centers. Both SPS and

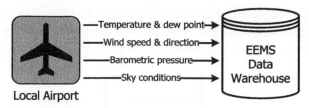

Figure 4-12. Weather data from a local airport should be used within an EEMS.

EEMS have distinct, complementary roles. Space Planning's role is to maintain the space relationships (since occupancy and cost centers change) and to transfer cost center information into the general ledger. The role of the EEMS system is to deliver accurate energy costs down to the space level where the SPS can roll up the costs by cost center.

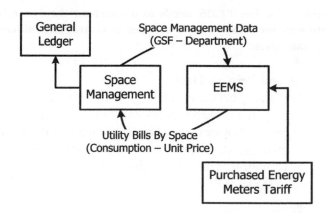

Figure 4-13. The EEMS and SPS work to combine energy and space data and write costs to the General Ledger.

Maintaining space planning data is an important discipline so that campus utility organizations, facility operations, planning and engineering functions can access common reference information. Today each of these distinct groups creates their own variation of a standard building name (for example) which makes their information useable only to themselves rather than many other departments.

Organizational Information

Organizational information is required for the EEMS so that it can roll up cost center information. Once the EEMS allocates costs to each space, organizational information is required that can relate space to department (or cost center). It is incumbent upon the EEMS to adapt to any hierarchical structure and to the constantly changing organizational structure of the institution. The EEMS system should not burden the space management system nor the personnel maintaining space planning data with this task. In essence, the EEMS functions increase the value of the investment already made in existing space planning systems.

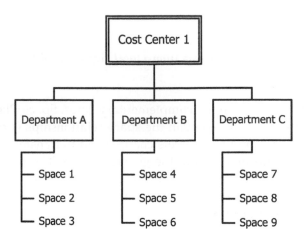

Figure 4-14. The EEMS needs to understand the dynamic nature of the relationships between organizational hierarchy and space.

Market and Pricing Data

For any purchaser of large amounts of energy, purchasing at favorable rates can make a significant impact on the bottom line. This is difficult to do since prices, pricing structures, and regulations are constantly changing. The role of the EEMS in purchasing energy is threefold:

• To display past and future energy usage patterns
• To convert and present utility billing and consumption usage into an equivalent real-time price
• To present real-time pricing information

All of this needs to be accomplished within the same interface.

Easy access to actual and predicted usage patterns enables the organization to make more informed utility purchasing decisions than ever before. For organizations that have secured the services of a third party to assist in making utility purchasing decisions, access to the EEMS must be made available to those individuals.

Don't Throw Data Away

Most large facilities and campuses spend millions of dollars annually on energy, millions on systems like building automation systems, and tens of millions for HVAC equipment. Yet the data that they generate is thrown away largely because there is so much of it, and it is difficult to access, manipulate, and interpret.

An EEMS captures this data and leverages it to provide a complete picture of utility consumption and an organization's energy infrastructure over time so that the most expeditious analysis can take place to reduce the total cost of ownership and operation.

EEMS DESIGN PRINCIPLE 2: NORMALIZE AND STRUCTURE DATA

EEMS Design Principle 1 focused on the importance and requirements of getting data into the system. EEMS Design Principle 2 constructs the data warehouse of historical trend information based upon the least common data denominator, a standard time interval. Once all data are normalized to this common standard, one can utilize it for a variety of applications.

Optimal Time Interval

There are a number of considerations in selecting a standard time series interval used by the EEMS, the most significant of which is ensuring that it is able to display sufficient data to identify transitions—this means that there should be enough data points gathered to discern performance fluctuations across transition time periods such as between day and night, 'office hours' and non-working hours, etc.

With this information, behavior patterns and problems become apparent quickly. For this reason, and because electricity is frequently metered within the same time interval, 15 minutes is an appropriate time series upon which an EEMS can be normalized. Longer intervals do not provide sufficient data granularity to always see behavioral changes. Shorter intervals can increase the data storage and processing requirements for the EEMS by 300% (or more) while increasing the information value very little.

Additionally, because capturing data at 15-minute intervals does not require a great deal of storage space to house the data records, data can be stored for the lifetime of a piece of equipment—20 years or more.

Normalized Time Series Data

Normalizing the data means that data from all sources are stored in the warehouse in the same time series interval. As discussed earlier, not all data sources provide data in the same time series, so an EEMS system must reconcile these differences. In some cases the actual time interval of the data source is one per day so the EEMS system must automatically convert this data into the normalized times series data. This problem isn't always due to the time series specified by the utility provider; it may occur for a number of reasons, including equipment failure.

The problem of time series inconsistencies is a fact that must be acknowledged and addressed. A well-implemented EEMS accepts these inconsistencies and "fills the gaps" with estimates that, when totaled, account for the total energy consumed and represent the pattern of that energy usage in a precise and accurate manner.

Data Calculations

Part of the function of any data warehouse is to provide access to the data in the most efficient means possible. This includes calculating and storing certain commonly needed values.

Once the data have been normalized, you can calculate additional trend data at each interval. For example, a Delta T calculation, simply the difference between supply and return temperatures, can be computed and stored as another monitoring point at each interval, making it available to the user as a Trend Line.

Storing calculated values is important to the overall data warehouse structure because it dramatically speeds up access by the user. The small amount of storage space used by the calculated data is more than made up for by the performance gains. (See more on calculations in a later section.)

Naming Conventions

There is a complete lack of uniformity in how buildings and systems are labeled within an institution. Today's facility organizations use building automation systems, store utility bills and meter readings in Excel or some utility system, generate their own utility bills for steam and chilled water, create campus maps and engineering drawings, maintain space planning systems, and work with outside engineering and construction firms. Each group has their own systems for specific tasks, each with a different nuance to the same information.

This is unlikely to be prevented or be brought under control—hence the EEMS should be able to present to each specific user group the naming convention they are familiar with, while providing the cross reference information required.

Data Warehouse Structure and Hierarchy

It is clear that an EEMS handles and stores an enormous amount of data from many disparate systems. In order to derive value from such a large data set, a defined structure and hierarchy must be implemented to make the data readily consumable. The structure of the data must be flexible too, since physical configurations are constantly changing. Buildings may be added, equipment may fail unexpectedly, and space may be modified to accommodate organizational changes. The design of the EEMS system must be flexible to adapt and keep pace with this dynamic environment.

Warehouse Objects

Structuring of data should mirror the manner in which those data are to be used to gain insight. For example, it should parallel the physical facility so that data can be viewed either in aggregate or in isolation when focusing on an individual building or piece of equipment.

The EEMS warehouse needs to support logical objects—sets of information and relationship hierarchies—that allow for this structure. The list on the following page shows six warehouse objects and the hierarchy that needs to exist within them. The warehouse structure also must support the way different elements within the object interrelate, i.e., the way meter interval data must connect to billing rate data and physical building data.

EEMS DESIGN PRINCIPLE 3: PROVIDE INTERACTIVE ACCESS TO ACTIONABLE INFORMATION

Defining Interactive and Actionable

Interactive access allows the users to work with the data in a dynamic fashion, moving seamlessly through the data with tools that provide near instantaneous response. This allows users to "work the way they think" rather than being limited to a series of static queries and reports. The EEMS must address both usability and performance issues to successfully provide interactive access to the data.

Actionable information can be used as the basis and rationale for effective decision-making, as opposed to merely indicating "status." To create actionable data the EEMS must collect *both* monitoring and control points. With monitoring data alone (without control information) it is difficult to verify or quantify the savings opportunity. Monitoring data and control data should be viewed in tandem to become "qualified" as actionable information.

Time Matters

It is crucial when collecting tens of millions of records per month that insight can be gained within minutes rather than days. The use and operation of an EEMS cannot be an arduous and time-consuming task.

Facility Objects
• Site
•   Zones
•   Buildings
•   Floors
•   Rooms
• HEGIS group
• HEGIS classification

Issue Objects
• Issues

Organizational Objects
• Departments

User Objects
• User privileges
➢ Users

Interval Data Collection Objects
• BAS & OPC point data
• Calculations
➢ Balance efficiency calculation
➢ Cooling efficiency
➢ Cost calculations
➢ Delta T
➢ Theoretical water loss
➢ Etc.
• Weather data

Meter Objects
• Utility type
•   Utility information
•   Account information
•   Meter information
•   Rate information
• Billing information

Users must be able to derive insight about the data (useful, actionable information) in a very short period of time if the system is to become embedded within the facility's management operations.

To accomplish this, data must be accessed and presented to the user at the *speed of thought*—able to view hundreds of thousands of data intervals in 15 or 20 minutes. In essence, if the user has to wait for information, their thoughts will wander. If they continually have to wait—if they are not constantly engaged—then their reaction will be that the EEMS is wasting their time. Once that point is reached, the EEMS will not be used and any potential value will be lost.

Currently, most facility personnel waste an enormous amount of time collecting and distributing data. Due to the historical lack of availability, this wasted time is culturally accepted as "part of the job," when in fact, instead of spending days manually gathering and piecing together data, a well-implemented EEMS can deliver it in seconds. When you have immediate access to the data, staff is freed up to devote time to the real engineering work of diagnostics, analysis, and planning.

Usability Matters

The amount of time required to gain insight is directly related to the usability of the system. With such vast amounts of data stored in the EEMS warehouse, it is essential that data can be assembled dynamically by the user via a simple, intuitive interface. The elements of the interface include the data organization and presentation, data display, and the program interface itself. The EEMS also needs to support other aspects of the facility staff's workflow, such as tracking identified issues or interacting with external analysis tools.

Data Organization and Presentation

As discussed in "Design Principle 2," the data must be structured in a manner that is inherently useful. This needs to happen not only at the data structure level, but also at the user presentation level.

With such large volumes of data, the EEMS must allow them to be categorized in meaningful ways, such as organizing within the facility hierarchy (zones, buildings, floors, rooms), organization (departments, rooms), or systems (chiller system, air handling system, etc.). In many cases data must be accessible through multiple views so that, for example, a building manager has access to the building information while a plant engineer can look at chiller operations facility-wide.

The EEMS must make the presentation simple by providing a master organizational structure that offers a mechanism for users to select which data they see.

Users must be able to "drive through the campus" from their desktop viewing thousands of data intervals per minute. The EEMS must also provide a method for users to define their own organized views.

Trend Lines

An earlier section discussed the importance of collecting trend data as the basis of diagnostics, monitoring, M&V, and more. The EEMS must use the trend data to provide *trend lines* as the primary mechanism to display and interact with the EEMS data. These are very different from BAS trend logs—a difference that elevates, by orders of magnitude, the effectiveness of an EEMS for diagnostics, monitoring and other applications.

Trend lines provide insight into operational data over extended time periods and permit the expeditious identification of problems—problems spanning multiple BASs and inefficiencies that arise suddenly due to changing circumstances such as weather, equipment degradation over time, or system configuration adjustments. In contrast to Trend Logs, Trend Lines:

- Contain all the data from every monitoring and control point
- Do not impact BAS control performance at all
- Combine data from multiple BASs and other data sources such as meters, utilities, and weather data
- Capture all data from the moment the EEMS is turned on, so it is constantly available
- Create a historical record of building operations—both monitoring points and control settings

BAS trend logs do have their place in providing needed information—where their ability to collect real-time data is useful—as shown in Figure 4-15.

Calculated Data

Calculations are often done ahead of time and stored in the data warehouse. The efficiency gains in having co-mmonly desired calculations available for monitoring and diagnostics are tremendous-having a dramatic impact on usability. A visual display of an ongoing trend built on a calculation can provide insight instantly that would otherwise take hours of number crunching and charting in Excel.

A Delta T is a simple calculation example commonly viewed as a basic performance measure of a chiller system. A Delta T is even more effective when the EEMS displays the supply and return temperatures used in the calculation at the same time. This way, if the Delta T fluctuates, the user can immediately see if the change was affected by a supply or return temperature rise or drop.

More complicated calculations can provide users with an overall operational efficiency rating, or, applying billing rate data, even a Trend Line that shows at every 15-minute interval the total energy cost for that 15-minute period. Calculations of this complexity rely on the EEMS's ability to fully integrate data from all sources and present it in a normalized fashion.

The EEMS should support calculations of any complexity, although it is appropriate to restrict the creation of calculated points to a system administrator who understands how calculations fit into the underlying data warehouse.

A sample of desired calculations—calculated every 15 minutes—include:

- Balance efficiency
- Chiller efficiency
- Chiller total cost
- Chiller plant total hourly cost of operations
- Cooling tower cost & efficiency
- Cooling tower make-up water cost & efficiency
- Delta T (primary & secondary)
- Pump brake horsepower
- Pump efficiency
- Pump kW
- Theoretical water loss
- Tons output
- Tower total cost

Application Interface

All of the usability factors mentioned must come together in the software user interface (UI). It is how end users interact with the data-through mouse clicks, menu selections, etc.

The EEMS should use interface conventions already familiar to its users, such as expandable data trees to display available points and hierarchies, tabs for organization, contextual menus, drag and drop, etc.

Users should be able to take an iterative approach with each action building on the last for diagnostic purposes. Monitoring should be fast and efficient, allowing the user to quickly cycle through hundreds of Trend Lines—hundreds of thousands of intervals—in minutes. Trend Lines are the source of actionable information, but it is the UI and system performance of the EEMS that allow the user to work at the speed of thought.

Tracking Issues

As an expert reviews the data within the Trend Lines, they are able to identify areas of concern or high-

Application of BAS Trend Logs and EEMS Trend Lines		
Criteria	BAS Trend Logs	EEMS Trend Lines
When to Use		
Diagnosing operations	After problem has been identified as under the control of the BAS and further data is needed for final diagnosis	Always—far superior tool for nearly al diagnostics and all cases where historical data or data outside the BAS must be considered
Monitoring operations	When real-time data for a small number of BAS points needs to be watched	Always—provides the ability to monitor hundreds of trends in minutes, combining data from any and all sources
Typically used by	BAS control engineers and technicians	BAS control engineers & technicians, energy engineers, facility managers, performance monitoring contractors, commissioning agents, HVAC design engineers
Technology Perspective		
Data storage	Stores point data for trends defined	Stores data for all points from all systems
Time interval of data	Captures data in increments from milliseconds to minutes	Captures data from all systems every 15 minutes
Displays data from	Native BAS	Multiple BASs, metering systems, utilities, weather, billing
Display time period	Typically a few days or weeks	Between a day and a year, with historical data going back years
Data storage	Up to a few months and data is discarded	Up to 20 years

Figure 4-15. Some of the uses, users, and differences between BAS Trend Logs and EEMS Trend Lines.

light known problems. For each problem discovered, the EEMS should take a snapshot of the Trend Line(s) at that point in time, which can be annotated by the expert and sent to the appropriate building control technician. That technician is then able to take action and modify settings within the BAS as appropriate. The technician can then log the fix while the expert can verify that the change is working appropriately before closing out the issue.

Issues must be accessible by a variety of personnel that need the information to put together the action plan and ultimately resolve the issue. It is appropriate, however, to have controls that limit some users' scope of what they can see or modify.

Interact with External Analysis Tools

As good of a diagnostics tool as an EEMS is, there are many types of data analysis that are best performed by other tools created for that purpose. For example, Excel can do curve fitting, regression analysis, and many other calculations that would be wasteful to duplicate within the EEMS.

To make using external applications an easy process, the data from the EEMS should be easily exported into Excel or other analysis software. This allows engineers who have built their own analysis routines in Excel (or other packages) to continue to use them.

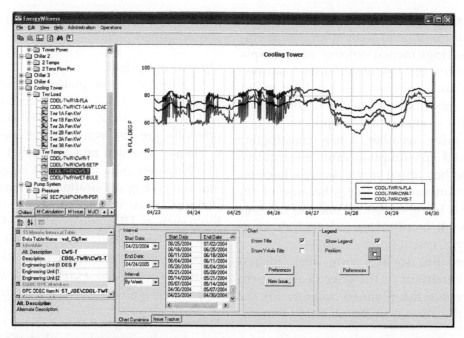

Figure 4-16. An EEMS should allow interaction with data, such as drag & drop capabilities to add more Trend Lines onto an existing chart, one-click scanning back through time, and zooming in or out.

Present User Specific Information

An EEMS must support a variety of applications and users. Data must be pertinent for each user while providing the ability to conduct additional investigation and access data not typically made available via other systems. Users of an EEMS may be BAS engineers and technicians, energy engineers, HVAC design engineers, facilities engineers and managers, commissioning agents, energy purchasers, and/or performance contractors.

Users of information from an EEMS extend well beyond that group to building managers, zone maintenance groups, department heads, finance personnel, and anyone who has occasional needs for some facilities information.

Information Publishing

In many cases, this second tier of users do not need interactive access to the data at the individual point level. Their interests are mostly static and may be better served by periodic reports.

An EEMS needs to provide a variety of output options that enable everything from detailed reports with data tables and charts of multiple Trend Lines, to summary reports that roll up information into cost breakdowns and overall operating efficiency ratings. The output needs the flexibility to be delivered via paper, electronic documents, or the Web.

EEMS DESIGN PRINCIPLE 4: MEASURE & VERIFY RESULTS

All too often performance measurement and verification (M&V) ends up neither measuring nor verifying performance. It is a simple case of not having access to data to do M&V properly[1]—a problem an EEMS solves.

In measuring and verifying results it is first important to define terms often used to justify and quantify the impact of investments in utility operations.

Real Savings versus Stipulated Savings

Real savings is proving cost savings via actual dollar savings, while stipulated savings is based upon savings of equipment. For example, new lighting fixtures may consume 40% less electricity while generating the same amount of lumens and, while the savings per lumen are real, the heating costs may have increased because the new lighting now produces less heat. An EEMS is required to view both the data for the electricity consumed by equipment type and for the utility consumed to heat (or cool) the area, all within the context of a specific space if a real savings assessment is to be made.

[1]IPMVP Volume I: Concepts and Options for Determining Energy and Water Savings, 2001, Sect. 5.6

Energy Savings versus Dollar Savings

The EEMS must be able to account for both energy savings and cost savings (actual dollars). For example, just because the utility bill has dropped does not necessarily indicate that money/energy has been saved; rather it confirms only that less was spent. It confirms nothing about reduced energy consumption.

Factors like price, weather, a new construction coming on line, and consumption rates are required to understand whether actual dollar savings have occurred. Consumption savings occurs because of the way energy usage is being controlled and this can only be validated through monitoring and control points that highlight energy consumption changes.

To understand the dollar savings realized, one must account for variations in the price of energy, the actual weather versus the planned weather for the time period, and the change in the amount of energy used.

To realize true energy savings, which in turn lead to dollar savings, consumption must be reduced independent of the factors above and this can only be achieved when energy usage is being controlled more efficiently.

Use Life Cycle Costing

An important philosophical concept to adopt is "Life Cycle Costs." It is the most appropriate way to assess equipment and building costs. The initial purchase of HVAC equipment is a significant capital investment, but its true costs lie in this number plus the cost of its operation, service, maintenance, and total life span. The dramatic impact of an EEMS on life cycle costs is discussed later, but it is important to highlight this "real cost" assessment as a true measure of the total cost of equipment ownership.

EEMS DESIGN PRINCIPLE 5: A PLATFORM THAT EMBRACES INDUSTRY STANDARDS

Using industry standards and an open architecture is the right way to build an enterprise-class application. This has been proven repeatedly at all levels of technology and business where broad support and interoperability are significant benefits. A platform architecture and the use of standards protects the organization by minimizing dependency on any single vendor, even allowing functionality to be added outside the vendor's development cycle.

There are standards in several areas that an EEMS should adhere to:

Operating System

There are three platforms, sufficiently open and standardized, that an EEMS could run on. The first, and by far the most popular, is Microsoft Windows. It offers the greatest availability of tools and options, and is already installed and supported nearly everywhere. The Microsoft .NET platform is excellent for developing and integrating application components. Other options include Linux, which has strong server support and tools such as J2EE, but is a limited end-user platform; or a Web-based platform, which is typically made up of Windows or Linux servers using the Web for network communications and a browser for the application front end (which instills limitations on the UI).

Database Management

Equally important is that the data stored within an EEMS is housed in a manner that is open, accessible, and interoperable with other systems such as a standard relational database (i.e. SQL Server, Oracle, etc.). Proprietary data managers will handcuff users to rely on the vendor for everything.

Data Collection

As discussed within "Design Principle 1," data are the lifeblood of an EEMS, and access to this data is a complex and arduous task. The mechanism for ensuring that this data extraction/transfer takes place can be both cost prohibitive and extremely difficult without adherence to standards such as OLE for Process Control (OPC), BACnet, LonWorks, and Modbus.

Analysis

Data analysis tools range from the most general purpose and broadly available, Microsoft Excel, to highly specialized analytics. Ideally the EEMS will provide direct support for Excel, allowing users to take advantage of analysis routines already developed. Minimally the EEMS will allow data to be exported into any analysis program through cut and paste or by using an intermediate file.

Space Planning and Classification

The EEMS should support HEGIS groups and classifications for space planning. It should also interface with the leading space planning system, FAMIS.

Publishing

Making information available throughout the organization is an important function that saves significant time. There are four document formats an EEMS can consider publishing to—Microsoft Office, PDF, HTML, and XML. Office, which includes Word, Excel, and PowerPoint, is ubiquitous in business settings. The other formats are also nearly universally readable, although they require less common tools or special skills to edit.

BUSINESS APPLICATIONS OF AN EEMS

The EEMS has many business applications, each with different benefits to the campus. Some of these applications deliver benefits that are operational savings (reduced energy consumption) and some of the applications enhance the infrastructure. The most common applications of an EEMS are:

- Operational diagnostics & monitoring
- Empowering efficient building control strategies
- Enabling continuous commissioning
- Chiller plant efficiency calculations
- Controlling building comfort
- Accurate energy cost allocation
- Capital request justification
- Information publishing
- Providing more accurate budgeting & forecasting
- Purchased utility accounting
- Vastly improved performance measurement & verification

Operational Diagnostics & Monitoring

Operational diagnostics is the process of reviewing operational energy data and identifying targets for energy savings, highlighting engineering design deficiencies, and alerting staff to malfunctioning equipment, to name just a few. While these tasks sound familiar, it is important to contrast the manner in which an EEMS performs this function with that of today's technology and processes.

In short, an EEMS streamlines the process. It provides a complete picture that can be explored quickly and easily to identify problems, validate improvements, and test hypotheses—all without the technical literacy and timescales required by today's approaches. With an EEMS, this process simply involves a straightforward visual analysis of the data, quickly digging deeper into the information where anomalies are evident to uncover problems and gain real insight.

An engineer diagnosing a problem using the data in an EEMS can typically resolve the issue to its root cause in one fifth to one tenth the person-hours than with standard approaches. Also, the elapsed time from the first indication that something is wrong to resolution can be reduced by a factor of 100 or even more. Appendix A shows a case study where this was the situation.

Once an issue has been resolved, the results should be verified and measured. This is to ensure that any actions taken had the intended result, and also to be sure that no unintended side effects occurred. Trend Lines provide the ability to start verifying results as soon as 15 minutes after the change has been made.

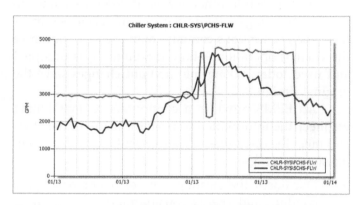

Figure 4-17. An example of a chart used in the process of identifying a problem. Here the diagnostician has identified significant differences between the Btu output of different chillers even though they should be producing equal amounts.

When not diagnosing specific issues, an EEMS is the ideal tool for ongoing monitoring of operations. The data organization, performance, and Trend Lines allow engineers to monitor as many as a half million interval data points in 20 minutes. Instead of watching unconnected real-time monitors and waiting for BAS alarms to sound, an EEMS's monitoring capability provides far more in-depth information about how systems are operating and reveals issues at their earliest stage.

Efficient Building Control Strategies

A great deal of effort is expended programming building automation systems to meet the disparate comfort needs of building occupants. Since seasons and weather change constantly, the strategy used to control the building will need to change accordingly.

A key role in the development of efficient building control strategies is the EEMS Trend Lines function. They show how the change in the control strategy is working, or highlight areas where the strategy should be adjusted. Importantly, by using a data collection interval of just 15 minutes, control strategy change requirements and the results of implemented changes will be apparent almost immediately.

Continuous Commissioning[SM]

Continuous Commissioning[2] is an ongoing process for monitoring systems, diagnosing and resolving issues, and making energy consumption as efficient as possible while maintaining or improving building comfort. It includes anything from physical maintenance, to control strategies, to prioritizing and implementing retrofits. While other forms of commissioning on existing buildings have initial design specifications as their goal, continuous commissioning seeks to optimize the current operations—how the building is occupied and used today, taking into account changes since the original design.

An EEMS provides a continuous commissioning engineer the information to perform many of the steps and meet the objectives of a successful continuous commissioning program:

• Document the existing state of operations, space comfort, and energy consumption

• Locate system issues, diagnose, and develop a plan to resolve them

• Create new recommendations for control settings, setpoints, etc.

• Identify and prioritize retrofit projects that will have the greatest impact

• Measure and document improvements in system performance and energy consumption

• Monitor ongoing operations to ensure all benefits are sustained

Chiller Plant Efficiency Calculations

A true EEMS must enable holistic utility performance measurement and management, including the frequently neglected chiller plant data. Chiller plant efficiency calculations provide simple to understand graphs that show the cost of running the chiller plant; accounting for energy usage by each chiller, cooling tower, and primary and secondary distribution loops. As the control strategies change, operators can assess the true overall impact on costs.

Building Comfort

As well as operational efficiency and energy cost related applications of an EEMS, it is important to remember that the real purpose of HVAC systems is building comfort. Well-maintained systems improve comfort and health of building occupants through properly controlled temperature, humidity, and ventilation.

Improving comfort levels and maintaining their consistency can be accomplished while lowering energy consumption and its associated costs. This was proven by the Energy Systems Laboratory at Texas A&M University where their Continuous Commissioning efforts resolved major comfort issues while reducing energy consumption by 15-30%[3].

An EEMS provides the data to see exactly how building or room conditions change and what the control settings are to try to maintain comfort. For example, a large meeting room filled to capacity will typically experience a rise in temperature during a long meeting. The solution of "cranking up the AC" halfway through does not work very well, yet is often what is done in response to occupant complaints. With an EEMS, engineers can use the data to determine the proper control settings to manage temperature, relative humidity, supply air static pressure, terminal damper position, etc. This is a more complex control, but maintains comfort better and is often more energy efficient.

More Accurate Energy Costs Allocation

By integrating all energy sources and consumers across the campus along with space planning information, organizations can identify specific energy consumption by space. Once space consumption is determined, the rates can be applied to generate accurate costs for each space. These can then be rolled up to assign costs by department or other organization structure.

Today, energy costs per space are typically allocated by a combination of approximation and averaging, producing an inaccurate cost allocation. In essence, the high energy consumers are being subsidized by the low energy consumers, resulting in cost centers being billed inaccurately.

Capital Request Justification

Large capital requests tend towards two categories: replacing old worn-out or out-dated equipment, and doing major overhauls resulting from design flaws

---

[2]Energy Systems Laboratory, Texas A&M University, Continuous Commissioning In Energy Conservation Programs, *www-esl.tamu.edu/cc*

[3]U.S. Department of Energy, 2002, "continuous Commissioning Guidebook, Maximizing Building Energy Efficiency and Comfort," ch. 2, pp. 1.

or changing requirements that render the design obsolete. Replacing old equipment is part of a normal cycle, even if only done every 20 years. Proving the need for an overhaul is much more complicated and requires proof to show that the investment is needed.

Today, engineers design systems and specify how they should be operated. However, they receive very little feedback as to how a particular piece of equipment performs in the field, under load, supporting usage patterns that may be different from when the original design was done. EEMS Trend Lines can be used to "instrument" the equipment and provide verification of air handling system performance, distribution loop efficiency, and overall system performance under various loading conditions, to name just a few.

Many design engineering firms build various computer models to simulate operations; but now, Trend Lines provide real feedback, based upon live data for the first time. The information is based on actual operations, not original design intent.

While a building automation system cannot provide this feedback due to the large volume of data created, and the inability to look outside itself, an effectively implemented EEMS can perform this function with ease.

In short, an EEMS can be used to identify and verify design flaws, whether from bad initial design or due to changing requirements or equipment upgrades. This capability gives facilities personnel the information needed to justify the capital budget requests needed to fund system overhauls.

Information Publishing

The current approaches to publishing operational data fall into two categories: the ad hoc, "that information should be here somewhere" approach, and the comprehensive reporting approach. Both consume vast amounts of time from engineers, technicians, and other facilities personnel because the data are not readily available, and what is available is not organized. It is a giant exercise in manually compiling and distributing spreadsheets.

The EEMS provides the starting point where all data are present. It is organized so that building-specific, departmental, or system-centric views of data can easily be reported. Reports can be financial or consumption based, or correlate the two with ease. The EEMS makes more information available at a fraction of the effort and associated cost.

More Accurate Budgeting and Forecasting

An EEMS should also show deviations from budget in terms of causes (consumption, weather, price and new construction).

Through the EEMS, individuals can view daily updates versus budget on a graph, thereby preventing month-end budget shortfalls. By providing this insight concerning current status on an ongoing basis, the EEMS provides opportunities to prevent budgetary overages before they blossom, while at the same time providing the vehicle through which the most effective cost savings can be realized should a budget overrun actually occur.

Purchased Utility Accounting

Today, at their own significant expense, many organizations install their own meters in parallel with those of the utility company for the purpose of verifying the utility bills. However, there are other techniques to verify utility billing without requiring additional meters.

An EEMS can summarize energy usage from all of the consumers of energy, identify losses, and provide supporting documentation to reconcile discrepancies with the utility. Furthermore, an EEMS can do this in a fraction of the time and cost.

Vastly Improved Performance Measurement & Verification

In the past, the first phase of a performance contract required that the contractor collect baseline information regarding the status of the building. This labor intensive and expensive task invariably only provided small samples of data from locations across the campus. Attempts to collect more data through BAS Trend Logs will have a negative impact on the BAS's ability to control operations. With such a random data collection process, it is simple to understand why this form of baseline information is inadequate for use to make significant investment decisions.

EEMS Trend Lines, discussed earlier, can significantly improve performance contracting for both the customer and contractor. With an EEMS, all the point data can be collected (with zero impact on the BAS) to provide a complete and accurate baseline as a matter of course, rather than as an expensive manual task, and the subjectivity can be removed from performance contracting. As improvements are implemented, their impact can be measured and verified quickly and easily.

FINANCIAL BENEFITS OF EEMS

It is clear that an EEMS is able to provide the infrastructure to support many different business purposes. Importantly, these applications of EEMS technology are also able to deliver a rapid return on investment in a number of areas including:

- Lowering operational costs
- Positive cash flow
- More effective staff deployment
- Greater indirect cost recovery
- Reducing equipment maintenance costs and increasing equipment life
- Improving the efficiency of energy purchasing

Lowering Operational Costs

EEMS Trend Lines provide the insight necessary to identify and achieve operational savings. Today, organizations typically identify operational savings targets by engaging a seasoned HVAC engineer to tour the campus, every building, and each piece of equipment to determine what is happening. Decisions are made based on a tiny fraction (often just one or two percent) of the data and many problems are not found until long after they appeared or are completely missed.

With an EEMS, this process can be accomplished within days instead of months, without leaving the office, through the rapid visual analysis of data. Unlike a manual campus tour, an EEMS uses all of the operational data to quickly identify common operational problems that, when addressed, lead to a reduction in operational costs. Below are samples:

- Fans running constantly when they do not need to
- Chillers and air handling systems unable to keep up with setpoints
- Over controlling of the building
- Operation of the equipment that deviates from the instructions of supervisors

Furthermore, using an EEMS, the operational changes and the financial benefits derived from them are instantly measurable and verifiable through simple monitoring of the data.

Positive Cash Flow

A typical retrofit project involves replacing dated components with new, efficient equipment. Such projects require a huge up-front investment to fund a few months of design and planning, followed by 6-24 months of installation/construction. It can be multiple years of negative cash flow before the project is complete and stipulated savings begin, and much longer before the return on investment is fully realized.

With an EEMS, savings start after just a few weeks. The initial investment is typically recovered in 5-6 months, and thereafter generates a positive cash flow. The realized savings and additional cash could even fund other projects once the operational issues are resolved.

More Effective Staff Deployment

Most facilities staffs are operating on guesswork due to the lack of data. Energy and control engineers may study real-time monitors and run BAS Trend Logs, often taking weeks to correct an operational defect.

With the data provided through the EEMS, facilities staff prioritizes the right problems and resolves them in far less time. Less time is spent on fire fighting and more on improvements that upgrade campus/facility physical assets in line with the master plan.

Equipment Maintenance and Life

Access to Trend Lines and comparison data between the weather, like-equipment, and facilities can serve as a powerful tool to reduce maintenance costs and prolong the life of equipment.

For example, cooling tower fans are often run at 100% capacity unnecessarily when the dew point is too

Figure 4-18. Electrical consumption in six cooling tower fans is reduced by (1) employing new control strategies for cooling tower approach offset, and further reduced by (2) setting fans to run in unison at lower rpm.

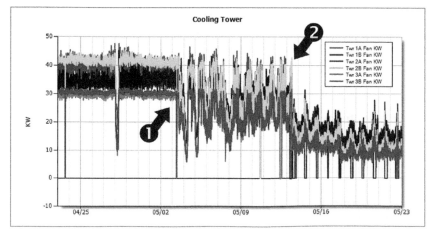

high for the environment to accept the heat that the towers are attempting to transfer. In this example, running the fans at 50% or less would be equally beneficial as running at 100%. Access to this data via an EEMS can extend the life of the fans, reduce the service call frequency, and reduce overall costs dramatically. The same approach can be applied across other equipment types.

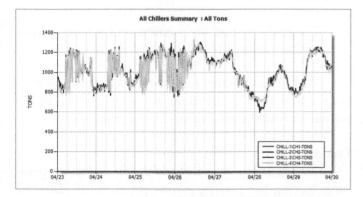

Figure 4-19. Equipment trashing was stopped when the chiller setpoint was set to a fixed value, resulting in an expected increase of life and reduction in repair/maintenance.

More Efficient Energy Purchasing

Whether an organization makes energy purchasing decisions alone, or with the help of consultants, an EEMS can have a positive and profound impact on the energy purchase process.

Purchasing energy is a complex task that requires a thorough understanding of demand. In addition, if energy costs are to be minimized, it requires an insight into how to manipulate the timing and height of peaks in demand.

Also, because energy pricing structures change regularly, effective purchasing requires a sound knowledge of the energy market's price drivers so as to "lock in" the best price. Organizations frequently buy extra energy, locking in prices via futures contracts. When decisions concerning contracts of this magnitude are made, organizations, like people, make better choices when presented with all of the facts. Without a robust knowledge of an institution's consumption patterns, the drivers of these patterns, and the impact of the weather on operations, an energy purchase is being made without visibility of the whole story.

Indirect Cost Recovery
Allocation

By accurately identifying energy cost for research space, rather than using an estimated allocation approach that invariably falls below the actual energy cost of the space, institutions are neglecting access to government funds.

Through the use of an EEMS, energy cost for space can be accurately and verifiably attributed to ensure that all of the institution's revenue opportunities are maximized and an institution's low energy users do not necessarily have to subsidize the high energy consuming departments.

Productivity

When examining financial benefits and focusing on cost savings, it is easy to overlook the impact of having space comfort. Studies have shown that maintaining a consistent, comfortable environment has a positive impact on attention span and productivity, and lowers time lost due to illness. Also, fewer comfort complaints mean that facilities engineers and technicians are not forced to spend time servicing comfort complaints.

CONCLUSIONS

When implementing a true EEMS, the following criteria must be specified:

- The EEMS must collect data from all monitoring and control points from all sources—building automation systems, utilities, metering systems, weather, and space planning systems. It must gather consumption data (in instantaneous, totalized, and average forms as appropriate), control settings, billing and rate information to provide a holistic view.

- It should structure the data within a data warehouse in a manner that provides the flexibility to handle complex relationships, hierarchies and calculations, and adjust to meet evolving requirements. Data should be normalized so that it may be more easily compared and contrasted, and so that problems can be pinpointed more precisely.

- It must present actionable information and Trend Lines so that data are represented in an informative manner. Performance and user interface must combine to provide interactive access to the data that can then be manipulated, supporting detailed investigation and efficient monitoring.

- The EEMS must deliver real savings that can be measured and verified, and demonstrate these returns over time.

- The EEMS should be more than an application—it should be a platform for energy management and other facilities operations that is expandable. It must conform to open standards so that integration and interoperability between related systems is a seamless and straightforward process.

An effectively implemented EEMS provides unparalleled insight into the day-to-day, week-to-week, and month-to-month operation of an institution's utilities. It nearly eliminates the time wasted by facilities staff gathering needed data. It provides individuals with the ability to rapidly find and address inefficiencies (fixing the root cause, not just treating symptoms) that can result in immediate cost savings and an ongoing financial return—often when these problems have gone undetected for many months or even years.

An EEMS also reduces the total cost of ownership of equipment by reducing maintenance costs and extending its life, all while providing the most actionable data to effectively secure lower utility rates and both improve and expedite master planning. It makes it easy to publish operational data to constituents within and outside the facilities staff.

An Enterprise Energy Management System presents the opportunity to span all existing building automation systems and energy related data so that assessments can be made in the context of the whole facility, environment, and billing climate. Complete information leads to better decisions—decisions that address building comfort, energy consumption, operational costs, capital investments, and stewardship of the assets.

APPENDIX A: SHORT CASE STUDY DIAGNOSTIC PROCESS USING EEMS

This short case study is from a hospital in the southeastern United States. The data shown is real, taken from a live EEMS.

Problem

The water temperature leaving the cooling tower was not meeting setpoint.

Figure A-1. Bumps in trends show where setpoint is not held, causing BAS alarm to sound.

Based on traditional data and investigation, all that was known was:

- The first indication of the issue from the BAS was an intermittent alarm

- Alarms came from cells 2 and 3 of the 3-cell system

- The problem had been going on for over a week without being able to identify the cause

With no prior knowledge of the systems or operations at this facility, the diagnostician using the EEMS set out to find the root cause of the problem.

Process

The first step was to look at the water temperatures leaving the cooling towers for each cell. Tower 3 had the most difficult time achieving setpoint (80°F).

A correlation with weather was investigated by overlaying wetbulb with Tower 3's water temperature.

Scanning back through time verified that the behavior was related to weather. This was the first time in several months that wetbulb reached that high.

The next area of interest was the power consumption of the two fans in the tower. The EEMS immediately showed that two identical fans were using very different amounts of power. Adding control information to the monitoring data showed the exact same signals being sent to each variable speed drive (VSD).

Another scan back in time shows the fan behavior starting about the same time the BAS alarms started. Earlier, the fans run in unison (although wetbulb was not as high).

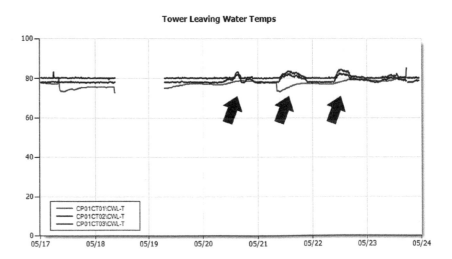

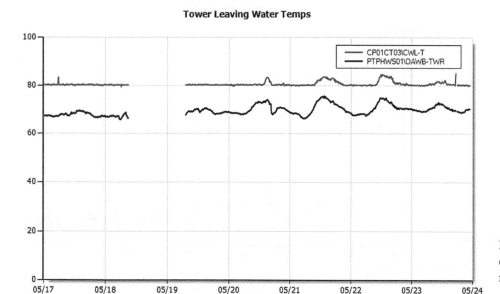

Figure A-2. Each time the wetbulb exceeded 73.4°F, the tower could not maintain setpoint.

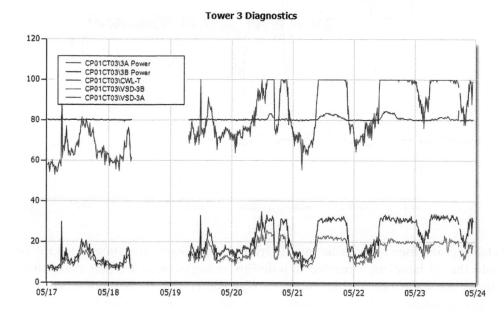

Figure A-3. The bottom two lines show the separation in power when wetbulb rises. The upper line that follows the same shape is actually two lines—the control signals.

Diagnostics Conclusion

The conclusion was that a fan belt was likely slipping because it was a VSD, causing the problem. It only took about 30 minutes to complete the diagnosis.

The fan belt was checked and found to be loose. It was tightened and the EEMS verified the result.

Although the problem was properly resolved, its return suggests that the equipment ultimately needs replacing, not adjusting. The EEMS provided the facilities manager with the data to prove that this was a necessary expense.

APPENDIX B: HOSPITAL CASE STUDY

Introduction

This case study shows the value of using interval data from the building automation system (BAS) as a tool to diagnose and monitor facility operations. You will get a close-up view of the first six months of using EnergyWitness™—an Enterprise Energy Management System (EEMS) from Interval Data Systems—from installation and initial data collection to discovering, fixing, and validating a variety of issues, including operations, comfort, and engineering design. The case

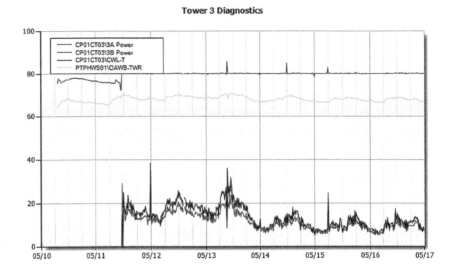

Figure A-4. The fans first started to split a week earlier.

Figure A-5. After the belt tightening, the fans ran properly again, but monitoring showed slippage again a month later.

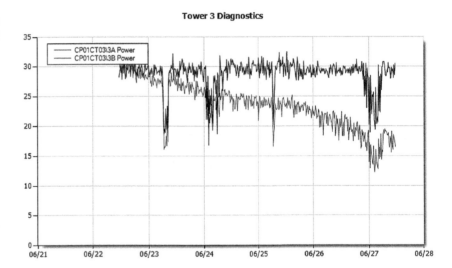

study traces issues starting in the chiller plant, then out through the distribution system, into the air handlers, and finally to the terminal boxes. In all, we'll see more than $260,000 of annual cost savings quantified and verified through the data.

The site is a large hospital complex in the southeastern U.S. with nearly 1,000 beds. To meet the demanding comfort requirements of a southern hospital, there are four chillers and three cooling towers capable of producing over 5,000 tons of output at the heart of their HVAC system, which contribute to a multi-million dollar annual energy cost.

Situation

The hospital had come to suspect that there were significant issues within its chiller plant. They believed it was operating far from optimally, and that there were perhaps engineering design issues at the root of the problems. If true, a significant investment would be needed to redesign, re-engineer, and repair the system. Additionally it was known that there were space comfort issues in several areas of the hospital, including the operating rooms.

Before seeking funding, the facilities staff knew they needed to prove that engineering design problems did indeed exist. Further, if they could quantify the ongoing wasted cost of current operations, it would provide the justification to proceed with the anticipated redesign.

Getting Started

Hospital staff found that they were unable to extract enough information from their BAS to analyze the problem. Engineering studies were also inconclusive. They needed a system that could help diagnose the issues beyond the obvious symptoms they were already

Hospital chilled water pumping system.

aware of and find the source of the problems.

Selecting an EEMS

It was their BAS vendor, Johnson Controls, who knew of a company that provided a system with the level of comprehensive data gathering and analysis tools needed. Interval Data System (IDS) was brought in and put EnergyWitness to work. IDS also provided expert diagnostic services to work with the hospital staff interpreting the data, identifying the issues and making recommendations for an action plan.

Installing the System

Installing EnergyWitness required addressing server hardware, software, networking, security, and remote access components. Once installed, BAS monitoring and control points had to be mapped into a relational database. The challenges associated with the installation fell into two categories—technical and project management. It took four to five days of labor time over a three-week period to get to a point where data were being collected 24/7.

Technical Installation

There were three main components to the hardware and software installation:

- An OPC (OLE for Process Control) server: a software component that resides on the BAS server hardware, typically supplied by the BAS manufacturer, used to collect BAS point data.

- OPC client: the software that EnergyWitness uses to communicate with the OPC server to collect data.

- The EnergyWitness server: a Windows PC running the EnergyWitness server application and Microsoft SQL Server for database management.

In order to provide the necessary diagnostic services, as well as providing software support (technical support, software updates, backup of data), IDS staff needs remote access to the EnergyWitness server. As with any site, it was critical for the installation and ongoing support to follow the policies set forth by the IT department. In this case, the hospital set up a VPN (virtual private network) line to provide secured access.

The final phase of installation was data collection. Since the initial focus was the chiller plant, all of the BAS point data for the chillers, cooling tower, and chilled water loop were defined in EnergyWitness as collection points. This was the most time-consuming part of installation, as it is important to not just identify the points, but also determine which need instantaneous, averaged, and/or totalized readings. All of the interval data are collected every 15 minutes. At this point the system was monitored, and given a couple weeks to collect data before data configuration and diagnostics would begin.

Project Management

The project management related activities (coordination with IT, BAS managers, and project leaders) required far more time and effort than did the technical issues. Most times the management issues were the critical path to *allow* the technical work to happen. This is normal, but often overlooked, for any enterprise-scale software installation—especially when it involves network connections and a need for remote access. Communication between the vendor, facilities team, and IT department is the foundation to a smooth, successful installation.

Data Configuration

Data configuration is an ongoing process. EnergyWitness uses a combination of a tabbed interface for major grouping of information, and a tree structure to organize and group the points to be meaningful and easily accessible to the user. Points were gathered from the BAS into a master list—the raw data coming into the system. The tree in back shows the start of the master list, including all of the chiller plant points and points for some of the air handling systems. The master tab categorizes approximately 3,000 BAS points.

Phase two of data configuration organized points from the BAS into more navigable structures. The front list shows a chillers tab that further organized the data into a tree structure. The structure makes locating the right data easier and trend lines for all points within a folder can be displayed as a group for fast monitoring. When configuring the data within the chillers tab, weather data were also included for convenient access. You can also see two types of trend lines: the first, as can be seen in the CHW tons folder, are BAS points; the second, seen in the tower power folder, are calculations.

Over 60 calculations, based on the chiller plant data, were added to the structure to provide a higher level of information. These range in complexity from Delta Ts, to balance efficiencies and theoretical water loss, to overall chiller and plant efficiency and operational cost calculations. All calculations are updated at every 15-minute interval.

DIAGNOSTICS PHASE I: OPERATIONS AND CONTROL ISSUES

The data capture and configuration was phased in, as mentioned above. At the start, with only 700 BAS points from the chiller system and weather data being collected, the IDS diagnosticians started to review the data. In just three days of diagnostics, 45 preliminary issues were identified and documented (Figure B-2). The issues spanned control problems, sensors that needed calibration, and the suspected engineering design issues began to surface.

Using the collected data, IDS and the hospital staff were able to categorize each issue, define the implications to the overall operation, and identify (quantification comes later) the likely financial benefit of correction. This classification effort helped prioritize the plan of action.

Diagnostics Reveal Instability

Before the analysis on the engineering design problem could begin, it became clear that there were operational problems that could be resolved first. The system was showing a significant amount of instability that needed to be addressed to identify the root causes of problems and inefficiencies. Much of the instability was a result of hospital staff addressing symptoms—they simply did not have the data to get to the source of the issues.

The next three charts show some of the initial concerns found by studying the data for literally just a few hours. The issues shown revealed sensor problems,

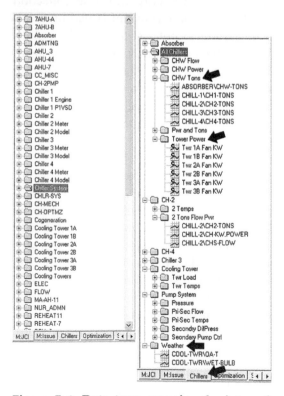

Figure B-1. Data trees organize the interval data for instant access.

Hospital Chiller Plant Preliminary Investigation Summary					
Issue Num	Issue Description	Category	Implications	Comment	Financial Benefit of Correcting Issue
1	CW set point adjust every 10 minutes causing CW & CHW systems to hunt	Over Control	Cost of Operations, Equipment Life		Straight forward change to BAS will pay for itself very quickly
2	Cooling tower fans (all) controlled in unison	Inadequate number control points and control sequence	Cost of Operations, Equipment Life	Each pump will need own control points added so that flow can be controlled	Eligible for rebate with significant energy cost savings
3	Although systems may be new, they do not seem to take advantage of high CW delta T for increased chiller efficiencies and reduced pumping rate.	Engineering design	Cost of operations (significant)	More effort required to determine savings benefit as costs could be high	TBD
4	Secondary pumping systems seem to operate from the same pressure controller thus causing increased pump hp operating costs.	Inadequate number control points and control sequence	Cost of Operations, Equipment Life		Straight forward change to BAS will pay for itself very quickly
5	Current CHW operation causes flow in path of least resistance through idle chillers/absorber.	Suspect of engineering design	Large cost of operations, Equipment Life	Needs further investigation to determine course of action. Possible piping change req.	Candidate for rebate
7	Current CHW operation causes mixing of 42° F water to higher temperature which will reduce design cooling load because control valves will be open trying to dehumidify.	Engineering design	Cost of operations (significant)		
8	Current CW operation causes flow in path of least resistance through idle chillers	Engineering design	Large cost of operations	Likely addition of smaller in series valve will fix situation	TBD
9	Current CW low ΔT causes chillers to work less efficiently and increases hp energy.	Chiller Control	Cost of operations		
10	Primary CHW pumps are controlled through circuit setters instead of VFDs which can cause high annual operating costs. i.e.. 10ft head at 3000 gpm x $0.08/kWh ≅ $5K/year pumping hp	Engineering design	Cost of operation		Eligible for rebate with significant energy cost savings

Figure B-2. The beginning of the initial diagnostics report—the first ten of 45 preliminary findings.

outputs cycling by as much as 500 tons every 15 minutes (Figure B-3), temperatures within the primary and secondary chilled water loops that were not as they should be (Figure B-4), and bypass flow issues (Figure B-5).

Disable Chiller Reset Program—Fix Setpoint at 42°F

In order to stabilize the overall chiller plant operations, the first recommendation was to discontinue using the reset program that was in place and fix the setpoint at 42°F for all chillers. This suggestion was made to the hospital during meeting on 4/26 at about noon. The control change was made immediately afterwards, and you can see in Figure B-6 the cycling of output that was happening immediately settled down. The stabilization that occurred was immediately verifiable because the interval data are constantly collected, showing the impact of the control change in setpoint.

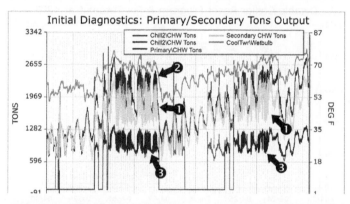

Figure B-3. The chiller operations were unstable. Tons output measured from the primary and secondary sensors (❶-green and ❷-dark purple lines) and the chillers (❸-blue & dark green lines) showed cycling by as much as 500 tons every 15 minutes.

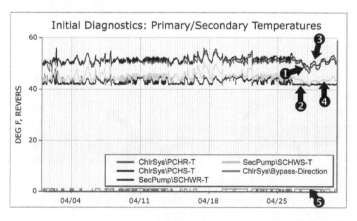

Figure B-4. The primary supply and return temperatures (❶-dark green and ❷-blue lines) should not be lower than the secondary temperatures (❸-purple and ❹-bright green lines) when the flow is reversed (❺-magenta line at bottom).

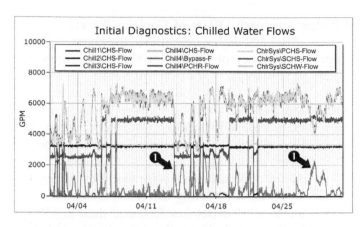

Figure B-5. There is often too much flow through the bypass (❶-magenta line across bottom). This causes the secondary supply temperature to rise, reducing cooling capacity.

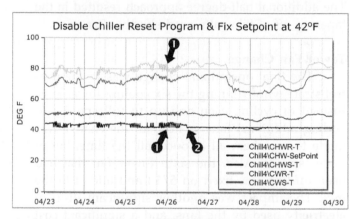

Figure B-6. The tons output load was cycling widely (❶-green line) until the setpoint (❷-magenta line) was fixed at 42°F on 4/26.

Looking at the impact of chilled water and condenser water temperatures, the change had a similar positive impact. The chilled water supply temperature, which had been oscillating between 42° and 45°, stabilized at the 42° setpoint (Figure B-7). The vacillating chilled water return temperature and the condenser supply and return temperatures also smoothed out with the fixed setpoint.

After any operational change it is important to verify that the intended impact occurred, as we saw above. It is also critical to check that there are no unexpected side effects. Because all of the chiller plant BAS points are available, it was trivial to look at cooling tower data and check the impact of the setpoint change. Similar positive results had occurred. As seen in Figure B-8 the condenser water temperatures smooth out and the full load amps percentage (which had been cycling 20 percentage points) stabilized significantly.

Change Approach Offset of Cooling Towers

After stabilizing operations, the goal was to improve efficiency of the chiller plant operations. Attempting to improve the efficiency of the chillers, the hospital had set the cooling towers to achieve the lowest possible condenser water temperature. However, this led to the cooling tower fans running at full load all the time, negating any savings the chillers were achieving.

A series of adjustments were made to the approach offset for the cooling towers. A change was first made to a 3°F offset (on 5/3) and then to 3.5° (on 5/7). Figure B-9 shows the impact, where the first change immediately allowed the fans to draw less power at night—20 to 30 kW less power for each fan—especially between midnight and 7:00 a.m. The additional half-degree approach resulted in the tapering off of kilowatts used shortly after noon.

Tower Fan Change

The final operational change was to the control strategy for running the six cooling tower fans. Instead of cycling each one up to full speed and adding additional fans as needed, the change was made to run them all at once, all the time, but at lower speeds. Also, the sixth tower fan that had been off-line was fixed on 5/13 and added to the mix. The impact was another significant drop in electricity used by the fans, and a significant cost savings for the hospital as a result.

Verification of Results

The reason for the drop is that the power used by the fans is related to the cube of the fan speed. So, as fans were pushed to 100%, the power consumption was going up exponentially. Keeping the fan speeds down keeps the kW down.

Further analysis was done in Excel, plotting the kilowatts used versus cooling tons output. Figure B-11 shows that running only five fans uses more power at all load levels. As the load increases, the penalty of only running five fans increases. At 1,600 tons output, the difference is about 35 kW, but at 2,400 tons the difference is nearly 60 kW.

Savings Estimate

The final stage of this operational change was to determine the annual cost savings resulting from the improvements. The bottom line was a calculated $40,000 annual savings. The calculation is summarized in Table B-1.

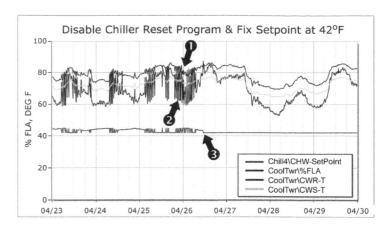

Figure B-7. Temperatures of chilled and condenser water stabilized (**❶**) immediately when the chiller setpoints were set to 42° (**❷**).

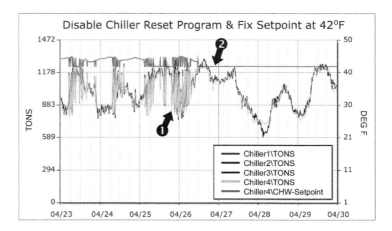

Figure B-8. The effects of the chiller setpoint change on the cooling tower show condenser water temperature (**❶**-purple line at top) and % full load amps (**❷**-blue line) smooth out considerably with setpoint (**❸**-green line at bottom) change.

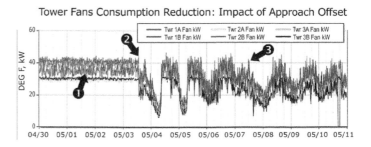

Figure B-9. Electrical consumption of the six cooling tower fans (**❶**-fan kW lines across the bottom) was reduced by setting the approach offset to 3° on 5/3 (**❷**) and then 3.5° on 5/7 (**❸**).

Ongoing monitoring in the weeks and months following this change have further confirmed that the fans continue to run at a lower rate, validating that the projected cost savings are being realized.

Also during this initial phase, because the data from the initial diagnostics showed that many pressure and temperature sensors needed calibration, the hospital reprioritized their work plan and fixed them to ensure the accuracy of incoming data.

DIAGNOSTICS PHASE II: ENGINEERING DESIGN ISSUES

With the chiller plant operating in a much more stable manner, the focus could finally turn to the

Table B-1. The data show an average savings of 78 kW for all six fans. The calculation assumes that there are 100 days/year where (in the south) the wetbulb isn't high enough to cause this condition.

A.	Fan energy savings (all fans):	78 kW
B.	Electricity rate:	$0.08/kWh
C.	Daily savings (A*B*24):	$150
D.	Days/year affected by change:	265
E.	Annual savings (C*D):	$39,800

larger set of issues due to the engineering design of the chilled water piping. The belief was that a significant amount of money was being wasted and staff resources were spending considerable time controlling around the deficiencies—treating the symptoms, as they didn't have enough information to identify and tackle the root problems.

Given that the expected cost of a piping redesign is in the millions of dollars, the hospital needed quantitative proof of the issues to justify the investment in a redesign.

In order to assist the diagnostics process of determining the redesign requirements needed to optimize the chiller performance, a series of calculations were put into the system. These calculations are applied at each 15-minute interval and made available to the system users as trend lines, exactly in the same way the individual point data are available.

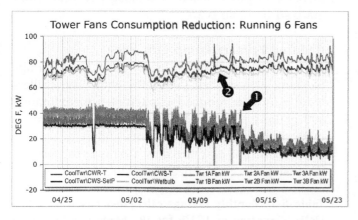

Figure B-10. In this monthly view, you see the impact of the approach offset changes also seen in Figure B-9, then an equally dramatic impact of the fan repair and control change on 5/13 (❶) to run all six fans together at lower speeds. The control change benefit occurs despite higher wetbulb temperatures (❷-bright green line).

Underutilized Low-Cost Chiller

The chiller plant consists of three electric chillers, each with a 1,500-ton capacity, and a gas-fired chiller with an 800-ton capacity. The electric chillers need to maintain minimum loads to prevent surging and resulting damage. Because of that, the gas-fired chiller is only used during peak hours. Despite its 800-ton capacity, the gas chiller rarely outputs more then 600 tons, and its output trails off during the course of the day. That leaves, on average, 236 tons of unused capacity.

The unused capacity is so important because this chiller operates at a much lower cost per ton than the electric chillers do. As the analysis in Figure B-13 shows, the gas chiller (chiller 1) costs an average of $0.07 per ton to operate, while during on-peak hours the electric chillers cost $0.22 per hour. By running below 75% of capacity, there is an unrealized savings of $83,000 per year (Table B-2).

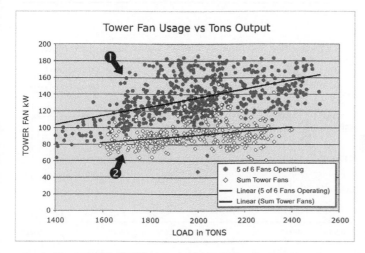

Figure B-11. Using only 5 of 6 fans (❷-purple dots) uses more kW at all load levels than using all 6 (❷-green diamonds), and the difference gets greater the higher the output load.

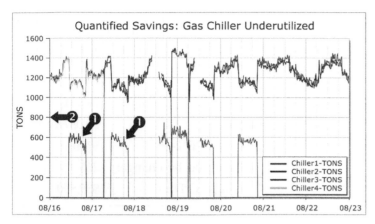

Figure B-12. Chiller 1, the gas-fired chiller (❶-dark green line), operates well below its 800-ton capacity (❷).

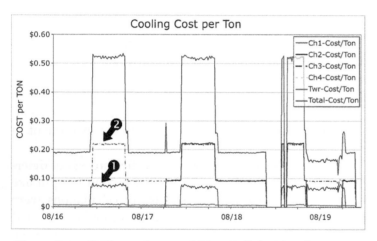

Figure B-13. Chiller 1, the gas chiller (❶-light blue line), operates at a third of the cost of the electric chillers (❷) during on-peak hours.

Reduced Chiller Efficiency

Issues in the chilled water flow of the electric chiller systems (chillers 2, 3, and 4) were discovered that caused additional inefficiencies. At times the primary chilled water flow was higher than the secondary flow, resulting in secondary chilled water mixing with primary chilled water supply (note the bypass loop in Figure B-14). This, in turn, caused the temperature of the water returning to the chiller to be lower than it should (a low Delta T). The result was a decrease in chiller efficiency and a corresponding increase in operating costs. Figure B-14 shows the increased primary flow during peak hours, and Figure B-15 shows the matching drop in return temperature. The picture is completed in Figure B-16, which shows the kilowatts per ton increasing each time the flow problem occurs.

Table B-2. The gas chiller is 14.73 cents/ton cheaper to operate during the 9-10 peak hours per day.

A.	Electric chiller on-peak:	$0.2182
B.	Gas chiller on-peak:	$0.0709
C.	Savings/ton-hour (A-B):	$0.1473
D.	Unused gas tons:	236
E.	Hours per day:	9
F.	Days per year:	265
G.	Unused ton-hours (D*E*F):	562,860
H.	Lost savings (C*U):	$82,900

The operational cost associated with the increase in power used is $38,500 per year, summarized in Table B-3.

Reduced Primary Pump Efficiency

The three electric chillers also have pumps working at less than peak efficiency. The chillers were run at a reduced capacity—approximately 300 gallons per minute flow reduction—to minimize differential flow problems and help balance the system. The way the flow was reduced was by increasing the pump head pressure, resulting in a

Table B-3. Monthly savings calculated based on a lost Delta T average of 1.73°F for 197 hours/month, differential kW/ton of 0.027, on-peak rate of $0.124/kW, and off-peak rate of $0.053/kW.

A.	Total monthly savings:	$5,500
B.	Months savings will occur:	7
C.	Lost savings (A*B):	$38,500

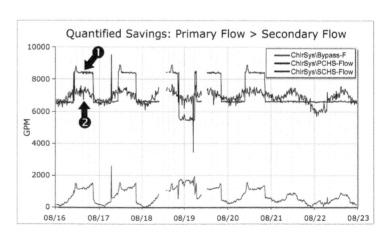

Figure B-14. Primary chilled water flow (❶-blue upper line) often exceeds secondary flow (❷-purple line) during peak hours.

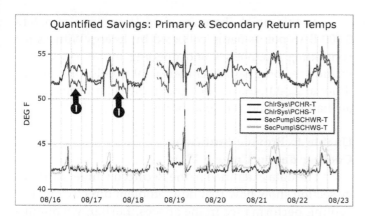

Figure B-15. Lower chiller water return temperatures (❶) result when the primary flow exceeds secondary flow (B-14).

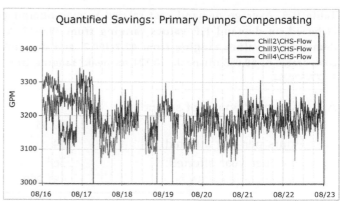

Figure B-17. The three electric chillers running below capacity.

Figure B-16. Power consumption (❶) increases when the primary flow exceeds secondary as well, resulting from a higher supply temperature and corresponding lower Delta T.

75 hp variable speed pumps. The control for the header is managed by the worst-case pressure across all seven zones. The system was working the pumps too hard, providing too much pressure to five zones in order to meet the needs of zones 4 and 6—the worst-case zones. Figure B-18 shows that the pressure in five of the seven pumps is well above setpoint nearly all of the time.

The average demand placed on all of the pumps causes an hourly cost excess of $9.35 per hour. Since this condition is constant, the annual savings possible is nearly $82,000.

Summary of Quantified Lost Savings

Using the interval data collected by the BAS, EnergyWitness provided the means to diagnose several problems with the chilled water system. The data enabled IDS and the hospital staff to quantify the excess expenditure these issues are currently costing on an

15 kilowatt power consumption increase per pump.

The annual cost of the primary pump inefficiencies is nearly $23,000.

Reduced Secondary Pump Efficiency

A much bigger problem, as it turns out, were efficiency issues with the secondary pumps. There are seven zones served by the secondary header with four

Table B-4. Daily savings based on a 6" water column differential pressure, on-peak rate of $0.124/kW, and off-peak rate of $0.053/kW.

A.	Total daily savings for all three electric chillers:	$85.50
B.	Days/year savings will occur:	265
C.	Lost savings (A*B):	$22,700

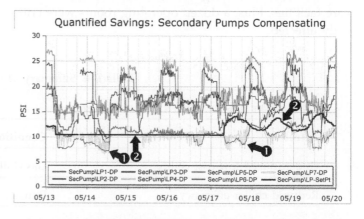

Figure B-18. Controlling all seven zones to meet the worst-case pressure demands of zones 4 & 6 (❶-bright green & teal lines) causes the other five zones to exceed setpoint (❷-gray line), forcing the pumps to work harder.

Table B-5. Average demand cost related to over-pumping is based on demand HP values ranging from 39.9-177.8, demand kW values ranging from 42.7-190.2, and demand costs (rates of $0.053 off-peak, $0.124 on-peak) ranging from $3.77-$22.54.

A.	Average demand of secondary pumping system:	$9.35
B.	Hours per day affected:	24
C.	Days per year affected:	365
D.	Lost savings (A*B*C):	$81,900

annual basis. All of this analysis was done by a small team of diagnosticians and facility engineers, requiring just a week of time from each team member, spread over about two months. The final compilation of results speaks for itself (see Table B-6).

DIAGNOSTICS PHASE III:
SPACE COMFORT ISSUES

A most important goal of the pending engineering redesign effort is to correct space comfort deficiencies. Although not quantifiable in the same manner as the operational cost issues in the previous section, it is mandatory that the hospital be able to reach required comfort levels throughout the facility as a completely separate concern from the cost of doing so.

The current design was simply making it impossible to deliver the desired air quality to some areas. Appropriate cooling levels were not achievable or could not be maintained throughout the day. Priority areas to examine included certain floors of the main building, floors in the north wing, and the children's and cancer care facilities. Plus, a dozen operating rooms were critically affected.

At this point there are data from approximately 3,000 BAS points being collected by EnergyWitness for continuing diagnostics and ongoing monitoring. They include the original 700 chiller plant points, plus an additional 2,300 points throughout the distribution loop, air handlers, and terminal boxes.

Unable to Provide Sufficient Cooling due to Uneven Flow

The data examined in the Reduced Chiller Efficiency section show how that flow problem directly impacts cooling performance in the spaces. Earlier, we quantified the costs associated when primary flow exceeded secondary flow. Cooling issues also existed when secondary flow exceeded primary flow (Figure B-19). This resulted in higher chilled water supply temperatures delivered to the air handling units, reducing cooling efficiency and space comfort. Figure B-20 shows the rise in supply temperature caused by the flow problem.

Inadequate Pressure for Cooling to Reach Some Zones

During peak cooling periods (e.g. summer) some zones are unable to meet their pressure setpoint. This indicates inadequate flow and insufficient cooling capacity, allowing increased humidity and temperature in those areas. Although the cooling capacity exists, the pump differential pressure issues keep it from being delivered to all of the hospital zones that need it.

Doctors "Requesting" Colder Temperatures

Additional evidence of space comfort problems were found by examining the operating room controls. The temperatures in the operating rooms were seldom as low as the doctors wanted them, as evidenced by looking at the warm-cold slider control data. Doctors

Table B-6. Summary of quantified lost savings.

Quantifiable Cost of Chilled Water Problems

Affect	Condition	Cost per Year
Underutilized low-cost chiller	Gas chiller unable to reach full capacity	$82,900
Reduced chiller efficiency	Primary flow greater than secondary flow	38,500
Reduced primary pumping system efficiency	Operating water pumps to compensate	22,700
Reduced secondary pumping system efficiency	Operating water pumps to compensate	81,900
Total		$226,000

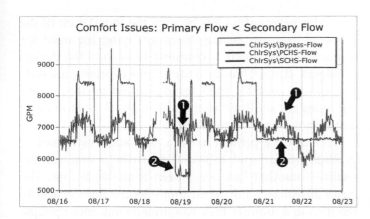

Figure B-19. Secondary chilled water flow (❶-purple line) exceeds primary flow (❷-blue line).

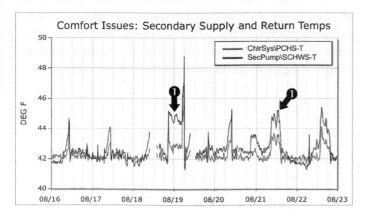

Figure B-20. Supply water temperatures (❶-blue line) rose when secondary flow exceeded primary flow (Figure B-19).

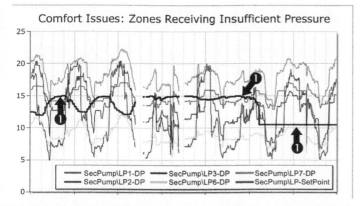

Figure B-21. Pump pressures are unable to meet setpoint (❶-thick teal line) in many of the zones.

most often push the sliders down to get the rooms colder, as seen in Figure B-22.

Space Comfort Summary

The information gathered through the interval data gave the hospital facilities director with ability to:

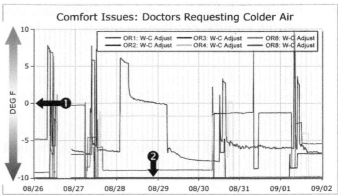

Figure B-22. Colder temperatures are requested most of the time in each OR as shown by the sliders in the negative position (❶-below the zero setting). Note that three of these six ORs sat with the sliders at -10 all weekend (❷).

- Document (for management) the affects on comfort and air quality resulting from the current HVAC design
- Define the requirements for the engineering design work to ensure that all known comfort problems get resolved

SUMMARY

This case study represents only the beginning of the impact that using an EEMS, EnergyWitness, has had on the hospital. The information that has been extracted from the interval data, and the quality of the decisions and planning that have come from it, was not possible before.

Earlier efforts relying on the BAS trend logs took substantially more time to garner much less information even though most of the data used in this case came from the hospital's BAS. In short, the hospital personnel determined that he BAS is unequipped to supply the level of diagnostic capabilities the EEMS has provided, as are the tools used by the engineering company to determine the redesign. This is why even the BAS vendor suggested EnergyWitness.

So far EnergyWitness has been used to:

- Improve chiller plant operations that are saving the hospital $40,000 a year in electric costs, plus anticipated maintenance reductions and extended equipment life due to smoothing out operations, which are not quantified. (More operational savings to come.)

- Identify and document piping design issues that cost the hospital nearly a quarter-million dollars ($226,000) per year.

- Identify and document space comfort problems, showing the root cause, which should disappear with the right engineering redesign.

- Reprioritize staff assignments to address the most important issues.

- Focus the right resources on the right problems—stop addressing symptoms with technician time and fix the real problems with engineering work.

- Provide the facilities staff with a tool to monitor operations, continue driving down operational costs, and document needed equipment and services.

- Prove the necessity of instant access to all the data in a visual form in order to achieve goals.

This information has become the foundation of the proposal to upper management justifying the need for a multi-million dollar investment to correct the design problems. It also has provided the hospital with data that can be used by the engineers hired to do the piping redesign, and then enable the hospital to validate or question their recommendations.

Data have proven to be a new and viable tool to provide operational information to help manage the facility—one that has never been available before. The hospital realized that when more people have access to the data in a manner they can understand, more people can participate in discussions and strategy decisions. An Enterprise Energy Management System such as Energy-Witness is the only tool capable of delivering this level of operational diagnostics. Facilities directors now have the information—and the credibility that goes along with documented facts—to make the right decisions, fund the right projects, and fix the root problems.

Chapter 5

The Case for Energy Information

Jim Lewis

This chapter discusses the benefits to a facility of gathering energy use information and explains some of the costs involved in collecting the information. It includes a discussion of the importance of submetering and describes submeter applications including cost allocation, accountability metering, efficiency monitoring, and measurement and verification. Examples of applications covered in the chapter are: accountability metering, tenant submetering, load curtailment/demand response, and web display of existing meter data.

INTRODUCTION

There's an old acronym in the business world that in order to ensure that goals are met, they must be SMART, meaning that they need to have:

- **S**implicity—the goals and tasks to achieve the goals must be easy to understand
- **M**easurability—if you can't measure it, how will you know if you are successful
- **A**uthority—control to affect the outcomes and achieve the goals
- **R**esponsibility—goes hand in hand with authority and makes someone accountable for the success or failure
- **T**imeliness—access to timely and accurate information is crucial to making changes to ensure success

Clearly, these same parameters can be applied to many energy management strategies, but in many cases key elements are left out of the process. Many building owners and managers embark on expensive and time consuming projects that are doomed to failure because some or all of the SMART requirements are either not met or are left to the last minute. How many energy managers have implemented energy projects based on elaborate (or back of the napkin) calculations for projected savings, only to see utility bills that are the same or higher than before the project began? The problem,

of course, is that relying solely on utility bill analysis means the process:

- Is Not **S**imple—complex calculations must be applied to the data to try to massage meaningful information
- Is Not **M**easurable—most projects impact only part of the buildings systems while utility bills measure the entire building
- Lacks **A**uthority —total building energy consumption is impacted by almost every user to some extent
- Lacks **R**esponsibility—when everyone is responsible, no one is responsible
- Is Not **T**imely—typically analyzed weeks or months after the usage has occurred

So, why are so many energy managers spending thousands (if not millions) of dollars on projects without any means of measuring and verifying the results? For once, energy managers and energy service providers (ESPs) are all singing from the same hymnal: *measurement and verification of energy savings is too expensive*! The view espoused by many in the industry is that installing and managing hardware and software for data acquisition is simply an added expense and wastes dollars that could otherwise be used to buy more cool energy savings toys.

On the surface, this sounds like a reasonable response, but let's look at this argument from a slightly different perspective: imagine that a Fortune 500 company plans to invest $100 million in research and development of new products. The plan is to put $50 million into two separate technologies with good opportunities for growth, but some degree of uncertainty as to whether the technologies will work and be accepted by the company's customers. In other words, the outcome looks promising for both based on the projections and calculations of the business development people, but the outcomes are not certain. Now imagine that the same company commits to the projects, but decides not to have the accounting department track the progress of the projects to determine whether there is any return on investment. They are content to assume that the business development people did their calculations correctly, that the engineering team met the cost requirements, that manufacturing doesn't encounter any problems in building the new products and that the sales force meets the target sales plans.

Company executives might argue that measuring each project would require hiring some new accountants and maybe investing in some new software or hardware and if the projects meet the projections, these additional investments would be a waste of resources that could be spent on future R&D.

If you were on the board of directors of this company, how much money would you give them? How about when they come back next year with requests for two more projects, but can't give you any results from the prior years' investment?

More importantly, how much of your 401(K) would you like to risk on the management team of this company? Carrying it one step further, how would you feel if the manager of your retirement fund told you that she was investing in a number of different companies, but wouldn't be able to tell you how each was doing and that at the end of the year, hopefully you'd have more money. Unfortunately, if you did lose money, there wouldn't be any way to figure out which investments were good and which were bad so you'd just have to guess and hope things went better next year.

Providing timely and accurate feedback to the success of an investment is such an integral part of our personal and professional lives that it seems very inconsistent to accept anything less in energy information and projects. Near real time information about energy usage "behind the meter" serves to provide accountability for operations and allow for corrections to be made where appropriate.

Another major benefit to gathering energy information is that making someone accountable for energy usage almost always results in a reduction in that usage. If an employee or tenant feels that no one is watching or cares about energy use, the tendency is for that employee to become lax about turning off lights, shutting down computers, etc. Many studies have shown that energy consumption reductions of 5 to 10% commonly occur simply through submetering energy usage and making individuals responsible and accountable for energy.

If we agree on the value of timely and accurate information, we are still confronted with the issue of the cost of acquiring, storing and reporting the information.

Historically, this process has been expensive primarily because there has not been hardware and software specifically designed to perform the data acquisition functions, so each installation has required a great deal of specialized integration of sensors and meters to accomplish the task of getting data.

In many cases, energy managers and providers have had to rely on building automation systems (BAS) or programmable logic controllers (PLC) to gather interval energy information. While these systems are capable of gathering the data, their primary focus is to accept inputs from a variety of sensors, calculate appropriate output parameters and implement those commands in a timely manner. Gathering and trending of status points (e.g., electrical meters) is, at best, a secondary function for these systems and consequently the cost of setting up these trending programs is high and the timeliness of the information is suspect.

Recently, a number of companies (including Obvius LLC in Portland, OR) have begun to develop solutions that are focused on the data acquisition needs of commercial and industrial customers. The AcquiSuite from Obvius is a stand-alone server that is preconfigured for many common inputs such as Modbus, plus analog and pulse inputs. This pre-configuration of drivers means that the system recognizes and configures many popular Modbus meters automatically with the only user input required being the naming of the point (i.e., no scaling or multipliers is required). This reduces the installation from hours to minutes and provides a much higher level of accuracy as input errors are minimized.

The second cost driver for energy information is the cost of communications from the local data acquisition server (DAS) to a host server. Since the DAS is a stand-alone web server, all the setup is accomplished by filling out forms using a standard web browser. There is no need for expensive software to configure the system or to set up or change communication parameters (upload times, etc.). The DAS comes with both a modem and an ethernet port and is capable of connection to a remote server via LANs or phone lines with options for direct dialing or connection to an ISP.

The last cost component for energy information is converting the raw data from the DAS into useful information. In the past, getting this information required the purchase, installation and support of proprietary software on a new or existing server. Sharing the costs of supporting a data warehouse was not possible and as a result, the cost of getting useful information was prohibitive. Several companies (including Obvius) offer web based services that provide access to energy information reports using standard web browsers for less than a dollar a day per building.

SUB-METERING FOR ENERGY MANAGERS

Managers of almost any kind of commercial or industrial facility have as all or part of their goals the implementation of an energy strategy. While these strategies take a variety of forms depending on the organization, the overriding purpose is to maximize the efficiency of operations so that the most productivity and profit is realized for the least amount of energy input. For many (if not most) managers, the first challenge in developing an energy policy is to determine how and where energy is being used within the building as most facilities simply rely on monthly energy bills as the measure of performance. The information gained from a monthly bill is generally not timely, nor does it provide indications of where within the building energy is being used and thus it is difficult if not impossible to even know where to start.

Historically, submetering of energy usage within the building has been an expensive and time consuming process due to the lack of hardware and software targeted to this application, so most users have simply avoided looking "behind the meter." Fortunately, recent developments in the energy information field have greatly reduced the time and expense required to gather and analyze data from one or many locations. This chapter will examine some of the ways managers with responsibility for energy can leverage existing systems and technologies to provide cost effective measurement of energy usage.

WHY SHOULD WE SUBMETER AT ALL?

The traditional view of submetering of energy has been that it is, at best, a necessary evil that accompanies investments in energy saving products and services. Today's energy managers realize that simply monitoring energy usage within the building and reporting that usage to the occupants can provide a significant return on investment. Typical submeter applications include:

Cost Allocation
Submeters are installed to provide more accurate information for charging tenants or departments for their energy usage. Cost allocation scenarios can range from very simple programs (e.g., dividing the total bill

based on consumption) to very complex programs calculating coincident demand and power factor penalties in addition to consumption.

"Accountability Metering"
Similar to cost allocation applications, accountability meters are used not for billing or accounting purposes, but to identify and impact behaviors of the building occupants in meeting goals for energy usage

Efficiency Monitoring
Submeters and other sensors are installed to monitor the efficiency of specific systems such as industrial processes in plants or chiller plants in commercial buildings

Measurement and Verification
These applications are generally associated with energy improvement strategies and submeters are used to benchmark usage before changes are made and then to measure and verify the actual savings

COST ALLOCATION

In general, cost allocation can be used where a single meter serves a campus or large facility and there is a benefit to assigning the actual costs to tenants or departments or product lines. Typical customer uses include:

Industrial/Manufacturing
Many plants produce a number of different products and one key component of the cost of goods sold is energy used to produce the specific products. Cost of sales numbers and gross profit can be greatly distorted if one or more products require heavy energy input and the costs are simply divided by square footage or sales volume. In order to allocate energy costs accurately and make better operational decisions, manufacturers can install submeters for specific lines and provide a more accurate cost accounting

Retail
Many retail centers have multiple tenants that share common spaces or centralized services such as HVAC. Some users, such as restaurants and bars may have different operational hours and energy needs that are not adequately reflected in the base rental agreement and submetering can be a very effective tool for measuring energy use and for allocating the costs of shared services such as HVAC.

Hospitality

Hospitality providers (including hotels, theme parks, etc.) frequently have space that is rented out to independent businesses such as restaurants or retailers. In many cases, the rental agreement provides a fixed cost per square foot for energy that may be either too high or too low. If the cost is too low, the owner will be subsidizing the costs for the renters and potentially overcharging other tenants. In this case, submetering not only produces a more accurate means of allocating costs, but also provides significant information on the profit provided from these non-owned users.

Commercial

Most commercial real estate properties provide utilities and conditioned spaces during fixed operating hours and have no adequate means of allocating these costs to specific tenants. This means that the building owner has little ability to provide flexible scheduling for tenants who have either long term or short term needs for changes in schedule. It also means that all tenants are treated as having the same energy density and the operations of some tenants (e.g., restaurants, bars and data centers) may be subsidized by other tenants, resulting in a long term competitive disadvantage in the market for the owner of the building.

Campuses/Schools

Most colleges and universities have a single primary feed meter that supplies power to all the buildings on campus. Campuses typically also have one or more central plants providing heating and cooling to multiple buildings via underground lines. This type of installation makes it difficult to assign costs of operations to specific departments or to outside groups using the facilities on a rental basis. Installing electrical submeters and Btu meters on the individual buildings allows for costs to be allocated to groups using the campus facilities

ACCOUNTABILITY METERING

In accountability metering, the submeters are used not for accounting purposes, but to reinforce desired behaviors by providing "report cards" on the activities of the users of the facility. If there is no means of measuring the impact of energy savings activities, it is likely that employees will be less conscientious about managing energy wisely. A large hospitality group found that energy costs were reduced significantly (>5%) simply by implement-

ing a program to submeter the rides and operational areas of their parks so that they could, for instance, determine whether lights and motors were being turned off when the park was not in operation. The results for various areas of the park are published monthly to allow managers to see which areas of their operations are meeting their objectives and which are not. The system also provides the capability to "drill down" into functional areas to isolate particular offenders.

EFFICIENCY MONITORING

Monitoring the efficiency of energy consuming systems such as chiller plants can not only provide valuable feedback about energy conservation and operational practices, but can also be used as an early warning system for maintenance and repairs. For example, if the tubes on a HVAC chiller plant start to plug, the amount of energy required to produce the same amount of cooling (typically measured in kW/ton) will go up. The system will continue to produce chilled water at the desired temperature, but because the efficiency of the heat exchanger is reduced due to buildup in the tubes, the compressor will have to run longer to produce the desired amount of cooling.

Monitoring large motors with current sensors can provide valuable information about bearings going out, insulation breakdowns or other mechanical deficiencies which result in more current being required for motor operation. In addition, a simple current switch can be installed on one leg of the motor to monitor runtime on the motor to determine maintenance schedules, filter changes, etc.

MEASUREMENT AND VERIFICATION

The majority of energy saving retrofit projects are, quite reasonably, implemented based on engineering calculations of the projected return on investment. As with any projections of ROI, much of what goes into these calculations are assumptions and estimates that ultimately form the basis for implementation. As the folks at IBM used to say, "garbage in—garbage out," which in the case of energy retrofits means that if any of the assumptions about parameters (run times, setpoints, etc.) are wrong, the expected payback can be dramatically in error. The establishment of good baselines (measures of current operations) is the best way to determine the actual payback from investments in energy and submetering is a key element in a baselining program.

Just as important as building an accurate picture of the current operation is measuring the actual savings realized from an investment. If there is no effective means of isolating the energy used by the modified systems, it may be impossible to determine the value of the investment made. Using monthly utility bills for this analysis is problematic at best since actual savings which may be achieved can be masked by excess consumption in non-modified systems.

Consider for example, a commercial office building with a central chiller plant with an aging mechanical and control structure that provides limited capability for adjusting chiller water temperature. To improve efficiency, the building owner plans to retrofit the system to provide variable speed drives on pumps for the chilled water and condenser water systems along with control upgrades to allow for chilled water setpoint changes based on building loads. In the absence of baseline information, all calculations for savings are based on "snapshots" of the system operation and require a variety of assumptions. Once the retrofit is completed, the same process of gathering snapshot data is repeated and hopefully the savings projected are actually realized. If the building tenants either add loads or increase operational hours, it is difficult if not impossible to use utility bills to evaluate the actual savings.

In contrast, the same project could be evaluated with a high degree of accuracy by installing cost-effective monitoring equipment prior to the retrofit to establish a baseline and measure the actual savings. While each installation is necessarily unique, building a good monitoring system would typically require:

- *Data acquisition server* (DAS) such as the AcquiSuite from Obvius to collect the data, store them and communicate them to a remote or local host.

- *Electrical submeter(s)*—the number of meters would vary depending on the electrical wiring configuration, but could be as simple as a single submeter (e.g., Enercept meter from Veris Industries) installed on the primary feeds to the chiller plant. If desired, the individual feeds to the cooling tower, compressors, chilled water pumps, etc. could be monitored to provide an even better picture of system performance and payback.

- *Temperature sensors (optional)*—in most installations, this could be accomplished by the installation of two sensors, one for chilled water supply and the

other for chilled water return. These sensors do not provide measurement of energy usage, but instead are primarily designed to provide feedback on system performance and efficiency.

- *Flow meter (optional)*—a new or existing meter can be used to measure the GPM and calculate chiller efficiency.

This benefits of a system for actually measuring the savings from a retrofit project (as opposed to calculated or stipulated savings) are many:

- Establishing a baseline over a period of time (as opposed to "snapshots") provides a far more accurate picture of system operation over time and help to focus the project.

- Once the baseline is established, ongoing measurement provides a highly accurate picture of the savings under a variety of conditions and the return on investment can be produced regardless of other ancillary operations in the building.

- The presence of monitoring equipment not only provides a better picture of ROI, but also provides ongoing feedback on the system operation and will provide for greater savings as efficiency can be fine-tuned.

VIEWING AND USING THE DATA

Historically, much of the expense of gathering and using submetered data has been in the hardware and software required and the ongoing cost of labor to produce useful reports. Many companies (such as Obvius) are leveraging existing technologies and systems to dramatically reduce the cost of gathering, displaying and analyzing data from commercial and industrial buildings. Using a combination of application specific hardware and software, the AcquiSuite data acquisition server provides user interface using only a standard web browser such as Microsoft Internet Explorer.

The AcquiSuite DAS automatically recognizes devices such as meters from Power Measurement Ltd and Veris Industries, which makes installation cost effective. The installer simply plugs the meters into the DAS and all configuration and setup is done automatically with the only input required being the name of the device and the location of the remote server. Data from the meters is gathered on user-selected intervals (e.g., 15 minutes) and transmitted via phone line or LAN connection to a remote host where it is stored in a database for access via the internet.

To view the data from one or more buildings, the user simply logs onto a web site (e.g. www.obvius.com) and selects the data to view (see Figure 5-1).

INSTALLATION AND COSTS

It is, of course, difficult to generalize on the costs of submetering as factors such as the amperage of the service and wiring runs will vary greatly from building to building. A couple of examples may provide some rough estimates of typical installations.

The example illustrated below is for a single submeter on a 400 amp service panel.

The installation outlined in Figure 5-2 is for a project with these costs:

- One Obvius AcquiSuite data acquisition server (DAS)
- One 400 amp submeter (Veris model H8035)

- Labor to install the meter and DAS
- Wiring labor to connect the DAS to the internet

Assuming no extraordinary costs of installation, this project could be completed for less than $2,000.

If the data are sent to a remote server via the internet for display, the annual cost for the single AcquiSuite would be approximately $240 (note: the annual cost listed is per AcquiSuite, not per meter, so up to 32 meters can be monitored for the same $240 annual cost).

EXAMPLE ONE:
ACCOUNTABILITY METERING

Accountability metering is a term used to describe the use of meters, sensors and software in commercial and industrial facilities to ensure that users are using energy efficiently. This process can be used to compare total usage for a number of different facilities or to measure energy usage within a single facility (a.k.a. behind-the-meter monitoring) or both. The purpose is to make users accountable for conforming to best practices in energy usage.

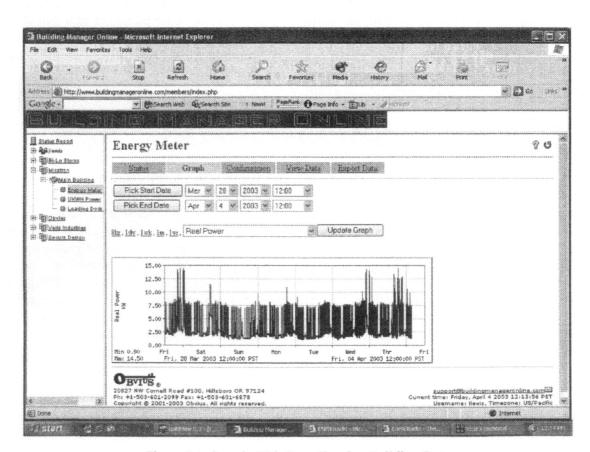

Figure 5-1. Sample Web Page Showing Building Data

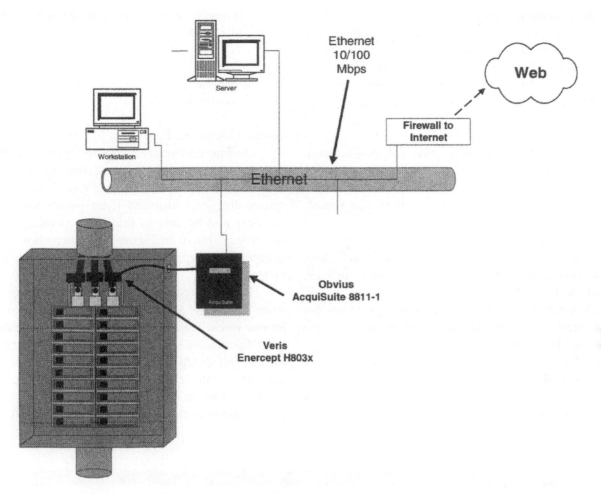

Figure 5-2. Typical Single Submeter Installation

Background

Most building owners and managers have established procedures for employees, tenants and other users that are designed to minimize the energy usage in a building (e.g., turn off lights in unoccupied spaces, keep thermostats set to reasonable levels), but lack the ability to monitor whether these procedures are followed. In most cases, users who are not held accountable (financially or otherwise) will, over time, fail to follow through on the procedures and energy will be wasted.

How Does it Work?

The building owner or manager installs equipment (meters, sensors, etc.) that monitors energy usage either for the entire building or parts of the building. The equipment gathers the data on a regular basis and uploads the information to a local or remote database server. This database server can then produce regular or custom reports that provide feedback on the performance of the building occupants in meeting energy goals and following procedures.

Benefits

Many studies have shown that simply metering and sub-metering the energy use and providing accountability for the users will produce savings of 5% to 10% annually. Once building users are aware that the owner has the means to verify that procedures are being documented, they will adjust their behavior to conform to the energy saving goals of the owner. In many cases it has been demonstrated that not only will the users change their behavior to meet expectations, but in fact will actively seek opportunities for additional savings, particularly when these activities are rewarded with incentives for the additional savings (reduced rent, bonuses, etc.).

Drawbacks

Accountability metering requires some investment in hardware, software and time. It may also be viewed as intrusive by some employees or tenants as it may be perceived as "Big Brother" invading their privacy.

Installation Requirements

As with most energy information applications, the specific hardware and software required for any project will vary depending on the systems to be monitored and the level of detail required from the software. At a minimum, the installation in each building to be monitored will be:

- *AcquiSuite data acquisition server (DAS)*—a stand-alone web server located on the building site that communicates with the sensor(s), stores interval information and communicates with the remote server
- *Pulse output* from an existing primary energy meter (electric, gas, water or steam) or a "shadow" meter installed behind the primary meter(s) to provide data to the DAS
- *Sub-meters (optional)* to monitor energy usage for physical areas of the building (departments, operations, HVAC, lighting, etc.)
- *Sensors (optional)* to monitor runtime or energy consumption for specific equipment
- *Phone line or local area network (LAN) connection* for communication with the remote server

- *Software or services* to provide standard and customized reports

Reports

In general, the reports for accountability metering fall into two categories:

On-line Customized Reports

On-line customized reports provide daily information on consumption and demand to users and managers. The purpose of these reports is so that those responsible can monitor their performance in near real-time and be able to fine tune operations. A typical example is shown in Figure 5-3 which shows an actual internet page depicting a week's demand profile for one meter.

This page can be viewed with any web browser (such as Internet Explorer) and shows the peak demand for one week. This information can be used to determine what operations within the facility are contributing to the overall energy cost. In the case of buildingmanageronline.com, clicking on the graph "zooms" in to allow the user to determine precisely what time critical power consumption occurred and relate it directly to

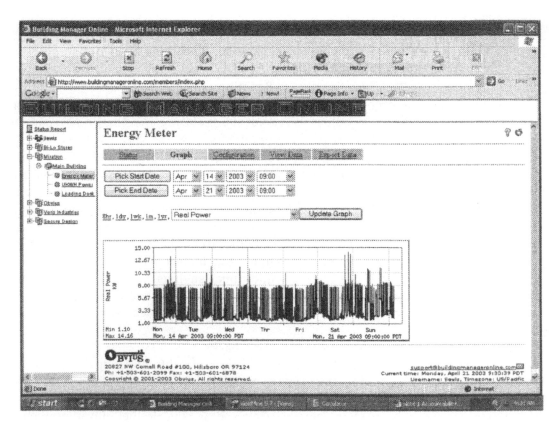

Figure 5-3. Sample web page showing a week's demand profile.

activities and systems in the building (see Figure 5-4).

Clicking on a point in the graph zooms in to a single day.

Regular Monthly Reports

Regular monthly reports provide a more standardized mechanism for reporting usage and the success (or lack thereof) of individual users or facilities at meeting stated objectives. These monthly reports might be published as a web page or distributed via email to provide accountability. For example, assume that XYZ company has established a corporate goal of 5% reduction in energy this year. A typical monthly report to company managers is shown in Figure 5-5.

Actions

On receiving the report displayed in Figure 5-3, the western regional manager has a great deal of information on which to act. First, the manager knows that while her region is ahead of the 5.0 % goal for the year at 5.26%, the month was only slightly over 3.0% and a repeat for another month will likely push her below plan (and cost her bonus money). More importantly, she can instantly determine that the cause of the excess energy usage for July 2002 is a significant increase in Seattle, a location that has performed well year-to-date. This could signal a change in operating hours, significantly higher temperatures, or failure in the control or mechanical system. Clicking on the hyperlink to the Seattle branch would provide additional insight on run times for equipment and average outside air temperature for the month, potentially valuable data for further analysis. Either way, a call to the manager of the Seattle location is probably in order.

Over the longer term, the regional manager (in conjunction with the corporate energy manager), will not only be able to change the behavior of the employees, but will also be able to identify targets for energy studies and investment. If a location has a consistently higher cost per square foot than other locations, it is a likely candidate for further study and potential installation of energy saving equipment.

Costs

As indicated above, the costs for this service will vary widely, depending both on the nature of the installation and the level of detail desired. If we consider a simple installation:

* AcquiSuite server for data acquisition
* Connection to an existing utility pulse output for

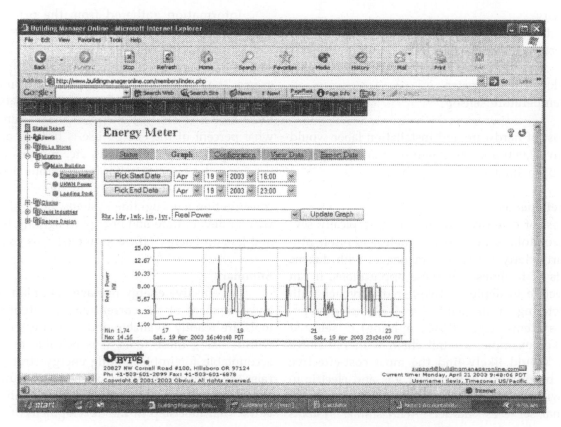

Figure 5-4. Sample web page showing a week's demand profile.

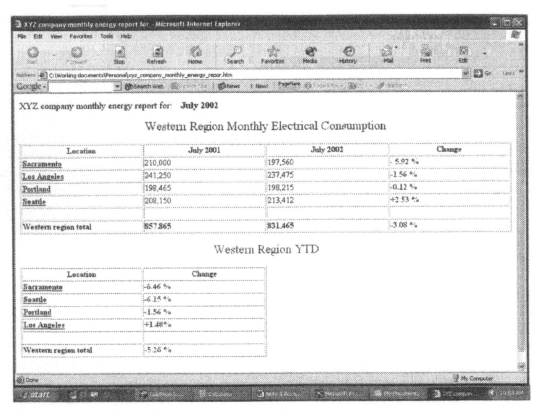

Figure 5-5. Sample web page showing a monthly report.

electricity
- Connection to an existing utility pulse output for gas
- Connection to an existing utility pulse output for water
- One electrical sub-meter
- One temperature sensor

The cost of hardware and installation labor would be less than $2,000 and the ongoing monitoring cost would be approximately $20 per month.

Notes/miscellaneous

The cost of extensive sub-metering (typically several hundred dollars per electric meter) is often prohibitive, particularly for energy managers with dozens or hundreds of facilities. For these applications, it can be cost effective to deploy a two-tiered approach with primary metering in all facilities and extensive metering of sub-systems (e.g., HVAC, lighting) in selected buildings. For example, a retail chain with 150 stores nationwide would likely find the most cost-effective solution to be primary metering of gas and electric in all stores, with extensive sub-metering in perhaps 5 or 10 stores that serve as "models" with similar systems and operations.

Information gathered from the primary meters would be useful in determining how particular buildings perform (for example, energy density and cost per square foot) and for benchmarking usage for later programs. The model stores would provide valuable insight into not just how much was used, but where. Using these model stores, the energy manager can determine what percent of the energy use is heating and air conditioning, lighting, operations, etc. and this information can form the basis for allocating retrofit dollars. If the manager finds that, say, 45% of the electrical energy consumption is in lighting systems, he or she would be wise to devote time and energy to lighting technologies and less on energy efficient motors, fans etc.

Summary

An effective energy management plan begins with an assessment of the current energy situation and the tools are readily available and cost effective in today's world. Using these tools, it is possible to make users accountable for implementing energy strategies and to raise the level of awareness of energy consumption as it relates to operational activities. Accountability metering is the vital first step in the development of an effective energy management program.

EXAMPLE TWO: TENANT SUB-METERING

Tenant sub-metering is a broad term applied to the use of hardware and software to bill tenants in commercial facilities for their actual usage of energy. The goals of tenant sub-metering are: 1) to ensure that the owner recovers the cost of energy from tenants, and 2) to make sure that tenants with high energy usage are not subsidized by those with lower usage.

Background

Many buildings are equipped with only a primary metering system for measuring and billing energy consumption for the entire building. In buildings with this configuration, the tenants are typically billed for energy usage on either a fixed rate (cost per square foot) that is built into the lease, or the bill is allocated to tenants based on their square footage. Each of these methods have inherent flaws, but both share the common problem that energy costs are unlikely to be accurately charged to the tenants. Under a cost per square foot arrangement, the owner will almost certainly collect more or less than the actual bill, and the discrepancy will be even greater during times of energy volatility or low occupancy rates. If the bill is simply divided amongst the tenants on a per square foot basis, tenants with lower energy density (Btu per square foot) will subsidize the space costs for those with higher energy density (e.g., data centers).

These errors become particularly acute when there is a wide variance in occupancy schedules (retail vs. office space) and the building provides central services such as chilled water or conditioned air. Additional complexity is added when the building owner must make decisions in advance on the cost to add when the rate structures for commercial buildings are taken into account.

Residential electrical customers typically pay a flat rate for electrical consumption in the form of cost per kilowatt-hour ($/kWh) which makes calculating a bill relatively simple: read the kWh from the meter and multiply by the $/kWh to get a cost. Commercial rate structures are much more complex. Commercial rate structures typically have the following components:

- Consumption (kWh) charge—this part of the bill is basically the same as the residential charge, but usually has multiple tiers so there is not a single fixed cost. The owner pays at different rates for the amount consumed (e.g., the first 100,000 kWh is billed at $0.07, the next 100,000 kWh is billed at $0.065). These costs will also generally vary by season, so there will commonly be a winter rate and summer rate depending on the supply and demand for electricity that the utility experiences. For purposes of sub-metering, these costs can generally be blended into an average cost per kWh.

- Demand (kW) charge—since the utility has to be certain that adequate supplies of power are available, large customers such as commercial properties are billed not only for the total energy consumed during the month, but also for the maximum power used during a short interval (typically 15 minutes). The demand charge is used to help pay for the costs associated with having generating capacity to meet the highest period of demand (hot summer days, for example) that is not required during lower periods of demand. The demand charge is also applied because the utility will typically have to bring on less-efficient generating plants to meet the peak loads and thus the cost of generating goes up.

- Power factor charge—In a perfect world, the electrical energy provided to a device (e.g., a motor) would be converted to mechanical energy with 100% efficiency. Unfortunately, with very rare exceptions, this is not the case and the inefficiencies associjated with this energy transfer mean that the utility must provide more power than it can actually bill for (the actual math is quite complicated and beyond the scope of this chapter, but it's true). This inefficiency is usually just bundled into the other charges along with things like line loss, etc., but many utilities will bill customers with a power factor penalty if the power factor (the measure of efficiency) falls below a certain level (commonly 92% to 95%).

- Other charges—In addition to things like taxes and surcharges, most utilities are studying or have implemented rate structures that are intended to more directly reflect the actual cost of generation and have large users bear more of the burden during peak times. Things such as time of use metering, load curtailment penalties, etc. will only serve to complicate the average commercial electrical bill even more as time goes on.

How Does it Work?

In the simplest sense, the owner installs meters to monitor the consumption of electricity, gas, water and steam by individual tenants. These meters are connected to a data acquisition server (DAS) like the AcquiSuite from Obvius. The DAS gathers data from the meters on the same schedule as the utility supplying the building and then communicates this data to a local or remote database such as *http://www.buildingmanageronline.com* (BMO). The tenant is then billed at the end of the month at the same rate the building owner pays for the building as a whole and thus the owner recovers the cost of energy from each tenant.

While this process seems very straightforward, as with most things, the devil is in the details. Using the Obvius hardware and BMO service, the gathering, storing and reporting of the data is relatively simple and in most cases the information can be available with just a few hours of installation time. Analyzing the data and producing accurate billings for tenants can be considerably more challenging due to the different rate structures and billing components in a typical commercial setting as outline above. Options for allocating these costs will be considered in the "Actions" section below.

Benefits

The most obvious benefit is that tenants pay their fair share of the energy costs and the owner does not get stuck with unrecoverable costs. An often overlooked, but extremely important benefit is that the building owner is not placed at a competitive disadvantage in the marketplace. If the cost of serving high energy density

tenants is spread over all the tenants, the total cost per square foot of leased space goes up and the owner may lose new or existing tenants to lower cost competition.

Drawbacks

There are two key issues to consider before implementing a tenant sub-metering program:

*   **Costs**—depending on the layout of the services in the building, the cost of sub-metering may be high, particularly for utilities like water and gas where pipe cutting and threading may be involved.

*   **Regulatory agencies**—in many jurisdictions, the state Public Utilities Commission (PUC) regulates the ability of building owners to charge tenants for energy consumption to prevent the owners from overcharging "captive" users. It may be difficult for owners to implement tenant sub-metering programs and to recover the costs of setting up and managing these programs.

Installation Requirements

Details and costs of installation will naturally be heavily dependent on the layout of the building and utility services, but in general most applications can be met with the following hardware:

*   *AcquiSuite data acquisition server (DAS)*—a stand-alone web server located on the building site that

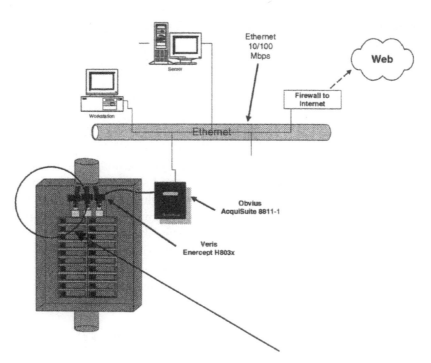

Figure 5-6. AcquiSuite Data Acquisition Server (DAS)

communicates with the sensor(s), stores interval information and communicates with the remote server.

- *Electrical sub-meter(s)*—several companies (including Power Measurement, Ltd. and Veris Industries) produce electrical meters designed for sub-metering applications. These meters can be simple pulse output devices or can provide information using serial communications to provide additional information such as power factor, current and harmonics.

- *Flow meter(s)*—flow meters are used to measure the volume of flow of gas, water and steam. There are a variety of technologies that can be employed, but typically these meters produce either pulse or analog signals that can be read by the AcquiSuite and converted to billable units of measure (gallons, therms, etc.)

- *Btu meter(s)*—Btu meters combine flow meters with temperature sensors to measure the actual energy usage for chilled or hot water. These meters can be useful in cases where a central chilled or hot water plant serves multiple tenants as the tenant is billed not only for the water consumed, but also for the input energy required to produce the conditioned water. These meters can provide simple pulse or analog output to the AcquiSuite or can provide more sophisticated analysis via serial connections

Reports

The level of complexity of reports depends on the method chosen for the tenant billing. The data from the BMO site provides all the information necessary

for calculating tenant bills except for the rate structure information. In the simplest scenario, the owner simply downloads the consumption data from BMO for each tenant and allocates the total bill cost to each tenant based on his or her proportionate consumption (kWh). For most applications, this simple process provides the most cost-effective solution that distributes the cost fairly without creating a complex and expensive process for administration.

The standard reports from BMO provide all the information necessary to do a more thorough allocation that incorporates the demand (kW) charge and power factor penalty (if applicable). The back end processing required to accurately allocate demand charges can be significant as the owner and tenants (and potentially the PUC and utility) must agree on the mechanism used for allocating demand costs. While it is relatively simple to determine the peak interval for the billing period and compare the demand for each tenant for that same interval, the actual allocation of this cost (known as coincident demand) can be very difficult and time-consuming (see "Analysis/Actions" section below).

Analysis/Actions

In the case of simple allocation based on consumption outlined above, the owner imports data from BMO into a spreadsheet (or other cost allocation software) and the software generates a bill for the tenant that is added to the monthly rent.

Billing for coincident demand and time or use charges becomes more complex because there are judgment issues involved as well as simple quantitative analysis. Does the tenant with a flat constant demand (e.g., data center) have to absorb the additional penalties

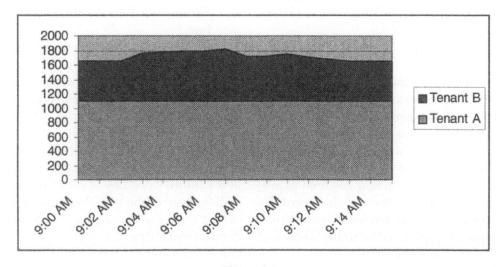

Figure 5-7

for tenants with highly variable rates? Does the tenant whose use is relatively low, but has incremental demand that pushes the total building into a higher demand charge have to absorb all the additional costs or are those costs spread among all tenants?

Notes/Miscellaneous

Simple tenant sub-metering is a relatively easy process that can be implemented by the building owner with the assistance of the providers of the hardware and software. More complex analysis is best left to consultants and resellers who specialize in rate engines and billing analysis.

Summary

Tenant sub-metering is a valuable tool for commercial property owners who want to accurately allocate the costs of energy to tenants and occupants, but it is extremely important to define the scope of the program up-front and to do the homework necessary to ensure compliance with leases and regulatory agencies such as the PUC.

EXAMPLE THREE: LOAD CURTAILMENT/ DEMAND RESPONSE

Load curtailment (or demand response) programs offered by utilities provide commercial and industrial building owners with reduced electrical rates in exchange for an agreement to curtail energy use at the request of the utility. Typically, these requests come during periods of high load such as hot summer afternoons. Building owners or managers who have the ability to reduce loads by turning off equipment or using alternative sources of energy can realize significant savings under these programs.

Background

Utility rate structures typically provide residential, commercial and industrial customers with fixed rates for energy regardless of the generating cost. Not surprisingly, these utilities use the most efficient (lowest cost) generating plants (e.g., nuclear and hydroelectric plants) for the bulk of their load and only bring on less efficient generation (e.g., older coal and gas-fired plants) as load requirements increase. Because of the essentially fixed price of energy to the customer, using less efficient resources has a negative impact on the utility's earnings and they would like to have alternatives.

For the utility, the best option at certain cost levels is to not bring on additional inefficient generating capacity and many utilities find that it is more cost effective to pay customers to curtail loads. If enough customers reduce their usage, the utility does not have to add generation (or purchase additional supplies on the spot market). This compensation can take several forms, but it generally is reflected in a lower overall rate schedule for the owner throughout the year.

How Does it Work?

Load curtailment can take a variety of forms depending on the severity of the shortfall in supply and the type of agreement between the utility and the end user and the equipment in place to implement the reductions. In the simplest form, the utility notifies the owner of a curtailment request (typically a day in advance) and it is up to the customer to voluntarily meet the requested load reduction. Options for the end user range from adjusting temperatures to shutting off lights to closing facilities to meet the requested reduction levels.

The future of demand response is likely to contain more options for automatic, real-time reductions in load, triggered directly by the utility with little involvement of the owner. This option allows for the matching of loads much more closely to actual demand levels in real time, but obviously requires much higher levels of automation and investment. In this scenario, the owner and utility agree in advance what steps can be taken to lower the energy usage in the facility and the utility can initiate load reduction measures remotely using the customer's control system or additional controls installed in the building.

Benefits

For the utility, the primary benefits include:

- Eliminating the cost of bringing another plant on line
- Providing more cost-effective generating sources (i.e., more profit)
- Minimizing the environmental impact of generating plants with poor emissions records (fossil fuel plants)

For the customer, the clear benefit is a reduced cost of energy in the near term and avoiding cost increases in the future since the utility is at least theoretically operating more efficiently. There may also be an additional benefit from the installation of controls and equipment in facilities that provide the user with more information and control over the operation of the facilities during periods when load curtailment is not in effect.

Drawbacks

From a financial perspective, both the utility and customer are likely to incur costs to add or retrofit controls and equipment in the customer's facility. Both must also commit ongoing resources to track and manage the operation of the load curtailment and to provide reports. The customer is also likely to experience inconvenience in the form of less comfortable space temperatures (i.e., higher in summer, lower in winter) than desired if HVAC equipment is shut off, or reduced lighting levels. This kind of program will clearly impact the customer in some respects and these effects need to be maintained within acceptable limits.

Installation Requirements

Most facilities with installed building automation systems (BAS) already have the equipment in place to meet day-ahead requests by using the BAS to initiate new or pre-programmed operational strategies to limit energy use. The primary installation of new equipment could include additional metering and also likely some form of remote monitoring equipment to allow the utility to monitor the success of the program at reducing loads in the building.

Customers without existing BAS systems or those participating in real time demand response programs will have to make additional investment in monitoring and control systems (for example, remote setpoint thermostats). In most cases, the most cost effective way to implement a real time program is to use the internet and web-enabled data acquisition servers (DAS) like the AcquiSuite from Obvius to provide real time feedback to the owner and the utility of the load before and after curtailment. The DAS can also function as the conduit for the utility to provide a supervisory signal to the BAS or to the systems directly.

Reports

The information needed to implement and evaluate the effectiveness of a load curtailment program is near real-time interval data (kW). This report (either for a single facility/meter or an aggregated load from multiple locations) might look like the following figure (Figure 5-8).

Analysis/Actions

The utility determines that there will be a shortage of available power (or that there will be a need to bring additional generating capacity on line) and informs users with demand response contracts that they will be expected to reduce their demand to the contracted levels. Users are expected to meet these requirements through

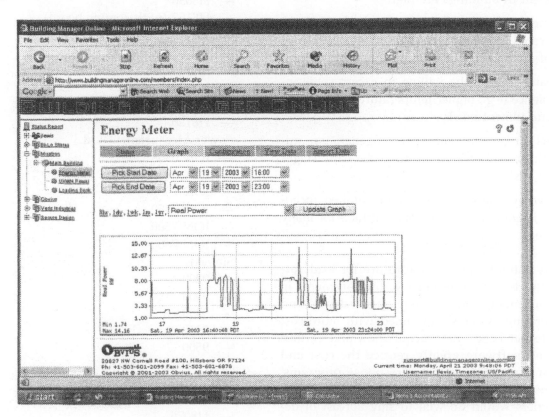

Figure 5-8. Near Real-time Interval Data (kW)

some combination of automatic or manual shutdown of equipment, temperature adjustment or closing down some or all of their operations. The DAS is used to verify that the user has met the required load curtailment and that the utility has achieved its objectives for taking load off the grid.

Costs

As with all the application notes in this series, it is very difficult to estimate costs due to a variety of factors (wiring distances, communications issues, scheduled shutdowns, etc.), but some general guidelines for costs (hardware and installation) are:

- AcquiSuite™ data acquisition server—$1,200 to $1,800

- Electrical sub-meter (3 phase)—$600 to $1,000

- Data storage and reports—$20 per month per AcquiSuite

Notes/Miscellaneous

To date, implementation of demand response programs has been limited for a variety of reasons:

- Costs to monitor and control energy consuming equipment in buildings are high

- Implementation of these types of programs can be complicated, particularly if the utility desires some form of automated control via the internet

- Occupants have to be willing to except some inconvenience (e.g., higher temperatures) in order to meet curtailment needs

- Options for on-site generation (cogeneration, microturbines, fuel cells, etc.) are in the early stages and not cost effective for many owners

- Low overall energy costs provide limited incentive for negotiating curtailment contracts for many customers

All this notwithstanding, the future for load curtailment contracts is very promising. Improvements in building equipment (e.g., variable speed drives and on-site generating systems) combined with more cost effective internet based data acquisition and control hardware and software greatly reduce the cost and impact of implementing demand response programs. DR programs are likely to become more prevalent and building owners would do well to stay abreast of developments in this area.

EXAMPLE FOUR: WEB DISPLAY OF EXISTING METER DATA

Using either the AcquiSuite™ or AcquiLite™ data acquisition servers (DAS) from Obvius to connect existing electrical, gas or flow meters to the web. Once web-enabled, meter data are available for viewing from any web browser at www.buildingmanageronline.com.

Background

Many owners of commercial and industrial (C&I) buildings have meters installed in their buildings that either provide or could provide outputs (either pulse or serial) that will allow users to see energy information on the web. Many of these meters were installed by the utility for primary metering or as submeters within the building to monitor usage. Examples of meters that are compatible with the DAS from Obvius include:

- Pulse output from any utility meter with a pulse output sub-base;
- Pulse output from submeter;
- E-MON—any submeter with a pulse;
- Veris Industries
  - H8053
  - H8035
  - H8036
  - H8163 (with comms board)
  - H8075
  - H8076
  - H8238
  - H663
  - H704
  - H8126
- Power Measurement Ltd.
  - ION 6200
  - ION 7300
  - ION 7330
  - ION 7350
  - ION 7500
  - ION 7600
- Siemens
  - 9200
  - 9300
  - 9330
  - 9350
  - 9500
  - 9600
- Square D
  - H8163-CB
  - H8076
  - H8075

Other meters using Modbus RTU may be compatible with the DAS, contact the factory for information or questions.

How Does it Work?

Each AcquiSuite DAS can support up to 32 Modbus meters and 128 pulse meters and each AcquiLite can support up to 4 pulse meters. There are two basic connection approaches, depending on the type of output available:

Pulse Output Meters

The two wires from the pulse output from the meter (electric, gas, water, steam) are connected to one of the four pulse inputs on the AcquiSuite (A8811-1). The installer uses a web browser to name the meter and add the appropriate multiplier to convert each pulse to valid engineering units. The DAS is then connected to the internet (either via phone line or LAN connection) and data are pushed to the web site for viewing. (See Figure 5-9.)

Modbus RTU Meters

Devices from the list above with a serial output will be automatically recognized by the AcquiSuite as soon as they are connected. The DAS has drivers for these devices that will recognize and configure the device in unit, so the installer only needs to give the meter a name and setup the parameters for reading interval data from the meters and uploading to the remote server.

Regardless of whether the input is pulse or Modbus, the DAS gathers data on user-selected intervals (from 1 to 60 minutes) and stores the data until it is uploaded to the BMO server (typically daily). Once uploaded, the data from all the meters is available for viewing from any web browser (see "Reports" below for sample reports).

Benefits

Many C&I building owners have installed submeters with local display options that are also capable of providing pulse or serial outputs, but the meters have never been connected to a local or remote server due to cost or other constraints. Having only local display means that someone must physically read the meter, record the values and input this information into a spreadsheet or database for calculation. This approach is not only inefficient, but is also prone to error. It is virtually impossible to synchronize readings with the utility bills, which means that accurate accounting is unlikely.

Because the AcquiSuite automatically recognizes supported Modbus meters, installation can be done by any electrician or local building personnel without the need for expensive software and integration.

Using a DAS provides many benefits, including:

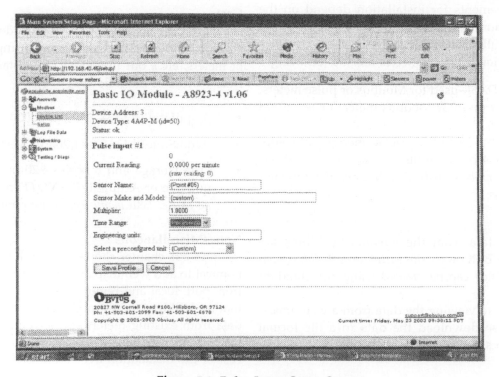

Figure 5-9. Pulse Input Setup Screen

- Continuous interval reading makes synchronizing to utility bills simple
- Data from multiple, geographically dispersed buildings are automatic as all data come to a single site for viewing
- Information can be viewed and downloaded from any web browser in spreadsheet and database compatible file formats
- All data are stored at a secure site and record-keeping i10s minimized
- The DAS can be programmed to call out on alarms via email or pager in the event of a problem

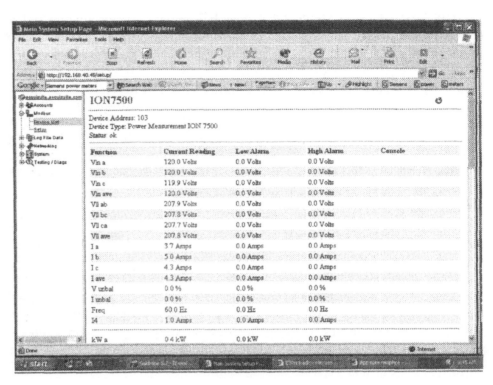

Figure 5-10. Modbus RTU Setup Screen

Drawbacks

The major obstacle to this approach is that it requires some investment of time and materials to connect the meter(s) to the DAS and to provide phone line or LAN connection for communications.

Installation Requirements

The requirements for installation depend on the type of installation and whether new meters are being installed. Generally the only requirements for connection to one or more existing meters are the following:

- *AcquiSuite DAS*—used for Modbus or pulse meters, can support up to 32 Modbus meters or up to 128 pulse meters
- *AcquiLite DAS*—used for pulse meters only, supports up to 4 pulse inputs
- *Phone line (can be shared) or LAN connection* for communications

Reports

Once the data from the various buildings are uploaded to the BMO web site (*http://www.buildingmanageronline.com*), they can be viewed using any standard web browser.

In addition to viewing the data from a web browser, users can also download the data in a file format compatible with spreadsheets or databases:

Analysis/Actions

Once the data are exported from BMO to a local spreadsheet, the submeter energy usage can be allocated to the tenant or department.

Costs

Typical installed costs will vary depending on the specific requirements of the job (wiring runs, number of meters, etc.), but in general the installed cost for the DAS will be in the following range:

- AcquiLite™ DAS—$500 to $600
- AcquiSuite™ DAS—$1,200 to $1,800
- Data storage and reports—$20 per month per AcquiSuite or AcquiLite™ (*NOTE: the cost is the same no matter how many meters are connected to a DAS*)

Notes/Miscellaneous

As this chapter shows, it is both practical and economical to add web display capability to existing meters from both local and remote sites. It is important to note that the building owner or manager who wants to gather data from existing meters can also add new submeters to existing buildings at the same time and spread the cost of the installation over more points.

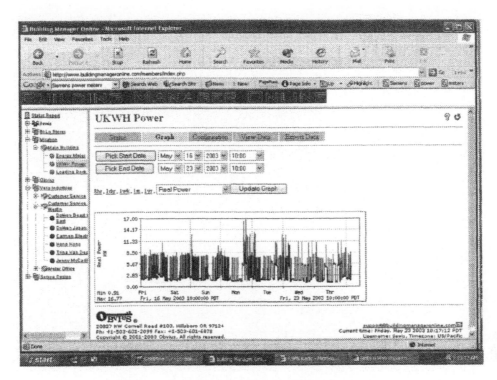

Figure 5-11. Sample kW Report from BMO Site

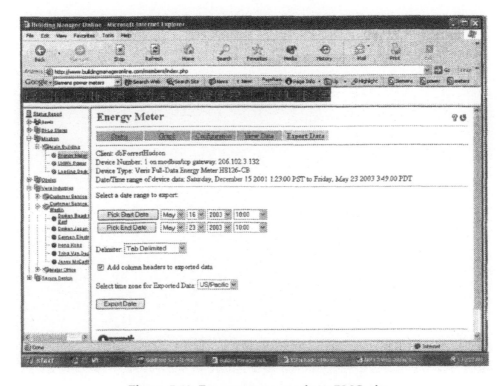

Figure 5-12. Export setup page from BMO site

Chapter 6

Web Based Building Automation Controls and Energy Information Systems

Paul J. Allen, David C. Green, Steve Tom, and *Jim Lewis*

INTRODUCTION

A successful energy management program (EMP) has three main components. The first system is the building automation system (BAS) that provides controls for air conditioning, lighting and other systems in each building or facility. The second system is the Energy Information System (EIS), which is a suite of information technologies that work with the EMS to provide data and information to energy managers and other stakeholders. The final key component is a commitment from both management and staff that collectively participate in the EMP. It is the combination of technology and people that makes an EMP successful and sustainable.]

This chapter will focus on the technical systems, the BAS and the EIS, and will get "under-the-hood" to show how these systems work together to effectively control and measure energy usage.

BUILDING AUTOMATION SYSTEMS

The combination of low cost, high performance microcomputers together with the emergence of high-capacity communication lines, networks and the internet has produced explosive growth in the use of web-based technology for direct digital control building automation systems (BAS) [1]. Many of these current BAS systems use a proprietary information structure and communications protocol that greatly limits the plug and play application and addition of interchangeable components in the system. Control solutions such as BACnet and LonWorks have helped this situation somewhat, but they have also introduced their own levels of difficulties. The BAS of the future will integrate state of the art Information Technology (IT) standards used widely on the internet today. These new IT based systems are rapidly overtaking the older BAS systems. Most of the established BAS companies are quickly developing ways to interface their systems using IT standards to allow the use of web browsers such as Internet Explorer and Firefox.

This section will examine all facets of a BAS, from field controllers to the front-end interface. The emphasis is on understanding the basic BAS components and protocols first, then examining how a BAS has changed based on the influence of IT standards. Finally, this section will also discuss upgrade options for legacy BAS systems and BAS design strategies.

Even though we will be referring exclusively to the term BAS in this chapter, the building automation controls industry also uses the following terms interchangeably with BAS: direct digital control (DDC), energy management system (EMS), energy management and control system (EMCS), building automation and control system (BACS) and building management system (BMS).

The Basics of Today's BAS

At a minimum, a BAS is used to control functions of a heating, ventilating, and air conditioning (HVAC) system, including temperature and ventilation, as well as equipment scheduling. Even basic BAS are generally expected to perform control functions that include demand limiting and duty cycling of equipment. Additional basic features recording utility demand and energy use, building conditions, climatic data, and equipment status. BAS report outputs can show the facility utility load profiles, trends and operation logs of equipment, and generation of maintenance schedules. Another basic feature of a BAS is to notify service personnel of defective equipment and of sensors out of normal range. Using email or text pages, these alarms notify those responsible individuals so that repairs can be implemented quickly.

More elaborate BAS can integrate additional building systems, such as video surveillance, access control, lighting control and interfacing with the fire and security systems. A BAS might provide a status summary from multiple dedicated systems through a single operator interface, and sometimes to share at least a limited amount of data between controllers. A BAS may not control access to a building or generate security alarms, but it may show the status of door and window switches on the floor plan. Similarly, a single occupancy sensor may control both the HVAC and the lighting systems for a zone. However, in large organizations and campuses today, it is still more common to see dedicated systems for these additional building systems due to divisions in management functional responsibility, code issues, and features/performance of dedicated systems.

Today's BAS can receive and process more sophisticated data on equipment operation and status from such sensors as vibration sensors on motors, ultrasonic sensors on steam traps, infrared sensors in equipment rooms, and differential pressure sensors for filters. Top of the line BAS today also have additional capabilities, such as chiller/boiler plant optimization, time schedule/setpoint management, alarm management and tenant billing to name a few. Most BAS manufacturers today have started to offer some form of web-based access to their existing control systems and are actively developing web-based capability for their future products.

Controller-level Hardware

BAS controllers are used to provide the inputs, outputs and global functions required to control mechanical and electrical equipment. Most BAS manufacturers provide a variety of controllers tailored to suit the specific need. Shown below is a list of the most common BAS controllers:

Communications Interface

Provides the communication interface between the operator workstation and lower-tier controller network. On a polling controller network, a communications interface is used to transfer data between the controllers.

Primary Controller

Provides global functions for the BAS control network that can include, real-time clock, trend data storage, alarms, and other higher-level programming support. Some BAS manufacturers combine all these functions into one primary controller while other manufacturers have separate controllers that are dedicated to each global function.

Secondary Controller

Contains the control logic and programs for the control application. Secondary controllers usually include some on-board I/O and may interface to expansion modules for additional I/O. Inputs include temperatures, relative humidity, pressures, and fan & pump status. Outputs include on/off, and valve/damper control. Also included in this group are application specific controllers that have limited capability and are designed for a specific task. Examples include controllers for VAV boxes, fan coil

units or multistage cooling and heating direct expansion (DX) air conditioning systems.

For further reference, the Iowa Energy Center has an excellent web site (http://www.ddc-online.org) that shows a complete overview of the designs, installations, operation and maintenance of most BAS on the market today.

Controller-level Programming

BAS controllers typically contain software that can control output devices to maintain temperature, relative humidity, pressure, and flow to a desired setpoint. The software programming can also adjust equipment on-off times based on a time-of-day and day-of-week schedule to operate only when needed.

The software used to program the controllers varies by BAS manufacturer and basically falls into three categories:

1. Fill-in-the-blank standard algorithms
2. Line-by-line custom programming
3. Graphical custom programming.

Fill-in-the-blank

Uses pre-coded software algorithms that operate in a consistent, standard way. The user fills in the algorithm configuration parameters by entering the appropriate numbers in a table. Typically, smaller control devices, like those that control a fan coil or VAV box controller use this type of programming. These devices all work the same way and have the same inputs and outputs.

A few manufacturers have used fill-in-the-blank programming for more complex devices such as air handlers where a variety of configurations can exist. Standard algorithms use the same program algorithm for each individual air handler device. As an example, the chilled water valve for an air-handling unit is programmed using a standard algorithm with only the configuration parameters adjusted to customize it for the particular type of valve output and sensor inputs. The rest of the air-handler components (supply fan, heating coil, outside air damper, etc.) are programmed using the appropriate standard algorithm.

The advantage of fill-in-the-blank standard algorithms is that they are easy to program and are standard. The downside is that if the standard algorithm does not function the way you want, or there is not a standard algorithm available, then the system requires development of a custom program.

Line-by-line Custom Programming

Control programs are developed from scratch and

are customized to the specific application using the BAS manufacturer's controls programming language. In most cases, programs can be re-used for similar systems with modifications as needed to fit the particular application.

The advantage of the line-by-line custom programs is that technicians can customize the programs to fit any controls application. The disadvantage is that each program is unique and trouble-shooting control problems can be tedious since each program must be interrogated line-by-line.

Graphical Custom Programming

BAS manufacturers developed this method to show the control unit programs in a flow chart style, thus making the programming tasks easier to follow and troubleshoot.

Below are some additional issues to consider regarding control unit programming:

- Can technicians program the control units remotely (either network or modem dial-in) or must they connect directly to the control unit network at the site?
- Does the BAS manufacturer provide the programming tools needed to program the control units?
- Is training available to learn how to program the control units? How difficult is it to learn?
- How difficult is it to troubleshoot control programs for proper operation?

Controller-level Communications Network

The BAS controller network varies depending on the manufacturer. Several of the most common BAS controller networks used today include RS-485, Ethernet, ARCNET and LonWorks.

RS-485

Developed in 1983 by the Electronic Industries Association (EIA) and the Telecommunications Industry Association (TIA). The EIA once labeled all of its standards with the prefix "RS" (Recommended Standard). An RS-485 network is a half-duplex, multi-drop network, which means that multiple transmitters and receivers can exist on the network.

Ethernet

The Xerox Palo Alto Research Center (PARC) developed the first experimental Ethernet system in the early 1970s. Today, Ethernet is the most widely used local area network (LAN) technology. The original and most popular version of Ethernet supports a data transmission rate of 10 Mb/s. Newer versions of Ethernet called "Fast Ethernet" and "Gigabit Ethernet" support data rates of 100 Mb/s

and 1 Gb/s (1000 Mb/s). Ten-gigabit Ethernet provides up to 10 billion bits per second.

ARCNET

A company called Datapoint originally developed an office automation network in the late 1970's. The industry referred to this system as ARC (attached resource computer) and the network that connected these resources as ARCNET. Datapoint envisioned a network with distributed computing power operating as one larger computer.

LonWorks

Developed by the Echelon Corporation in the 1990s. A typical LonWorks control network contains a number of nodes that perform simple tasks. Devices such as proximity sensors, switches, motion detectors, relays, motor drives, and instruments, may all be nodes on the network. Complex control algorithms are performed through the LonWorks network, such as running a manufacturing line or automating a building.

Controller-level Communications Protocol

A communications protocol is a set of rules or standards governing the exchange of data between BAS controllers over a digital communications network. This section describes the most common protocols used in a BAS.

BACnet

Building Automation Control Network is a standard communication protocol developed by ASHRAE specifically for the building controls industry. It defines how applications package information for communication between different building automation systems. The American National Standards Institute has adopted it as a standard (ASHRAE/ANSI 135-2008). The International Organization for Standardization adopted it as ISO 16484-5 in 2003.

LonTalk

An interoperable protocol developed by the Echelon Corporation and named as a standard by the Electronics Industries Alliance (ANSI/EIA-709.1-A-1999). Echelon packages LonTalk on their "Neuron chip" which is embedded in control devices used in a LonWorks network.

Proprietary RS-485

The protocol implemented on the RS-485 network is usually proprietary and varies from vendor to vendor. The Carrier Comfort Network (CCN) is an example of a proprietary RS-485 communications protocol.

Modbus

In 1978, Modicon developed the Modbus protocol for industrial control systems. Modbus variations include Modbus ASCII, Modbus RTU, Intel® Modbus RTU, Modbus Plus, and Modbus/IP. Modbus protocol is the single most supported protocol in the industrial controls environment.

The Internet Protocol Suite

This is a set of communications protocols used for the internet and other similar networks. It is commonly also known as TCP/IP, named from two of the most important protocols in it: the transmission control protocol (TCP) and the internet protocol (IP), which were the first two networking protocols defined in this standard. Most BAS today use communications devices that take advantage of these standards to extend the physical distance of their networks. For further reference, http://www.protocols.com/pbook/tcpip1.htm provides an excellent source of information on this subject.

Enterprise-level Client Hardware/Software

Normally, a PC workstation provides operator interface into the BAS. The PC workstation may or may not connect to a LAN. If a server is part of the BAS, the PC workstation would need LAN access to the server data files and graphics. Some smaller BAS use stand-alone PCs that have all the BAS software and configuration data loaded on each PC. Keeping the configuration data and graphics in-sync on each PC becomes problematic with this design.

A graphical user interface (GUI) is one of the client-side software applications that provides a window into the BAS. The GUI usually includes facility floor plans that link to detailed schematic representations and real-time control points of the building systems monitored by the BAS. The GUI allows technicians to change control parameters such as setpoints, time schedules, or temporarily override equipment operation. Other client-side software applications include:

- Alarm monitoring
- Password administration
- System setup configuration
- Report generation
- Control Unit programming and configuration

Enterprise-level Server Hardware/Software

Servers provide scalability, centralized global functions, data warehousing, multi-user access and protocol translations for a mid to large size BAS. Servers have become more prominent in the BAS architecture as the need has grown to integrate multi-vendor systems, publish and analyze data over an intranet or extranet and provide multi-user access to the BAS. While having a central server on a distributed BAS may seem contradictory, in reality a server does not take away from the stand-alone nature of a distributed control system. Servers enhance a distributed control system by providing functions that applications cannot perform at the controller level. In fact, a BAS may have several servers distributing tasks such as web publishing, database storage, and control system communication.

Servers provide the ability to globally control a BAS. Facility-wide time scheduling, load-shedding, or setpoint resets are examples of global functions a BAS server can perform. Since these types of functions are overrides to the standard BAS controller-level programs, having them reside in the server requires that steps be taken to insure continued control system operation should the server go down for any length of time. The distributed BAS should have the ability to "time out" of a server override if communications with the server is lost. When the server comes back on line, the BAS should have rules that govern whether the override should still be in effect, start over, or cancel.

BAS Design Issues

Aside from the impact that IT will have on future EMS, there are some fundamental characteristics that owners have always desired and will continue to desire from a new BAS:

- Single-seat user interface
- Compatible with existing BAS
- Easy-to-use
- Easily expandable
- Competitive and low-cost
- Owner maintainable

There have been several changes made by the BAS industry to help satisfy some of these desires. The creation of open protocols such as LonWorks and BACnet has made field panel interoperability plausible. The development of overlay systems that communicate to multiple BAS vendor systems has made a single-seat operation possible.

There are two strategies available for the design and specification of a BAS for new or existing facilities:

1. Specify a multi-vendor interoperable BAS.
2. Standard on one BAS manufacturers system.

Specifying a multi-vendor interoperable BAS is

probably the most popular choice of the facility design community. Using this approach, the engineer's controls design is more schematic and the specifications more performance-based. The engineer delegates the responsibility of the detailed BAS design to the temperature controls contractor since the engineer does not actually know which BAS vendor will be selected. Therefore, the resulting BAS design is by nature somewhat vague and entirely performance-based. The key to making this approach successful is in the details of the performance specification, which is not a trivial task. Competition results from multiple BAS vendors bidding on the entire BAS controls installation. The engineer may further specify that the BAS protocol be open (LonWorks or BACnet) to have the systems be interoperable. Unfortunately, there is currently little commonality between different BAS vendors' low-level panel programming, and different service tool software is needed for each BAS vendors' system regardless of the open protocol. In the end, even though the BAS meets the original specification, it might not be the same as or even compatible with the existing facility BAS. Thus, the owner might operate this new system as just one of many disparate BAS systems in their BAS portfolio.

Another alternative to the multi-vendor BAS would be to have the owner pre-qualify two or three BAS vendors all using the same standard protocol (generally BACnet) and then allow these selected vendors to competitively bid projects. One vendor is chosen to provide the server front-end software, and the others integrate the projects they win into that front-end. This allows competitive bidding, while keeping the number of service tools and systems the maintenance crews must understand to a manageable number.

The second approach is based on standardizing on one BAS manufacturer's system. To create competition and keep installation cost low, the engineer must create the BAS design as part of the design documents and prescriptively specify all components of the BAS. This allows multiple temperature control contractors to bid on the BAS installation (wire/conduit/sensor actuators)—everything outside of the BAS field panel. Everything inside the BAS field panel is owner furnished. Contractors familiar with the owners' BAS, or the owners' own technicians perform the controller wire termination, programming and startup. This approach is successful when all parties work together. The design engineer must produce a good BAS design. The temperature controls contractor must install the field wire, conduit, sensors and actuators properly. Finally, the BAS contractor must terminate and program the BAS panel correctly. A successful project is a system that integrates seamlessly with the owners' existing BAS.

Upgrading an Existing BAS

Most users already own and operate a legacy BAS that they might desire to upgrade from a stand-alone BAS to a network-based system [2]. The benefits of a network-based BAS include better standard operational practices and procedures, opportunities to share cost-savings programs and strategies, and wider access to building control processes. The keys to justifying the costs associated with networking a BAS are that it can be done at a reasonable cost and it is relatively simple to implement and operate.

There are three main strategies available when upgrading a BAS from a stand-alone system to a network-based system:

1. Remove existing BAS and replace with new network-based BAS.

2. Update existing BAS with the same manufacturer's latest network-based system.

3. Install a BAS interface product that networks an existing BAS.

The first upgrade strategy is to simply replace the existing BAS with a newer network-based BAS that has been established as a standard within your company. The cost for this option is solely dependent on the size of the BAS that will be replaced. However, this approach might be justified if the existing BAS requires high annual maintenance costs or has become functionally obsolete.

The second upgrade strategy available is to request a proposal from the original BAS manufacturer for their upgrade options. Most BAS manufacturers have developed some form of Ethernet network connectivity. Typically, some additional hardware and software is required to make the system work on an Ethernet network. The cost for this might be very reasonable, or it could be very expensive. It all depends on how much change is required and the associated hardware, software and labor cost to make it all work.

The third upgrade strategy involves the installation of a new network-based system that is specifically designed to interface to different BAS systems [EDITOR: insert chapter reference here]. These systems typically have dedicated hardware that connects to the BAS network and software drivers that communicate to the existing BAS controllers. The new BAS interface controllers also have an Ethernet connection so they can communicate on the corporate LAN. Users view the BAS real-time data using web browser software on their PC. The advantage of this strategy is that a multitude of different BAS systems can be interfaced together. The disadvantage is that users must still use the existing BAS software to edit or add new control programs in the existing BAS field panels. Furthermore, software license fees must be maintained for all original BAS along with the new integration software.

Enhancing an Existing BAS with Custom Programs

Users might find they needing their BAS to meet certain functions that are not fully integrated into their BAS. The user could request the manufacturer make the changes to the BAS software, but this is a slow and potentially expensive process. A quicker solution is to use the existing BAS software to obtain the data needed which is then exported to your own custom programs that are designed to perform the desired requirement. Shown below are a few BAS enhancements that use this technique:

Alarm Notification System

When mechanical systems become defective or operate out normal design range, the existing BAS software can be programmed to create an alarm. In most cases, this alarm is either printed on the BAS alarm printer or displayed on the BAS terminal. To enhance the alarm notification beyond the existing BAS functionality, a separate program could be created to read all of the incoming BAS alarms and then notify all responsible service personnel automatically via email or text page. For example, suppose the chilled water temperature exceeded 55F on an operating chiller. The BAS would generate an alarm that would be read by the alarm notification program and automatically emailed/text paged to responsible maintenance department personnel. If desired, the alarm notification program could be designed to interface to the users preventative maintenance program to create a work order for repair. A web-based interface program could also be developed to display a history of the alarms by time and date to let users see the frequency of the alarm occurrence.

Interval Data Collection

Most BAS have the ability to collect data from the sensors and utility meters connected to their system. Typically, the data collection report is stored on the BAS front-end PC for the user to display using a spreadsheet program. An enhancement to this data function would be to create a separate program to automatically read all of the data from these reports into a relational database or to read from the BAS database directly. Another web-based

program could be developed to graphically display the collected data quickly an easily. This web-based program could be part of a larger energy information system (EIS), which is covered in detail later in this chapter.

Time and Setpoint Reset

A BAS uses time schedules to turn the equipment on/off at scheduled intervals. Likewise, setpoints are used to modulate control devices to maintain specific temperature, relative humidity, flow and other requirements. The time and setpoint schedules will eventually get change from their optimal settings resulting from periodic too hot/too cold calls received by the maintenance department. Adjustments to the time and setpoint schedules are done manually through the BAS software and once changed will stay at those values indefinitely. A better solution is to create a separate program to keep the time and setpoint schedules at their optimal state by automatically resetting them on a daily basis. The program is design to provide the user with a method to manage the time and setpoint schedules for a large campus facility in a master schedule database.

Future Trends in BAS

The future BAS can be found on the web. Most all BAS manufacturers see the need to use web-based system using IT standards. Tremendous economies of scale and synergies can be found there. Manufacturers no longer have to create the transport mechanisms for data to flow within a building or campus. They just need to make sure their equipment can utilize the network data paths already installed or designed for a facility. Likewise, with the software to display data to users, manufacturers that take advantage of presentation layer standards such as HTML, Java and Flash can provide the end user with a rich, graphical and intuitive interface to their BAS using a standard web browser.

Standards help contain costs when new products are developed. While there is a risk of stagnation or at least uninspired creativity using standards, internet standards have yet to fall into this category, due to the large consumer demand for rich content on the internet. A BAS, even at its most extensive implementation, will only use a tiny subset of the tools available for creating content on the internet.

When a BAS manufacturer does not have to concentrate on the transport mechanism of data or the presentation of that data, new products can be created at a lower cost and more quickly. When the user interface is a web browser, building owners can foster competition among manufacturers since each BAS system is inherently compatible with any competitors at the presentation level. All that separates one BAS from another on a web browser is a hyperlink.

Another area where costs will continue to fall in using internet standards is the hardware required to transport data within a building or a campus. Off-the-shelf products such as routers, switches, WiFi, WiMAX, general packet radio service (GPRS), Zigbee and server computers make the BAS just another node of the IT infrastructure. Standard IT tools can be used to diagnose the BAS network, generate reports of BAS bandwidth on the intranet, and backup the BAS database.

Owners will reap the benefits of internet standards through a richer user interface, more competition among BAS providers, and the ability to use there IT infrastructure to leverage the cost of transporting data within a facility.

The Enterprise

Extensible Markup Language (XML)

XML is an internet standard that organizes data into a predefined format for the main purpose of sharing between or within computer systems. What makes XML unique is that data tags within the XML document can be custom, or created on the fly; and unlike HTML, are not formatted for presenting the data graphically. This makes XML a great choice for machine-to-machine (M2M) communication. XML data exchanges often utilize SOAP, or "Simple Object Access Protocol" to carry the data. The combination of XML/SOAP is commonly referred to as "web services."

Why is M2M so important? Because the next wave of BAS products will include "hooks" into other internet based systems. BAS systems have done a great job of integrating building-related components together. BACnet, LonWorks, and Modbus provide the capability of connecting together disparate building components made from different manufacturers, so that a lighting control panel can receive a photocell input from a rooftop building controller, or a variable frequency drive can communicate an alarm on the BAS when a failure occurs.

The future will require a BAS to connect to enterprise level systems, not just building level systems. This is where M2M and web services come into play. Web services can be thought of as plug-ins for your BAS to communicate with a web based system or server. An example of this would be time synchronization. The internet has many time servers that can provide the exact local as well as GMT time. A BAS can have a web service that would plug-in to the BAS, synchronizing all of the time clocks within a facility to the atomic clock in Boulder, Colorado. Another example would be obtaining the outside air

temperature from the local weather service. Instead of the BAS just measuring the outside air temperature at a local controller, a web service could provide the outside air temperature, humidity, barometric pressure, and any other weather related data. Now the BAS can make more intelligent decisions on using outdoor air for comfort cooling, determining wet bulb setpoints for cooling towers, or even announcing an alert that a storm is imminent.

More enticing than connecting to weather and time servers is the promise of connecting to a facility's enterprise data. The BAS of the future must become an integral part of the decision making for allocating personnel, budgeting maintenance and upgrades, purchasing energy, and billing those that use the energy. Most larger facilities have departments that provide these types of services, yet the BAS has always stood alone, providing input through exported reports, system alarms, or human analysis. Enterprise level integration would create web services to connect directly to these systems, providing the data necessary to make informed decisions about capital investments, energy, or personnel. See figure 1 for what a BAS might look like in the future.

The good news is that XML and web services have gained the market acceptance to become the standard for enterprise level connectivity. The bad news is that this is still in its infancy for most BAS vendors. It is a very costly effort to create an enterprise level web service today. Even though web services are supported by Microsoft, Apple, Oracle, and others, they can still be custom solutions, tailored to a specific accounting, maintenance management, or energy procurement system. For web services to become mainstream in the BAS world, common services will need to be created that can be used by all BAS vendors. In addition, for web services to be properly implemented the skill set for BAS programmers and installers will need to include XML and a basic understanding of IP. If facility managers and technicians are to be able to make changes, adjustments and enhancements to their enterprise system, they too will require this skill set.

The future will also need to better define the decision logic and troubleshooting tools when implementing web services. When the BAS sends duplicate alerts to a maintenance management system, where does the logic reside to send only one technician to the trouble call? This is currently undefined. Standard tools for testing scenarios online and offline need to be developed. Even though web services typically rely on XML, which is a self-documenting standard, it can be very verbose. Tools are available to validate XML documents. When a facility decides to change their accounting system to a newer version or a different vendor, will the BAS be able to adapt?

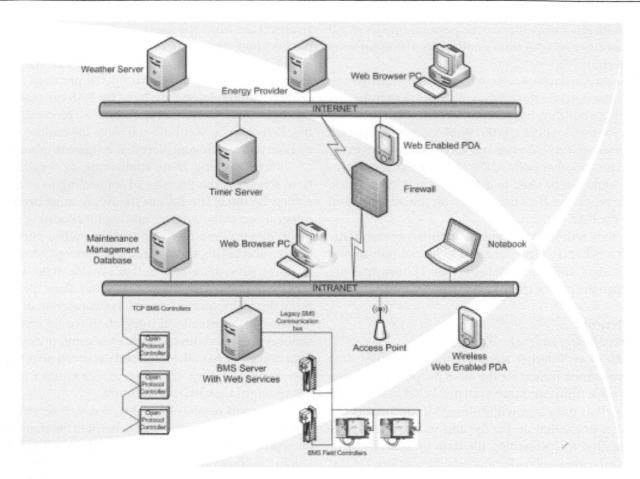

Figure 6-1. BAS Network Schematic

Conversion and upgrade tools also need to be considered when defining BAS web services.

Even without all the tools identified, enterprise level connectivity is moving ahead rapidly. The benefits of integrating BAS data within the facility's other systems can outweigh the immediate need for a complete set of tools. Web services through XML place the BAS directly into the facility data infrastructure. That is a good place to be for an energy manager wanting to maximize the investment in their BAS.

The BAS of old relied heavily on a collection of separate systems that operated independently, and often with proprietary communication protocols that made expansion, modification, updating and integration with other building or plant information and control systems very cumbersome, if not impossible. Today the BAS is not only expected to handle all of the energy and equipment related tasks, but also to provide operating information and control interfaces to other facility systems, including the total facility or enterprise management system.

ENERGY INFORMATION SYSTEMS

The philosophy, "If you can measure it, you can manage it," is critical to a sustainable energy management program. Continuous feedback on utility performance is the backbone of an energy information system[3]. A basic definition of an energy information system is:

Energy Information System (EIS)

Equipment and computer programs that let users measure, monitor and quantify energy usage of their facilities and help identify energy conservation opportunities.

Everyone has witnessed the continuing growth and development of the internet—the largest computer communications network in the world. Using a web browser, one can access data around the world with a click of a mouse. An EIS should take full advantage of these new tools.

EIS Process

There are two main parts to an EIS: (1) data collection and (2) web publishing. Figure 6-2 shows these two processes in a flow chart format.

The first task in establishing an EIS is to determine the best sources of the energy data. Utility meters monitored by an energy management system or other dedicated utility-monitoring systems are a good source. The metering equipment collects the raw utility data for electric, chilled & hot water, domestic water, natural gas and compressed air. The utility meters communicate to local data storage devices by pre-processed pulse outputs, 0-10V or 4-20ma analog connections, or by digital, network-based protocols. Meters are either hardwired directly to the local data storage devices or might use some form of wireless connectivity, such as cell phone technology.

Data gathered from all of the local data storage devices at a predefined interval (usually on a daily basis) are stored on a server in a relational database (the "data warehouse"). Examples of relational databases are MS SQL Server and Oracle*.

---

*Any reference to specific products or name brands of equipment, software or systems in this chapter is for illustrative purposes and does not necessarily constitute an endorsement implicitly or explicitly by the authors of this chapter or the others in this book.

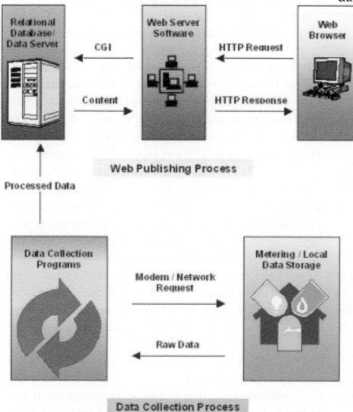

Figure 6-2. Energy Information System Schematic

Data Collection

Identifying and organizing the energy data sources is the first step in establishing an EIS. Since utility meters can be connected to several disparate data collection systems, there can be several methods used for data collection. Transferring this data in a common format and storing the data in a common EIS relational database is a simple way to pull all this together for further analysis by a web publishing program.

A BAS can provide an excellent utility data collection system. Utility meters that provide a pulse output for each fixed consumption value (i.e. 1 kWh/pulse for electric meters) can be read into a discrete pulsed input point and totalized to record the utility meter consumption. Most BAS have a daily report feature that can automatically generate reports that show the utility meter data at the desired interval (15 minutes, 1 hour, etc.). Once the reports are created and stored on the BAS Server, a separate program can be used to read each report and extract the data into the EIS relational database. On some newer BAS, utility meters are read and the data stored directly into a database that is part of the BAS. The data can be used by the BAS analysis tools or by other enterprise level systems by means of web services or API calls. However, if multiple data collection systems are being used, these data can be copied into a common EIS relational database so that all data from all data collection systems can be stored in one location.

Another approach to collecting utility data is to use a dedicated data acquisition server (DAS). The DAS allows users to collect utility data from existing and new meters and sensors. On a daily basis the DAS uploads the stored data to the EIS Server. Once the data has been transferred to the EIS Server, a program reads the DAS data files and updates the data in the EIS relational database for use by the web publishing program.

The AcquiSuite system from Obvius is typical of the emerging solutions and is a Linux based web server which provides three basic functions:

- Communications with existing meters and sensors to allow for data collection on user-selected intervals
- Non-volatile storage of collected information for several weeks
- Communication with external server(s) via phone or internet to allow conversion of raw data into graphical information

The backbone of the system is a specially designed web server. The DAS provides connectivity to new and existing devices either via the on-board analog and digital inputs or the RS 485 port using a Modbus protocol. The analog inputs permit connection to industry standard sensors for temperature, humidity, pressure, etc and the digital inputs provide the ability to connect utility meters with pulse outputs. The serial port communicates with Modbus RTU devices such as electrical meters from Veris, Square D and Power Measurement Ltd.

Web Publishing

The internet, with the world wide web—or web—has become accessible to all. It has allowed the development of many new opportunities for facility managers to quickly and effectively control and manage their operations. There is no doubt that web-based systems are the wave of the future. The EIS web publishing programs should take full advantage of these web-based technologies.

To publish energy data on the internet or an Intranet (a private network that acts like the internet but is only accessible by the organization members or employees), client/server programming is used. The energy data are stored on the EIS server, and wait passively until a user, the client, makes a request for information using a web browser. A web-publishing program retrieves the information from the EIS relational database and sends it to the web server, which then sends it to the client's web-browser that requested the information.

There are many software choices available for the web-publishing process. One method uses a server-side common gateway interface (CGI) program to coordinate the activity between the web-server and the web-publishing program. Using CGI enables conventional programs to run through a web browser.

The web-publishing client/server process for an EIS uses the steps below (See Figure 6-2). This entire process takes only milliseconds depending on the connection speed of the client's computer to the web.

1.   A user requests energy information by using their web browser (client) to send an HTTP (hypertext transfer protocol) request to the web server.
2.   The web server activates the CGI program. The CGI program then starts up the web-publishing program.
3.   The web-publishing program retrieves the information from the relational database, formats the data in HTML (hypertext markup language) and returns it to the CGI program.
4.   The CGI program sends the data as HTML to the web

server, which sends the HTML to the web browser requesting the information.

Web-publishing Programming Options

Although there are many web servers available to choose from, two are the most popular by far. Microsoft's Internet Information Services (IIS) comes with Windows server software. Apache web server is a good choice for other operating systems. Any web server needs some configuration to produce web content, especially if it is querying a database. The web-publishing task will likely require custom folders, special access permissions and a default page.

There are many programming alternatives available other than the CGI approach described above. Active Server Pages (ASP and ASP.NET), JavaScript, Java Applets, Java Server Pages, Java Servlets, ActiveX controls, and PHP are a few of the more popular choices available today. Some of these are easier to implement than others. ASP for instance, is a part of IIS so no installation is required. PHP require installation of their respective programs on the web server machine to run. Javascript and VBScript are somewhat limited in that they are just a subset of the other full fledged programming languages. Most browsers interpret them correctly so no installation is required. Java Server Pages and Java Servlets run on the web server in the same way as ASP but may require some installation depending on the web server used.

After installing the web server, the web-publishing administrator must put a default page in the root directory of the web server. This is the first page users will see in their browser when they type in the web site's internet address. The pages are usually named "default.htm" or "index.htm" but can be anything as long as the web server is configured to treat them as the default page.

Next, if CGI is used, the administrator creates a special folder to store the scripts. This is usually called "cgi-bin" or just "scripts." This folder must have permissions specifically allowing the files in the folder to be "executable." In some cases, "write" permissions are required for the folder if the CGI programs write temporary files to it. Other custom folders may be required to organize the web publishing content. Once the web-publishing administrator configures the web server he or she can install and test custom CGI programs and pages. If the CGI program or pages accurately return data from the database then the task of creating custom reports for the energy data can begin.

Alternatively, if a scripting language such as PHP is used, a separate program must be installed and configured to work with the web server. http://us.php.net

is an excellent resource for installing and configuring PHP on many operating systems. After PHP is installed and configured on the web server any file stored in the web directory ending with ."php" will automatically be parsed by the web server. Other scripting languages are built into certain web servers and do not require any separate installation of software. ASP is built into internet information server (IIS) but not Apache web server. http://httpd.apache.org is a good source for information on using scripts with Apache web server.

EIS Implementation Options

Deciding which web server and programming method to use along with configuring and implementing it to create a web publishing system can be quite a task. It really requires an expert in these areas to do a reliable job. Three approaches have evolved to satisfy web-publishing requirements.

1. Use internal resources to accomplish this task. This works well if there are already experienced web programmers available and they have time to work on the project. This makes it easy to customize the web publishing content as needed quickly and cost effectively. Finding time for internal personnel to focus on the project is usually the problem with this option.

2. Hire an outside consultant to do the configuration and programming as needed. This works well if the consultant has a good working relationship with someone internally to facilitate access to the protected systems and help with understanding the data. The consultant must be willing to work for a reasonable rate for this approach to be cost effective. The consultant must also be responsive to requests for support.

3. Purchase and install a somewhat "canned" version of the web publishing software and then customize it to fit the energy data as required. This approach has many possible problems in that the software is usually quite expensive and requires a great deal of customization and support from the outside to make it work well. However, for small simple projects this may be a good fit.

For users, who do not want to invest the time and effort required for this "do-it-yourself" approach, numerous companies provide a complete EIS service for an on-going monthly service fee. The EIS service company provides all of the IT-related functions, including the energy data collection/storage and the web-publishing program. The user accesses the EIS service web site by using a web browser, enters a user ID and password and then uses the available reports/graphs to analyze energy data.

The advantage of this approach is that the user does not get involved with the details and operation of the EIS, but instead is able to work with the EIS service provider to develop the utility data reports most helpful to their operation. The downside to this approach is the on-going monthly service fee that is a function of the amount of data processed—the more meters or bills processed the higher the monthly fee. There may also be additional costs to customize any reporting from the standard reports already created by the EIS service provider. The Building Manager Online service from Obvius* is one of the many choices available to users today.

EIS Web-publishing Example

The utility report cards (URC) is an example of a web-based energy information system that reports and graphs monthly utility data for schools. Each month, a web-based report is automatically generated and e-mailed to school principals and staff as encouragement to examine their school's electricity usage (energy efficiency) and to identify schools with high-energy consumption needing further investigation. The URC also is intended for teachers and students to use as an instructional tool to learn about school energy use as a complement to the energy-education materials available through the U.S. Department of Energy's EnergySmart Schools program (ESS). To see how the URC operates, go to http://www.utilityreportcards.com and click on "URC Live."

The URC was created to help the school staff understand and, therefore, manage their utility consumption and associated costs. The URC allows school principals to become aware of how their school is performing relative to a projected benchmark and to other schools of similar design and capacity. Giving recognition to schools that improve performance from prior-year levels creates a spirit of competition with the opportunity to recognize success. Those schools identified as high-energy users become the focus of attention to determine the reasons for their consumption level and ultimately to decrease the energy used. All of this is done by using the monthly utility data that is provided electronically at minimal or no cost to the schools by the utilities.

Turning Data into Useful Information

The installation of sub-metering is a positive step toward understanding the dynamics of the building systems being measured. However, energy savings are only

---

*Jim Lewis, one of the authors of this chapter, is the CEO of Obvius, LLC.

achieved when the sub-meter data are turned in to actionable information. An EIS that continuously collects the sub-meter data and displays them in easy-to-understand charts and graphs is the key. By continuously "shining a light" on utility usage at each facility, utility costs are minimized by the actions of those who receive these timely and informative reports. Continuous feedback on utility performance pinpoints problems that result in energy waste which are corrected though BAS programming changes or repair of defective equipment.

Sub-metering utility usage is the most direct method for energy saving measurement. The majority of energy saving retrofit projects are implemented based on engineering calculations of the projected return on investment [4]. As with any projections of ROI, much of what goes into these calculations are assumptions and estimates that ultimately form the basis for implementation. As the folks at IBM used to say, "garbage in—garbage out," which in the case of energy retrofits means that if any of the assumptions about parameters (run times, setpoints, etc.) are wrong, the expected payback can be dramatically in error. The establishment of good baselines (measures of current operations) is the best way to determine the actual payback from investments in energy and sub-metering.

Just as important as building an accurate picture of the current operation is measuring the actual savings realized from an investment. If there is no effective means of isolating the energy used by the modified systems, it may be impossible to determine the value of the investment made. Using monthly utility bills for this analysis is problematic at best since the actual savings achieved can be masked by excessive consumption in non-modified systems.

Consider, for example, a commercial office building whose central chiller plant has an aging mechanical and control structure that provides limited capability for adjusting chilled water temperature. To improve efficiency, the building owner plans to retrofit the system to provide variable speed drives on pumps for the chilled water and condenser water systems along with control upgrades to allow for chilled water setpoint changes based on building loads. In the absence of baseline information, all calculations for savings are based on "snap-shots" of the system operation and require a variety of assumptions. Once the retrofit is completed, the same process of gathering snapshot data is repeated and hopefully the savings projected are actually realized. If the building tenants either add loads or increase operational hours, it is difficult if not impossible to use utility bills to evaluate the actual savings.

In contrast, the same project could be evaluated with a high degree of accuracy by installing cost-effective monitoring equipment prior to the retrofit to establish a baseline and measure the actual savings. While each installation is necessarily unique, building a good monitoring system would typically require:

• Data acquisition server (DAS) such as the AcquiSuite from Obvius to collect the data, store them and communicate them to a remote file server.

• Electric submeter(s)—the number of meters would vary depending on the electric wiring configuration, but could be as simple as a single submeter (e.g., Enercept meter from Veris Industries) installed on the primary feeds to the chiller plant. If desired, the individual feeds to the cooling tower, compressors, chilled water pumps, etc. could be monitored to provide an even better picture of system performance and payback.

• Temperature Sensors (optional): in most installations, this could be accomplished by the installation of two sensors, one for chilled water supply temperature and the other for chilled water return temperature. These sensors do not provide measurement of energy usage, but instead are primarily designed to provide feedback on system performance and efficiency.

• Flow Meter (optional)—a new or existing meter can be used to measure the gallons per minute (gpm). By measuring both the chiller input (kW) and the chiller output (tons) the chiller efficiency can be calculated in kW/ton.

The benefits of a system for actually measuring the savings from a retrofit project (as opposed to calculated or stipulated savings) are many:

• The establishment of a baseline over a period of time (as opposed to "snapshots") provides a far more accurate picture of system operation over time.

• Once the baseline is established, ongoing measurement can provide a highly accurate picture of the savings under a variety of conditions and establish a basis for calculating the return on investment (ROI) regardless of other ancillary operations in the building.

• The presence of monitoring equipment not only provides a better picture of ROI, but also provides ongoing feedback on the system operation and will provide for greater savings as efficiency can be fine-tuned.

Case Study—Retail Store Lighting

A retail store chain in the Northeast was approached by an energy services company about converting some of their lighting circuits to a more efficient design. On paper, the retrofit looked very attractive and the company elected to do a pilot project on one store with a goal to implementing the change throughout the entire chain if it proved successful. The retailer decided to implement a measurement and verification (M&V) program to measure the actual savings generated by comparing the usage before the retrofit (the baseline) and after.

The store had 12 very similar lighting circuits, all of which were operated on a time schedule from a central control panel in the store. Since the circuits were very similar, it was decided that measuring the impact on one circuit would provide a good indication of the savings from the other circuits. The sub-metering equipment consisted of the following:

- An electrical sub-meter was installed on the power lines feeding the lighting circuit.

- A data acquisition server was installed in the store to record, store and upload time-stamped interval data to a remote server for storage and display. The DAS provides plug and play connectivity to the sub-meter and uses an existing phone line or LAN to send data from the store to a remote server on a daily basis.

- The remote server was used to monitor consumption before the retrofit and to measure the actual savings.

Figure 6-5 shows the actual kW usage over roughly 24 days. The left side of the chart shows the kW usage for the first 11 days before the retrofit and the average usage is fairly constant at around 1.45 kW. On Feb. 11, the retrofit was performed, as indicated by the drop to zero kW in the center of the chart. Immediately after the retrofit (the period from Feb. 11 to Feb. 15, the kW load dropped to around 0.4 kW, a reduction of over 70% from the baseline load in the left of the graph.

The good news for the retailer was that the retrofit performed exactly as expected and the M&V information obtained from monitoring the energy on this circuit provided clear evidence that the paybacks were excellent. The initial good news, however, was tempered somewhat

after looking at the chart. It was immediately evident that this lighting circuit (and the other 11 identical circuits) were operating 24 hours per day, seven days a week. The store, however, operated from 10 AM to 9 PM each day and the lighting panel was supposed to be shutting off the circuits during non-operating hours.

The electrical contractor was called in to look at the system and determined that a contactor in the panel had burned out resulting in continuous operation of the lighting circuits throughout the store. Once the contactor was replaced, the operation of the lighting panel was restored so that the lights were only on during operating hours and shut off during the night, as indicated by the right side of the chart.

This simple chart of energy usage provides an excellent example of two uses of energy information:

1. *Measurement and verification of energy savings*—The chart clearly shows the actual energy reduction from the lighting retrofit and the data provided can be used to extrapolate the payback if this same retrofit is applied throughout the chain.

2. *Use of energy information to fine-tune building operations*—In addition to the M&V benefits of energy information, this example also shows how a very simple review of energy usage can be used to make sure that building systems are operating properly.

CONCLUSION

The BAS is used for real-time *control* of building systems. An EIS is used to *measure* the buildings energy usage. The information from the EIS provides feedback to the building operator to make sure the BAS is working properly.

The web provides the means to share information easier, quicker, and cheaper than ever before. There is no doubt that the web is having a huge impact on the BAS industry. The BAS of tomorrow will rely heavily on the web, TCP/IP, high-speed data networks, and enterprise level connectivity. Improving facility operations in all areas, through enterprise information and control functions is fast becoming an equally important function of the overall BAS or facility management system.

Historically, hardware, software and installation of EIS has been prohibitively expensive and has limited implementation to those commercial and industrial facilities that could afford to pay for custom systems integration services. These costs have fallen dramati-

Figure 6-3. Retrofit Electric Sub-Meter

Figure 6-4. Data Acquisition Server

cally as companies leverage the enormous investment in the internet to provide the building owner with tools that make do-it-yourself data acquisition a cost effective reality. Hardware and software designed specifically for data acquisition and using available tools such as TCP/IP, HTTP and Modbus put valuable energy information literally at the fingertips of today's facility owners and provide an excellent method for measurement and verification of energy saving projects.

Web integration of BAS and EIS are inevitable, so if you have not done so already, it is a good time for Energy Managers to know their IT counterparts. Getting a good handle on the technical-side of things can be a daunting task. A successful Energy Manager will find a way to master their BAS and EIS.

At the same time, it is important to remember that commitment from people (management and staff) is the most important aspect of a successful energy management program. Once all three components are working together, the energy-saving results are significant and sustainable.

References

[1]   Barney Capehart, Paul Allen, Rich Remke, David Green, Klaus Pawlik, *IT Basics for Energy Managers—The Evolution of Building Automation Systems Toward the Web*, Information Technology for Energy Managers, The Fairmont Press, Inc., 2004

[2]   Paul Allen, Rich Remke, Steve Tom, *Upgrade Options for Networking Energy Management Systems,*, Information Technology for Energy Managers—Vol II, The Fairmont Press, Inc., 2005

[3]   Barney Capehart, Paul Allen, Klaus Pawlik, David Green, *How a Web-based Energy Information System Works*, Information Technology for Energy Managers, The Fairmont Press, Inc., 2004

[4]   Jim Lewis, *The Case for Energy Information, Information Technology for Energy Managers*, The Fairmont Press, Inc., 2004

Software References

RS-485, http://www.engineerbob.com/articles/rs485.pdf

Ethernet, http://www.techfest.com/networking/lan/ethernet.htm

ARCNET, http://www.arcnet.com

LonWorks & LonTalk, http://www.echelon.com/products/Core/default.htm

BACnet, http://www.bacnet.org/

Modbus, http://www.modbus.org/default.htm

XML, http://www.xml.com

Iowa Energy Office, http://www.ddc-online.org

TCP/IP, http://www.protocols.com/pbook/tcpip1.htm

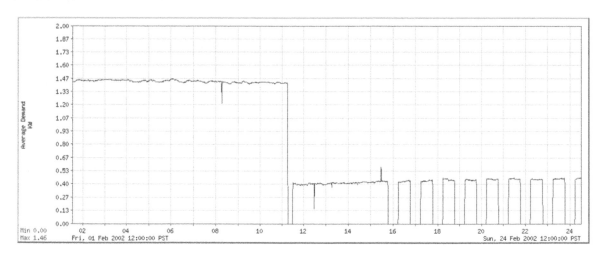

Figure 6-5. kW Loads for a 24-day Period

Chapter 7

Turning Energy Information Into $$$$

Jim Lewis

INTRODUCTION

This chapter examines the use of energy information as a valuable tool in finding and correcting costly operational deficiencies and identifying high value retrofit opportunities in commercial, educational and government facilities.

Most owners and managers of commercial and industrial facilities are familiar with and recognize the value of submetering within their buildings, particularly as it relates to the use of energy consumption and costs for traditional purposes such as:

- Cost allocation to departments or projects;
- Tenant submetering to assign costs to new or existing third party tenants;
- "Shadow" metering of utility meters to verify the accuracy of bills and the quality of the power being delivered;

Many owners are not, however, aware of the potential for using energy information to monitor the performance of both automated and non-automated energy-consuming systems and equipment within the building. Energy consumption information gathered from a variety of sensors and meters within one or more buildings provides not only verification of the efficiency of the equipment, but also the management of the systems by the users.

This chapter examines the practical use of energy information and the hardware and software needed to apply the information to everyday use.

SOME BASIC CONCEPTS

The term "energy information" as it is used in this chapter refers to data gathered from a variety of sources within the building that relates to:

1. Quantity of energy and/or water consumed during a particular interval
2. The time period during which the energy was con-

sumed by one or more systems
3. The relationship between the energy consumed, the time it was used and the operation of the building's systems

Measurement of energy information consumed during a particular interval, whether from primary meters or from secondary metering and sensing devices is the first line of defense for the building owner wishing to monitor operations. Proper selection and installation of sensors and meters in critical areas provides the most valuable and timely source of operations verification.

THE VALUE OF TIMELY AND ACCURATE INFORMATION

Remember, **if you don't measure it, you can't manage it**. Using utility bills that arrive weeks after the energy was used and lump all of the building's systems into one account is hardly a timely and accurate measure of building performance and efficiency.

When your August electric bill is 20% higher than your budget, how do you determine the cause? Was it lights left on, HVAC systems operating inefficiently, new equipment installed in a tenant space, hotter weather than normal or ___ ?

Demand charges make up half of your electric bill. When does the demand occur? Is it a short term spike or is the demand profile relatively flat? Is it something you can manager to limit the cost?

When your November gas bill is 30% higher than last year, was it because of the colder weather or did the hot water heater not shut off at night? Is there a leak in the system, or are you simultaneously heating and cooling occupied spaces?

How much of the steam your central plant produces is lost to leaky steam traps? How much of the water your building uses goes to irrigation and landscaping and how much is used for makeup water to the boiler?

Most municipalities bill for sewer charges based on the water bill. As sewer charges become a greater part of the energy cost for many buildings, how much of the water you bring in actually reaches the sewer? How much goes to landscaping or cooling towers or evaporation from water features?

HOW DOES ENERGY INFORMATION HELP IN OPERATIONS?

The previous section of this chapter touched on several operations related issues that many building owners and managers are concerned about, but how does gathering information about energy and other utilities contribute to more efficient building operation? We will look at a couple of examples from real buildings being monitored by Obvius.

The first example is a retail store located in the Northeastern United States. This example provides a very clear case for the value of submetering and timely monitoring of electricity usage. The customer became concerned about electric bills that were higher than historical usage and called in a consultant to review the building's operations and make recommendations.

After looking at the operations, one of the primary opportunities identified for saving electricity was in the lighting systems for the warehouse/operations area of the store. The lighting consisted of a mix of incandescent and fluorescent tubes and it was determined that based on the operating hours of the store, a conversion to more efficient lighting would generate significant savings and an attractive payback.

The owner of the store and the energy consultant determined that since this retrofit would likely be used as a pilot project for all of the company's stores, it would be a good idea to provide some measurement and verification of the savings realized. The energy consultant provided two alternatives:

1.   A snapshot view of the consumption before and after using simple handheld tools like a multi-meter and an amp clamp; or

2.   Installation of a monitoring system to measure the actual power consumed by each of the circuits on 15 minute intervals. This option, while more expensive than the first option, would obviously provide much more accurate feedback on the success of the installation and would also have the added benefit of providing near real-time access to the data using a web browser.

After reviewing the two options, the customer decided to install electrical submeters on the lighting circuits before the retrofit to establish a baseline and then to leave the submeters in place for a period of time after the installation to verify the exact savings realized from the changes. The meters would be monitored and time stamped interval data recorded using an AcquiSuite Data Acquisition Server (DAS) from Obvius (see Figure 7-1).

The data gathered by the DAS would be sent each night to the Building manager online (http://www.buildingmanageronline.com) website hosted by Obvius so that the data would be available the next morning via the internet.

The installation of the monitoring required the following:

*   AcquiSuite DAS (A8811-1) from Obvius to monitor and record the data from all the submeters;
*   Enercept submeters (H8035) from Veris Industries (see Figure 7-2) to connect to each of the 12 lighting circuits to be monitored;
*   Ethernet connection to the existing store LAN to provide a path for sending the data to the host server;
*   Electrical installation labor for the devices and the wiring

The meters were installed and connected via an RS485 serial cable to the DAS. The DAS gathered data from each meter and stored the kW information on 15-minute intervals in nonvolatile memory. Every night, the data were uploaded to the BMO server site where they was

Figure 7-1. AcquiSuite Data Acquisition Server

Figure 7-2. Retrofit electrical submeters

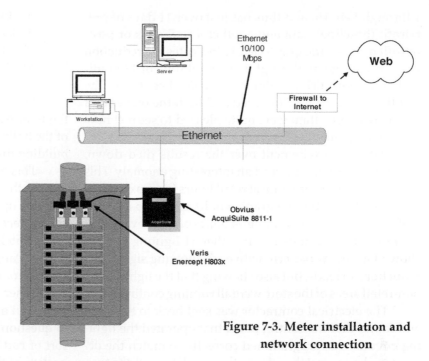

Figure 7-3. Meter installation and network connection

automatically stored in a MySQL database. Once the time-stamped data were stored in the database, they could be accessed by any authorized user with a web browser such as Internet Explorer.

The monitoring system ran for several weeks to establish a baseline for energy usage before the lighting retrofit was done, and then ran for a period of time after the installation to verify the savings from the installation. To begin the verification process, the energy consultant and the store management reviewed the kW data for the month that included the benchmark period, the installation period and a few days after the installation. The data they saw are shown graphically in Figure 7-4.

This graph represents approximately 15 days (Feb.

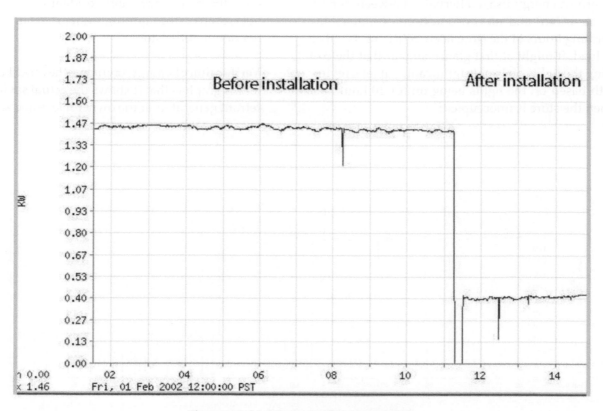

Figure 7-4. Lighting retrofit energy graph

1 through Feb. 15) and thus has just over 11 days of pre-retrofit (baseline) data and another 4 to 5 days of post-retrofit data. As the graph clearly indicates the reduction in energy usage by this lighting circuit (one of 12 modified in this retrofit) was almost 70%. Needless to say, the results were very well received and both the owner and the energy consultant were quite pleased to see just how much energy had been saved.

After the excitement over the results died down, however, the owner noticed an interesting anomaly. This was a retail store with typical retail hours (approximately 10 AM to 9 PM), but the graph would appear to indicate that the lights were operating 24 hours per day every day. Reviews of the graphs for the other 11 lighting circuits showed a similar pattern, with each showing significant reductions in loads, but also showing that the lights in the non-retail areas of the store were all running continuously.

The electrical contractor was sent back to the store and found that the control panel that operated the lighting circuits was programmed correctly to match the occupancy hours, but the relay in the panel designed to turn the circuits on and off had failed and was continuously on. The lighting relay was replaced and the next read of the graph showed the following pattern:

As this second graph clearly indicates, once the relay was replaced and the control panel functioned properly, the pattern of energy usage alternates between normal consumption (on) and no consumption (off), matching the operating hours of the store itself. Since this is only a 1.5 kW load, it might at first glance appear that the cost of leaving the lights on is not significant, but let's review the math based on the lights being on for 10 hours per day when the store is unoccupied:

(1.5 kW) x (10 hrs/day) = 15 kWh per day
(15 kWh/day) x ($0.10/kWh) = $1.50/day
($1.50/day) x (365 days/yr) = $547.50/yr
($547.50/yr/circuit) x (12 circuits) = **$6,570/yr**

The wasted energy in this example is only part of the total waste since it does not reflect the reduced life of the bulbs and ballasts or the added cost of cooling the building due to the heat from the lights.

This example provides a very clear indication not only of the value of monitoring energy consumption to monitor operations, but also just how easy it is to spot and correct malfunctions. Anyone reading this graph can immediately spot the problem (although not necessarily the root cause) without the need for sophisticated analytical tools or experience in energy analysis or engineering. The owner knew that his store did not operate 24 hours a day, but a quick glance at the data in this chart led him to question why the lights were on all the time. It's this sort of rudimentary analysis that provides a significant portion of the savings on energy.

This example highlights the value of energy monitoring on several levels, any or all of which may be important to the building owner:

1. **Highlighting of inconsistencies**—as we see in this case, the owner was able to identify inconsistent operation of the lighting systems relative to the operation of the store

2. **Verification of energy savings**—the other key point to this graph is that it shows the actual savings this retrofit generated. If the owner had relied solely on

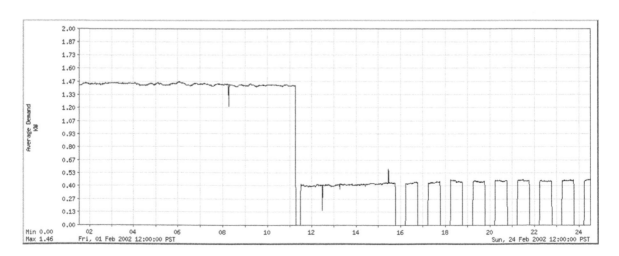

Figure 7-5. Lighting energy after relay replacement

the utility bills to verify savings, the savings from the retrofit would have been overstated by more than $6,000. In this case, the owner was planning to use the data from this pilot project to determine the payback of rolling out a similar retrofit to hundreds of stores and the overstated savings would have been impossible to duplicate in other locations.

3. **Supervisory monitoring of control systems**—in this case, the lighting control panel appeared to be functioning as designed, and the cursory review by the electrical contractor prior to the retrofit gave no indication of any problem. It was only after the issue was made clear by the energy data that the problem was identified and corrected.

In the example of this store, everyone assumed that since there was a properly programmed lighting control panel in the store, that the system was operating properly and no one was concerned (or accountable) for the lights being on when they left. The only way this problem was identified and corrected was through the use of proper monitoring equipment.

EXAMPLE 2: United States Coast Guard Yard

The United States Coast Guard Yard located in Baltimore, MD, is a full-service ship repair facility and the primary provider of repairs and upgrades to a wide variety of fleet vessels. As a full service shipyard, the yard has a significant number of very large energy consuming systems including welders, compressors, cranes plus all the support systems including HVAC and lighting.

The yard was the first Coast Guard facility to employ submetering to monitor the operation of both operational and environmental systems and to identify opportunities for reducing energy consumption. The submetering

program has since been expanded to a number of other Coast Guard facilities around the world is presently being implemented at all of the bases to meet the requirements of EPACT 05 and other federal mandates for reduction of electricity, gas, water and steam.

Deployment of the submetering system began with a single meter and data acquisition server to monitor the operation of two large air compressors and has since been expanded to include many other mechanical and lighting systems throughout the base. This initial installation was designed to monitor the provide 15-minute interval data on the primary electrical feeds providing power to two large compressors. The operational plan calls for both compressors to be operating during the weekday when the primary work is being done at the yard, with the compressors turned off in the evening and on weekends as there is no requirement for compressed air..

The purpose of the monitoring was to determine how much the load was reduced at night and whether the operators were actually turning off one of the compressors in the evening as expected. The results of the monitoring are shown in Figure 7-6.

In the initial week of monitoring (May 20-27) it is clear that the compressors are being turned off on the weekends, but the compressors are operating at night. Once this pattern was detected, the operational team began turning the compressors off as required and the energy consumption was dramatically reduced as seen in the ensuing weeks above.

HOW DOES IT WORK?

In order to get the information from the building and make it available in a user-friendly web format, there are two major pieces that must be in place:

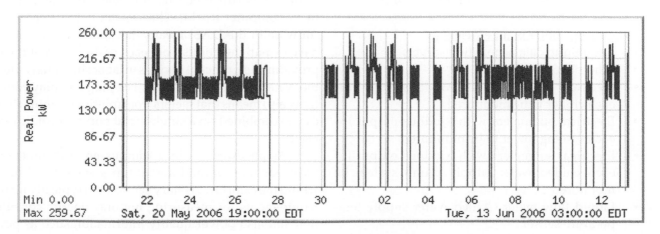

Figure 7-6. Compressor operation

1. On-site hardware and software—each facility being monitored needs to have the meters and other sensors to actually measure the desired parameters and a data acquisition device to gather and store the data and to communicate it to a remote web server

Application service provider (ASP)—the ASP provides two primary functions: first, an internet accessible connection to each of the locations being monitored for uploading the interval data; and second, a web based user interface for viewing information from all the locations using a web browser such as Internet Explorer or Netscape.

ON-SITE HARDWARE AND SOFTWARE

The equipment required for specific monitoring projects will obviously vary depending on the application. Each job will require some type of data acquisition server or logger to log data from the meters and sensors and to communicate with a local or remote host server to upload the interval data gathered. Some representative applications and the hardware required:

1. Monitoring existing utility meters with pulse outputs
    * AcquiSuite or AcquiLite data acquisition server;
    * Ethernet or phone line

2. Monitoring submeters
    * AcquiSuite data acquisition server;
    * Ethernet or phone line
    * Electrical submeters with Modbus output for each circuit or system to be submetered
    * Gas meters with pulse or analog (4 to 20 mA) output
    * Flow meters with pulse or analog output

3. Monitoring chiller plant efficiency
    * AcquiSuite data acquisition server;
    * Ethernet or phone line
    * Electrical submeters with Modbus output for each chiller to be monitored
    * Electrical submeters with Modbus output for each cooling tower to be monitored
    * Electrical submeters with Modbus output for each chilled water supply pump system be monitored
    * Liquid flow meter Chilled water supply temperature sensor
    * Chilled water return temperature sensor

4. Monitoring indoor air quality
    * AcquiSuite data acquisition server;
    * Ethernet or phone line
    * Analog (4 to 20 mA) output sensors to monitor:
      — Temperature
      — Humidity
      — NO_x
      — CO
      — CO_2

Since the DAS is capable of monitoring any analog output device, the installation can be readily customized to meet the needs of a particular application including:

* Submetering of loads within the building for cost allocation, verification of operations or tenant billing;
* Runtime monitoring to verify that loads are shutting down when scheduled
* Monitoring flows such as chilled water to allocate costs from central heating and cooling plants to individual buildings
* Monitoring flows to sewer services to verify sewer charges based on water usage
* Measurement and verification of energy saving retrofits
* Supervisory monitoring of facilities to benchmark energy usage and provide feedback on operations

The on-site software required is typically contained in the DAS and should require little if any customization. Since the DAS is a Linux based web server, interface requires only a standard PC on the network with the DAS. Any changes or modifications are made using only a web browser (see Figure 7-7).

One of the major changes in installation of submeters today is the ability of the DAS to automatically recognize most meters (the metering equivalent of plug and play in the computer world). This capability (written into the software in the DAS) means that the installer does not have to map the points from each of the different meters, providing a significant savings in labor since many meters today have 60 or more points. The plug and play connectivity also means that meters from multiple manufacturers can be combined on a single serial port to meet different needs.

As an example, in a campus environment, there are needs for several different types of electrical meters within the campus. Primary service will typically include requirements for measuring not only energy consumption, but also power quality information such as power factor, phase imbalance and harmonic distortion. The DAS

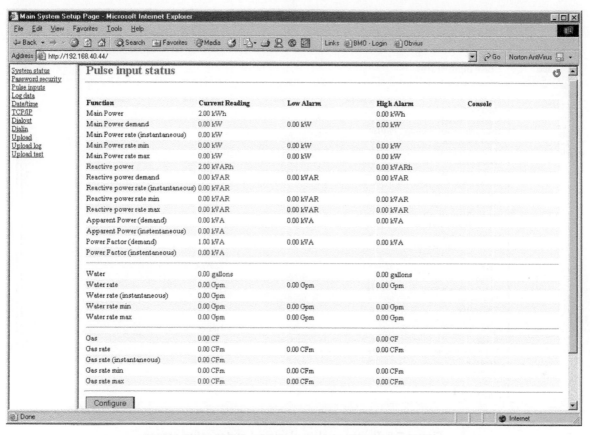

Figure 7-7. Browser based setup and configuration

has the flexibility to automatically recognize a variety of meters, from high-end power quality meters to simple submeters for measuring kW and kWh.

For example, the primary service at a campus might have a meter with a variety of inputs such as shown in Figure 7-8.

In contrast, the same campus would likely deploy a number of less expensive energy only meters where the purpose is primarily to monitor energy consumption and provide cost allocation to departments and third-party occupants (see Figure 7-9).

Once the setup and configuration of the DAS is complete, the unit logs energy or power information on user-selected intervals that is stored by the DAS until it is uploaded to a remote or local server for viewing and interpretation. Upload options include automatic upload via modem or LAN or manual upload using a web browser on the network with the DAS. Locally uploaded data can be stored and viewed with a variety of standard programs including any ODBC database or spreadsheet programs such as Excel. In Figure 7-10, interval data for one month is displayed in an Excel chart.

The information contained in the chart above can be used in a variety of ways:

- Cost allocation of consumption and/or demand charges if this is a submeter for a tenant or department

- Comparison to historical demand profiles to indicate changes over time that might indicate operational inefficiencies

- Prioritization of capital expenditures to reduce demand profiles and costs

PUTTING IT ALL TO WORK

Now that we have seen how data can be gathered and displayed, the key question is: what do we do to turn this information into actions that will save energy?

For most facility managers, the answer is best illustrated in reviewing four strategic approaches to the use of energy information. It is important to note that each of these strategies builds on the prior strategies and thus can be implemented in turn as part of an overall energy strategy:

1. **Installation of submetering equipment**—the key here is that not only are the meters installed, but the

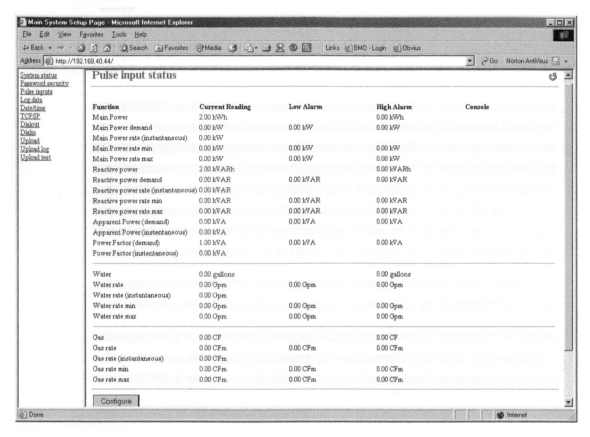

Figure 7-8. Power quality electrical meter setup screen

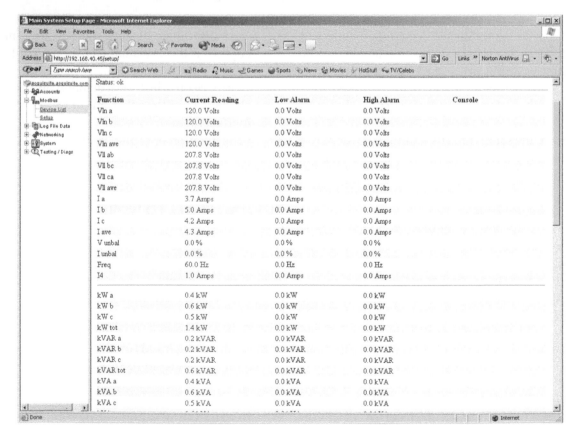

Figure 7-9. Energy only submeter (demand and consump tion)

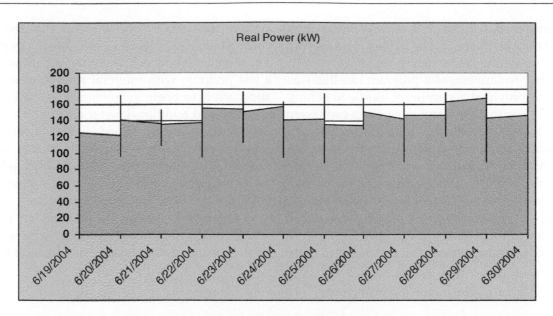

Figure 7-10. Excel chart of kW usage

information gained from these meters is reviewed and communicated to building users responsible for turning off lights, HVAC, computers, etc. It has been conclusively shown that if occupants know that the usage is being tracked, energy reductions of up to 2.5% will be realized (this is generally known as the Hawthorne effect).

2. **Allocation of costs to tenants and other users**—The next logical step is to add a reporting component to the monitoring and begin to hold tenants and other occupants accountable for the energy use in their space. Whether this cost allocation is to in-house departments or third-party tenants, the presentation of a "bill" for energy usage will prompt further reductions in energy usage as occupants can associate a cost to their activities. The expected savings from this approach are typically 2.5 to 5.0%.

3. **Operational analysis and performance reviews (the Building Tune-up Process)**—The next level of implementation involves a regular, comprehensive review of the performance of the equipment in the building, with particular emphasis on scheduling and occupancy. Questions to asked and studied include whether the control systems and HVAC equipment are functioning as designed, whether the schedules for occupancy match the actual usage, and whether there are unusual loads or demands on the system. Using this information, the operation of the facility can be fine-tuned to provide occupant comfort with a minimum of waste. These data can also be used

to identify areas for potential investment in energy retrofits and to pinpoint opportunities for maintenance and repair in a timely manner. Savings from the Btu process are generally in the 5.0 to 15.0% with a very small investment in time and money.

4. **Continuous Commissioning**—Originally developed by the Energy Studies Lab at Texas A&M, the concept of continuous commissioning carries the use of energy information to the next level. This strategy involves the use of energy information as a tool for continuously refining the operations of building systems for maximum efficiency. Many (if not most) systems are installed and commissioned in a less than optimal manner, but are generally not fine-tuned to meet the needs of the occupants with the best efficiency. Continuous commissioning uses energy information in conjunction with specialized software to identify and correct deficiencies in operations on a regular ongoing basis with the involvement of building personnel and occupants a key element. Studies have shown that Continuous Commissioning provides savings of up to 45% of a building's energy usage as compared to buildings where no monitoring is performed.

WHERE DO I START?

Hopefully this chapter has provided at least a prima facie case for the value of gathering and using energy information for operational analysis. For the building owner or manager contemplating an energy information program, but not sure where to begin, there's good news on the technology front. Historically, the investment in time and money to implement a program like this was substantial and provided a significant hurdle to getting started as design, installation and integration costs were prohibitive. Changes in technology today allow the building owner to take a "do-it-yourself" approach to energy information and to use existing building resources for implementation. Highly scalable hardware and internet-based data hosting make installation of submetering products a project that can be accomplished by in-house personnel or any local electrical contractor.

For most building owners, the scalability of these systems means that they can start with the most valuable metering projects and expand the system as needed and the minimal investment required allows them to do a test program with minimal risk. In general, the best approach is to begin submetering at the highest levels and then add additional metering equipment as savings opportunities are identified. For example, the facility manager of a campus would likely start by submetering each building, identifying those facilities where energy use is highest and adding additional submeters to those buildings to isolate and correct problems.

Energy information is a valuable tool and the most important first step in any energy strategy. As stated earlier, **"If you don't measure it, you can't manage it."**

Chapter 8

Why Can't a Building Think Like a Car? Information and Control Systems Opportunities in New Buildings

Barney L. Capehart, University of Florida
Lynne C. Capehart, Consultant

ABSTRACT

This chapter examines the information and control technology used in new vehicles and points out the potential for using similar information and control technology in new buildings. The authors draw on their knowledge of new cars and new buildings to present a list of information and control functions, together with the available sensors, computers, controls and displays used in new cars that can provide significant opportunities for our new buildings. Methods for integrating this new technology into new buildings are also discussed. The use of information and control technology in new cars should serve as a model for new building technology. This potential for new buildings should be recognized, and similar technological improvements should be implemented.

INTRODUCTION

A great deal of new technology is available for buildings. The labels "Smart Buildings" and "Intelligent Buildings" have been around for years. Unfortunately, this wealth of new technology for buildings only exists in pieces and as products from many different companies; virtually no building constructed today utilizes a significant amount of this new technology. Most new buildings operate just like the buildings of the 1970s. Even though new materials, new design and construction methods, and new ASHRAE building codes have greatly improved new buildings, these buildings still look and function much as they did 20 years ago. While most new buildings do have new equipment and better insulation, there is little in the way of new controls and display technology for the building occupants to see and use. Individuals seldom have the ability to control personal comfort and preferences.

In contrast, every new automobile—regardless of its price—is filled with new technology compared to the automobile of the 1970s. A new car typically comes with about fifty separate computers or microprocessors, has around forty to fifty sensors, and provides about twenty electronic display and control functions. It does this for as little as $20,000. This automotive information and control system commonly requires little or no maintenance or repair for a period of three to five years. The technology is often visible, it can be used by the driver and passengers, it is generally standard on all new cars, and it is inexpensive and reliable. There is much fancier technology available if you want to pay for it (Lincoln Navigators, 7-Series BMWs and S-Class Mercedes have around 100 processors on-board), but the majority of new automotive technology is found on every new car.

With all this new technology, today's cars are much more reliable and have significantly reduced maintenance requirements. In the 1970s, an automobile needed a tune up every 10,000 miles. Today, a typical new car does not need a tune up for 100,000 miles. Older cars needed new brakes about every 20,000 miles. Now it's every 50,000 miles. The authors bought a new mini van in 1998, and did not have to take it back to the dealer for any service for 40,000 miles! The vehicle had several oil changes in that period, but it needed no mechanical or electrical work.

In comparison, our buildings need maintenance people from the moment we start using them. We're not talking about janitorial work, but about maintenance of lights, air conditioners, switches, controls, doors and windows. This is like the old days with our cars when we started making a list of things to be fixed as soon as we drove the car off the dealer's lot. We are paying extra for building commissioning just to make sure everything in the building is operating correctly and is fixed if it is not. Why can't a new building operate for six months, a year, or even several years without needing any maintenance? Our cars do.

What is the potential for using reliable, comprehensive, integrated, and inexpensive components in our new buildings to create a transparent and efficient information and control system? And what should we do in terms of buying new buildings? Clearly, progress in adapting and implementing technology for new buildings has a long way to go. Nonetheless, we should demand more technology—a lot more. Technological improvements should be standard features that come with every new building without question rather than options that add significant cost to the building. The only question should be where do we draw the line between standard features and additional new technology that we will pay extra for?

FEATURES OF AUTOMOBILES THAT WE COULD USE IN BUILDINGS

Individual Control Systems

One of the most noticeable features of new automobile technology is how it provides the driver and often the passengers with individual control systems. Compared to a building, a new car has far more sensors, controls and displays for a much smaller space. There are individually controllable air supplies for the driver and the front passenger. Large vehicles often have air controls for the rear seat passengers too. Temperature ranges for heating or air conditioning are individually controllable, often for the front passenger as well as the driver. The air velocity is controllable with a multi-speed fan. The outlet vents are easily reached and can be moved to direct the airflow onto or away from the person. The amount of outside air can be controlled by selecting fresh air or recirculation. Some lights such as headlights and interior dome lights are activated by sensors. Other lights are individually controllable. The driver or the passenger can turn on selected interior lights, can often dim these lights, and can direct the light to the area where it is needed. The moon roof can be opened or closed by the driver or front passenger. Both front seats are individually adjustable for horizontal position, height, tilt, and back support; and many are heated, too. In addition, in some cars, these individual settings or preferences for functions like HVAC and seat positions are provided through a memory setting tied to an electronic key, and settings for more than one person can be stored in memory.

Compare this technology to the control systems currently available in a common new building. A typical room in a new building may have a thermostat with a control setpoint and a temperature display at that location. It also usually has an unseen VAV control function, and in a few instances a humidistat with a setpoint control and a display of the relative humidity at that location. Lighting is controlled with a single light switch or possibly a single occupancy sensor for lighting. Otherwise, the occupants usually have no other sensors, controls or displays in that room.

An example of a new technology that is currently available and that achieves some of the goals of individual control over personal space within a building comes from Johnson Controls. Their Personal Environments system is an easy-to-use, desktop control unit that gives each person the flexibility to adjust temperature, lighting, air flow and acoustic characteristics as often as necessary to maintain personal comfort levels. Individuals can adjust the air temperature and control the amount and direction of air flow at their desktop. They have a heating panel under the desk to adjust the temperature to their legs and feet. The Personal Environments system also allows an individual to control task lighting and to mask background noise. The system has a sensor that turns off all functions when the workstation is unoccupied for more than 10 to 15 minutes. Although this system is being used by a number of companies, it is the exception rather than the rule.

Operational Controls

In addition to personal comfort controls, the new car also has a large number of automatic control systems to optimize and control its own operation. Engine control systems insure fuel efficiency and reduce air pollutants from the combustion process. Sensors for inlet air temperature and relative humidity allow optimum fuel flow control and optimum combustion. System computer modules also control the ABS, transmission, cruise control, and body controller. These microprocessor systems are standard on new vehicles, but new buildings are not built the same way. Operational controls are available for new buildings, but they require special design criteria. No one considers the possibility that they should be standard equipment.

Display Systems

New cars tell the owner about much of the maintenance and repair that needs to be done, and certainly notify the driver whenever one of the major systems is in need of attention. A new car has sensors that report tire pressure, unclosed doors, lights or other controls left on, unfastened seat belts, brake fluid status, and many other operational features related to the safety of the car and the occupants. Even a cursory comparison shows that our new buildings lag very far behind the present use of technology in new cars.

Much of the information on car maintenance and safety is aimed at the driver. What comparable information does a building operator get about the maintenance needs of the building or the various rooms in a building? Things that would be helpful to know include whether the air handling system filters are dirty, whether the refrigerant is at the proper level, whether sensors are working properly, whether lights are burned out, or whether the doors have been left open.

The present system in buildings is essentially a manual system. Filters are checked by maintenance personnel on a time schedule. Maintenance workers often depend on "human" sensors to notify them of burned-out lights, improperly functioning photosensors, or temperature problems in individual rooms.

Options

New cars have options, and new buildings have options—but these mean very different things. An option for a new car is an item or function that is already available and can be installed on the car, but at extra cost. For a building, an option is an item or function that an owner wants to add at extra cost, but expensive additional design, engineering integration and testing work must usually be performed before it can be installed and operated.

Table 8-1 summarizes many of the sensor control and display functions of new cars, and provides a model for desired technology in new buildings.

HOW DID THE AUTOMOTIVE INDUSTRY DO THIS?

We must understand how new automobiles can have so much new technology at such a low cost, and why they are so reliable in order to know how to utilize similar innovations in the building industry.

Engineering Analysis and Design

A significant amount of engineering analysis and design goes into both the structural and operational features of a new car. In addition, significant engineering analysis and design also goes into the manufacturing and production processes for assembling the new cars. A ma-

jor benefit of this approach is that the car's entire system and subsystems, as well as each of the car's components, are carefully engineered. For example, the electrical power consumption of the components and systems in a new car are carefully analyzed, built and selected to make sure that the total power demand is not greater than the capacity of the electrical power supply system, i.e., the 12-volt battery. Thus, with cars, the need for energy efficient electrical systems is built in from the start.

When a building is designed, the electrical load is specified first, and then a power supply system is specified that is big enough to handle the load of the building. Little or no thought is given to minimizing the electrical load itself because there are generally no constraints on the amount of power a utility will supply to the building.

Overall Quality Control Programs

A new car is reliable because a significant amount of engineering goes into both the car design and its manufacturing process. Quality and quality control start with the engineering design, and are strongly emphasized throughout the manufacturing and assembly of the car. Individual components are designed and made with quality and reliability as major goals. Subsystems and final systems—including the entire car—are similarly produced. Ordinary and accelerated life testing are conducted on the car's components, subsystems and systems. These extensive tests include the effects of temperature, moisture, mechanical and thermal stress, and other factors. As a result, most of the car's components and systems will last at least three years or 36,000 miles. Warranties on some new cars are now available for seven years or 70,000 miles.

Quality control and warranties in building design and construction are very different. Auto manufacturers provide the warranty for the entire vehicle (with the possible exception of the tires); the systems in new buildings are likely to be under several different warranties. HVAC manufacturers cover the HVAC system; flooring manufacturers guarantee the carpet/flooring; plumbing manufacturers guarantee plumbing fixtures; etc. There is usually no centralized quality control or warranty for a new building as there is with cars.

Widespread Use of Microprocessors and Computers

Much of the technology and operational features of our new cars comes from the use of microprocessors and microcomputers. A new car may have as many as 50 separate microprocessors and 11 major computer-based systems. Some new luxury cars have up to 90

Table 8-1. Sensor, Control and Display Comparison for Cars and Buildings

S=Sensor   C=Control   DI=Display
D=Driver   FP=Front Passenger   RP=Rear Passengers
CBO=Controllable by Occupant   NCBO=Not CBO

Function	Cars			Buildings	
	All	**Mid-cost**	**Luxury**	**All**	**Some**
I. Comfort and convenience					
A. Climate Control (HVAC)				Yes	
1. Zone of Control					
Single Zone	D	D	D	Few	
Dual Zone		D, FP	D, FP	Very Few	
Multi-Zone			D, FP, RP	Most	Individual Zone, CBO
2. Temperature				Yes	
Lever setting (C, DI)	D	D, FP	D, FP	Some	
Thermostat (S, C, DI)	–	D, FP	D, FP	1 per zone	Individual Zone, CBO
3. Air Supply					
Directional Vents (C, DI)	D, FP	D, FP, RP	D, FP, RP	Partial	
Multi-Speed Fan (C, DI)	D	D, FP	D, FP, RP	No	
Ventilation (S, C, DI)	D	D, FP	D, FP	Yes, NCBO	
Recirculation (C, DI)	D	D	D	Yes, NCBO	
4. Humidity (S, C)	–	–	Yes	Some, NCBO	Yes, CBO
5. Air Quality (S, C) (CO, NO$_2$)	–	–	Yes	Yes, NCBO	Yes, On/Off
6. Advanced Features					
Reheat Operation			Some	Most, NCBO	Yes, CBO
Window Fog Control			Some	No	
Air Filters			Some	Yes	
Sun Sensors			Some	No	
B. Seating BT=Back Tilt					
1. Basic – Mech or Elec (C)	D,	D, FP	D, FP, RP	Yes	–
2. Horiz Position + Back Tilt (BT)	FP		(BT)		
3. Six-Way (C)	–	D, FP	D, FP		Yes
4. Back Support (C)	–	D, FP	D, FP		Yes
5. Heated (C)	–	D	D, FP		No
6. Memory Function (C)		D	D, FP	No	No
C. Visual (Inside Lighting)					
1. Dome Light (C)	Yes	Yes	Yes	Yes	
With Occupancy Sensor (S,C)	Yes	Yes	Yes	No	Yes
Delayed Dimming (S, C)	No	Yes	Yes	No	No
2. Overhead/Task (C)	No	Yes	Yes	Some	Yes
Directional (C)				Few	Yes
3. General					
Door, Glove Box (S, C)	Yes	Yes	Yes	No	?
Visor, Map (C)	No	Yes	Yes	No	?
D. Windows					
Power Windows	No	Yes	Yes	No	
Power Sunroof	No	Yes	Yes	No	Yes
II. Normal Operation					
A. Speedometer (S, D)	Yes	Yes	Yes	No	*
Cruise Control (S, C, DI)	No	Yes	Yes	No	*
B. Odometer (S, C, DI)	Yes	Yes	Yes	No	*
C. Tachometer (S, DI)	No	Yes	Yes	No	*
D. Fuel (S, DI)	Yes	Yes	Yes	No	*

Table 8-1. (*Continued*)

III. Safety and Maintenance					
A. Engine					

1. Engine Control Module (S, C)
     Temperature, Oil Pressure, Check
     Engine
	Yes	Yes	Yes		

Status Light (S, DI)	Yes	Yes	Yes		
Gauges (S, DI)	No	Yes	Yes		
B. Auxiliary Systems					
1. Electrical					
Generator (Charge)					
Status Light (S, DI)	Yes	Yes	Yes		
Voltage Gauge (S, DI)	No	Yes	Yes		
Lights					
Headlights (C)	Yes	Yes	Yes		
Backup Lights	Yes	Yes	Yes		
2. Brakes					
ABS (S,C)	No	Yes	Yes		
Status Light (S, DI)	Yes	Yes	Yes		
Brake Light Out (S, DI)	No	No	Yes		
3. Others					
Seat Belt (S, DI)	Yes	Yes	Yes		
Turn Signals On (S, DI)	No	Yes	Yes		
Headlights On (S, DI)	No	Yes	Yes		
Low WW Fluid (S, DI)	No	Yes	Yes		
Door Not Closed (S, DI)	No	Yes	Yes		
Exterior Temp (S, DI)	No	Yes	Yes		
Tire Pressure Sensor					
Parking/Backup Sensor					
Automatic Day/Night Mirrors					
IV. Pleasure & Entertainment					
A. Audio System					
Radio	Yes	Yes	Yes		
Satellite Radio			YES		
CD Player	No	Yes	Yes		
B. Video System					
TV	No	No	Yes		
VCR, DVD	No	No	Yes		
C. Computer	No	No	Some		
Internet	No	No	Option		
D. Communications					
Cell phone	No	Yes	Yes		
Internet	No	No	Option		
V. Advanced Systems					
A. Navigation Systems	No	No	Yes		
B. Collision Avoidance Systems	No	No	Option		
C. Rain Sensing Wipers	No	No	Option		
D. DewPoint & Glass Temp Sensors	No	No	Option		
E. Voice Commands	No	No	Option		

microprocessors. It is often said that a new car has more computer power in it than our first manned space capsule. Computer-based systems are found in the system modules for new cars, and account for much of the engine performance, reduced emissions, sophisticated diagnostics, and many of our comfort and convenience features. The engine control unit (ECU) is the most powerful computer in the car, and it has the demanding job of controlling fuel economy, emissions from the engine and the catalytic converter, and determining optimum ignition timing and fuel injection parameters. These computers, microprocessors and system modules greatly simplify the diagnostic job of finding problems with the car, and providing information on what kind of repair or replacement work is needed.

While a large new building with a sophisticated BAS or building automation system may well contain 50 or more microprocessors, this does not match the new car in terms of having equal computing power per room or per group of rooms with 2 to 4 occupants. The rooms and offices in our buildings do not have monitoring and self-diagnostic features. They could, because the technology, equipment and systems exist, but they are not supplied as a standard item, and they are not available in the same way that options are available on new cars.

System Modules

As discussed above, the system modules are where the computer-based systems reside in new cars. These system modules are highly complex, and highly important systems in new cars. Many of our highly desirable performance and comfort features are provided by system modules. Typical system modules in a new car are: the engine control unit, the instrument panel module, the climate control module, the transmission control module, the power distribution box module, the airbag module, the driver's door module, the ABS module, the body controller module, and the cruise control module. These are the system modules on every basic car. Additional system modules are options for lower priced cars, or standard features of higher priced cars. These include navigation control modules, entertainment system modules, advanced comfort control modules and communication control modules for computers, cell phones and internet access.

Communications Buses

Using standardized communications buses with these system modules makes both designing and building new cars much easier than it was in the old days.

Two major services must be accessible to every area of a new car—electric power and the communications bus. All of a car's system modules must be able to communicate with each other, receive signals from most of the sensors in the car, and send signals to the control components, systems and actuators. Using a communications bus greatly simplifies the wiring, reduces the number of data sensors, and implements additional features at very little additional cost. Without the communications bus, the job of wiring up a car during the assembly operation would simply be too labor and time consuming to have a reasonable cost product. Also, the speed of communications is so important now that only a digital bus has the speed and capacity to handle the data collection and data transfer load for a new car.

The communications bus and the system modules work together to make the design and building of the car much easier. Information is sent over the communications bus in a standard communications protocol—usually the SAE J1850 standard, or the controller-area network (CAN) standard, although some manufacturers are using FlexRay, which is a faster and more sophisticated communications bus. Data are sent in packets with a standard structure—a label and some data. For example, an information packet with Speed for the label and 52.5 for the speed data in MPH is picked up by the instrument control module, which refreshes the indication on the speedometer with this new data. The standard communications bus makes the design of the various system modules much more straightforward. In addition, the sensors in the car only need to send packets of data to the communications bus; therefore, the carmaker does not have to deal with the problem of a particular sensor putting out a strange voltage or current signal that must be converted somewhere into a true physical parameter of the car's operation. In our example, the alternative is to tell the instrument panel module maker that the signal for speed was going to be a 4-20 mA current loop value, and that 10 mA was equivalent to 40 MPH.

The use of the standardized communications bus also makes it easy to use outside suppliers and sources for many of the components and systems in a new car. The carmakers do not have to worry about how a specific sensor or module works internally; they only need to know that the data will be transmitted in a known, standardized manner, and that it will have a known, standardized structure. Much of the success with using modern technology in cars, and much of the reliability of that technology comes from using the simplified approach of a standardized communications bus.

This same type of technology is essentially available for our new buildings. BACnet, LONWorks, and TCP/IP are the most common standard communication protocols. TCP/IP may be the ultimate answer, but another level of standardization is also needed to insure that data that comes across TCP/IP means the same thing to each different piece of equipment in a facility. Most buildings are being wired for a Local Area Network (LAN) with either coaxial cable or fiber optic cable. Thus, the hardware and software are available, but there is no organization responsible for requiring or enforcing the standardized interconnection of all of the building components, subsystems and systems like the automakers have. Without a standardized communications bus running through the entire facility—together with accessible electric power—buildings will never have the kind of technology that cars have, and we will never have the cost benefit or the reliability that this kind of technology can bring to our buildings.

Smart Sensors

Most of the basic automobile sensors that were used in the past to read continuous physical parameters such as temperatures, pressures, flows and levels operated on the principle of producing a voltage or current output proportional to the real value of the parameter. The output of these sensors was almost always nonlinear, and also varied with the temperature or other physical parameters. This resulted in poor measurements, or required using more equipment and processing power to correct the sensor reading for the nonlinearity and to provide temperature compensation curves to get accurate readings. Today, smart sensors are used to provide these functions and to output data to a microprocessor or system module. The sensor output is input to the microprocessor, and the sensor reading is digitized, corrected, temperature compensated and sent out over the standardized communications bus.

These smart sensors interface directly to the communications bus, and provide fast and accurate measurements. Since the sensor package contains a microprocessor, much of the load is removed from the system module that the smart sensor is supporting. Designed and built as an integrated package, the smart sensor fulfills its mission reliably with a low initial cost.

The sensors for buildings are expensive, and many of them are not very reliable. They are certainly not reliable in comparison to those in cars. In particular, the relative humidity sensors and CO_2 sensors are notoriously unreliable, and require frequent cleaning, calibration and general maintenance. That level of performance

would be unacceptable for these sensors in a car. Why shouldn't the sensors in buildings work reliably for a period of three to five years before they need any significant attention?

Wiring Harnesses and Standard Connectors

The use of pre-assembled wiring harnesses and standard connectors has made the task of wiring up a new car much easier. It is important to use stranded, not solid, wire cable. Each length of stranded wire consists of a twisted bundle of very thin thread-like wires. Solid wire, on the other hand, is a single thick wire segment. The advantage of stranded wire is that it is much more flexible than solid wire, and also less susceptible to breakage. One thread of a stranded wire can break without affecting the performance of the connection, but if a solid wire breaks the connection is lost. Also, if there is one weak link in the reliable performance of any electrical or electronic system, it is the connectors. With this in mind, the importance of carefully and correctly built standardized connectors cannot be overemphasized.

Use of Skilled Assembly Workers

The auto industry has a large supply of skilled workers at its design, engineering and assembly operations. These skilled workers receive training in their specific jobs as well as training in quality control and process improvement techniques. Many of the manufacturing and design improvements in new cars have come from the production workers themselves. In addition, skilled workers have made a great improvement in the overall reliability and quality of the new cars. Auto workers are usually paid more than those working in other industries or services.

Problems with the construction of new buildings often come from the use of workers with minimal or insufficient skills for the job. Finding skilled workers may be difficult, and is certainly expensive. The nature of building construction often impedes the retention of skilled workers. As a result there may not be a large pool of highly qualified building construction workers available when a particular building is being built.

One of the most common problems in building structures is the roof, which is the subject of the greatest number of lawsuits in building construction. Most roofs leak, and leak from the day the building is occupied. Roof leaks are the result of poor installation and construction rather than problems with roofing technology and materials. When a roof leaks, water leaks into the walls and may not be noticed until mildew and rot are visible; by then the building may be significantly damaged. Mold, mildew and IAQ problems in the building will require more time and money to fix. Using sensors in new buildings to identify roof and wall leaks when they occur is a critical application of automotive type technology in our new buildings. New cars use infrared reflectance sensors to identify rainfall on windshields, and automatically start up the windshield wipers. These sensors, or other types of moisture sensors, if installed throughout our new buildings, would quickly identify leaks and moisture buildup and alert building operational people to this serious problem.

Poor workmanship can cause many other problems in buildings. Even the HVAC system can be affected since random testing has shown that many air conditioning systems are installed with an improper charge of refrigerant. In economic terms, the problem of workers with insufficient skills and quality training results in the need to commission buildings to check and see if the building components and systems work as they should. (See discussion on Commissioning below.) This expense is clearly attributable to lack of adequate engineering, lack of quality control measures, and especially lack of highly trained workers.

Why Doesn't New Building Construction Include More New Technology As Standard Equipment And Systems?

Automobiles are built according to a standard plan; building architects on the other hand reinvent the wheel each time they design another building. This lack of standardization in buildings impedes the introduction of new technology in new building construction. Other factors also influence this difference in approach.

Unlike new cars, most new buildings are site built, and are built to "cookie cutter" specifications that emphasize lowest first cost of construction.

Even "custom built" buildings are held hostage to the lowest first cost syndrome. Thousands of different construction companies build residential and commercial buildings. Hundreds of different companies build fairly large commercial buildings. These companies range in size from small businesses to major architectural and engineering firms and major construction firms. It is extremely difficult to implement standards of technology when this many individual companies are involved.

The fact that most buildings are site built impedes the assembly line and systems approach to installing new technology that is used in the auto business. One area of building construction that is immediately ame-

nable to the assembly line approach of the carmakers is the construction of prefabricated or modular buildings. This manufacturing sector could easily incorporate the knowledge from the automotive assembly sector to produce buildings with the same level of technology and reliability as new cars. The engineering and quality control functions are much more cost effective in this sector. This sector could easily use more computers, more microprocessors, more system modules, more Smart Sensors, and a standardized communications bus.

Cars are constructed in a factory assembly line and moved to their ultimate market and user.

The factory environment makes it easier to train workers in installing the equipment in new cars as well as training them in quality control procedures. Buildings, however, are constructed at the point of use. Construction workers may work for a long time on a single building doing all types of work. Their training is not likely to be technology specific. Auto assembly workers typically specialize in some part of the assembly process, and therefore can be trained on this more limited work task. In addition, they become quite knowledgeable on this part of the assembly operation, and soon become able to add value to the company by suggesting improved methods of designing and constructing components and systems that they assemble. Quality control is more easily stressed in this environment, and many of the workers actually see the final result of their work drive off the assembly line, which serves to positively reinforce the need for a high skill level and the need to perform high quality work. In fact, these workers often own and drive cars produced by the company they work for. They are more likely to reject poor quality parts, components, systems and assembly procedures.

More new cars are sold each year than new buildings, so there is a larger market for the technology, and the price can be reduced due to bulk purchase of the equipment.

This is certainly true at face value, but when the scale of use of technology for buildings is considered, the numerical superiority of the cars goes away. If we consider that the unit of interest in buildings is rooms, and that we are interested in having the same technology level in each room that we have in a car, we now have a very different perspective. There may very well be more rooms than cars built each year. Thus, the comparison of a room to the car, rather than a building to a car, will lead to a much greater economy of scale for new building construction, and should provide a strong economic incentive to move in this direction for buildings.

Cars have a shorter lifetime than buildings, so new technology can be introduced faster, and the customers can develop a faster appreciation for what it does.

Cars do have a shorter lifetime than buildings, but most buildings end up being refurbished, or equipment and systems retrofitted, so there is still a lot of opportunity to use new technology in older buildings. Sensors, controls, system modules and many of the other features of new car technology can be added to older buildings when they are needed. In general, the most cost effective way to build an energy-efficient and functionally superior building is to do it right the first time, rather than retrofit it later. However, new equipment, and especially new information technology, can be added to rooms and to the entire building. It would have been easier and cheaper to install coaxial or fiber optic cable in a building when it was built, but we still have managed to find a way to get the LAN cable and connections into our rooms and offices so we could network our PCs.

Purchasers of new cars are influenced by features they have seen on other cars. Therefore, consumer demand is important in increasing the marketability of new technology options.

This is one reason we need to start installing some of this new technology in buildings. Once building owners, managers and occupants start seeing what has been done in other buildings, and how much more enjoyable and productive it is to work in buildings with this advanced technology, they will start to demand more technology as a result. It is somewhat amazing that the people who drive cars with all this new technology will go happily to work in buildings that do not come close to providing similar comfort and operational features of automobile technology!

Cars are designed for use by individuals; buildings are designed for use by companies.

The motivation of the designers and the manufacturers of cars is frequently different from that of people who design and build buildings. Car manufacturers build a car to attract a buyer; they add bells and whistles to make their car different. They encourage innovation and thinking outside the box. Architects and construction companies are building a box, so their thinking often stays in the box. They may work on the exterior design; they may make the interior appearance pleasing; but they do not think very hard about what goes on inside the box, and they don't consider the needs of the individuals living and working in the box. A car designer should consider safety when drawing up plans for a new car; beyond putting in emergency exits and sprinkler systems, a building designer may not think about how to make the building safer because that is not part of the job. Among the questions that building designers should be asking are "How can this building be made more comfortable, safer, and more user-friendly?" "How can occupants interact with this building to increase their comfort and safety levels?" "How can we make this a building of the future?" With a little imagination and an increased use of information and controls technology, building designers can make significant changes in the comfort level of the occupants.

What Does The Building Construction Industry Need To Do?

Establish an integrated design-and-build engineering and management structure. The amount of engineering work that goes into a new building must increase significantly. The building structure should be designed with high technology use in mind, and should utilize new technology to deliver the performance and comfort features that we want in our new buildings. In addition, quality control and reliability should be designed and engineered into the building from the start of the project. Then, quality management techniques should be employed so that the building is actually constructed to provide the quality and reliability features that we expect.

Use equipment and system modules in new buildings.

This approach has facilitated the use of most new technology in new cars at a reasonable cost, and with extremely good reliability. However, the standardized communications bus has made the most dramatic difference. By using a standardized communications bus and system modules, car technology could be transferred to buildings relatively easily. Individual HVAC modules for occupants, individual lighting modules, other comfort modules such as for seating, and building operation and maintenance modules could all be used to greatly increase the performance and reliability of new buildings and yet allow us to build them at reasonable costs. Certain sectors such as the residential manufactured housing sector, the hotel/motel sector, and many office buildings could easily adopt this approach.

Even site-built homes could incorporate some of these features. Residences are often pre-wired for intercoms, telephones, security systems, cable, and high-speed internet connections. Designing a central integrated system for monitoring and controlling the performance and comfort of a home and pre-wiring the house for such a system is well within the realm of feasibility. It is possible to envision a home with a central control panel that was accessible from the internet. Homeowners could monitor their homes from work. They could receive security or fire alarms. They could make changes to thermostat settings if they knew they were going to be early or late getting home. They could get alarms if there was a water leak or if the refrigerator stopped running.

Build more modular buildings.

The solutions to providing greater use of technology in new buildings and providing quality and reliable buildings are much easier for modular buildings with significant pre-site construction performed in a factory or controlled environment. High-tech components and equipment can be installed more easily in prefabricated and modular buildings within a controlled environment and with highly skilled and quality control trained workers.

Impose standards on equipment and system suppliers.

Most major construction companies are already in a position to do this. They have the financial leverage to specify components and equipment that meet their exact requirements. The residential manufactured housing sector in particular could do this quite easily. The federal sector, states and large companies also have excellent opportunities to set these standards. One of the most important standards is to require a standardized communications bus in a building with all sensors and controls interfacing directly with that communications bus.

Support codes, standards, or legislation to increase the use of new technology in buildings.

Building codes and standards have been responsible for many of the improvements in standard buildings. With minimum equipment efficiencies, minimum thermal transfer levels, and minimum structural standards in place, companies that construct buildings must meet these minimum standards—regardless of whether it increases the first cost of the building. Without minimum standards such as the ASHRAE 90.1 standard, many buildings would still have inferior equipment and poor insulation, because it was cheaper to put in initially. Other programs like LEED and EnergyStar could incorporate requirements for adding new comfort and control technology in buildings. The standards for utilizing new technology could be set voluntarily by large companies and big purchasers of buildings like the federal sector,

states, schools, and the hotel/motel sector. The auto industry has certainly incorporated many of the new technological features without needing government intervention.

Integrate new building technology with the desktop computers and BAS (Building Automation Systems) that are already being installed in new buildings.

The types of smart sensors, system modules and standardized communications buses that the authors have been recommending for use in new buildings should be considered an integral part of the overall Building Automation System. All of these components, systems and equipment must work together seamlessly to provide the expected level of performance and comfort and all the desktop computers should be tied in to these systems through a Local Area Network.

The desktop computer could be the equivalent of the car dashboard or instrument panel, and it should be the personal interface to an expanded BAS. It could tell what the space temperature is and how much ventilation is being provided. It should allow occupants to set their personal preferences for lighting levels, seat positions, window or skylight openings, etc. It should also let them enter new desired values of these space parameters.

Benefits of Standardized Commissioning of Buildings

Commissioning a building is defined in ASHRAE Guideline 1—1996 as: The processes of ensuring that building systems are designed, installed, functionally tested over a full range, and capable of being operated and maintained to perform in conformity with the design intent (meaning the design requirements of the building). Commissioning starts with planning, and includes design, construction, start-up, acceptance and training, and can be applied throughout the life of the building.

Commissioning a building involves inspection, testing, measurement, and verification of all building functions and operations. It is expensive and time consuming, but it is necessary to insure that all building systems and functions operate according to the original design intent of the building. Commissioning studies on new buildings routinely find problems such as: control switches wired backwards; valves installed backwards; control setpoints incorrectly entered; time schedules entered incorrectly; bypass valves permanently open; ventilation fans wired permanently on; simultaneous heating and cooling occurring; building pressurization actually negative; incorrect lighting ballasts installed; pumps running backwards; variable speed drives bypassed; hot and cold water lines connected backwards; and control dampers permanently fully open. And this is only a short list!

The process of commissioning a building constructed like a new car, and using the new car-type technology would be far quicker and simpler, as well as much less expensive. The use of standardized components, subsystems and systems could actually eliminate the need to check and test these items each time they are used in a new building. A factory or laboratory, standardized commissioning test could well determine their acceptability with a one-time procedure. The use of a standardized communications bus would dramatically shorten the time and effort of on-site testing of the building components, subsystems and systems. Data from all sensors and controls would be accessible on the communications bus, and would allow a significant amount of automated testing of basic functions and complex control actions and responses in the building. A commissioning module could also be added to the building systems, and would even further automate and speed up the commissioning process. This commissioning module would remain as a permanent building system, and would not only aid in the initial commissioning process, but also the recommissioning process, and the continuous commissioning process.

Presently, the cost of commissioning a new building is around 2 to 5 percent of the original cost of construction. The use of standardized commissioning tests, and the use of a commissioning module, would greatly reduce this cost. Although commissioning is a cost effective process—usually having a payback time of one to two years—many building owners do not want to spend the additional money for the commissioning effort. A prevailing attitude is "I have already paid to have the job done correctly. Why should I have to be the one to pay to check to see that it has actually been done correctly?" This is a difficult attitude to overcome, and it is often a hard sell to convince new building owners that they will actually come out ahead by paying to verify that their building does work as it was designed to work.

One final note on commissioning is that from one of the author's energy audit experience. Many problems found when conducting audits of existing buildings are clearly ones where the problem has been there since the building was constructed. For example, in the audit of a newspaper publishing company it was found that the cost of air conditioning was excessive. Further checking showed that the heating coil

and the cooling coil of the major air handling unit were both active during the hottest part of the summer. The control specifications specifically called for that simultaneous heating and then cooling! Once that original problem was corrected, not only did the air conditioning bill go down dramatically, but the building occupants reported that they thought the air conditioning system was working much better since they were much more comfortable.

DO NEW BUILDINGS NEED "DASHBOARDS?"

The dashboard and instrument panel is the heart of the driver—car interface. Status information on the car's operation is presented there in easily understood form. A similar feature in a new building would make sense. Not the complex HMI or GUI from a BAS, but a simplified display for average building occupants, and maybe even one for the building engineer or maintenance supervisor. Each floor of a building could have a "dashboard" type of display. It could be located in a visible place, and occupants could see the status of energy use in terms of peak cost or off-peak cost, daily use of kWh and therms of gas. They could also see temperature and RH conditions at a glance, and could get red light/green light indicators for energy use and maintenance actions. Several of these "dashboards" could be provided to the operation and maintenance staff. These simplified "dashboard" type of displays could also be available on the PCs of the occupants and operating personnel. Cars provide a powerful model to use in many building technology applications.

CONCLUSION

New buildings have not kept up with technological advances, especially when compared to automobiles. All we need to do is to make one trip in a typical new car, and then make one visit to a typical new building to see this for ourselves. Comfort levels, safety levels, reliability levels, quality control levels and automation levels are all much higher in new cars than in buildings. The imagination and creativity that goes into new car technology and manufacture should be harnessed for our new buildings as well. We really do need to start building our new buildings like we build our new cars.

ACKNOWLEDGMENT

An earlier version of the material in this chapter appeared in a paper titled, "If Buildings Were Built Like Cars—The Potential for Information and Control System Technology in New Buildings," by Barney L. Capehart, Harry Indig, and Lynne C. Capehart, *Strategic Planning for Energy and the Environment*, Fall 2004.

Bibliography

Argonne National Laboratory program on sensor development, *www.transportation.anl.gov/ttrdc/sensors/gassensors.html*

Automated Buildings website, *www.automatedbuildings.com*

Court Manager's Information Display System, *www.ncsc.dni.us/bulletin/V09n01.htm*

Delphi Automotive Electronics Website, *www.delphi.com/automotive/electronics*

How Car Computers Work, *www.howstuffworks.com/car-computer.htm*

"Motoring with Microprocessors," by Jim Turley, *http://www.embedded.com/showArticle.jhtml?articleID=13000166*

New car features, *www.autoweb.com.au*

"Sensors," Automotive Engineering International Online, *http://www.sae.org/automag/sensors/*

Smart Energy Distribution and Consumption in Buildings, CITRIS—Center for Information Technology Research in the Interest of Society, *www.citris.berkeley.edu/SmartEnergy/SmartEnergy.html*

"Today's Automobile: A Computer on Wheels," Alliance of Automobile Manufacturers, *http://www.autoalliance.org/archives/000131.html*

Chapter 9

Web Resources for Web Based Energy Information and Control Systems

Ken Sinclair

INTRODUCTION

This chapter is intended to provide additional perspective on the topics covered in this book. In particular, through the website www.automatedbuildings.com there is a considerable literature on the topics of web based energy information and control systems, as well as other areas of intelligent buildings. Since 1999 AutomatedBuildings.com has been an on line magazine and web resource. We provide the news and connection to the community of change agents that are creating our present definition of smart, intelligent, integrated, connected, green, and converged large buildings. Our virtual magazine and web resource provides a searchable platform for discussion and exchange while creating opportunities for B2B for all new and existing stakeholders.

The web based media is the message and we hope that providing web access to this information will amplify the power this book with access to related information that has only been published electronically to date. Here we provide four articles with web links related to longer articles published on our web site where you can gain more information.

PREDICTIONS FOR SMART BUILDINGS IN 2011
Jim Sinopoli, PE

Expect a few small and medium size companies to exponentially grow, some to be acquired by large companies, but some culling will take place.

> *"Never make predictions, especially about the future."*
> Casey Stengel, American Baseball Player and
> Manager, 1891-1975

Traditional building management systems (BMS) from major international manufacturers will become obsolete. The major systems are slow to adopt the latest IT software and applications, their interfaces are not the customizable dashboards users are looking for, their applications suites aren't broad enough and they will have to be completely retooled to handle demand response applications. Third-party developers are already tapping into the BMS databases or using open communication protocols to read and write to each point and taking the BMS functions to a new level. Without a quick response major manufacturers may be left with just software configuration tools for their controllers and field devices.

There will be a major shakeout in the "energy management software" sector. There is some great energy software available, much of it developed by small and medium sized companies that have agility and speed not seen in larger companies. However, developing software is one thing; it is straightforward to read energy data from Modbus meters and create graphs, charts and tables. Ongoing support and additional development is different and will take deeper pockets and larger companies. Expect a few small and medium size companies to exponentially grow, some to be acquired by large companies, but some culling will take place.

Internships for facility technicians, engineers and managers will dramatically grow. The building operations industry is short on qualified people. The skill set and knowledge base to operate and maintain a building is rapidly changing. Young men and women can go to an academic institution to gain knowledge of the technical systems in buildings but there's nothing like working in real world building operations to ground and grow that expertise. Large organizations with significant needs for facility technicians, engineers and managers will move to team up with technical institutions and offer internships to find and develop the talent.

Telecom carriers will enter the energy management and building operations industry. Yes, your cell phone or cable television service provider may be looking at facility and energy management as a new opportunity in their "network connectivity" or "managed services" business.

We tend to think of managed services as primarily dealing with information technology, where a company such as IBM manages the everyday operation of an enterprise client's IT hardware and software. However several service providers have figured out that it's not only IT systems they can manage for their clients but any sensor or device on any system or network. From a business standpoint the service providers are just leveraging their existing assets and expanding their offerings. The idea is to expand telemetry services where data from remote devices and sensors can be collected and communicated to a central point and analyzed for meaningful information, something that could bring value to building owners and managers. Major carriers are already deploying command and operations facilities and "collaboration centers" for the development of M2M applications. Their new slogan may be "Check your minutes, order a movie, and manage your building!"

Except for California, the trading of carbon emissions is pretty much dead. The business approach to capping CO_2 amounts and letting companies sell or buy tons of CO_2 sounded good but fraud in trading, the recession, and even the infamous "scientific" emails of Climategate have managed to sink the idea. The largest exchange in North America, the Chicago Climate Exchange (COX), was once estimated to eventually handle a $10 trillion trading market in CO_2, but in October the COX announced it is closing shop. On top of that are revelations of fraud in the Danish emissions trading registry where corrupt traders have walked away with an estimated $7 billion in the last two years. Similar fraud allegations are being investigated in other European countries; Germany, Britain, France, Spain, Norway and the Netherlands.

Expect a few new energy companies who started in the last few years and are involved with the smart grid and buildings to flourish after the recession is over. Recessions are transformative periods. Some of the world's largest companies started during a recession and managed to grow and evolve into industry leaders. This recession we've had large government funding for the smart grid, keen focus on energy, and a whole new market and industry created. Demand response is one example. In 2009 only 11% of the Demand response market was captured, with the remaining 89% being nothing except a great business opportunity. Expect companies such as Enernoc and Converge to grow and become industry stalwarts.

The importance of certification and accreditation of designers, contractors and operators for integrated building systems will increase rapidly. When we're undergoing major transformation of how buildings are designed, constructed and operated; when the complexity of the buildings is rising steeply; when the skill set and knowledge base to operate a building is shifting, the marketplace will sort through the qualified and unqualified by accreditation. The certification and accreditation process has been and is likely to continue to be a messy process, involving many organizations with fragmented or overlapping focus, and a knowledge base covering everything from BIM and LEED to integrated systems, security systems, energy and much more.

"White-space" networking will start to develop into the next major wireless standard. When television broadcasting was exclusively analog the regulators of radio frequency bands required "white-space" or "guard bands" between the frequencies so as to "guard" against radio interference. Given broadcasting has gone primarily digital and is also "compressing" more signal into less frequency, the guard bands are no longer needed, thus opening up radio frequencies for "white-space networking." Why the excitement and anticipation? Network speeds should be around 50-100MBps and above, thus easily accommodating streaming video and eliminating problems with VOIP quality. Seeing opportunities in new markets and new devices (called "white space devices" or WAD) all the big technology players are backing the endeavor: Google, Microsoft, Dell, Intel, Hewlett-Packard, etc.

For more information, write us at info@smart-buildings.com

BUILDING AUTOMATION SYSTEMS

Review of Chapter 15 on Building Automation Systems in the New Third Edition of the http://www.ashrae.org/greenguideASHRAE GreenGuide
Ken Sinclair

This is one of the best documentations of our Building Automation Industry's role in Green Building Design that I have seen.

Introduction

The ASHRAE GreenGuide is a guide for the design, construction, and operation of sustainable buildings. The ASHRAE GreenGuide aims to help you answer your biggest question—"What do I do now?" Using an integrated, building systems perspective, it gives you the need-to-know information on what to do, where to turn, what to suggest, and how to interact with other members of the design team in a productive way. Information is provided on each stage of the building process, from planning to operation and maintenance of a facility, with emphasis on teamwork and close coordination among interested parties.

Why is Chapter 15 Important to Our Industry?

Building control systems play an important part in the operation of a building and determine whether many of the green design aspects included in the original plan actually function as intended. Controls for HVAC and related systems have evolved over the years, but in general, they can be described as either distributed (local) or centralized. Local controls are generally packaged devices that are provided with the equipment. A building automation system (BAS), on the other hand, is a form of central control capable of coordinating local control operation and controlling HVAC and other systems (e.g., life-safety, lighting, water distribution, and security from a central location). Control systems are at the core of building performance. When they work well, the indoor environment promotes productivity with the lighting, comfort, and ventilation people need to carry out their tasks effectively and efficiently. When they break down, the results are higher utility bills, loss of productivity, and discomfort. In modern buildings, direct digital control systems operate lights, chilled- or hot-water plants, ventilation, space temperature and humidity control, plumbing systems, electrical systems, life-safety systems, and other building systems. These control systems can assist in conserving resources through the scheduling, staging, modulation, and optimization of equipment to meet the needs of the occupants and systems that they are designed to serve. The control system can assist with operation and maintenance through the accumulation of equipment runtimes, display of trend logs, use of part-load performance modeling equations, and automated alarms. Finally, the control system can interface with a central repository for building maintenance information where operation and maintenance manuals or equipment ratings, such as pump curves, are stored as electronic documents available through a hyperlink on the control system graphic for the appropriate system.

Outline of Chapter 15 on Building Automation Systems

This chapter presents the key issues to designing, commissioning, and maintaining control systems for optimal performance, and is divided into seven sections as follows:

- *Control System Role in Delivering Energy Efficiency.* Through scheduling, optimal loading and unloading, optimal setpoint determination, and fault detection, controls have the capability of reducing building energy usage by up to 20% (or sometimes even more) in a typical commercial building.

- *Control System Role in Delivering Water Efficiency.* Used primarily in landscape irrigation and leak detection, controls can significantly reduce water usage compared to systems with simplistic control (such as time clock-based irrigation controllers). Building controls can also provide trending and alarming for potable and no potable water usage.

- *Control System Role in Delivering Indoor Environmental Quality (IEQ).* In most commercial buildings, controls play a crucial role in providing IEQ. Controls can regulate the quantities of outdoor air brought into the building based on occupancy levels, zone ventilation, zone temperature, and relative humidity, and can monitor the loading of air filters.

- *Control System Commissioning Process.* Of all the building systems, controls are the most susceptible to problems in installation. These can be addressed by a thorough process of commissioning and post-commissioning performance verification.

- *Control System Role in Attaining Leadership in Energy and Environmental Design (LEED®) Certification.* This section describes the elements of LEED certification that can be addressed by control system design and implementation.

- *Designing for Sustained Efficiency.* Control systems help ensure continued efficient building operation by enabling measurement and verification (M&V) of building performance and serving as a repository of maintenance procedures.

Control System Role in Attaining LEED Certification

Chapter 15 provides a great overview of the control systems' roles in green building design while providing connection to how to obtain LEED credits.

In Section 7, the LEED® and other green building rating programs are discussed. This section explicitly discusses how controls can be used in various sections of the LEED 2009 Green Building Operations and Maintenance (USGBC 2009) (for existing buildings) and the LEED 2009 for New Construction and Major Renovations (USGBC 2009). (These are the latest versions in effect as of this writing.) A BAS or building control system can be of great assistance with the certification and maintaining certification for existing buildings under the LEED-

Existing Building program, but the impact is dependent on the type of control system available within the building. This section on LEED and controls will connect control methods discussed earlier in this chapter with either of the two LEED rating systems cited above. Be sure to get your copy of this important guide and work with your consultants and clients to insure that their projects are automatically green forever.

How Can Non ASHRAE Members Get This Guide?

The guide is available at ASHRAE's online bookstore at member and non-member pricing. The book also has a student price and is encouraged for use in classroom instruction in engineering, construction and architectural curricula.

ASHRAE has vast resources available to the building community to reduce the environmental footprint of buildings. Standards 189.1 and 90.1 come to immediate mind.

THE NEXT BIG THINGS
Ken Sinclair

Introduction

One of the advantages of a life spanning over four decades in an industry is being able to recognize trends that will radically redefine that industry. In 1975 I worked on a direct digital control "DDC" project that saw the first buildings operate without physical controls. That was the first big thing I saw in the Building Automation Industry. The problem was the cost was too high and the system too complex for the masses.

The next big thing I saw was the rapid evolution of these DDC concepts into low cost stand alone panels "SAP" with simple operator control languages "OCL." These concepts were pioneered by the likes of Delta Controls and Reliable Controls in my area of the world. This SAP concept was then deployed at a card level as microprocessor costs radically dropped. This was the early 1980's and the rest of the world was still installing pneumatic devices while we had moved to virtual devices controlled by powerful OCL. This was truly the next big thing that redefined our industry. By the early 1990s DDC became a way of life as the major control companies were forced to abandon their costly pneumatic controls. DDC allowed many new players into the industry which forced the big three to change.

Early DDC system communication protocols were a Proprietary Babel and in the early 90s the concept of open protocols such as Lon and BACnet started to gain

traction, truly the next big thing. The open protocols were quickly adapted by the new DDC companies who finally provided enough market pressure to force major control companies to embrace.

The next big thing in the late 1990s in our industry was the internet and early adopters the likes of Automated Logic Controls, Andover, and Enflex blazed a trail of how the internet would become an integral part of our industry.

How the internet would become an integral part of our industry was the next big thing of interest to me, and in May of 1999 we started AutomatedBuildings.com to provide an online saga of the evolution of our industry as part of the world wide web. Working with the web and its pioneers allowed me insight to this next big thing and it was huge. Even in the early years I heard of the coming of powerful web services.

The Next Big Thing

More evolution, however, was required by the web and our industry to unleash the web deployment. From Andover grew a new company—Tridium that more than any at that time provided close coupling of our industry to the web. Truly the next big thing. This step was very significant because it allowed our industry to grow up into the information technology "IT" industry but more important it allowed the IT folks to grow down to our industry.

All of this scene setting is necessary so that I can tell you that the next big thing is "Web Deployment" and have you believe me.

Although my peers cringe at my loose definition of web deployment, which I have chosen to include any services provided by the web automatically or even utilizing manual intervention, web deployment of all our services is redefining our industry. Because we are now part of the IT industry and our industry is just viewed as a data stream, we are being swept at an unbelievable rate into web ways defined not so much by our industry but by the incredible daily evolution of the web by the IT industry.

The Latest New Thing

A recent editorial of mine contained this quote: *The past decade has been an extraordinary adventure in discovering new social models on the Web—ways to work, create and organize outside of the traditional institutions of companies, governments and academia. But the next decade will be all about applying these models to the real world.*

The collective potential of a million garage tinkerers is now about to be unleashed on the global markets, as ideas go straight into entrepreneurship, no tooling required.

Web was just the proof of concept. Now the revolution gets real. These are words from Chris Anderson Editor in Chief of *Wired Magazine.* Understand we are part of this web revolution and this is the next big thing.

So can my old mind grasp what might be the next big thing after web deployment? I am betting that rapidly evolving low cost wireless BAS devices that actually work well with open protocols with self discovery, self healing, cell phone like networks closely coupled with web deployment will completely redefine our industry.

The Dynamic Duo

This dynamic duo will eliminate complex installations, provide; self set up, connection, and commissioning while self populating the needed information for web deployment. As industry equipment such as chillers, boilers, air handlers, etc are moved into place and powered up they will wirelessly connect to their creator and powerful web services will take over complex commissioning. Their identity, an IP address will be a part of the manufacturers, owners, designers, and operators web deployment, it is a new world and these are the next big things.

CONTINUOUSLY CONNECTED OPEN INFORMATION FOR BAS

Ken Sinclair

Introduction

This chapter is also the focus of our 12th consecutive year of providing free education sessions at the AHR Expo 2011 Education Session Las Vegas. The agenda for those sessions is used as an outline and provides linkage to online resources that will help you better understand what the heck we are talking about. All of our sessions focus on the topic of continuously connected open information for BAS. Details of these sessions are posted on the automatedbuildings.com web site.

The five AHR Expo Education sessions are:

1. Connecting Building Automation to Everything

Session discussed managing building systems all the time from anywhere with a continuously connected open web environment to allow the user to complete all tasks including engineering, commissioning and facility maintenance. The role of evolving technologies such as smart phones and tablets for users and facility managers interface was discussed plus how to sorting through and present data. Information technology "IT" is changing everything while connecting everything, and it's this combination of changes that is dramatically changing

the BAS industry. The driver is not "IT," but the way forward is enabled by IT. How we prepare our building automation information for continuous connection will be the focus of session. This session will also covered a few predictions for 2011 plus the new products, innovations and developments since AHRExpo Orlando.

Buildings connected with open protocols to the powerful internet cloud and its web services are redefining the building automation industry, with the result that the reach and the visibility of the industry have never been greater nor has change been so rapid. Our clouded future includes new virtual connections to buildings from the communities they are part of with both physical and social interactions. An example is digitally displayed energy / environmental dashboards to inform all of the building's impact in real-time energy use, plus the percentage generated from renewable sources. And connections to the smart grid make buildings a physical part of their supply energy infrastructure.

The ability to operate buildings efficiently via the internet cloud from anywhere allows the building automation industry to be better managed and appear greatly simplified. Web services, or software as a service (SaaS) as it is sometimes called, coupled with powerful browser presentation are changing how we appear and interact with clients.

Building information model (BIM) software allows the power of visual relational databases to improve decisions throughout the building design. And new visualization and simulation tools reveal the effects of decisions made prior to the commitment of funds. In similar fashion, cloud computing provides a collaborative process that leverages web-based BIM capabilities and traditional document management to improve coordination.

The data cloud for our industry has become real. As we see applications and services moved "off-site," you can imagine the opportunities for managing real estate, reducing energy and providing value-added applications for buildings.

We must unhinge our minds and find new pivot points from which to build our future. We must embrace the power of the cloud while increasing our comfort level in using the solutions within. In an ideal world, we will be able to be vendor and protocol independent. Everything will talk to everything.

What is the Ontology?

Since ontology allows us to represent specific knowledge in an abstract and organized way it could be used to create a common layer between different application protocol paradigms.

Creation of such technology where the devices and networks comply, even if they are produced by different vendors, represents a major move to really open systems. Today within one technology the devices are able to communicate with each other. This vendor-independence we call *interoperability*. Interoperability does not guarantee just communication, it guarantees distributed application processing among products from different vendors within one system and one technology domain. Since, all technologies have their data model as well an application model, we cannot mix devices from different technology domains. We can do it through gateway translations, but that makes integration very complex and expensive. We see that integrating dissimilar technologies is not a trivial task. Within different networks (BACnet, LonWorks, ZigBee) particular application models use different communication services as well as different data structures to store application data. A mapping between different models is the only way for integration of dissimilar systems. Commonly, we use gateways for such applications. Usage of ontology principles in integration of heterogeneous networks brings major benefits. First we are able to co-mmission and configure an automation system centrally through ontology changes only. The central management approach guarantees system consistency. Since there is no translation between protocols, there is always translation between the protocol and ontology model, we don't have the data overhead we have normally in most of today's heterogeneous systems. Generally, what we do is shift interconnection on top of the application layer. That significantly simplifies information management and processing. Simply, that brings us to a possibility to integrate variables (information) from various protocols as input parameters of a particular function, do the processing, and the output parameters could again go to a different protocol. To be able to do so, we have to separate generic information from a dependent installation using an abstract model.

2. Continuous Commissioning and Today's Aggressive Energy Standards

Session discussed the need for a systematic approach to tracking energy utilization that detects problems early long before they lead to tenant comfort complaints, high energy costs, or unexpected equipment failure. Today's aggressive energy standards are greatly increasing the need to insure all technologies in place work. Once successful operation has been achieved continuous commissioning is the only way to maintain and improve aggressive energy standards.

Persistent monitoring and diagnostics of system operations directly impacts sustainable energy efficiency in commercial buildings. Examples include everything from detecting heating and air conditioning programming errors to identifying out-of-adjustment settings on control systems, improperly balanced parallel chillers that cause unwanted surges, high head pressure on rooftop unit compressors, oscillating controls that cause unnecessary heating and cooling run times, and incorrect refrigerant charge.

In today's complex buildings, even small problems can have big impacts on building performance. Lighting, heating, ventilating and air conditioning systems need continuous performance tracking to ensure optimal energy efficiency. Yet, a formal process for data gathering and analysis is not commonplace in the nation's building stock. Plus, there's often a disconnect between the energy modeling done in isolated, one-time re-commissioning or energy audit projects, and what really happens in day-to-day building operations.

What's needed is a systematic approach to tracking energy utilization that helps detect problems early, before they lead to tenant comfort complaints, high energy costs, or unexpected equipment failure. That's why new robust energy monitoring technologies and Monitoring-based Commissioning (MBCx) techniques are now at the fore-front in building energy management.

MBCx has the potential to keep buildings running at peak efficiency by addressing the "performance drift" which occurs when building systems fall out of calibration or fail altogether. A sensor network gathers discrete data measurements and with analysis capabilities identifies trends, detects leaks and alerts building engineers to hidden problems that waste energy.

A recent Lawrence Berkeley National Laboratory study revealed that MBCx is "a highly cost-effective means of obtaining significant energy savings across a variety of building types." The program combined persistent monitoring with standard retro-commissioning (RCx) practices with the aim of providing substantial, persistent, energy savings. There were three primary streams of energy savings from the MBCx project:

3. Key Technologies for our Connected Future

This session carried on from the "Connecting Building Automation to Everything" session discussing and providing the details of the key technologies used to connect all types of data and services to the information cloud. New sensors, video analytics, wireless, SaaS, artificial intelligence, ownership of metering are all changing how we connect to the future. In addition we will discuss building system analytic software, operations centers, micro video cameras, facial recognition security, plug load control and more. Our new found graphically technologies allow us to demonstrate and tell the world about our ability as an industry to reduce environmental impact and energy.

An energy dashboard provides much more technical information than an energy education dashboard. It consists of a series of gauges, graphs, and live display values that provide a building operator with a summary of the important energy metrics within their facility. These dashboards are very useful and are typically utilized by the operation management side. In fact, one "dashboard" definition even defines it as a reporting tool that presents key indicators on a single screen, which includes measurements, metrics, and scorecards.

While an organization's management team can proactively manage a building's resources, it's just as important that the occupants are aware and involved in this effort. The second dashboard to emerge is the energy education dashboard, a dashboard that is used to provide education and facilitate the widespread understanding of sustainable building. Educating building occupants on measures they can take to be more energy efficient can actually reduce the resource consumption within a building, and proper education on sustainability efforts can ensure those efforts continue into the future.

Energy education dashboards focus more on education about a building's efficiency and sustainable features than the technical information that operation management would utilize. Real-time resource use is provided to give occupants insight on how their actions directly impact the building, and the data is presented in an easy to understand format, at a more simplified level than what a building operator might expect to see. Another important aspect of displaying this building data for occupants and the public is to benchmark that data versus a baseline energy model, an ASHRAE standard, or even against another company or governmental goal. Seeing how a building compares to others is very informative, especially when that energy savings can be translated into monetary savings. Even though the data are simplified, it is still helpful for facility managers and building owners to review these data; mainly, because the information is so easy to access. Many energy education dashboards are presented on a touch screen, located in a building's main lobby for example, as well as through the organization's intranet or internet for the public to see.

At the top of the list of benefits of an operations center is the capability of operators and technicians having a comprehensive and common image of events or a situation. The common operating picture among operators

breeds communications, collaboration and often some degree of cross-training and workforce flexibility among the operators. If you have building technology systems that are integrated and interacting, this collaborative and holistic understanding of how the total building is performing is critical to managing the building.

The operations center is where technicians, engineers and management monitor, manage and troubleshoot issues. The operations center monitors building performance, systems configurations, policy implementation, scheduling, report generation and documentation.

At the heart of an operations center are the "human factors." "Human factors" sounds like some mushy soft science, but it is a well-recognized scientific discipline called human factor engineering. It is utilized to address the environmental design of an operations center, ergonomics, reengineering of operational processes, and the human interface to the technology. There is a tendency to focus on the technology in the operations center rather than the human factors (who isn't wowed by a video wall of high-def plasma displays?), however, the focus on the bells and whistles misses the underlying premise that technology is simply an enabler, and should be used to change the behavior and operations of the people using it.

4. *A Panel discussion. Incentives to Motivate and Connect our Industry*

A panel discussion of what is currently motivating the industry to significantly improve building systems, i.e. the green building movement, deep design analysis, continuous commissioning, sustainability, LEED, etc. The panel explored issues like, "Do utility energy efficiency incentives and rebates work?" and "What is the best vehicle to motivate our building automation industry?" Panel will also explore a variety of tools in the marketplace used for benchmarking and monitoring. Will energy performance BEPIs become part of the due diligence? Become part of this discussion of our how to incent our future. Join us and provide your input and ask questions of the panel.

"Does owning the meters make you smarter?" Answer: Only if you do smart things with them. As you see by the above rant we are now spending billions of dollars without a plan. If we own the smart meter or collection of sub meters we no longer need to deal with other people politics. We can decide if we wish to open up energy information to all; we can feed energy information to an energy dashboard in the lobby; we can bring on and monitor the performance of renewables; run a micro grid generation behind the wall. Our options are unlimited and unrestricted driven by only what is right for the building. You would not run your car without a gas gauge, why would you run a building without dynamic energy reporting? In fact car fuel gauges have gotten very sophisticated including: instant consumption readings; projections; histories because they can and at a very low cost. Why not have all this, plus more at a building level, with monthly tank full costs that greatly exceeds our car?

I see a movement to better metering at a building level—the first step in creating the necessary changes. If we are to achieve improved energy efficiencies we must be able to measure it. To understand the value of continuous commissioning we must measure it. All of the changes we are able to make at a building level need to be validated. Information is power, information that has been freed can be used for many purposes and its power is greatly increased.

What else can we do with our new smart meters? We can provide much better information than the utility can as to the amount of energy we are using because for the most part they have 20 to 30 year old meters connected to oversized Current Transformers based on the total power requirements for the complex not actual usage. New smart meters provide higher accuracy and amazing information such as harmonic analysis.

Do not be surprised if your installation of new smart meters on your side of the wall causes the utility to replace their meters on their side of the wall.

Now that we have the utilities' attention we can start meaningful dialog while we demonstrate to them how we can shift loads and generally what a B2G relationship could provide. Although utilities are very political there is a real side at a local level to keep running with available distribution equipment and any ideas that avoids them increasing sub station size or distribution are still of interest.

If all this is not of interest to the local utility it is possible that a well documented energy profile will be of interest to an energy aggregator to add to his stable. If we are to go shopping for the best way to buy our power we had better know how much we use and how controllable it is.

What we have achieved in owning our own smart meters is control of our future and its' options.

Smart metering and sub-metering can be added to most buildings for relatively low cost and almost imme-diately. It is an investment that will keep paying for itself over and over and even when the smart grid does become a reality it will be a second source of opinion, and when the utility will not share their information, you can.

Some building owners have been seduced by the concept that the utility is going to pay for smart meters and this has prevented them from investing in their own smart meters. It is clear that whatever the utility company calls a smart meter will be limited by their imagination and motives and will be less useful than owning your own accounting device for the millions of utility dollars you spend.

Use our site search to read our many articles on smart metering and how to implement.

Leverage the smartness in your own smart meters unrestricted by other's cluttered thoughts and motives of what they think a smart meter may be.

Yes owning the meters does make you smarter!

5. *A Panel discussion. Creating Budget for Implementing Information Management*
"How much should a facility management organization set aside from their annual budget for information management?" A fascinating question not generally discussed. The "convergence" of IT and BMS solutions, the evolution of BIM and related technologies, and the emergence of a new facility function "Facility Information Officer" suggests that the time is ripe to get some discussion on this question by our panel. Discussions will include new facilities with a BIM as well as older ones with a lot of data needing connection and convergence. "Connectivity has reached much further than we originally imagined and this session will explore and provide an update on evolving standards and trends in industry and there effect on this budget. Budgeting for the cost of information beyond the building connecting to the collaborative Connected Communities for Building Systems will also be discussed. Join us and provide your input and ask questions of the panel.

You can go to the AutomatedBuildings.com site to get more answers as this is an evolving discussion.

Reference

AutomatedBuildings.com

Section III

Building Information Systems

Chapter 10

Building Energy Information Systems: State of the Technology and User Case Studies

Jessica Granderson
Mary Ann Piette
Girish Ghatikar
Phillip Price

EXECUTIVE SUMMARY

The focus of this chapter is energy information systems (EIS), broadly defined as performance monitoring software, data acquisition hardware, and communication systems used to store, analyze, and display building energy data. At a minimum, an EIS provides hourly whole-building electric data that are web-accessible, with analytical and graphical capabilities [Motegi 2003a]. Time series data from meters, sensors, and external data streams are used to perform analyses such as baselining, benchmarking, building level anomaly detection, and energy performance tracking.

Energy information systems are viewed as a promising technology for a number of reasons. There is widespread recognition that there is often a large gap between building energy performance *as designed* and measured post-occupancy energy consumption, and a growing body of evidence indicates the value of permanent metering and monitoring [Brown et al. 2006; Mills et al. 2005; Mills 2009; Piette et al. 2001b]. Energy information systems are also well aligned with current trends toward benchmarking and performance reporting requirements, as in recent federal and state mandates.

Dozens of EIS are commercially available, yet public domain information is often vague, and demonstration software may not be available. In addition, a lack of common terminology across vendors, and a significant degree of salesmanship, makes it difficult to discern *exactly* what functionality the tools offer, what the hardware requirements are, or what makes one product more effective than another. This study was designed to extend and update an earlier report [Motegi and Piette 2003], and it is guided by three high-level objectives:

To define a characterization framework of EIS features that provides a common terminology and can be used to understand what EIS are and what they do.

To apply the framework to EIS products to achieve a better understanding of the state of the technology, its distinguishing capabilities, and its leading-edge functionality.

To conduct case studies, to begin to understand the interplay between common features, diagnostics, and energy-saving actions.

EIS State of the Technology

The EIS characterization framework was developed iteratively, beginning with the features identified in prior work and a scoping of current technologies. In its final form the framework comprises eight categories with five to ten features each. This framework was then applied to characterize approximately 30 EIS. Key findings that are related to distinguishing capabilities, leading edge functionality, and the general state of EIS technology are presented in the following list, grouped by major feature category.

Business models (General)
- EIS are most commonly offered through an Application Service Provider (ASP) with no hardware, or optional hardware based on client needs.
- Optional or bundled services are nearly universally offered.

Display and visualization
- Features have converged to a near common set. Data can be viewed over user-defined intervals of time, trended variables can be aggregated into totals, and the user can overlay multiple data sets on a single plot.
- X-y scatter plotting is offered in only half of today's EIS solutions.

Energy analysis

- Two-thirds of the EIS feature greenhouse gas analysis, or provide custom or configurable options to do so. Most apply a simple energy/carbon dioxide (CO_2) relationship, but almost half account for regional differences in generation or other standards.

- Nearly every EIS permits the user to quantify an energy consumption baseline, however weather normalization is rare.

- Every tool that was evaluated supports (or will soon support) multi-site benchmarking. Distinguishing aspects include:
  - Composition of the comparative cohort: buildings within the user's enterprise; comparison to buildings from the vendor's database; or less commonly, national data sets.
  - Display of results: static reports versus dynamically accessible functions; results depicted in tables, plots, or charts.

Advanced analysis

- About three-quarters of the EIS address data quality, and they do so via three principal means: flagging or summative reporting, cleansing and/or correction, and linking to external or third-party software packages.

- Anomaly detection is typically trend-based and accomplished by identifying departures from normal energy consumption patterns.

- More than half of the EIS forecast near-future loads, usually by coupling historic trends and weather data; very few provide model-based capabilities.

- The large majority of EIS accommodate some form of measurement and verification (M&V) or the ability to track the impact of operational changes.

Financial analysis

- Energy costing is supported in nearly all of the EIS, and more than half have implemented model- or tariff-based calculations.

Demand response

- Demand response (DR) capabilities have advanced since early 2000 and have converged to a common set of features.

- Automated response to DR signals is supported in all but three of the DR systems that were characterized.

Remote control and management

- Just over half of the EIS surveyed report the ability to control according to a program, and just under half report internet-capable direct remote control.

The EIS product evaluations indicated that, overall, visualization and analytical features are distinguished by the degree to which they accommodate dynamic user-defined selections versus statically defined reporting, calculation, and plotting parameters. Rigorous energy analyses that include normalization, standards-based calculations, anomaly detection, and forecasting are robustly integrated in some EIS products, but less so in others.

EIS User Case Studies

The case studies included in the scope of this study attempted to answer questions related to energy savings and actions attributable to EIS use, performance monitoring challenges, and successful implementation models. Wal-Mart, Sysco, the University of California (UC) Berkeley, and UC Merced were selected, representing commercial enterprises and campuses with a diversity of performance-monitoring technologies, commercial building types, and portfolio sizes, as described in Table 10-1. These cases encompass buildings that range from Wal-Mart and Sysco's relatively repeatable warehouse and retail designs, to UC Berkeley's legacy and historic sites, to UC Merced's very-low energy new construction.

UC Merced

The UC Merced case illustrated the challenges in using a web-based energy management and control system (EMCS) as an EIS, the web-EMCS as enabling critical information links, and realization of the campus as a living laboratory. Typically, WebCTRL use at UC Merced is dominated by operational EMCS investigations, however, WebCTRL meter data are used annually to track energy performance. Gas, electricity, hot water, and chilled water consumption are quantified at the campus level and for critical buildings. On a monthly basis, the campus energy manager uses the web-EMCS data to determine utility recharges for non-state buildings, and he reports a high level of satisfaction with WebCTRL. He emphasizes that UC Merced trends extremely large volumes of data and that intensive monitoring needs to be undertaken deliberately, with close attention to a spectrum of issues including wiring, system programming, network architecture, and hardware selection.

Sysco

The Sysco case highlighted: (1) enterprise-wide EIS use and information sharing, both vertically and horizontally throughout the corporation, (2) limited, yet powerful,

Table 10-1. Characteristics of case study sites

Case	Type, size (square feet)	Controls	Performance Monitoring
UC Merced	Campus (800,000)	Automated Logic Corporation WebCTRL	Automated Logic Corporation WebCTRL Utility bills
UC Berkeley	Campus (15.9M)	Barrington Some ALC, Siemens	Obvius Utility bills
Sysco: Stockton, California Sygma site	Refrigerated/dry warehouse (52 M, Stockton 95,000)	DOS-based refrigeration control	NorthWrite Energy WorkSite Utility bills
Wal-Mart	Retail/grocery (675M)	Novar Danfoss Emerson CPC	Energy ICT EIServer Utility bills

on-site use of the EIS, and (3) use of EIS technology to ensure persistence in savings and energy accountability. Sysco adopted a three-part approach to achieve portfolio savings of 28% in under three years: expert site visits to conduct tune-ups and identify low-/no-cost energy-saving measures; customization of the EIS to accommodate and map to Sysco's goals; and continuous communication and collaboration between corporate managers, energy services contractors, and on-site "energy champions." Sysco performs both site-specific and portfolio analyses on a monthly basis. Managers coordinate monthly group reviews with each site's "energy champion," who is accountable for energy use. The energy champion who was interviewed reports that the EIS is most highly valued for its role in supporting and encouraging accountability and staff motivation, so that efficiency gains might persist over time.

Wal-Mart

Wal-Mart is a case of "siloed" EIS use by specific groups or individuals for a few key purposes. A group of internal supporters champion the use of the EIS technology and maintain a vision for how its use might be expanded throughout the organization, yet regular operational analytics are not yet widespread vertically or horizontally within the enterprise. The EIS features a custom module for M&V tasks that has been used extensively, although it has been used on an ad-hoc basis, to determine the effectiveness of energy efficiency improvements. The wholesale power procurement and demand response group also uses the EIS intensively, making considerable use of forecasting and normalization. The EIS is also used to gauge the performance of new designs, particularly at "High Efficiency" supercenters. Each month, the benchmarking analyst identifies the twenty poorest-performing

sites; however, custom benchmark models and downloading constraints in the interface require that EIS data be exported to conduct this portfolio tracking.

UC Berkeley

There is no central EIS at UC Berkeley; it is a contrasting case that is included to illustrate the challenges that are encountered in the absence of a campus-wide performance monitoring system. Although there is no campus EIS, there is a large volume of energy and system performance data, yet it comes from disparate sources and is used by different staff groups. The utility group uses utility bills and monthly manual meter reads to manage the purchase and billing of all campus energy, performing reviews for approximately 200 utility accounts. The EMCS group uses a web-accessible interface to oversee the campus Barrington control systems. Independently, a number of efficiency and commissioning interventions have implemented remotely accessible electric interval metering at approximately 30 buildings, totaling 11 million gross square feet. UC Berkeley's energy manager identified several energy management priorities including: more remote-access metering to reduce the resources dedicated to manual meter reads, submetering beyond the whole-building level, and access-controlled public data for researchers and special projects.

Conclusions

Resources and staffing were a significant constraint in every case studied, and clearly affect the extent to which energy data are successfully used to identify energy-saving opportunities. They also directly affect a site's ability to make meaningful use of submetered data. With the exception of Sysco, where current levels of engagement with the EIS are viewed as sufficient to meet efficiency goals, each organization expressed a strong desire to engage more with measured data in order to improve efficiency.

Reliable, high-quality data are a critical aspect in automated analysis of building energy performance, and can have a significant impact on EIS usability. The Merced case shows that particular attention must be paid to wiring and hardware integration, system programming, and network communications. In contrast, Wal-Mart and Sysco did *not* report significant data quality issues, probably for two reasons: EIServer has embedded validation estimation error checking (VEE) routines, and data quality is usually a concern only in cases of submetering and energy sources other than electric. In the four EIS cases that were studied, the most common energy-saving actions related to fixing incorrect load scheduling, performing measurement and verification (M&V) tasks, and identifying and fixing inefficient operations. Reported savings resulting from these improvements were on the order of 20%–30% for measures applied at the end-use and whole-building level.

The degree to which a site uses embedded analytical capabilities depends on the particular performance metrics and benchmarking data that are utilized. Our cases showed that the more tailor-made the calculations, the more likely it is that the data will be exported for analysis in third-party modeling or computational software. Although EIS offer a wide range of features, actual use of these features can be very limited, and it is not clear that users are always aware of how to use the capabilities of the technology to generate energy-saving information.

Future Needs

Future research needs concern four key areas:
1. Features and usability
2. Anomaly detection and physical models
3. Technology definitions and scalability
4. Successful use and deployment models

Questions concerning the most useful features, potentially useful but underutilized features, and energy savings attributable to EIS use merit further attention. For instance, a more extensive set of typical actions and associated energy savings, as well as documented records of building consumption before and after EIS implementation, would enable stronger conclusions on the range of expected savings from EIS use. Closely related to features and usability, there is considerable analytical potential in linking EIS anomaly detection methods to physical models. Today's EIS algorithms rely purely on empirical historic performance data to detect *abnormal* energy consumption. However, they do not provide a means to identify *exces-*

sive energy consumption relative to the design intent, or to realize model-predictive control strategies. Standardizing the format and structure of information at the data warehouse level could encourage such advancements, as could the development of features to configure exported data files into formats that can be used by modeling tools such as Energy Plus or DOE-2. Standard formatting of EIS data would also facilitate the transfer of energy information from the building to outside entities, supporting and aligning with current developments in demand side management, and the smart grid.

From a technology standpoint, definitions and scalability require further study. The question of whether a given system is or is not an EIS is not trivial. This study defines EIS broadly, stipulating whole-building energy analyses, graphical capabilities, and web accessibility. Therefore, many technologies that were included in the study are EMCS or DR tools that are less immediately thought of as EIS, but that *can be used* as an EIS. Scalability is a concern that may provide insights as to where to draw the line between EIS and related technologies. In the future it will be necessary to understand the tradeoffs between diagnostic capabilities, trend volume and number of points monitored, and the resulting burden on the system's underlying hardware and communication networks.

Finally, there remains much to learn about effective EIS use within organizations. A common view is that EIS are primarily the domain of in-house staff, and that services are used to a minimal degree during installation and configuration. However, the general prevalence of staffing constraints, Sysco's successful efficiency gains through partnership with service providers, and the number of EIS vendors that offer analytical services indicate the potential for alternate models of successful EIS use. Additional research is needed to understand the full spectrum of approaches to data-centered energy management. Large enterprises and campuses have cost-effectively implemented EIS, yet for other organizational sizes, commercial segments, and building ownership models the appropriate balance between on-site analysis, technology sophistication, and expert services is not well understood.

INTRODUCTION

The focus of this chapter is energy information systems (EIS), broadly defined as performance monitoring software, data acquisition hardware, and communication systems used to store, analyze, and display building energy data. Time-series data from meters, sensors, and external data streams are used to perform analyses such as baselining, benchmarking, building-level anomaly detection, and energy performance tracking. Newly adopted initiatives such as the Energy Information and Security Act, the zero-energy Commercial Building Initiative, and the Smart Grid have brought building energy performance to the forefront of the national energy dialogue. At the same time, national energy use intensities across the commercial sector increased 11% between 1992 and 2003 [CBECS 1992, 2003], marking a trend that must be quickly reversed in order to meet national net-zero building energy goals. It is clear that a multiplicity of solutions will be required to effect deep efficiency gains throughout the nation's building stock, and analogous to home energy displays, building EIS have received significant attention as a technology with the potential to support substantial energy savings.

Energy information systems are viewed as a promising technology for a number of reasons. There is widespread recognition that there is often a large gap between building energy performance *as designed* and measured post-occupancy energy consumption. A growing body of evidence indicates the value of permanent metering and monitoring [Piette et al. 2001], particularly in the context of monitoring-based and continuous or retro commissioning [Brown et al. 2006; Mills et al. 2005; Mills 2009]. Also pointing to the value of monitoring, researchers have increasingly documented the positive behavioral impacts of making energy consumption visible to building occupants and residents [Darby 2006; Petersen et al. 2007]. Energy information systems are also well aligned with current trends toward benchmarking and performance reporting requirements. For example, recent federal and state mandates require benchmarking of public buildings, and many corporations now participate in greenhouse gas (GHG) emissions reporting. While these requirements can be met through utility bill tracking, EIS can certainly simplify the process through increased levels of automation.

This work is motivated by two closely related, yet unproven concepts. First is the idea that buildings are complex, dynamic systems, and that realizing optimal energy performance requires higher-granularity data and more timely analysis than can be gained from monthly utility bills. Second is the notion that EIS are critically important because they can process data into *actionable information*, and thereby serve as the informational link between the primary actors who affect building energy efficiency. This concept is illustrated in Figure 10-1, using the following example. Time-series data from electric interval meters

and weather information services are analyzed by the EIS, which displays information in the form of weekend versus weekday energy consumption. The EIS user is then able to take action based on this information, for example, ensuring that weekend schedules are properly implemented. Further, since the EIS is implemented in software, the energy manager who might detect the mis-scheduling is able to share this information with the operators who are responsible for equipment settings and controls, and with owners or other decision makers who might need to authorize such changes, or to track energy costs. Clearly, as one transitions from the whole-building focus of EIS to component or system level fault diagnostics, there is a spectrum of what is considered "actionable information." For example, EIS do not typically generate information as specific as, "third-floor damper stuck open." Rather, the current state of the technology is such that a knowledgeable operator can use the visualization and analysis features to derive information that can be acted upon.

There is not an extensive body of prior work or literature from which to draw an understanding of contemporary EIS technology or the energy savings that they might enable. Dozens of EIS are commercially available, yet public domain information is often vague, and demonstration software may not be available. In addition, a lack of common terminology across vendors and a significant degree of salesmanship makes it difficult to discern *exactly* what functionality the tools offer, what the hardware requirements are, or what makes one product more effective than another. These questions must be

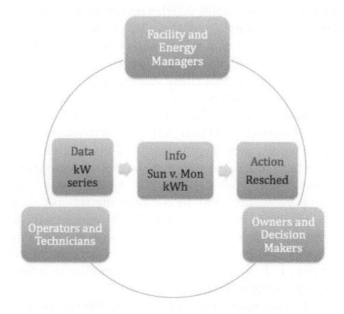

Figure 10-1. EIS translate data into actionable information and link the actors who impact building energy

better understood before it is possible to evaluate the energy saving potential of EIS. What is the full spectrum of analyses and diagnostics that EIS support? Which capabilities are standard in EIS, and which denote more sophisticated functionality? What are users' experiences with EIS, and how do they leverage embedded features to improve energy performance?

Correspondingly, this study is guided by three high-level objectives:

1. To define a characterization framework of EIS features that provides a common terminology and can be used to understand what EIS are and what they do.

2. To apply the framework to EIS products to understand the state of the technology, distinguishing capabilities, and leading-edge functionality.

3. To conduct case studies to reveal critical aspects of EIS usability and begin to understand the interplay between common features, diagnostics, and energy-saving actions.

While the body of prior work dedicated to EIS is sparse compared to other aspects of building control and diagnostics, there are several studies and key articles that merit attention. Two books published in 2005 and 2007 contain editors' compilations of articles that document the implementation of web-based building control and automation systems and their use for enterprise or site energy analysis [Capehart and Capehart 2005, 2007]. This year, at the request of the U.S. Environmental Protection Agency (EPA), the New Buildings Institute published a report that considers EIS in the context of advanced metering technologies [NBI 2009]. The Lawrence Berkeley National Laboratory (LBNL) has a long history of research addressing EIS, as well as system-specific performance monitoring and diagnostics [Motegi and Piette 2003; Piette et al. 2001, 2001b]. This study extends and updates the outcomes of research published by LBNL in 2003, which comprised a smaller-scale evaluation of features and EIS products [Motegi and Piette 2003]. Finally, a substantial body of work is dedicated to the use of building automation systems (BAS) and energy management and control systems (EMCS). However, it tends to focus on leveraging heating, ventilating, and air conditioning (HVAC) data for applications external to the EMCS, and on HVAC performance diagnostics [Friedman and Piette 2001; Heinemeier 1994; Webster 2005]. In contrast, this work considers EMCS only in terms of their utility in whole building energy monitoring.

The remainder of the report details the content and structure of the characterization framework and findings from our review of commercial EIS. In its totality, the frame-work represents the full range of analytical, diagnostic, and visualization features that EIS support. In addition, each major feature category is discussed with a focus on typical offerings versus more sophisticated or more rare ones. It is important to emphasize that all reported findings are based on vendor-supplied information at the time of the study. Current capabilities are subject to change, and readers are encouraged to confirm information based on their specific needs. Moreover, the EIS that were selected for evaluation are representative of the market but not comprehensive, and inclusion in the study does not imply endorsement.

The case studies are presented first, followed by conclusions and future work. The appendices contain the characterization framework, EIS evaluations, a technical discussion of baseline methods, and case study narratives.

EIS CHARACTERIZATION FRAMEWORK AND EVALUATIONS

As depicted in Figure 10-2, EIS are defined as products that combine software, data acquisition and storage hardware, and communication systems to store, analyze, and display building energy information. At a minimum an EIS provides hourly whole-building electric data that are web-accessible, with analytical and graphical capabilities [Motegi and Piette 2003]. Data types commonly processed by EIS include energy consumption data; weather data; energy price signals; and demand response (DR) information. These data are processed for analyses such as forecasting, load profiling, and multi-site and historic benchmarking. Energy information systems may also provide submeter, subsystem, or component-level data, as well as corresponding analyses such as system efficiencies or analysis of end uses, yet these are not requirements.

Four general types of EIS were identified in prior work: (1) utility EIS, (2) DR systems, (3) web-based energy management and control systems (web-EMCS), and (4) enterprise energy management (EEM) tools [Motegi and Piette 2003]. As indicated in Figure 10-3, EIS consist of the intersection of support tools from a number of domains.

The distinction between what is and what is not an EIS is better understood using EMCS as an example. While their traditional design intent is to monitor and control building systems, EMCS *can* integrate whole-building utility meters and weather sensors. In turn, these data can be used to define energy performance metrics that can be included in plots, calculations, and reports. In addition, some EMCS are web-accessible. If the monitoring-focused

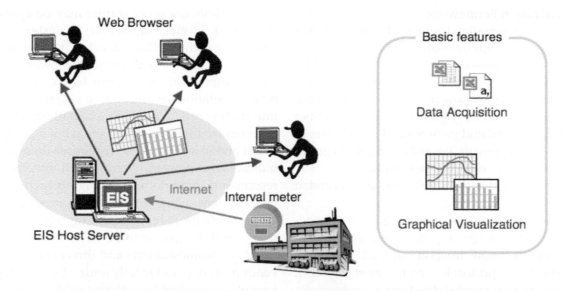

Figure 10-2. Basic Energy Information System [Motegi and Piette 2003]

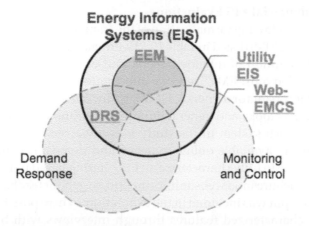

Figure 10-3. Types of EIS and overlapping functional intent [Motegi and Piette 2003]

features of an EMCS are implemented and used in this manner, the web-based EMCS can be considered an EIS. That is, the functionality of some EMCS can be applied to whole-building data in such a way that the software serves as an EIS, although scaling issues for data management and storage may be encountered in large enterprises. On the other hand, conventional EIS may not have control capability or subsystem data, but rather embody a design intent to understand patterns of whole-building energy use. Energy information systems provide support for benchmarking, baselining, anomaly detection, off-hours energy use, load shape optimization, energy rate analysis, and retrofit and retro-commissioning savings. In this way, traditional building automation or control systems, and equipment specific diagnostic software tools do not fall within the scope of EIS.

In contrast to EIS software, we treat information "dashboards" according to the traditional definition: single-screen graphical displays of the most critical information necessary for a job or task, commonly used to communicate business information [Few 2006]. Dashboards have recently gained popularity in energy applications, because of their ability to distill a large volume of complex data into a summative set of graphics that can be interpreted at a glance. Common graphical elements in dashboards include gauges and dials evocative of a vehicle dashboard, as well as graphs and charts that are often color-coded to map quantitative measures to qualitative terms. There is clearly overlap between the two technologies—for example, EIS may include dashboard views or layouts—however, we consider EIS to be full-featured software offerings with a variety of menu, display, and analytical options.

It is tempting to attempt to provide a more constrained definition of EIS that goes beyond a general set of use contexts and an accompanying set of technology capabilities. For example, one may seek a specific minimal set of features that must be offered in order for a specific technology to qualify as an EIS. This study targets technologies that are commonly considered EIS, that are used as EIS, or that could arguably be considered EIS. The immediate objective is to identify the full set of features that are supported, in order to provide a common framework for understanding and discussing this diverse set of technologies. This framework could be used in the future by an industry standards group to then determine by consensus an appropriate set of capabilities that could serve as the criteria for a given technology to qualify as an EIS.

EIS Characterization Framework

The EIS characterization framework was developed iteratively, beginning with the features that were relevant in 2003. That set of features was augmented to better fit today's systems based on preliminary knowledge of industry advances and a cursory scoping of current systems. Feedback from a technical advisory group and a small number of vendors was solicited and incorporated in revisions. In its final form, the framework consists of eight categories with five to ten features each (see Appendix A).

The categories within the framework (and associated features) include the following:

- Data collection, transmission, storage, and security
  — Accepted energy inputs, storage capacity, minimum trend interval, upload frequency, supported protocols and interoperability, archived and exported data formats, and security measures
- Display and visualization
  — Daily, summary, or calendar plotting intervals, daily and trend display overlays, three-dimensional plotting, DR status and reduction, and x-y plotting
- Energy analysis
  — Averages, high/lows, efficiencies, normalization, carbon tracking, multi-site, historical, and standards-based benchmarking
- Advanced analysis
  — Forecasting, fault detection and diagnostics (FDD), data gaps, statistics, on-site generation, renewables, and load shape analysis
- Financial analysis
  — Simple and tariff-based energy costing, meter/bill verification, estimation of savings from capital or operational changes, bill processing/payment, and end use allocation
- Demand response
  — Signal notification, event response recording, manual vs. automated response, opt out, blackout, test dates, response analysis, and quantification
- Remote control and management
- General information
  — Browser support, purchase and subscription costs, intended user, number of users, vendor description, traditional and newly targeted markets

This framework characterizes standard out-of-the-box functionality across a broad spectrum of EIS technologies. Depending on the specific software under consideration, not every feature may be applicable. The framework is most applicable to systems that target end users at the facilities level, with a minimum level of bundled or optional services. However, even tools with a number of options can be characterized with a bit of annotation beyond simple yes/no assignments. In interpreting product-specific evaluations, it is important to recognize that within the context of a given product's target and objectives, "no" responses do not necessarily indicate a less-powerful overall solution; conversely "yes" responses do not automatically signify increased usability or effectiveness. In terms of specific products, the framework should be understood as a high-level starting point from which to gain an understanding of any particular offering. Demonstrations and direct conversations with vendors are required to fully understand the appropriateness of any one tool for a given facility and its associated energy management needs.

Commercial EIS Evaluations

Following formalization of the framework, approximately 30 EIS (listed in Table 10-2) were characterized, with a description of intended users. Out-of-scope products included most EMCS, energy information "dashboards" for occupants or owners, GHG footprint calculators, batch analysis tools, and general building environment tools.

Each system in the study was reviewed based on publicly available online material and demos. It is not possible to fully characterize an EIS offering based purely on brochures and website information, so vendor feedback and input was included in the evaluation. Where possible we characterized features through interviews with the vendor, although in some cases the vendors preferred to evaluate their offering independently, and they then provided us with their evaluations.

General findings concerning the state of the technology are presented in the remainder of Chapter 10, with product-specific evaluations provided in Appendix B. It is clear that product-specific yes/no responses taken over a family of capabilities do not directly lead to an understanding of key differentiators and driving trends. To better understand those differentiators and trends, the body of EIS that were characterized is analyzed from a number of perspectives, corresponding to primary feature categories in the framework. Specific products are referenced only to illustrate the conclusions that are drawn.

Business Models

It is quite difficult to map the diversity of EIS offerings to traditional software business models. The array of optional services, varying degrees of customization or

Table 10-2. EIS evaluated according to the characterization framework

Vendor	EIS	Intended Users or Facility Types
Agilewaves	The Resource Monitor	Energy managers, operators
Apogee Interactive	Commercial Energy Suite	Facility managers
Automated Energy		Commercial, enterprise, utility customers
Automated Logic	Web-CTRL	Data center, commercial
Chevron Energy Solutions	Utility Vision	Energy managers
Energy Connect	Web Connect	DR participants, energy and facility managers
EnergyICT	EIServer and modules	Enterprises, utilities, multi-site
EnerNOC	Power/CarbonTrak	Internal use, commercial and government DR participants
Envinta	ENTERPRIZE.EM	Enterprises, utilities
FactoryIQ	eMetrics	Large commercial, industrial
	Green Energy Management System (GEMS)	
Gridlogix	Automated Enterprise Management	Enterprise
Interval Data Systems	EnergyWitness	Enterprises, facility managers
Itron	EEM Suite	Energy managers
Matrikon	Operational Insight	Enterprise
NorthWrite	Energy WorkSite	Commercial, industrial, utility customers
Novar		Internal use, big-box retail enterprise
Noveda	Facilimetrix	Facility managers
Powerit Solutions	Spara EMS	Facility managers
PowerLogic	Energy Profiler Online	Commercial
PowerLogic	Ion EEM	Enterprise, industrial
Richards Zeta	Mediator	Commercial
SAIC	Enterprise Energy Dashboard (E2D)	Enterprise and industrial facility, energy managers
Small Energy Group	Pulse Energy	Managers, owners, occupants
Stonewater Controls	InSpire	Enterprise, utilities, government
Tridium	Vykon Energy Suite	Facility and energy managers, owners, energy service providers
Ziphany	Energy operation, energy information, and DR platforms	Energy service and DR providers

configuration, and alternatives for data and IT management and pricing quickly blur the lines that define common software models. Nevertheless, some of the familiar structures are useful in attempting to understand the EIS market.

Standard software products are typically purchased with a one-time fee, are licensed according to number of installations, and include limited support with no additional services. **Enterprise client-server applications** are commonly licensed based on the number of users, and include one-time fees as well as support and upgrade subscriptions. **Application Service Providers** (ASP) offer solutions in which the ASP owns operates and maintains the software and servers for web-based applications that are usually priced according to monthly/annual fees. **Turnkey solution providers** offer fully packaged solutions that include pre-installed software, hardware, and accessories in a single "bundle."

Although it is rare to find an EIS vendor that cleanly fits into a single model, EIS offerings and providers can be differentiated according to the following considerations:

- ASP or traditional ownership: who houses, owns, and maintains the servers and software application?
- Bundled or optional services: data and IT management, interface customization, and energy-specific data analysis
- Intended end user: energy service providers, aggregators, operators, facilities managers, corporate enterprise managers, utilities, and systems integrators
- Hardware requirements: does the offering include specific or proprietary hardware, no hardware, or hardware only as necessary for the clients' objectives?
- Payment options: per site, per user, billing frequency, subscription or one-time fee

A minimum number of tools included in this study are offered as traditional enterprise client-server applications, with the user responsible for on-site IT management (e.g., Energy Witness). More commonly, EIS are offered via ASP with no hardware, or optional hardware as might be dictated by client needs (e.g., Facilimetrix, Energy Work-Site, EEM Suite, Pulse Energy). Just as frequently, EIS are offered via ASP with optional or bundled services (e.g., Automated Energy, Ziphany, E2D). In a limited number

of cases the EIS software is offered free of charge (e.g., PowerTrak, Novar, Web Connect), as its primary end users are service providers.

Solutions that feature software bundled with hardware tend to include web-EMCS by definition, in addition to some of the DR tools (Web-CTRL, The Resource Monitor, Spara EMS). Finally, it is important to understand that EIS can be intended for diverse user groups. The tool may be intended directly for the on-site or enterprise end users or for third parties to offer to their own clients. For example utilities, aggregators, energy consultants and service providers, and systems integrators may develop or customize applications for on-site end users.

EIS Architectures

The discussion of business models naturally leads to a review of the architectures underlying common EIS tools and services. Figure 10-4 illustrates the hardware, subsystems, and software that comprise or are utilized in a typical EIS.

From left to right in the figure, the three hierarchical levels underlying the data acquisition and controls, storage and analysis, and display functionality of EIS are:

1. **Facility End-Use Meter and Control Systems:** These systems measure and monitor using variety of communication protocols such as BACnet, and Modbus.

2. **Facility or Third-party Data Center:** This is typically a data warehouse within a facility or third-party (service provider) location.

3. **EIS Web Interface and Client Access:** The front-end

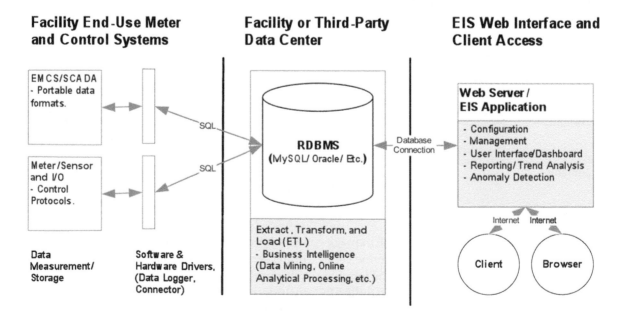

Figure 10-4. Hardware, subsystems, and software that comprise or are utilized in typical EIS

Note: SCADA = supervisory control and data acquisition; SQL = Structured Query Language; RDBMS = relational database management system

application is used to configure, manage, and display EIS data. Remote internet access is provided via web browsers or other clients such as mobile devices.

At the data center or facilities storage level detailed in Figure 10-5, monitored information is posted to a data warehouse. Typically, a relational database management system (RDBMS) stores and archives the data, although online analytical processing (OLAP) is sometimes used. The RDBMS might follow a variety of database offerings, including those such as MySQL, Microsoft SQL Server, or Oracle, as well as proprietary solutions. Structured Query Language (SQL) or variations such as Procedural Language SQL (PL/SQL) are standard communication languages to query and post information between meter sources and databases.

The EIS data warehouse can be a standalone server or a high-volume cluster, and can be physically located at the site or at the EIS provider's (third-party) data center. For the purposes of EIS, data are processed in three major steps: transmission to the data center, data cleansing or filtering (if provided), and database archiving for post-processing. Archived data are the basis of facility-specific analyses, including energy, finances, weather modeling, and others. Algorithms for baselining, load forecasting, fault detection, energy costing, are applied to processed data. Finally, for front-end web interfaces to display and report information, EIS application programmers make use of database connection drivers such as Java Database Connectivity or Open Database Connectivity.

Display and Visualization

Since 2003 there have not been significant changes in display and visualization features. Across all of the

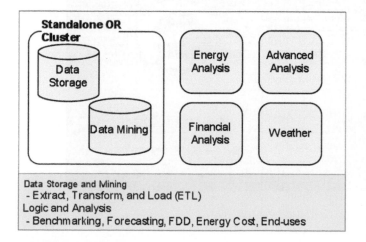

Figure 10-5. Detailed view of the data center level of EIS architectures

EIS that were evaluated, load profiling and point overlay display capabilities are largely accommodated. With a few exceptions trends can be viewed over user-defined intervals of time, be it years, months, or minutes of data; trended variables can be aggregated into totals (e.g., kilowatt-hours [kWh] last week); and features that allow the user to overlay multiple trends on a single plot are nearly universal. Slightly less common, but still standard is the ability to overlay trends for different time periods on a single plot (e.g., Monday kW and Saturday kW). Flexibility is one aspect of visualization that is found to vary from tool to tool. Display parameters might be dynamically altered "on-demand" as user need arises, or more statically defined within configurable options. For example all tools will display a plot with multiple trend overlays, but in some implementations these trends must be predefined in reports settings, while others allow the user to plot any value on the fly.

Similar to trend display and overlay features, the ability to show DR event status and reduction levels is almost universally supported, as it was in 2003. Three-dimensional surface plots in contrast are *not* common, and were encountered in just a handful of the tools that were reviewed. This finding is not surprising, as it is unclear how additional dimensionality enhances the ability to process, understand, or analyze energy information. X-y scatter plotting was not a common or standard visualization capability in 2003, and while it has grown some, it remains an under-supported feature in today's EIS solutions. Given their power in facilitating diagnostic troubleshooting, it is discouraging that only half the tools surveyed include x-y plotting. Those tools that do include it usually accommodate the feature through correlation analyses. The EIS that originated in the industrial sector are especially likely to support x-y or correlational plotting, due to the historic demand for site-specific key performance indicators. As for the more general display features discussed above, an important distinction in evaluating x-y plotting is whether it is dynamically defined by the user or statically defined in configured graphics.

Energy Analysis

Features related to GHG analysis did not appear in the 2003 study, but they are an element of the EIS framework. Two-thirds of the EIS that were reviewed feature carbon tracking and analysis as a standard capability or provide custom or configurable options to do so. The majority of analyses apply a simple energy/carbon dioxide (CO_2) relationship; however, about half account for regional differences in generation or other standards. For example:

• PowerTrak uses EPA's eGRID (emissions and generation resource integration) database paired with client zip codes.

- Ion EEM determines emissions factors based on Scope 1 and 2 of the GHG Protocol, a GHG accounting framework used in standards and programs such as the International Standards Organization and the Climate Registry.[1]
- Automated Energy, EIServer, and Energy Witness apply knowledge of utility-specific fuel mixes.
- Energy WorkSite uses Department of Energy values for state-by-state emissions.

Time-varying GHG intensities are not yet frequently addressed. Time variance is expected to be a useful feature for sites that perform load shifting, for example via thermal energy storage systems. The few exceptions that were encountered include EPO and ION EEM, both of which permit the definition of multiple emissions factors, and Commercial Energy Suite, which cites time-variance as an optional feature. Energy Witness reports that the feature is under development for upcoming releases.

Normalization is an important feature of energy analysis that is widely accommodated, although at diverse levels of rigor. Only a handful of tools report that they offer no means of normalization (e.g., Utility Vision, EEM Suite, The Resource Monitor, Web Connect, InSpire), or require that data be exported to third-party software such as Excel to do so. Normalization capabilities may be offered via reporting options or definable arithmetic calculations (monthly kWh divided by monthly degree days) or plot-

table trend points created from other trends (e.g., ION EEM, Operational Insight). Weather normalization may make use of environmental sensors that are integrated into the EIS database, external sources of weather data (e.g., Automated Energy uses Accuweather), or manual entry within calculation functions.

Quantification of a building's historic energy performance baseline is supported in nearly every EIS in the study. See Figure 10-7. The majority implement trend-based or report-based solutions, while weather-normalized baseline models or implementation of standard methodologies are far less prevalent. Some exceptions include the ION EEM energy modeling module, as well as Powerit Solutions and Novar, who integrate expert knowledge and heuristics.

Multi-site benchmarking is used to relate one building's energy performance to that of other buildings, for comparative purposes. Every tool that was evaluated for this study supports some form of benchmarking, currently or in upcoming version releases. Distinguishing aspects of EIS benchmarking functionality include the following:

- Composition of the comparative cohort: buildings within the end user's enterprise, other clients from the vendors databases, or data sets such as the Commercial Buildings Energy Consumption Survey (CBECS)
- User access: embedded in static reports or dynamically accessible functions
- Display of results: numerically in tables or graphically in plots or charts

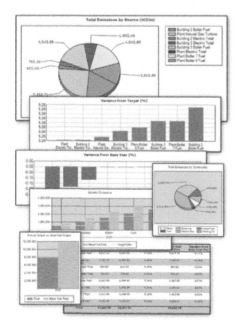
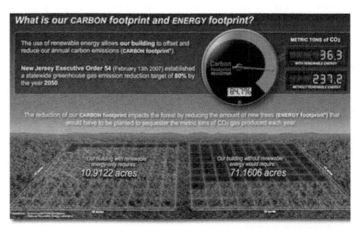

Figure 10-6. ION EEM emissions reporting module[2] and Noveda Carbon Footprint Monitor[3]

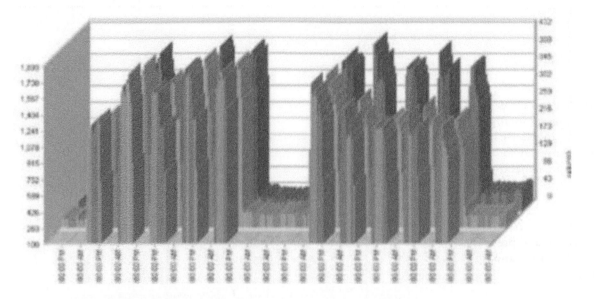

Figure 10-7. EEM Suite[4] baseline and metered consumption, with total production

Two examples of benchmarking against national data sets include Energy Witness' use of CBECS data and Energy WorkSite's calculation of Energy Star rankings.

Financial and Advanced Energy Analyses

The ability to identify corrupted or missing data is critical in EIS, given the number of performance calculations that are automated based on trended historic data, as well as the large volumes of data that are stored. Three-quarters of the systems that were evaluated accommodate this capability, via three principal means: identification through flagging or summative reporting; actual cleansing and/or correction; and linking to external or third-party software.

- Utility Vision, Automated Energy, Energy Witness, and ENTERPRIZE.EM identify gaps/corruption by flagging, reporting, or e-mail notification
- Energy WorkSite automates error checking, data cleansing, and interpolation
- Energy ICT, PowerTrak, and Ziphany make use of validation editing and estimation standards (VEE)
- Vykon offers configurable cleansing options in reports and documents communication faults to identify potentially corrupted data sets
- eMetrics provides data cleansing as a service

Depending on the tool and the extent to which the vendor offers services, data filtering and correction is purchased for additional fees, custom-defined, or out-of-the-box.

Some EIS provide building-level anomaly detection, or departures from normal consumption or trend patterns, however as expected based on the whole-building emphasis of EIS, automated fault detection and diagnostics at the component level is not typical. The exceptions include a few tools that link to external software packages or to dedicated, compatible FDD modules. For example EEM Suite recommends linking to Metrix IDR to identify corrupted data and to perform FDD, and Operational Insight links to an FDD module separate from the EIS.

Over three-quarters of the EIS that were reviewed are able to provide simple estimates of the energy cost of operating the building, and the majority of those that do so also handle model-based or tariff-based costing. It is not surprising that the DR tools tend to offer the most robust energy cost estimates. In addition to estimating energy costs, more than half of the tools evaluated report the ability to forecast near future load profiles, typically by coupling historic trends with weather data and perhaps pricing or cost data (e.g., Automated Energy, Energy Witness, Facilimetrix). In those tools that are bundled with services, the level of forecasting sophistication is largely dependent upon the needs of the client (e.g., E2D, Novar). Few solutions feature model-based or algorithmic forecasting, although Energy ICT applies neural networks, Energy WorkSite employs a bin methodology, and Pulse Energy uses a proprietary method of weighted averaging. Although forecasting, anomaly detection, and benchmarking are separate features in the framework, it is important to recognize that these functions often rely

Figure 10-8. PowerTrak's[5] departure from normal/programmed schedule (left); ION EEM[6] trend overlay to compare typical and actual trends (right)

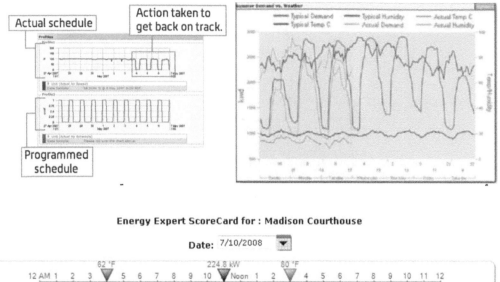

Figure 10-9. NorthWrite display of actual and predicted use, costing, and forecasting

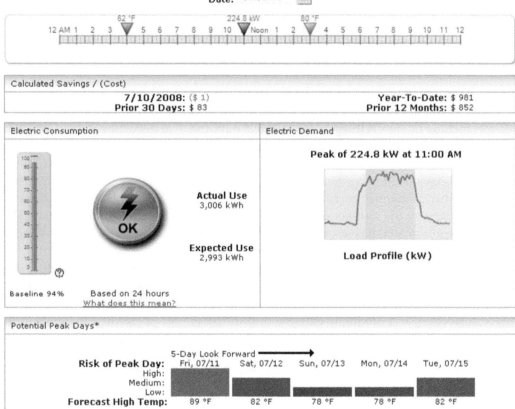

upon a single underlying baseline method. Appendix C contains a technical overview of several approaches to baseline calculation that are found in EIS, and discusses how baselines are used for prediction, M&V, benchmarking, and anomaly detection. It was not possible to learn the precise baseline methods used in each and every EIS in the study, therefore this discussion details general approaches, with three specific examples.

Closely related to energy costing and forecasting is the ability to calculate or predict savings from retrofit operational strategies, or EIS use. As with energy costing, roughly three quarters of the EIS that were reviewed support this feature in some form, and EIS that do not support costing do not tend to calculate or predict savings from operational changes. Determination of savings from retrofits is one of the more common applications of M&V efforts, and in theory it is possible to facilitate M&V through the baselining, normalization, user-defined arithmetic, and tariff-based costing in an EIS. However it may be difficult to configure an EIS to conform to specific M&V

protocols, e.g., the International Performance Measurement and Verification Protocol (IPMVP), that may require baseline or routine adjustments via regression modeling of independent variables or minimum monitoring periods. In fact, at least one large utility reports that many EIS vendors who offer comprehensive feature sets are unable to configure their systems to provide an acceptable M&V methodology.

Approximately half of the EIS accommodate the comparison of meter readings and utility bills to verify accuracy, but the feature may not be fully automated. As one might expect, the EEM systems often provide more sophisticated or robust financial analyses, as they are designed to address corporate/executive needs in addition to energy monitoring.

Control and Demand Response

Energy information systems control and management capabilities commonly appear in two varieties: (1) control according to a program via gateway or EMCS, or (2) remote control over the internet [Motegi and Piette 2003]. Just over half of the EIS surveyed report the ability to control according to a program, and just under half report internet-capable remote control. Remote control is intimately related to demand response capabilities, which have advanced since 2003, converging to a common set of features.

* Automated response is possible in all but three of the DR systems that were evaluated (Commercial Energy Suite, PowerTrak, and EPO are limited to manual DR.
* E-mail, phone, pager, and alarm notifications of DR event status are all standard, although not every tool implements all four contact methods.
  All of the tools surveyed calculate baselines accordi*ng

to utility program formulas, allow testing events, and support response recording/documentation. Recording may be formalized and structured or simply captured in historic trend logs.
* All of the systems evaluated permit selection of opt out and blackout dates. One exception is PowerTrak; as with the automation feature, this is an artifact of the specific service they offer, rather than a software limitation.

While the features detailed above are standard across all DR systems, the ability to predict savings from a given response is a key distinguisher of EIS capability. The near uniformity in features offered in today's DR systems begs the question of what would expand today's response capabilities? One potential advance is to allow for several increasingly severe DR strategies that could be implemented if the primary strategy were not effecting large enough demand reductions to meet the target. In addition, calculation of DR potential or expected savings might be enhanced through model-predictive or intelligent algorithms. Ultimately, as automated DR becomes commonplace in commercial buildings, post-event rebound will become more critical, and DR systems that address rebound will be an advantage.

EIS USER CASE STUDIES

While exceptionally useful in building an understanding of the state of the technology, individual product characterizations and conclusions regarding software capabilities do not answer questions of usability and real-world utilization. Correspondingly, four case stud-

Figure 10-10. Ziphany's load curtailment platform for utilities[7]

ies were conducted to answer questions such as: Which features have proved most useful in attaining energy savings? What actions are taken based on the information provided via an EIS? How much of a building's low energy use or energy savings can be attributed to the use of an EIS? What are common challenges encountered in whole building performance monitoring? What are successful, realistic EIS implementation and use models?

The existing body of case studies documenting EIS use is modest, and is typically comprised of vendor-authored publications or literature from the commercial building energy community. Vendor-authored case studies are typically written to publicize successful implementation of a specific EIS, and as such are inherently biased to emphasize positive aspects of the technology that the vendor wishes to advertise. Vendor case studies are usually posted on the website, as is true of a number of the EIS that were evaluated in this chapter.[8,9,10,11] These profiles tend to emphasize cost savings over energy savings, although a number do include both metrics. The case studies conducted for this project are markedly different, in that they present users' technology challenges as well as successful savings. Further, they document EIS use based on the user's perspective, rather than the vendor's.

Case studies from the building energy community are more objective in their assessments and more varied in content and level of detail. Integration and installation, and the use of EMCS data, tend to be more frequently addressed [Capehart and Capehart 2005, 2007; Webster 2005] than the relationship between software features, actions taken, and resulting energy savings [Motegi et al. 2003]. When features, actions, and savings are addressed in the literature, the overall topic is usually not whole-building EIS diagnostics, but rather equipment and system-level operational diagnostics. The cases presented in this chapter offer several unique contributions. First,

they are coupled to the framework and EIS evaluations, providing a structured context from which to relate overall technology capabilities to real-world uses. Second, they cover a range of commercial building sectors and types. Finally, they target the less commonly explored aspects of EIS use at a high level of detail.

Wal-Mart, Sysco, the University of California (UC) Berkeley, and UC Merced were selected for case study based on the following criteria: users with a high level of engagement with energy data and a role in energy management; aggressive savings or high-efficiency performance; and willingness to participate in three to four hours of interviews and site visits, location permitting. The UC Merced case study was conducted with the campus energy manager, and the UC Berkeley case included the associate director of sustainability and engineering services, and members of her staff groups. The Sysco case study was informed by the energy services provider and the person who is accountable for energy performance at a Northern California warehouse site. The Wal-Mart case combined discussions with a benchmarking analyst from the Energy Department, the Electrical Engineering Manager from Prototypical Design/Construction Standards, and the Senior Manger of Energy Systems and Technology Development. As summarized in Table 10-3, the four cases that were chosen represent commercial enterprises and campuses with a diversity of performance-monitoring technologies, commercial building types, and portfolio sizes. These cases encompass buildings that range from Wal-Mart and Sysco's relatively repeatable warehouse and retail designs, to UC Berkeley's legacy and historic sites, to UC Merced's very-low energy new construction.

The following four bold headings detail case-specific findings and the particular research themes that each case illustrates. For example, UC Merced exemplifies the challenges of using a Web-EMCS as an EIS, and the

Table 10-3. Case study sites and characteristics

Case	Type, (square feet)	Controls	Performance Monitoring
UC Merced	Campus (800,000)	ALC WebCTRL	ALC WebCTRL Utility bills
UC Berkeley	Campus (15.9M)	Barrington Some ALC, Siemens	Obvius Utility bills
Sysco: Stockton, California Sygma site	Refrigerated/dry warehouse (52M, Stockton 95,000)	DOS-based refrigeration control	NorthWrite Energy WorkSite Utility bills
Wal-Mart	Retail/grocery (675M)	Novar Danfoss Emerson CPC	Energy ICT EIServer Utility bills

value of a web-accessible dense data set. Sysco, on the other hand, is a case of classic enterprise-wide EIS use, and use of the EIS to ensure persistence of savings and corporate accountability for energy performance. The final section in the chapter summarizes the energy savings and challenges that were documented for each case. The UC Merced and Sysco cases are presented in deeper detail in the narratives in appendices D and E.

UC Merced

Opened in 2005, UC Merced is the newest University of California campus. Prior to opening, the campus made a strong commitment to energy-efficient building design, and energy conservation plays a fundamental role in campus objectives. Campus energy activities focus on three areas:[12] building energy performance targets; ongoing monitoring of energy use; and climate neutrality. The University of California at Merced uses custom benchmarks for UC/California State University (CSU) campuses [Brown 2002] and has set targets ramping over time from 80% to 50% of average performance.

In support of these efficiency requirements and three focal activity areas, standardization of the campus control systems was made to be a priority during the design and construction phases, and Automated Logic Corporation's WebCTRL was selected. The University of California at Merced features a uniquely dense metering and monitoring infrastructure, already trending over 10,000 points at three academic buildings, the central plant, and smaller auxiliary buildings. Custom benchmarks and deep monitoring capability with a sophisticated web-EMCS are central to the main themes embodied at the campus: (1) the challenges in using a web-EMCS as an EIS, (2) the web-EMCS as enabling critical information links, and (3) realization of the campus as a living laboratory. These themes are visible in typical uses of the web-EMCS, site-specific data and technology challenges, and energy-saving opportunities identified in the data.

UC Merced Web-EMCS Uses

Typically, WebCTRL use at UC Merced is dominated by operational EMCS investigations, rather than EIS energy performance diagnostics. The technology is used most extensively to respond to trouble calls. In addition to troubleshooting problems that have already been brought to attention, WebCTRL is also regularly used to verify that individual buildings are operating as expected. Use of the web-EMCS for more traditional EIS analyses directed at campus and whole-building energy performance has been complicated because the EMCS and monitoring instrumentation was not explicitly commissioned for EIS diagnostics. The metrics used to track energy performance are more complicated than simple energy use intensities [Brown 2002], and the logic-based arithmetic in WebCTRL was not configured to perform the associated calculations. For example, allocations from the central plant, based on chilled water consumption, are added to the whole-building electric meter data, as are allocations for campus road lighting. Therefore, building and campus energy data are commonly exported to spreadsheet software for additional computation, cleansing, and computation. WebCTRL meter data are used annually to track energy performance; gas, electricity, hot water, and chilled water consumption are measured at the campus level and for critical buildings, as summarized in Table 10-4. On a monthly basis, the campus energy manager uses the web-EMCS data to determine utility recharges for non-state buildings (that is, buildings that are located on campus but are differently financed, requiring that they "reimburse" the campus for utilities).

The campus steam system provides two examples of the use of the web-EMCS to inform operational changes leading to energy-savings. Gas trends at the central steam plant showed significant gas use throughout the night, when the system was not intended to operate; at the same time, steam trends at the central plant revealed non-zero operating pressures at night. The energy manager shared

Table 4. UC Merced metrics, benchmarks, and data sources

Web-EMCS	Annual Building Metrics	Data Sources	Benchmarks
Automated Logic Corporation WebCTRL	Peak electric demand Total electric use Peak chilled water demand Total gas use (incl. steam and HW)	Building electric meter Central plant electric submeters Central plant gas meters Building gas meters Building chilled/hot water flow, and supply/return temperature Utility bills	UC/CSU weather and building-type normalized energy use intensity [Brown 2002]

the data with the superintendent, who returned the system to true zero overnight pressure, securing 30% reduction in average daily gas consumption (therms/day) at the steam plant and an estimated $4,500 monthly savings. In the lower portion of Figure 10-11 the change to zero overnight pressure is plotted; in the upper portion the resulting drop in overnight gas use is shown. In addition, the energy manager is in the process of combining gas trends from the steam plant with temporary steam use logging at the building level to confirm the efficiency of the steam plant. Knowledge of the plant efficiency will direct a decision to continue centralized steam production or move to a distributed supply.

In addition to providing a rich set of operational and energy consumption data, the web-EMCS has also facilitated the realization of the campus as a *living laboratory*. The campus energy manager emphasizes that this has been a tangible benefit of the WebCTRL system, and that the living laboratory concept is of critical importance, particularly at an academic institution. To this end the web-EMCS data are used in engineering thermodynam-

ics course modules that the energy manager teaches; to inform student and faculty research efforts; and in short- and long-term research and demonstration collaborations with the external buildings research community. For example, the U.S. Department of Energy and the California Energy Commission have sponsored research at Lawrence Berkeley National Laboratory that focuses on using EMCS data to: develop and pilot model predictive control strategies at the chilled water plant; design a real-time diagnostic tool based on comparing meter data to calibrated building energy models; and analyze how the campus demand response potential is affected by the thermal energy storage system.

UC Merced Challenges and Needs

Data quality issues arise in a number of contexts at UC Merced, further challenging the use of the EMCS for automated analyses. Networking and connectivity problems have led to dropped or miscommunicated values that generate errors, lock out equipment, and cause large volumes of false data and cascading false alarms. This

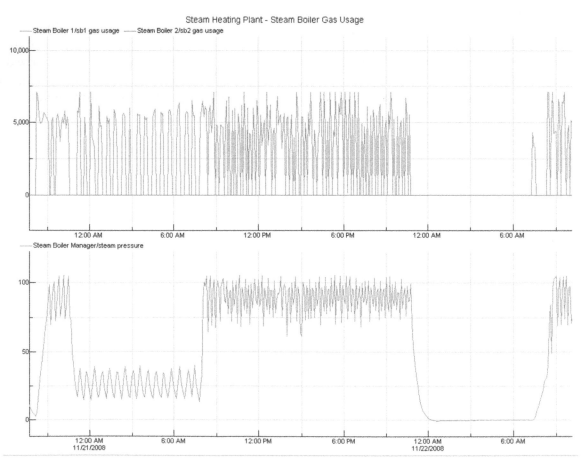

Figure 10-11. Web-EMCS trends and energy-saving operational change at UC Merced, showing overnight drop in gas use from zero overnight pressure."

has been a significant problem in using and maintaining WebCTRL at UC Merced, however network communications are viewed as affecting operations more than energy monitoring, and over time many of these challenges have been addressed.

While not attributable to the capabilities of the EMCS, meter or sensor calibration and configuration errors also affect data quality, thereby affecting the ability to use the EMCS as an EIS. With the exception of whole-building electric data, significant resources were required to manually validate the EMCS data quality and to quantify the campus energy performance relative to benchmark. Manual validation included inspections to trace the physical meter point to its representation in WebCTRL, as well as energy and mass balances to confirm accuracy of logged data and interpolation or estimation of missing data. To date, manual validation has affected building science researchers more than WebCTRL users at UC Merced, but it does have implications for advanced use of the data within the living laboratory context.

Staffing and resources are a recurrent theme that arises in the case studies. At UC Merced, the energy manager has not been able to investigate building and submeter trends to the full extent desired, and campus-wide it has taken some effort to transition from reactive to proactive use of the data. For example, the central plant operators have begun performing hourly reviews of WebCTRL trends according to a defined check-sheet, and the reviews are documented and commented. This process was implemented as a structured way for the operators to be able to leverage the web-EMCS technology. Analogously, more routine campus and building-level energy diagnostics based on web-EMCS trends has been somewhat hindered by constraints on the energy manager's time. Note that prior research in the use of building management systems at government buildings identified similar challenges in proactivity, resources, and energy management [Webster 2005].

There are no embedded features in the software that are unused at UC Merced or that are considered superfluous or too time-consuming or difficult to learn. Although addressing trouble calls may dominate use of the web-EMCS, these operational efforts have allowed the energy manager to maintain exceptional energy performance. Were the metering infrastructure better calibrated and commissioned, and were the EMCS configured to track key performance metrics, it might be used more easily for EIS-like analyses.

In spite of outstanding needs and imperfections with the technology, the energy manager reports a high level of satisfaction with WebCTRL and what it has en-abled him to accomplish. He emphasizes that UC Merced trends extremely large volumes of data and that intensive monitoring needs to be undertaken deliberately, with close attention to a spectrum of issues including wiring, system programming, network architecture and hardware selection. Further, Automated Logic Corporation has been particularly accommodating, working with UC Merced and other large institutions to develop a revised network and hardware infrastructure.

Sysco (Sygma)

Sysco has implemented a three-year corporate-wide energy efficiency program that targets a 25% reduction in energy consumption across a portfolio of 143 distribution centers in the United States and Canada. Sysco has a long-standing energy services and consulting contract with Cascade Energy Engineering, with whom a collaborative three-part approach was adopted: (1) site visits by expert refrigeration engineers and technicians to perform tune-ups and identify low-/no-cost energy-saving measures; (2) customization of NorthWrite's Energy WorkSite EIS to accommodate Sysco's goals; and (3) continuous communication and collaboration between corporate managers, Cascade Energy Engineering, and on-site "energy champions." This approach has enabled Sysco to outpace its goal, reaching 28% savings in kilowatt-hours per thousand square feet (kWh/ksf) before the end of the program period. This amounts to roughly 18,000,000 kWh savings each month.

While the UC Merced case revolved around the particular constraints and power of a densely populated, sophisticated web-EMCS platform, the Sysco case highlights the following themes: (1) classic enterprise-wide EIS use and information sharing; (2) limited, yet powerful, on-site use of the EIS; (3) use of EIS technology to ensure persistence in savings and energy accountability. These themes are reflected throughout the organization in typical uses of the EIS and in the ways in which the 28% energy reductions were achieved. The first 12–18 months of Sysco's efficiency program were dedicated to site visits, control tune-ups, and installation of the EIS meters and software. A combination of EIS data, expert assessments, and on-site staff insights was used to gain 18% savings from no-/low-cost measures. Over the remainder of the program a further 10% savings in total energy use were gained through capital improvements such as variable frequency drives (VFDs), lighting retrofits, and HVAC upgrades.

Sysco EIS Uses and Challenges

Sysco performs both site-specific and portfolio analyses on a monthly basis, using Energy WorkSite's embedded reporting capabilities. Cascade Energy Engineering inputs utility billing invoices into the data warehouse, and portfolio benchmark rankings are generated as listed in Table 10-5. Managers coordinate monthly group reviews with each site's "energy champion," who is accountable for energy use. Monthly rankings are compared based on a metric called the *efficiency factor,* which takes into account wet bulb temperature, the total volume of frozen and refrigerated space, total and daily energy consumption, and weather predicted energy performance. While not deeply understood by energy champions and managers, the efficiency factor is a metric that was custom defined for Sysco's portfolio of refrigerated warehouses, and preconfigured within Energy WorkSite reporting options, as in Figure 10-12. In addition to serving as the basis for portfolio rankings, each site's efficiency factor is tracked over time as a means of ensuring accountability for performance and persistence of savings. Although the predictive algorithms that form the basis of efficiency factor are not well understood, it is understood that the metric is a unit-less number and that larger magnitudes indicate excessive use.

The Sysco site visit was conducted at the Stockton,

California, Sygma distribution center. The visit was based on Cascade Energy Engineering's experience that the Stockton energy champion is one of the most highly engaged EIS users, with one of the higher-performing sites. Stockton ranks highly in the Sysco portfolio, and has reduced site energy 36% since the start of the efficiency program. In this case, daily use of the EIS was limited but inarguably powerful. The energy champion makes near-exclusive use of the "meter monitor" view for his most energy intensive building's utility meter. As shown in Figure 10-13, this view contains a two-point overlay comparing the current week's or day's kilowatt time series to that of the prior week, a summary of cumulative kilowatt-hour for both time periods, the average ambient temperature, and the percent change in consumption and temperature.

Use of the EIS to monitor the meter dedicated to refrigeration loads has allowed the energy champion to implement a powerful daily energy efficiency strategy, with data to confirm its effectiveness. The existing controls do not permit it; however, the frozen goods can tolerate fluctuations in temperature between -5°F to 10°F for short periods of time without compromising quality. In

Table 10-5. Sysco metrics, benchmarks, and data sources

EIS	Performance Metrics	Data Sources	Benchmarks
NorthWrite Energy WorkSite	Unit-less efficiency factor kWh/ksf	Electric utility meter Utility invoices Weather feed	Portfolio rankings based on efficiency factor Pre-program consumption [kWh/sf]

Site	kWh	Wet-Bulb Temp	kWh / day	Total Frozen cu-ft	Total 28-55°F cu-ft	Total Dry cu-ft	Weighted Volume Cu-ft	Space Weighted Eff Factor	Weather Predicted Eff Factor	New Effiency Factor
Sygma - Denver	1,902,060	41.9	10,337	891	971			2.66	4.52	.666
Sygma - Southern California	2,587,710	48.4	14,064	920	1,142			3.01	4.76	.732
Sygma - Northern California	1,855,050	52.3	10,082	748	595			3.01	4.9	.756
Sygma - Kansas City	1,077,360	47.7	5,855	340	641			3.43	4.74	.775
Sygma - Illinois	3,558,070	46.4	19,337	1,680	1,043			3.66	4.69	.835
Sygma - Oklahoma	1,391,650	52.8	7,563	498	392			3.61	4.92	.844
Sygma - Portland	1,277,340	48.1	6,942	425	442			3.62	4.75	.863
Sygma - Detroit	1,567,380	45.2	8,518	384	407			3.32	4.65	.927
Sygma - Dallas	3,397,560	56.6	18,465	919	1,083			4	5.06	.930
Sygma - Boston	1,390,940	43.7	7,559	445	383			3.85	4.59	1.037
Sygma - Georgia	2,341,800	55.6	12,727	388	584			4.67	5.02	1.043
Sygma - Florida	3,333,500	64.5	18,117	794	824			5.09	5.35	1.110
Sygma - Carolina	2,899,140	53.7	15,756	685	592			4.86	4.95	1.111
Sygma - Pennsylvania	2,689,120	47.3	14,615	597	884			4.36	4.72	1.152
Sygma - San Antonio	5,479,980	59.7	29,783	924	936			7.43	5.17	1.532
Sygma - Columbus	5,816,070	45.3	31,609	1,006	942			7.09	4.65	1.677

Figure 10-12. Efficiency factor report for Sygma distribution centers

Electric Demand

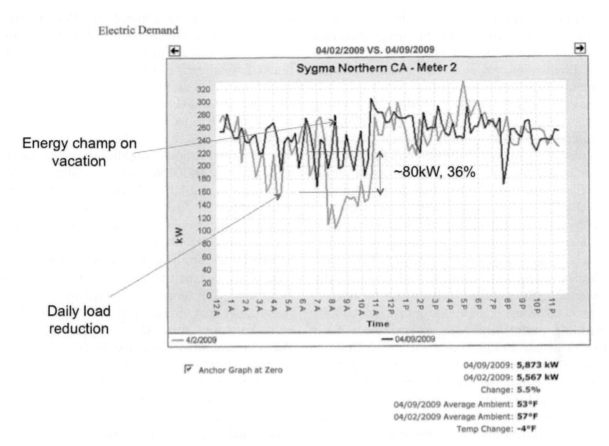

Figure 10-13. Energy WorkSite's "meter monitor" and the Stockton Sygma efficiency strategy

response, upon arriving in the morning, he accesses the DOS-based control programs for ten refrigeration units via dial-up modem and manually raises the setpoints to force the compressors to shut down. Throughout the morning he monitors the temperature of the conditioned spaces and the metered power consumption, returning the setpoints to their original levels around 11 a.m. The lighter-colored trend in Figure 10-13 reflects an instance of this daily strategy, whereas the darker line reflects a day in which the energy champion was on vacation and the strategy was not implemented. In spite of a four-degree temperature increase, the energy champion effected an average load reduction of approximately 35% throughout the morning, relative to a day in which the strategy was not implemented.

The Stockton Sygma site contains five utility meters and accounts, and while the meter dedicated to refrigeration loads is the primary focus of EIS use, minor energy management tasks are performed with the remaining four meters. Unanticipated, unexplained spikes in consumption are plotted and shared with equipment technicians, and deviations from expected profiles are investigated and remedied. For example, the energy champion has noted

instances in which loads did not decrease as expected after the final shift of the day, and based on knowledge of the building end uses was able to determine that lights were not being turned off. Staff reminders were sufficient to correct the situation. Over time, the EIS has played an especially useful role identifying such behavioral impacts on site energy consumption, and it has served as a motivational benefit to prevent backsliding performance. The energy champion perceives that staff behavior is now well aligned with site efficiency goals.

End users of the EIS and Cascade Energy Engineering did not bring up challenges in data acquisition and quality until explicitly asked to do so during the case study interviews. Sysco monitors electric utility meters and has not pursued submetering, with the result that aside from infrequent minor glitches in cellular communications, data quality has not been a critical challenge. Timely entry of utility invoices into the EIS data warehouse is a challenging aspect of the services contract, since the provider manually inputs the billing data for storage in the central data warehouse. As a result the Stockton site relies on personal spreadsheets, forgoing Energy WorkSite's comprehensive embedded utility modules. In fact, the Stockton site visit revealed that much of the EIS functionality was unused and unexplored. It was difficult for the energy champion to navigate outside of the default meter monitor view; for example, to identify the previous year's total consumption or the previous year's peak demand.

Although NorthWrite offers on-demand training, their clients do not commonly request it, revealing one of the more compelling case study findings. The highly customized implementation of the EIS configured to meet Sysco's needs and the collaboration with expert service providers has resulted in a notion that deep diagnostics from on-site energy champions are not necessary to attain energy savings. Successful measures implemented during the initial stages of the program, accountability based on monthly reporting, and an emergent corporate culture of competition have precluded the perceived need to use the more powerful features of the EIS. It may be that refrigerated warehouses pose limited opportunities for extensive whole-building performance diagnostics, and as such present a special case for EIS. In contrast to other building types, a full 50% of the load is dedicated to refrigeration, and another 20% to lighting. At the Stockton site the EIS is most highly valued for its role in supporting and encouraging accountability and staff motivation, so that efficiency gains might persist over time. However, it is possible that additional energy savings have gone unidentified because energy champions have not seen the value in the full set of EIS capabilities. For example, what added savings could be gained at the Stockton site if the energy champion made use of the "daily scorecard" to compare predicted to actual consumption, or to view month-long load profiles to identify historic trends? How might forecasting feature be leveraged to optimize the daily efficiency strategy that is currently based on implicit heuristic knowledge?

Wal-Mart

Wal-Mart maintains a portfolio of 67 million square feet of commercial retail space, and uses Energy ICT's EIServer to collect and monitor energy consumption data. Wal-Mart's decision to implement an EIS was motivated by an overarching business philosophy that holds that with billion dollar utility expenses, energy information limited to 60- or 90-day billing cycles is wholly insufficient. Wal-Mart's Energy Systems and Technology Development manager and building design engineers analogize that they would never base retail decisions on sixty-day old sales data, and that energy considerations are just as critical. Motivated by this viewpoint, Wal-Mart determined that the organization required access to real-time data at the electric submeter level, and issued a request for EIS implementation proposals in which functionality and

cost were prioritized. Ultimately, EIServer was selected for the ability to forecast near-future time series using neural networks. At that time, around 2003, Wal-Mart found that competing technologies either did not provide model-based forecasting, or they were far less willing to share the details behind their specific methodology. Further, Energy ICT was willing to customize applications for Wal-Mart, and their final quotes were lower in price.

The central themes highlighted in the Wal-Mart case contrast markedly to those at Sysco. Rather than integrated EIS use throughout the enterprise to meet portfolio goals, as at Sysco, Wal-Mart is a case of "siloed" use by specific groups or individuals for a few key purposes, among various departments and teams in the enterprise. A group of internal supporters champion the use of the EIS technology and maintain a vision for how its use might be expanded throughout the organization, yet regular operational analytics are not yet widespread vertically or horizontally within the enterprise. In addition, the Wal-Mart case illustrates that even the more-sophisticated EIS may not satisfy all of an organization's analytical and energy performance monitoring needs. For uses such as measurement and verification (M&V), EIServer's embedded functionality is well suited to user needs, while for others such as portfolio benchmarking, the EIS data are exported to third-party software for analysis.

Wal-Mart EIS Uses and Challenges

EIServer features a custom module for M&V tasks that has been used extensively at Wal-Mart, although on an ad-hoc basis, to determine the effectiveness of energy efficiency improvements. "Project Tracking" is used at a given site or group of stores to quantify the savings associated with efficiency measures. Regression analyses establish weather-normalized baseline forecasts against which actual measured consumption data are compared. Wal-Mart does not have a dedicated M&V analysis team, although the software tool is available to any project. The wholesale power procurement and demand response group also uses the EIS intensively. This group makes considerable use of EIServer's forecasting and normalization features, with experience indicating that the technology is sufficiently accurate for week-ahead predictions, and accurate to within to within 1% for hourly time intervals.

Wal-Mart's EIS data come from independent meters that "stand alone" from the building management systems. HVAC, lighting, and refrigeration mains are the most metered, however some stores do monitor gas and water as well. Real-time data from a subscription weather feed are imported into the EIS. Store and portfolio performance metrics are summarized in Table 10-6.

At the individual store level, the EIS is used to gauge the performance of new designs, particularly at "High Efficiency" supercenters. Beginning in 2007 four series of high-efficiency prototype designs have been constructed, targeting 20%-45% savings compared to the typical Wal-Mart store [Wal-Mart 2009]. New stores are tracked to ensure that the design performance is met. One user reports that High Efficiency stores are best analyzed by exporting EIS data for use in Virtual Environment models, because of the ability to run computational fluid dynamics, solar thermal, and daylighting simulation modules. Due to usability constraints and the use of custom benchmark models, EIS data are also exported to for portfolio tracking. From a usability standpoint it is too cumbersome for the analyst to select trend data meter by meter, for the entire portfolio. More critically, Wal-Mart applies a custom model-based approach to calculate weather and sales-normalized energy use intensities. Each month, the benchmarking analyst identifies the twenty poorest performing sites, and refers them for further investigation at the operations and maintenance level. In some cases the benchmarking analyst delves into the data for

an individual store; however, she does not rely upon the EIS normalization capability, preferring to ensure validity by comparing stores from similar climates.

Measurement and verification and benchmarking activities provide two examples of energy savings attributable to Wal-Mart's use of the EIS. Non-functional dimming is one of the more common problems that are detected with the EIS. As shown in Figure 10-14, high energy consumption at a store in Texas was traced back to a 225 kW static lighting load due to a failed dimming control module. The benchmarking analyst identified the problem, corrected it, and avoided thirty-five thousand dollars of additional energy costs. Avoided waste due to failed hardware also arose in a VFD retrofit program. There, the EIS Project Tracker module was used to identify several sites in which a failed or incorrectly installed VFD prevented actual energy savings.

Wal-Mart's EIS challenges are largely independent of the EIS technology itself. Submetering has been difficult because it has not been financially feasible to meter each store to the degree desired by the corporation's internal EIS champions. Given that the average supercenter con-

Table 10-6. Wal-Mart metrics, benchmarks, and data sources

EIS	Performance Metrics	Data Sources	Benchmarks
Energy ICT EIServer	Weather and sales normalized kWh/sf M&V for energy saving measures	Building and submetered electric Some gas and water Subscription weather feed ICT project tracking	Portfolio rankings Pre-measure baseline

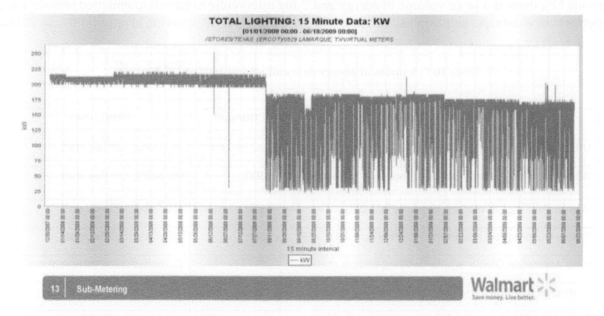

Figure 10-14. Non-functional dimming module at Wal-Mart identified with EIServer, and then fixed

tains a dozen submeters, consistency in the quality of contracted installations has also been a concern. More central to understanding real-world EIS use, Wal-Mart has faced difficulty integrating regular EIS use into standard daily activities, particularly during the current economic downturn. For example, believers in the power of the EIS technology would like to see, at a minimum, that all staff have access to the system through web-based executive reporting. Similarly, one person currently performs benchmarking tasks every thirty days, whereas the vision is to support a benchmarking group that would engage with the data on a daily basis.

UC Berkeley

The University of California at Berkeley (UC Berkeley) is a 140-year-old, 15.9 million square-foot campus with a wide diversity of building ages, types, and sizes. This accounting includes off-campus buildings and non-state buildings such as the health center. Campus energy performance has been prioritized to differing degrees throughout the last decade, and Berkeley is currently experiencing a period of renewed attention to efficiency. Following a two- to three-year gap, the campus energy management position has been re-staffed. There is no central EIS at UC Berkeley; it is a contrasting case that is included to illustrate the challenges that are encountered in the absence of a campus-wide performance monitoring system. It also provides insights as to the information needs and energy management desires of a specific energy manager, when a large, aging campus is tasked with reducing its climate impact. Although there is no campus EIS, there is a large volume of energy and system performance data. As summarized in Table 10-7

however, the data come from disparate sources and are used by different staff groups.

UC Berkeley Data Uses

The utility group uses utility bills and monthly manual meter reads to manage the purchase and billing of all campus energy. They process all invoices, and perform accounting reviews for approximately 200 utility accounts, including water, electric, gas, and steam. UC Berkeley uses an in-house DOS-based database program to store manual meter reads, which are exported to spreadsheet software for analysis. In addition to utility recharges, manual meter records are maintained to provide data for building energy analysis. Although there are not dedicated energy analysts on the energy manager's staff, from time to time the group receives external requests for building data, for example from staff who are responsible for cohorts of buildings, students conducting research projects, and developers of the campus Strategic Energy Plan.

The EMCS group at UC Berkeley uses Broadwin's WebAccess Project Manager to remotely access and oversee the campus' Barrington control systems. Fifteen to twenty servers are managed exclusively by the EMCS group to monitor sixty-one buildings, with approximately forty thousand trend points. Each day eight person hours are dedicated to building-by-building HVAC equipment checks. Beginning with the graphics screen pictured in Figure 10-15, appropriate on/off status and setpoints are verified. When problems are detected, the staff delves further into time-series plots of relevant trend data.

A number of campus efficiency and commissioning interventions have implemented remotely accessible electric interval metering at approximately 20 buildings,

Table 10-7. Sources of energy data and user groups at UC Berkeley

Data Sources	Number	Users, uses
Utility bills	Gas, electric	Invoicing, utilities staff
Whole-building electric meters, monthly manual reads	>200	Invoicing, utilities staff
Whole-building gas meters, monthly PGE bills	<100	Invoicing, utilities staff
Whole-building steam meters, monthly manual reads	<50	Invoicing, utilities staff
Web-accessible Obvius whole-building electric meters	20	Commissioning interventions
Prototype building performance monitoring website		
Barrington EMCS – control settings, states, equipment energy consumption	61 bldgs. 40K points	Four-person EMCS staff

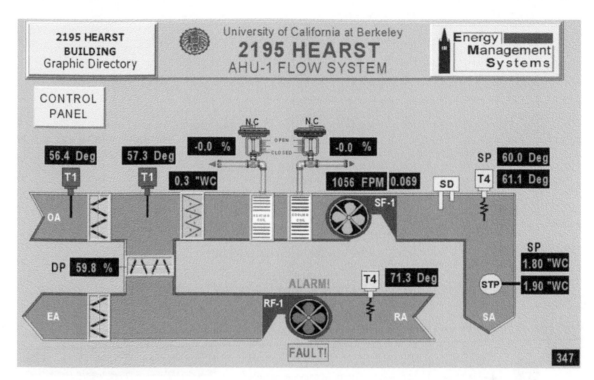

Figure 10-15. Air handler graphic from UC Berkeley's Broadwin Web-EMCS

totaling 11 million gross square feet. Obvius field devices acquire 15-minute pulse outputs and upload the data to an off-site data warehouse daily. Meter data can be visualized, plotted, or exported via a web application maintained by Obvius. While the data are continuously acquired and constantly available, it has been used most extensively for specific projects on short-term bases. It is worth noting that a potentially useful tool is under development in a student-funded research project that pairs Obvius meter data with monthly utility data. The *Building Energy Dashboard* includes monthly representations of energy, water, and steam, as well as real-time displays of meter data from Obvius devices. Although it is still under development and the final version may be quite different, a prototype was made accessible for the purpose of this report. Figure 10-16 shows a "live data plot," which contains a zoom-able representation of the most recently uploaded data from Obvius field devices; Figure 10-17 shows a "detailed building plot" in which this week's consumption is plotted against the previous week, with minimum maximum and average demand.

The *Building Energy Dashboard* targets occupants, and it is primarily intended to inform faculty, staff, and students [Berkeley Campus Dashboard 2009]. While the software is still under development, anecdotes of student trials revealed an instance in which excessive operation of the ventilation system and over-illumination in the

architecture building were identified. Based on these observations, the ventilation schedule was reduced by six hours per day, and a lighting retrofit was conducted, resulting in a 30% reduction in total energy use. The two trends in Figure 10-17 show the whole-building power before and after these changes were made. Because the dashboard combines data from the utilities group with interval data that are currently used only on a limited basis, the application might be useful for the campus energy management team, as well as for building occupants.

UC Berkeley Energy Management Needs

Similar to the other cases in the study, resources were cited as a challenge at UC Berkeley. In particular, the energy manager prioritizes tracking performance at the building level and providing feedback to building coordinators, EMCS and HVAC staff, and technicians. The energy manager also emphasizes that continuous maintenance is a critical element of any efficiency program, noting that healthy equipment is a precursor to optimal energy performance.

Regarding energy information and data, Berkeley's energy manager identified several priorities. More remote-access interval metering, with near-real time (as opposed to daily) uploads would reduce the resources dedicated to manual meter reads and increase the resolution of existing building data. Submetering beyond the whole-

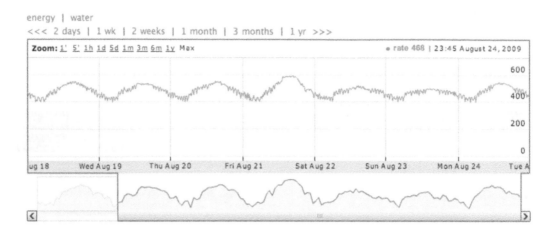

Figure 10-16. UC Berkeley Building Energy Dashboard prototype, "live data" view

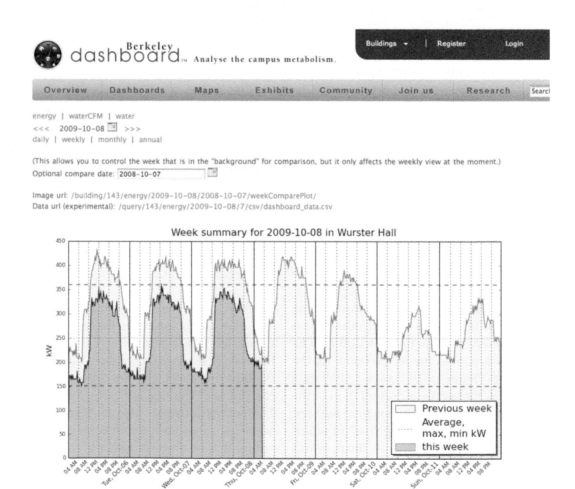

Figure 10-17. UC Berkeley Building Energy Dashboard prototype, "detailed building plot"

building level is desired to support improved decision making related to building technology, operations, and proposed use or space changes. Finally, access-controlled public data would simplify the process of satisfying data requests from researchers and special projects. While she did not cite an EIS as an outstanding need, the manager's challenge in processing the existing data, her desire for remote-access permission-based meter data, and increased density of electric metering does imply the need for an analysis-rich EIS.

Summary of Energy Savings and Challenges

Table 10-8 summarizes actions that were taken based on building energy data in each of the cases studied, and where available, the associated energy impacts. The most common actions and observations that were encountered concerned incorrect implementation of scheduled loads,

M&V, and inefficient or excessive operations. Table 10-9 summarizes the challenges, needs, and successes that were found. Note that in this respect each case truly is different, and that one case's success may represent another's challenge.

CONCLUSIONS AND FUTURE WORK

Energy information systems encompass a diverse set of technologies that are sold under an array of business models, with a complicated mix of features, architectures, and optional or required services. The sheer number and variety of options, in combination with rapidly advancing analytical and IT capabilities makes it difficult to distinguish one product from another or to understand the general state of the technology. Vendors' public domain

Table 10-8. Summary of actions taken based on building energy information

Site	Observation/Action	EIS Data Points	Energy impact
UC Merced	Excessive overnight gas use due to non-zero pressure at steam boilers	Steam plant pressure, gas	30% reduction in average daily gas use, $4,500/mo avoided costs
UC Merced	False peaks in observed chilled water demand at buildings, due to central plant operations	Building chilled water flow, supply and return temperature Central plant chilled water supply temperature	
Sysco	Lights left on after hours at Stockton Sygma	Building electricity	
Sysco	Multi-hour daily energy efficiency strategy at Stockton Sygma	Building electricity, control system setpoints and temperatures	35% demand reduction *Single observation
Sysco	Identification of low-/no-cost savings opportunities, e.g., retro-commissioning and refrigeration tune-ups	Warehouse electric meters	18% reduction in portfolio energy use 36% reduction in Stockton site energy
UC Berkeley	Excessive ventilation and over illumination identified, leading to lighting retrofit and ventilation schedule change	Whole-building electric meter	30% reduction in whole building energy use
UC Berkeley	Multi-week chiller lockout that prevented shut-down	Control system setpoints	
Wal-Mart	Static 225kW load at dimming control submeter	Submeter electricity	$35,000/yr avoided costs
Wal-Mart	Failed or disconnected VFDs used in retrofit programs		Avoided zero savings at program sites

Table 10-9. Summary of performance monitoring challenges, needs, and successes

Site	Challenges/Needs	Successes
UC Merced	Network communications – largely resolved Meter/sensor configuration, and commissioning Resources, staff, proactive use of data Commissioning the Web-EMCS as an EIS Metering aligned w/ metrics – e.g., sampling, vs. totalization	Living laboratory realization Dense instrumentation and data Meeting energy performance goals
Sysco, Stockton Sygma	Timely integration of utility data On-site knowledge and use of EIS features	Network reliability and data quality Portfolio-wide energy reductions Persistence of energy savings Accountability in energy performance Enterprise information sharing
Wal-Mart	Resources, staffing for more intensive EIS use Enterprise-wide use of EIS Portfolio benchmarking within the EIS Ensuring quality of submetering installs	EIS forecasting for DR and purchasing M&V of energy saving measures EIS for large portfolio data acquisition
UC Berkeley	Resources, staff to use energy data Central energy information system External requests for energy data Continuous maintenance	In-house EMCS IT management Utility invoicing, whole-building meters Volume of distributed building data Efficiency intervention meter monitoring

information is typically vague, demonstration software is often not available, and vendor-documented use cases tend not to critically evaluate the technology usefulness. In response, a framework to characterize today's EIS market was developed and applied to several dozen commercial products. The framework provides common nomenclature, as well as a structured classification of existing functionality, while the evaluations permit characterization of the state of today's technology. In addition, four case studies were conducted to explore how the various features and technologies in the framework and evaluations are actually used to achieve energy savings.

EIS Characterization Framework and Technology Evaluations

The categories in the framework comprise the highest-level functions and uses of the technology, such as graphics and visualization or energy and financial analysis. They also include aspects related to purchase and implementation, such as data transmission, storage and security, and general business and licensing models. The sets of features associated with each category are based on typical capabilities as well as leading edge functions that may not yet be widely implemented, for example time-varying analysis of GHG emissions. These findings represent a snapshot of the state of the technology in a

quickly changing field with frequent shifts in offerings and ownership, and they should be interpreted in this context.

The EIS product evaluations indicated that overall, visualization, and analytical features are distinguished by the degree to which they accommodate dynamic, user-defined selections versus statically defined reporting, calculation, and plotting parameters. Rigorous energy analyses that include normalization, standards-based calculations, actionable anomaly detection, and forecasting are either more or less robustly integrated, depending on the specific product. The fact that EIS capabilities are largely distinguished by flexibility in parameter selection, dynamic versus static options, and robustness of analyses reveals the single most difficult aspect of the EIS evaluations. Although out-of-the-box capabilities were stressed as the focus of the study, vendors were quite reluctant to differentiate between embedded "clickable" functionality and actions that the user *conceivably could* perform based on the software features. For example, one EIS might have dedicated modules specifically for M&V investigations, whereas another might report that M&V is supported through no-limit trend storage, aggregate totaling functions, and configurable arithmetic.

The following summarizes specific evaluation findings according to the different categories in the framework:

Business models (General)
- EIS are most commonly offered via application service provider (ASP) or software as a service (SaaS), with no or optional hardware based on client needs
- Optional or bundled services are nearly universal across EIS technology solutions

Display and visualization
- Supported features have converged to a near-common set, including the ability to display load profiles, point overlays, aggregation into totals, etc.
- X-y scatter plotting remains under-supported and relatively uncommon, given the potential for powerful diagnostics
- 3-D surface plotting is among the least common features

Energy analysis
- GHG analysis is a newly emergent feature in EIS; the majority apply a simple energy/carbon relationship, but just under half include knowledge of regional generation or other standards
- Nearly every EIS permits the user to quantify an energy consumption baseline, but weather-normalization is rare
- Benchmarking is widely supported, provided that a portfolio of meters is included in the historic data warehouse; only two EIS in the study used national data sets for comparison

Financial analysis
- Energy costing is supported in nearly all of the EIS, and more than half have implemented model or tariff-based calculations

Advanced analysis
- About three-quarters of the EIS handle corrupted or missing data, and do so via three principal means: flagging or summative reporting; actual cleansing and/or correction; and linking to external or third-party software packages.
- Anomaly detection is typically trend-based, and is accomplished by identifying departures from normal energy consumption patterns
- More than half of the EIS forecast near-future loads, usually by coupling historic trends and weather data; very few provide model-based capabilities
- The large majority of EIS accommodate some form of M&V or the ability to track the impact of operational changes

Demand response
- DR capabilities have advanced since early 2000 and have converged to a common set of features
- Automated response to DR signals is supported in all but three of the DR systems that were characterized.

Remote control and management
- Just over half of the EIS surveyed report the ability to control according to a program, and just under half report internet-capable direct remote control.

The EIS that supports the most features is not necessarily the most powerful solution for a given building. Identifying the most suitable EIS for a commercial implementation must begin with a purposeful consideration of the site's operational and energy goals. Once the immediate and longer-term needs are understood, high-priority features and functionality can help narrow the options, and the most appropriate technology can be selected. For example, an organization that uses custom benchmark models to gauge performance might prioritize flexible definition of metrics and calculations over a dynamic configuration; a geographically diverse enterprise that requires proof of savings from large retrofit initiatives may require robust baselining, data cleansing, and tariff-specific energy costing. Similarly, a business with a history of energy awareness that has implemented a phased, multi-year energy plan is likely to have different needs than a business that has just begun to consider building energy performance.

EIS Case Studies

While exceptionally helpful in gaining an understanding of the state of the technology, individual product characterizations and conclusions regarding software capabilities do not answer questions of usability and real-world utilization. The case studies included in this study attempted to answer questions related to energy savings and actions attributable to EIS use, performance monitoring challenges, and successful implementation models. Because the associated findings overlap considerably, they are grouped into organizational impacts and success factors, and usability and analysis.

Organizational Impacts and Success Factors

The existence of data or performance monitoring software does not guarantee shared knowledge or actionable information. Enterprise-wide EIS use at Sysco has encouraged persistent savings and a corporate culture of energy accountability, awareness, and competition. Similarly, extensive use and sharing of energy data at UC

Merced has contributed to highly efficient operations and energy performance, and it has supported the realization of the living laboratory concept. On the other hand, Wal-Mart and UC Berkeley are both working toward more extensive use of data to reduce energy consumption.

Resources and staffing were a significant constraint in every case studied, and those factors clearly limit the extent to which energy data are successfully used to identify energy-saving opportunities. They also directly affect a site's ability to make meaningful use of submetered data. With the exception of Sysco, where current levels of engagement with the EIS are viewed as sufficient to meet efficiency goals, each organization expressed a strong desire to engage more with measured data in order to improve efficiency.

A common view is that EIS are primarily the domain of in-house staff, and that services are used to a minimal degree during installation and configuration. At the alternate end of the spectrum, EIS may be primarily intended for use by third-party energy service consultants and providers. However, the general prevalence of staffing constraints, Sysco's successful efficiency gains, and the number of EIS vendors that offer analytical services indicate the potential for alternate models of successful EIS use. For example, Cascade Energy Engineering is seeking opportunities for inclusion in utility energy efficiency programs, confident that careful application of engineering expertise, services, and software-based performance tracking will prove a guaranteed pathway to deep energy savings for enterprises. The varying degree to which these cases were successful in leveraging energy data emphasizes that factors such as organizational resources, commercial subsector, size, and resources have a critical impact on the most effective balance between on-site analysis and expert services.

Usability and Analysis

Reliable high quality-data are a critical aspect in automated analysis of building energy performance, and those data significantly affect EIS usability. At UC Merced, failure to commission the instrumentation and web-EMCS for EIS analytics has impacted the ability to track and diagnose building performance. More generally, usability at UC Merced is affected by a number of challenges specific to implementation of an intensive monitoring infrastructure and the acquisition and storage of extreme volumes of trend data. The UC Merced case shows that particular attention must be paid to wiring and hardware integration, system programming, and network communications—not all of which lies wholly in the domain of the EMCS developer. In contrast, Wal-Mart

and Sysco did *not* report significant data quality issues, which is likely for two reasons: EIServer has embedded validation estimation error checking (VEE) routines, and data quality is usually a concern only in cases of submetering and energy sources other than electric.

The degree to which a site uses embedded analytical capabilities depends on the particular performance metrics and benchmarking data that are utilized. Our cases showed that the more tailor-made the calculations, the more likely it is that the data will be exported for analysis in third-party modeling or computational software. In addition, users may develop personal analyses or spreadsheets that prefer to the EIS, even when the EIS provides similar or more powerful functionality. These cases indicated that sophisticated EIS normalization and forecasting methods are not universally understood across users and technology champions. Even so, these methods are commonly used to great success, in a "black-box" manner.

Finally, although EIS offer a wide range of features, actual use of these features can be very limited, and it is not clear that users are always aware of how to use the capabilities of the technology to generate energy-saving information. As evidenced in the Sysco case, partial use of analytical features can result in very powerful outcomes; however, it is possible that further potential savings have gone undetected. In the four EIS cases that were studied, the most common energy-saving actions were related to incorrect load scheduling, M&V, and inefficient operations. The actual savings attributable to these actions are expressed in a number of ways (if at all) depending on a site's standard performance tracking procedures and metrics. Avoided costs or energy consumption, percent reductions in component or end-use loads, reductions in portfolio consumption, and total site energy or power reductions are examples of the diverse measures that each site used to quantify EIS savings.

Future Research

Taken together, the EIS characterization framework, technology evaluations, and user cases studies have resulted in a complementary set of findings, to be extended in future research. These findings and future research needs concern four key areas:

1. Features and usability
2. Anomaly detection and physical models
3. Technology definitions and scalability
4. Successful use and deployment models

While the four case investigations generated useful insights as to the value of EIS, questions concerning the most useful features, potentially useful but underutilized features, and energy savings attributable to EIS use merit further attention. For example, a more extensive set of typical actions and associated energy savings, as well as documented records of building consumption before and after EIS implementation, would enable stronger conclusions on the range of expected savings from EIS use. In addition, typical EIS actions and associated features can be linked to a classification of standard EIS uses such as M&V for retrofit support, continuous building-level anomaly detection, or GHG emissions reporting. Specific building ownership models may also affect these standard uses, as geographically diverse enterprises likely have different organizational objectives than do medium-sized tenanted offices or government-owned buildings.

Closely related to features and usability, there is considerable analytical potential in linking EIS anomaly detection methods to physical models. Today's EIS algorithms rely purely on empirical historic performance data to detect *abnormal* energy consumption. However, they do not provide a means to identify *excessive* energy consumption relative to the design intent, or to realize model-predictive control strategies. Standardizing the format and structure of information at the data warehouse level could encourage such advancements, as could the development of features to configure exported data files into formats that can be used by modeling tools such as Energy Plus or DOE-2. Standard formatting of EIS data would also facilitate the transfer of energy information from the building to outside entities, supporting and aligning with current developments in demand side management and the smart grid.

From a technology standpoint, definitions and scalability require further study. The question of whether a given system is or is not an EIS, is not trivial. This study defines EIS broadly, stipulating whole-building energy analyses, graphical capabilities, and web accessibility. Therefore, many technologies that were included in the study are EMCS or DR tools that are less immediately thought of as EIS, but that *can be used* as an EIS. The UC Merced case illustrated some of the challenges in using an EMCS as an EIS, indicating an outstanding research question: can an EMCS serve as a robust EIS, reliably adding whole-building energy analyses to management and control functionality? Scalability is a concern that may provide insights as to where to draw the line between EIS and related technologies. In the future it will be necessary to understand the tradeoffs between diagnostic capabilities, trend volume and number of points monitored, and the resulting burden on the system's underlying hardware and communication networks. These considerations become especially relevant as a campus or owner's portfolio of buildings grows, or as a user moves to increased levels of submetering or subsystem monitoring and analysis.

Finally, there remains much to learn about effective EIS deployment and use models within organizations. The Sysco case reveals a potentially powerful approach in which in-house use and expert services are combined. This is critical when facility managers have limited time to devote to energy analysis. Additional research is needed to better understand where this approach is most useful and to determine alternate success models that are appropriate to a diversity of organizational sizes, commercial segments, and building ownership models. Neither the EIS evaluations nor the case studies delved very deeply into the costs of EIS. Not surprisingly, vendors were very reluctant to provide price details, and the case studies were primarily focused on the use of EIS features to achieve energy savings. Future investigations into successful EIS use models will be most informative if they are able to link features, whole-building energy savings, the role of services, and EIS cost. The outcomes of the work in this report and prior research will provide the foundation for a broader set of case studies sponsored by the Department of Energy. They will be pursued in collaboration with the New Buildings Institute, beginning in 2010.

Acknowledgment

Vendor participation was critical to the success of this study, and the authors wish to acknowledge their generosity and willingness to be included in this work. This work was supported by the California Energy Commission and the California Institute for Energy and Environment under Contract No. MUC-08-04.

References

Berkeley Campus Dashboard. Berkeley dashboard™; Analyse the campus metabolism. Berkeley Dashboards; 2009 [cited July 27, 2009]. Available from: http://dashboard.berkeley.edu/building/.

Brown, K. Setting enhanced performance targets for a new university campus: Benchmarks vs. energy standards as a reference? ACEEE Summer Study Proceedings. 2002: 4.29–40. Washington, D.C.: American Council for an Energy-Efficient Economy.

Brown, K., Anderson, M., and Harris, J. How monitoring-based commissioning contributes to energy efficiency for commercial buildings. ACEEE Summer Study Proceedings. 2006: 3.27–40. Washington, D.C.: American Council for an Energy-Efficient Economy.

Capehart, B., and Capehart, L., editors; Allen, P., and Green, D., associate editors. *Web based enterprise energy and building automation systems.* Fairmont Press, Inc. 2007.

Capehart, B., and Capehart, L., editors. *Web based energy information and control systems: case studies and applications.* Fairmont Press, Inc. 2005.

Commercial Buildings Energy Consumption Survey (CBECS). Energy information Administration; 1992. Available from: http://www.eia.doe.gov/emeu/cbecs/detailed_tables_1992.html.

Commercial Buildings Energy Consumption Survey (CBECS). Energy information Administration; 2003. Available from: http://www.eia.doe.gov/emeu/cbecs/cbecs2003/detailed_tables_2003/detailed_tables_2003.html#consumexpen03.

Darby, S. *The effectiveness of feedback on energy consumption: A review for DE-FRA of the literature on metering, billing, and direct displays.* April, 2006. Environmental Change Institute, University of Oxford. Available from: http://www.eci.ox.ac.uk/research/energy/downloads/smart-metering-report.pdf.

Few, S. *Information dashboard design: The effective visual communication of data.* O'Reilly Media, Inc. 2006.

Friedman, H., and Piette, M.A. Comparison of emerging diagnostic tools for HVAC systems. April 6, 2001. Lawrence Berkeley National Laboratory. Paper LBNL-47698. Available from: http://repositories.cdlib.org/lbnl/LBNL-47698/.

Granderson, J., Piette, M.A., Ghatikar, G., Price, P. Preliminary findings from an analysis of building Energy Information System technologies. Proceedings of the 2009 National Conference on Building Commissioning, Seattle, WA, June 3–5, 2009. Lawrence Berkley National Laboratory. Paper LBNL-2224E. Available from: http://eis.lbl.gov/pubs/lbnl-2224e.pdf.

Heinemeier, K. The use of energy management and control systems to manage the energy performance of commercial buildings. 1994. Ph.D. Thesis: Department of Architecture, UC Berkeley, and Energy and Environment Division, Lawrence Berkeley National Laboratory. Paper LBL-36119.

Mills, E., Bourassa, N., Piette, M.A., Friedman, H., Haasl, T., Powell, T., and Claridge, D. The cost-effectiveness of commissioning new and existing commercial buildings: Lessons from 224 buildings. 2005. Proceedings of the 2005 National Conference on Building Commissioning; Portland Energy Conservation, Inc., New York, New York. Lawrence Berkeley National Laboratory Report No. 56637. Available from: http://eetd.lbl.gov/emills/PUBS/PDF/NCBC_Mills_6Apr05.pdf

Mills, E., and Mathew, P. *Monitoring-based commissioning: Benchmarking analysis of 24 UC/CSU/IOU projects.* June 2009. Report Prepared for: California Energy Commission Public Interest Energy Research (PIER) Technology Demonstration Program. Lawrence Berkeley National Laboratory. Paper LBNL-1972E.

Motegi, N., and Piette, M.A. Web-based energy information systems for large commercial buildings. March 29, 2003. Lawrence Berkeley National Laboratory. Paper LBNL-49977. Available from: http://repositories.cdlib.org/lbnl/LBNL-49977.

Motegi, N., Piette, M.A., Kinney, S., and Dewey, J. Case studies of energy information systems and related technology: operational practices, costs, and benefits. Proceedings of the Third International Conference for Enhanced Building Operations, Berkeley, California, October 13-15, 2003. Available from: http://txspace.tamu.edu/handle/1969.1/5195.

New Buildings Institute (NBI). Advanced metering and energy information systems. White Salmon, Washington: New Buildings Institute, 2009. For the U.S. Environmental Protection Agency, Grant 83378201.

NorthWrite. Energy Expert: a technical basis. EnergyWorkSite; [cited September 1, 2009]. Available from: http://www.myworksite.com/energyworksiteMBS/htmlArea/files/documents/244_eetechdesc.pdf.

Petersen, J., Shunturov, V., Janda, K., Platt, G., and Weinberger, K. "Dormitory residents reduce electricity consumption when exposed to real-time visual feedback and incentives." *International Journal of Sustainability in Higher Education* 2007; 8 (1): 16–33.

Piette, M.A., Kinney, S., and Friedman, H. EMCS and time-series energy data analysis in a large government office building. April, 2001. Lawrence Berkeley National Laboratory. Paper LBNL-47699. Available from: http://www.osti.gov/bridge/servlets/purl/787118-Qg4Jfr/native/787118.pdf.

Piette, M.A., Kinney, S., and Haves, P. "Analysis of an information monitoring and diagnostic system to improve building operations." *Energy and Buildings.* 2001; 33(8): 783–791.

Wal-Mart. Sustainable buildings. Wal-Mart Stores, Inc. – Sustainable Buildings. Wal-Mart Stores; 2009 [cited July 27, 2009]. Available from: http://Wal-Martstores.com/Sustainability/9124.aspx.

Webster, T. Trends in Energy Management Technologies - Part 5: Effectiveness of Energy Management Systems: What the experts say and case studies reveal. November 20, 2005. Lawrence Berkeley National Laboratory. Paper LBNL-57772. Available from: http://repositories.cdlib.org/lbnl/LBNL-57772.

APPENDIX A:
EIS CHARACTERIZATION FRAMEWORK

Category	Main Feature	Feature Details	Feature Description
	Accepted energy inputs	Electricity	Does the EIS accept metered electricity data?
		Water	Does the EIS accept metered water data?
		Hot water	Does the EIS accept metered how water data?
		Natural gas	Does the EIS accept metered natural gas data?
		Oil	Does the EIS accept metered oil data?
		Steam	Does the EIS accept metered steam data?
		Chilled water	Does the EIS accept metered chilled water data?
		Liquefied petroleum gas	Does the EIS accept metered LPG data?
		Utility billing data	Does the EIS accept utility billing data?

Category	Main Feature	Feature Details	Feature Description
Data Collection, Transmission, Storage and Security	Storage capacity	Months, years, memory limit, duration	What are the storage limits?
	Manual data entry		Can the user manually input the collected data (via GUI)?
	Minimum trend interval	Daily, hourly, near real-time, real-time	What is the minimum resolution of interval data?
	Upload frequency	Daily, hourly, near real-time, real-time	How often does the EIS retrieve data?
	Upload type/connectivity	Phone, cellular, internet	Does the EIS use internet or telecommunication?
	Data sources	Interval meter (building), submeter	Does the EIS provide submeter in addition to building data?
	Data transmission standards, protocols, interoperability	BACnet, LonMark, MV 90, IP, OPC	What transmission protocols or standards does the EIS use/interoperate with?
	Archived data	SQL, .net, XML, CSV, .xls	How is data archived (relational database, flat file, binary proprietary)?
	Exported data	ASCII delimited (ex. CSV, TDL), XML	What export formats are supported for archived data?
	Security	Https encryption, VPN, pgp, authentication	What security protocols/procedures does the EIS use?
Energy Analysis	Performance indicators/metrics	Averages	Does the EIS calculate hourly, daily, weekly, or monthly average consumption?
		Highs/lows	Does the EIS calculate the highest/lowest hourly, daily, or weekly consumption?
		Efficiency	Does the EIS calculate system (plant) or component (chiller) efficiencies?
		Load duration	Does the EIS calculate load duration: number of hours at a set of demand levels, usually annually
		End-use breakdown	Does the EIS estimate energy consumption by end use? (w/o submeter data)
	Normalization	Cooling degree days (CDD)	Does the EIS normalize consumption by CDD?
		Heating degree days (HDD)	Does the EIS normalize consumption by HDD?
		Outside air temperature (OAT)	Does the EIS normalize consumption by OAT?
		Square feet (sf)	Does the EIS normalize consumption by sf?

Category	Main Feature	Feature Details	Feature Description
	Carbon	Standards-based, relational	Is carbon analysis standards-based (CA AB 32), or is an energy/CO_2 relationship applied?
		Time-varying intensity	Does the carbon analysis account for time-varying intensity?
	Benchmarking	Multi-site comparison	Is it possible to comparatively analyze one building's use with respect to another?
		Historical	Is it possible to analyze a building's use with respect to a historic benchmark?
		Standards-based	Does the EIS benchmarking analysis rely upon standards such as Energy Star or Labs 21?
Advanced Analysis	Forecasting	Algorithm, trend-based, model-based, neural net	Does the EIS forecast near-future load profiles?
	Fault detection and diagnostics (FDD)		Does the EIS perform FDD?
	Data gaps/faults		Does the EIS identify corrupted data or gaps in trends?
	Statistics	Regression	Does the EIS perform regression analysis?
		Percentiles	Does the EIS calculate percentiles within a defined cohort?
		Deviation	Does the EIS calculate standard deviation and/or variance?
	Renewables	Solar, wind	Does the EIS provide analysis modules/functions for data from renewable energy sources?
	On-site	Cogeneration	Does the EIS provide analysis modules/functions for data from on-site energy generation?
	Load shape		Does the EIS identify base load, peak demand, and other (weather-based) consumption patterns?

Category	Main Feature	Feature Details	Feature Description
Financial Analysis	Simple energy cost prediction	Estimation	Does the EIS perform simple energy cost estimates?
	Model-based costing	Rate tariff, dynamic (online), or static	Does the EIS include specific rate tariffs in energy cost analyses?
	Bill and meter verification		Does the EIS compare meter readings to utility bills to validate billing and metering accuracy?
	Savings estimation		Does the EIS calculate/predict savings from retrofits, operational strategies, EIS use
	Bill outsourcing		Does the EIS transmit data sufficient to outsource bill processing/payment? (campus recharges)
Demand Response	Automation type	Manual, semi-auto, auto	How does the system respond to DR signals - manual initiation of load-shed, or automatic based on utility signal?
	DR signal notification	Pager, e-mail, phone, fax	How is the operator notified of DR events?
	Real-time DR response		Is the load-shed quantified in real-time?
	Analyze DR		Does the product calculate energy and/or $ savings due to event response?
	Baseline		Does the product calculate a DR baseline according to a utility program formula?
	Savings		Does the product predict expected savings from a response?
	Opt-out		Can the operator choose to ignore a DR event signal?
	Black-out dates		Can the operator pre-specify dates to ignore DR signals?
	Test		Can the operator test DR events (simulate DR signals)?
	DR recording		Does the product record DR data: time received, actions performed?

Category	Main Feature	Feature Details	Feature Description
Control, Management	Automated		Is the EIS capable of controlling building systems according to a program (either through gateways or EMCS)?
	Internet remote Management		Does the product offer remote control and/or management of buildings via the internet?
General, Meta	Browser support	Internet Explorer, Mozilla Firefox, Safari	Are all three current major web browsers supported?
	Upgradeability		Are upgrade modules available (vs. full version purchases)?
	Purchase cost	Hardware	What are the associated hardware costs?
		Software	What is the product software cost?
	Ongoing costs (if any)	Software fee	What is the cost of annual or monthly licensing fees?
		System usage fee	What is the cost of annual or monthly system usage fees?
		Service and maintenance	What are the annual or monthly service and maintenance costs?
	Lifespan		What is the expected product lifespan before major upgrades are recommended?
	Target market	Commercial, industrial, school, etc...	Which market segments does the product/company traditionally target (historic versus newly targeted)?
	Intended user	Building manager, facility manager, enterprise/multi-site	Who are the intended users (most common, as opposed to all-that-apply)?
	Number of users		Approximately how many customers currently use the product?
	Company profile	System integrator, control vendor, software, energy service provider, utility	Which descriptor best characterizes the company?
	EIS product type	Software only, software and hardware	Does the product offering include hardware in addition to software?

APPENDIX B:
EIS TECHNOLOGY EVALUATIONS

This appendix contains the specific vendor evaluations that were used to inform the state of the technology findings presented in this study. The appendix comprises a spreadsheet that can be downloaded from: http://eis.lbl.gov.

All reported findings are based on vendor-supplied information at the time of the study (November 2008–April 2009). Current capabilities are subject to change, and readers are encouraged to confirm information based on their specific needs. The EIS that were selected for evaluation are representative of the market, but not comprehensive, and inclusion in the study does not imply endorsement.

APPENDIX C:
SELECTED EIS BASELINE METHODS

Energy information systems use baseline energy consumption models to perform measurement and verification (M&V) or savings tracking, historic performance tracking, multi-site benchmarking, anomaly detection, and near-future load forecasting. In this context, the term *baseline* refers the typical or standard energy consumption. To ensure fair comparisons and consistency across time, climate, and buildings, baselines should be normalized to account for weather, time of day or week, and other factors.

- M&V analyses compare post-measure energy consumption to the baseline. Similarly, historic performance tracking compares recent or current consumption to the baseline.

- Multi-site benchmarking is accomplished by comparing one building's appropriately normalized baseline to that of a cohort of buildings. The cohort might be other sites in a portfolio, other sites in a vendor's databases, or national or state databases such as the Commercial End Use Survey (CEUS) or Commercial Buildings Energy Consumption Survey (CBECS) or ENERGY STAR.

- Anomaly detection is accomplished by predicting would-be consumption by inputting current/recent conditions into baseline models, and then comparing the predicted and actual consumption.

- Near-future load forecasting is accomplished by inputting current/recent and forecasted conditions into baseline models.

Linear regression and non-linear estimation techniques are common approaches to quantifying baselines in EIS. The following susbsections describe these methods, and their relative strengths and weaknesses. Linear regression pairs historic energy trends and weather data to determine a functional relationship between the two. In regression models, explanatory variables (e.g., humidity, air temperature and day of week) are used to determine the value of the dependent variable (e.g., demand). Baseline models differ according to:

- the number of explanatory variables included
- the resolution of weather data, e.g. daily high, or hourly air temperature
- the resolution of the baseline, e.g. daily peak, or hourly peak demand
- the goodness of fit between the model and the data

Three non-linear estimation techniques that were evaluated in the study are used in the EIS. Energy Work-Site uses a bin methodology, Pulse Energy uses weighted averaging, and EIServer uses neural networks.

Bin method used in Energy WorkSite

The bin method predicts the energy consumption at a given time to be equal to the average consumption at times when conditions were similar. To understand the bin method, consider the case in which air temperature, relative humidity, and time of week are the explanatory variables and are used to estimate energy. The three-dimensional space of explanatory variables is "binned," or broken into mutually exclusive volumes. For example, temperature might be binned into five-degree intervals, time of week into weekend and weekday, and relative humidity into five percent intervals. Energy consumption data are placed into the appropriate bins, as in Figure 10C-1.

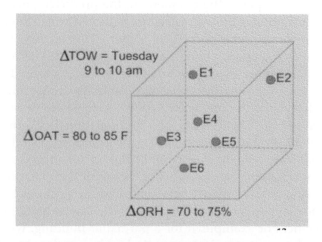

Figure 10C-1. Energy Worksite bin methodology[13]

Once the bins are sufficiently populated with historic data, the explanatory variables are used to identify which bin corresponds to the current conditions. The predicted energy consumption for the current conditions is then taken as the average consumption across the historic data in the bin. The bin method has proven effective in building energy analysis, as well as component systems, but it breaks down as the number of explanatory variables grows: beyond 3–5 variables, the number of bins becomes too large.[14] With a very large number of bins, the current state of the system will often correspond to an empty bin, or one with few data points, and in these cases averaging the energy consumption of the points that are in the bin will often not yield good predictions. For example, there may never have been a previous period in which, at 10:30 a.m. on a Tuesday morning in summer, the outdoor temperature was between 75°F and 80°F, the relative humidity was between 55% and 60%, the sky was cloudy, and the wind was strong from the south.

Advantages of the bin approach are that it is simple to explain and understand, and it works well when only a few input variables are important. Disadvantages are that the approach can only handle a small number of input variables, and that bin models may be unable to predict, or may provide inaccurate predictions for conditions that have occurred rarely or never before. Also, some bin implementations use fixed bin boundaries; in these cases there can be problems when current conditions are closely aligned to bin boundaries, e.g., if outdoor air temperature is 84.9°F and bins are defined by 80–85°F and 85–90°F. (Energy Worksite's implemention does not have this problem: it dinamically redefines bin coundaries so that current conditions are always in the middle of the bin.)

Pulse Energy

The method used by Pulse Energy applies the same basic principle as the bin method: the predicted energy consumption is the average consumption during similar periods. The method creates a metric to describe the degree of similarity between the current conditions and similar conditions at other times. It then takes a weighted average of the energy performance at these similar times to determine the predicted Typical Performance for the current time. The weights used in calculating the average depend on the degree of similarity, with highly similar conditions receiving a high weight. Pulse uses a proprietary patent-pending method to define the metric that quantifies the degree of similarity between current conditions and conditions in the database. This metric can be building-specific. For example, if for a particular building, wind speed turns out to be unusually highly correlated with energy consumption, the metric for this building will be more sensitive to wind speed than it is in other buildings.

As with bin-based methods, a method based on a weighted average of values during similar conditions can suffer if the current conditions have rarely or never been encountered before. However, the predictions are bounded by the lowest and highest values that have been historically recorded so the predicted values will always be physically possible values, which is not true for all other methods. Pulse Energy developers are working on extrapolation methods to be able to make predictions for conditions that have not occurred before.

Advantages of the Pulse Energy approach are that the basic principle is easy to understand, and that there is no limit on the number of input variables that can be used effectively. In contrast to bin methods, the Pulse Energy approach allows input variables to be differently weighted, potentially improving accuracy. Similar to bin methods, a disadvantage of the approach is that it may provide poor predictions for conditions that have rarely or never occurred before.

Neural Network method used in EIServer

Artificial neural networks are so named because they simulate some of the behavior of neurons in the central nervous system. Input variables such as outdoor temperature and humidity are mathematically processed to create a potentially large number of secondary, or "hidden," values. These hidden values are then processed to generate a (usually small) number of output values, such as predicted energy consumption (see Figure 10C-2.) The mathematical functions that process the input values and the hidden values have adjustable parameters known as *weights*, so that the effect of every input value on every hidden value is adjustable, as is the effect of every hidden value on every output value. Neural networks "learn" by adjusting the weights so that the outputs are as close as possible to their desired values, for a large set of "training" data. For example, data from several weeks or months

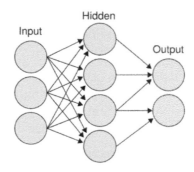

Figure 10C-2. Graphical representation of a neural network[15]

of building operation can be used to train the network to predict energy consumption, given input data such as temperature, humidity, and time of day.

EIServer beings with a simple model, adding additional input variables only if the network generates inaccurate outputs. Once the initial training is complete, the system can perform energy forecasts. EIServer features a built-in scheduling system to retrain the model occasionally as more data become available. Neural networks have the advantage of being able to handle a large number of input variables, and the large number of automatically adjusted parameters can provide accurate predictions. However, the concept is difficult to understand, and if the network behaves poorly even experts can be challenged in identifying improvements. Finally they may not perform well for conditions that differ greatly from those in the "training set."

Outstanding questions

Each of the three systems discussed above has advantages and disadvantages compared to the others. Unfortunately we do not have enough information to judge which system or systems work best, and it is even possible that some methods will work well in some buildings and poorly in others. As far as we know, there has never been a comparison of how the different approaches work in the same set of buildings.

All of the methods discussed above may make poor predictions when conditions differ substantially from those in their database, which can be a problem for periods on the order of one year after the system begins operating: all of the systems need to accumulate at least a few weeks of data in the cooling season and the heating season, and may still have problems making accurate predictions when conditions are extreme (such as the coldest or hottest weeks of the year, the most humid week of the year, and the cloudiest week of the year). Some of the methods may perform better than others when conditions are outside those in the historical data, but we are not aware of any studies that have investigated this issue.

Many systems for estimating baseline energy consumption or for recognizing anomalous behavior can fail to recognize a slow creep or shift in use, in which at any given moment the energy consumption is not greatly different from normal, but over time the consumption creeps up or down. Since the systems continue to incorporate new data as they become available, the baseline can slowly shift with time, without any particular data point appearing anomalous. Some of the EIS provide approaches to recognizing or quantifying this issue, for example by comparing predictions using the system's

current model to the predictions that would have been generated using last year's model. The effectiveness of these approaches is not known.

APPENDIX D:
UC MERCED EIS CASE STUDY NARRATIVE

Case Background and Introduction

Opened in 2005, UC Merced is the newest University of California campus. Prior to opening, the campus made a strong commitment to energy-efficient building design, and energy plays a fundamental role in campus objectives targeting environmental stewardship and high-quality, affordable instruction, research, and employee working environments. At UC Merced, campus efficiency requirements have been developed to:[16]

- reduce operating costs toward minimizing life-cycle cost of campus facilities,

- achieve maximum subsidies for energy efficiency,

- contribute as many points as practical to facility LEED™ ratings,

- minimize infrastructure costs,

- minimize impact of the campus on the environment and on the energy infrastructure, and

- maintain high-quality energy services in campus facilities.

Campus energy activities focus on three areas:[17]

- **Building energy performance targets,** to ensure that new buildings are significantly more efficient than required by code or compared to other university buildings in California.

- **Ongoing monitoring of energy use,** to facilitate continuous improvement in campus operational efficiency and design, as well as serve as a primary component of UC Merced's "living laboratory" for the study of engineering and resource conservation.

- **Climate neutrality,** to pursue use of renewable energy resources and other strategies to reduce and offset greenhouse gas emissions with an eventual goal of climate neutrality.

In support of these efficiency requirements and three focal activity areas, standardization of the campus control systems was made to be a priority during design/construction, and the campus control was bid as a full package. This is in contrast to many UC campuses in which building controls are bid on an individual basis, and it is common to encounter a diversity of solutions. Automated Logic Corporation's WebCTRL was selected, largely for the internet/intranet connectivity and control capabilities. The UC Merced staff and the engineering company both found it especially useful to log in to the system remotely throughout the campus design and commissioning process as new buildings were constructed and opened for full-time use.

Installation and Configuration

Because UC Merced was newly constructed, integration with existing systems was not an issue, as is often the case when an EIS is purchased. Automated Logic distributors are responsible for system installation, configuration, programming control sequences, and desired monitoring points. In addition, UC Merced holds a small ongoing maintenance contract with their distributor.

Energy Savings and web-EMCS Use

Approximately twenty people use the WebCTRL system at UC Merced, including:
- central plant operators,
- HVAC technicians,
- building superintendents,
- the campus energy manager, and
- electricians, on occasion.

In addition, twenty internal and external researchers use the software for building energy research projects. Relative to other professional and technical software applications, UC Merced WebCTRL users have found the system easy to learn and to use. For example, in contrast to the maintenance management software, the campus energy manager does not feel that there are capabilities that he does not understand or not know how to use. Only 2–3 other users at Merced understand the system at the same level of detail as the energy manager, however that is more an artifact of job structuring and responsibilities than of usability.

Specific Web-EMCS uses

At UC Merced WebCTRL is most extensively used as a troubleshooting tool in response to trouble calls. During the initial year of operation, problems involving the reliability of the air handler units (AHUs) arose 2–3 times a week. The AHUs would commonly trip off, causing a severe rise in buildings and IT room temperatures. The facilities staff and energy manager found that the only way to consistently solve the problems was to build up diagnostic trends that would permit identification of the source.

In addition to troubleshooting problems that have already been brought to attention, WebCTRL is also regularly used to verify that the individual buildings are operating as expected. Roughly 10 multiple-trend "opera-

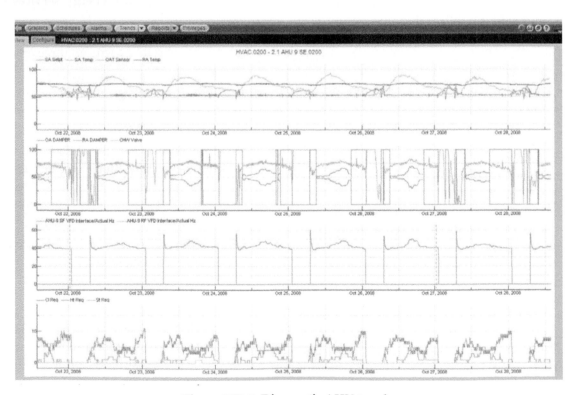

Figure 10D-1. Diagnostic AHU trends

tional plots" were defined, for example, for the hot and chilled water bridges, AHUs, and a representative array of individual zones. Figure 10D-1 contains an example of the AHU plot, including trends for supply air temperature and setpoint, outside and return air temperatures, damper and valve positions, supply and return VFD power, and other factors.

The central plant operators have begun performing hourly reviews of WebCTRL trends according to a defined check-sheet, and the reviews are documented and commented. This process was implemented as a structured way for the operators to be able to leverage the tool and be more proactive. The campus energy manager views this as their first successful step in moving beyond reactive use of the EIS which is limited to responding to trouble calls and alarms.

On a monthly basis WebCTRL meter data are used to determine utility recharges. Twelve-month snapshots are also compiled for annual analyses of campus and building-level energy performance with respect to California campus benchmarks. These include hot water, chilled water, electricity, and gas. In addition laboratory $/cfm (cubic feet per minute) and cooling plant $/ton are calculated on annual bases. Finally, WebCTRL data are used on an as-needed basis, to generate diagnostic variable air volume (VAV) summaries in the form of trends or reports. For example, there has been some difficulty keeping the dining facilities warm in the winter. In response, the energy manager used a summative report of every VAV in the space in combination with trends of temperatures and flows relative to setpoints, to characterize the number of zones not meeting setpoint. From that knowledge he was able to isolate problem areas to be serviced by technicians. These reports and plots are shown in Figures 10D-2 through 10D-4.

UC Merced was designed for low energy use and energy-efficient performance. Manual diagnostics based on EMCS data have been used to manage energy use over time as the campus grows. For example, actual operating data from the EMCS were used to verify that campus energy performance has in fact exceeded ambitious targets. For the 2007-2008 fiscal year, UC Merced used only 48%-73% percent of the energy used at other campuses. That is, it was operated 27%-52% more efficiently than average.

Data quality

Data quality issues arise in a number of contexts at UC Merced, further challenging the use of the EMCS for automated analyses. Networking and connectivity prob-

VAV Summary Report

Location: University of California / Merced Campus / Kolligian Library: 0201 / HVAC: 0201   Run Date: 10/28/2008 2:24:22 PM

Name	Zone Temp	Htg Setpoint	Clg Setpoint	Setpt Adj	Air Flow	Airflow Stpt	% of Flow Setpoint	Cool Req	Heat Req	HW Valve	CO2 Lev
VAV-130 MDF	73.8	65.0	74.0		440.0	410.0	107.3	0.0	0.0		
VAV-122 MDF M1-2	61.6	63.0	72.0		1,480.0	1,500.0	98.7	3.0	0.0		
VAV-131 U1-2,3	72.0	72.0	76.0		530.0	520.0	101.9	0.0	0.0	52.0	
VAV-101 RM-101	72.1	72.0	76.0		280.0	300.0	93.3	0.0	0.0	0.0	
VAV-102 RM-107A,B,E	73.4	74.0	78.0		70.0	60.0	116.7	0.0	1.0		
VAV-103 RM-107,C,D	73.4	72.0	76.0		120.0	100.0	120.0	0.0	0.0	0.0	
VAV-104 RM-108	72.0	68.0	72.0		1,550.0	1,600.0	96.9	1.0	0.0		
VAV-105 RM-108	72.6	68.0	72.0		950.0	1,000.0	95.0	1.0	0.0	0.0	
VAV-106 RM-106A	71.2	71.0	74.0		100.0	100.0	100.0	0.0	0.0		
VAV-107 RM-106B	72.3	69.0	73.0		100.0	100.0	100.0	0.0	0.0	0.0	737.0
VAV-108 RM 108-D,E	71.0	68.0	71.0		130.0	130.0	100.0	0.0	0.0		
VAV-109 RM-109	72.3	70.0	74.0		220.0	230.0	95.7	0.0	0.0	0.0	
VAV-110 RM-113	69.1	67.0	73.0		30.0	60.0	50.0	0.0	0.0	0.0	
VAV-111 RM-117	70.9	69.0	73.0		20.0	20.0	100.0	0.0	0.0	0.0	619.7
VAV-112 RM-119	72.4	72.0	76.0		0.0	40.0	0.0	0.0	0.0	0.0	807.3
VAV-113 RM-122	75.9	71.0	75.0		1,410.0	1,500.0	94.0	1.0	0.0	0.0	
VAV-114 RM-122	72.5	70.0	74.0		180.0	290.0	62.1	0.0	0.0	0.0	
VAV-115 RM-122A	71.8	70.0	73.0		100.0	100.0	100.0	0.0	0.0		
VAV-116 RM-122B	73.6	70.0	74.0		120.0	110.0	109.1	0.0	0.0	0.0	666.6
VAV-117 RM-122C,D,E,F	73.6	70.0	74.0		90.0	100.0	90.0	0.0	0.0		
VAV-118 RM-122G	73.6	72.0	79.0		120.0	120.0	100.0	0.0	0.0		

Page 1 of 10

Figure 10D-2. Diagnostic VAV summary report

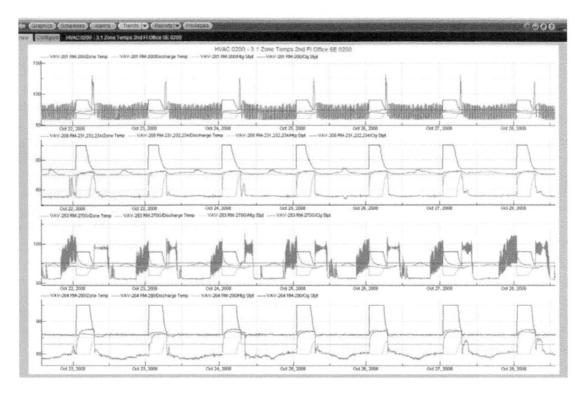

Figure 10D-3. Diagnostic trends of VAV zone temperatures

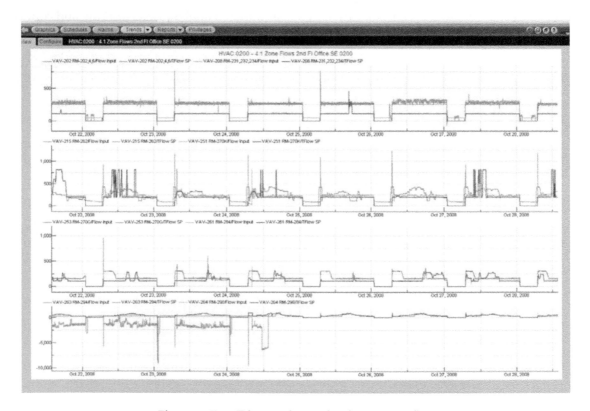

Figure 10D-4. Diagnostic trends of VAV zone flows

Table 10D-1. UC Merced campus energy performance, 2007/2008 fiscal year

	Performance vs. Benchmark
Peak Electric Demand*	48%–63%
Annual Electric Use	68% source, 69% site
Annual Gas Use	73%

*With and without use of the Thermal Energy Storage system

lems have led to dropped or miscommunicated values that generate errors, lock out equipment, and cause large volumes of false data and cascading false alarms. This has been a significant problem in using and maintaining WebCTRL at UC Merced; however, network communications are viewed as affecting operations more than energy monitoring, and over time many of these challenges have been addressed.

While not attributable to the capabilities of the EMCS, meter or sensor calibration and configuration errors also affect data quality, thereby affecting the ability to use the EMCS as an EIS. With the exception of whole-building electric data, significant resources were required to manually validate the EMCS data quality and to quantify the campus energy performance relative to benchmark. Manual validation included inspections to trace the physical meter point to its representation in WebCTRL, as well as energy and mass balances to confirm accuracy of logged data and interpolation or estimation of missing data. To date, manual validation has affected building science researchers more than WebCTRL users at UC Merced, but it does have implications for advanced use of the data within the living laboratory context.

Future web-EMCS use

Looking to the future, there are several measures that the energy manager would like to implement. Currently the data are in the form of single-point samples acquired every 15 minutes, however the energy manager would also like to make use of 15-minute averages to more accurately reflect standard monitoring protocols. Similarly, in an effort to reduce the volume of data to be processed, he would like to identify the minimum sampling frequency necessary to accurately reflect the energy parameters that are continuously tracked. The energy manager would also like to enhance the operational plots and fully integrate them into the daily routines of the HVAC technicians. The ability to review a standard set of plots and data each time there is a problem has proved to be a valuable time saver, but it is not yet a habit.

The metrics that quantify energy performance with respect to benchmark are currently calculated annually, by exporting WebCTRL data to third-party software for computation. Ideally, these metrics would be calculated directly within WebCTRL. For instance, building-level metrics could be combined with a range of Central Plant efficiencies (actual vs. best-practice) and a basic annual load shape. This would make it possible to determine for example, if three months into the year the campus was on-track to meet the annual performance targets. The energy manager expects that these calculations can be defined using the WebCTRL's logic, but that actually programming the logical sequences will require the expertise of the distributor.

In addition to enhanced metrics and calculations and performance tracking, the energy manager would like to delve further into the building electrical submeters to better understand building end uses, and to inform and justify proposed changes. For example, the exterior zones of the library and science buildings currently feature banks of lights that are switched on/off according to daylight, in addition to scheduled on/off operation. The energy manager reports significant hassle with the setup and maintenance of such controls, and suspects that increased personal control options combined with vacancy sensing may be more effective and more efficient. Regular tracking of end use data, which is currently acquired and stored but configured for display, would permit quantitative comparison of different conditioning strategies.

In addition to facilities staff, the energy manager would like to make end-use data available to building occupants. In the summer of 2008 the campus participated in a single-day manual demand response event that relied in part on building occupants to reduce their electric demand. The most valuable result of that event, beyond the savings that were achieved, was that the campus community became engaged and began to think about building energy in new ways and ask questions of facilities. As a result, the energy manager has expressed interest in making the WebCTRL data publically available so that during the next event occupants can view the load reductions in real-time and assist in participation at their building. Analogously, there is a desire to use WebCTRL data for a newly constructed 1-megawatt solar panel, in combination with a front-end panel graphical user interface to encourage public awareness and engagement with efficiency measures.

Usability and Enhancement

Overall, UC Merced WebCTRL users are quite satisfied with the system's plotting and graphical capabilities.

In contrast to some tools such as Excel, WebCTRL offers a simple, clean way to graph and zoom in and out over very large sets of data. However, it was noted that it would be useful to have an easy way to run basic statistical analyses (un-accommodated in logical programming blocks) and identify gaps in historic trend logs.

There are no embedded features in the software that are unused at UC Merced or considered superfluous by the facilities' end users. There are however, features that are not included that would be very useful for enhanced energy analysis and performance monitoring. The ability to create x-y scatter plots was highlighted as the single-most useful, yet absent, feature in WebCTRL. The University of California at Merced has experienced significant seasonal difficulty in tuning nested proportional-integral-derivative (PID) control loops, and x-y capability would permit visual and numerical troubleshooting that is not possible today. Expanded data analysis options would also be useful; there are limits to the calculations that can be automated via logic, such as identification, filtering, or interpolation of gappy meter data.

At UC Merced, the energy manager did not have decision-making power over what type of performance monitoring to use, because WebCTRL was pre-selected as the campus-wide operational and control tool. Despite not being involved in the selection, the energy manager feels that the overwhelming majority of analytical and operational tasks he would like to implement are easily accommodated within WebCTRL. Furthermore, WebCTRL's capabilities have been instrumental in complementing the realization of UC Merced as a *living laboratory*.

Early in the design process, campus stakeholders opted to heavily instrument the campus to support the link between research, instruction, and facilities operations. This concept is very highly valued by the energy manager, who asserts that every academic institution should support such connections. WebCTRL data have been used to facilitate several conversations between facilities and faculty in the department of Engineering, resulting in research proposals to the California Energy Commission, a thermodynamics curriculum that includes a module to quantify the performance of the chilling plant, student employment or project work, and collaborative research projects with the U.S. Department of Energy and Lawrence Berkeley National Laboratory.

General EIS Perspectives

Regarding general perspectives independent of vendor-specific solutions, the energy manager at UC Merced has a strong preference for multi-option, user-configurable designs over pre-configured quick-access

displays of the variety commonly see in information "dashboards." In terms of specific features, the energy manager believes that carbon tracking and alignment with benchmarking and other standards will be of increased importance in the future, though especially challenging due to the need to anticipate how people will use software to comply with reporting requirements. It was noted that the campus reports to the California Climate Registry are based on monthly utility bills, but that it would be ideal to use WebCTRL data to automate the reporting process. Data quality, filtering, and fault detection were also highlighted as critical features that will remain critical to any EIS, meriting increased levels of sophistication.

Understandably, the features and capabilities of a single EIS are not likely to support each and every diagnostic or analytical procedure that especially engaged operators and managers may wish to conduct. Ultimately, it may be necessary to export data to third-party software with more robust graphical and visualization or data processing and manipulation capabilities. At UC Merced, the preference is to rely upon a single system in spite of its inherent constraints, rather than attempting to leverage the capabilities of a suite of software tools. For example, Pacific Gas and Electric's Universal Translator offers a convenient means to synchronize, filter, and analyze data from loggers and energy management systems,[18] and it could prove quite valuable to UC Merced, given the history of non-uniform sampling configurations and data corruption. Ultimately however, the prospect of adding another step to the WebCTRL-based monitoring procedures outweighed the potential advantages. Moreover, Web-CTRL has recently developed an integrated plotting and visualization module that is expected to enhance use of the EMCS for monitoring purposes.

In terms of powerful features that are not accommodated in contemporary EIS, UC Merced would benefit most significantly from embedded functionality to link performance analysis and maintenance. The energy manager emphasized that rather than deeper analysis he would like to see current analytical capabilities merged with knowledge of operations. At UC Merced, efficiency is not perceived as a stand-alone goal in and of itself, but rather as an aspect of an ongoing need to ensure that the campus operates and performs as it should. Therefore the energy manager believes that the EIS should enable facilities to conduct decision-making that will protect the campus efficiency investment. For example, reports would indicate when a technician should be dispatched in order to maximize system performance and minimize costs. That is, what should be done today to improve tomorrow's performance? It is important to note that the energy manager believes that this absence of EIS capability is rooted in a lack of understanding within the industry, rather than in software development challenges. Finally, although less critical, the ability to better detail IT capacity and energy demands to equipment performance and long-term growth planning would also be of great use to enterprises or campuses such as UC Merced.

APPENDIX E
SYSCO EIS CASE STUDY NARRATIVE

Case Background and Introduction

Sysco has implemented a corporate-wide energy efficiency program that targets a 25% reduction in energy consumption across a portfolio of over one hundred distribution centers in the United States and Canada. Two-and-a-half years into the three-year program Sysco has exceeded its goal, achieving 28% energy savings. Sysco has a long-standing energy services contract with Cascade Energy Engineering, but did not make use of an EIS prior to beginning the efficiency program. When the energy targets were determined, a three-part approach was adopted in collaboration with Cascade Energy Engineering: (1) site visits by expert refrigeration engineers and technicians to perform tune-ups and identify low/no cost energy-saving measures, (2) implementation of an EIS to accommodate Sysco's performance monitoring need and energy savings goals, and (3) continuous communication and collaboration between corporate managers, Cascade Energy Engineering, and on-site "energy champions."

The first 12-18 months of the program were dedicated to site visits, tune-ups, and installation of the EIS meters and software. A combination of EIS data, expert assessments, and on-site staff insights was used to gain 18% savings from no-/low-cost measures. Over the remainder of the program period, a further 10% savings were gained through capital improvements such as variable-speed drives, lighting retrofits, and HVAC upgrades. In addition, Sysco experienced significant growth over the program period and was able to successfully apply energy-saving recommendations to new facilities. Throughout the enterprise, the performance tracking metric is daily savings per thousand square feet. Current monthly savings with respect to the program baseline amount to nearly 18,000,000 kWh. The success of the initial three-year energy efficiency program has encouraged adoption of a second phase in which underperforming sites will form the focus to achieve enterprise savings of 30%–35%.

Sysco uses NorthWrite's Energy WorkSite EIS, and it serves as an interesting example of an EIS that was at least in part developed to support the specific needs of a large client with a complementary vision of EIS technology, use, and design. With the exception of the Energy Expert module, key configurations in reporting, benchmarking, and utility billing utility modules were defined based on the needs of the Sysco efficiency project. In addition to a willingness to collaboratively define the EIS information content, look, and feel, the NorthWrite system was selected for usability and relatively low cost. While it did not perform an extensive screening process, Cascade was able to determine that relative to competing technologies, NorthWrite was intuitive, learnable, and presented a sufficient but not overwhelming number of configurations and user-selected options.

Installation and Configuration

Interoperability between the EIS and preexisting systems and controls was not a notable challenge in the case of the Sysco implementation. This was largely due to the nature of refrigerated warehouse energy consumption and Sysco's specific program needs. Gas is not a significant portion of total energy use, and with approximately 50% of energy consumption devoted to refrigeration, and 20% to lighting, even minor operational changes are reflected in whole-building electric meters. Because most Sysco warehouse sites contain multiple utility meters but do not feature submetering beyond the whole-building level, the monitoring aspect of the efficiency program did not require extensive integration with existing control systems or equipment-level metering. Several sites expressed interest in NorthWrite's ability to integrate submetering, but ultimately they were unable to justify the additional associated costs. Across the enterprise, 15-minute interval pulse outputs are uploaded to the NorthWrite central data server via cellular communication.

Energy Savings and EIS Use

The NorthWrite EIS is used throughout the Sysco organization. The energy champion at each site interacts with the EIS to varying degrees, depending on individual work styles and site-specific operational concerns. Additionally, site energy champions attend monthly meetings to discuss their site ranking relative to others in the portfolio and to share successes and ensure accountability. At the executive level, monthly reports that aggregate site performance into portfolio savings are regularly reviewed.

Sysco's Northern California Stockton SYGMA affiliate was studied for this case, including a site visit and interview with the energy champion. Typically, the energy champion is the only staff member that regularly uses the EIS, as is the case at the Stockton distribution center. The title of "energy champion" is not a dedicated assignment, but rather a responsibility that is assumed in addition to the traditional aspects of their role in the organization. In Stockton, and throughout the enterprise, Sysco's contract with Cascade Energy Engineering precluded significant involvement of on-site staff in the identification of reporting options, trend resolution, tracked performance metrics, and other configurable options within the software.

Specific EIS Uses

The Stockton site was selected for this case study because of the reported degree to which the energy champion engages with the EIS, and because of his energy performance relative to his peers in the organization. Stockton ranks highly in the Sysco portfolio, and it has reduced site energy 36% since the start of the efficiency program. The energy champion makes extensive daily use of the "meter monitor" view that contains a two-point overlay comparing the current week's or day's kilowatt time series to that of the prior week. In addition to time series overlays this view contains a summary of cumulative kilowatt-hours for both time periods, the average ambient temperature, and the percent change in consumption and temperature between the two overlaid time periods. This view is illustrated below in Figure 10E-1.

The Stockton Sygma site contains five utility meters, and while the meter dedicated to refrigeration loads is the primary focus of EIS use, minor energy management tasks are performed with the remaining meters. Unantici-

pated or unexplained spikes in consumption are plotted and shared with equipment technicians, and deviations from expected profiles are investigated and remedied. For example, the energy champion has noted instances in which the lights were not shut off following the last shift of the day, and has responded with staff reminders. Over time, the EIS has played an especially useful role identifying such behavioral impacts on site energy consumption, and has served as a motivational benefit to prevent backsliding performance. The energy champion perceives that have staff behavior is now well aligned with site efficiency goals.

In addition to the analyses embedded in the meter monitor, which are utilized daily, several analyses are performed in monthly reporting runs. Cascade Energy Engineering inputs utility billing invoices into the data warehouse, and monthly reports are used to generate portfolio benchmark rankings. Within the network of SYGMA affiliates, the Ohio-based project manager coordinates monthly group reviews with each site's energy champion. Each energy champion's access is limited to their own site; however, executive level staff have portfolio-wide account permissions. Comparative benchmark rankings are based on a metric called the *efficiency factor,* which takes into account wet bulb temperature, the total volume of frozen and refrigerated space, total and daily energy consumption, and weather predicted energy performance. Report-generated ranking tables and efficiency factors are shown in Figures 10E-2 and 10E-3.

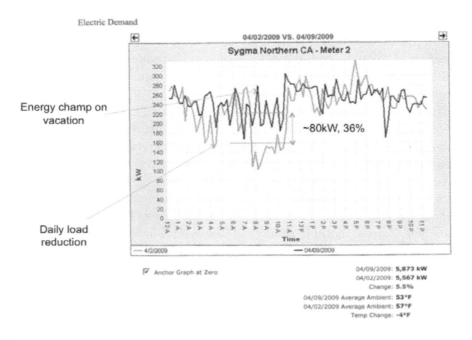

Figure 10E-1. NorthWrite meter monitor and Stockton Sygma daily efficiency strategy

Ranking Table Report - January 2009

North Central

Ranking Table

Rank	Site	Efficiency Factor
1	GRAND RAPIDS	0.521
2	KNOXVILLE	0.525
3	CINCINNATI	0.577
4	BARABOO	0.676
5	CLEVELAND	0.687
6	ROBERT ORR	0.712
7	ASIAN FOODS - Chicago	0.738
8	E. WISCONSIN	0.742
9	INDIANAPOLIS	0.763
10	KANSAS CITY	0.811
11	MINNESOTA	0.829
12	ROBERTS	0.849
13	HARDINS	0.858
14	NORTH DAKOTA	0.864
15	DETROIT	0.865
16	IOWA	0.867
17	ST. LOUIS	0.920
18	LOUISVILLE	0.956
19	PEGLER	0.959
20	CHICAGO	1.118
21	CENTRAL OHIO	1.500
22	ASIAN FOODS - St. Paul	1.665
	AVERAGE	0.864

SYGMA

Ranking Table

Rank	Site	Efficiency Factor
1	Sygma - Denver	0.625
2	Sygma - Southern California	0.663
3	Sygma - Northern California	0.671
4	Sygma - Portland	0.822
5	Sygma - Oklahoma	0.864
6	Sygma - Dallas	0.867
7	Sygma - Kansas City	0.926
8	Sygma - Detroit	0.982
9	Sygma - Illinois	0.983
10	Sygma - Georgia	1.015
11	Sygma - Boston	1.059
12	Sygma - Pennsylvania	1.070
13	Sygma - Florida	1.097
14	Sygma - Carolina	1.125
15	Sygma - San Antonio	1.555
16	Sygma - Columbus	2.209
	AVERAGE	0.955

Figure 10E-2. Energy performance ranking tables for Sysco North Central and Sygma affiliate distribution centers

are to be entered into the system by Cascade Energy Engineering. However, the Stockton site has experienced several months lag in the data entry process, perhaps due in part to the larger number of meters at the site—most Sysco sites do not have multiple meters. Therefore, the Stockton energy champion uses a personally designed spreadsheet to track energy expenditures. He also uses this personal tracking to produce documentation in support of his annual employee performance review. An example is provided in Figure 10-5.

Daily use of the EIS to monitor the meter dedicated to refrigeration loads has encouraged and confirmed the effectiveness of operational changes implemented by the energy manager. The existing controls do not permit it; however, the frozen goods can

Site	kWh	Wet-Bulb Temp	kWh / day	Total Frozen cu-ft	Total 28-55°F cu-ft	Total Dry cu-ft	Weighted Volume Cu-ft	Space Weighted Eff Factor	Weather Predicted Eff Factor	New Effiency Factor
Sygma - Denver	1,902,060	41.9	10,337	891	971			2.66	4.52	.666
Sygma - Southern California	2,587,710	48.4	14,064	920	1,142			3.01	4.76	.732
Sygma - Northern California	1,855,050	52.3	10,082	748	595			3.01	4.9	.756
Sygma - Kansas City	1,077,360	47.7	5,855	340	641			3.43	4.74	.775
Sygma - Illinois	3,558,070	46.4	19,337	1,680	1,043			3.66	4.69	.835
Sygma - Oklahoma	1,391,650	52.8	7,563	498	392			3.61	4.92	.844
Sygma - Portland	1,277,340	48.1	6,942	425	442			3.62	4.75	.863
Sygma - Detroit	1,567,380	45.2	8,518	384	407			3.32	4.65	.927
Sygma - Dallas	3,397,560	56.6	18,465	919	1,083			4	5.06	.930
Sygma - Boston	1,390,940	43.7	7,559	445	383			3.85	4.59	1.037
Sygma - Georgia	2,341,800	55.6	12,727	388	584			4.67	5.02	1.043
Sygma - Florida	3,333,500	64.5	18,117	794	824			5.09	5.35	1.110
Sygma - Carolina	2,899,140	53.7	15,756	685	592			4.86	4.95	1.111
Sygma - Pennsylvania	2,689,120	47.3	14,615	597	884			4.36	4.72	1.152
Sygma - San Antonio	5,479,980	59.7	29,783	924	936			7.43	5.17	1.532
Sygma - Columbus	5,816,070	45.3	31,609	1,006	942			7.09	4.65	1.677

Figure 10E-3. Efficiency factor report for Sygma distribution centers

In addition to use in determination of monthly site rankings, efficiency factors are tracked over time. As reflected in Figure 10-4, the Ohio project manager generates tabular reports that show monthly efficiency factors for each site, over a rolling period of more than a year. Each cell is color-coded to indicate increases and decreases relative to the previous month. Changes in efficiency factor between the previous year and year-to-date are also carried in this table.

The NorthWrite EIS includes a module called *Utility Bill Manager,* and under Sysco's contract all utility invoices

tolerate fluctuations in temperature between -5°F to 10°F for short periods of time without compromising quality. In response, the energy champion implements a daily energy-efficiency strategy. Upon arriving in the morning, he uses a dial-up modem to access DOS-based control programs for ten freezer units and manually raises the setpoints to force the compressors to shut off. The energy champion observes the temperatures and metered power consumption throughout the morning, and reduces the setpoints to their original levels around 11 a.m. The lighter trend in Figure 10-1 reflects this daily strategy; whereas, the darker

SYGMA Operations Tracking Report

Occupancy

FY 09   Period 9

Energy Efficiency Factor (EEF) - through Period   7

	LY thru Period 7	YTD thru Period 7	% Change P/06 P/09	% Change LY - YTD	May-07	Jun-07	Jul-07	Aug-07	Sep-07	Oct-07	Nov-07	Dec-07	Jan-08	Feb-08	Mar-08	Apr-08	May-08	Jun-08	Jul-08	Aug-08	Sep-08	Oct-08	Nov-08	Dec-08	Jan-09
Boston	1.005	1.037	-23.3%	3.2%	0.892			1.044	1.070	0.958	0.927		1.020	1.023				1.042		1.107				0.980	
Carolina	1.246	1.111	-59.6%	-30.8%		1.286	1.299		1.265	1.217	1.237						1.182	1.197				1.110			
Columbus	1.671	1.677	-19.9%	0.4%				1.590	1.587	1.809			1.797			1.574		1.485	1.497						
Dallas / Dallas North	0.940	0.930	-12.0%	-1.1%		0.943	0.953	0.964		0.958	0.910	0.905			0.857	0.873			1.004						
Denver	0.797	0.666	-21.0%	-36.4%	0.760	0.700		0.796	0.785	0.779	0.770		0.834	0.639	0.797	0.750					0.641				
Detroit	1.135	0.927	-36.9%	-18.3%	1.183		1.297		1.292	1.211	0.872			1.076								0.680		
Florida	1.059	1.110	-15.9%	4.8%				1.163	1.084	0.962		0.900	1.038		1.074		1.148	1.169	1.189						
Georgia	0.902	1.043	-9.3%	15.6%				0.933		0.887	0.898	0.836	0.854						1.035	1.071	1.104				
Illinois	1.051	0.835	-5.6%	-20.6%					0.836						0.799	0.766	0.737	0.753	0.764	0.767		0.862	0.897	
Kansas City	0.665	0.774		13.0%		0.672	0.896		0.602					0.674		0.770	0.795							
Northern Cal	0.935	0.758	-35.5%	-20.6%		1.141								0.821	0.851		0.875							
Oklahoma	1.070	0.844	-33.2%	-21.1%	1.021	1.050	1.089		1.092	1.047		1.016	1.090				0.890	0.932				0.795		0.864
Pennsylvania	1.208	1.152	-21.6%	-4.6%				1.207	1.256			1.043						1.275	1.291						1.070
Portland	1.131	0.863	-32.9%	-23.7%	1.130	1.144		1.139	1.150		1.143				0.989			0.964						0.852	0.622
San Antonio	1.601	1.533	-13.8%	-4.3%		1.532			1.613		1.567				1.535		1.554	1.562			1.433			
Southern Cal	0.914	0.732	-	-19.9%				0.898	0.888	0.913	0.952		0.965		0.910		0.833	0.865					0.696	
OVERALL	1.118	1.022	-22.9%	-8.2%	1.105	1.111	1.112	1.122	1.130	1.047	1.056	1.080	1.050	1.030	1.000	0.972	1.027	1.012	1.005	1.045	0.944	0.929	0.908	0.931	0.955

Improvement from previous month

Increased usage by 0.050 or more

Figure 10E-4. Efficiency factor tracking over time

2009 / 2008 BLENDED kWh RATES COMPARISON (P-9)

Meter #	Provider	2009 kWh used	2009 Total $ billed	2009 True kWh cost	2008 kWh used	2008 Total $ billed	2008 True kWh cost
OR2235	PG&E	3,491	$ 237.78	$ 0.0681	3,397	$ 247.77	$ 0.0729
OR2235	Sempra	3,491	$ 431.27	$ 0.1235	3,397	$ 294.84	$ 0.0868
PG&E / Sempra Combined >		3,491	$ 669.05	$ 0.1916	3,397	$ 542.61	$ 0.1597
C38928	PG&E	124,681	$ 4,808.34	$ 0.0386	164,841	$ 7,109.32	$ 0.0431
C38928	Sempra	124,681	$ 15,450.39	$ 0.1239	164,841	$ 14,277.79	$ 0.0866
PG&E / Sempra Combined >		124,681	$ 20,258.73	$ 0.1625	164,841	$ 21,387.11	$ 0.1297
Bldg A Total >		128,172	$ 20,927.78	$ 0.1633	168,238	$ 21,929.72	$ 0.1303
63P886	PG&E	2,830	$ 379.24	$ 0.1340	2,822	$ 362.57	$ 0.1285
6M9291	PG&E	19,760	$ 2,678.15	$ 0.1355	17,040	$ 2,121.14	$ 0.1245
9M4491	PG&E	15,418	$ 1,789.34	$ 0.1161	14,309	$ 1,557.09	$ 0.1088
Bldg B Total >		38,008	$ 4,846.73	$ 0.1275	34,171	$ 4,040.80	$ 0.1183
36P844	PG&E	2,294	$ 310.56	$ 0.1354	2,787	$ 358.58	$ 0.1287
6M9304	PG&E	29,000	$ 3,653.23	$ 0.1260	22,993	$ 5,576.94	$ 0.2425
28P847	PG&E	3,222	$ 739.29	$ 0.2295	10,569	$ 1,376.47	$ 0.1302
Bldg C Total >		34,516	$ 4,703.08	$ 0.1363	36,349	$ 7,311.99	$ 0.2012
Total ALL		200,696	$ 30,477.59	$ 0.1519	238,758	$ 33,282.51	$ 0.1394

PERIOD 9	kWh usage	Invoice Amounts	Blended Rate
2009	200,696	$ 30,477.59	$ 0.1519
2008	238,758	$ 33,282.51	$ 0.1394
$ Increase from last year			$ 0.0125
% Rate increase from last year			8.94%
Variance due to rate increase from prior year			$ 2,500.87

Variance in kWh useage from last year   (38,062)

% Change from last year   -15.94%

Figure 10E-5. Analysis of utility billing data based on exported meter data and utility invoices

line reflects a day in which the energy champion was on vacation and the strategy was not implemented. In spite of a four-degree temperature increase, the data for these days show approximately 35% reduction in load when the energy manager was present to execute the strategy.

In addition to site-specific uses, the NorthWrite EIS was an integral component in the identification and pursuit of the low- and no-cost measures that resulted in 18% energy savings across Sysco's portfolio. To begin the three-year program, Cascade Energy Engineering and expert refrigeration engineers and technicians conducted three-day site visits to over 100 distribution warehouse centers. They used the NorthWrite EIS information to support retro-commissioning and tune-up activities, and to support the implementation of low-cost measures. For example, occupancy sensors were installed at the Stockton site for bathroom and break room lighting, and locked-out digital thermostats were placed in conditioned staff areas. The "Projects and Tasks" tool within the EIS was used to track these measures for savings, cost, and persistence, and to provide administrative task checklists.

Across the portfolio, Cascade Energy Engineering reports that approximately one-third to one-half of energy champions engage with the EIS data on a daily or weekly basis, typically making use of the meter monitoring view, as at the Stockton site. Also similar to the Stockton case, a typical use of the EIS is to verify that consumption dips during off or sleep-mode hours of operation. In addition, most Sysco sites have a refrigeration operator who uses whole-building trends to optimize setpoints and number of active compressors if site performance should slip.

EIS Data Quality

Data quality is managed by Cascade Energy Engineering, who report that in general NorthWrite's pulse acquisition and cellular relay hardware is quite reliable. Further, the Sysco sites monitor at the utility meters only, removing the quality issues commonly encountered in submetered installations. When data feeds do drop out, Cascade Energy Engineering receives the alarm notification, and notifies the specific site to service the acquisition devices. There are also occasional glitches in cellular transmission of the data; however, these are perceived more as annoyances than critical problems, particularly given that cellular solutions are quick and relatively straightforward to install.

EIS Usability and Enhancement

As might be expected, given his emphasis on daily energy efficiency in refrigerated spaces, the Stockton energy champion reports that the most useful feature in the EIS is the ability to monitor meter trends and changes in total electric use, as illustrated in Figure 10E-1. In his experience, this is the only analytical support he requires to maintain good performance at his site. After the initial site tune-ups and low-cost measures were implemented, the EIS software has proven most useful at the site for motivation, awareness, and accounting and verification of persistence in savings. In this sense, the Stockton energy champion has also found monthly comparison rankings and corporate accountability meetings especially valuable. The identification of an energy champion, provision of the EIS software to prevent backsliding, and accountability for performance have resulted in a corporate culture of energy awareness and competitiveness relative to energy efficiency.

In contrast to other commercial implementations of EIS, the Stockton Sygma case revealed limited exploration most of the analytical features offered. While the energy champion does not feel that he could not manage energy performance as successfully *without* the NorthWrite technology, and while he uses it to implement a powerful efficiency strategy, he had difficulty navigating beyond the meter monitoring view that he accesses as a default. For example, it was a challenge to locate performance indicators such as total kWh last year, an entire month of time series, or the annual peak for the most critical refrigeration meter. Similarly, analyses beyond those automatically included in monthly reports are largely unused and in some cases misinterpreted.

Throughout the enterprise, the energy champions have identified a set of recommended improvements to the NorthWrite EIS, as currently configured for the Sysco portfolio. Utility billing graphs will be modified to allow extrapolated data points to be displayed for projected energy use and costs based on month-to-date data. In addition, a real-time metering graph will be added to show this week versus last year, with an option to display today versus last year. A monthly report addressing underperformers has also been requested. This report is to include the two least efficient sites in each benchmark group; the ten sites portfolio-wide that have improved the least, relative to pre-program baselines; and sites that have backslid more than four percent relative to the prior fiscal year. Backsliding sites may ultimately have the option to undergo recommissioning with Cascade Energy Engineering.

General EIS Perspectives

Sysco has achieved significant energy savings by coupling corporate goals and accountability methods with Cascade Energy Engineering's expertise and the performance-tracking capabilities of the NorthWrite technology. In addition to the Sysco program, Cascade has recently initiated a program with Super Value centers, reaching 9% energy savings in the first nine months. Cascade Energy Engineering therefore views this as a compelling model that promises widespread traction for enterprise energy-saving initiatives, and it has begun to seek opportunities for formal inclusion in utility programs. They are optimistic that careful application of engineering expertise and energy services, combined with software-based tracking and performance documentation within a context of corporate promotion of efficiency goals, will prove a reliable pathway to secure, low-cost, deep energy savings.

Footnotes

[1] http://www.ghgprotocol.org/
[2] http://www.powerlogic.com/literature/3000HO0603R1108_IONDemand.pdf
[3] http://www.noveda.com/en/page/105?l1=3&l2=5&l3=0
[4] www.**itron**.com/asset.asp?region=sam&lang=en&path=produ cts/specsheets/itr_008021.pdf
[5] http://www.enernoc.com/pdf/brochures/enernoc-mbcx-brochure.pdf
[6] http://www.powerlogic.com/literature/3000HO0603R1108_IONDemand.pdf
[7] http://www.ziphany.com/Files/drp-utilities.pdf
[8] http://noveda.com/en/page/130?l1=5&l2=0
[9] http://www.enernoc.com/customers/case-studies.php
[10] http://www.intdatsys.com/pdfs/EnergyWitness-Hospital_Case_Study.pdf
[11] http://www.pulseenergy.com/category/case-studies
[12] UC Merced, http://administration.ucmerced.edu/environmental-sustainability/energy [13] Image from Energy Expert: A Technical Basis, available from http://www.myworksite.com/energyworksiteMBS/htmlArea/files/documents/244_eetechdesc.pdf.
[14] Energy Expert: A Technical Basis, p.5
[15] Image from: http://en.wikipedia.org/wiki/File:Artificial_neural_network.svg
[16] UC Merced, http://administration.ucmerced.edu/environmental-sustainability/energy
[17] Ibid.
[18] http://www.pge.com/mybusiness/edusafety/training/pec/toolbox/tll/software.shtml

Section IV

Data Analysis
and
Analytical Tools

Chapter 11

Data Quality Issues and Solutions for Enterprise Energy Management Applications

Greg Thompson, Jeff Yeo, and Terrence Tobin

ABSTRACT

Web-based enterprise energy management ("EEM") systems are delivering the information and control capabilities businesses need to effectively lower energy costs and increase productivity by avoiding power-related disruptions. However, the quality of energy decisions is directly affected by the quality of the data they are based on. Just as with CRM, ERP and other business intelligence systems, EEM systems have data quality issues, issues that can seriously limit the return on investment made in energy management initiatives.

Data quality problems result from a number of conditions, including the reliability and accuracy of the input method or device, the robustness of the data collection and communication topology, and the challenges with integrating large amounts of energy-related data of different types from different sources. This chapter describes how dedicated data quality tools now available for EEM applications can be used to help ensure that the intelligence on which an enterprise is basing its important energy decisions is as sound, accurate, and timely as possible.

THE IMPORTANCE OF MANAGING ENERGY WITH RELIABLE DATA

Under growing competitive pressures, and spurred by recently introduced energy policies and mandates, businesses are becoming increasingly aware of the need for, and advantages of, proactively managing the energy they consume. Industrial plants, commercial facilities, universities and government institutions are looking for ways to lower energy costs. For some operations, the quality and reliability of power is also critical, as it can negatively affect sensitive computer or automation equipment, product quality or research results, provision of 24/7 service, and ultimately revenues.

In response, facility management and engineering groups have been tasked with finding the latest available technology capable of delivering and managing the energy commodities that have such a significant affect on their bottom line. A number of options have emerged in recent years, the most comprehensive of which are enterprise energy management ("EEM") systems.

An EEM system typically comprises a network of web-enabled software and intelligent metering and control devices, as well as other inputs (Figure 11-1). A system can track all forms of utilities consumed or generated, including electricity and gas, as well as water, compressed air and steam. Data can be gathered from the utility billing meters or other meters positioned at each service entrance, from tenant or departmental sub-meters, and from instruments that are monitoring the conditions of equipment such as generators, transformers, breakers, and power quality mitigation equipment. Other inputs can include weather information, real-time pricing information, occupancy rates, emissions data, consumption and condition data from building automation systems, production data from enterprise resource planning (ERP) systems, and other energy-related data.

Based on these diverse inputs, EEM systems deliver analysis and reporting for energy, power quality and reliability information. Armed with this intelligence, managers can verify utility billing, sub-bill tenants, aggregate energy use for multiple locations, compare the affect of utility rate choices by running "what if" scenarios, procure energy or manage loads and generators in real-time based on pricing inputs, identify power conditions that could potentially cause downtime, and perform a variety of other tasks that

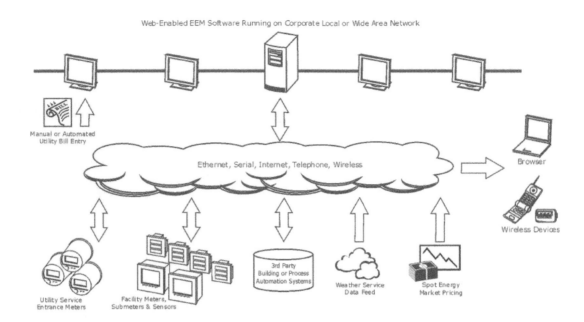

Figure 11-1. Enterprise energy management ("EEM") system inputs

can improve energy performance.

Though EEM technology effectively gathers, stores, processes and delivers customized information from key points across an enterprise to the people that need it, the important economic and operational decisions based on that information can be negatively impacted by poor data quality. EEM systems, like other business intelligence systems such as customer relationship management (CRM) and enterprise resource management (ERP), are susceptible to the effects of poor data quality; for any system it is only a matter of degree.

THE IMPACT OF POOR DATA QUALITY ON ENERGY DECISIONS

A recent global data management survey by Pricewaterhouse Coopers of 600 companies across the United States, Australia and Britain showed 75% reporting significant problems as a result of data quality issues, with 33% of those saying the problems resulted in delays in getting new business intelligence systems running, or in having to scrap them altogether. Industry analysts at the Gartner Group estimate that over 50% of business intelligence projects will suffer limited acceptance, if not outright failure, due to lack of attention to data quality issues.

Data are the foundation of strategy and success, and sound business decisions depend on their quality. But data can only truly be an asset when they are of high quality; otherwise, they're a liability. As the old saying goes, "garbage in, garbage out."

For example, repeated data cleansing for CRM systems is common to ensure customer and sales prospect name and address information is up-to-date and accurate. Professionals depending on that data are aware of the many pitfalls of poor data quality, as the impacts can be serious if parts of information are missing, invalid or inconsistent. The impact can be worse if the data needed are late, inaccurate, or irrelevant to the decisions being made.

For these systems, the costs of poor quality data can be high, in terms of bad decisions, lost opportunities or lost business, damaged relations with partners, suppliers, or customers due to overcharging or underpayment, or even regulatory noncompliance due to faulty indicators. Managers need to have confidence in the reports they are using. Instead of wasting time wondering if an anomaly is the result of a problem with the supporting data, they should be identifying what caused the anomaly.

In the context of energy management, businesses using EEM systems will also have data quality problems, but many might not realize they do (Figure 11-2). The costs of low quality information can mean an inability to take advantage of better real-time pricing, not identifying energy waste, missing a large discrepancy on a utility bill, incorrect sub-billing, incurring an expensive utility demand or power factor penalty, or being issued a fine by a regulating authority for

Overpayment, under-billing, or wasted energy due to inability to validate consumption

Disagreements over inaccurate or unreliable reports

Tenants angry about over-billing due to data inaccuracies

Incurring penalties or fines due to unreliable tracking of real-time conditions

Figure 11-2. Impacts of poor energy data quality

exceeding an emissions standard.

Some specific examples of problems caused by poor data quality in EEM systems, in order of increasing impact:

1. **Misleading energy trend reports**: Collected energy data are used to generate a monthly report showing consumption trends for all utilities – if the data are inaccurate or incomplete, you may fail to identify a serious trend toward over-burdening your system, or miss an opportunity to save energy by rescheduling a process or shutting off a load.

2. **Inaccurate billing reports:** Data quality can affect the accuracy of revenue-related applications. Data from "shadow meters" (installed in parallel with the utility's billing meters) are used to generate a "shadow bill" to help verify if the bill you receive

from your utility each month is accurate. Poor data quality from the shadow meters can mean potentially missing a large billing error, or falsely accusing your energy provider of making one. If you use sub-meters to bill your tenants for the energy they consume, poor data quality can cause you to under bill or over bill your tenants, either of which can cause problems.

In the two scenarios above, if your utility billing tariff includes a real-time pricing (RTP) component, you will need to integrate an RTP data feed from the *independent system operator* that is responsible for setting energy pricing for a given region. If there are missing data in that feed, your shadow bills will be inaccurate and so will your tenant's bills. The potential for problems is compounded with each additional input from different energy metering systems (electricity, gas, steam, etc.), building automation systems, and ERP systems.

3. **Reduced confidence in critical business decisions:** Effective long-term energy management can include dynamic procurement strategies and contract negotiation. These require comprehensive modeling and projections of energy requirements based on a depth and breadth of information that includes all of the data described in the preceding examples as well as other data inputs such as weather, occupancy, etc. With this increased complexity and greater potential impact of the resulting decisions on your profit margins, data quality can seriously affect confidence.

All of these examples in turn represent the overall effects on the return on investment you achieve from your EEM system and the energy management program it supports.

DATA QUALITY PROBLEMS AND WHERE THEY COME FROM

To address data quality problems it is first important to understand what is meant by data quality. As each category of business intelligence system has its own data types, and in turn its own data quality criteria, this discussion will be restricted to the quality of data in EEM systems. However, data quality concepts, in general, are applicable across all business systems.

Data quality can be considered in terms of three main categories of criteria:

Validity. Not only does each data location need to contain the information meant to be there, but data also need to be scrutinized in terms of whether they are reasonable compared to established patterns. The data must be within the allowable range expected for that parameter. For example, if a monthly total energy value is being viewed for a facility, there will be a maximum to minimum range that one would expect the usage to fall within, even under the most extreme conditions. If a value is "out of bounds," it probably indicates unreliable data.

Accuracy. The data gathered and stored need to be of high enough accuracy to base effective decisions on. This not only requires that metering

equipment be rated adequately for its accuracy, but that every internal and external input to the system is considered in terms of its accuracy, including third-party data feeds. For enterprise-wide systems, it is also important to accurately record the time at which each measurement is taken. When an aggregate load profile is being developed for multiple facilities across geographically dispersed locations, the measurements need to be tightly time-aligned. This is also true for sequence-of-events data being used to trace the propagation of a power disturbance.

Completeness. For any business intelligence system, incomplete data can seriously compromise the precision of trends and projections. There must be a complete data set; each recorded channel of information must contain all the records and fields of data necessary for the business needs of that information. For example, if interval energy data are being read from a tenant submeter, there can be no empty records. Such gaps might be mistakenly interpreted as zero usage, and in turn the tenant could be under-billed for energy that month.

In EEM systems, the above types of data quality issues can come from a number of sources, and for a number of reasons (Figure 11-3). For example, data that are out of range might be the result of an energy meter being improperly configured when it was installed, or a meter that has been improperly wired to the circuit it is measuring. There may also be inconsistencies between how a number of meters on similar circuits are configured, or differences between how the meters and the head-end software are set up.

Another source might be the "rollover" characteristic of registers inside most energy meters. Most energy meters have a specific maximum energy value they can reach, for example 999,999,999 kilowatt-hours. The registers will then rollover and start incrementing again from a count of zero (000,000,000). A system reading the information from the meter may not recognize this behavior and instead interpret values as being in error, or worse, interpret it as a negative value which produces large errors in subsequent calculations.

When there are gaps in data records, the source might be a loss of communications with a remote meter or other device or system due to electrical interference, cable integrity, a power outage, equipment damage or other reasons. Some communication methods are inherently less reliable than others; for example, a dial-up

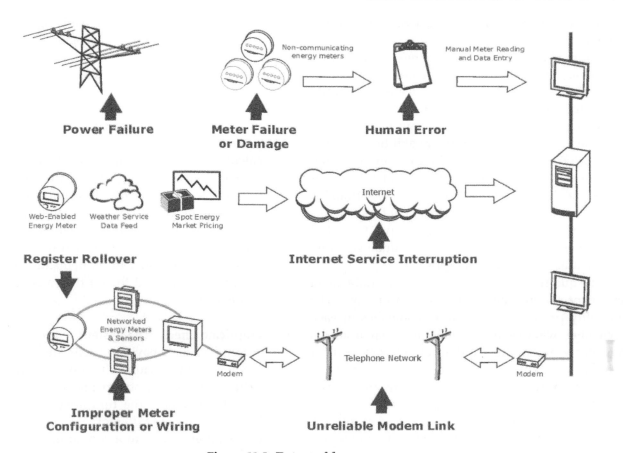

Figure 11-3. Data problem sources

modem connection over a public telephone network will likely be less reliable than a permanently hard-wired Ethernet connection. Some meters offer onboard data logging that allows saved data to be uploaded after a connection has been restored, reducing the possibility of gaps. But an extended communication loss can still cause problems.

Other breaks in communication can include the interruption of an Internet connection over which weather or utility rate information is being imported, or the failure of the network feeding information from a third-party building or process automation system. As additional, diverse sources of real-time and historical information are integrated into an EEM system, the possibility of communications problems increases.

A remote meter, sensor, or other instrument may operate incorrectly or fail altogether, the latter condition causing a continuous interruption in data flow until the device is repaired or replaced. Finally, in cases where some remote meters are not permanently connected by a communications link, their data might be collected manually with a dedicated meter reading device or laptop computer, and then manually entered into the head-end system. Anytime this kind of human

intervention is required there is room for error.

Ultimately, users judge the reliability of a system by the delivery point of information; they do not know or care where the data originated. If incorrect information is being displayed they simply consider the entire system to be at fault. Thus, the success of the EEM system as a whole is very dependent upon the quality of its data.

NEW DATA QUALITY TOOLS AND HOW THEY HELP

Identifying and correcting data quality problems takes the right tools, as well as the people to use them and a regular, repeatable process to ensure maximum effectiveness.

Electric and gas utilities have traditionally used tools to compensate for data quality problems in their metering and billing systems. The need for these tools was driven by the fact that the revenue billing meter and its data represent a utility's "cash register." Without accurate, reliable and complete information an energy supplier cannot be sure it is being properly compensated for the energy product delivered to the customer.

Utilities use what is commonly referred to as *validation, editing, and estimation tools*, collectively known as *VEE*. Until recently, these tools were not readily available for use by commercial and industrial businesses doing their own internal energy metering and management. New data quality capabilities are now emerging, specifically customized for energy consumer applications and expanding the concept of VEE to encompass *total data quality*. Though they borrow from the capabilities of utility VEE, new data quality tools take into account the wide range of input types that EEM systems leverage to develop a complete understanding of energy usage across an enterprise.

Data quality tools help ensure that data meet specific expectations and are fit for each purpose. The data quality process first identifies anomalies in each data channel, whether that channel is a remote meter, a third party automation system or Internet feed, or data input manually. Specific tools are then provided to correct for errors or gaps so that data are *cleansed* before they are analyzed, reported or otherwise put into action. The entire process is achieved through a sequence of validation, estimation and editing steps.

Validation

To validate that data are of high quality, a set of internal *standards* is created that defines the level of data quality required for each purpose. For example, a property manager may decide that data for sub-billing is acceptable with lower quality than the data used for utility bill verification.

Based on the quality standards, a set of rules is constructed that the data quality tools use to automatically check the quality of energy-related data coming into the EEM system (Figure 11-4). Examples of these rules include the following:

- **Constraints checks**. As mentioned, incoming data representing a particular parameter must meet a *bounds check* to see if they fall within reasonable values, such as between an acceptable minimum to maximum range. Those that fall outside predefined constraints are flagged as "out of bounds." In the case of meter energy register rollovers, a *delta check* can be done to see if the previous and currently read values differ by too much to be reasonable. Similarly, checks can be run to verify data correspond to an established set of allowable values. For example, a record indicating alarm status should either show

an active or inactive state, possibly represented as a 1 or a 0. No other value is acceptable.

Note: Though not an error check, tests will also be run on some data to find which sub-range a measurement falls in within the acceptable boundaries of values. This is needed when the value of a measurement determines how it is used in subsequent calculations. For example, some utilities charge for energy by applying different tariff charges to different levels of energy demand measured over specific demand intervals (usually 15 minutes), or to energy consumed at different times of the day, week or year. If a recorded demand level is greater than or equal to "x demand," a different tariff may be applied than if the value is less than "x demand."

- **Duplicate check**. The system will check for consecutive records having exactly the same data in them. This can include situations where both the value and the timestamp are the same for both records, or where the timestamps are the same but the values are different. Normally this will indicate an error due to a communication problem, improper system settings or metering logging configuration, time jitter, or other issue.

- **Completeness (gap) check**. When verifying interval energy data, where records are expected at specific time intervals (e.g. every 15 minutes), gaps in data are flagged. These can be due to message transfer issues, power outages, communication issues, etc. Missing records can then be compensated for in a number of ways, as described in the following section.

- **Dead source detection.** If a gap in data is long enough the system can flag it as a dead source so that appropriate steps can be taken to investigate the cause.

- **Zero, null and negative detection.** If an energy meter is showing "zero" energy usage for a particular facility, load or other metered location, when that condition is not expected, it will be flagged as a possible error. This can include either a zero reading for an individual interval, or a delta (difference) check between two consecutive readings of a totalizing register. It can then be investigated to see if there may have been a major power out-

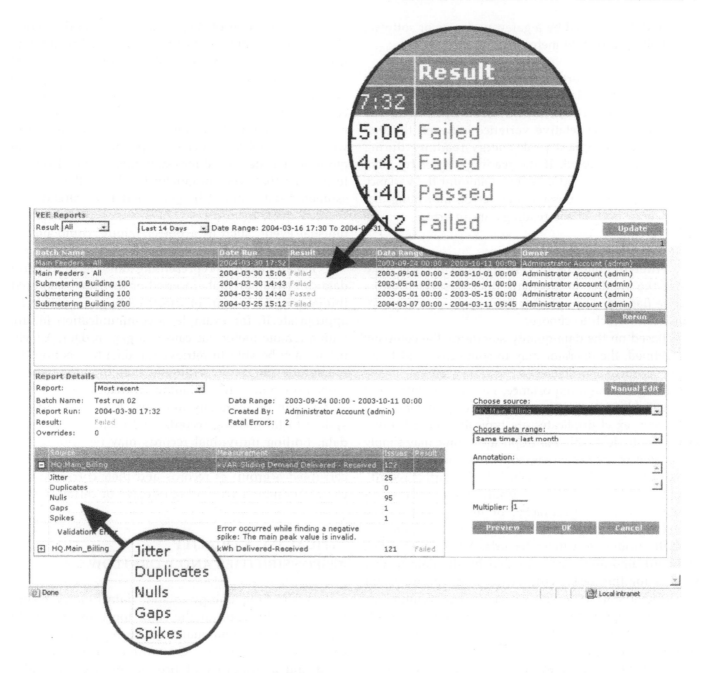

Figure 11-4. Data quality validation screen. Highlights show a series of records for passed and failed tests, and a total count for each error type found when the selected test was run.

age event. If so, the error indication can be manually *overridden*. Null readings (e.g. no value) can also indicate a problem. Finally, negative checks are done to ensure consecutive readings in a cumulative register are not decrementing instead of incrementing. This can catch conditions such as meter resets, register rollovers, and other issues.

• **Timestamp check.** As mentioned earlier, it is important that measured values being aggregated together are accurately time-aligned. The data quality tools will verify if all timestamps for all values being summed are within an acceptable proximity of each other, often referred to as *time jitter*. Excessive jitter can sometimes be the result of delays caused by a gateway device or software polling a remote meter.

- **Other tests.** Further tests can be applied to help determine if data input is reasonable. For example, a *spike check* (commonly done by utilities) will compare the relative variance between the first and third highest peak energy readings during a specific period. If the readings differ by more than a predefined acceptable amount the variance is flagged as a possible data error, indicating that one or both of the readings are in error.

Estimation & Editing

After the data quality system has been used to verify the validity of incoming data, suspected errors will be flagged. The operator then has a variety of options from which to choose.

Based on the data quality standards the company has defined, the operator may in some cases opt to ignore a particular problem with a data element, if it is not of high enough importance. For critical data, tools are available to correct or compensate for errors.

First, exact duplicates are typically deleted. Rules can be set to deal with *near* duplicates; some may simply be deleted in the same way, some may need to be analyzed further to determine which is the correct record.

Second, automated estimation tools allow erroneous data to be replaced, or missing data to be completed, by "best guess" calculated values that essentially bridge over those records. A variety of preset standard algorithms are provided by the data quality system for this task, with each being optimized for the specific data type and situation. For example, an estimation algorithm for kilowatt-hour measurements will be different than the treatment for humidity or real-time pricing data.

The data quality system will make recommendations as to how the data should be corrected, and may incorporate exogenous factors, such as weather, to make those recommendations more feasible.

One of the most common and simple examples of estimation is straight-line *averaging*. In this case, a bad data point for a particular energy interval reading is replaced with a value representing the straight-line average of the data point values on either side of it. This kind of point-to-point linear interpolation can be applied to multiple contiguous data points that are either missing or otherwise in error. Rules can be set defining the maximum time span allowable for interpolation to be applied.

If a time span of suspect data exceeds the allowable duration, estimation can be performed using data from other similar days. Typically, a number of selected reference days are chosen and their data averaged to produce the replacement data. Reference days need to closely represent the day whose data are being estimated, for example by being the same day of the week, weekend, or holiday as close as possible to the day in question. The data used for estimating would also need to be data that were originally valid; in other words, estimated data cannot be generated from already estimated data. In addition, days that experienced an unusual event, such as a power failure, could not be used for this purpose.

Finally, records representing missing or corrupt data can have new data inserted or their data replaced through manual input or direct editing. This may be appropriate if, for example, a communication failure with a remote meter has caused a gap in data. A technician may be able to retrieve the data by visiting the remote site and downloading the data from the meter's on-board memory into a laptop or meter reader, which in turn can be manually imported into the head-end system to fill missing records, or to replace estimated data. Editing individual records may be appropriate where rules allow, such as when a known event has corrupted a group of records and their correct values can be presumed with a high degree of certainty.

TYPICAL DATA QUALITY RESPONSIBILITIES AND WORKFLOW

Addressing energy-related data quality issues takes a combination of the right tools and the right process. Using a data quality software application can solve specific problems. Commitment from management and availability of proper resources are also needed to ensure that data quality assurance is an ongoing process.

Data quality needs a champion to drive the program and one or more data stewards to execute the necessary steps. Given the importance of energy as a key commodity for an organization, and in turn its impact on profits, the champion can be anyone from an executive through middle management level, including corporate energy managers, operations or facilities managers, or engineering managers.

In terms of day-to-day execution, data cleansing tasks are typically assigned to one or more people within a facility management group, someone with a

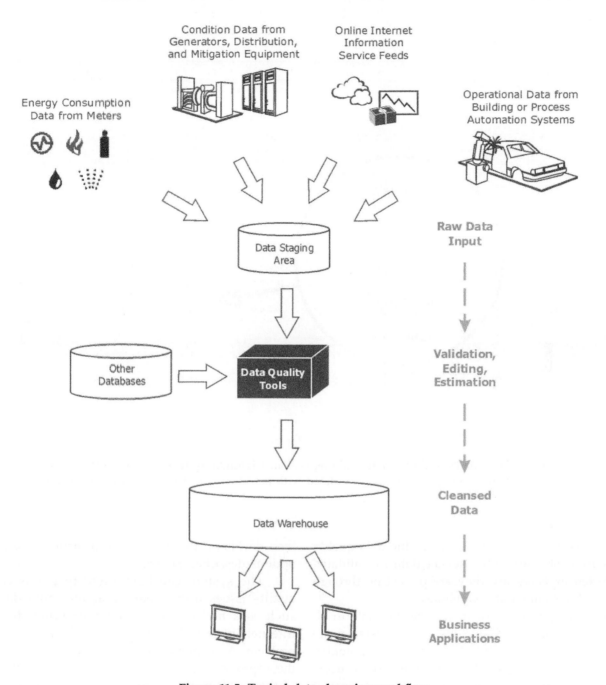

Figure 11-5. Typical data cleansing workflow

title such as *data administrator*, or clerical staff specially trained in the data quality tools and rules. Often, it makes more sense for a business to concentrate on their core competencies and outsource the data quality function to an energy management services company.

In general it is always best to fix data problems up front rather than later. That is why the data cleansing process should be positioned at the point where collected data first enter the enterprise energy management system, before they makes it through to where data are ultimately stored in a data warehouse on a data server (Figure 11-5). If this is not done, data problems within the data warehouse start to affect critical calculations and decisions, and can propagate further problems before they are isolated and corrected.

To be most effective, an EEM system is configured to include a front-end data staging area. Data in this area have already been broken out from combined data packets from remote devices or other data steams and translated into the proper units as necessary. The stag-

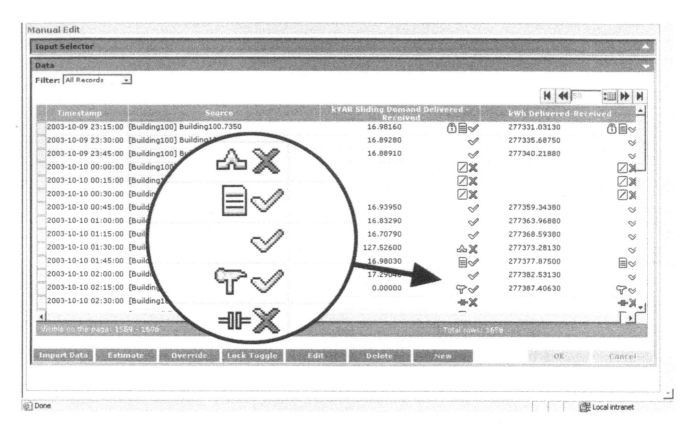

Figure 11-6. Typical data quality estimation and editing screen. Highlighting (from top to bottom) test result indicators for: spike test failure, edited and passed, validated with no errors, failed test override, and gap test failure.

ing area acts as the *raw data* input to the data quality process. In most cases, only after the data are validated or corrected as necessary they are passed on through to the EEM system's data warehouse.

In some special cases, it may be desirable to allow data entering the staging area to be passed on to the data mart without validation, despite the potential for data errors or gaps. This will allow some users that require near "real-time" data to benefit, even if there is an issue with a few readings. In this case, the data quality process can be run on the data at the next scheduled time. If data issues are identified at that point, they will be addressed, and the cleansed data will then propagate to the data mart.

Beyond the real-time inputs to the EEM system, the data quality tools can also be used to validate and cleanse data in previously stored databases before it is integrated within the EEM system database.

During the data quality validation process data problems will be highlighted visually on screen and, if desired, through alarm annunciations. The user can

then decide on the best course of action based on the options described above.

The system can help identify persistent data quality issues from a particular incoming data feed, such as a faulty remote meter or other device, an Internet interruption, or a communications network problem. A maintenance protocol can be set up to flag the appropriate technical staff to investigate the source of the problem. If a meter or other data source exhibits an intermittent problem, a decision can be made on whether to repair or replace by comparing that cost to the ongoing man-hours and cost of repeated error correction using data quality tools. The data quality system may also uncover recurring problems with a particular data entry method or other process.

How often the data quality tools need to be used to cleanse data depends on a number of system conditions and the workflow preferences of the user:

• **System size.** The greater the number of data sources (e.g. number of metering points), the higher the probability of data problems, and the harder it is to identify and correct problems. Data problems are often compounded due to sheer size of the data. The user may wish to run the data quality process more often (daily instead of weekly) to keep on top of the workload.

- **Real-time data requirements.** Monthly data cleansing, well before the billing date, may be sufficient for tenant billing. For real-time applications such as load management, energy procurement, etc., data may be required more frequently and thus the data will need to be cleansed often to deliver up-to-date reliable data.

- **System topology.** Some types of communications (e.g. modem links), may be inherently less reliable than others (e.g. hardwired). The geographical breadth of a system may also affect the reliability of data collected from remote points. Both of these issues may create higher frequencies of data problems that need to be addressed more often.

- **Variety of data types.** The number of distinctly different sources of data (electric meter, gas meters, weather feed, RTP feed, etc.) will add to the complexity of the EEM system and, in turn, influence the expected rate of data errors or gaps.

- **Efficiency.** As mentioned above, the sooner a problem is discovered the easier it is to fix; therefore, the data quality process should be run more often rather than less.

CONSIDERATIONS WHEN CHOOSING A DATA QUALITY SOLUTION

Data quality tools can be effective in addressing data quality problems, but only if they are well designed. If a business intends to have in-house responsibility for the data cleansing process, a number of criteria should be considered when choosing a solution:

- **Flexible and modular.** The data quality system should be flexible enough to align with utility standards to support bill verification and energy procurement. It should also be able to adapt to

evolving internal business standards. A modular architecture is an additional advantage, as it allows for sub-components or features to be added or engaged as required for testing of different kinds of problems.

- **Applicable to** *all* EEM data sources. Data quality tools should be designed for a "whole system approach," available to cleanse not only metered electrical energy data, but also data representing other energy sources like gas, steam, etc. Further, they should be able to validate external data feeds such as weather, real-time pricing (RTP) rate forecasts from the ISO/RTO, etc.

- **Notification system.** The system should allow data administrators to subscribe to desired information and be notified when necessary, for example when a dead source is detected, or when a data quality report has been run and shows data issues. Notification methods should include email, pager, and other convenient options.

- **Report generation.** Reporting tools should provide scheduled or on-demand reporting, listing details on data quality problems and summary "roll-ups" of data quality performance metrics. The data quality reports and, ideally, the bills generated by the EEM system should both reflect valid data statistics.

- **Audit trails and raw data backup.** A complete audit trail should be provided for any data that have been edited. This should indicate the user that executed the change, what was changed and how. For data that have been changed, the complete raw data set should be retained in a backup file and be accessible in case a particular data cleansing step needs to be reversed.

- **Security.** Different password-protected access levels should be provided. This can include "view only" access for some users, while "administrator" access allows viewing and the ability to make changes to the data.

- **Override capability.** The system should allow an administrator to override an error indication for what first appears as an error but may be valid data (e.g. a meter is showing zero energy usage due to a known power outage.)

- **Ease of use.** The data quality process must be cost effective, so the tools must be efficient and easy to use. A number of features can help in this regard. For example, error indicators on data quality screens should provide quick links to view the supporting data. Ideally, this should also be a feature of bills generated by the EEM system. Data should be clearly marked to differentiate between valid, estimated or corrected data.

CONCLUSIONS

Enterprise energy management systems represent the key to energy-related savings and productivity improvements, but their effectiveness is significantly influenced by the quality of data they deliver. As with all business intelligence systems, the right tools and processes must be in place to avoid data quality issues that could otherwise seriously affect business decisions, tenant relations and return on investment. New data quality tools are available for industrial, commercial and institutional energy consumers help ensure the intelligence delivered by EEM systems is accurate, complete and timely.

Whether businesses choose to dedicate in-house staff to the data quality process or outsource it, the design features of the data quality application are critical. Due to typical EEM system breadth, the variety of networking methods, and the number and types of data sources, a comprehensive set of data quality tools is needed to identify and compensate for all potential data quality problems. The data quality solution chosen should provide the flexibility and modularity needed to adapt to evolving business rules and needs. It should also be applicable to all EEM data sources beyond energy metering, including external feeds such as weather and real-time energy pricing. Finally, to be cost-effective, data quality tools must be easy and efficient to use.

Chapter 12

Using Standard Benchmarks in an Energy Information System

Gerald R. Mimno
Jason Toy

ABSTRACT

Once interval data from an electric meter are captured, the question soon arises "What does my load profile data tell me?" There are some common benchmarks that can answer this question. These are: 1) The Energy Star building rating system. 2) Utility industry class load profiles and 3) Weather normalization based on regression analysis which reveals the baseline energy consumption of a facility Benchmarks give an owner or manager a frame of reference on how efficient a facility is; where, when, and how energy is being wasted; and what action items might reduce the monthly energy bill.

INTRODUCTION

For the last several years we have been engineering, manufacturing, and installing Internet based Energy Information Systems (EISs) for very typical high schools, factories, hospitals, retirement homes, quarries, municipal offices, recreation facilities, and many others. The EIS provides real time data every 15 minutes on a web browser. The owner or manager can select a wide variety of reports showing daily, weekly, or monthly load profiles. The cost of equipment to get an on-line signal is about $1,000, roughly half for metering and half for the wireless equipment to get the signal on the Internet. The equipment is often installed and managed by Energy Service Companies (ESCOs) who charge $75 to $90 month to provide a live interval data service.

Alternatively, many utilities offer a monthly load profile service. They pick data up from smart commercial electric meters such as the GE kV2 which has digital output and can store and download 15 minute data during the monthly read into a hand-held device. Data are available a few days later on the Internet for $30 per month. If you want to connect your phone to the meter, the utility will supply a meter modem for a few hundred dollars. The utility will call your meter at midnight and provide the previous day's interval data on the web for about $50 a month. In many systems, you can initiate a call and get a reading in about half an hour. We recommend making the effort to get "near real time" data on the web continuously. If your meter is less than ten years old, your utility can likely install a plug-in card for Internet service costing about $500 or your utility might provide a new meter with an Ethernet jack for about $1200.

Figure 12-1 shows the amount of energy used in a small office every 15 minutes over four days. Note that the area under the curve represents Kilowatt Hours (kWh) or the amount of energy used during a day or month. This is represented by the kWh charge on the monthly bill. To reduce the kilowatt hour charge, you need to reduce the area under the curve.

The peak of the curve represents the kilowatt demand charge (kW) or the highest amount of power used in a fifteen minute period. A mark of efficiency is a low night load. A mark of inefficiency is a high peak load. Another consideration is "coincident peak demand." Does your facility draw its peak power at the same time as the electric grid system peaks? There will be advantages and financial incentives to moving your peak off the system peak.

Residential customers do not pay demand charges, but virtually all commercial customers do. Demand represents the peak capacity of the electric system. The utility says, "You need to pay a demand charge because I had to build the system large enough to meet your peak need." Some utilities use a ratchet system. Even if you only reach a peak for 15 minutes, you pay the demand charge every month for a year. Other utilities charge only for the demand you reached in the highest 15 minutes of the month. In New England for example,

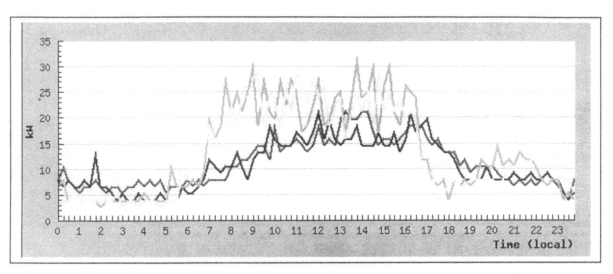

Figure 12-1. Load profiles from an Energy Information System or Utility Load Profile Service

a typical kilowatt hour costs $0.12. A typical kilowatt demand costs $12.00 per month. While it is not listed on the bill, a suburban house will have a peak of about 3 kW. In many utilities, small commercial customers have a demand less than 200 kW; large customers are over 200 kW. Very large users may have a demand of 1,000 kW or one megawatt. Generally about a third of the commercial electric bill is in the demand charge.

Many factories want to see their load profile so they can tell how to reduce energy cost. One method to accomplish this is to create a single facility system including meter pulses, Ethernet connection, computer, and software for about $3,500. The Independent System Operator (ISO) in New England offers such a package to participants in its summer load management program. Another approach is to collect data over cellular or traditional phones and send the data through the phone network to an off-site energy management network and then to the facility's web browser. The public telephone company charge to convey interval data can be quite costly depending on the type of service contract. Monitoring one meter at 15-minute intervals represents 2880 calls per month. We recommend you connect your meters to the Internet to eliminate recurring interval charges. The Automatic Meter Reading (AMR) industry offers many suitable wireless and power line carrier technologies you can use to connect to the Internet.

Your monthly utility bill includes your meter reading and energy use but this information has limited value in managing your operations. ESCOs can find ways to save five, ten, fifteen percent or more of the energy used in any facility, but facility managers seldom do so on their own. To date, the market pen-

etration for commercial interval data is less than one percent, but the Internet has recently provided the means to make obtaining interval data practical and we expect the use of interval data will expand

Why don't more facility managers find value in interval data? What they need is a frame of reference. It is hard to look at a load profile and understand what it means. It is like a map without a scale or north arrow. The reaction of many of our customers is, "Well that looks interesting, but what does it mean?" They also question whether they should pay for a load profile service since they can't attribute a value to the data. Utilities offering load profile services at $90, $50, or even $30 per month have had very few customers sign up.

In our experience, there are three stages required to manage facilities better, reduce waste, and conserve energy. These stages and technologies needed to manage energy and reduce waste are shown in Table 12-1.

This chapter discusses how to meet the second requirement. If you have successfully completed the arduous journey placing real time interval data on the web, the next step is to find a frame of reference or **benchmark** for that data. Benchmarks will show how your facility compares to similar properties. They can also help identify where your waste and inefficiency is located, which will help you find opportunities to save money. In the past, this analysis was available on a custom basis or through expert outside analysis. Our interest is to find ways to automate this analysis and make it available at little or no cost beyond the basic monthly cost of an interval data service.

Table 12-1. Three stages and technologies needed to manage energy and reduce waste.

Stage 1
Show me my load.   Automated Meter Reading System

Stage 2
Benchmark me.   Energy Information System

Stage 3
Save me money and
do it for me.   Energy Control System

BENCHMARKS

There are several useful benchmarks you should consider. The Environmental Protection Administration (EPA) has expanded the Energy Star rating program from residential products to commercial buildings and has developed a web based program for ranking the efficiency of facilities. [3] Another benchmark is provided by power companies. For decades, utilities have studied and published "Class Load Profiles." Class load profiles show the typical consumption of residences, small and large commercial buildings, industrial facilities, and even street lighting. Utilities recognize that within a class, many customers operate in a similar way. Therefore the class load profile can be used for forecasting demand, setting tariffs, and settling payment of wholesale power contracts. Comparing your own load profile to the class load can offer a valuable benchmark.

Another reference point is the extensive studies made on the relation between weather and the consumption of electricity. An industry rule of thumb is that weather explains 30 % of the changing demand for power. Without taking account of the weather, it is difficult to compare this month's energy use to last month or last year. Removing the transient effects of weather reveals the baseline energy consumption of a facility.

As one vendor, we have incorporated these benchmarks in our Energy Information software. We use an XML link from our server to the Energy Star database. We have mined the web both for class loads in the public domain and for weather analysis. With a few key strokes, users can bring up a load profile of their own facility, use weather data to normalize the load profile, and then benchmark this data through Energy Star or their own utility's class load profile. We describe the details below. Our purpose is to make an initial pass at analyzing a load profile and to set the stage for a more detailed design of energy management and control systems to mitigate waste, reduce monthly bills, and conserve energy.

Benchmarking Architecture In An EIS

An EIS and benchmarking system has the following parts:

1) Automatic Metering System. If your utility does not provide interval data, have an electrician install a shadow meter downstream from the revenue meter. The shadow meter will use Current Transformers (CTs) and a transducer which outputs Watt Hour pulses or Kilowatt hours. We have had very good experience with shadow metering using the Square D Veris Hawkeye [1] and Ohio Semitronics Inc. WL55 [2]. The Veris is quick to install and costs about $600. The Ohio Semitronics unit is exceptionally accurate and also economical. The CT's and transducer for a 3 phase system total about $275. A benchmarking system needs some additional data inputs. Vendors offer software which contains links to a variety of databases including EPA's Energy Star database [3], libraries of utility class load profiles, and weather data. Live hourly weather data are available from the bigger airports. [4] The Energy Information Administration (EIA) [5] offers tables with 30 years of weather normalization data for every region in the US. You can also get useful temperature data from a building management system (BMS).

2) *Networking*. You should use the Internet to collect data for your benchmarking. A good benchmarking process needs access to websites which have historical data, libraries, and live weather information. Just as benchmarking is a continual process, so is the need to continually update the benchmarks. There are always additional good sites to find and link to your benchmarking process.

You should also use the Internet to distribute the results of your benchmarking to all the relevant parties in your energy management organization. These include facilities managers, budget managers, operations people, power suppliers, and interested users such as teachers and administrators.

We have found that the first and heaviest users of interval data are operations people who manage the performance of retirement homes, hospitals, factories, commercial rental property, and quarries. They use live interval data to monitor their operations and like the convenience of checking on the facility from wherever they may be.

3) *Server and Database.* Interval data from hundreds of points in an operation comes in over the Internet as packets with a header, contents, and checksum. The server sorts the incoming data by such categories as packet type, business unit, ID, date & time stamp, and inserts data in the proper form into the EIS database.

The database is the heart of the benchmarking system. The databases frequently chosen for Energy Information Systems are Oracle, Microsoft SQL Server, and the open source MySQL. The advantages of open source (including the Linux operating system) are that a user group with talented software people can extend the features of the software on their own and make this available to other members of the group. This is particularly helpful for users, such as universities, electric coops, or schools, who share many common problems.

Alarm automation software monitoring the database can trigger alarms and make telephone calls using audio files, send emails, send Short Messaging Service (SMS) messages over cell phones, and chase recipients to their after hours and weekend addresses. Alarm incidents are logged in the database with a record of who responded to or canceled an alarm. The alarm automation program also monitors conditions in the field and

can issue an alert, for example when a meter or device exceeds specified ranges such as the run time or start-stop cycles on a pump.

4) *The Browser.* Data from the data base is presented to users in reports they see on their web browsers. The browser presents a dashboard with a few key functions displayed and access to much more information available by drilling deeper into the database. There are many advantages to browser-based software. Upgrades to the browser software can be made in the server and are then immediately available to all users. Administrative and support functions can also be accessed by any authorized user anywhere. New meters can be enrolled and addresses and telephone numbers entered from the field without returning to the office. Tech support personnel can diagnose and fix many problems remotely without field calls. There are many EIS products on the market. Most present some form of live data on a running basis. This data can then be printed as reports or downloaded into Excel for further processing by the user.

In a benchmarking system, the customer brings up their own facility on a charting window, and then adds additional information from the benchmark features. For example, the benchmarks for School A will typically show School A's load profile compared to a population of other similar schools, the class load profile for an average school, and School A's load profile with variations due to temperature stripped out.

5) *The Control System.* Some vendors incorporate control systems into the EIS while others maintain a control system in the facility and have a separate EIS to reflect the performance of the control system remotely. Historically the facility control industry is a Tower of Babel but progress is being made on common protocols and standardized interfaces. Some of the simplest technologies are already a standard such as pulse outputs or the 4-20 MA industrial control protocol. Anyone desiring to tie an EIS to the control system is going to have to contend with different products that have incompatible outputs. The usual way to cope with this is to attach a relatively smart and flexible digital communications device to a relatively dumb and inflexible control or instrument. The

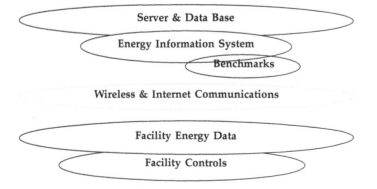

Figure 12-2. Data architecture in an Energy Information System

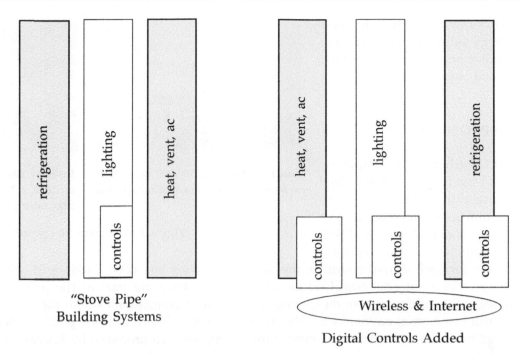

Figure 12-3. Adding digital controls to a facility

digital device mimics the signal the instrument expects and packages the results in a communications packet in TCP/IP which can be sent or received over the Internet.

Benchmarking is not going to tell you how to set or reset your controls. The benchmark information is too general. But a benchmark will tell you where to look and where to make more detailed analyses. Is your inefficiency problem in night load? Is it in peak load? Are you way off from a standard or just a little off? These are the answers you can expect to find in your benchmarks.

ENERGY STAR BENCHMARK

The first benchmark we want to discuss is a *score* from the Energy Star National Energy Performance Rating System. Energy Star is an EPA program long associated with the sticker rating on refrigerators and boilers. Five years ago, Energy Star rolled out a program for benchmarking whole buildings including offices, schools, supermarkets, hospitals, and hotels. Energy Star has benchmarked 15,000 buildings representing 2.5 billion square feet. This is 14% of the eligible market. [6] The system is gradually being extended to more building types and will include higher education, residence halls, warehouses, banks, retail, and additional commercial building types.

You can get a score for your building by filling out a web form at *www.energystar.gov/benchmark.* Energy Star has made detailed analyses of thousands of buildings and the program uses this data to evaluate some relatively simple data you supply about your own facility. You provide figures from the building's utility bills, the building size and type of construction, the year it was built, and some operational data such as hours of operation and number of employees and computers. After comparing your building against the national stock of similar buildings, the Energy Star program returns an efficiency score between 1 and 100. The program takes account of weather, hours of use, occupant density, and plug load. The EPA has found the distribution of buildings along the efficiency scale is pear shaped. The majority of buildings is clustered at low efficiency while about ten percent reach out to the high efficiency "stem" of the pear. The best buildings are 4 times more efficient than the worst.

The Energy Star score is an excellent example of a benchmark. The single number could be described as an "efficiency motivation index." EPA makes recommendations for utilizing the score (see Table 12-2).

The National Energy Performance Rating System is only the first step in an ongoing Energy Star support network. Once a building has been rated, the building manager can create an on-line portfolio to track improvements and performance over the years. EPA also has a Financial Value Calculator to assess alternative energy improvement strategies on financial indicators including payback, market value, and earnings per share.

CLASS LOAD PROFILES

The second benchmark for comparison is Class Load. A class load is a 24 hour load profile for a typical utility customer in a specific category. Utilities conduct load research to determine patterns of utilization and then construct class loads. Class loads represent the pattern of use for each different tariff in the utility's rate structure. The residential class load for large apartment houses will be different from the commercial class load for small commercial customers and the commercial class load for offices. The class load shows how your facility compares to the average customer in your class. The class load does not show the absolute amount of kWh but rather the pattern of use.

Often the class load is described in a table of percentages. The percentage of power used between midnight and 12:15, the percentage between 12:15 and 12:30, etc. Multiplying your total use by the percentages will result in a load profile for your facility. The EIS will then chart your actual use and compare it with the class load rendered to a common scale. We recommend a 15 minute profile giving 96 intervals per day. An hourly profile does not show enough detail and five minutes shows unnecessary detail for most people. While we are talking about electric loads, class loads also apply to gas, steam, fuel oil, and water.

Built in 1969, this 500,000 sq.ft. high school has an Energy Star rating of 8.

Built in 2001, this 190,000 sq.ft. middle school has an Energy Star rating of 32.

Figure 12-4. Energy Star scores

Who has class loads and where can we find them? They are part of the public record at Public Utility Commissions where they are used in rate proceedings. Many utilities publish their load research. They are also produced by Energy Service Companies, power marketers, large users, and in wholesale power settlement. Another good source is the US Government Energy Information Administration. [5] A search of the web is the most practical source. Excel is the standard software used for organizing, downloading, and charting a class load.

Loads change for a number of reasons. Daily activity varies. Weather varies. Processes and uses vary in a facility. However, there are also common patterns among users. These are represented in class loads and are typically divided into residential, small commercial, large commercial, and industrial categories. In addition, class loads are provided for weekdays, weekends, and sometimes holidays. Class loads are accurate enough that a lot of wholesale power is purchased and paid for

Table 12-2. Benchmark scores and strategies.

SCORES BELOW 50: INVESTMENT STRATEGIES	— Buildings in this range need new equipment. — Replacement equipment can be amortized by substantial savings in monthly utility costs. — New operational practices will also have a substantial impact on the bottom line.
SCORES BETWEEN 50 AND 75: ADJUSTMENT STRATEGIES	— Concentrate on simple, low cost measures such as improved operations and maintenance practices. — Upgrade equipment for additional savings.
SCORES BETWEEN 75 AND 100: MAINTAIN BEST PRACTICES	— Buildings with these scores represent best practices of design, operations and maintenance. — Slacking off will lower the building's score.

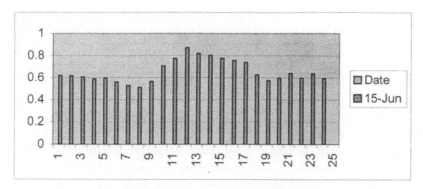

Figure 12-5. Sample Class Load Profile

on the basis of class load. The class load shows the timing of the delivery of power and charges for capacity or peak power.

In examining a facility load and comparing it to a class load, we want to look at a number of items. These are night load, day load, seasonal load, peak summer load, and the spring and summer shoulder months. In comparing your load profile to the class load, look for three things:

Base load percent (night load/day load)
Peak to base ratio (day load/night load)
Coincident peak

The base load is the amount of power always on. It is the area of the rectangle at the base of the load profile. Typically in a single facility your base load will be about 33% of your day load. The area of the base load is also defined as your "load factor." It is the

percent of power you use 24/7. You should compare your facility base load with the class base load.

The peak-to-base ratio is the number of times the height of your day load exceeds the height of your base load. A small facility will use three times more power at the height of the day than the minimum used at night. A large facility will use 30% more at the daily peak than at night.

The third factor to note is the coincident peak. The entire state of California peaks at 3:30 p.m. Your facility will show a peak at some time during the 24 hours of the day. Does your facility peak at the same time as the class load? If so, you have a coincident peak.

Comparing these numbers to the class load will begin to show you if you should look for inefficiencies (i.e., if you are wasting kWh), and if you should try to reduce your peak demand (i.e., if you are wasting kW).

If your facility has a relatively high base load, you will typically find that people in your facility are unnecessarily leaving a lot of lights, fans, computers, and other equipment on. Note that the class load is an aggregation of many facilities and will look relatively high and flat. The smaller your facility, the smaller you would expect your base load to be when compared to the class load. Schools are often found to have a relatively high base load. School utility bills may be paid by the superintendent's office, custodians have little information about cost, teachers have their own priorities, the whole building is left on to accommodate a few night meetings,

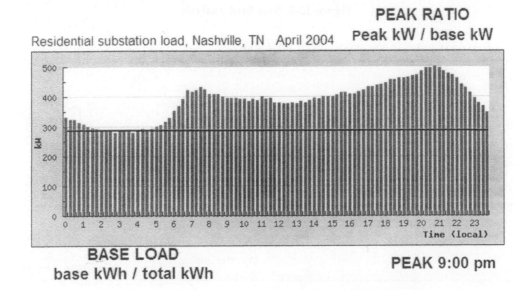

Figure 12-6. Ratios important in analyzing a load profile

and the operating cost of the building drifts up. Even in buildings with an energy management system, controls and time clocks can be overridden or defeated, negating the advantages of the system. A comparison with class load may alert you to this.

The peak to base ratio is a measure of what is turned on during the day. The smaller the facility, the steeper the rise is expected. A comparison with class load will suggest some things to look for. The expected range of variation may be between 130 percent and three times, (day to night load). The class load (an aggregation) should show less variation than your facility. The steeper the facility peak, the more the small facility will be penalized in peak demand charges. A very sharp peak indicates more equipment and HVAC may be turned on at the peak than necessary and that HVAC might be coasted through the peak to reduce peak demand charges.

The coincident demand measure is the degree to which your facility peaks at the same time as the class load. A coincident peak is a sign of expensive power. Increasingly, commercial customers are going to be exposed to the real time price of power. The more power you use on-peak, the higher the price you will pay. On the flip side, if your peak is not coincident, you have the opportunity to use a Time of Use (TOU) tariff to lower your cost. There is a conflict between consumers and generators on peak power. Consumers want cheap power. Generators want a flat load. The mediator between these two is in many parts of the US the Independent System Operator, (ISO). The ISO takes bids from generators usually in a day ahead and hour ahead market and builds a load stack taking the cheapest generator first and then adding higher bids as the demand increases. A few times a year we can expect a Critical Peak Pricing Incident in which

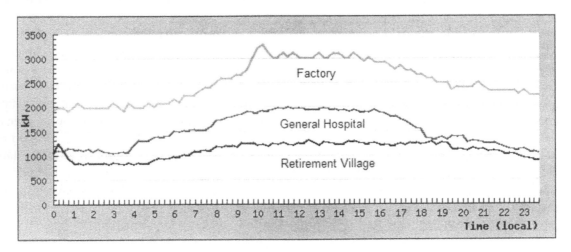

Figure 12-7. Base load analysis

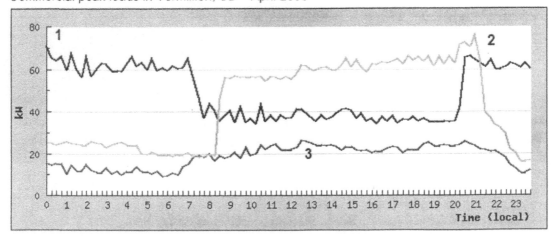

Figure 12-8. Peak load analysis

demand outstrips supply and the price of wholesale power may shoot from $50/MWh to $500/MWh. If your peak demand can be moved, you may want to participate in one of the many new load management programs which pay the market price for reducing load. New York and California both offer about a dozen new load management programs. They also support "demand response aggregators" who can take your small response and combine it with the efforts of many others to reduce megawatts of demand. Peak reduction is the equivalent of building more generation and demand response aggregations will be treated by system operators as if they were generators bidding a price into the load stack. Your ability to participate in demand response programs will be increasingly rewarded. Comparing your facility to your class load should give you an opinion of how well you might be able to participate in load management. If your peak is very high, it is likely you have uncontrolled operations and you can add controls to mitigate this.

WEATHER NORMALIZATION

Many electric bills now show consumption for the past 13 months. A few even provide the average monthly temperature. This gives consumers a crude measure of the annual pattern of consumption. But is this month's bill comparable to the same month last year? The hours of daylight will be the same, but the economy can change, the utilization of a facility can change, and the weather can change. Energy managers need a means to factor out the variables and determine the baseline energy use. The baseline then becomes the standard against which efficiency improvements can be measured. Why is this needed? We can't do much about sunrise or the economy, but we can try to remove the effects of weather from the monthly bill and develop a measurement scale that shows us if we can make changes that will save energy and if so, how much can we save. Every business is used to paying the monthly utility bill but many are reluctant to spend "additional" money on conservation. Knowing the baseline will offer a true measure of energy savings and help calculate what efficiency measures are worth buying and what savings are possible on the monthly bill. Without baseline information, businesses are not likely to implement any conservation measures.

A single bill is useless in determining an underlying pattern of energy use. Since the bill is virtually the only information most people have, the majority of energy users are in the dark. With a little work, they can chart quarterly and annual patterns and get a sense of their annual pattern of consumption. Typically this is represented by load profiles showing a band of consumption representing the four seasons. The wider the band, the greater the weather effect. The Energy Information Administration (EIA) goes a step further by publishing a 30 year data set of temperature information [5]. The data are presented as Heating Degree Days (HDD), and Cooling Degree Days (CDD). By convention the data sets sum the deviation in Fahrenheit and duration in hours from 68 degrees (HDD) and 72 degrees (CDD)

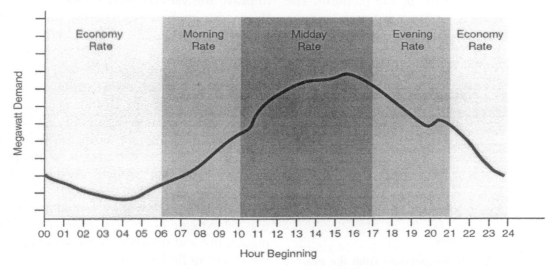

Figure 12-9. Coincident load analysis

Weather normalization is the statistical process of removing the variability of weather from energy consumption. Regression analysis is used to relate energy use to temperature; multiple regression analysis is used to relate energy use to multiple variables such as changes in day/night occupancy and activities. Regression analysis commonly requires highly paid consultants or expensive and data intensive software. Our interest was in producing a simple and automated process which would take the "first cut" at removing the effects of weather. This is the process we have implemented in our live Energy Information System.

The data we work with are 15-minute interval data collected by wireless from a commercial electric meter and temperature data provided hourly on the Internet from the nearest airport. Both sets of data are channeled through the Internet into a web server running an SQL data base. The raw material is normally plotted over the 24 hour period.

The next step is to plot the two variables against each other and to use a simple Microsoft Excel™ regression program to determine a linear relation between the two variables.

The third step is to use the equation developed in the regression to replot the load profile both in its original form and as it has been modified by the regression equation. Analysis of weekend data would require an additional data set because the utilization of the building is different. This may seem like a lot of computation, but it is accomplished quickly, automatically, and out of sight in the data base. To normalize a load profile, the user clicks on "Show Normal Plot." This compares the raw data plot with the processed data on the screen or in a report. We provide the normalized data at no additional charge in our EIS platform. The automated process does not cost us any more and we want to encourage our users to plan and implement conservation strategies.

In using this technique we have made several simplifying assumptions. The first is to view the load profile as 96 independent readings (one for each 15 minutes in the 24 hours). This means that our process uses 96 regression equations, one for each 15 minute interval in the day. In this way we regard each interval of the day as a fixed point which varies only according to the temperature, and we assume this variation has a linear relationship. For example, if the actual summer temperature is higher than the normal summer temperature at a specific time of day, then we assume that more cooling energy will be used at that time. At some point a user may want to hire a consultant to do a more complex analysis. This type of analysis would represent the load profile as a quadratic equation with additional variables for each inflection point and would require a year or more of data and powerful analytic tools to give confidence to the results.

We use our simple system because the data loading is automatic and it can give useful results within days. The longer it runs, the more the scatter plots will yield a discernible pattern. Effectively we are measuring the number of lights turned on, the number of computers running, and the processes and machinery running in the building at any given time in the 24 hour cycle. What we remove is the run time of the heating or air conditioning equipment. We can then look at both plots—the "normal load" and the variable load—and devise different strategies to address each. To improve the normal load we need to consider how much lighting or process equipment such as heaters or air compressors may be left idling when not needed because no one takes the time to turn them off. To improve the variable load we need to consider when this load is on, how efficient it is, and whether it could be re-timed to operate off peak. When there is a lack of insulation, consumption will vary more in response to weather. It is also possible to see how a change in operations will affect cost. Holding to a narrow comfort range despite noticeable heat or cold will require more energy. Allowing more variation in response to changes in weather will be more economical. The normalized data will also let us compare what happens on average days and what happens on days of extreme heat or cold. One strategy may be appropriate for the 1900 regular hours of the work year and another strategy for the 100 extreme hours. The comparison between your regular use and normalized use will begin to point out the places to find savings.

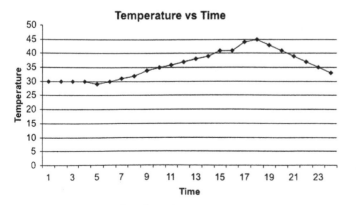

Figure 12-10. Charting hourly temperature from the nearest airport

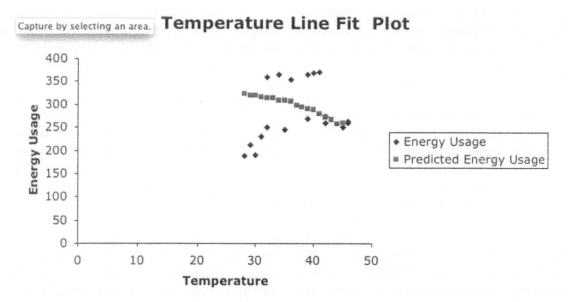

Figure 12-11. Finding the regression equation

Using Microsoft Excel™ for Regression Analysis

Microsoft Excel™ has some easy to use tools for plotting linear regressions. First you need to define your independent variable and your dependent variable. Put the independent variable (temperature) in the left column of an Excel spreadsheet. Put the dependent variable (kilowatt hours used) in the right column. Select the charting function, select the two data columns with their headings, and use the chart wizard to plot the data. Next select the chart, click on a data point, and then choose Chart > Add Trend line. Excel will produce the trend line and give you the equation for the linear regression. You can experiment with several different trend lines according to whether your data are best represented by a linear, a logarithmic, exponential, or moving average trend line.

In normalizing the utility data based on weather, look for a linear relationship in the data. If your historical data set runs between 30°F and 50°F, you cannot use this data to normalize energy consumption at 60°F. You must stay within the bounds of the data you have collected. If the temperature rises to 60°F, collect data for a few days and make a new plot and use linear regression for that part of the season. We do not try

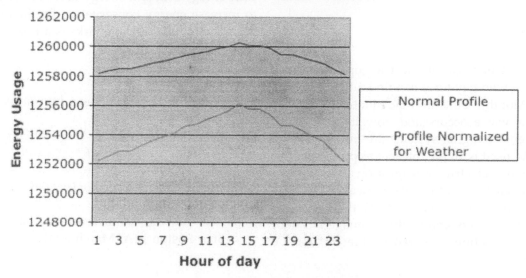

Figure 12-12. The normalized load profile

to make one equation fit all circumstances. Rather we apply many different equations over small ranges of our normalization: one equation for weekdays, one for weekends, others for heating season, cooling season, and shoulder months. The software knows what day of the week and season you are normalizing and picks an appropriate linear relation for that period.

Excel also calculates R^2 values which measure the fit between the plotted points and the linear equation. The closer all the points lie to the line, the larger the R^2 value. An R^2 value of 0.0680 means that 68% of the variation in the data is explained by the linear relationship. 32% is due to some other factor plus random error. The R^2 value also shows the "Standard Error of Regression." An R^2 of 0.068 indicates that 68% of values plotted lie within one standard error of regression. Looking at the plots outside this range—the outliers—may also prove valuable as something in the building may be causing unusually high use when the temperature changes. For example, in the shoulder months, the heating and cooling systems may both be running at the same time.

To normalize an interval reading, the software inserts the kW and the temperature into the appropriate equation and returns a value for the "normalized" kW. This is then plotted along with the raw meter data. The narrower the range between the meter data and the normalized data, the less weather effects the operation of the building, and conversely, the wider the range, the greater the effect of the weather. The normalized load profile gives you the energy use baseline for the building. You will know you are saving energy when you make a change and you see a reduced baseline.

CONCLUSION

The amount of waste in electric consumption is large, and those who pay for it can find a better use for their money. Monthly bills are inadequate in pointing out where this waste occurs and how it can be prevented. Inexpensive digital systems are now available, and they can be used with the Internet to provide better information about electric consumption especially in the form of the facility load profile. However the mass market for EISs has never developed. We think that one reason is that bill payers and facility managers cannot interpret the information in a load profile and ascribe little value to it.

We have made a first step in answering this misperception by automating benchmarks and including them as an added feature in an EIS. The user can display the facility load profile and then click down to add additional frames of reference from which to view the data. The Energy Star score shows how the user's facility compares to other facilities of the same type. If your facility receives a percentile score of 30, you know that 70 percent of other facilities of the same type are more efficient then yours. You also know that a lot of energy efficiency capital improvements could be funded from savings in your monthly bill.

Class load profiles offer another perspective on your consumption with regards to the size of your base load, peak load, and coincident demand. A careful comparison of the facility load profile with the class load profile should give you some insight into operational changes that can reduce your facility demand.

Finally, normalization reduces the ever changing variations caused by weather. These changes prevent any meaningful comparison of the May bill to the June bill, or to the May bill for the previous year. Using regression analysis, weather factors can be removed revealing a normalized load. A wide spread between normalized load and the actual load suggests looking for more efficient heating and cooling. A narrow spread means the focus should be on how occupants use the building and whether wasteful practices, policies, or habits can be changed.

The idea of automated benchmarking in the EIS is in its early stages. As facility managers begin to recognize the value of the information they get from automated benchmarking, they will be able to reduce their energy cost and energy use. The technology exists to increase energy conservation by double digits. Providing automated benchmarks is a first step in starting that process.

References
[1]   Square D Veris Hawkeye *http://www.veris.com/ products/pwr/8000.html*
[2]   Ohio Semitronics Inc. WL55 *http://www.ohiosemi-tronics.com/pdf/wl55.pdf*
[3]   *www.energystar.gov*
[4]   *http://weather.noaa.gov/index.html*
[5]   *www.EIA.gov*
[6]   AEE Globalcon Proceedings, Chapter 14, Energy Star, Boston MA, March 2004

Chapter 13

An Energy Manager's Introduction to Weather Normalization of Utility Bills

John Avina

UTILITY BILL TRACKING: THE REPORT CARD FOR FACILITIES AND ENERGY MANAGERS

Energy managers all too often have to justify their existence to management. They may be asked: "How much did we save last year?"; "Did your recommendations give reasonable paybacks?"; "Since the last project didn't save any money, why would we expect the next one to?"

Since over the reign of an energy manager, many energy conservation projects, control strategies, and operation and maintenance procedures may be employed, the simplest method to report on the energy manager's complete performance is to look at the utility bills. Management often sees it quite simply—it is all about the utility bills, since the bills reflect how much you are paying. Did the energy manager save us money or not?[1]

Since most energy managers are already tracking their utility bills, it should only take an additional step to see whether you have saved any energy and costs from your energy management program. In theory, you could just compare the prior year's bills to the current year's bills and see if you have saved.

But if it is so easy, why write a chapter on this? Well, it isn't so easy. Let's find out why.

Suppose an energy manager replaced the existing chilled water system in a building with a more efficient system. He likely would expect to see energy and cost savings from this retrofit. Figure 13-1 presents results the energy manager might expect.

But what if, instead, the bills presented the disaster shown in Figure 13-2?

Imagine showing management these results after you have invested a quarter-million dollars. It is hard to inspire confidence in your abilities with results like this.

How should the energy manager present these data to management? Do you think the energy manager is feeling confident about his decisions and about get-

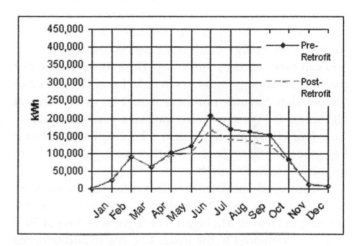

Figure 13-1. Expected Pre- and Post-Retrofit Usage for Chilled Water System Retrofit.

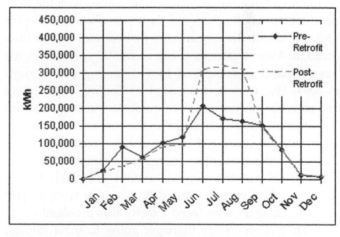

Figure 13-2. A Disaster of a Project? Comparison of Pre-Retrofit and Post-Retrofit Data

ting funding for future energy savings projects? Probably not. Management may simply look at the figures and, since figures don't lie, conclude they have hired the wrong energy manager!

There are many reasons the retrofit may not have delivered the expected savings. One possibility is that the project is delivering savings, but the summer after the retrofit was much hotter than the summer before the retrofit. Hotter summers translate into higher air conditioning loads, which typically result in higher utility bills.

Hotter Summer ➡ Higher Air Conditioning Load ➡ Higher Summer Utility Bills

In our example, we are claiming that because the post-retrofit weather was hotter, the chiller project looked like it didn't save any energy, even though it really did. Imagine explaining that to management!

If the weather really was the cause of the higher usage, then how could you ever use utility bills to measure savings from energy efficiency projects (especially when you can make excuses for poor performance, like we just did)? Your savings numbers would be at the mercy of the weather. Savings numbers would be of no value at all (unless the weather were the same year after year).

Our example may appear a bit exaggerated. But it begs the question: Could weather really have such an impact on savings numbers?

It can, but usually not to this extreme. The summer of 2005 was the hottest summer in a century of record-keeping in Detroit, Michigan. There were 18 days at 90°F or above compared to the usual 12 days. In addition, the average temperature in Detroit was 74.8°F compared to the normal 71.4°F. At first thought, 3 degrees doesn't seem like all that much, however, if you convert the temperatures to cooling degree days[2], as shown in Figure 13-3, the results look dramatic. Just comparing the June through August period, there were 909 cooling degree days in 2005 as compared to 442 cooling degree days in 2004. That is more than double! Cooling degree days are roughly proportional to relative building cooling requirements. For Detroit then, one can infer that an average building required (and possibly consumed) more than twice the amount of energy for cooling in the summer of 2005 than the summer of 2004. It is likely that in the upper Midwestern United States there were several energy managers who faced exactly this problem!

How is an energy manager going to show savings from a chilled water system retrofit under these circumstances? A simple comparison of utility bills will not work, as the expected savings will get buried beneath the increased cooling load. The solution would be to somehow apply the same weather data to the pre- and post-retrofit bills, and then there would be no penalty for extreme weather. This is exactly what weather normalization does. To show savings from a retrofit (or other energy management practice), and to avoid our disastrous example, an energy manager should normalize

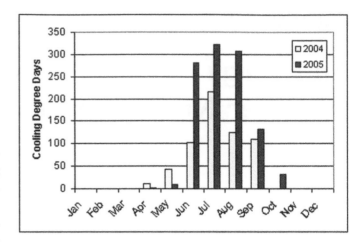

Figure 13-3. Cooling Degree Days in Detroit, Michigan for 2004 and 2005

the utility bills for weather so that changes in weather conditions will not compromise the savings numbers.

More and more energy managers are now normalizing their utility bills for weather because they want to be able to prove that they are actually saving energy from their energy management efforts. This process has many names: weather correction, weather normalization, tuning to weather, tuning or weather regression.

HOW WEATHER NORMALIZATION WORKS

Rather than compare last year's usage to this year's usage, when we use weather normalization, we compare *how much energy we would have used* this year to *how much energy we did use* this year. Many in our industry do not call the result of this comparison, "savings," but rather "usage avoidance" or "cost avoidance" (if comparing costs). Since we are trying to keep this treatment at an introductory level, we will simply use the word *savings*.

When we tried to compare last year's usage to this year's usage, we saw Figure 13-2, and a disastrous project. We used the equation:

Savings = Last year's usage – This year's usage

When we normalize for weather, the same data results in Figure 13-4 and uses the equation:

Savings = How much energy we would have used this year – This year's usage

The next question is how to figure out *how much energy we would have used this year*? This is where weath-

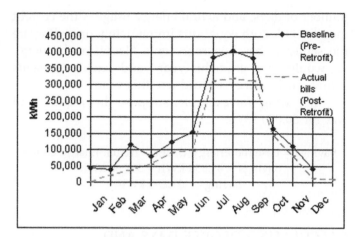

Figure 13-4. Comparison of Baseline and Actual (Post-Retrofit) Data with Weather Correction

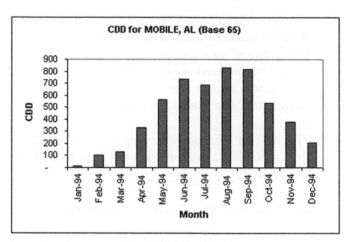

Figure 13-5. Cooling Degree Days

er normalization comes in.

First, we select a year of utility bills[3] to which we want to compare future usage. This would typically be the year before you started your energy efficiency program, the year before you installed a retrofit, the year before you, the new energy manager, were hired, or just some year in the past that you want to compare current usage to. In this example, we would select the year of utility data before the installation of the chilled water system. We will call this year the **base year**[4].

Next, we calculate degree days for the base year billing periods. Because this example is only concerned with cooling, we need only gather cooling degree days (not heating degree days). A section on calculating degree days follows later in the chapter. For now, recognize that only cooling degree days need to be gathered at this step.[5] Figure 13-5 presents cooling degree days over two years.

Base year bills and cooling degree days are then normalized by number of days, as shown in Figure 13-6. Normalizing by number of days (in this case, merely, dividing by number of days) removes any noise associated with different bill period lengths. This is done automatically by canned software and would need to be performed by hand if other means were employed.

To establish the relationship between usage and weather, we find the line that comes closest to all the bills. This line, the **best fit line**, is found using statistical regression techniques available in canned utility bill tracking software and in spreadsheets.

The next step is to ensure that the best fit line is good enough to use. The quality of the best fit line is represented by statistical indicators, the most common of which, is the R^2 value. The R^2 value represents the goodness of fit, and in energy engineering circles, an

$R^2 > 0.75$ is considered an acceptable fit. Some meters have little or no sensitivity to weather or may have other unknown variables that have a greater influence on usage than weather. These meters may have a low R^2 value. You can generate R^2 values for the fit line in Excel or other canned utility bill tracking software.[6]

This best fit line has an equation, which we call the **fit line equation**, or in this case the **baseline equation**.[7] The fit line equation from Figure 13-6 might be:

$$\text{Baseline kWh} = (5 \text{ kWh/Day} * \text{\#Days}) + (417 \text{ kWh/CDD} * \text{\#CDD})^8$$

Once we have this equation, we are done with this regression process.

Let's recap what we have done:

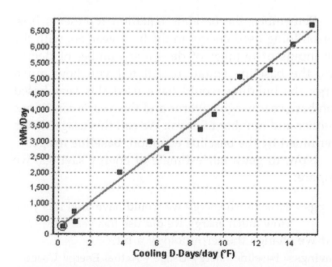

Figure 13-6. Finding the relationship between usage and weather data. The dots represent the utility bills. The line is the best fit line.

1. We normalized base year utility bills and weather data for number of days in the bill.
2. We graphed normalized base year utility data versus normalized weather data.
3. We found a best fit line through the data. The best fit line then represents the utility bills for the base year.
4. The best fit line equation represents the best fit line, which in turn represents the base year of utility data.

Base Year bills ≈ Best Fit Line = Fit Line Equation

The fit line equation represents how your facility used energy during the base year, and would continue to use energy in the future (in response to changing weather conditions) assuming no significant changes occurred in building consumption patterns.

Once you have the baseline equation, you can determine if you saved any energy.

How? You take a bill from some billing period after the base year. You (or your software) plug in the number of days from your bill and the number of cooling degree days from the billing period into your baseline equation.

Suppose for a current month's bill, there were 30 days and 100 CDD associated with the billing period.

$$Baseline\ kWh = (5\ kWh/Day * \#Days) + (417\ kWh/CDD * \#CDD)$$

$$Baseline\ kWh = (5\ kWh/Day * 30) + (417\ kWh/CDD * 100)$$

$$Baseline\ kWh = 41,850\ kWh$$

Remember, the baseline equation represents how your building used energy in the base year. So, with the new inputs of number of days and number of degree days, the baseline equation will tell you how much energy the building would have used this year based upon base year usage patterns and this year's conditions (weather and number of days). We call this usage that is determined by the baseline equation, **baseline usage**.

Now, to get a fair estimate of energy savings, we compare:

Savings = How much energy we would have used this year
How much energy we did use this year
Or if we change the terminology a bit:
Savings = Baseline Energy Usage – Actual Energy Usage

where baseline energy usage is calculated by the baseline equation, using current month's weather and

number of days, and actual energy usage is the current month's bill. Both equations immediately preceding are the same, as baseline represents "How much energy we would have used this year," and actual represents "How much energy we did use this year."

So, using our example, suppose this month's bill was for 30,000 kWh:

Savings = Baseline Energy Usage – Actual Energy Usage
Savings = 41,850 kWh – 30,000 kWh
Savings = 11,850 kWh

CALCULATING DEGREE DAYS AND FINDING THE BALANCE POINT

Cooling degree days (CDD) are roughly proportional to the energy used for cooling a building, while heating degree days (HDD) are roughly proportional to the energy used for heating a building. Degree days, although simply calculated, are quite useful in energy calculations. They are calculated for each day, and are then summed over some period of time (months, a year, etc.).[9]

In general, daily degree days are the difference between the building's balance point and the average outside temperature. To understand degree days then, we first need to understand the concept of balance points.

Buildings have their own set of **balance points** for heating and for cooling – and they may not be the same. The **heating balance point** can be defined as the outdoor temperature at which the building starts to

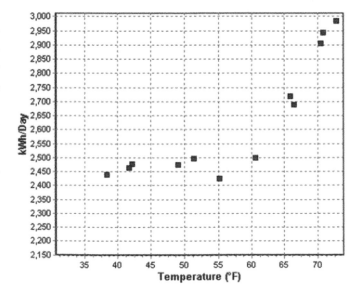

Figure 13-7. Determining the Balance Point using a kWh/day vs. Outdoor Temperature Graph

heat. In other words, when the outdoor temperature drops below the heating balance point, the building's heating system kicks in. Conversely, when the outdoor temperature rises above the **cooling balance point**, the building's cooling system starts to cool.[10] A building's balance point is determined by nearly everything associated with it, since nearly every component associated with a building has some effect on the heating of the building: building envelope construction (insulation values, shading, windows, etc.), temperature set points, thermostat set back schedules if any, the amount of heat producing equipment (and people) in the building, lighting intensity, ventilation, HVAC system type, HVAC system schedule, lighting and miscellaneous equipment schedules among other factors.

In the past, before energy professionals used computers in their everyday tasks, degree day analysis was simplified by assuming balance points of 65°F for both heating and cooling. As a result, it was easy to publish and distribute degree days, since everyone calculated them using that same standard. It is more accurate, however, to recognize that every building has its own balance points and to calculate degree days accordingly. Consequently, you are less likely to see degree days available, as more sophisticated analysis requires you to calculate your own degree days based upon your own building's balance points.[11]

A way to find the balance point temperature of a building is to graph the usage/day against average outdoor temperature (of the billing period) as shown in Figure 13-7. Notice that Figure 13-7 presents two trends. One trend is flat, and the other trend slopes up and to the right. We have drawn lines signifying the two trends in Figure 13-8. (Ignore the vertical line for now.) The flat trend represents **non-temperature sensitive consumption**, which is electrical consumption that is not related to weather. In Figure 13-7, non-temperature sensitive consumption is roughly the same every month, about 2450 kWh per day. Examples of non-temperature sensitive consumption include lighting, computers, miscellaneous plug load, industrial equipment and well pumps. Any usage above the horizontal line is called **temperature sensitive consumption**, which represents electrical usage associated with the building's cooling system. Notice in Figure 13-8, the temperature sensitive consumption only occurs at temperatures greater than 61°F. The intersection of the two trends is called the **balance point** or balance point temperature, which is 61°F in this example.

Notice also that, in Figure 13-8, as the outdoor temperature increases, consumption increases. As it gets hotter outside, the building uses more energy, thus the meter is used for cooling, but not heating. The balance point temperature we found is the cooling balance point temperature (not the heating balance point temperature).

We can view the same type of graph for natural gas usage in Figure 13-9. Notice that the major difference between the two graphs (electric and gas), is that the temperature sensitive trend slopes up and to the left (rather than up and the right). As it gets cooler outside, they use more gas, therefore, they use gas to

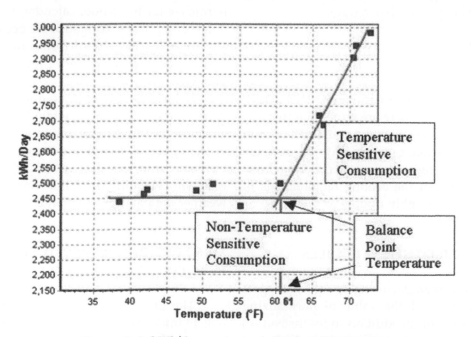

Figure 13-8. kWh/day vs. Average Outdoor Temperature

heat the building.

Now that we have established our balance point temperature, we have all the information required to calculate degree days. If your graph resembles Figures 9, you will be using heating degree days. If your graph resembles Figure 13-8, you will be using cooling degree days. If you calculate degree days by hand, or using a spreadsheet, you would use the following formulae for your calculations. Of course, commercially available software that performs weather normalization handles this automatically.

For each day,

$$HDD_i = [\ T_{BP} - (T_{hi} + T_{lo})/2 \ ] \times 1 \ Day^+$$
$$CDD_i = [\ (T_{hi} + T_{lo})/2 - T_{BP} \ ] \times 1 \ Day^+$$

Where:

HDD_i = Heating Degree Days for one day

CDD_i = Heating Degree Days for one day

T_{BP} = Balance Point Temperature,

T_{hi} = Daily High Temperature

T_{lo} = Daily Low Temperature

$^+$ signifies that you can never have negative degree days. If the HDD_i or CDD_i calculation yields a negative number, then the result is 0 degree days for that day.

Heating and cooling degree days can be summed, respectively, over several days, a month, a billing period, a year, or any interval greater than a day. For a billing period (or any period greater than a day),

$$HDD = \Sigma HDD_i$$
$$CDD = \Sigma CDD_i$$

Now, let's take a look back to Figure 13-3, where you may have noticed that there are more than twice as many cooling degree days (CDD) in August 2005 than in August 2004. Because cooling degree days are roughly proportional to a building's cooling energy usage, one could rightly assume that the cooling requirements of the building would be roughly double as well.

NORMALIZING FOR OTHER VARIABLES

More and more manufacturing energy managers are coming to understand the value of normalizing utility data for production in addition to (or instead of) weather. This works if you have a simple variable that

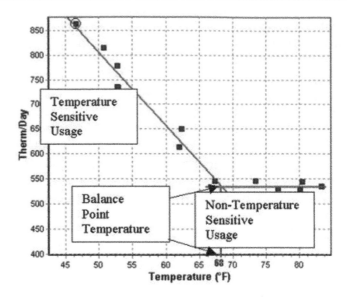

Figure 13-9. Therm/day vs. Average Outdoor Temperature

quantifies your production. For example, a computer assembly plant can track number of computers produced. If your factory manufactures several different products, for example, disk drives, desktop computers and printers, it may be difficult to come up with a single variable that could be used to represent production for the entire plant (i.e. tons of product). However, since analysis is performed on a meter level, rather than a plant level, if you have meters (or submeters) that serve just one production line, then you can normalize usage from one meter with the product produced from that production line.

School districts, colleges, and universities often normalize for the school calendar. Real estate concerns, hotels and prisons normalize for occupancy. Essentially any variable can be used for normalization, as long as it is an accurate, consistent predictor of energy usage patterns.

Figure 13-10 presents normalized daily usage versus production for a widget factory. The baseline equation for this normalization is shown at the bottom of the figure. Notice that units of production (UPr) as well as cooling degree days (CDD) are included in the equation, meaning that this normalization included weather data and production data.

MANAGING UNEXPECTED CHANGES IN ENERGY USAGE PATTERNS

The greatest difficulty involved in using utility bills to track savings occurs when there are large, unexpected and unrelated changes to a facility. For example,

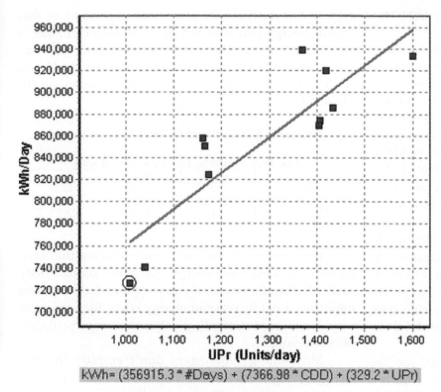

kWh= (356915.3 * #Days) + (7366.98 * CDD) + (329.2 * UPr)

Figure 13-10. Daily Usage Normalized to Production and Weather. The Baseline Equation is Shown at the Bottom of the Figure.[12]

suppose an energy manager was normalizing usage to weather for a building in order to successfully determine energy savings, and then the building was enlarged by several thousand square feet. The comparison of baseline and actual usage would no longer make any sense, as the baseline number would continue to be determined based upon usage patterns before the building addition, whereas the actual bills would include the

addition. Now we would be comparing two different facilities, one with the addition and one without. If there were any energy savings, they might be buried in the additional usage from the new addition. Figure 13-11 presents our hypothetical case in which the new addition came online in August.

Notice that in Figure 13-11 the actual usage has increased while the baseline did not. As a result, savings are hidden by the increase in usage from the building addition. Here, the energy manager would need to make a baseline adjustment (also known as baseline modification) to handle the increase in usage due to the building addition (since the actual bills already include it). The energy manager would make a reasonable estimate of the additional usage and add that onto the baseline. Our earlier equation, then, becomes:

Baseline kWh = (5 kWh/Day * #Days) + (417 kWh/CDD * #CDD) + Adjustment

where the adjustment represents the additional usage due to the building addition. Figure 13-12 presents an example with the addition of the baseline adjustment.

Baseline adjustments are the most troublesome part of using utility bills to analyze building usage. Buildings continue to change their usage patterns regardless of the energy managers' efforts. To maintain usefulness, baseline adjustments must be added to the analysis.

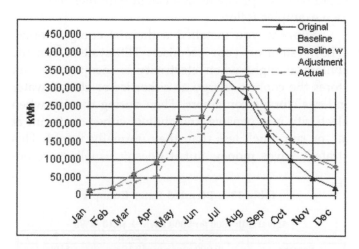

Figure 13-11. Example of increase in energy usage due to increase in square footage starting in August.

Figure 13-12. Baseline now Adjusted to Account for Increase in Usage Due to Building Addition

APPLYING COSTS TO THE SAVINGS EQUATION

Energy managers often need to present their savings numbers to management in a form that managers can comprehend, which means showing savings in cost, rather than energy or demand units. Transforming energy savings into cost savings can be done quite simply, and there are several methods by which this can be done. As in most things, the simplest methods yield the most inaccurate results. The methods investigated here are blended rates and modeled rates. There are some variations on these themes, but they will not be covered here.

In many areas, utility rates may be difficult to understand and model. Once the energy manager understands the rate, he might have to explain it to management, which can be even more difficult. Simplicity is always worth striving for, as many energy managers don't have the time to learn their rates and model them explicitly.

Blended rates (also called average costs) are the simplest way to apply costs to energy units. Suppose for a billing period that baseline usage was 10,000 kWh and the current usage was 8,000 kWh, and current total cost was $800. It doesn't matter how complex the rate is, to apply blended rates, we just consider total cost. The simplest application of blended rates would be to determine the average $/kWh of the *current bill*. In this case, we have $800/8000 kWh = $0.10/kWh. So, the blended rate ($0.10/kWh) would be applied to both the baseline usage and the actual or current usage, as shown in Table 13-1.

This may seem like the best solution, and many energy managers use blended rates as it does simplify what could be unnecessarily complex. However, there could be some problems associated with blended rates. Two examples follow.

Suppose you installed a thermal energy storage (TES) system on your premises. TES systems run the chillers at night when electricity is inexpensive and stores the cooling energy as either ice or chilled water in large storage containers. Then during the day when electricity is expensive, the chillers either don't run at all, or run much less than they normally would. This strategy saves money, but it doesn't usually save energy. In fact it often uses more energy, as some of that cooling energy that is stored in the storage container is lost through the walls of the container, and the extra pump that runs the system consumes energy. If you applied a blended rate to the TES system you might see the numbers in Table 13-2.

Table 13-2. Where Blended Rates Can Go Wrong

Actual Bill: $10,000 for 1000 kWh → $0.10 /kWh

	KWh	$/kWh	Cost
Baseline	9,500	$0.10	$950
Actual	10,000	$0.10	$1000
Savings	-500	$0.10	-$50

If you modeled the rates, you would see that even though you used more energy, you saved on electricity costs. On the other hand, if you used blended rates, you might see a net increase in energy costs. Blended rates would deliver a dramatically incorrect representation of cost savings.

Most energy managers don't employ thermal energy storage, but they may shift demand to the evening. Suppose a facility is on a time-of-use rate and there is a small net increase in energy usag, combined with a significant shift in energy usage to off peak (less expensive) periods. What happens then? Since less energy is consumed during the more expensive on peak period and more is consumed during the less expensive off peak period, the total costs might decline (in real life). But if the usage increases, your blended rate strategy will show a net increase in costs (in your analysis), which is exactly wrong. Again you can refer to Table 13-2.

Another example demonstrates a weakness in the blended rate approach. Suppose you installed a new energy efficient boiler and boiler controls in a building that is mostly vacant in the summer. Suppose the utility charges a $25 monthly charge plus $1.30/therm.

A January bill, with 100 therms usage, is presented in Table 13-3.

If our baseline usage for January was 120 therms, then savings would be calculated using the blended rate, as shown in Table 13-4.

That seems to work well. Now try July, in which the current bill might have had 1 therm usage, the bill is presented in Table 13-5.

And suppose baseline usage in July was 4 therms, then savings would be calculated as in Table 13-6.

The blended rate calculation told us that the customer saved $78.90, whereas the actual rate calculation would have told us that the customer saved 3 therms * $1.30/therm = $3.90. This problem is not unusual.

Table 13-1. Blended Rates Example

Actual Bill: $800 for 8000 kWh → $0.10 /kWh

	KWh	$/kWh	$
Baseline	10,000	$0.10	$1000
Actual	8,000	$0.10	$800
Savings	2,000	$0.10	$200

Table 13-3. A Hypothetical Winter Gas Heating Bill

Charge	Usage	Rate	Cost
Monthly Charge	N/A	$25	$25
Usage Charge	100 therms	$1.30/therm	$130
Total Bill			$155
Blended Rate $/therm = $155 / 100 therms = $1.55/therm			

Table 13-4. Savings Calculations Using a Blended Gas Rate

	Therms	$/therm	Cost
Baseline	120	$1.55	$186
Actual	100	$1.55	$155
Savings	20	$1.55	$31

Table 13-5. Problematic Hypothetical Summer Gas Heating Bill

Charge	Usage	Rate	Cost
Monthly Charge	1	$25	$25
Usage Charge	1	$1.30/therm	$1.30
Total Bill			$26.30
Blended Rate $/therm = $26.30 / 1 therms = $26.30/therm			

Table 13-6. Hypothetical Gas Heating Savings Problems

	Therms	$/therm	Cost
Baseline	4	$26.30	$105.2
Actual	1	$26.30	$26.30
Savings	3	$26.30	$78.90

Often, this type of overstatement of savings occurs without anyone noticing. Blended rates can simplify the calculations and on the surface may return seemingly correct savings numbers. However, upon further analysis, it can usually be found that using blended rates introduces inaccuracies that can, at times, prove embarrassing. The whole point of weather normalization was to reduce the error (due to weather and other factors) in the savings calculations. What is the point of going through the weather normalization procedure if you are only going to reintroduce a potentially even greater error when you apply costs to the savings equation?

If you want to get more accurate cost savings numbers then you would elect to model the rates, which unfortunately means that you will have to understand them.

This would involve retrieving the rate tariff from the utility (usually, they are on the utility's website), and then entering all the different charges into your software or spreadsheet. There are a few difficulties associated with this approach:

1. Many rates are very difficult to understand
2. Some tariff sheets do not explain all the charges associated with a rate.
3. Some software packages have limitations and can model most but not all of the different charges, or even worse, some packages don't model rates at all.
4. Rates change often, which means you will have to continually keep updating the rates. The good news on this front is that once the rate is modeled, the changes are usually very minor.

As mentioned before, if you are modeling rates, then usually the same rate is applied to both baseline and actual usage and demand. There are exceptions of course. If you changed your facility's rate or changed utility providers, then you should apply your old rate to the baseline, and your new rate to your actual usage. To understand which rate should be used for the baseline, answer the same question: "How much would we have spent if I had not run the energy management program?" The answer is, you would still be on the old rate, therefore, baseline gets the old rate, and actual gets the new rate.

Regardless of how you apply costs to your savings equation, good utility bill tracking software can handle all of these situations.

WEATHER NORMALIZATION IN EXCEL VS. SPECIALIZED UTILITY BILL TRACKING SOFTWARE

Weather normalization can be done in Excel; however, it can be laborious and oftentimes may not be as rigorous as when done using specialized software. Excel will give regressions, fit line equations and statistical indicators which show how well your usage is represented by the fit line.

However, it is difficult to find the best balance point in Excel, as you can in specialized software.[13] If you use Excel, the steps we outlined in this paper will have to be done manually, whereas with canned software, most of it is done for you automatically. In addition, in Excel, if you want to achieve a good fit to your data, you may have to iterate these manual steps for different balance points. The most tedious process in Excel is matching up the daily weather to the billing periods. Try it and you will see. Assuming the weather and bill data are already present, it should take less than two minutes in canned software to perform weather normalization, versus at least 30 minutes in Excel.

AVAILABLE WEATHER NORMALIZATION DESKTOP SOFTWARE

All of the major desktop utility bill tracking software packages will now normalize for weather data. All of them will correct for your own variables as well; however, only some of them will normalize for weather in addition to your own variables. The major desktop programs are *Energy CAP®, Metrix™* and *Utility Manager™ Pro*. You can find information on all of them online.

AVAILABLE WEATHER NORMALIZATION IN WEB SOFTWARE

At the time of this writing, only one of the above desktop programs is also offered on a web platform, though a web front end is available from some of the other providers which allow users to enter bill data, perform diagnostic tests and make reports online.

WEATHER NORMALIZATION IN INTERVAL DATA WEB SOFTWARE

There are some interval data programs that perform weather normalization as well, but for these packages, weather normalization is done primarily for forecasting applications, not for verifying energy savings. The method is more complex as the data are in finer increments. Weather forecasts are downloaded and then projected usage is then calculated. At least one of the programs uses weather normalization, or any of a handful of other techniques to forecast energy usage. Energy managers can use these forecasts to adjust their energy consuming activities to prevent high peak demands.

CONCLUSION

Weather varies from year to year. As a result, it becomes difficult to know whether the change in your utility bills is due to fluctuations in weather, your energy management program, or both. If you wish to use utility bills to determine energy savings from your energy management efforts with any degree of accuracy, it is vital that you remove the variability of weather from your energy savings equation. This is done using the weather normalization techniques described in this chapter. You may adjust your usage for other variables as well, such as occupancy or production. You may have to make baseline adjustments to further "correct" the energy savings equation for unexpected changes in energy usage patterns such as new additions. Finally, the method in which you apply costs to your energy savings calculations is very important. Blended rates, although simple, can result in inaccurate cost savings numbers, while more difficult modeling rates, are always right.

Footnotes

1. What are the alternatives? The most common might involve determining savings for each of the energy conservation activities using a spreadsheet or perhaps a building model. Both of these alternative strategies could require much additional work, as the energy manager likely has employed several strategies over his tenure. One other drawback of spreadsheets is that energy conservation strategies may interact with each other, so that total savings may not be the sum of the different strategies. Finally, spreadsheets are often projections of energy savings, not measurements.

2. Cooling degree days are defined in detail later in the chapter; however, a simplified meaning is given here. Warmer weather will result in more cooling degree days; whereas a colder day may have no cooling degree days. Double the amount of cooling degrees should result in roughly double the cooling requirements for a building. Cooling degree days are calculated individually for each day. Cooling degree days over a month or billing period are merely a summation of the cooling degree days of the individual days. The inverse is true for heating degree days.

3. Some energy professionals select 2 years of bills rather than one. Good reasons can be argued for either case. Do not choose periods of time that are not in intervals of 12 months (for example, 15 months, or 8 months could lead to inaccuracy).

4. Please do not confuse base year with baseline. Base year is a time period, from which bills were used to determine the building's energy usage patterns with respect to weather data, whereas baseline, as will be described later, represents how much energy we would have used this month, based upon base year energy usage patterns and current month conditions (i.e. weather and number of days in the bill).

5. Canned software does this automatically for you, while in spreadsheets, this step can be tedious.

6. The statistical calculations behind the R2 value and a treatment of three other useful indicators, T-Statistic, Mean Bias Error, and CVRMSE are not treated in this chapter. For more information on these statistical concepts, consult any college statistics textbook. (For energy managers, a combination of R2 values and T-Statistics is usually enough.)

7. Baseline equation = fit line equation +/- baseline modifications. We cover baseline modifications later in this chapter.

8. The generic form of the equation is:
Baseline kWh = (constant * #days) + (coefficient * #CDD)

where the constant and coefficient (in our example) are 5 and 417.

9. Summing or averaging high or low temperatures for a period of time is not very useful. (Remember the Detroit example mentioned earlier.) However, you can sum degree days, and the result remains useful, as it is proportional to the heating or cooling requirements of a building.

10. If you think about it, you don't have to treat this at the building level, but rather can view it at a meter level. (To simplify the presentation, we are speaking in terms of a building, as it is less abstract.) Some buildings have many meters, some of which may be associated with different central plants. In such a building, it is likely that the disparate central plants would have different balance points, as conditions associated with the different parts of the building may be different.

11. Some analysts had separate tables of degree days based upon a range of balance points (65, 60, 55, etc.), and analyzed their data painstakingly with several balance points until they found the best balance point temperature for their building. On the other hand, other analysts believe that all degree days are calculated assuming the standard balance point of 65 °F.

12. A better presentation of the data would be in 3 dimensions (with Units Produced in X axis, Weather in Y axis, and kWh in Z axis), however due to limitations of a printed page, a 2 dimensional image is shown here, which is just one slice (or plane) in the actual relationship. This explains why the line may not look so close to the bills.

13. It is not necessary to find the best balance point, and you might choose instead to use published tables of degree days, which are often based on a 65-degree balance point. Using these standard degree days will in most cases lead to decreased accuracy and poorer fits. Using the base 65-degree balance point, many meters will not have an acceptable fit (R2 > 75%) at all.

Chapter 14

Data Analysis and Decision Making: Using Spreadsheets and "Pivot Tables" To Get A Read On Energy Numbers

Partha Raghunathan

In recent years, the computing world has witnessed a large-scale adoption of spreadsheet technology by business users worldwide to gather and manage numeric information. Today, spreadsheets have become the de-facto standard for business data analysis. Spreadsheet applications like Microsoft® Excel© are very easy to use, easy to share, have many rich in-built analytic and statistical functions and can handle large volumes of data while still providing sub-second performances and, finally, are easily integrated to web pages on the internet.

This chapter's objective is to take a deeper look at data analysis features in spreadsheets and understand how they can be applied to common tasks performed by an energy manager. Using a simple tutorial, we will attempt to show how spreadsheets can be a very effective, cheap and simple solution for most data analysis needs of an energy manager.

INTRODUCTION

The last decade has seen a massive proliferation of software applications and databases in almost every conceivable business process. From buying a book on the internet to paying bills from a bank account to tracking a FedEx® package, web-based software systems are dramatically changing the very way we live our lives. These systems allow companies to electronically capture vast amounts of information about their customers, their suppliers and their competitors. Consequently, these companies that are now capturing vast amounts of information, are demanding sophisticated analysis tools to deal efficiently with this information and make profitable decisions. Mainly driven by this commercial segment, the data analysis and decision support industry market has seen a huge boom in the past decade—technologies like Data Warehousing, Business Intelligence, On-line Analytical Processing (OLAP), Data Mining etc. have seen massive advances—as early adopters of these technologies attempt to get a competitive advantage through intelligent analysis of data culled from consumers, competitors and business partners.

As a few elite business analysts ride this wave of leading-edge, decision-support technology on the lagging edge, business users worldwide are very quietly beginning to use everyday spreadsheet technology in very imaginative ways, stretching its limits to accomplish the most sophisticated of analyses. Hundreds of software vendors are using more advanced features of spreadsheets such as the open application programmable interfaces (APIs) to develop complex business applications for budgeting, forecasting, manufacturing and even strategic operations planning as well as consumer applications such as personal financial planning.

Today's spreadsheet applications, like Microsoft® Excel©, are very easy to use, easy to share, have many rich, in-built analytical functions and can handle large volumes of data while still providing sub-second performances and, finally, are easily integrated to web pages on the internet.

CASE STUDY FOR TUTORIAL

This chapter contains a tutorial on using spreadsheets, showing how they can be an effective solution for most data analysis needs of an energy manager.

This case study shows how an energy manager, Bob Watts, uses spreadsheet technology to gather quick insight into the energy consumption patterns of a small commercial customer. This example will also show how this insight leads to decisions resulting in reduced energy usage and costs. Further, we will illustrate how Bob can easily share his findings with his customer and his peers over the web. In our estimate, this whole analysis would take Bob, a skilled spreadsheet user, less than 2 hours to complete from start to finish.

Meet Bob Watts, an account manager at a leading utility company in Atlanta, Georgia. Bob owns a few large commercial customer accounts and is responsible for providing services such as energy cost reduction programs to his customers, not only to keep customers happy but also to make sure the utility company doesn't run into an energy shortfall.

Bob recently visited Plastico Inc., a manufacturer of small plastic widgets, to assess their energy usage patterns and suggest ways to reduce consumption and costs. Plastico is headquartered in small town, south of Atlanta, in Georgia and has sales offices around the country. The headquarters in Georgia has 3 major locations:

• Office: Offices for sales & marketing staff, executive management staff.

• Warehouse: Holding area for finished product inventory to be loaded onto trucks.

• Plant: Facility for manufacturing, testing and packing operations.

While he was at Plastico, Bob collected information about the various electric and gas equipment in the 3 major areas, along with estimated hours of operation. Table 14-1 shows a sample snapshot of the types of data Bob was collecting. The column "Efficiency" is Bob's qualitative assessment of the relative efficiency of the equipment (for example, a standard incandescent

would rate "Low," a standard fluorescent would rate "Medium" and a high-efficiency compact fluorescent lamp would rate "High").

Bob now wants to analyze the data he has gathered. The next section will describe how Bob converts this data into meaningful information. First, he enters all the data into an Excel© spreadsheet. Table 14-1 shows sample data in a spreadsheet. In this tutorial, for the sake of simplicity, we will only consider the information for lighting appliances, although Plastico also has several pieces of production equipment, gas heaters and furnaces, and the HVAC systems. This example can easily be extended to all kinds of energy equipment.

Step 1. Preparing the data for Analysis

The first step involves performing all basic computations. In this case, Bob uses "Annual Usage (hours)," "Rating (Watts)" and "Quantity" to compute estimated "Annual Usage (kWh)" and "Demand Usage (kW)" using the formula:

Annual Usage (kWh) = Annual Usage (hours) (Rating (watts) (Quantity/1,000 (Wh/kWh)

Demand Usage (kW) = Rating (watts) (Quantity/1,000 (Wh/kWh)

Table 14-2 shows the computed column along with the collected data. It is possible to have several other computed columns like "Average Hours of Operations Per Day," "kWh consumed per day" etc. using simple spreadsheet formulas.

Step 2: Creating a "Pivot Table"

Once the data are complete and ready for analysis, Bob creates a "pivot table" for his analysis. A "pivot table" is an interactive table that Bob can use to quickly summarize large amounts of data. He can rotate its rows and columns to see different summaries of the

Table 14-1. Sample energy Consumption Data Collected by Bob at Plastico

Location	Area	Category	Sub Category	Quantity	Efficiency	Rating (Watts)	Annual Usage (hours)	Comments
Plant	Shopfloor	Lighting	HPS	316	Low	300	8760	
Warehouse	Working Area	Lighting	Fluorescent	240	Medium	48	8,760	
Office	Cafeteria	Lighting	Fluorescent	156	Medium	48	2,480	
Warehouse	Main Storage Area	Lighting	Metal Halide	150	Low	480	8760	
Warehouse	Cafeteria	Lighting	Fluorescent	128	Medium	48	8760	
Warehouse	Back-end Storage	Lighting	Metal Halide	100	Low	480	3744	
Plant	Cafeteria	Lighting	Fluorescent	96	Medium	48	8760	
Office	Laboratory	Lighting	Fluorescent	92	Medium	28	8760	
Office	Back Office	Lighting	Fluorescent	76	Medium	48	2,480	

source data, filter the data by displaying different pages, or display the details for areas of interest.

> *Tip: For more information on "pivot tables," refer to Microsoft® Excel© Help on PivotTable"*

To create a pivot table, Bob highlights all the data in the spreadsheet and selects from the Excel© Menu, Data–> PivotTable and PivotChart Report.

In the window titled, "PivotTable and PivotChart Wizard - Step 1 of 3," Bob clicks "Finish." This results in the creation of a new worksheet titled "Sheet2" that has a blank table and a floating menu titled "Pivot-Table."

Step 3: Some basic reports

Once the pivot table has been created, Bob wants to generate some very basic reports to understand where to begin looking for opportunities. There are three reports that Bob wants to create to understand these basic trends in Plastico's energy consumption patterns:

- #1 - Annual Usage (kWh) by Location—i.e. office, warehouse, plant

- #2 - Annual Usage (kWh) and Demand (kW) by sub-category—i.e. fluorescent, incandescent etc.
- #3 - Annual Usage (kWh) and Demand (kW) by efficiency—i.e. high, low, medium

Report #1: Annual Usage (kWh) by Location

To create this report, Bob starts with the empty pivot table created in Step 2.

1. Bob selects the "Location" field from the "pivot table" menu and drags it onto the "Drop Row Fields Here" section in the empty pivot table.

2. Next, he selects the "Annual Usage (kWh)" field from the "pivot table" menu and drags that onto the "Drop Data Fields Here" section in empty pivot table.

3. Bob then double-clicks on the pivot table title "Count of Annual Usage (kWh)." In the "PivotTable Field" window, he selects "Sum" instead of "Count" (Figure 14-2).

4. In the same window, Bob also clicks on the button "Number," formats the numeric display to "0 decimals," and checks the box "Use 1000 separator (,)." He then clicks on "OK" in the PivotTable Field window. Table 14-3 shows a finished report "Annual Usage (kWh) by Location."

Table 14-2. Fully Computed Data

Location	Area	Category	Sub Category	Quantity	Efficiency	Rating (Watts)	Annual Usage (hours)	Annual Usage (kWh)	Demand Usage (kW)	Comments
Plant	Shopfloor	Lighting	HPS	316	Low	300	8760	830,448	94.8	
Warehouse	Working Area	Lighting	Fluorescent	240	Medium	48	8,760	100,915	11.5	
Office	Cafeteria	Lighting	Fluorescent	156	Medium	48	2,480	18,570	7.5	
Warehouse	Main Storage Area	Lighting	Metal Halide	150	Low	480	8760	630,720	72.0	
Warehouse	Cafeteria	Lighting	Fluorescent	128	Medium	48	8760	53,821	6.1	
Warehouse	Back-end Storage	Lighting	Metal Halide	100	Low	480	3744	179,712	48.0	
Plant	Cafeteria	Lighting	Fluorescent	96	Medium	48	8760	40,366	4.6	
Office	Laboratory	Lighting	Fluorescent	92	Medium	28	8760	22,566	2.6	
Office	Back Office	Lighting	Fluorescent	76	Medium	48	2,480	9,047	3.6	

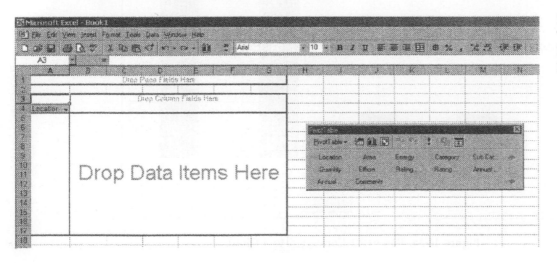

Figure 14-1. A Pivot Table (left) with the Floating Pivot Table Menu (right)

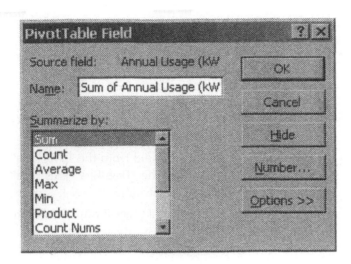

Figure 14-2. Changing the Summarize Option of a Pivot Table Field

(This report has also been formatted. To apply a format to a finished report, you can right-click anywhere on the pivot table and select "Format Report" and choose one of the several standard formats. The format selected below is "Table 7.")

**Table 14-3. Report #1—
Annual Usage (kWh) by Location**

Location	Annual Usage (kWh)
Office	162,183
Outdoor	159,782
Plant	970,377
Warehouse	1,015,070
Grand Total	2,307,413

Tip: Unlike the normal formatting in a spreadsheet, pivot table formats are dynamic. By applying the formats once, the fonts, colors, etc. are preserved automatically as the content in the table changes. For example, if Bob adds more lighting entries in the main spreadsheet, all he has to do is refresh the pivot table and the results are automatically recomputed.

5. This report can easily be converted into a chart by clicking on the "Chart" icon on the Excel© toolbar OR by selecting Insert->Chart from the menu (Figure 14-3a). Since Bob is going to use the same

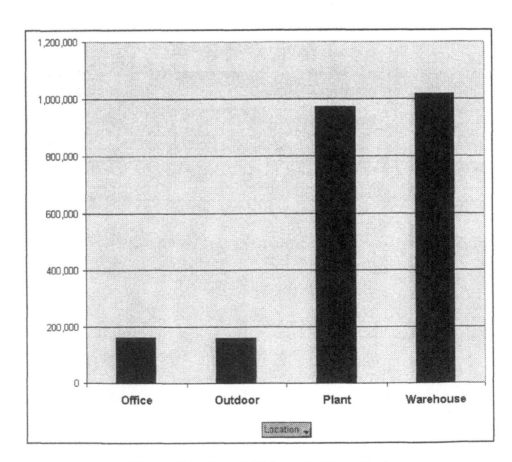

Figure 14-3a. Report #1 Converted to a Chart

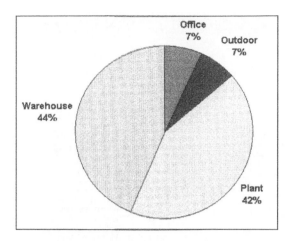

Figure 14-3b. Pie-chart View of Figure 3a

pivot table to create other reports and continue with his analysis, he now takes a snapshot of this report and copies it over to anotheer file. To do this, he selects the entire results table, copies it (Edit->Copy) and pastes it into a new spreadsheet file.

Report #1: Analysis

This report gives Bob a good feel for the break-up of energy usage by area. From Figure 3b, since the Plant and Warehouse areas account for 86% of overall kWh consumption, one straightforward conclusion is that Bob should first investigate these two areas before the other two. Lets see what the other reports reveal...

Report #2: Annual Usage (kWh) and Annual Demand (kW) by sub-category—i.e. fluorescent, incandescent etc.

To create this report, Bob starts with Report #1:

1. First, Bob needs to replace "Location" with "Sub-Category" in the Row field. To do this, Bob opens the "Layout" window by right clicking anywhere on the pivot table and selecting "Wizard" (Figure 14-4). From the resulting screen, he selects the "Layout" button.

2. In the "Layout" Window, Bob re-arranges the report view. He first drags "Location" from the ROW area onto the PAGE area on the left. Next, he drags the "Sub Category" field from the palette onto the ROW area. He also drags the "Demand Usage (kW)" onto the DATA area along with "Annual Usage (kWh)." Finally, he formats the two data measures to show the "Sum of" values instead of the default "Count of." (See Step 3 of Report #1 above). Figure 14-5 is an illustration of the finished layout.

3. He then clicks "OK" and in the Wizard window, selects "Finish."

4. This results in a report that shows Annual Usage and Demand Usage by lighting sub-category. The resulting report is the completed report as shown in Table 14-4.

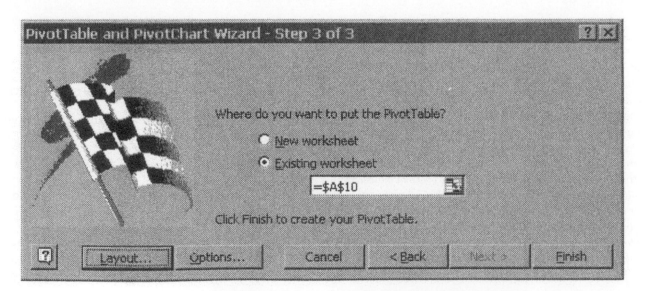

Figure 14-4. Modifying Existing Reports using the PivotTable Wizard

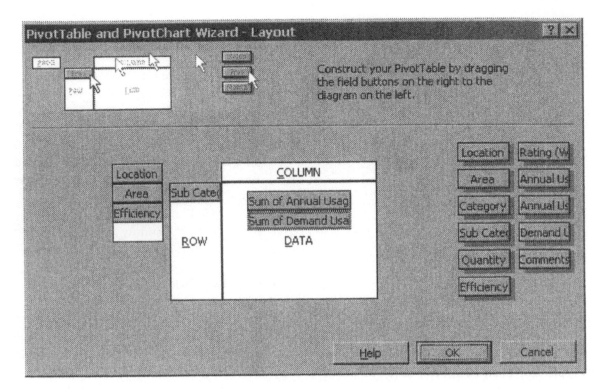

Figure 14-5. PivotTable Layout Tool

Table 14-4. Report #2: Annual Usage (kWh) and Demand Usage (kW) by Sub-Category

Sub Category	Annual Usage (kWh)	Demand Usage (kW)
Fluorescent	366,658	62
HPS	830,448	95
Incandescent	140,093	19
Metal Halide	970,214	156
Grand Total	**2,307,413**	**332**

Tip: You might have figured out by now that it is possible to create dynamic computations or formulas within pivot tables (right-click on any DATA measure title and select Formulas-Calculated Field). For instance, Bob could just as easily have computed demand usage (kWh) using a pivot table formula instead of computing it in the datasheet. These are two ways to approach the same problem, each with its pros and cons. Computing formulas in pivot tables keeps the original datasheet small and simple but at the same time, it takes up computation time each time you pull up a pivot report as the computations are being done dynamically. As a general guideline, it is better to perform commonly used computations up-front and use pivot tables for formulas that are only used in a few reports.

Report #3: Annual Usage (kWh) and Demand Usage (kW) by efficiency— i.e. high, low, medium.

To create this report, Bob starts from scratch.

1. First, from the "Layout" window, he removes all items from the ROW area.

2. Next, he adds "Efficiency" to the ROW area.

3. He makes sure Annual Usage (kWh) and Demand Usage (kW) are in the DATA area and properly formatted.

4. Bob now wants to drill deeper into the Low-efficiency devices, and get a prioritized list of the top 5 energy saving opportunities from the low-

Report #2: Analysis

Looking at the report #2, it is clear that metal halides and HPS lamps account for a bulk of the kWh usage. Also, metal halides account for nearly half of the total annual demand (kW). If reducing annual kWh usage is a big priority, obviously, these two lamp types are the best targets. To show the power of pivot tables, with a single-click, Bob can drill deeper into metal halide and HPS usage. First from the row drop down, Bob deselects all other lamp types except metal halide and HPS. Then, he drags "Location," "Area" and "Rating (Watts)" fields onto the Row area, which explodes Report #2 into detailed usage. Figure 14-6 shows a chart view of this report (remember: clicking the chart icon automatically creates a chart of the pivot table report).

From Figure 14-6, it is quite clear that the two big areas of kWh usage are the 150 480-Watt MH lamps in the "Main Storage Area" and 316 300-Watt HPS lamps in the Shop floor. Together, they account for 1.5 Million kWh (approx. 75% of the total 2.3 million kWh usage by Plastico). A 10% reduction in this usage either through wattage reduction or usage reduction will result in savings of 150,000 kWh—Bob quickly estimates that to be worth nearly $15,000 at 10 cents/kWh, which does not include any savings from demand (kW) usage reduction. Also, a 20% reduction in metal halide wattages will lead to approximately 30kW monthly demand reduction (actual peak demand reduction is typically less and depends on usage patterns). At $10/kW/month, this represents roughly $3,600 of annual savings. Bob jots this away for later and continues analyzing further trends.

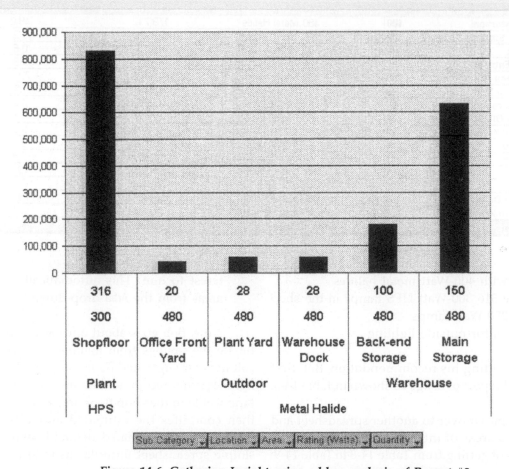

Figure 14-6. Gathering Insight using ad-hoc analysis of Report #2

efficiency devices. To do this, Bob takes Report #3 and first adjusts the layout.

5. First he drags Area onto the ROW section and moves Efficiency to the PAGE section. After clicking "Finish" in the layout window, from the pivot table, he selects only "Low" from the "Efficiency" drop-down box in the PAGE section of the report. In the resulting report, Bob adds "Rating (Watts)" and "Sub Category" to the ROW section which leads to a listing of all the low-efficiency devices along with their kWh and kW usages. (Table 14-6)

6. The next step is to narrow down this list to the top

Table 14-5. Report #3: Annual Usage (kWh) and Demand Usage (kW) by Efficiency

Sub Category	(All)	
Location	(All)	
Area	(All)	

Efficiency	Annual Usage (kWh)	Demand Usage (kW)
Low	1,940,755	270
Medium	366,658	62
Grand Total	2,307,413	332

Table 14-6. All Low-efficiency Devices

Efficiency	Low				

Area	Quantity	Rating (Watts)	Sub Category	Annual Usage (kWh)	Demand Usage (kW)
Back-end Storage	100	480	Metal Halide	179,712	48.0
Cardboard Storag	36	90	Incandescent	28,382	3.2
Entryway	8	60	Incandescent	4,205	0.5
Mailroom Entry O	32	60	Incandescent	4,762	1.9
Main Storage Are	150	480	Metal Halide	630,720	72.0
Maintenance Hall	28	60	Incandescent	3,931	1.7
Mens Lockeroom	54	60	Incandescent	28,382	3.2
Mezzanine Office	68	60	Incandescent	35,741	4.1
Office Front Yard	20	480	Metal Halide	42,048	9.6
Paint Office	4	60	Incandescent	2,102	0.2
Paint Room	22	60	Incandescent	11,563	1.3
Plant Yard	28	480	Metal Halide	58,867	13.4
Shopfloor	316	300	HPS	830,448	94.8
Warehouse Dock	28	480	Metal Halide	58,867	13.4
Womens Lockero	40	60	Incandescent	21,024	2.4
Grand Total				1,940,755	269.9

age areas) with 400-Watt metal halides
2. Replace the 316 300-Watt HPS lamps in the shop floor with 250 Watt lamps
3. Install timers for outside lighting

To begin building his recommendation, Bob first creates a detailed report of usage as shown in Table 14-8.

Next, he copies it over to another spreadsheet and only retains the 3 areas of interest to him (Table 14-9). There are 2 ways of getting from Table 14-8 to Table 14-9:

Method 1: From the Area dropdown, Bob can deselect all areas of no interest

Method 2: This is a lot easier (and is obviously the one Bob prefers!) than Method 1. In the pivot table, Bob deletes the rows that contain areas of no in-

terest to him. This automatically deselects these areas from the Area drop-down.

Now, Bob goes about adding columns to help estimate the savings from making the changes. He adds columns for suggested hours (applicable to the outdoor lighting timer recommendation) and suggested wattage (applicable to the shop floor and warehouse lamps). He then computes the estimated annual energy savings (kWh/yr) and estimated demand savings (kW) using simple spreadsheet formulas as follows:

For Warehouse and Plant areas:

Estimated Demand Savings (kW) = Estimated Wattage Reduction (Watts) (0.001 kW/Watt

Estimated Annual Energy Savings (kWh/yr) = Estimated Demand Savings (kW) (Annual Usage (hours/yr)

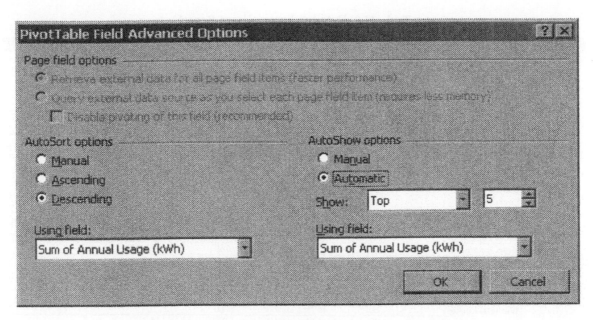

Figure 14-7. Sorting and Filtering Results in a Pivot Table

Table 14-7. Top 5 Low-efficiency Areas (by Annual kWh Usage)

Efficiency	Low			
Area	**Rating (Watts)**	**Sub Category**	**Annual Usage (kWh)**	**Demand Usage (kW)**
Shopfloor	300	HPS	830,448	94.8
Main Storage Are	480	Metal Halide	630,720	72.0
Back-end Storage	480	Metal Halide	179,712	48.0
Warehouse Dock	480	Metal Halide	58,867	13.4
Plant Yard	480	Metal Halide	58,867	13.4
Grand Total			1,758,614	241.7

Report #3: Analysis

Report #3 suggests that a high-efficiency lighting upgrade will greatly benefit Plastico. There are primarily 3 opportunities that stand out:
— The 300 watt HPS Lamps in the shop floor
— The 480 watt Metal Halide Lamps in the warehouse (2 storage areas)
— The 60 to 90 watt Incandescent Lamps all over the Offices

Looking at Figure 14-8, the first two account for nearly 80% of all low-efficiency kWh usage. Therefore, these two areas are immediate priority areas. The office upgrades are also important because this has minimal impact on operations.

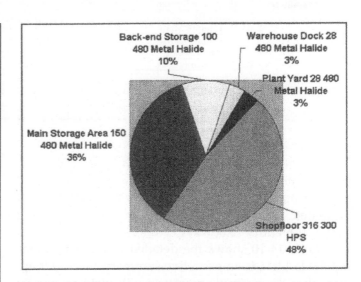

Figure 14-8. Top 5 Low-efficiency Areas (by Annual kWh Usage)

Table 14-8. Detailed Energy Usage Report

Location	Area	Rating (Watts)	Quantity	Sub Category	Ann	Annual Usage (kWh)	Demand Usage (kW)
Office	Back Office	48	76	Fluorescent	2480	9,047	3.6
	Break Room	48	14	Fluorescent	2480	1,667	0.7
	Cafeteria	48	156	Fluorescent	2480	18,570	7.5
	Closet room	48	4	Fluorescent	2480	476	0.2
	Conference Roor	28	24	Fluorescent	2480	1,667	0.7
	Conference Roor	28	12	Fluorescent	2480	833	0.3
	Conference Roor	48	8	Fluorescent	2480	952	0.4
	Entryway	60	8	Incandescent	8760	4,205	0.5
	H Room	48	24	Fluorescent	3968	4,571	1.2
	Laboratory	28	92	Fluorescent	8760	22,566	2.6
	Literature room	48	12	Fluorescent	2480	1,428	0.6
	Lobby	48	56	Fluorescent	2480	6,666	2.7
	Locker Room	48	28	Fluorescent	2480	3,333	1.3
	Mailroom Entry C	60	32	Incandescent	2480	4,762	1.9
	Main Personel O	28	76	Fluorescent	2480	5,277	2.1
	Mens Lockeroom	60	54	Incandescent	8760	28,382	3.2
	Mens Restroom	48	4	Fluorescent	8760	1,682	0.2
	Mens Restroom E	28	8	Fluorescent	8760	1,962	0.2
	Press Office	109	16	Fluorescent	8760	15,277	1.7
	Reception	48	12	Fluorescent	3968	2,286	0.6
	Snack Room	48	16	Fluorescent	2480	1,905	0.8
	Womens Lockerc	60	40	Incandescent	8760	21,024	2.4
	Womens Restroo	48	4	Fluorescent	8760	1,682	0.2
	Womens Restroo	28	8	Fluorescent	8760	1,962	0.2
Office Total						**162,183**	**35.8**
Outdoor	Office Front Yard	480	20	Metal Halide	4380	42,048	9.6
	Plant Yard	480	28	Metal Halide	4380	58,867	13.4
	Warehouse Dock	480	28	Metal Halide	4380	58,867	13.4
Outdoor Total						**159,782**	**36.5**
Plant	Cafeteria	48	96	Fluorescent	8760	40,366	4.6
	Employee Devel	28	44	Fluorescent	2480	3,055	1.2
	Main Maintenan	28	20	Fluorescent	2340	1,835	0.8
	Maintenance Hal	60	28	Incandescent	2340	3,931	1.7
	Maintenance Off	28	64	Fluorescent	2340	4,193	1.8
	Mezzanine Office	60	68	Incandescent	8760	35,741	4.1
	MIS Center	28	28	Fluorescent	8760	6,868	0.8
	Offices	48	72	Fluorescent	8760	30,275	3.5
	Paint Office	60	4	Incandescent	8760	2,102	0.2
	Paint Room	60	22	Incandescent	8760	11,563	1.3
	Shopfloor	300	316	HPS	8760	830,448	94.8
Plant Total						**970,377**	**114.8**
Warehouse	Back-end Storag	480	100	Metal Halide	3744	179,712	48.0
	Cafeteria	48	128	Fluorescent	8760	53,821	6.1
	Cardboard Stora	90	36	Incandescent	8760	28,382	3.2
	Individual Office	48	60	Fluorescent	3968	11,428	2.9
	Loading Area	48	24	Fluorescent	8760	10,092	1.2
	Main Storage Are	480	150	Metal Halide	8760	630,720	72.0
	Working Area	48	240	Fluorescent	8760	100,915	11.5
Warehouse Total						**1,015,070**	**144.9**
Grand Total						**2,307,413**	**332.0**

For Outdoor area:

Estimated Annual Energy Savings (kWh/yr) = Estimated Annual Usage Reduction (hrs/yr) (Rating (Watts) (Quantity (0.001 kW/Watt

Table 14-10 shows the detailed recommendations and Table 14-11 is an executive summary. Overall it appears that these simple measures will yield an annual savings of over $34,000.

Step 5: Reporting and Collaboration

This section will describe how Bob quickly communicates his findings and analyses with his team members in the utility company as well as with his client, Plastico. Bob does not have extensive web-site authoring skills and will again depend on the basic capabilities in MS Excel© spreadsheets to report and collaborate his results.

Table 14-9. Key Areas for Recommendations

Area	Location	Rating (Watts)	Quantity	Sub Category	Annual Usage (hours)	Annual Usage (kWh)	Demand Usage (kW)
Shopfloor	Plant	300	316	HPS	8760	830,448	94.8
Main Storage Area	Warehouse	480	150	Metal Halide	8760	630,720	72.0
Back-end Storage	Warehouse	480	100	Metal Halide	3744	179,712	48.0
Warehouse Dock	Outdoor	480	28	Metal Halide	4380	58,867	13.4
Plant Yard	Outdoor	480	28	Metal Halide	4380	58,867	13.4
Office Front Yard	Outdoor	480	20	Metal Halide	4380	42,048	9.6
Grand Total						1,800,662	251.3

Table 14-10. Energy Recommendation Details

Area	Location	Rating (Watts)	Quantity	Sub Category	Annual Usage (hours)	Annual Usage (kWh)	Demand Usage (kW)	Suggested Hours	Suggested Wattage	Energy Savings	Demand Savings	$ Savings
Shopfloor	Plant	300	316	HPS	8760	830,448	94.8	same	250	138,408	15.8	$ 15,737
Main Storage Area	Warehouse	480	150	Metal Halide	8760	630,720	72.0	same	400	105,120	12.0	$ 11,952
Back-end Storage	Warehouse	480	100	Metal Halide	3744	179,712	48.0	same	400	29,952	8.0	$ 3,955
Warehouse Dock	Outdoor	480	28	Metal Halide	4380	58,867	13.4	3,650	same	9,811	0	$ 981
Plant Yard	Outdoor	480	28	Metal Halide	4380	58,867	13.4	3,650	same	9,811	0	$ 981
Office Front Yard	Outdoor	480	20	Metal Halide	4380	42,048	9.6	3,650	same	7,008	0	$ 701
Grand Total						1,800,662	251.3			300,110	35.8	$ 34,307

Table 14-11. Energy Recommendation Summary

Area	Location	Sub Category	Rating (Watts)	Recommendation	Quantity	Energy Savings	Demand Savings	$ Savings
Shopfloor	Plant	HPS	300	Reduce to 250 Watts	316	138,408	15.8	$ 15,737
Main Storage Area	Warehouse	Metal Halide	480	Reduce to 400 Watts	150	105,120	12.0	$ 11,952
Back-end Storage	Warehouse	Metal Halide	480	Reduce to 400 Watts	100	29,952	8.0	$ 3,955
Warehouse Dock	Outdoor	Metal Halide	480	Install Timer and reduce usage	28	9,811	0	$ 981
Plant Yard	Outdoor	Metal Halide	480	Install Timer and reduce usage	28	9,811	0	$ 981
Office Front Yard	Outdoor	Metal Halide	480	Install Timer and reduce usage	20	7,008	0	$ 701
Grand Total						300,110	35.8	$ 34,307

First, Bob decides what he wants to share with whom. In this case, Bob has a very simple collaboration need:

- Post the final recommendations (Table 14-10 and Table 14-11) as well as Figure 14-6 (Top 5 low-efficiency usage areas) in an internet web site and send a link to Plastico so that they can view the results

Given recent advances in internet publishing, there are several ways of achieving the above two objectives—starting with email (directly to the concerned people) and all the way to the other extreme end which involves creating a sophisticated interactive web site that requires a user-id and password to access the files.

In this exercise, we will show how Bob converts all the charts/tables he wants to share with partners, to "HTML format." In doing so, Bob will have the information in a format that is compatible with every operating system and will be easily rendered in a browser on any machine. It is also possible to link these HTML files within a web site easily using simple "URL tags" (Please talk to your web administrator to help you post your HTML files on your web site or on the company intranet) and a basic editor like Microsoft® FrontPage™.

For the first task, Bob takes snapshots of the three reports in Table 14-10, Table 14-11 and Figure 14-6 and saves them as separate Excel© spreadsheets, say, "Plastico_Detail.xls," "Plastico_Summary.xls" and "Plastico_Usage.xls" respectively.

Next, he opens "Plastico_Detail.xls" and then selects File->Save As. In the resultant window, he selects "Web Page" in the field "Save As type" (Figure 14-9). He leaves the file name as the same. He also clicks on the button "Change Title" and renames the web page as "Plastico Detailed Recommendations" and then clicks "OK" in the "Set Page Title" window (Figure 14-10). He also clicks OK on the "Save As" window.

This creates a file named "Plastico_Details.htm." This file can now be easily embedded in any web page.

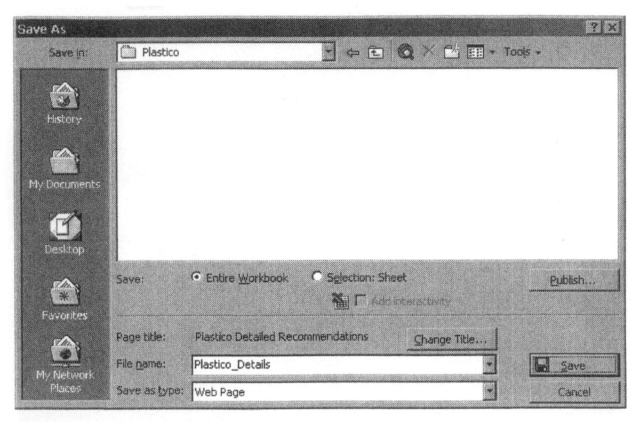

Figure 14-9. Saving Spreadsheets as "Web Pages"

Figure 14-11 shows an example of this report that has been linked into a web site.

Similarly, Bob creates HTML versions of the other reports he wants to share and links them up to the same web site to with his peers and with Plastico.

CONCLUSION

This tutorial has explained how to use Excel for easy energy use calculations. The tutorial is available on the web at: http://www.utilityreporting.com/spreadsheets/IT_2003_Plastico_Sample_Data.xls

Using the techniques shown in the tutorial, Bob was able to slice and dice a large spreadsheet of data, perform quick analyses and narrow down to the top energy saving opportunities in two hours. Further he was able to instantly compute the estimated savings and share that via the internet with his customer and his peers. Bob is no technology wizard and yet he was able to use everyday spreadsheet technology to perform such a comprehensive technical analysis.

This oversimplified case study was easily handled by Excel and it is quite possible that a larger or more sophisticated audit would require more advanced decision support tools and maybe a full fledged database to hold the data. However, the features contained in spreadsheets can easily handle a large majority of energy management analysis. Also, as spreadsheets continue to mature and advance, newer versions will allow users to handle larger data volumes, provide more computational algorithms and finally, provide more internet publishing capabilities.

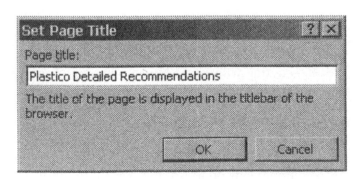

Figure 14-10. Renaming a Web Page Title

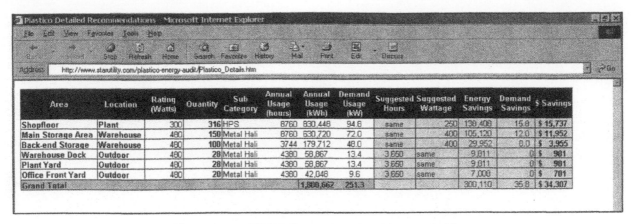

Figure 14-11. HTML Report Rendered in a Web Site

Chapter 15

Fundamentals of Database Technology and Database-Driven Web Applications

Fangxing Li

Database technology involves the access and manipulation of information. It is critical to the development of highly efficient information systems. It also plays an important role in the development of web-based applications that require information processing and distribution between web-browsers and web-servers. This chapter provides a quick guide to database technology and illustrates common structures of database-driven web applications.

INTRODUCTION

Like many other information systems, utility information systems contain and process a large amount of data. Without the ability to manage that data efficiently, it is difficult for utilities to provide satisfactory services to customers. The development of information technology has answered this challenge. Databases and database management systems (DBMS) have been broadly deployed to manage bulk data in enterprise information systems.

Database

A database is a self-describing collection of data. The term "self-describing" implies that the database contains not only the actual data, but also the structure of the data (or the meta-data).[1-3]. A database may achieve high integrity because the meta-data typically describes the relationships among different tables. This feature of "self-describing" is the main difference between a database and a flat file that was used in the early age of computing. A flat file does not contain any information about the structure and relationships among different pieces of data, and therefore is less integrated than a database.

DBMS

A DBMS is a software tool that helps users define, access, and maintain the underlying data contained within a database. As the definition shows, a database is the collection of structured data, and a DBMS is a software program that helps users manage the database efficiently. Figure 15-1 describes the logic interaction between a user, a DBMS and a physical database.

There are many commercial DBMS products available from different vendors. Microsoft's Access is a popular example of a desktop DBMS. Microsoft's SQL Server is an example of an enterprise DBMS that works across a network for multiple users. Other popular DBMSs are IBM's DB2, Oracle's series of database management products, and Sybase's products.

DATABASE TABLES

Typically, a database is organized as a collection of multiple two-dimensional tables. Each entry in a table is single-valued. Each row in a table represents an instance of a real world object. Each column in a table represents a single piece of data of each row.

Figure 15-2 shows two tables in a database of a utility information system. Each row in the first table represents the information about a different metering device. Each row in the second table represents the amount measured by a specific water meter at a given time.

In the first table, there is a specific column ID, which uniquely identifies a metering device in the real world. This column is the primary key of this table. The primary key of a table may be a combination of several columns. For example, the primary key of the second

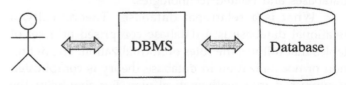

Figure 15-1: Information flows among a user, a DBMS and a physical database

235

Metering_ Equipment Table

ID	Name	Utility	Facility	Manufacturer
X1001	GasA1	GAS	BLDG1	M101
X1002	ElecA1	ELECTRIC	BLDG1	M201
X1003	WaterA1	WATER	BLDG1	M301
X1011	GasB1	GAS	BLDG2	M101
X1012	ElecB1	ELECTRIC	BLDG2	M201
X1013	WaterB1	WATER	BLDG2	M301

Water_Consumption Table

ID	Measurement_Date	Measurement_Time	Gallons
X1003	2/1/2003	12:00:00	341.23
X1003	2/1/2003	12:15:00	355.68
X1003	2/1/2003	12:30:00	362.42
X1003	2/1/2003	12:45:00	377.81

Figure 15-2. Two tables in a utility information system

table is the combination of ID, Measurement_Date and Measurement_Time. That is, the remaining columns (in this case only one column—'Gallons') are uniquely determined by the combination of the first three columns.

RELATIONAL DATABASES

As the previous definition shows, a database contains structural information about the data as well as the actual data. How is the structure defined? Or, what is the structural model? The most popular model over the past 20 years is the relational model. Other models include the hierarchy model and the network model that existed for some time but did not gain considerable market share. There are also some emerging models like the object-oriented model and the object-relational model, both of which are gaining some market share. However, in general the relational model is still the dominant force. Hence, this chapter focuses on relational databases and related technologies.

What is a relational database? Theoretically, a relational database is a database comprised of a set of tables that follows the rules of normalization. The definition of normalization in database theory is complicated if expressed in a mathematical way. For simplicity, the primary principles of normalization can be roughly interpreted as the following guidelines:

1. When you design a table in a database, your table should avoid repeating groups.

2. The columns in a table depend on the primary key only.

3. There is no column depending on another column that is not part of the primary key.

The "repeating groups" problem is illustrated by the following example. A table is created to store all purchase orders, while each order may have different items. The first table in Figure 15-3 shows an un-normalized design, which can handle orders with no more than two items. If the maximum number of items is 30, then the columns must be expanded to contain 30 groups of item and quantity, i.e., from {Item1, Qty1} to {Item30, Qty30}. This could cause a serious waste of space if most of the other orders have less than 5 items.

The second table in Figure 15-3 shows a normalized design, which involves only four columns. The column Sub_Order_ID is used together with Order_ID to avoid the repeated grouping problem in the first table. In other words, if two rows have the same Order_ID, then the items in these two rows are associated with the same order. The column Sub_Order_ID can be used to identify different items within the order. With this design, there is no limit on the number of items within

Purchase_Order Table: Un-normalized Design

Order_ID	Item1	Qty1	Item2	Qty2
1	Circuit Breaker	4	Transformer	1
2	Sectionalizer	2	Distributed Generator	1

Purchase_Order Table: Normalized Design

Order_ID	Sub_Order_ID	Item	Qty
1	1	Circuit Breaker	4
1	2	Transformer	1
2	1	Sectionalizer	2
2	2	Distributed Generator	1

Figure 15-3. Two designs for a purchase order table

one order. Further, there is no redundant information stored and efficiency is achieved.

The term "depending on" mentioned in the above guidelines can be interpreted as "being uniquely determined by." That is, if the column A depends on the column B, then the value of A is uniquely determined by the value of B, but not vice versa. For example, if we know the email of a person, then we know his or her name, title, address, etc. Thus, the column of name, title, or address depends on the column email. However, if we know the name of a person, it is possible that we cannot identify his or her email since people may have the same names. In the second table in Figure 15-3, the columns Item and Qty each depend on the combination of the columns Order_ID and Sub_Order_ID.

In order to achieve efficient database systems, the normalization rules or the above rough guidelines should be followed. Practical tests show that an un-normalized database may result in much poorer performance (5+ times slower) and need much more programming involvement. Although database designers may consider normalization by intuition without knowing it, they should be required to explicitly follow the rules to ensure high performance and efficiency. The performance and efficiency issue is particularly important for web-based database-driven applications, since users of web-based applications may experience longer delays than users of stand-alone applications. The longer delays may be attributed to the following features of web applications:

- There may be many clients (users) concurrently accessing the server.
- The users may be geographically distributed across the country.

SQL

SQL is a standard to create, retrieve and manipulate a relational database. Originally, SQL stood for the acronym of "Structured Query Language," but now it has become generally accepted as a non-acronym standard to access the internal data of a database, usually a table-based relational database.

Unlike full-featured programming languages such as C/C++, VB and Java, SQL is not a full-fledged programming language. It may be considered as a sublanguage to create, retrieve, and manipulate information contained in databases. It can be dynamically coded into high-level languages like C/C++, Java or VB to facilitate the control of the underlying databases.

SQL consists of a set of text-based commands or queries to control data. The SQL commands can be classified into two major categories, Data Definition Language (DDL) and Data Manipulation Language (DML). DDL is used to create tables or change table structures, while DML is used to insert, delete or modify rows of tables. The DDL commands include statements of CREATE, ADD, DROP, ALTER and others. The DML commands include statements of SELECT, INSERT, UPDATE, DELETE, JOIN, UNION and others.

The following brief examples are given as a quick

guide to explain how the SQL commands work. The examples are based on the needs of a utility information system administrator who wants to create a database table to host information, populate the table, manipulate the table, etc.

1.   Create a Table
The following command creates a blank table with the similar schema as the second table in Figure 15-2.

    CREATE TABLE Water_Consumption(ID TEXT(20), Measurement_Date DATE, Measurement_Time TIME, Gallons DOUBLE)

The above CREATE command creates a table named Water_Consumption. The table has four columns called ID, Measurement_Date, Measurement_Time and Gallons. The data types of these four columns are a text string of 20 characters, Date, Time and Double, respectively. The Date data type is usually input in the format of "MM/DD/YEAR" or "YEAR/MM/DD." The Time data type is usually input in the format of HOUR:MINUTE:SECOND with the HOUR filed using a 24-hour clock. For example, "18:45:00" is input for 6pm, 45 minutes and 0 seconds.

2.   Populate a Table
The following command adds a row into the table Water_Consumption. It should be noted that a text column is enclosed by opening and closing quotes (single or double). Columns in Date or Time data type should be enclosed in quotes as well.

    INSERT INTO Water_Consumption VALUES ('X1003', '1/1/2003', '18:45:00', 165.82)

To add many rows into the table, users may use the command in the format like "INSERT INTO target SELECT ...†FROM source," in which the "SELECT... FROM" statement will be mentioned next.

3.   Select Data
The most popular command used in SQL probably is the SELECT statement. The following command selects all rows and all columns from a table.

    SELECT * FROM Water_Consumption

The * represents all columns in the selected table. Users may select partial columns by specifying the actual column names. For example, the following SQL command selects all rows but only 'ID' and 'Gallons' columns.

    SELECT ID, Gallons FROM Water_Consumption

There are also various clauses that can be appended after the above SELECT statements to filter some rows. For example, the following command selects the information only related to Meter X1003 using WHERE clause.

    SELECT ID, Gallons FROM Water_Consumption WHERE ID='X1003'

4.   Delete Data
The following command deletes all rows from the table Water_Consumption.

    DELETE * FROM Water_Consumption

The WHERE clause can be used as a filter for DELETE statement. The following command deletes the rows from meter X1003 and with a date no later than 12/31/2001.

    DELETE * FROM Water_Consumption WHERE ID='X1003' AND Measurement_Date<'01/01/2002'

Since this chapter is not a detailed SQL guide, the above examples do not cover all aspects of SQL commands. For details about SQL, users may check references 1 and 2. Despite its simplicity, this section is expected to serve as a quick start for further SQL studies.

SQL CODED IN OTHER PROGRAMMING LANGUAGES

Although SQL is a powerful tool specifically designed for database access, it is not a full-featured programming language. Hence, to maximize the benefit of SQL in applications, embedded SQL is used. That is, SQL is coded into a programming language like C/C++, VB or Java in a database-driven application. The programming language is employed to perform the common "programming" tasks while the embedded SQL queries are utilized to access the database. This interactive process can be described as follows:

1.   The application creates a connection to a database so that the host application can "talk" with the database and its tables.
2.   The application generates a SQL query to obtain a table in the database.

3. The content in the table is retrieved and then mapped to the internal data structure of the application.

4. Operations on the data are carried out. This could be very simple or complicated depending on the application's requirement.

5. The updated data may be saved back into the database and output may be generated.

The VB code shown in Figure 15-4 illustrates a sample process of the five steps above, including how to set up a database connection, retrieve data from a database, identify rows with gallon amount over a predefined threshold, and generate a warning report. The report file contains all metering records with a gallon amount over the threshold.

In short, SQL queries could not do much but access the database. Programming languages are more flexible and powerful in many other functions, but not in direct database access. Hence, the combination of these two is a good choice to create fast, efficient, and easy-to-program applications involving underlying databases.

DATA ACCESS INTERFACES

After reading the previous example, readers may have this question: "How does the database receive the SQL query, interpret it and then send the response back to the VB application?" To answer this question, the mechanism of database access interfaces is explained.

Database access interfaces are software modules that provide connections between an application and a specific database. They play a key role in implementing database-driven applications. Different vendors may have different database drivers. At times, this could cause portability and extensibility problems since users may have to deal with different vendors and even different platforms. An early solution to this problem was the ODBC (Open Database Connectivity)[4] technique provided by Microsoft. The ODBC module sits between

```
Dim dbs As Database, rst As Recordset
Dim OutputFileName As String
Dim k As Long
Dim TheID As Long, TheDate As String, TheTime As String, TheGallons As Double

'Connect to a database
Set dbs = OpenDatabase("MyTestEIS.mdb")
'Create a SQL query to obtain the database table Water_Consumption
Set rst = dbs.OpenRecordset("SELECT * FROM Water_Consumption")
If rst.RecordCount = 0 Then
     MsgBox "No record in the table."
     Exit Sub
End If

OutputFileName = "OutputTest.csv" 'Output to a CSV spreadsheet file
Write #1, "ID", "DATE", "TIME", "GALLONS" 'Output a header

Open OutputFileName For Output As #1 'Open an output report file
For k = 1 To rst.RecordCount
     'Retrieve the field and store it in VB's internal data structures
     TheID = rst("ID")
     TheDate = CStr(rst("Measurement_Date")) 'Get date in string format
     TheTime = CStr(rst("Measurement_Time")) 'Get time in string format
     TheGallons = rst("Gallons")

     'Perform operations on the extracted data. Here, GetThresholdFor()
     'is a function to obtain the warning threshold of a metering device
     Threshold = GetThresholdFor( TheID )

     Write #1, TheID, TheDate, TheTime, TheGallons
     rst.MoveNext
Next k
Close #1

'Close the database connection
rst.Close
dbs.Close
```

Figure 15-4. An Example of VB and SQL Queries

applications and vendor-specific databases to provide the necessary connectivities.

Microsoft has recently replaced ODBC with Universal Data Access (UDA), which provides access to all kinds of data sources like ODBC databases, traditional SQL data, non-SQL data like spreadsheets, etc. UDA is the database access part of Microsoft's Component Object Model (COM), which is an overall framework for creating and distributing object-oriented programs in a network. UDA consists mainly of the high-level application program interface (API) called ActiveX Data Objects (ADO) and the lower-level services called OLE DB. SQL queries are sent to ADO interfaces from applications and then forwarded to OLE-DB interfaces. OLE-DB communicates with vendor-specific data providers to retrieve information from physical databases. The retrieved information is then sent back to the applications. This database access strategy is shown in Figure 15-5.

Sun Microsystems presents another data access technology, Java Database Connectivity (JDBC)[5], which is an API that can be used to access almost any tabular data source from the Java programming language. The JDBC interface provides cross-DBMS connectivity to a wide range of SQL databases. The latest JDBC API also provides access to other tabular data sources, such as spreadsheets.

JDBC architecture contains a driver manager and database-specific drivers to provide transparent connectivity to databases from different vendors. This is shown in Figure 15-6. A JDBC driver translates standard JDBC calls into a protocol that the underlying database can understand. This translation function makes JDBC applications able to access many different databases. There are four distinct types of JDBC drivers. Details about the four drivers and their mechanisms are beyond this chapter but can be found in reference 5.

An interesting and noteworthy point is that Sun's JDBC technology supports multiple operating systems but is restricted to the Java programming language. As opposed to JDBC, Microsoft's UDA technology is restricted to the Windows platforms but supports multiple languages like VB, C/C++, J++ and the latest .NET technology.

Although a thorough discussion of the technologies on database access interfaces is beyond the scope of this chapter, it is helpful to readers to understand the basic concepts. With the assistance of these interfaces, applications can essentially "talk" with physical databases and can efficiently retrieve the information contained in various databases. Also, database access interfaces like ADO and JDBC can minimize application developers' efforts, because developers only need to deal with the ADO or JDBC rather than do the detailed work that the interfaces perform.

DATABASES AND WEB APPLICATIONS

Like stand-alone applications, web applications[6-8] may rely on database technologies for efficiency. The architecture of a database-driven web application is illustrated in Figure 15-7. The information flows and activities are also illustrated with the arrow-lines. Here is the description of the activities in the figure.

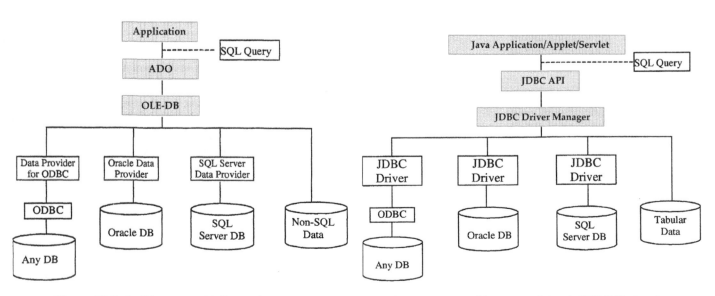

Figure 15-5. Architecture of Microsoft's UDA   Figure 15-6. Architecture of Sun's JDBC Technology

1. A client browser sends an HTTP request to a web server for a specific web page that may contains regular HTML code as well as code written in server-side script (SSS).

2. The web-server receives the request. If the requested web page involves SSS code, the web server invokes a SSS engine to process the SSS code.

3. If the SSS code involves database operations, the SSS engine queries the database through database interfaces to obtain the needed results, and may generate part of the returning HTML code based on query results.

4. The web server creates a response HTML page that is a combination of the returning HTML code from the SSS engine and the regular HTML code in the original web page.

5. The response HTML page is sent back to the client browser and the browser displays it for users.

The server-side script (SSS) is employed to generate dynamic web pages, which may require database manipulations. Since HTML is a markup language designed mainly for information displaying, it cannot handle complicated computation and database access. To make a web-server more powerful, some scripts are usually embedded into an HTML page to direct the web-server to perform some specific tasks. The web-server invokes the SSS engine to handle complicated tasks like database access. The SSS engine passes the database queries to the database interface/driver that communicates with the physical databases.

Typically, the associated database interface/driver is located at the server side. This makes many database-related operations at the server-side transparent to the client side. This is an advantage of web-based applications, because users do not need to worry about the complicated database access and set-up processes. Once everything at the server side is set up, users from any place with internet access can benefit from the web application.

SERVER-SIDE SCRIPT TECHNOLOGIES FOR DATABASE-DRIVEN WEB APPLICATIONS

There are several server-side script technologies that can assist developers to implement database-driven web applications. Two of them, Active Server Pages (ASP) and FoxWeb, are briefly reviewed in this section.

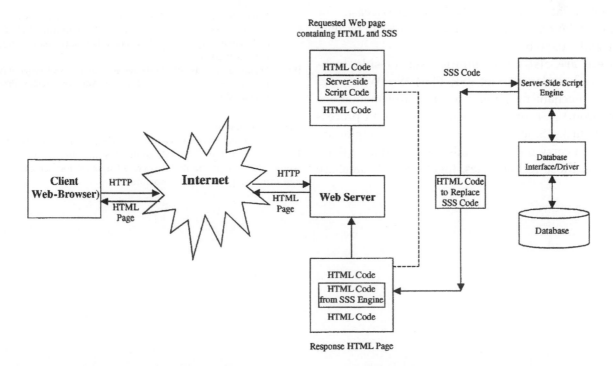

Figure 15-7. Generic Architecture of a Database-driven Web-based Application

ASP

ASP is the Microsoft's technology for building dynamic and interactive web pages. The overall architecture of ASP-centered web applications is similar to the generic architecture depicted in Figure 15-7. The main difference is that the so-called ASP script engine replaces the SSS engine in Figure 15-7. The ASP script engine can handle various requests including intensive database manipulations. The requests are coded in ASP scripts that are usually written in the VB Script language. An ASP web page is a text file with the extension of *.asp that contains HTML code and ASP scripts.

FoxWeb

FoxWeb is another technology that enables developers to create dynamic web pages, especially if the pages involve underlying FoxPro databases. The overall architecture of FoxWeb-centered web applications is also similar to that depicted in Figure 15-7. The FoxWeb script engine is a specific SSS engine. Like the ASP script engine, the FoxWeb script engine can handle complicated database manipulations, especially for FoxPro databases. This makes FoxWeb very attractive to developers who need to convert legacy desktop applications powered by FoxPro databases to web-based, FoxPro-driven applications. Similar to an ASP web page, a FoxWeb web page is essentially a text file with extension of *.fwx that contains HTML code as well as FoxWeb scripts. Also, the latest FoxWeb scripting object is compatible with Microsoft's Active Server Pages (ASP) objects. This makes it easier for developers who are already familiar with ASP to become familiar with FoxWeb.

There are also other similar server-side technologies such as JavaServlet, JSP, PHP, WestWind, etc., which use mechanisms similar to (but implementations different from) ASP or FoxWeb technologies to process server-side tasks. Since all of the above technologies are developed to carry out server-side tasks including database access, those technologies, together with database technologies, are the driving force of the evolution of web applications. For details of the server-side technologies, readers may refer to other chapters in this book.

CONCLUSION

Like many other information systems, utility information systems usually employ database technology to store and retrieve data to achieve high performance and efficiency. Database technology is especially important to web-based information applications since a large amount of data needs to be processed at the server-side and distributed to geographically remote clients. As a quick tutorial and guide, this chapter reviews the basics of database technology such as relational databases, SQL, and database access interfaces. The chapter also provides an illustration about common architecture of database-driven web applications.

References

[1] Raghu Ramakrishnan, *Database Management Systems*, McGraw-Hill, 1997.

[2] Jesse Feiler, *Database-Driven Web Sites*, Morgan Kaufmann Publishers, Inc., 1999.

[3] Paul Dorsey and Joseph. R. Hudicka, *Oracle 8—Design Using UML Object*, Oracle Press, 1999.

[4] Kyle Geiger, *Inside ODBC*, Microsoft Press, 1995.

[5] Cay Horstmann and Gary Cornell, *Core Java*, vol. 2, The Sun Microsystems Press, 2001.

[6] Chris Ullman, et al, *Beginning ASP 3.0*, Wrox Press Ltd., January 2000.

[7] Fangxing Li, Lavelle A.A. Freeman, Richard E. Brown, "Web-Enable Applications for Outsourced Computing," *IEEE Power and Energy Magazine*, vol. 1, no. 1, (Premier issue) January-February 2003.

[8] Dustin R. Callaway, *Inside Servlets: Server-Side Programming for the Java® Platform*, Second Edition, Addison Wesley Professional, May 2001.

Acknowledgment

The author would like to thank Mr. David Green and Dr. Barney L. Capehart for their valuable comments and suggestions. The author would also like to thank Ms. Lynne Capehart for her careful editing and formatting of the final document.

Chapter 16

BCS Integration Technologies— Open Communications Networking*

Tom Webster, P.E.

The purpose of this chapter is to provide energy managers with some basic informational tools to assist their decision making process relative to energy management systems design, specification, procurement, and energy savings potential. It is important for energy practitioners to have a high level of knowledge and understanding of these complex systems.

This chapter will focus on building control system (BCS) networking fundamentals and an assessment of current approaches to open communications protocols. Networking is a complex subject and the networks form the basic infrastructure for energy management functions and for integrating a wide variety of OEM equipment into a complete EMCIS.

Note: Please refer to the Glossary of this book for a complete listing of all acronyms and their definitions.

INTRODUCTION

The two primary driving forces behind the vast changes that are occurring in BCS communications networking technology are: 1) technological change, and 2) the open systems movement. Technological change was discussed in Chapter 8; this chapter will focus on open systems.

In general, open systems embrace three major concepts: open source software, open communications protocols, and open data exchange. Open data exchange includes standardization of databases, data objects, and data presentation software (e.g., browsers). While a major driver for open systems derives from the user's desire for simplification, interoperability, and low cost, one of the strongest drivers comes from the internet

and the move to "web-enable" virtually all applications. While all of these categories of open systems will have an impact on BCS development, we will focus here on open protocols and open data exchange. Furthermore, since a discussion of open systems cannot be divorced from a discussion of standards, standards issues will be interwoven into the open systems discussions.

The central focus of open protocol efforts in the BCS industry is the standardization efforts by BACnet and LonMark, and the corresponding changes the BCS vendors are making in their proprietary offerings to accommodate openness. A similar process is occurring in other industries, most notably industrial process control and information systems (IS), which will also affect BCS development.

In assembling this assessment, we have relied heavily upon building control system experts, product literature, white papers, technical papers, and news and journal articles. Our intent is to provide an impartial and accurate portrayal of the state of practice with emphasis on evolutionary trends and emerging technologies.

NETWORKING FUNDAMENTALS

It is essential to have a grasp of basic networking concepts to be able to clearly understand the issues as well as to interpret the information being provided by vendors and consultants. This chapter covers the following concepts: (1) network architectures as they apply to the BCS industry, (2) networking fundamentals, including a short primer on protocols, and (3) the contending approaches to open protocols.

Network Architectures

Although product and technical literature contains descriptions of various BCS network architectures, it is the evolution of these architectures and how they fit into

*This chapter was originally published as a report for the Federal Energy Management Program, New Technology Demonstration Program, Report No. LBNL—47358

a broader perspective that is most important. The on-going convergence of technologies (voice, data, video) and the increasing internetworking of communications infrastructures are the hallmarks of the information age.

Figure 16-1 shows a generalized view of the inter-relationship between network types and how BCS networks relate to others. This diagram illustrates the most common arrangements being developed today but does not indicate the vast array of legacy systems that make up the bulk of the installed base.

Networks can be broken down into two fundamental types:
1. **Point-to-point**—store and forward, or switched WANs that pass messages through a network node by node, and
2. **Broadcast**—multiple access or multi-drop LANs that (typically) use baseband* signaling with various access methods to share a single channel; i.e., each node sees packets sent to other nodes.[1]

BCSs can be characterized by the four-level architecture shown in Figure 16-1. While this hierarchical structure predominates in the buildings industry, it is being "flattened" by merging levels together as the technology evolves; e.g., sensor and terminal bus merged together so sensors and terminal controllers co-exist on the same sub-network, or field panels and terminal unit controllers on the same bus.** However, in general, the sensor bus is not yet implemented as a discrete layer, although it is being developed in the industrial process industry. For BCS networks the trend is toward more internetworking just as it is for IS networks. In addition, as IS protocols become more standardized, and components less expensive, they are migrating further down the network hierarchy (see discussion below about Opto 22). One reason this is important is because of the impact it is having on the development of EMCIS standards.

As indicated in Figure 16-1, the sensor bus connects sensors and actuators together and interfaces to the *terminal bus* level (referring to HVAC terminal equipment unit controllers such as VAV boxes, fan coils, etc.), which in turn connects to field panels at the BCS *backbone* level.

Above the backbone level are various levels of EIS networks that ultimately connect to WANs. Typically EISs are client-server based, which distinguishes them from real-time peer-to-peer control networks. Client-server functions are contained in a set of protocols that sit on top of the networking protocols that are most familiar to the building control industry. Integration with the EIS is important in distributing control system information to higher-level EIS applications. In fact, one of the major drivers for change in BCS technology is the ability to port control system information into the EIS environment and thus service a much more diverse set of enterprise information needs than previously was possible.

Table 16-1 is a "roadmap" showing the relationship between the typical BCS network architecture and the various protocols used. This list includes common protocols that are or will be candidates for the architectural level shown. These are included to provide the reader with a basic framework for tracking the evolution of protocol development efforts that affect the BCS industry.

Networking Protocols & Standards, Basic Concepts
Protocol

A protocol is a detailed, structured method of communicating certain types of information. When we use the term "protocol" in reality we are referring to a "protocol stack," i.e., a series of protocols that are used to send messages between devices (nodes) on a network or between different networks. When discussing protocols it is useful to use the OSI multi-layer model for a protocol stack that was created by ISO and is outlined in Table 16-2. The message, or packet, consists of various headers that are significant to the services being performed at each layer as well as for interfacing between the layers (i.e., addressing, contention resolution, etc.), and a data element that contains the application information. The lower four layers are considered "connectivity" layers, and the upper three "application" layers. For BCS applications, only the top application layer is generally used.

The workhorse protocol for WANs has been the TCP/IP stack that was originally developed for the internet but now is used routinely for intranets (LANs within enterprises) as well. Of course, there are many other protocols being used in various types of LANs and WANs but there appears to be a steady migration in newer LANs towards the TCP/IP/ethernet* standards. Although this makes internetworking easier it still does

---

*Baseband refers to digital signaling whereas broadband refers to analog signaling. LANs and WANs use both of these techniques.

**In actual practice flattening depends more on the functions of a system, its size, and the desire to segregate communications traffic than on technology availability. Merging these networks lowers costs and allows more direct integration with enterprise wide applications and WAN access and is changing the administration of these infrastructures away from the facility departments to the MIS departments.

---

*The slash between acronyms is synonymous with "over" and denotes the hierarchical relationship of the OSI model; i.e., TCP/IP means TCP (Layer 4) over IP (Layer 3).

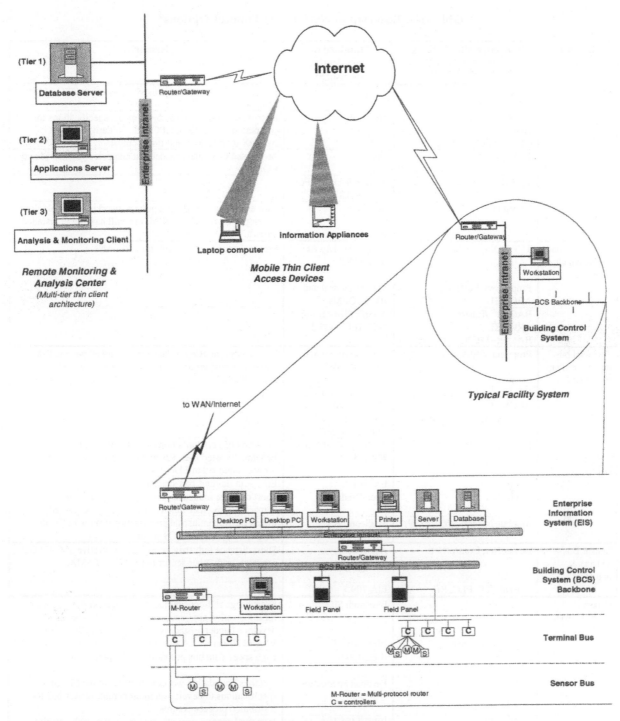

Figure 16-1. Network Architectures

not make it possible to connect LANs and WANs together without some sort of interface device. For TCP/IP intranets it is usually a "multi-protocol router" since the signaling protocols for the two networks are different (see discussion below about networking protocols). All of the networks can be classified under the terms "distributed processing" or "distributed intelligence" networks, as opposed to the central processing systems of old.

Standard

More care should be exercised in the industry when this term is used. We reserve this term for "true" or *de jure* standards, i.e., those promulgated by recognized standards organizations such as ANSI, ASHRAE, IEEE, EIA, ISO, and ITU. To apply this term to propri-

Table 16-1. Roadmap of Networking Protocol Options*

Network Level	Protocol Suite Options* Note: / indicates 'on top of' or 'over' data link/physical layers.	Standard or Specification Sponsor**	Remarks
Enterprise/ Intranet	CORBA IIOP	OMG	Client-server object standards via consortium//SIG of 800 companies
	COM, DCOM DDE OLE,ActiveX,OPC	Microsoft	Microsoft's client-server object model specifications that compete with CORBA. OPC is the initiative to adapt ActiveX to industrial real-time control.
	Java RMI	Sun	Java based C-S object model that competes with COM and CORBA
	TCP-IP/Ethernet	Internet Society RFCs/ IEEE 803.2	
	Novell IPX SNA	Novell IBM	Two examples of early computer networking technologies developed by Novel and IBM that still have large installed bases
BCS Backbone - Field panels (Control)	TCP,UDP-IP/Ethernet	Internet Society RFCs/ IEEE 803.2	Protocol suite of choice for intranet LANs
	Proprietary/Ethernet, ARCNET	Various control and HVAC OEMs	
	BACnet® /Ethernet, ARCNET BACnet-IP/Ethernet	ANSI/ASHRAE −135-1995, IEEE-803.2	
Terminal bus - unit controllers (Fieldbus)	Proprietary/EIA-485	Various control and HVAC OEMs	Typically proprietary master-slave or token passing low speed, low cost protocols used ubiquitously by BCS vendors since 1980's.
	Proprietary/Ethernet, ARCNET	Various control and HVAC OEMs	
	BACnet MS,TP/EIA-485	ANSI/ASHRAE −135-1995, IEEE-803.2	
	LonTalk/TP, PLC, RF	ANSI/EIA-709.1 and 709.2,.3	Adoption of Echelon's LonTalk. Requires a license from Echelon. Twisted pair, RF, and powerline carrier standards for are contained in EIA-709.2,.3
	Profibus	EN 50 170	European fieldbus bus standard
	DeviceNet	Allen Bradley	Short distance industrial sensor bus derivative of CAN that competes with other field & sensor buses. Uses CSMA.
	Modbus/Modbus+//TCP//IP	Modicon	Token passing industrial process protocol for PLCs now supported by several BCS vendors
Sensor Bus – sensors & actuators	DeviceNet, ControlNet	Allen Bradley, Rockwell	Short distance industrial sensor bus derivative of CAN that competes with other sensor buses. Uses CSMA.
	LonTalk/TP, PLC, RF	EIA-709.1	
Others (Emerging protocols to watch)	Bluetooth	Bluetooth SIG	Short-range self-configuring wireless networking protocol for information appliances (e.g., PDAs, etc.) being developed under auspices of 1200 member Bluetooth SIG
	Wireless Ethernet	IEEE 803.11	Wireless Ethernet LAN protocols being developed by IEEE.
	Firewire	IEEE 1394	High speed data bus primarily for PC peripherals
	Fieldbus	Fieldbus Foundation	Fieldbus Foundation works to implement the ISA SP50 specifications to develop an interoperable sensor bus for distributed process control.
	IEEE 1451	IEEE 1451.1-.4	Industrial process sensor/actuator interoperable interface standards.

* We do not attempt to distinguish individual OSI layers here, only indicate the various protocol suites that are used or under development for the particular architecture level shown. Parentheses indicate typical industrial process level designations.

** In this table we include both *de jure* standards as well as "specifications" being developed by SIGs and private companies.

Table 16-2. OSI Protocol Model

An easy way to conceptualize protocols is by using a layered model. Each layer performs specific functions and interfaces to the layers above and below. While the generalized model used by ISO (called the OSI model) consists of 7 layers, most HVAC protocols utilize only 3-5 layers (typically Layers 1-3,4 and 7).

Layer	Description	Function	Examples
Layer 7	Application	Interface to applications logic; data object interfaces	HTTP (HTML/XML web pages); SMTP, SNMP (email), Telnet, FTP BACnet
Layer 6	Presentation		
Layer 5	Session		
Layer 4	Transport	End to end security; assembly and ordering of message fragments	TCP UDP
Layer 3	Network	Addressing and Routing	IP BACnet LonTalk
Layer 2*	Data Link (LLC+MAC**)	Contention resolution, medium access, channel allocation	TDM FDM WDM (DWDM) ATM CDMA (wireless) CSMA (Ethernet) PPP, SLIP (for IP protocols) Token passing (ARCNET) Master/Slave
Layer 1*	Physical	Signaling, topology and media	EIA 232,485 10,100BaseT SONET (fiber) RF,FHSS, DSSS (wireless) ISDN

*Note that many LAN standards combine these two layers into one specification; i.e, Layer 2 and Layer 1 are not necessarily "mix and match."
** The MAC sub-layer is a very important component that specifies the packet structure and the rules for how a device accesses the network.

etary or *de facto* protocols tends to confuse and mislead and usually results from overzealous marketing efforts. Just as important as standardization is the availability of multiple suppliers; a standard that does not capture enough of a market to warrant multiple suppliers fails to achieve one of the primary goals of standardization—low cost from competition and thus easy access to a uniform implementation.

Even worse is the overuse of the designation "open system." Open can have just about any meaning one wants to put on it. As generally applied in the BCS industry, "open system" usually refers to those systems that allow connection by alternative protocols used by competing vendors. This definition has little to do with standards, proprietary systems can be interconnected by virtue of an agreement between two parties (see discussion below). "Open" implies the willingness to cooperate but the devil is in the details. For example, when a license is required to use a protocol, it represents a barrier to use and may compromise the concept of "free and open access to any and all parties." The author believes that a protocol cannot be truly "open" unless it is a standard as defined here. The process of standardization makes the specifications available to all comers without restriction or preconditions at virtually no cost. Even this definition is not completely adequate, since for example, implementing ANSI/EIA-709.1* requires a license from Echelon Corp.**

LAN Characteristics and Performance

A detailed discussion of networking technology is beyond the scope of this chapter but a basic understanding of some key characteristics of LANs typically used in BCSs is important to frame the following discussions about BACnet and LonMark. There are three data link protocol technologies that are important to BCS networking: Master-slave (M-S), token passing (TP), and carrier sense multiple access (CSMA).

The master-slave protocol has been the most popular choice for lower levels in the network due to its simplicity and flexibility. This technique relies on a master node to orchestrate traffic on the network; the master initiates all transactions usually by polling the slaves. Although this is not technically a peer-to-peer

solution, in practice it can be made to work like one.

Token passing schemes are "multi-master" and operate by passing a token from node to node. Only when a node holds the token can it transmit a message.

CSMA is the technique used in ethernet (IEEE 802.3) and LonTalk. CSMA methods are fundamentally different than the two above. M-S and TP are orderly non-contention schemes while CSMA is a contention method where each node begins to transmit a message at will unless it detects* a collision with a message being sent at the same time by another node. Various types of probabilistic backoff algorithms are used to schedule retries.

Each of these has advantages and drawbacks. There is no perfect solution, and each was invented to solve certain problems and/or to overcome limitations from previous developments. The pros and cons can be evaluated to some extent from the following key factors:**

- **Efficiency**—ratio of message data bits to total bits in a packet; accounts for overhead in the protocol.
- **Throughput**—ratio of actual packet transmission rate relative to line speed signaling rate; depends on how efficiently network traffic is managed.
- **Response time**—the time it takes to respond to a request for data; important for time critical operations.
- **Determinacy**—the consistency in delivery of message packets; are there varying or known delays.***
- **Peer-to-peer**—the ability of nodes to talk directly to one another without a master being involved.
- **Signaling method**—baseband vs. broadband; i.e., digital vs. analog signaling; also determines cost of implementation.
- **Data rate**—raw signaling rate, i.e., bandwidth.
- **Complexity**—relates to difficulty of implementation in terms of initial cost, memory size, and maintenance.
- **Topology**—the geometric layout of the nodes and wiring. Bus, star, and ring are the configurations of most importance to BCS networks.
- **Bus length and number of nodes**—the combination of length of wire and number of nodes that can be supported without repeaters.

All of these factors influence one another in de-

---

*ANSI/EIA-709.1-A-1999 is the ANSI/EIA standard that incorporates Echelon's LonTalk protocol.[3]

**When a license is required there is another critical concern—who determines interoperability. To preserve openness, a separate, independent certification organization is required to allow competing vendors a certification method that does not require the "blessing" of the company granting the license.

---

*Technically this protocol is called CSMA, CD where the CD denotes collision detection..

*Most of these factors result from the MAC and physical layer specifications, but upper layers also will have an affect.

***This issue is being overcome for CSMA schemes by advances in the technology; see the discussion under the BACnet section.

termining overall performance. For example, for CSMA networks, speed and bus length determines contention slot timing and therefore the minimum packet size required. Bus length is inversely proportional to speed; the lower the speed the longer the bus can be. Table 16-3 provides an overview of the three data link protocols of interest in terms of these key factors.

When considering the pros and cons it is good to bear in mind Tannenbaum's view. He points out that numerous studies of these different access schemes have been conducted and have not identified any clear winner in terms of performance (i.e., throughput, efficiency, etc); they all work reasonably well so frequently factors other than performance drive the choice of one over the other.[3]

Interoperability

This is a complex subject. However, for our present purposes, we assume that the Holy Grail of interoperability is "plug and play"—the ability to substitute devices of equivalent *functionality* for one another without special tools and configuration. This allows end users to enjoy the benefits of easily interchanging devices from multiple vendors that work the same way as one another.

Communications connectivity (meaning the lower four layers of the protocol stack) is assumed to be compatible within this concept. However, compatible connectivity itself is no assurance of interoperability. Interoperability is facilitated by the adherence to standards in the upper layers, primarily the application layer (Layer 7). Object standards, are application level methods for facilitating interoperability. Interoperability requires that all disparate applications adhere to the same object standards. Furthermore, use of these objects may be necessary for interoperability, but in and of themselves they do not guarantee equivalent *functionality* which is in the domain of the control logic.

Integration

Closely allied with interoperability, integration connotes the interfacing of multiple systems of distinct

Table 16-3. Data Link Protocol Comparison

	CSMA	TP	M/S
Peer-to-peer	Yes	Yes	No
Complexity (relative)	Medium	Medium -High	Low
Signaling method (typical for BCS industry)	Baseband IEEE 802.3, EIA-485 (and derivatives)	Baseband IEEE 802.4,5 or EIA-485	Baseband EIA-485
Data rates (typical bandwidths for BCS)	39 kps – 100 Mbps	19.4 Kbps – 10 Mbps	9.6 kps – 78 Kbps
Determinate	No	Yes	Yes
Efficiency	Decreases with speed increase since minimum packet size increases. Low (10-15%) for short messages due to minimum packet size.	Medium to high - Increases with load and message size	Medium to high - Increases with message size
Throughput	Excellent at low to medium loads, saturates at high loads due to increase in number of collisions	Excellent at high load	Low to medium
Response time	Fast for low traffic loads, degrades with load	Slow relative to CSMA for low loads due to token passing time, but no degradation for high loads	Slow
Topology	Star, bus	Bus, Ring, Star	Bus
Bus Length	500 m – 2.5 km	~600 m	~1200 m
Other	Variations in implementation have potential to improve on determinacy and performance	Susceptible to lost token which complicates protocol	Requires a master node which makes the network susceptible to its failure and compromises response time and throughput.

functionality such as HVAC, lighting, security, access, fire and life safety. While interoperability is useful to accomplish system integration it is not sufficient; many other issues must be resolved for these complex system to interact in a seamless and synergistic way. For example, deciding which data to exchange between the systems and what an appropriate system response should be to data input from other systems.

Networking Components
Gateway, Routers, Bridges and Repeaters

There is much confusion (as well as marketing hype) generated around these concepts. In an attempt to avoid being bogged down in technical nuances inherent in these concepts, we believe HVAC practitioners and energy managers will benefit from using the conceptual framework illustrated in Table 16-4. All of the devices listed in Table 16-4 are considered at minimum two-port devices that support interfacing functions between two message streams. Remember that there is a "protocol" associated with each layer of the OSI "protocol stack" and a message (or packet) uses all of the available layers.

The five component types listed in Table 16-4 perform a service (action or "translation") of some kind between the two message streams at the protocol layer(s) indicated.* It is only the gateway, however, that actually performs a translation of the *data* portion of the packet. This is why the gateway generally has limited functionality, is a customized device, and requires support and maintenance.** As such it can be an expensive undertaking. Implementation of a robust gateway is estimated to cost $20-50K and take 3-6 months to design, program and test. However, for simple data objects it can be considerably cheaper and in fact has been routinely done in the form of "drivers" in the industrial process industry for a number of years at costs as low as $2-5K.

Likewise in the BCS industry this gateway function is also routinely supported in field panel level controllers that support sub-networks of various other controllers (see Figure 16-1). Some vendors such as Johnson Controls (JCI), have made gateways a fundamental part of their business model; e.g. JCI's Metasys Connectivity Partners program which claims to support protocols from over 100 other companies. Newer web-

based product companies such as Silicon Energy also rely heavily on gateways to allow access to legacy networks to support their energy monitoring and analysis software products. An excellent discussion of gateways can be found in[4].

Tunneling routers obviate the need for a gateway to pass messages through different types of networks; e.g., using the internet to collect data from remote buildings. In tunneling, the entire message (data, addressing, etc.) of a given protocol is "wrapped" (i.e., contained in the *data* portion of the packet) in a secondary protocol as opposed to making a translation. The message travels between two nodes of the tunneling (secondary) network that are each in turn connected to primary networks. At the destination the message is unwrapped and placed on the primary network port untouched. The interface devices between the two network types are called "multi-protocol routers" since they are performing a routing function between two different networks but also generally include support for different data link and signaling formats. Tunneling is being used extensively to support interconnection between networks via the internet, and within enterprise ethernet based intranet LANs.

Driver

Driver is a term frequently used as a catchall phrase for code used to interface a protocol to a device. It is similar to an application-programming interface (API) in that it provides a means for interfacing the protocol to a platform's computing resources/OS and thereby performing a gateway like function of translation into the platform's native schema.

OPEN COMMUNICATIONS APPROACHES

Proprietary Networks

Proprietary protocols have been the workhorses of earlier generations of BCS networks. They have become robust due to continual upgrading over long periods of time and because the vendors had a vested interest in ensuring a reliable infrastructure for their control devices. Since one vendor provided virtually everything, users had a single point of responsibility to address problems. Many users prefer to work with a single providers system simply to reduce complexity.

The Present

Proprietary networks use the same protocol layers and techniques described earlier. Generally, they have evolved from early generations that used simple col-

---

*Layers 5 and 6, (Session and Presentation) are not listed here because they are generally not used in BCS protocol stacks.
**Gateways can become complex because they must link domains that may not share the same ideal of what objects are and how their associated methods perform; the gateway has to know a lot about both object domains to successfully bridge between them. To simplify this process, the object translations can be done at low levels in the network thereby reducing the burden on higher level objects.

Table 16-4. Networking Components

Protocol				Gateway	Tunneling or Multi-protocol Router	"Pure" Router	Bridge	Repeater
Layer	Description	Function	Device Function	Route/forward and translates data frames of packets between dissimilar networks	Route/forward packets between dissimilar networks	Route/forward packets between similar networks	Store and forward data link frames	Copy bits between cable segments; regenerate weak signals
7	Application	Data object interfaces		X*				
4	Transport	End to end reliability		X				
3	Network	Addressing and routing		X	X	X		
2	Data Link	Contention resolution		X	X		X	
1	Physical	Signaling		X	X			X
			Examples	• LonMark to BACnet • Proprietary to BACnet or LonTalk	• LAN-WAN router • LonTalk to IP/Ethernet • BACnet/Ethernet to IP/LAN, WAN (Annex H tunneling) • LonTalk iLON** • LonTalk to BACnet	• BACnet routers • LonTalk Routers • LAN-LAN and WAN-WAN routers		• Media extenders • LonTalk repeaters • EIA-485 repeaters

X indicates which layers are involved in providing the services of the indicated device; the functionality of the other layers is the same for both ports.

* Although generally a gateway device supports two entirely different protocols and therefore supports routing functions, it's the application layer "translations" that are key to the gateway concept. A special form of gateway may involve only the applications layer; where all other functions remain unchanged except a translation is made for the application data objects. This could occur, for example, if BACnet objects were used in a LonTalk network.

** The iLON now being developed by Echelon performs both tunneling router functions and gateway functions. The latter is inherent in supporting a web-server thus requiring a translation between a BCS protocol and HTTP for HTML support.

lapsed three layer structures. For the lower layers, many vendors have used specifications very similar to one another and/or used older *defacto* standards such as Modbus, Opto22, or simple EIA-485 master-slave protocols running at 1.2 to 9.6 kbps. Most of these protocols have been upgraded to higher speeds due to the availability of better transceivers and embedded processors. Also, IT developments have fostered the upgrading of the lower layers of these proprietary networks to more modern LAN protocols such as ethernet and therefore provide much better support for greater data transfer demands. However, the proprietary nature of these protocols is derived not so much from the lower layer specifications but from the Applications layer implementations that were and still are custom solutions.

Openness with this approach is based on vendor-supported access; i.e., agreements between vendors to support each other's needs. This has worked well and can be considered an alternative to a standards-based approach. As mentioned previously, this approach follows the model that the industrial process industry has used for many years.

There are four basic ways that proprietary networks support openness:

1. **Open access protocol**—Equipment OEMs provide specifications for an access protocol to their equipment controllers that allows network providers to integrate the equipment into a BCS network usually using a gateway device or integration module of some sort that is interfaced to the proprietary network. JCI supports at least 75 third-party vendors using this approach.

2. **Open network protocol**—BCS providers supply open access via published bus protocol specifications (usually upon request and at lower levels of the network hierarchy) to allow implementation by others directly into the BCS network. This then becomes a third party vendor supported gateway to third party devices. JCI's Open N2 offering is a good example of this approach.

3. **Gateways**—Gateway support for the BACnet standard or LonMark (see section below). Most of the major BCS vendors now support both technologies.

4. **Front ends**—System integration provided by support of third party protocols at the front-end workstation. JCIs Unity workstation product is an example of this approach.

The Future

Complications have arisen with the proprietary openness approach because users are now demanding that these traditional networks support integration with other vendors' equipment, legacy systems, and other vendor sub-networks in a seamless way. Since the rate of adoption of new standards is slow and there is a large installed base of proprietary networks, and because the major BCS providers still want to retain control over the supply of BCSs, there will always be a mix of proprietary and standards-based approaches to networking similar to what exists today. However, the lower layers are rapidly becoming standardized either via IT developments or by HVAC industry efforts such as BACnet. The Application layers however, will most likely remain proprietary in the core of the major vendors networks until there is more widespread adoption of either BACnet or LonMark. However, support for these two approaches alongside proprietary offerings has, and will continue to, become more widespread. It is also likely that BCS vendors will continue to provide an alternative integration mechanism by supporting a wide number of third party protocols.

BACnet and LonMark*

BACnet, the standard protocol suite that ASHRAE has developed, and LonMark, a protocol suite developed by Echelon Corp. based on their LonTalk protocol, currently represent the two main contenders for BCS standardization. Although there is competition between these two technologies, one thing is clear—there will be no "winner."[5] These two technologies (and others) will share in being options for EMCIS specifications. This competition is most intense at the terminal and sensor bus levels of the network where the control devices are located. BACnet and LonMark devices located on the same bus are incompatible with one another. Thus the more established one protocol becomes, the greater the potential for revenues based on it. Of course both of these protocols are competing with established proprietary offerings by BCS vendors.

This is somewhat analogous to the situation in other industries where newly developed protocols and standards are continually competing with older ones that make up a large installed base. In fact we contend that all of the BCS protocols will be impacted to a large extent by ongoing advancements in information technology. The market for IT is orders of magnitude

---

*Henceforth in this chapter when we refer to LonMark we assume that it represents the LON protocol suite that is comprised of LonMark application level objects and the LonTalk protocol. See Appendix A for a more complete description of the LON Technology.

larger than those for BCS and industrial processes so component prices are low, standardization efforts are greater, and capabilities are ever increasing. However, although IT dominates the landscape, each industry still needs its own standard objects and services to integrate with IT standards in order to support industry specific requirements. It is this set of industry specific objects and services that defines the real value to efforts such as BACnet.

Major efforts have been mounted in the industrial process arena to adapt IT to real-time control applications. PC based SCADA, and OPC, are examples of these efforts, but the most notable is the continuing development of ethernet. Vast changes in the ethernet standards are being made to make it more suitable for real-time control and thus a candidate for the lower layers for virtually all levels of the network.[7-11] Opto 22 offers a digital I/O product today that uses TCP/UDP-IP over ethernet for transmitting sensor data. It even includes a web server so that this device can be accessed over the internet with a web browser. Cisco and GE have recently formed an alliance to pursue factory automation based primarily on the realization that ethernet is now ready for widespread use in real-time networks.*[13]

BACnet

BACnet is a good example of a "true" or *de jure* standards based technology. The development of BACnet has been long and difficult, but significant progress has been made. Most BCS vendors now support BACnet to a greater or lesser degree**, but only a few such as Alerton, Automated Logic, and Delta Controls have complete, native implementations at all levels of the network. However, there is still considerable confusion about BACnet's usefulness and impact. Some of the issues are discussed in the following comments.

The Present

BACnet was developed by a recognized standards body, ASHRAE, under a consensus process and is truly open and non-proprietary. Ignoring all arguments about technical issues and innovation impacts, this process is the best method for ensuring open standardization. This process has been, is being, and will be increasingly used throughout the world in virtually all industries in order

to level the playing field and ensure broad and uniform implementation of and access to technology.

In the original standard, BACnet developed Layer 7 applications objects and Layer 3 addressing conventions. For Layers 1 and 2, existing standards were specified; e.g., ethernet and ARCNET. Although not a standard at the time, LonTalk was also included as another data link option. BACnet also developed a version of the commonly employed master-slave and token passing schemes called MS/TP used over the popular EIA-485 signaling protocol. These latter protocols have long been the workhorses for terminal level devices in older systems. The BACnet specification also includes a point-to-point (PTP) protocol based on the ubiquitous EIA-232 physical layer. PTP is the basis for accessing networks over modems or direct connections of BACnet gateways to workstations.

With the approval of Addendum 135*a* in January 1999, BACnet fully conforms to IP standards. Technically BACnet/IP is a version of BACnet that consists of BACnet Layer 7 objects, UDP for transport, IP for addressing, and choices (typically ethernet) for Layer 2 data links. It also includes support for broadcast messages. This is a significant development in that BACnet/IP devices can operate on standard TCP/IP networks using widely available IT networking components. It also greatly reduces the need for gateways to access BACnet networks. BACnet internetworking options are summarized in Figure 16-2.

The Future

There are those that argue that TCP/IP is not appropriate for real-time control applications.[14] The argument is that since TCP/IP protocols were primarily devised for client-server networks that do not require the robust and deterministic two-way peer-to-peer communications capabilities of control networks, they are fundamentally unsuitable for control applications. Client-server applications generally are dedicated to transactions that are large, bursty and not time critical. Control applications are just the opposite. Furthermore, TCP (Layer 4) is a "connection oriented" protocol that establishes a virtual circuit between nodes during transactions and uses many sub-transactions for acknowledgments, packet sequencing, etc. that compromise the ability to perform real-time control. UDP on the other hand is connectionless with few accouterments and thus has become the basis for adaptations for real-time applications.

Many of these arguments are being overcome by advancements in IT and control technologies. This situation is analogous to the early arguments about using eth-

---

*For the upper layers, the trend in IT is for the applications to talk to databases and web servers using IT object standards (i.e., XML) and protocols thus obviating the need for a separate set of object standards.[12]

**Forty-four companies that offer one or more product types are listed on the BACnet web site.

1. **Basic BACnet** - Original specification for LANs that uses BACnet specifications for Layers 7 and 3, and choices for data link and physical layers.

2. **Tunneling BACnet (Annex H)** - A part of the original specification that supports BACnet messages over WANs via tunneling with TCP/IP, Routers carry the burden of managing IP.

3. **BACnet/IP (Annex J, Addendum 135a)** - Added in January 1999, this capability allows BACnet to support both LANs and WANs using "true" TCP/IP; i.e., network nodes are IP addressable.

Figure 16-2. Internetworking with BACnet

ernet for control networks in that ethernet timing is not deterministic (due to the use of contention techniques). This issue has largely been overcome by the brute force of high speeds (fast ethernet at 100 Mbps and, in the near future, gigabit ethernet at 1 Gps), switching technology (allows for full duplex transmissions and private channel communications), segmented network design, and advancements in ethernet technology (e.g., prioritize messages).[10,15,16] As a result, ethernet has supplanted ARCNET (ANSI Standard 878.1, a deterministic token passing protocol) as the data link of choice in today's BCS networks. Some BCS manufacturers have migrated to supporting ethernet after basing their backbone network on ARCNET for many years. In addition, it is likely that advances in industrial process technology will, as usual, filter down to the BCS industry. The industry is pushing the development of ethernet very hard, as noted previously.

As more emphasis is placed on open object based systems and on integrating control and enterprise information systems, it would appear that mapping BACnet objects in a way that is compatible with these trends is the preferred path. BACnet object services might even be augmented by more advanced and robust client-server implementations such as CORBA, DCOM, and Java RMI.* (Only CORBA, however, is an industry consensus specification.) This trend is currently being fueled by an alliance between Tridium, Inc. and Sun Microsystems with the development of the Building Automation Java Architecture (BAJA) standard. This effort is an attempt to standardize interoperability at the enterprise level using JAVA, XML and other standard internet proto-

cols.[17] In any event, developments like these leave the industry-specific object definitions and services themselves as the primary elements of significance that BACnet brings to the table.

The availability of BACnet/IP is a major step in facilitating the wider use of BACnet but the following additional efforts also will have a major impact:

*   Conformance classes are being replaced by new BIBB specifications.* A BIBB is a collection of BACnet services that support functions such as data sharing, alarm/event management, trending, scheduling, and device and network management. These BIBBS are in turn used in standard BACnet application profiles. All of the functionality of a BACnet device (both standard and proprietary) is required to be reported in the device PICS.

*   There is much discussion in BACnet circles about developing high level objects similar to LonMark profiles that would simplify configuring and programming. However, progress has been very slow and no clear consensus has emerged as to how to proceed.**

*   Conformance testing tools, procedures and testing agents are being established. Currently this capability is embodied in the open source VTS tools and procedures developed by NIST, which are the basis for a companion standard (Standard 135.1)

---

*Since higher level applications will most likely rely on IT object standards, it is important that BACnet objects do not interfere with implementation of services at this higher level; i.e., the focus should be on the behavior of objects, not on the ultimate implementation of them.[12]

**Conformance classes may have been a good idea but the particular way they were implemented was confusing.
**In this regard LonMark might be considered to be ahead of BACnet; i.e., BACnet has only recently (1999) started attempting to develop higher-level objects similar to the LonMark profiles. And, in fact, there has been some discussion of using the LonMark profiles as a model for BACnet profiles.

to BACnet currently (4/2001) under public review. These tools have formed the basis for current conformance testing activities by the BMA.[18]

- NIST has developed a DDC guide specification for BACnet systems.[19]

- Revisions and improvements in the standard are continually being made (e.g., Addendum 135*b* contains 17 changes and additions to the standard). The BACnet committee has been proactive in tracking and adapting to new technologies, as they become available.

Adoption

At first BACnet was being adopted slowly, but now it seems to be gaining momentum. The BMA has compiled the statistics shown in Table 16-5 reflecting the state of deployment of BACnet devices as of late 2000. [20]

Note that these numbers are based on reports from only six BACnet vendors but represent about 90% of the production of BACnet devices. Note also the small number of gateways and the large number of MS/TP devices relative to the others. It should also be pointed out, however, that one manufacturer who reports 15,000 installations dominates the number of installations. The change in these statistics over time will be of key importance in assessing the ultimate penetration of BACnet.

LonMark

As opposed to BACnet, LonMark exemplifies a *de facto* standards-based technology. The strategy with this approach is to create such a presence in the marketplace that users will be compelled to use it simply because

Table 16-5. BACnet Deployment

Item	Number
Installations	19,054
Gateways	2,410
Devices Network Type	
• ARCNET	95,567
• Ethernet	11,920
• MS/TP	248,500
• PTP	1,549
Workstations	15,807
Large controllers	53,391
Unit controllers	299,600

everybody else does and ultimately to have it adopted by a standards setting body. Most BCS vendors offer LON support and a few support it exclusively (e.g., ESUSA, Circon). Even more so than with BACnet, there is considerable confusion and controversy about the overall efficacy of this technology. Appendix A contains a detailed analysis of LON technology and its attributes and limitations. The material in this section is largely excerpted from the more complete analysis that appears in Appendix A.

The Present

The LonTalk communications protocol stack (a part of the LonMark protocol suite) is modeled after the full seven layer OSI stack contrary to many other BCS systems that use much simpler 3-4 layer structures. LonTalk consists of new protocols for each layer rather than implementing existing standards. In fact the lower layers are a derivative of the CSMA technique that ethernet uses. This approach was taken so LonTalk could address a wide variety of applications in various industries and operate over various media. To some extent these changes improve on low-load efficiency and high load saturation characteristics of IEEE 802.3 protocols. It also resulted in a maximum data rate of 1.25 Mbps, although most systems seem to use 78 Kbps.* While this scheme suffers from the same issues of non-determinacy as ethernet, it seems to work well for the lower levels in the BCS architecture as long as appropriate network design is followed.

These features (plus the "packaged" LonWorks technology) denote the major innovation that Echelon has brought to the BCS industry: peer-to-peer networking technology at the terminal and sensor bus level, using twisted pair (EIA-485 type) signaling.** The FTT polarity-free, twisted pair transceivers that Echelon has developed represent a major improvement over other EIA-485 implementations.

Another key feature of this technology is that it is *hardware based* in that the technology is imbedded in proprietary Neuron chips as opposed to software based solutions that can be used on any suitable hardware platform. LON technology originally derived its "openness" from the fact that multiple vendors implemented Echelon's proprietary technology.

LonTalk (not LonMark) is now a *standard* due to its adoption by ANSI and EIA. It is still not a *de jure* stan-

---

*As of 1999 the maximum rate was increased to 2.5 Mbps.
**The other media supported by LonTalk have not seen significant use in the BCS networks.

dard as is BACnet since it was not created by a standard setting body using a consensus process. The support of LonTalk by ASHRAE and EIA are fundamentally different. ASHRAE's BACnet adopts LonTalk as a *data link* specification only; none of LonTalk's upper layers are specified (nor any of LonMark's application level objects). Specification of LonTalk does not ensure BACnet conformance; it represents only one part of BACnet conformance—only the data-link and physical layers much like the MS/TP and ethernet specifications. For true compliance, BACnet objects and networking need to be implemented. LonMark's Functional Profiles are a competing object model to BACnet's Layer 7 objects; they are not compatible with one another.

EIA, on the other hand, has adopted LonTalk layers 2-7 in EIA-709.1 and Layer 1 options in EIA-709.2 and .3 but also does not include LonMark Functional Profiles in the standard: EIA-*709 standardizes LonTalk— not LonMark—profiles*. LonTalk does not support Layer 7 applications services other than the rudimentary SNVTs that can be used to facilitate sharing of variables over a network. The LonMark Profiles use these Layer 7 SNVTs to implement the interoperability guideline conventions. Thus the LON standardization effort falls short of being a complete standard since it is still missing an essential element—a full application layer object specification.

The LonMark organization was created in 1994 to further the cause of creating interoperable LonTalk based products for various applications. This was necessary to address the deficiencies in the LonTalk application layer for supporting interoperability. LonMark is a trade association sponsored and controlled by Echelon (i.e., Echelon owns the LonMark trademark) and therefore LonMark lacks the autonomy and neutrality of an independent industry organization or standards body. Furthermore, the LonMark guidelines are not subjected to public review, as is the BACnet standard. A degree of interoperability is obtained by the voluntary adherence of LonMark members to the LonMark guidelines (i.e., implementers' agreements). Vendors that do not have products certified are unlikely to be compatible with LonMark certified devices, despite having compatible connectivity.

Conformance is based on a review of conformance documentation submitted by the product manufacturer (.ixf interface files) for adherence to mandatory and optional variable definitions; it is not necessary to submit the product itself. "Testing" in the LonMark conformance process refers to the review process, not actual vendor-to-vendor compatibility testing. A new process was under development (slated for release in the second quarter of 2000) to allow self-certification using special testing devices that were to be used in-situ on each type of device offered by a vendor. As of October 2000 there was been no mention of this on the LonMark website.

A number of special tools and technologies have been developed to address the deficiencies inherent in LonTalk to service emerging requirements (i.e., LonMark for interoperability, LNS for client-server support, iLON and LNS for internet access). Although these are important for broader integration, it results in a cumbersome development, installation and maintenance process for what is ultimately a sub-network of a larger BCS network. For smaller systems that are solely LonTalk based, LonWorks technology may make more sense. For larger systems, offerings from providers such as Tridium that have well integrated support for LonMark products (as well as BACnet) in their web-enabled architecture obviate the need for LonWorks accouterments.

The Future

On one hand, LON technology has become well established in the buildings industry as evidenced by its wide support by BCS vendors. On the other hand, future potential is mixed as summarized in the following comments.

- The packaged concept of LonWorks as opposed to the protocol itself appears to be the most compelling reason for using LonMark devices. The LON technology is a fairly complete set of tools to build products around that includes most of the necessary micro-controller, programming, and networking components as well as network management and interfacing tools. The design and development tools were built around the "one size fits all" concept to offer developers a "universal" platform for control devices. This was to obviate the need to develop low level micro-controller capabilities from scratch for each new application; a basic micro-controller platform was made available that supported "typical" functions with communications built-in from inception. On the face of it, this "black box" concept allows designers a relatively easy path to build products without having to develop low level aspects from scratch. On the other hand, this approach results in some significant compromises as discussed in the *Other Limitations* section of Appendix A that limit its future potential.

- The adoption of LonTalk by EIA (EIA-709), has resulted in all layers of the LonTalk protocol now being "opened" so that the protocol can be implemented on alternative platforms. Although a license is still required from Echelon, developers are no longer required to buy Neurons or Echelon based workstation software to use the protocol. Although opened in 1996 via EIA-709 (and via ANSI acceptance in October 1999), very few alternative implementations can be found today. This suggests that there is not great incentive to "port" the protocol to other platforms most likely because it is so wedded to the Neuron processor structure and/or there is not enough market incentive to do so.*

• Neurons are computationally slow and relatively expensive. A better option might have been to develop a chip that implements the connectivity layers in firmware without the applications layers. In any event, even this approach would be challenged by the imminent rise of ethernet as a universal connectivity standard for all levels of the network.** In 1999 Toshiba introduced upgraded versions of the Neuron that included a 20 MHz clock speed (allowing 2.5 Mbps communications bit rates) and more on-board memory. This improves the raw processing capabilities but does nothing to improve the relatively old fundamental processor technology upon which the Neuron is based, or expand its computing capabilities; e.g., although promised years ago, there are still no 16-bit versions of the Neuron.

Since ethernet is now undergoing fundamental changes to improve its real-time performance, there is less and less reason to use other data link protocols such as LonTalk. On the other hand, third parties have provided ethernet support for LonTalk allowing for its use on high-speed networks. In addition, Echelon's iLON product, LNS network operating system software, network management tools, and development systems make up a complete development and operating suite of tools that cover at least the basic requirements for web-enabled systems.

• Echelon claims that they are finally on the verge

of major cost reductions due to new integrated chips being made by Cypress Semiconductor that combine the Neuron with the FTT transceiver and because of large orders derived from the adoption of LON technology by ENEL, the Italian utility as well as other non-building industries.* It remains to be seen if this in fact comes to pass.

• Although they are de facto standards, LonMark functional profiles appear to be the only high level objects available since BACnet has yet to develop them.

• The conformance process is weak and appears to be unfinished as discussed above. Also, there is no assurance in the current process that products of different types can be made compatible. This was supposed to be addressed by a new set of "system" certification procedures being developed by LonMark but there is no indication that these procedures have been adopted. In addition, there is no explicit control over future changes in the profiles since the modifications are voted on only by a select set of preferred members, i.e., "sponsors" that pay the greatest membership fees, of which Echelon is one. A further limitation is that development, configuration, and network management tools are based on proprietary technology that is only available from Echelon and a few select vendors.

Despite the drawbacks noted herein, it appears to be a significant ramp-up in vendor acceptance of LonMark technology that may ultimately have a major impact on the overall BCS market.

Adoption
Echelon has long claimed that LonTalk was a *de facto* protocol standard even before adoption by EIA. This, however, is questionable if the test is ubiquitous installation in the buildings industry (e.g., Windows OS is truly ubiquitous and therefore a *de facto* standard in the business environment; although some would argue that it is still proprietary because Microsoft drives the specification process). For example, Echelon estimates the following breakdown of Neuron uses as of June 2000.

• 13 million nodes sold
• 45% used for BCS, 25% for industrial process,* 20%

---

*Although the LonTalk reference implementation available from Echelon allows access to the protocol, the license agreement governing its use restricts commercial development. Commercial uses of LonTalk on other platforms are subject to additional license agreements governed by Echelon.

**The real competition to LonMark is not BACnet, but TCP-IP/ethernet based products that are likely to be the focus for the future.

*Pricing levels of $2 per Neuron long promised by Echelon have never been achieved.

transportation, and 10% in miscellaneous products.
- The split is roughly 50% US and 50% non-US.

Thus it appears that approximately 2.3 million Neurons were used in the US buildings industry over the past decade. Based on the analysis contained in Appendix A it appears that most of the ~2 million nodes are dedicated to a mixture of lighting, access, residential applications, and BCS vendor offerings.** This number of nodes is a small fraction of the BCS installed base.*** Since the major equipment OEMs currently offer very few or no LonMark based products most of the volume is provided by BCS vendors.

Echelon also claims that worldwide 3500 companies are involved in developing products and that 1400 products now exist. These numbers depend heavily on how the counting is done. Echelon literature suggests that 3500 represents the number of development systems sold, not the number of products being developed for sale. Moreover, many companies produce slightly different versions of the same product. If we use LonMark listed products as an example, we find that Leviton offers 7 types of occupancy sensors, and Siemens offers 29 versions of their DESIGO RX controller for fan coils and radiant heating and cooling systems.*** If *distinct* product types were counted, the number is likely to be far less as indicated by our estimates in Table A6;**** which shows that the total distinct products is about one-half of the total products listed on the LonMark website. Likewise, LonMark claims to have over 200 member companies. However, only about 50***** companies are listed on the LonMark product list.

SUMMARY AND CONCLUSIONS

*A study conducted by Venture Development Corp. disclosed the following facts about the industrial market and LonTalk penetration of it. (1) In 1998 the total *annual* device market consisted of 24 million control devices; Echelon's estimated 3 million nodes produced over 15 years is a very small fraction of the total. (2) Ethernet is estimated to increase to 22% of the industrial market in 2003 from 8.4% in 1998. Over 75% of the market is projected to be divided between just four basic protocols. Although not explicitly mentioned, LonTalk is assumed to be included in the "others" category that accounts for 24% market share.[7]*Lack of detailed and reliable data prevents a finer breakdown.
**If market growth had matched expectations projected at the 1995 LonUsers conference where annual volumes of 100 million (downgraded to 85 million at 1996 conference) Neurons were anticipated by year 2000, then claims of being a *de facto* standard may have been legitimized. In fact only a total of 10 million chips were sold in 12-15 years.[22]
***Ironically DESIGIO systems use BACnet for the BCS backbone.
****This product list grew by about 10% in a one-year period, mostly in lighting and I/O products categories.
*****As of mid-2000.

Our major conclusions regarding the evolution of open systems networks are summarized in the following:

General
- Information technology will drive the development of EMCIS and BCS communications networks; these technologies will augment and possibly displace elements of current protocol and/or object standards.

- Ethernet is likely to become the standard for the lower layers in all levels of the network.

- Networks will be "flatter" and less hierarchical.

- Despite the increased influence of new IT based technologies, there will still be a need for the industry-specific, application-level objects and services that LonMark and standards like BACnet provide.

Proprietary Networks
- Proprietary solutions have adapted to the demands for integration and interoperability by supporting third-party protocols and emerging standards.

- Proprietary communications network offerings represent an alternative to the pure standards based approach.

- Proprietary networks will continue to be an important part of the mix of solutions for the foreseeable future.

BACnet
BACnet is still very much a work in progress. However, BACnet has a number of attractive features:

- BACnet is truly an open and complete *de jure* standard allowing implementation on virtually any computing platform of choice without licensing requirements.

- BACnet/IP will facilitate the use of BACnet in TCP-IP/ethernet networks, and the emerging standard for higher levels in the BCS architecture. It will also foster integration with IS networks and the internet.

- The imminent approval of Standard 135.1p and the advent of conformance testing by the BMA will significantly improve the conformance certification process.*

- BACnet will continue migrating to lower levels in the network.

- BACnet is having difficulty moving beyond the primitive object level to create higher-level applications objects similar to LonMark's.

- Due to its inherent flexibility, software vs. hardware orientation, and scalability, BACnet is well suited to sophisticated solutions and adaptation to technological change.

LonMark

In terms of current availability, LonMark has an apparent edge over BACnet because it has a more complete offering including support and development tools and hardware components supplemented with the LonMark conformance certification procedure.

- Hardware dependence on Neurons will limit their long-term usefulness.

- The attractiveness of the LonTalk protocol will be challenged by the imminent rise of ethernet as a universal connectivity standard for all levels of the network and will compete with LonTalk and other similar protocols.**

- LonMark is a technology best suited for low-end applications for small systems (light commercial and residential) or for lower levels in large EMCIS networks.

- The LonMark profiles represent a significant contribution toward simplifying implementation of interoperability.

- For EMCIS specifiers (and developers), *caveat emptor* should be exercised when reviewing LonWorks marketing and promotional materials (see Appendix A).

- LonMark products are distributed broadly across low-end applications, are supported by many large BCS vendors, and the installed base is growing.

Federal Facilities Perspective

It is inevitable that energy practitioners will be drawn into the controversy surrounding protocols. This is especially true with regard to BACnet and LonMark because federal practitioners place greater emphasis on adherence to standards than their commercial counterparts.

However, one should bear in mind that the primary impact of standards will be on the configuration, procurement, and integration of systems and components. Although there is tremendous interest, lack of consensus, and even controversy surrounding protocol options, energy and O&M savings are derived primarily from *applications* and not communications technology or its infrastructure (except in so far as it might improve control dynamics). Ultimately, the applications are where the true intelligence of these systems resides.* None of the options (proprietary, BACnet nor LonMark) are total solutions or panaceas. If the goal is true interoperability and vendor independence, then BACnet and LonMark can be seen as one step in the process toward this goal, but they share the solutions landscape with proprietary offerings for the foreseeable future.

Although these protocols will have no significant direct impact on operations, control, and energy use, there may be an impact on reliability and on first cost (higher initially, lower later). The protocols represent esoteric details that manufacturers and implementers might care about, but end users are primarily interested in the functionality of the system and good reliability at low overall cost (including maintenance and upgrade cost).

Thus multi-vendor interoperability and interchangeability are important issues for the end user, over and above the subtleties of how they are achieved.

Standards help because they tend to cultivate uniformity, longevity, broad support, and reliability. However, standards need to be supplemented by an appropriate conformance certification and testing process and attention to equivalent functionality.

Trademark Notices:

- LON, LonTalk, LonWorks, LonMark, SNVT, Echelon, and Neuron are trademarks of Echelon Corp.
- ARCNET is a trademark of ARCNET Trade As-

---

*Critics have pointed to the BACnet conformance issue as evidence that BACnet is not really interoperable. However, most BACnet providers have either thoroughly tested their products with other vendors on their own initiative or through the NIST conformance testing standard development project. In addition, about 60% of BACnet vendors use the Cimetrics BACnet protocol stack which Cimetrics claims has been rigorously tested for interoperability.[31]

**The real competition to LonMark is not BACnet, but TCP-IP/ethernet based products that are likely to be the focus for the future.

---

*Considerable controversy surrounds projects such as 450 Golden Gate as to the impact that implementing a multi-vendor BACnet network has had on increasing energy savings. Any savings that have resulted have been due to the changes in control logic and equipment rather than overtly to the protocols themselves.[30]

sociation
- BACnet is a trademark of ASHRAE

All other products, trademarks, company or service names used are the property of their respective owners.

Acknowledgments
    The author would like to acknowledge the dedicated support and guidance provided by Bill Carroll of Lawrence Berkeley National Laboratory (LBNL) and DOE/FEMP/NTDP for providing funding for this work.

APPENDIX A

Part of the material in this appendix has been excerpted in the preceding chapter. However, several topics are covered here in much greater detail, and the editors of this book decided to include the appendix for the reader who desired more complete information on LON technology.

LON Technology
Definitions
One must distinguish carefully between some of the major elements of Echelon's technology and related terminology in order to avoid confusion.

- **LonWorks** refers to the overall technology developed by Echelon Corp.; it includes an array of hardware and software components and tools to develop and operate LonWorks based systems.

- **LonTalk** on the other hand, refers to the communications protocol part of the LonWorks technology; it is the only part that is standardized.

- **LonMark** refers to the trade organization that Echelon formed to develop implementers agreements to promote interoperability efforts. The LonMark organization has developed a series of Functional Profiles that represent the application level object definitions that promote interoperability between LonMark devices.

- **LON** stands for local operating network, an Echelon coinage of their LAN technology.

- A **neuron** is the fundamental building block of the LonWorks technology; it is a custom microcontroller now being manufactured by Toshiba and Cypress Semiconductor (Motorola, the original

maker of Neurons, has ceased production of these chips).

The LonTalk communications protocol stack is modeled after the full seven layer OSI stack contrary to many other BCS systems that use much simpler 3-4 layer structures. This was done so LonTalk could address a wide variety of applications in various industries and operate over various media. Unfortunately, this also introduces extra complexity and overhead not generally required in BCS control systems.* LonTalk consists of new protocols for each layer rather than implementing existing standards. In fact the lower layers are a derivative of the CSMA technique that ethernet uses. This was done to optimize its performance for lower speed networking typically found in low-end applications and to allow for consistent operation over multiple media. To some extent these changes improve on low-load efficiency and high-load saturation characteristics of IEEE 802.3 protocols. It also results in maximum data rates of 1.25 Mbps although most systems seem to use 78 Kbps.** Although this scheme suffers from the same issues of non-determinacy as ethernet, it still seems to work well for the lower levels in the BCS architecture as long as appropriate network design is followed.

These features (plus the "packaged" LonWorks technology) denote the major innovation that Echelon has brought to the BCS industry: peer-to-peer networking technology at the controller and sensor bus level, using twisted pair (EIA-485 type) signaling.* The FTT polarity-free, twisted-pair transceivers that Echelon has developed represent a major improvement over other EIA-485 implementations. What is less clear, however, is if there is any significant advantage to using LonTalk or CSMA for that matter at these levels of the network. As shown in Table 16-3 there are many tradeoffs that must be considered to determine whether there is a significant advantage for the particular applications being addressed. Moreover, since ethernet is now undergoing fundamental changes to improve its real-time performance, there is less and less reason to use other technologies like LonTalk. The choice to use LonTalk

---

*The interested reader may want to review the reasons why the full OSI protocol stack is not particularity good for actually implementing communications; it has been more useful as a model for discussing layered communications protocols (see Tannenbaum[1], Section 1.4.4 for an excellent discussion of this point).
*In 1999 Toshiba introduced upgraded versions of the Neuron that included a 20 MHz clock speed (allowing 2.5 Mbps communications bit rates) and more on-board memory.
***The other media supported by LonTalk have not seen significant use in the BCS networks.

frequently boils down to the "other factors" that Tannebaum denotes.

The packaged concept of LonWorks appears to be the most compelling reason for using LonMark devices. The LON technology is a fairly complete set of tools to build products around that includes most of the necessary micro-controller, programming, and networking components as well as network management and interfacing tools. The design and development tools were built around the "one size fits all" concept to offer developers a "universal" platform for control devices. This was to obviate the need to develop low level micro-controller capabilities from scratch for each new application; a basic micro-controller platform was made available that supported "typical" functions with communications built-in from inception.

On the face of it this "black box" concept allows designers a relatively easy path to build products without having to develop low-level aspects from scratch. On the other hand, this approach results in some significant compromises as indicated in the Other Limitations section below.

Standards

LonTalk (not LonMark) is now a standard due to its adoption by ANSI and EIA. It is still not a *de jure* standard as is BACnet since it was not created by a standard setting body using a consensus process. Echelon is attempting to follow the path of other proprietary protocol developments such as ethernet and more specifically ARCNET in becoming a standard, which is first to try to become a *de facto* standard by shear volume in the market.*

The real significance of the adoption of LonTalk by EIA (EIA-709), however, is the fact that all layers of the LonTalk protocol have now been "opened" and can be implemented on alternative platforms. Although a license is still required from Echelon, one is no longer required to buy Neurons or Echelon based workstation software to use the protocol. However, prior to EIA adoption, none of LonTalk protocol layers were standards; they were all proprietary. Another key feature of this technology is that it is hardware based in that the technology is imbedded in proprietary Neuron chips as opposed to software based solutions that can be used on any suitable hardware platform. LON technology derived its "openness" from the fact that multiple

vendors have implemented Echelon's proprietary technology. Although opened in 1996 via EIA-709 (and via ANSI acceptance in October 1999), very few alternative implementations can be found today. This suggests that there is not great incentive to "port" the protocol to other platforms most likely because it is so wedded to the Neuron processor structure and/or there is not enough market motivation to do so.*

The support of LonTalk by ASHRAE and EIA are fundamentally different. ASHRAE's BACnet adopts LonTalk as a *data link* specification only; none of LonTalk's upper layers are specified (nor any of LonMark's application level objects). EIA, on the other hand, has adopted LonTalk layers 2-7 in EIA-709.1 and Layer 1 options in EIA-709.2 and .3.

Specification of LonTalk does not ensure BACnet conformance; it represents only one part of BACnet conformance—only the data-link and physical layers much like the MS/TP and ethernet specifications. For true compliance BACnet objects and networking need to be implemented. LonMark's Functional Profiles are a competing object model to BACnet's Layer 7 objects; they are not compatible with one another. *EIA-709 standardizes LonTalk—not LonMark—profiles.* LonTalk does not support Layer 7 applications services other than the rudimentary SNVTs that can be used to facilitate sharing of variables over a network. The LonMark Profiles use these Layer 7 SNVTs to implement the interoperability guideline conventions. Thus the LON standardization effort falls short of being a complete standard since it is still missing an essential element—a full application layer object specification.

LonMark Products Penetration

Echelon has long claimed that LonTalk was a *de facto* protocol standard. This, however, is questionable if the test is ubiquitous installation in the buildings industry (e.g., Windows OS is truly ubiquitous and therefore a *de facto* standard in business the environment; although some would argue that it is still proprietary because Microsoft drives the specification process). For example, Echelon estimates the following breakdown of Neuron uses as of June 2000:**

• 13 Million nodes sold.

---

*Ethernet was created by Xerox/DEC/Intel and later adopted by IEEE as IEEE 802.3. ARCNET was a tightly controlled proprietary protocol (similar to LonTalk) for almost 20 years, finally standardized in 1992 but still has only two suppliers.

*Although the LonTalk reference implementation available from Echelon allows access to the protocol, the license agreement governing its use restricts commercial development. Commercial uses of LonTalk on other platforms are subject to additional license agreements governed by Echelon.

**As of June 2000 these numbers have changed somewhat: 13 million nodes worldwide, 40/60% US/other, 25% industrial.[21]

- 45% used for BCS, 25% for industrial process,* 20% transportation, and
- 10% in miscellaneous products.
- The split is roughly 50% US and 50% non-US.

Thus it appears that approximately 2.3 million Neurons were used in the US buildings industry over the past decade. This installed base is made up primarily of OEM factory-mounted control products and field-supplied products by BCS vendors. On the OEM side, we have estimated that a total of almost 9 million units of various types of commercial HVAC equipment** have been produced by equipment OEMs over the 10 year period of 1990 to 2000. Of these we estimate that about 3 million have digital controls. Roughly 150,000 of these units are likely to have LON based controls. Based on this analysis it appears that most of the 2 million nodes are dedicated to a mixture of lighting, access, residential applications, and BCS vendor offerings.*** This number of nodes is a small fraction of the BCS installed base.****

This analysis suggests two things; 1) a substantial fraction of the HVAC equipment production is still sold *without* factory mounted controls, and 2) BCS vendors are the primary purveyors of LonMark products. Since the trend is for more factory mounting of controls, it remains to be seen how this might change over time since the major equipment OEMs currently offer very few or no LonMark based products.

Echelon also claims that worldwide 3500 companies are involved in developing products and that 1400 products now exist. These numbers depend heavily on how the counting is done. Echelon literature suggests that 3500 represents the number of development systems sold, not the number of products being developed for sale. Moreover, many companies produce slightly dif-

ferent versions of the same product. If we use LonMark listed products as an example we find that Leviton offers 7 types of occupancy sensors, and Siemens offers 29 versions of their DESIGO RX controller for fan coils and radiant heating and cooling systems.* If *distinct* product types were counted, the number is likely to be far less as indicated by our estimates in Table 16A-1.** This list totals to about one-half of the total products listed on the LonMark website.

LonMark

The LonMark organization was created in 1994 to further the cause of creating interoperable LonTalk based products for various applications. This was necessary to address the deficiencies in the LonTalk application layer for supporting interoperability. LonMark is a trade association, sponsored and controlled by Echelon (i.e., Echelon owns the LonMark trademark); therefore LonMark lacks the autonomy and neutrality of an independent industry organization or standards body. Furthermore, the LonMark guidelines are not subjected to public review, as is the BACnet standard. A degree of interoperability is obtained by the voluntary adherence of LonMark members to the LonMark guidelines (i.e., implementers' agreements). Vendors that do not have products certified are unlikely to be compatible with LonMark certified devices, despite having compatible connectivity.

Conformance is based on a review of conformance documentation submitted by the product manufacturer (.ixf interface files) for adherence to mandatory and optional variable definitions; it is not necessary to submit the product itself. "Testing" in the LonMark conformance process refers to the review process, not actual vendor-to-vendor compatibility testing. A new process was under development (due for release in the second quarter of 2000) to allow self-certification using special testing devices that were to be used in-situ on each type of device offered by a vendor.*** However, there is no assurance in the current process that products of different types can be made compatible. This is being addressed by a new set of "system" certification procedures being developed by LonMark. In addition, there is no explicit control over future changes in the profiles since the modifications are voted on only by a

---

*A study conducted by Venture Development Corp. disclosed the following facts about the industrial market and LonTalk penetration of it. (1) In 1998 the total annual device market consisted of 24M control devices; Echelon's estimated 3M nodes produced over 15 years is a very small fraction of the total. (2) Ethernet is estimated to increase to 22% of the industrial market in 2003 from 8.4% in 1998. Over 75% of the market is projected to be divided between just four basic protocols. Although not explicitly mentioned, LonTalk is assumed to be included in the "others" category that accounts for 24% market share.[7]
**These consist primarily of VAV boxes, medium to large rooftops, water source heat pumps, fan coils, and packaged terminal air conditioners (used primarily in hotel and motels).
***Lack of detailed and reliable data prevents a finer breakdown.
****If market growth had matched expectations projected at the 1995 LonUsers conference where annual volumes of 100 million(downgraded to 85 million at 1996 conference) Neurons were anticipated by year 2000, then claims of being a *de facto* standard may have been legitimized. In fact only a total of 10 million chips were sold in 12-15 years.[22]

---

*Ironically DESIGIO systems use BACnet for the BCS backbone.
**This product list grew by about 10% in a one-year period, mostly in lighting and I/O products categories.
***As of October 2000 there was been no mention of this on the LonMark website.

Table 16A-1. LonMark Products

Product class	Distinct Products*
Access	1
Energy management	3
Fire	1
HVAC	
• Chilled Ceiling	2
• Fan coil	6
• Heat pump	2
• Damper actuator	5
• Equipment controller (e.g., AHU)	5
• Roof tops	4
• Thermostat	1
• Vav Box Controller	9
• I/O products	21
Industrial	4
Lighting	17
Motor controls	4
Networking	3
Sensors	16
Other	6

* Distinct products per company times the number of companies; e.g., two companies that make the same device are counted once each, but the same company that makes variants of a product for essentially the same application gets counted once only for each application, not each variant. [24]

cussion above). This means that communications with these systems is essentially proprietary and incompatible with newer LonMark based devices. Furthermore, these are virtually inaccessible to newer remote access technologies without a gateway.[23]

Another issue is device complexity. Apparently, LonMark objects are somewhat weak in terms of being able to inherit properties of other objects. Therefore, as complexity grows new objects have to be created rather than being a separate instance of a more robust single object. This could explain why there are so many versions of basically similar devices in the LonMark list. On the other hand, LonMark has succeeded in developing higher-level objects that make implementation easier, something that BACnet is still struggling with.[12]

LonMark Acceptance

LonMark claims to have over 200 member companies. However, only about 50* companies are listed on the LonMark product list. This plus the arguments presented earlier about distinct LonMark products suggests the following possibilities:

1. There are many LonMark products still in the pipeline awaiting agreements.
2. There is a lag in the commitment to interoperability in general.
3. LonMark is not being broadly accepted in the buildings industry.

Other Limitations

LonMark nodes are generally used for terminal and ancillary devices since there are limitations on the number of variables that can be shared on the network (64 network variables), in the amount of memory that can be supported with Neuron chips, and the bandwidth of the bus and therefore the amount of traffic that can be supported. Large applications need to be supported by a number of nodes with continuous interaction, a solution that has not been wholeheartedly embraced by the industry, or by using the Neuron as a communications coprocessor with another processor for applications— a better solution but one that increases cost.** Given these limitations, LonTalk has not been used as the sole EMCIS network protocol or as a backbone to any large extent in large building systems.* The vast majority of

select set of preferred members, i.e., "sponsors" that pay the greatest membership fees, of which Echelon is one. A further limitation is that development, configuration and network management tools are based on proprietary technology that is only available from Echelon and a few select vendors.

Moreover, interoperability between devices in legacy LonTalk networks is not assured since pre-LonMark systems (prior to 1994) relied heavily on the technique of "foreign frames" in which a proprietary protocol was embedded into a LonTalk frame (see the tunneling dis-

---

*As of mid-2000.
*Some practitioners believe that LonWorks does not scale well in large applications using multi-layer architecture due to the namespace limitations of the network variables.[12]

large system BCS vendors include LonTalk networks as sub-nets of larger systems for terminal or sensor bus level devices. Thus gateways are required at some point, usually at the field panels, in the network.

Some developers contend that it takes significant effort to get around the built-in limitations and roadblocks inherent in the LonWorks approach. For example, the SNVTs are actually quite limited resources that are mostly committed (i.e., bound) during configuration. If during later monitoring one wants to acquire unbound SNVTs data, one must resort to other more arcane methods to access them. These types of limitations are inherent in "packaged" or generic solutions. Packaging results in many tradeoffs and compromises and the broader the scope of applications to be addressed by a package the more compromises there are for any given application. Added to this is the fact that the basic Neuron processor technology is old—the basic design is now over 15 years old.* Many of the limitations arise from the need to protect against overwhelming the processor's capabilities. Unfortunately, the very high level of software integration that made the Neuron so attractive in the first place now makes it immune from improvement. Neuron software is not upgradeable, so the installed base of Neurons represents a non-upgradeable legacy product. Most current technologies such as system-on-a-chip solutions incorporate flash memory that allows quick, remote upgrades of OS, protocol and application codes.

The lack of effective network management and support tools has been a major impediment to easy deployment of LonMark systems. There are some alternative platforms available for network tools, including Echelon's LonWorks Network Services (LNS) and IEC's Peak Components. Performance (such as speed of discovery of networks) is an issue, as well as other features such as platforms that they can operate on, industry standard interfaces that they support, etc. For example, LNS is designed to operate on a PC platform, while Peak was designed to operate on smaller embedded platforms as well as PC platforms. LNS has it's own plug-in interface, which it markets as a "*de facto*" standard. The Consumer Electronics Association (CEA) has recently created an open device plug-in standard for network tools (EIA/CEA-860) that is independent of network management platforms like LNS and Peak. At the time of this writing, it is unknown how widely adopted EIA/CEA-860 will become in the future.[25]

These limitations result in LonWorks technol-

ogy, being relegated to lower end, simpler applications developed by lower skilled developers—precisely the way it is being played out in the market. LonWorks is not capable enough for high-end, custom, robust, high complexity systems. Going forward it will be at an increasing disadvantage compared to newer processing and communications technology currently being developed.[12,26] Implementation costs are another issue. Anecdotal comments suggest that building a product on LonWorks technology is not as simple as Echelon portrays—several projects required significantly more time and money than originally anticipated. These appear to result primarily from having to find ways around some of the limitations built into the technology as alluded to earlier.

Marketing and Promotion

A discussion of LonMark would not be complete without a comment about marketing and promotion.

In what may be a response to a competitive environment, Echelon's marketing strategies and methods appear to control information availability, and thus can make it difficult for a prospective client or specifier to make well-informed purchasing decisions.* Many of Echelon's marketing materials regarding their own products are heavily promotional in nature. From these materials it is often difficult to obtain a clear appraisal of the potential of LonWorks, LonTalk and LonMark in terms of acceptance, use, and capabilities, as we have indicated in the discussions above. This tends to cause confusion among developers, specifiers, and users. It also engenders lowered confidence about the overall merits of the technology in general. We therefore caution energy practitioners about accepting at face value statements in literature of this type. Additionally, Echelon statements regarding the merits of alternative approaches such as BACnet (see statements about BACnet in[24, 29]) should be viewed with caution and evaluated with care. We recommend using independent information sources and trusted, unbiased experts to evaluate functionality and performance claims before making purchasing decisions.

Because it is important to ensure availability of a sufficiently wide range of compatible products for future extensibility, practitioners should also be cautious when it comes to evaluating the penetration of both LonMark and BACnet technologies in the market. There is very

---

*See footnote 32.

*For example, IEC Intelligent Technologies has been repeatedly denied access to LonWorld to show their products that compete with Echelon's[28]

little solid data to back up claims being made. Data about types of products or, better yet, sales volumes of products by type are the only reasonable way to make definitive statements about penetration and growth rates. In terms of number of nodes being sold it appears the market is somewhat balanced between BACnet and LonMark with Honeywell and Siebe (Invensys) leading with LON devices and Alerton, Automated Logic, Delta Controls and Trane leading with BACnet products. It would be better if the types of products and their volumes were known.*

References

[1] Tannebaum, A.S., *Computer Networks*, 3rd ed., Prentice Hall, New Jersey, 1996.

[2] "Control Network Protocol Specification," EIA-709.1, Electronic Industries Alliance, April 1999.

[3] Tannebaum, p. 301.

[4] Bushby, S.T., "Communications Gateways: Friend or Foe," *ASHRAE Journal*, April 1998, p. 50.

[5] Hull, G.G., "Myths of LonWorks and BACnet," *ASHRAE Journal*, April 1999, p. 22.

[6] Internet Protocols[Website]. Cisco, November 8, 1999. Available from: www.cisco.com/univercd/cc/td/doc/cisintwk/ito_doc/ip.htm.

[7] "Control Network Shuffle Continues," *Control*, June/July 1999, p. 12.

[8] Caro, D., and R. Mullen, "Ethernet as a Control Network," *Control*, February 1998, p. 38.

[9] Merritt, R., "Technology Trends," *Control*, April 2000, p. 40.

[10] Waterbury, B., "Ethernet Ready to Strike," *Control*, August 1999.

[11] "Profibus Supporters Ponder Life with Ethernet," *Control Design*, October/November 1999, p. 14.

[12] McParland, C., Personal Communication, January 2001, *Computer Scientist*, Lawrence Berkeley National Laboratory.

[13] "GE, Cisco Systems Form Networks Unit for Factories," *Wall Street Journal*, June 6 2000.

[14] Knollman, R., Personal Communication, November 1999, Applications Engineer, CTI Products.

[15] Thomas, G., "Looking Deeper into Ethernet," *The Industrial Ethernet Book*, Spring 2000, p. 16.

[16] Feeley, J., R. Merritt, T. Ogden, P. Studebaker, and B. Waterbury, "100 Years of Porcess Automation," *Control*, December 1999, p. 17.

[17] Tridium, "White Paper: Baja: A Java-based Architecture Standard for the Building Automation Industry," Tridium Inc., 2000.

[18] "Method of Testing Conformance to BACnet," ASHRAE 135.1P, ASHRAE, 2000.

[29] Bushby, S.T., Newman, H.M., Applebaum, M.A., November, "GSA Guide to Specifying Interoperable Building Automation and Control Systems Using ANSI/ASHRAE Standard 135-1995, BACnet," NISTIR 6392, National Institute of Standards and Technology (NIST), 1999.

[20] BMA, "BMA 2000 Annual Report," BACnet Manufacturers Association (BMA), December 2000.

[21] Tennifoss, M., Personal Communication, February 27, 2001, Vice President, Marketing, Echelon Inc.

[22] "LonWorks Product Update," presented at LonUsers International, San Jose, CA, May 1995.

[23] Chervet, A., Personal Communication, January 2000, FAE, Echelon Corp.

[24] Current Listing of All LonMark Products[Website]. LonMark Interoperability Association, November 8 2000. Available from: www.lonmark.org/products/prod_list.cfm.

[25] Wittkowske, C., Personal Communication, May 31, 1999, Systems Integration Engineer, Trane Co.

[26] Robin, D., Personal Communication, February 2000, Senior Research Engineer, Automated Logic.

[27] DesBiens, D. "NOOO PEAKING!" at LonWorld®—AGAIN!

[28] IEC Denied Admission[Website]. IEC Intelligent Technologies, 2000. Available from: http://www.ieclon.com/News/20000928.html.

[29] LonTour99[Powerpoint]. Echelon Corp., November 1999.

[30] Diamond, R., T. Salsbury, G. Bell, J. Huang, and O. Sezgen, R. Mazzucchi, J. Romberger, "EMCS Reftofit Analysis Interim Report," LBNL-43256, Lawrence Berkeley National Laboratory, March 1999.

[31] Lee, J., Personal Communication, April 2001, President, Cimetrics.

Bibliography

Allen, E., Bishop, J., "The Niagara Framework," Tridium, Inc., 1998.

"BACnet Test Standard Recommended for Review," *ASHRAE Journal*, March 2000, p. 22.

"LonMark Product Conformance Review," LonMark Interoperability Association, December 1999.

"LonWorks Product Update," presented at LonUsers International, San Jose, CA, May 1995.

"The Next Wave," *Business Week*, August 31 1998, p. 80.

"Potential for Reducing Peak Demand with Energy Management and Control Systems," WCDSR-137-1, Wisconsin Center for Demand Side Research, 1995.

"Standard 135-1995—BACnet™—A Data Communication Protocol for Building Automation and Control Networks," ANSI/ASHRAE 135-1995, ASHRAE, 1995.

Bushby, S. T., Personal Communication, November 1999, Senior Program Manager, NIST.

Elyashiv, T., "Beneath the Surface: BACnet Data Link and Physical Layer Options," *ASHRAE Journal*, November 1994, p. 33.

Falk, H. S., "Overview of Plant Floor Protocols and their Impact on Enterprise Integration," presented at Autofact '94, Detroit, MI, 1994.

Feeley, J., "What Network Works for You," *Control Design*, February/March 2000, p. 30.

Hartman, T., "Practical Considerations for Protocol Standards," *Heating/Piping/Air Conditioning*, August 1994, p. 45.

Hougland, B., "Ethernet Advantages at I/O Level," *The Industrial Ethernet Book*, Spring 2000, p. 9.

Hess, M., Personal Communication, December 2000, Systems Marketing Engineer, Trane Co.

Kinsella, T., R. Hirschmann, "Ethernet in Industrial Automation—Today and Tomorrow," *The Industrial Ethernet Book*, Spring 2000, p. 29.

Klien, J., "Battle of the Protocols," *Energy User News*, February 1999, p. 1.

LeBlanc, C., "The Future of Industrial Networking and Connectivity," *The Industrial Ethernet Book*, Spring 2000, p. 6.

Merritt, R., "DCS Dead Yet?," *Control*, July 1999, p. 34.

Middaugh, K.M., "Industrial LANs: Sorting out your communications options," *I&CS*, November 1993, p. 45.

Mostia, W.L., "What's New in the Wireless World," *Control*, June 1999, p. 77.

Naylor, T., Personal Communication, December 1999, Application Engineer, Coactive Networks.

Newman, H.M., "BACnet Goes to Europe," *Heating/Piping/Air Condi-*

---

*Accumulating these numbers is a valuable contribution that an organization such as the BMA could make.

tioning, October 1998, p. 49.

Newman, H.M., Direct Digital Control of Building Systems, *Theory and Practice*, 1st ed., John Wiley & Sons, New York, 1994.

Newman, H.M., "Integrating Building Automation and Control Products Using the BACnet Protocol," *ASHRAE Journal*, November 1996, p. 26.

Putnam, A., Personal Communication, November 30, 1999, Development Director, Cimetrics Technology.

Roberts, J., Personal Communication, January 19, 2000, Principal Engineer, LonMark Interoperability Association.

Rosenbush, S., "Charge of the Light Brigade," *Business Week*, January 31, 2000, p. 62.

Sakamar, G., "FireWire Gets Ready for Control," *Control Design*, April/May 2000, p. 74.

Studebaker, P., "Object Technology Targets Process Control," *Control*, June 1998, p. 57.

Tatum, R., "Interoperability Marks new Generation of Commercial Building Products," Building Operating Management, July 1996, p. 23.

Turpin, J.R., "Getting to the BAS Point (and Counterpoint)," *Engineered Systems*, August 2000, p. 48.

Waterbury, B., "Network Computers are Dead: Long Live Thin Clients," *Control Design*, April/May 1999, p. 34.

Waterbury, B., "Platforms for Decentralized Intelligence," *Control*, March 2000, p. 97.

Waterbury, B., "Web Closes the Loop on Real-time Information," *Control Design*, April/May 2000, p. 37.

Weiss, G., "Smiling as Highfliers Blow Up," *Business Week, June 26 2000, p. 223.*

Chapter 17

ANSI/EIA 709.1, IP, and Web Services: The Keys to Open, Interoperable Building Control Systems

Michael R. Tennefoss
Alex Chervet

The advent of the ansi/eia 709 control standard, internet protocol (IP) based networks, and SOAP/XML web services has opened the door to a new generation of open, interoperable control systems. These systems leverage LANs, WANS, and the internet to deliver higher reliability, more vendor choices, lower life-cycle costs, and more flexibility. Seamlessly merging the worlds of data and control allows facility owners to gain deeper insights into the operation of their facilities, better analyze and manage their buildings, and better leverage their capital investments.

INTRODUCTION

Before the advent of solid-state electronics, control systems consisted of pneumatic controls or wire bundles connected to relays, switches, potentiometers, and actuators. Cabling was installed point-to-point between electrical panels, sensor inputs and actuator outputs. The functionality of these control systems was relatively rudimentary and inflexible, and changes often required the assistance of a controls engineer.

The arrival of the transistor provided a way to replace relays and pneumatics with logic circuits in direct digital controllers (DDCs), which were programmed or configured with a data terminal. As increasingly powerful microprocessors and more sophisticated control algorithms were made available, the control systems grew more complex as did the issues associated with making adds, moves, and changes. The system controllers and their associated software became more expensive, and susceptibility to a single point of failure grew as reliance on controllers increased (Figure 17-1). In order to ensure

a future revenue stream, manufacturers of DDCs developed them using proprietary internal architectures: the expansion of a DDC system required the use of components from the original manufacturer, making competitive bidding difficult if not impossible.

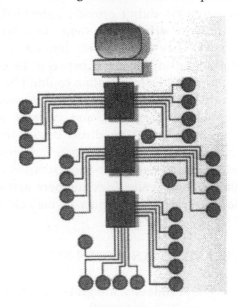

Figure 17-1.

The incompatibilities between products from different manufacturers were highlighted when customers attempted to interconnect DDC systems or devices from different manufacturers. The use of incompatible communication protocols, data formats, and electrical interconnections made it very difficult to exchange information. Seeking the communication equivalent of the "least common denominator," systems integrators and manufacturers turned to the use of gateways at the workstation level to tie together subsystems from different manufacturers (Figure 17-2).

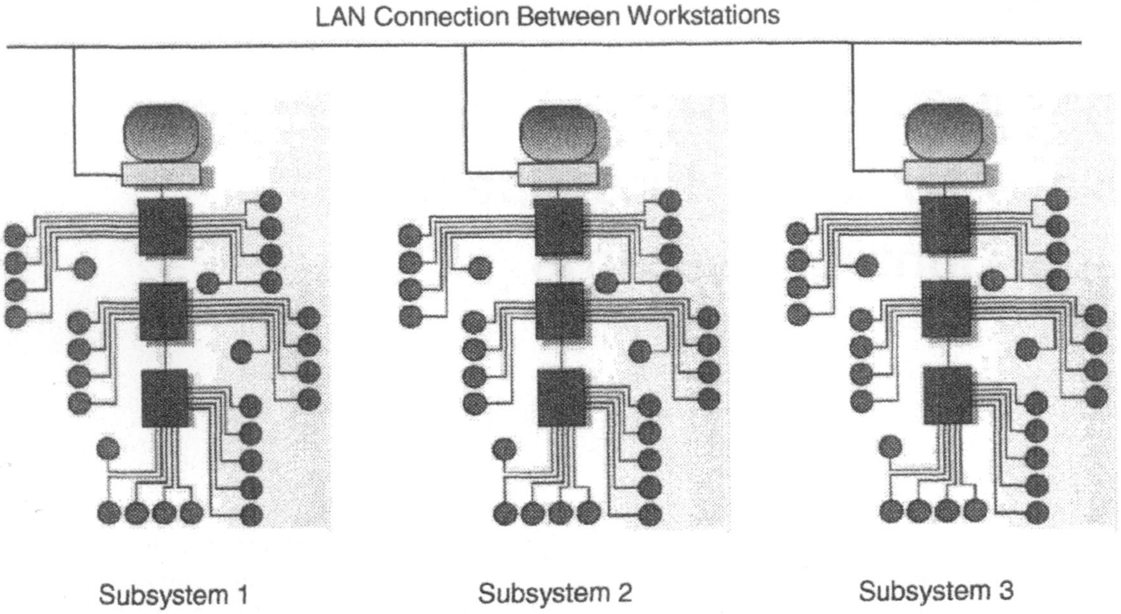

LAN Connection Between Workstations

Subsystem 1   Subsystem 2   Subsystem 3

Figure 17-2. Using Gateways in Workstations to Link Subsystems from Different Manufacturers

These gateways didn't provide a detailed, seamless view into the different systems to which they were connected. They allowed only limited status and control information to be passed between the different subsystems. Fault status information couldn't be shared, information from different sensors wasn't accessible for combinatorial logic programs, and systems couldn't adapt in real-time based on direct device-to-device communications. Furthermore, the gateways needed to be changed whenever one of the subsystems was modified, creating an open-ended development and support problem for integrators and facility owners alike.

INTEROPERABILITY AND OPEN SYSTEMS

Creating a seamlessly integrated control system requires interoperability among the components of that system, as well as other related systems that must exchange information (Figure 17-3). Interoperability is the process by which products from different manufacturers, including those in different industries, exchange information without the use of gateways, protocol converters, or other ancillary devices. Achieving interoperability requires a standardized means of communicating between the different devices and managing device commissioning and maintenance; it depends on a system level approach that includes a common communication protocol, communication transceivers of different types, media routing, object models, and management and troubleshooting tools.

The benefits made possible by interoperability are many. Since one sensor or control device can be shared among many different systems, fewer sensors/controls are needed and the overall cost of the control system drops appreciably. For example, in a building automation system, one interoperable motion sensor can share its status with the zone heating system for occupancy

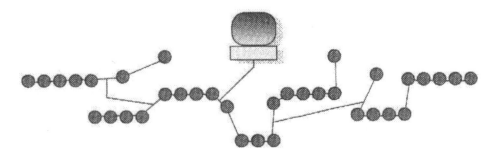

Figure 17-3. Open, Interoperable Control Network with Minimal Wiring

sensing, the access control system for request-to-exit purposes, the security system for intrusion detection, and the fire alarm system for occupancy sensing. The motion sensor still performs the same task—detecting motion—but it can share the information with the many subsystems that can make use of its status.

The ability to share more information between systems makes possible many long sought-after applications, including integrated energy control systems. For example, in response to access control reader data and daylight illumination sensors, the HVAC and lighting systems can automatically adjust the comfort and illumination levels in pertinent work areas based on individual preferences and energy costs. Lighting can be adjusted on a cubicle-by-cubicle basis for computer operators and occupants near windows—either automatically or through commands entered from a user's PC via the corporate LAN. Heating and air conditioning can be similarly tailored. Or, based on signals from smoke detectors, the HVAC system can create positive or negative air pressure of select areas to cause a fire to move away from occupied areas while the lighting system leads the way to the closest exit. The possibilities are limited only by the creativity of the designers.

For a facility owner, interoperable products offer the advantage that devices can be selected from among different manufacturers; the owner is no longer tied to any one manufacturer's closed technology. Aside from the cost savings achieved by open competition, the facility owner is safe in the knowledge that replacement products will be available if any one manufacturer goes out of business or discontinues products. Service contracts can be openly bid since no proprietary devices will be used, thereby avoiding single source service contracts.

Interoperability also benefits equipment manufacturers because their products will be assessed based on their quality and functionality—not on their ability to meet a closed, proprietary specification. Interoperability levels the playing field and increases competition, insuring that better devices will be built and the best devices

for the job will win.

Interoperability has been a driver in the growth of the internet, too. TCP/IP, the connectivity standard for internet communications, provided a common transport mechanism for connecting far-flung governmental and institutional computers. The availability of HTML and web browsers enabled the creation of interlinked pages that allowed anyone with a few basic tools to use the internet, fueling both its popularity and the development of new applications.

The next level of growth will be driven by web services, which are positioned to drive expansion in network connectivity by leveraging the huge infrastructure investment that created the World Wide Web. Web services (XML/SOAP) provide a standard means of allowing disparate computing systems to exchange information. Just as a web browser uses the internet infrastructure to find a specific computer URL, and the corresponding server responds with a stream of formatted text, a web service consumer allows a computer to request data via the internet and receive an intelligible response (Figure 17-4). By providing a standardized, interoperable platform for system-to-system communications, web services have the potential to drive new applications in the area of web interaction. The question is how web services can best be leveraged in the world of building controls.

THE DAWN OF BACNET

In an effort to create a standardized method of interconnecting heating, ventilation, and air conditioning (HVAC) subsystems from different manufacturers, the American Society of Heating Refrigeration and Air Conditioning Engineers (ASHRAE) set about to create an open standard called building automation and Control NETwork (BACnet). BACnet was originally intended to eliminate the need for proprietary gateways between workstations by defining a standardized means of communicating over a local area network (LAN) to which

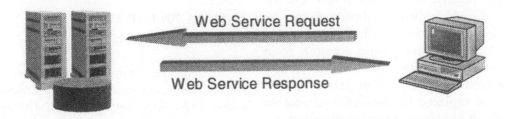

Figure 17-4. Web Services Communication Model

the workstations were connected. The workstations would, in turn, be connected to their respective control systems. Several different LANs were defined including point-to-point, master slave/token passing, ANSI/ATA 878.1, ethernet, and LonTalk® (an open control standard also known as ANSI/EIA 709.1).

One of the key features of BACnet is the use of an ethernet backbone running the BACnet protocol that was intended to improve overall system performance relative to the comparatively slow speed buses used by commercial control system vendors. The ethernet backbone was expected to be especially important in large systems at times of high network activity, such as an automatic restart following a power outage.

BACnet defined a messaging format that used objects (a logical representation of an input, output, or functional grouping of inputs and/or outputs), properties (the characteristics of an object through which it is monitored and controlled), and services (the means by which BACnet devices obtain information from one another). Since under BACnet, devices could have varying levels of functionality, even if they performed the same task from an end user's perspective, BACnet defined conformance classes that categorize the capabilities and functionality of devices.

Devices within a conformance class needed a minimum set of features, but optional features were permissible. All of the features of a device were presented in a device's protocol implementation conformance statement (PICS). A specifying engineer needed to know which objects and services were supported by which devices, since this varied from device to device, and the PICS provided most of this information. The PICS represented a point at which manufacturers could diverge in their implementation of BACnet: products which appeared to perform identically could vary considerably in terms of their functionality and the accessibility of data.

While the BACnet standard had the potential to be open and non-proprietary, its implementation varied considerably from manufacturer to manufacturer. This variability undermined the fundamental precepts of what BACnet was intended to be and do, and resulted in the creation of closed, proprietary devices flying the BACnet banner. Facility owners have the right to expect that BACnet workstations, sensors and actuators from different manufacturers may be used seamlessly in a common control network, and that devices from one manufacturer can be replaced by devices from another manufacturer without assistance from the manufacturer or the redesign of the control system. Due to variations

in the implementation of BACnet PICS by different manufacturers, however, neither scenario has been realized. In short, BACnet devices from different manufacturers are neither interoperable nor interchangeable. The truth of this point is driven home by the dearth of interoperable, multi-vendor BACnet systems that support device-to-device, peer-to-peer control.

In an effort to bridge the gap between existing control networks and BACnet networks, and to appease engineers who are specifying BACnet because of its promise, some manufacturers turned to BACnet gateways. A BACnet gateway converts data from the format used by one control network into the BACnet format. This allows the manufacturer to state that a product supports BACnet because BACnet packets can be received by the gateway and forwarded to the non-BACnet system, and vice versa.

The problem is that the use of a gateway violates the spirit of BACnet and fails to deliver interoperability to control systems. Why? As discussed earlier, in the process of converting information between two networks a gateway discards information, thereby limiting the scope of the tasks that can be performed across the gateway. Diagnostic information from nodes, network traffic statistics, network management messages—all are affected by the insertion of a gateway. While gateway manufacturers may have had the best of intentions in mind, gateways do little to realize the promise offered by BACnet and instead merely extend the life of closed, proprietary systems.

The vision of a unified BACnet network was transformed into the reality of islands of proprietary networks linked by workstations and gateways running the BACnet protocol over IP-based LANs. BACnet workstations and gateways impede the free flow of network information, making impossible peer-to-peer communication between sensors and actuators situated within different proprietary islands. BACnet did not deliver on its promise and has been sullied by hype that hasn't been matched in the execution of BACnet compliant products or systems.

ANSI/EIA 709.1, IP TUNNELING, AND WEB SERVICES

One of the control networks specified within BACnet for communications between sensors and actuators is the ANSI/EIA 709.1 protocol. The ANSI/EIA 709.1 protocol allows all manner of control devices to communicate with one another through a commonly shared protocol. Communication transceivers and transport mechanisms (ANSI/EIA 709.2 power line signaling, ANSI/EIA 709.3 twisted pair signaling, ANSI/EIA 709.4 fiber optics) are standardized, as are object models and programming/troubleshooting tools to enable the rapid design and implementation of interoperable, ANSI/EIA 709.1-based devices. Network management software, protocol analyzers, internet protocol (IP) servers, network interface cards, and development tools are available off-the-shelf to speed development and reduce time to market. In short, ANSI/EIA 709.x offers a system level approach to interoperability, and comprises a complete set of tools and products. Roughly 30,000,000 ANSI/EIA 709.1 compliant devices have been shipped to date by thousands of product manufacturers.

Ensuring the interoperability of these network communications is the responsibility of an organization called the LONMARK® Interoperability Association. Funded through member dues, the LONMARK Association defines the interoperability guidelines for devices based on the ANSI/EIA 709.1 protocol, including communication transceivers and object models. Products that bear the LONMARK logo are certified to adhere to the LONMARK interoperability guidelines and can be used with confidence in integrated control systems.

One of the underlying tenets of ANSI/EIA 709.1 is that every device in a network should have the ability to send packets to, and receive packets from, any other device in the network without an intermediate gateway that filters and modifies information. This capability is one of the cornerstones of any open, interoperable, peer-to-peer network.

Additionally, the ANSI/EIA 709.1 protocol is routable, and an internet server/tunneling router may be used with any IP-based network (including LANs, WANS, and the internet) as a seamless pathway to communicate on a peer-to-peer basis between ANSI/EIA 709.1-enabled workstations, sensors, actuators, and displays. Where BACnet workstations and gateways filter information, ANSI/EIA 709.1 compatible internet servers intelligently tunnel packets through the IP network on a peer-to-peer basis, making these packets available to any sensor, actuator, or workstation that needs them.

By bridging the data and control worlds, tunneling routers allow facility owners to leverage high speed information technology (IT) infrastructure as an extension of the control network. Peer-to-peer networking, network management, device diagnostics, and software downloading—all features of ANSI/EIA 709.1 are retained, but the overall installation costs are reduced because one common IT infrastructure supports from the control and data networks. From the IT manager's perspective, the two systems are seamlessly linked because the internet server behaves like typical IT gear (and includes a programmable packet aggregation feature which throttles the rate at which control packets are broadcast on the IP network, ensuring that the IP network is not overwhelmed with control-related packets). From the building automation System manager's perspective, the two systems are seamlessly linked because the internet server behaves like an ANSI/EIA 709.1-based device.

The ability to tunnel ANSI/EIA 709.1 packets through an IP network allows devices to communicate on a peer-to-peer basis across a LAN that connects floors of a building, a WAN that connects buildings within a city, or over the internet to facilities spread around the world. Eliminating the need to install a separate and dedicated data network for the control system, and instead using the existing LAN or WAN data infrastructure for control networking, saves both installation and on-going maintenance costs.

The use of an ethernet backbone was one of the claimed advantages of BACnet over other control systems. Today, however, ANSI/EIA 709.1-based systems have a significant advantage over BACnet because they support peer-to-peer networking, they share the IT infrastructure with existing data systems, and ANSI/EIA 709.1-based internet servers behave like standard IT equipment.

ANSI/EIA 709.1-based networks have one other significant advantage over BACnet—they can leverage the newest open web standard, web services. Using an internet server that exposes a SOAP/XML web services interface, enterprise applications from Oracle, PeopleSoft, Siebel, SAP, and others can draw data from, and send messages to, any ANSI/EIA 709.1-based networks. By leveraging existing enterprise applications and IT infrastructure, web services allow companies to derive greater utility from their existing software and capital investments, while providing unparalleled access to device-level information gathered from remote facilities.

From the perspective of a facility owner, web services can shorten the payback period of existing enterprise software by expanding its use into new areas such as remote facility monitoring, like-facility comparisons, preventive maintenance, predictive failure analysis, and support logistics optimization. These new tasks can be accomplished without replacing ANSI/EIA 709.1-based control devices since an internet server with SOAP/XML web services interface will be compatible with existing ANSI/EIA 709.1 devices. This is perhaps the best example of the power of using open, interoperable standards for building control—existing software and existing devices can perform new functions through the addition of a simple internet server operating over an existing IP network.

SUMMARY

The world of building control systems has come a long way technologically since the days of relay and pneumatic controls. The availability of ANSI/EIA 709.1-based devices has opened the door to a new generation of open, interoperable control systems that leverage the internet and SOAP/XML web services. These systems deliver on the benefits promised by those who espouse the benefits of open systems—higher reliability, greater vendor choices, lower life-cycle costs, great flexibility. Seamlessly merging the worlds of data and control allows facility owners to gain deeper insights into the operation of their facilities, better analyze and manage their buildings, and better leverage their capital investments. ANSI/EIA 709.1, IP, and web services are truly the wave of the future, so hitch a ride and enjoy the benefits.

Chapter 18

Network Security for EIS and ECS Systems

Joel Weber

As energy and facility managers increasingly utilize web-based information systems, they must also develop security for their information and their systems. This chapter describes some of the security nightmares that lurk in cyberspace, waiting like evil villains, to attack unsuspecting system operators and users. Methods for avoiding attacks, recovering from them, and minimizing their impact are covered.

INTRODUCTION

During the past two decades the energy sector and operations managers in charge of energy management in industrial sectors have become increasingly dependent on information technology (IT). At the operational level, IT has facilitated higher yields, more efficient quality control, and more effective inventory management. At the managerial level, IT has proven critical in customer relationship management, marketing, finance, accounting, and strategic planning. In short, IT systems are an essential element in maintaining and improving the integrity of any energy-focused or energy-dependent organization. Consequently, the security, reliability and availability of these systems are crucial to their successful implementation.

Many diverse IT systems are found within the energy sector, including energy marketing and trading, exploration and production, refining operations, power generation, and power and gas distribution. Such systems are also used by industrial, commercial and residential energy management applications. Energy IT systems can operate on desktop workstations, servers, laptops, portable devices, and embedded systems; they may be run on platforms such as Microsoft Windows 3.x/95/98/NT/2000/XP/CE/.NET, Linux, Unix, BSD, and Solaris.

Energy IT systems are subject to the same internal and external threats and vulnerabilities that threaten the integrity of other computer systems. Many high profile incidents in recent years such as the Melissa, Code Red and Nimda viruses, have underscored the need for increasing energy IT security awareness and making security a major feature of an energy manager's IT policy.

Types of Security Threats and Vulnerabilities

The following thirteen categories have been developed for classifying IT security violations:[1]

- Theft of proprietary information
- Sabotage of data or networks
- Telecom eavesdropping
- System penetration by outsider
- Insider abuse of network access
- Financial fraud
- Denial of service
- Spoofing
- Virus
- Unauthorized insider access
- Telecom fraud
- Active wiretapping
- Laptop theft

Each type of violation requires an intrusion or attack in order to succeed and the nature of such intrusions and attacks must be understood before energy IT assets can be secured and security policies can be implemented.

The Nature of Security Threats

To manage security threats we must understand the nature of threats, how intrusions and attacks are mounted and how they proceed. By definition a security intrusion requires some type of unauthorized access. An attack does not require entry into a system or network, but can cripple a system as fatally as an intrusion can. Intruders are either external or internal with respect to the organization. External intruders are those who are attempting to gain access to a system and do not have access authorization. Internal intruders are authorized users of the system who access data, resources or programs to which they are not entitled.[1] Whether or not an attacker is intruding upon a system, there are certain techniques he implements and processes he utilizes.

Security attacks are launched in either a direct or distributed manner. A *direct attack* is launched from a computer used by the attacker. A *distributed attack* makes use of other computer systems to cloak the origin of the attack and to potentially amplify the affect of the attack. Multiple systems can be enlisted to execute a distributed attack in a coordinated fashion. Consequently, the number of victims in this scenario can become quite large.

The most common form that distributed attacks take are denial of service (DoS) attacks. The purpose of a DoS attack is to render a network inaccessible by generating such a large amount of network traffic that the servers crash, the routers are overwhelmed, and the network's devices cannot function properly.[2] Many distributed security attacks are automated through the use of software attack tools. Such tools make launching attacks considerably easier by lowering the technical difficulty level and reducing the number of steps required to initiate an attack.

Until 2000, most automated distributed attacks would only execute one attack sequence. That year, the Nimda and Code Red attacks presented a new breed of automated distributed attack in which user interaction on a compromised system would initiate subsequent attacks and compromise more systems. The advent of such attacks underscores the difference between deliberate and accidental attacks. The parties that originally initiate such attacks clearly intend to harm other systems or users. However, the people who might accidentally perpetuate such attacks through otherwise normal computer interaction represent a source of "accidental" attacks. The success of automated distributed attacks is increasingly dependent upon accidental attacks made by unwitting users. IT security staff must remain aware of these types of attacks and must educate energy IT staff in order to mitigate the effect of such attacks on the organization's computer systems.

Phases of Attack

Each security attack has certain phases in common: Preparation, Initial Access, Full System Access, Establishing Future Access Opportunities, and Obfuscation. Understanding how security can be compromised and what steps should be taken to prevent security breaches is important for energy managers who are deploying and utilizing energy IT systems.

Preparation

The first step in mounting an attack on a system (the preparation stage) is to acquire information and develop a plan. This phase usually involves activities such as port scans and spoofing. A port is a service that runs on a computer and is a virtual door through which information enters and exits the system. There are 65,535 TCP (transmission control protocol) ports. Some ports are reserved for use by certain protocols and others are available for custom applications. See Appendix 1 for an abridged listing of these ports and their corresponding protocols and applications. A thorough listing of well-known port/protocol assignments can be found at *www. freesoft.org/CIE/RFC/1700/4.htm.*

Port scanning software tools are widely available on the internet as freeware. Such tools are not illegal in the United States because IT security personnel and law enforcement frequently use them to insure that only ports that are necessary for operations are open. However, these tools are frequently used and detected in the preliminary phases of attacks on computer systems. Much like weaponry in the physical world, port scanners can serve law-abiding citizens and heroes as much as they can serve criminals and villains. When a port scanner is used, it is roughly analogous to walking around the perimeter of a building and testing for whether the outer doors and windows are locked and/or open. For those preparing an attack, the scanner is used to locate and identify which ports (or doors and windows) are open and which protocol(s) is used by that port.

"Spoofing" is defined as the act of getting one computer on a network to pretend to have the identity of another computer, usually one with special access privileges, so as to obtain access to the other computers on the network.[3] There are three general subclasses of spoofing activity: ARP spoofing, IP spoofing, and DNS spoofing.

Address resolution protocol (ARP) is the communications protocol through which the network determines what transmissions should go to which computers. When a computer is logged onto a network, it receives an internet protocol (IP) address, which will serve as its address on the network, and all transmissions to or from that computer will be routed using that address. Each computer's network card has its own unique media access control (MAC) address, which serves as a physical address of the computer. The address resolution protocol maintains a table of IP-MAC address relationships called the ARP cache. ARP spoofing involves altering the ARP cache so that communications on the network are routed to a machine that the attacker chooses.

Internet protocol (IP) facilitates speedy, reliable and asynchronous communications between machines by dividing information into small packets that are easier to transmit across a network or between networks than if a dedicated communications channel were used. Each of these packets contains the IP address of the source of the transmission in a "header." Since the IP header can act as type of return address, many attackers will engage in IP spoofing in order to falsify the identity of the network and computer from which they are attacking a targeted system. IP spoofing involves changing the IP address in packet headers in order to impersonate a different machine. When an attacker knows the address of other trusted computers with which the targeted machine communicates, he can route his impending attack through one of those machines and conceal his identity and location.

The *domain name system* (DNS) is used to make the internet more efficient and user-friendly. This is the database system that keeps track of all of the domain names such as intel.com, and their associated static IP addresses. The DNS is by far the most critical single system for the operational integrity of the internet. Businesses, organizations, and individuals with an internet presence therefore depend on the DNS to insure that internet traffic intended for them is directed to their network. DNS spoofing is similar to ARP spoofing in that the address associations are altered. With DNS spoofing, the impact can be quite severe as web site traffic, file transfers, emails, etc. can be re-routed and web sites can be replaced with impostors. This method can even be used to con an unsuspecting individual into providing personal information through web forms.[4]

Initial Access

In order to gain initial access to a system that has been marked in the preparation phase, an attacker must either gain a password that yields access to a computer's operating system or application running on that platform, or exploit some known or newly discovered weakness in an operating system, application, or protocol.

Passwords can be compromised in many ways. The simplest method is that of gross enumeration or "brute force" in which the password is guessed until access to the target system is granted. The process of guessing words is tedious at best and requires a large helping of luck. To speed this process an attacker might perform a dictionary attack in which a software program automates the process of trying each and every word in a dictionary as a candidate password. This attack, however, is only successful against the least sophisticated computer users and those organizations without password policies. Password cracking has legitimate applications in the case of employees who forget their passwords or are terminated or deceased. In this context, however, password cracking is called "password recovery."

Operating system and application passwords are frequently stored on a machine. Some older operating systems store encrypted passwords in files but the encryption algorithms have proven to be quite weak. Newer operating systems such as Windows 2000/XP implement Kerberos authentication, which is currently one of the most robust encryption algorithms in existence, and store the encrypted passwords in a database that cannot be accessed through the file system. In the face of such safe password storage, an attacker must resort to intercepting a password as it is sent across a network in clear text format. This will require a network intrusion which, if appropriate network hardware and software solutions have been installed as part of the IT security strategy, will require significant resources, time and effort.

If a password cannot be decoded, an attacker may attempt to gain initial access to a system by exploiting a software weakness. Some operating systems allow for remote access, thus presenting an opportunity for unauthorized entry if these access points are not properly secured. One of the most common software vulnerabilities is called a buffer overflow. A buffer overflow occurs when input data exceed the size of their allocated program memory buffer because checks are lacking to ensure that the input data are not written beyond the buffer boundary.[5] Unfortunately, due to poor coding and quality assurance practices many operating systems and applications contain this vulnerability. An attacker exploiting a buffer overflow can remotely escalate his privileges on the target system. Through a complex process of analyzing memory addresses of applications with known vulnerabilities, the attacker can gain administrator privileges on the target system by either launching a command shell application or crashing an application or operating system such that the system restarts in 'Administrator' mode. The cost of a single compromise can be astronomical if the attacker is able to further infiltrate a system and access valuable information.[6] Exploitation of buffer overflows can generally be prevented by active maintenance of software patches as they become available and by improved software development processes within the organization.

Applications can also present security vulnerabilities if unexpected input from users leads to application failures. Protocols are often exploited via a denial of service (DoS) attack in which the protocol is overwhelmed with a high volume of requests or large single requests.

Full System Access

Once an attacker gains access to a targeted system he may engage in a variety of activities ranging from simply browsing the file system to introducing software or spyware to stealing data. Malicious software is commonly described as a Trojan (or Trojan horse), worm, or virus.

A Trojan is a software program that is introduced into a system under the pretense of serving a legitimate purpose when it actually aims to compromise system security and gain unauthorized access. It may be hidden in an application file, or disguised as a legitimate text file or email attachment. The 2000 "I Love You" Trojan appeared to be a text file attachment to an email when it was actually a visual basic script that forwarded the original email with the Trojan attachment in tow to every address in the user's Microsoft Outlook address book. The effects of this Trojan were felt worldwide as mail servers crashed from the overload. The end result of this attack was denial of service for email users worldwide.

Worms were the first incarnation of malicious software. A worm is a program that is capable of traveling across a network. Their original purpose was to distribute legitimate software across networks. In their malicious form, worms can make multiple copies of themselves and spread through a network. Their primary purpose is to replicate.[7]

A virus is an intrusive program that infects computer files by inserting copies of itself into those files. The virus program is usually executed when the infected file is loaded into memory. This allows it to infect other files, and so on.[8]

Viruses have several different morphologies. They can be logic bombs that "detonate" and execute their code when a series of conditions is met or a specific point in time occurs. The Michelangelo virus was the first incarnation of this virus class; it was programmed to erase the hard drives of the infected computers on the date of the artist's birthday. *Macro viruses* are another type that are embedded in documents and have been seen most commonly in Microsoft Word documents. *Application viruses* sometimes infect programs and execute their malicious code when the programs are run. *Boot sector viruses* are transmitted by portable media such as floppy diskettes. This type of virus inserts itself into the master boot record on permanent storage media like hard drives and is loaded into and executed in memory whenever a computer is booted. Fortunately Trojans, worms and viruses have caused so much upheaval in society that people are increasingly aware of the risks of opening unknown attachments and the means by which viruses are transmitted; therefore, they are implementing effective risk management practices including use of virus recognition software.

Upon gaining full access to a computer system, some attackers install *spyware* that allows the attacker to continually acquire data from the compromised system. In recent years, "key logger" software has become a tool of choice for attackers. Such software will actually log all of the keyboard keystrokes made by an unsuspecting user, thus giving the attacker complete information on the user's data, passwords, email, and other procedures.

An attacker with full access privileges can engage in many different activities. He may install or execute malicious software or steal privileged data or engage in other unanticipated misdeeds. Awareness of these major activity classes is a significant contribution to maintaining a credible IT security strategy.

Establishing Future Access Opportunities and Obfuscation

After an attacker has gained full access to a system, he is very likely to create difficult-to-detect doorways for his own use. This will make it much easier for him to use that same computer system again. He is likely to open additional ports or create separate user accounts for his use. Once his attack is complete, he will attempt to "cover his tracks" and obfuscate (conceal) his activities. To do this, he will alter and delete log files and eliminate all evidence of his unauthorized presence.

Information Security and Risk Management

The process of information security and risk management can be characterized by a modified version of the Deming cycle model of *Plan-Do-Check-Act* (Figure 18-1).

In this chapter, the hardening/securing and preparation phases of the information security assurance process will be our primary concern; the detection,

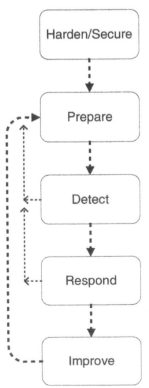

Figure 18-1. Deming Cycle for Securing Information Assets *CERT Guide to System and Network Security Practices.*

response and improvement activities are largely a function of a specific organization's resources, needs, and strategic objectives. There are many reputable resources for managing these phases of information security management. One of the best guides for information security management is the Operationally Critical Threat, Asset, and Vulnerability Evaluation (OCTAVE[SM]) framework for Information Security Assurance and Risk Management developed by the Software Engineering Institute at Carnegie Mellon University.

To be successful, an IT security strategy must have support at all levels of the organization and must be viewed as a process of continuous assessment and improvement. An organization that wishes to improve its security posture must be prepared to take the following steps:[9]

1. Change from a reactive, problem-based approach to proactive prevention of problems.
2. Consider security from multiple perspectives
3. Establish a flexible infrastructure at all levels of the organization capable of responding rapidly to changing technology and security needs.
4. Initiate an ongoing, continual effort to maintain and improve its security posture.

Effective definition and implementation of an organization's IT security strategy will require a forward-looking view, teamwork, senior management participation, vulnerability assessment, threat classification and prioritization, information systems audits, and information security risk analysis. Most importantly, ownership of the IT security strategy must be established at all levels as all personnel must bear a degree of responsibility for maintaining the integrity of the organization's information systems that is appropriate to their position and job within the organization.

In establishing an enterprise-wide IT security policy, the following elements must be addressed:

- Maintenance of the confidentiality of privileged digital records.
- Address risks associated with recently terminated and former employees.
- Email and internet usage policy.
- Document management and file storage and encryption policy.
- Password Maintenance and Change policy.
- Incident response methodology, policy and procedures.
- Information security audit and assessment meth-

odology, guidelines, and procedures.
- What security functions are internally addressed versus outsourced.
- Specific and general duties of employees within distinct functional areas.
- Workstation, server, and other digital device usage policies.
- Application use and maintenance policy.
- Backup and recovery policy.

To provide credibility to the IT security policy, the human resources department must be involved. This department should oversee the assurance that each employee, as a condition of employment, has signed an agreement to abide by the IT security policy as it pertains directly or indirectly to his or her job.

Securing and Hardening IT Assets and Security Preparation

Effective information security management can be modeled by the three legs of an equilateral triangle with the legs corresponding to the principles of confidentiality, integrity, and availability. The following recommendations will prove useful in meeting these persistent objectives. However, please note that the following discussion is not meant to be an exhaustive treatment of ways to secure and harden information system assets from intrusion and attack, and even if an organization follows all of these recommendations, they may not be sufficient to make it impervious to intrusion or attack.

Hardware and Software Management

Managing information security requires the proper installation, maintenance, and protection of numerous hardware and software assets. This section will discuss the securing and hardening of such assets that are common to nearly every business operation.

Updating Operating Systems and Applications

It is absolutely essential that operating systems and applications be kept up-to-date by technical administrators. Because the essence of software engineering is complexity, there will always be defects and weaknesses that a skilled attacker may exploit. Furthermore, with widespread use of certain operating systems and applications, the likelihood of a security flaw being discovered in such software is remarkably high.

Administrators must keep themselves apprised of current operating system and application upgrades and patches. Whenever an update is released, technical staff must evaluate it, determine if it is applicable to their organization's computer systems, and if so install it.[10] Failing to install relevant upgrades or patches can have disastrous consequences. For example, in January 2003, the SQL Slammer Worm caused worldwide chaos across corporate networks and the internet. This small but malicious program rapidly exploited a flaw in Microsoft SQL Server even though a patch had been available for six months, underscoring a dirty secret in the information technology industry: Software bugs are common, and administrators are slow to fix even widely publicized problems, said Johannes Ullrich, director of the security information site, Incidents.org.[11] Administrators must also weigh costs and benefits of installing upgrades, patches, and service packs as well as when and how to install them. Install an update can itself cause security problems, such as the following:[12]

- During the update process, the computer may temporarily become more vulnerable.
- If the update is scheduled inappropriately a computer or information asset may not be available when needed.
- If an update must be performed on a large number of computers, there may be a period when some computers on the network are using different, potentially incompatible versions of software, which might cause information to be lost or corrupted.
- The update may introduce new vulnerabilities.

Before installing software updates, system and user data must be backed up, as updates sometimes do not go as well as planned. Also, because viruses and other malicious software can sometimes be unintentionally introduced when deploying upgrades, it might be worthwhile to invest in integrity checking software such as Tripwire (http://www.tripwire.com). Integrity checking tools can identify changes made to files and directories when updates are installed.[13] When using such tools, a system baseline assay is established which can be used as a point of comparison in monitoring system changes and identifying those that might warrant investigation. For system maintenance to be performed in a precise and meaningful way, the maintenance methodology must be actively managed and thoroughly documented.

Configuring Operating Systems and Applications

The primary principle that should guide system administrators in configuring any operating system or application is: "deny first, then allow." That is, turn off as many services and applications as possible and then selectively turn on only those that are absolutely essential. [14] In the effort to monitor unexpected machine, network, or user behavior, an administrator must consider enabling the logging capability that is available with most modern operating systems. Logging will keep track of system, application, security, and internet browser events, errors, warnings and information that may prove quite valuable in enhancing and maintaining system security. On certain high traffic systems, this might require additional hardware. But should this need arise the cost of introducing more hardware to distribute computing will pale in comparison to the cost of a compromised system with no trail for conducting an analysis.

Workstations

The principle of "deny first, then allow" should govern workstation configuration activities. It is recommended that the most minimal operating system and application image that meets business requirements be installed on workstations. All unnecessary software should be removed or disabled on workstations also. Furthermore, users should be allowed only the minimal privileges that allow them to do their jobs on these workstations. Network services for each workstation should also be enabled frugally. In cases where extremely sensitive data resides on a workstation and that data should not leave the machine under any circumstances, portable media devices (floppy drives, CD-ROM writers, etc.) should be removed if system administrators and management consider such action appropriate. All organizations should have a workstation use policy that addresses such issues as inappropriate or private use of IT assets, locking operating systems and applications when machines are unattended, data backup, and user maintenance.

Servers

All of the recommendations for securing workstations apply to servers as well. However, special consideration must be given to the fact that servers usually serve multiple users and offer multiple services and applications, thus presenting a significantly larger security risk if servers are deployed or maintained haphazardly. For each network server, file systems, server maintenance methodologies, and protocols (and ports) offered must be determined. Adherence to the security principle of "deny first, then allow" is extremely important in the administration of servers. Nowhere is adhering to this principle more critical than in the configuration of services (ports). Depending upon the service (port), several configuration options can be considered[15]:

- Limit the network hosts that can access the service.
- Limit the users who can access the service.
- Configure the service to allow only authenticated connections. The authentication should not rely solely on network data such as IP addresses and DNS names, which can be spoofed, regardless of whether the host is trusted.
- Limit the degree of access (especially in cases where that would permit a user to change the configuration of network services).
- If applicable, limit the facilities and functions offered by the service only to necessary ones (e.g., if files will be shared via FTP, permit only file download and restrict file uploads).
- Isolate the service's files (configuration, data files, executable images, etc.) from those of other services and the rest of the system.

Backup and Recovery

File backups allow users and administrators to restore the availability and integrity of information assets following security breaches and accidents. Without a backup, organizations may be unable to restore a computer's data after system failures and security breaches. [16] A backup and recovery plan is a critical element of any successful information security policy. By implementing the principle of "deny first, then allow," which is also known as the principle of minimal privilege on workstations and servers and adequately documenting what software resides on each machine (the software "image"), the costs of the backup and recovery process will be better controlled. The backup and recovery procedures should be equally well documented, and users and administrators should be well aware of their responsibilities in the context of regular backup activities. When performing recovery operations users should watch for unexpected changes to their files or data.

For workstations there are two common approaches to backup processes:[17]

- Files are backed up locally at each workstation, often by the user(s) of that workstation. The advantage of this approach is that protected data does not have to traverse the network, which reduces the chances of its being monitored, intercepted, or corrupted. The disadvantage is that each workstation must have additional storage devices, which must be kept secure, and users must be trained to perform the backups.

- Backups are centrally administered, with data copied from workstations by a network-based backup

program. Encryption tools can be used to protect data passing from a user workstation to a central backup host server.

The server itself must have a backup system to protect its data and applications. Backup processes for servers generally involve establishing and maintaining a copy of the information content of the server on a separate secured server. Although implementation of encryption technology in backup processes is not commonplace, it would provide additional protection against intrusion, physical security failures, etc.

Malicious Software Protection

As mentioned earlier, malicious software such as Trojans, worms, and viruses represents a significant threat to system integrity. The best method for managing this risk involves both technology and human stewardship. Many anti-virus tools, such as McAfee VirusScan and Norton Anti-Virus, are available for both workstations and servers. These programs use sophisticated algorithms to recognize code patterns that are characteristic of malicious software. Anti-virus programs use a database of known malicious software and code patterns and the database must be updated frequently as new malicious software threats emerge literally every day. All levels of the organization should be involved in protecting the IT system from viruses and malicious software scanning on a regular basis. Users must therefore be trained to understand how malicious software works and how the available tools can be used to prevent damaging attacks. Administrators should actively discourage users from receiving, much less running, executable files received through email and should disallow such privileges if at all possible.

Secure Remote Administration

Energy management control systems (EMCSs) can be operated by a single computer at a central location within a facility or at another site belonging to the company. This provides significant cost benefits because many tasks can be automated, and the administrator does not physically have to visit each computer.[18] This convenience for the administrator can also create a significant security risk when the computers are network-based or linked to the internet. Authentication of user identity is the method for addressing this risk. User identity can be determined through public key authentication and use of secure protocols such as SSL and HTTPS. In addition, all remote data transmissions between administrator and computer must be made over an encrypted connection so that eavesdroppers cannot intercept confidential data.

Network Management
Routers and Firewalls

Networks by their very nature have several nodes, each of which may need to communicate with other nodes on the same network or with nodes on separate, outside networks. In networks with multiple segments using different protocols, the best device to use for efficient communications is a router.[19] A router is an intermediary device on a communications network that expedites message delivery. On a single network linking many computers through a mesh of possible connections, a router receives transmitted messages and forwards them to their correct destinations over the most efficient available route. On an interconnected set of local area networks using the same communications protocols, a router serves the somewhat different function of acting as a link between these local area networks, enabling messages to be sent from one network to another.[20] Because a router essentially handles all traffic across a network, it is a logical target for an attacker. Therefore, all routers used within an organization's computer network must have firewall capability.

Like the information security management cycle, implementation of firewalls is a Deming Plan-Do-Check-Act cycle; except here the process phases are Prepare, Configure, Test and Implement. See Table 18-1. There are numerous firewall topologies available that will significantly enhance network security. It is one of the network administrator's most critical responsibilities to be familiar with network and firewall topology and

to implement the one that best fits the organization's security needs.

As with all other hardware and software assets, routers and firewalls must be configured with minimal privileges and functions that meet the organization's needs. It is absolutely necessary that the firewall logging and alert mechanisms are enabled during the configuration phase so that the firewall will monitor and report attempted attacks in addition to enhancing network security.

Before the firewall system is implemented, it should be thoroughly tested. This will ensure that the design specifications and installation settings operate as intended and reveal any failures or weaknesses in the system. The features that must be tested include the following:[22]

• Hardware (processor, disk, memory, network interfaces, etc.)
• Operating system software (booting, console access, etc.)
• Firewall Software
• Network interconnection equipment (cables, switches, hubs, etc.)
• Firewall configuration software—including routing rules, packet filtering rules, and associated logging and alert options.

Testing the firewall system should also provide an opportunity for testing the network data backup and recovery processes and identifying strengths and weaknesses.

When installing a newly designed, configured, and

Table 18-1. Firewall Implementation Process Summary[21]

Process Phase	Practice
Prepare	Design the Firewall System
Configure	Acquire Firewall Hardware and Software
	Acquire Firewall Training, Documentation, and Support
	Install Firewall Hardware and Software
	Configure IP Routing
	Configure Firewall Packet Filtering
	Configure Firewall Logging and Alert Mechanisms
Test	Test the Firewall System
Implement	Install the Firewall System
	Phase the Firewall System into Operation

tested firewall system, administrators must consider the effects that such a system will have on the networks with which the firewall system will connect and how newly available connectivity will affect normal operations. Consequently, a firewall system should be phased in across the other network hosts that are connected to the new firewall system. Each host that is intended to send traffic through the firewall must be made aware of the new firewall's existence.[23] If they are not, the router will still function properly but no security benefits from the firewall will be realized. As for the hosts controlled by the new firewall system, those hosts must also be configured for sending data to and receiving data from the firewall system. Finally, all users should be notified of the firewall system's installation. The new network configuration will generally be transparent to users, but making users aware of the installation will expedite and ameliorate troubleshooting efforts.

Virtual Private Networks

Any organization with a network, and especially an organization that has significant data transmissions between internal and external networks, should use routers with Virtual Private Network (VPN) capability. A VPN is a set of nodes on a public network such as the internet that communicate among themselves using encryption technology. Their messages are as safe from being intercepted and understood by unauthorized users as if the nodes were connected by private lines. By installing a router with both firewall and VPN capability, the risk that data transmissions will be intercepted is markedly reduced. The VPN capability takes care of securing and encrypting transmissions and communications. This is especially necessary now that many computer users are highly mobile, and may travel to distant offices. A VPN can allow an organization with different regional offices to avoid the costly startup and maintenance of a Wide Area Network. As a general policy, employees should be required to use the company's VPN when remotely transmitting or receiving data over the organization's network.

Wireless Access

Since 2000 there has been a rush to implement wireless network access in the workplace. As a result, convenience has taken priority over security; in fact, as of September 2002, security was lacking in 80% of wireless networks.[24] Anyone with a wireless network card and a laptop can access a wireless network through an "access point" since the signal from nearly all "access points" spills onto the street. In fact, this spillover is so widespread that buildings are routinely "chalked"*[25] with information indicating known unsecured access points into corporate and private networks.[26] This behavior is known as "war chalking." There are many resources on the internet illustrating equipment that is helpful in addressing this security issue.[27]

A typical scenario might be the following. Unauthorized access starts when an intruder pulls his car into a public parking lot equipped with a laptop, wireless network card and free "sniffer" software. If the wireless network is not secured, the intruder changes the wireless card Service Set Identifier (SSID) on his laptop to match that of the wireless "access point." The intruder then requests an IP address from the network and opens Microsoft's Network Neighborhood to see what computers are unprotected. Any PC that has file and print sharing enabled could be accessed. From this point, an intruder can use advanced tools to probe into the network without the trouble of a firewall to prevent access. These access tools are available on the internet at no cost. Even if no other computers are exposed, the intruder has unfettered internet access and can impersonate a legitimate corporate computer or user. Any illegal computer activity will be traced back to the network owner who may not be able to determine the intruder's identity or location.

Currently, transmissions sent over a wireless IEEE 802.11b connection are not encrypted. Transmissions to pagers and between cellular telephones are also not encrypted. Consequently, the risk of eavesdroppers intercepting such transmissions is considerable. Moreover, the increasing popularity of "war-chalking" demonstrates the magnitude of the risk. Since many data transmissions are unencrypted, an intruder who acquires access through the wireless system could gain undetected and unfettered access to critical data transmissions and network resources. If the organization can afford to do so, establishing a sufficient number of static IP addresses to accommodate the needs of all devices will provide a significant barrier to intruders. If this is feasible for the organization, a network administrator would establish fixed relationships in the ARP (see Appendix B—Key Terms) cache between static IP addresses and individual device's network card MAC addresses. For this wireless security strategy to be complete, DHCP (see Appendix B—Key Terms) must be disabled at the host server level.

---

*Using chalk to place a special symbol on a sidewalk or other surface that indicates a nearby wireless network, especially one that offers internet access.

Wireless security can also be improved by making use of the wired equivalent privacy (WEP) protocol. Nearly all IEEE 802.11b-enabled cards and access points on the market implement the WEP standard, which makes it very difficult to use the wireless network without authorization. Since most access points are not protected by WEP, any organization that implements it will discourage all but the most dedicated intruders from gaining access to the network. After enabling WEP, businesses need to change the wireless network's default SSID to another character string that cannot be readily guessed. SSID broadcasting should also be disabled so that the SSID is not easily intercepted. At the hardware level, any wireless network should include or at least be contained within a network that has firewall and VPN capability.

Other Issues
Encryption of Critical Storage Media and E-mail

In order to mitigate the risk of eavesdroppers and interlopers gaining access to the exchange of operationally valuable data via email, the email should be encrypted using PGP, Kerberos, or some other public key encryption protocol. These methods implement significant barriers to decrypting ciphertext to plaintext. In addition to this technological solution for securing email, such communications should be explicitly marked as private, confidential, etc. and should contain an appropriate statement to that end in the footer of each and every email message. When documents are transmitted by email as attachments, they should be converted to portable document file (PDF) format or rich text format (RTF) before sending them. Microsoft Word and Corel WordPerfect file documents contain metadata (information that is hidden within the document). Such metadata may contain every modification or the change history of the document. PDF and RTF document formats contain minimal metadata about the document itself and therefore should be the only acceptable formats for transmission of formatted attachments. Finally, what little metadata can be extracted from the PDF and RTF file formats cannot be easily obtained by someone who has not had significant technical training.

Computer use is no longer limited to desktops located securely inside the walls of the buildings. Many computer users have laptop computers that are used for communications (email and peer-to-peer), word processing, document management, file storage, etc. Because these devices are highly mobile and are sometimes left unattended or are stolen, there is a significant risk of outside parties gaining access to privileged communi-

cations and data. Laptop users can easily protect their data by enabling the encryption capability on their permanent (hard drive) and local storage media (floppy, CD, etc.). While this solution is certainly no substitute for a user being cautious and protective of a laptop that contains privileged data, it will help prevent the data being compromised. As mentioned earlier, encryption protocols should be implemented wherever feasible, as doing so only enhances security by insuring greater data confidentiality and integrity.

In addition to data encryption, a user may incorporate the use of the new biometric devices that use fingerprints or other anthropocentric data for authentication purposes. Early devices allowed for only one possible user. This posed a risk that the data might be lost to other parties within the firm if the original user was not available to authenticate the data. Newer biometric device offerings allow multiple users to be able to authenticate a storage device. If an organization were to use such devices, the fingerprints of the primary user and a system administrator would be sufficient to provide a high level of security without excessively restricting access.

User Authentication Policy

One of the simplest methods of thwarting intruders and attackers is by having an effective user authentication policy. Such a policy should use hardware authentication features that are available on any computer's BIOS. User accounts should be actively managed at the administrator level with the principle of minimum privilege in mind. Unnecessary accounts and those belonging to recently terminated or deceased employees should be eliminated. Administrators should require users to re-authenticate themselves on machines that have been idle a significant period of time and should deny log-in to any device after a small number of failed attempts (three to five is common). Development of a robust password policy is probably the most significant element of user authentication policy administration; humans are notoriously weak protectors of information, and intruders and attackers frequently try to acquire a password in order to gain initial system access. A password policy should cover five elements:[28]

1. Length—passwords should have a minimum length of eight characters.
2. Complexity—passwords should contain a mix of characters, that is, both uppercase and lowercase letters and at least one non-alphabetic character.
3. Aging—users should change their passwords pe-

riodically (every 30-120 days). The policy should permit users to do so only through approved authentication mechanisms.

4. Reuse—administrators should decide whether a password may be reused. Some users try to defeat a password-aging requirement by changing the password to one they have used before.

5. Authority—administrators should decide who is allowed to change a user's passwords.

The password policy should be documented and communicated to users, and users should be trained and expected to follow the policy.

Redundancy

If a system is critical to a company's operations, there should be systems and storage redundancy. By having a redundancy strategy in place, low probability/high cost events are addressed. Major systems should have identical systems on-line that will seamlessly continue operations in the event of interruption for the primary system. The most critical area for implementing redundancy is in storage media. If a primary storage media systems failure occurs without having a redundant device immediately available, it can be disastrous. A daily backup policy alone is insufficient for addressing random failures between backups. This is true for both file systems and databases. Storage media are continually decreasing in cost per unit of storage and a redundant array of inexpensive disks, or RAID, technology is practical for addressing this need. RAID is a disk system that is comprised of an array of disk drives to provide greater reliability and storage capacity and better performance at a lower cost.

On a related note, redundancy is not only applicable to systems. It should also be considered when staffing system administrator positions. While one system admin may be cost effective in the short run, if that person leaves the firm, is incapacitated, or simply on vacation or sick, the firm could experience catastrophic results in the event of system failure. It is critical for any organization to have a redundancy strategy that addresses these issues.

Physical Access

Serious restrictions should be placed on physical access to the office facility, operations and computer facilities of any organization. Companies should use biometric or keycard systems as well as human security guards for maximum security. They should also employ or outsource an information security staff that is com-plementary to their regular IT staff. These individuals would conduct regular training as well as manage the daily physical security of all information technology employed by the organization. Security personnel should undergo strict background checks and regular controlled substance abuse testing. Active management of physical security, digital security and employee security will significantly reduce an organization's overall information security risk profile.

In addition to hiring security staff, IT assets should be used in secured facilities. Furthermore, the level of security associated with each IT asset should be commensurate with its importance to the continued operations of the business. System administrators should also limit the installation of unauthorized hardware as this is a frequently overlooked element of physical security. Such unauthorized hardware might include removable media storage devices, modems, or devices that might intentionally or unintentionally be used to bypass any security measures.

DETECTION, RESPONSE, AND IMPROVEMENT

The activities of detection, response, and improvement are important phases of the information security management process. Throughout the process, constant communication within the organization is absolutely necessary. Otherwise, any information security policy or strategy will fail to produce favorable results.

Detection

Intruders are always looking for new ways to break into networked computer systems. Even if an organization has implemented a number of information security protection measures, such as firewalls and intrusion detection systems, employees must closely monitor the organization's information assets and transactions involving these assets to check for signs of intrusion. [29] System administrators should regularly review and monitor the following:

- Network Alerts
- Network Error Reports
- Network Traffic
- Software image checksums where current software images on computers are compared to authoritative software images via complex mathematical algorithms.
- System performance statistics on all computers

— CPU, memory, storage media
— Message and print queues
— Changes in file system status or warnings
- Any unusual behavior by a system or by individual personnel

If a system administrator notes any anomalies during his reviewing and monitoring activities, he should ask himself the following questions:[30]

- Is the apparent anomaly the result of a legitimate new or updated characteristic of the system? (e.g., the unexpected process is executing a recently added administrative tool.)
- Can the anomaly be explained by the activities of an authorized user? (e.g., the user really was in Cairo last week and connected to the network; a legitimate user made a mistake.)
- Can the anomaly be explained by known system activity? (e.g., there was a power outage that caused the system to reboot)
- Can the anomaly be explained by authorized changes to programs? (e.g., the mail log showed abnormal behavior because the system programmer made a mistake when the software was modified)
- Did someone attempt to break into the system and fail?
- Did someone break in successfully? Does the administrator have the data that will tell him what the intruder did?

Response and Containment

In responding to any security incident, the administrator and management must remain disciplined in their procedures and use of documented processes and labor to generate sufficient documentation. This is critical to driving the continuous review and learning element of the information security process as well as collecting digital evidence of the anomalous activities.

At the outset of any incident response and containment process, administrators must assess the severity of the incident in terms of scope, impact and damage. From this analysis, it can be determined what subsequent actions must be taken. A worst-case scenario should be assumed in order to enumerate the major actions that must be taken when faced with such an incident.

Administrators and management should inform other sites, networks and organizations that might have been affected by the incident. Continuous communications with potentially concerned parties should be maintained until the incident is contained and normal system operations have resumed. Throughout the entire incident response process, information must be collected at all affected levels. The following data should be collected from all relevant system and network logs:[31]

- The name of the system
- The date and time of each incident affecting that system
- What actions were taken
- What was said during the individual system investigation
- Who was notified
- Who had access
- What data were collected
- What information was disseminated—to whom, by whom, when, and for what purpose
- What was submitted to legal counsel—to whom, by whom, and how it was verified (e.g., notarized)

Upon this investigation administrators and management might find it necessary to take further actions such as temporarily shutting down affected systems, disabling access, services and accounts, and contacting law enforcement agencies.

Once the root cause of the incident has been established, the damage has been contained, the system vulnerability addressed, and the appropriate protective and preventive measures taken, administrators and users must take actions to assure data and system integrity, particularly of backup and redundant systems. Also, every user on every affected system should be required to change his password(s). This should be done for all software at all affected levels too. Finally, because information security assurance is an ongoing process, system-wide monitoring should continue and lessons learned should be articulated and applied to the continuous improvement of the organization's information security stance.

Improvement

It is important to learn from the successful and unsuccessful actions taken in response to an intrusion. Capturing and disseminating what worked well and what did not will help reduce the likelihood of similar intrusions and will improve the security[32] of the organization. Organizations should also try to learn from actions taken during normal system activities and security assessments. If an organization fails to learn from its actions during both normal and abnormal operations and does not document and disseminate its discoveries, it will continue to maintain a reduced system security stance. One mechanism by which the learning process can be formalized is the postmortem review meeting. At such meetings, policies, procedures, system and administrative successes and failures should be evaluated, revised, and improved in order to strengthen the organization's security posture. By having event postmortem discussions the learning process will be served and vivid recent experiences from those involved will be captured and documented.

CONCLUSION

The importance of information of information security management for every energy professional cannot be underestimated. As the field of industrial energy management becomes increasingly automated and information technology becomes a more integral part of operational efficiency and strategic advantage, the breadth and depth of the risks that industrial organizations face will continue to expand. The diversity of threats and risks that the industrial energy manager faces must be managed at all levels of the organization. While it is impossible to manage all possible information security contingencies, continuous education about information security throughout the organization will facilitate the effective management of known and unknown security risks. In addition to education about information risk management and best practices, information security policies must be developed and continually reviewed. These policies must address the most probable security issues without constraining the organization to the extent that it fails to adapt to future challenges presented by the continually changing technological landscape. As any operations manager or energy manager remains ever vigilant about his physical inventories and resources, so he must also be about protecting his information systems assets. As the old Boy Scout motto advises: "Be Prepared."

Endnotes/References

[1] "2001 CSI/FBI Computer Crime and Security Survey," Computer Security Issues and Trends, vol. VII, no. 1. Computer Security Institute, Spring 2001.
[2] Cybercrime: Vandalizing the Information Society. Furnell, Steven. Pg. 25. 2002.
[3] Scene of the Cybercrime: Computer Forensics Handbook. Shinder, Debra Littlejohn; Tittel, Ed. Pg. 317. 2002.
[4] Information Assurance: Managing Organizational IT Security Risks. Boyce, Joseph G.; Jennings, Dan W. Pg. 207. 2002.
[5] Scene of the Cybercrime: Computer Forensics Handbook. Shinder, Debra Littlejohn; Tittel, Ed. Pg. 300. 2002.
[6] The CERT® Guide to System and Network Security Prac-

tices. Allen, Julia H. Pg. 101. 2001.
[7] Writing Secure Code. Howard, Michael; LeBlanc, David. Pg. 63. 2002.
[8] Scene of the Cybercrime: Computer Forensics Handbook. Shinder, Debra Littlejohn; Tittel, Ed. Pg. 338. 2002.
[9] Microsoft Press Computer Dictionary, Third Edition. 1997.
[10] Managing Information Security Risks: The OCTAVESM Approach. Alberts, Christopher; Dorofee, Audrey. Pg. 8-9. 2003.
[11] The CERT® Guide to System and Network Security Practices. Allen, Julia H. Pg. 39. 2001.
[12] "Work exposes apathy, Microsoft Flaws." Lemos, Robert, CNET News.com, January 26, 2003.
[13] The CERT® Guide to System and Network Security Practices. Allen, Julia H. Pg. 40. 2001.
[14] Ibid. Pg. 42.
[15] Ibid. Pg. 43.
[16] Ibid. Pg. 44.
[17] Ibid. Pg. 59.
[18] Ibid. Pg. 60.
[19] Ibid. Pg. 67.
[20] MCSE Networking Essentials. Sportack, Mark and Glenn, Walter J. SAMS Publishing. Pg. 194. 1998.
[21] Microsoft Developers Network, July 2000.
[22] The CERT® Guide to System and Network Security Practices. Allen, Julia H. Pg. 123. 2001.
[23] Ibid. Pg. 161-162.
[24] Ibid. Pg. 173.
[25] www.landfield.com/isn/mail-archive/2002/Sep/0046.html.
[26] www.warchalking.org.
[27] Equipment to "war-drive": www.bitshift.org/wardriving.shtml.
[28] The CERT® Guide to System and Network Security Practices. Allen, Julia H. Pg. 52. 2001.
[29] Ibid. Pg. 231.
[30] Ibid. Pg. 262.
[31] Ibid. Pg. 283.
[32] Ibid. Pg. 296.

APPENDIX A.
COMMON PORT/PROTOCOL ASSIGNMENTS

Port	Protocol
21	FTP
23	Telnet
25	SMTP
53	DNS
80	HTTP
88	Kerberos
110	POP3
119	NNTP
135	RPC
139	NetBIOS session service
194	IRC
389	LDAP
443	HTTPS
1,024—65,535	Open ports to be utilized by user processes or applications.

APPENDIX B. KEY TERMS

802.11—A family of specifications developed by the IEEE for wireless LAN technology. 802.11 specifies an over-the-air interface between a wireless client and a base station or between two wireless clients.**

ARP—Address Resolution Protocol. A TCP/IP protocol used to convert an IP address into a physical address, such as an ethernet address. A host wishing to obtain a physical address broadcasts an ARP request onto the TCP/IP network. The host on the network that has the IP address in the request then replies with its physical hardware address.**

BIOS—Basic Input/Output System. On PC-compatible computers, the set of essential software routines that test hardware at startup, start the operating system, and support the transfer of data among hardware devices. The BIOS is stored in read-only memory (ROM) so that it can be executed when the computer is turned on. Although critical to performance the BIOS is usually invisible to computer users.*

Ciphertext—Data that have been encrypted.***

DHCP—Dynamic Host Configuration Protocol. A TCP/IP protocol that enables a network connected to the internet to assign a temporary IP address to a host automatically when the host connects to the network.*

DNS—1. Acronym for Domain Name System. The system by which hosts on the internet have both domain name addresses and IP addresses. The domain name address is used by human users and is automatically translated into the numerical IP address, which is used by the packet-routing software. 2. Acronym for Domain Name Service. The internet utility that implements the Domain Name System (see definition 1). DNS servers, also called name servers, maintain databases containing the addresses and are accessed transparently to the user.*

Firewall—A security system intended to protect an organization's network against external threats coming from another network. A firewall prevents computers in the organization's network from communicating directly with computer external to the network and vice versa. Instead, all communication is routed through a proxy server outside of the organization's network, and the proxy server decides whether it is safe to let a particular message or file pass through the organization's network.*

FTP—File Transfer Protocol, the protocol used for copying files to and from remote computer systems on a network using TCP/IP, such as the internet. This protocol also allows users to use FTP commands to work with files, such as listing files and directories on the remote system.*

HTTP—HyperText Transfer Protocol. The client/server protocol used to access information on the World Wide Web.*

HTTPS—An extension to HTTP to support secure data transmission over the World Wide Web.**

IP—Internet Protocol. The protocol within TCP/IP that governs the breakup of data messages into packets, the routing of the packets from sender to destination network and station, and the reassembly of the packets into the original data messages at the destination.*

IRC—Internet Relay Chat. A service that enables an internet user to participate in a conversation on-line in real time with other users. An IRC channel, maintained by an IRC server, transmits the text typed by each user who has joined the channel to all other users who have joined the channel.*

Kerberos—A network authentication protocol developed by MIT. Kerberos authenticates the identity of users attempting to log on to a network and encrypts their communications through secret-key cryptography.*

LDAP—Lightweight Directory Access Protocol. A set of protocols for accessing information directories that supports TCP/IP, which is necessary for any type of internet access. LDAP makes it possible for almost any application running on virtually any computer platform to obtain directory information, such as email addresses and public keys. Because LDAP is an open protocol, applications need not worry about the type of server hosting the directory.**

MAC address—Media Access Control address. A hardware address that uniquely identifies each node of a network.**

NetBEUI—NetBIOS Enhanced User Interface. An enhanced NetBIOS protocol for network operating systems, originated by IBM for the LAN Manager server and now used with many other networks.*

NetBIOS—An application programming interface that can be used by application programs on a local area network consisting of IBM and compatible microcomputers running MS-DOS, OS/2, or some version of UNIX. Primarily of interest to programmers, NetBIOS provides application programs with a uniform set of commands for requesting the lower-level network services required to conduct sessions between nodes on a network and to transmit information back and forth.*

NNTP—Network News Transfer Protocol. The internet protocol that governs the transmission of newsgroups.*

NTP—Network Time Protocol. A protocol used for synchronizing the system time on a computer to that of a server or other reference source such as a radio, satellite receiver, or modem. NTP provides time accuracy within a millisecond on local area networks and a few tens of milliseconds on wide area networks.*

PGP—Pretty Good Privacy. A technique for encrypting messages that is one of the most common ways to protect messages on the internet because it is effective, easy to use, and free. PGP is based on the public-key method, which uses two keys—one is a public key that you disseminate to anyone from whom you want to receive a message. The other is a private key that you use to decrypt messages that you receive.**

Plaintext—Data that has not been encrypted.***

POP3—Post Office Protocol 3. A protocol for servers on the internet that receive, store, and transmit email and for clients on computers that connect to the servers to download and upload email.*

PPP—Point-to-Point Protocol. A data link protocol for dial-up telephone connections, such as between a computer and the internet.*

Public Key Encryption—An asymmetric scheme that uses a pair of keys for encryption: the public key encrypts data, and a corresponding secret key decrypts it. For digital signatures, the process is reversed: the sender uses the secret key to create a unique electronic number that can be read by anyone possessing the corresponding public key, which verifies that the message is truly from the sender.*

RPC—Remote Procedure Call. A type of protocol that allows a program on one computer to execute a program on a server computer. Using RPC, a client program sends a message to the server with appropriate arguments and the server returns a message containing the results of the program executed.*

SLIP—Serial Line Internet Protocol. A data link protocol that allows transmission of IP data packets over dial-up telephone connections, thus enabling a computer or a local area network to be connected to the internet or some other network.*

SMTP—Simple Mail Transfer Protocol. A TCP/IP protocol for sending messages from one computer to another on a network. This protocol is used on the internet to route email.*

SNMP—Simple Network Management Protocol. The network management protocol of TCP/IP. In SNMP, agents, which can be hardware as well as software, monitor the activity in the various devices on the network and report to the network console workstation.*

SSID—Service Set Identifier. A 32-character unique identifier attached to the header of packets sent over a wireless LAN that acts as a password when a mobile device tries to connect to the network. The SSID differentiates one wireless LAN from another, so all access points and all devices attempting to connect to a specific wireless LAN must use the same SSID. A device will not be permitted to join the network unless it can provide the unique SSID.**

SSL—Secure Sockets Layer, a protocol developed by Netscape for transmitting private documents via the internet. SSL works by using a public key to encrypt data that's transferred over the SSL connection. Both Netscape Navigator and Internet Explorer support SSL, and many web sites use the protocol to obtain confidential user information, such as credit card numbers. By convention, URLs that require an SSL connection start with https: instead of http:.**

TCP—Transmission Control Protocol. The protocol within TCP/IP that governs the breakup of data messages into packets to be sent via IP, and the reassembly and verification of the complete messages from packets received by IP.*

TCP/IP—Transmission Control Protocol/Internet Protocol—A protocol developed by the U.S. Department of

Defense for communications between computers. It is built into the UNIX operating system and has become the de facto standard for data transmission over networks, including the internet.*

Telnet—A protocol that enables an internet user to log on to and enter commands on a remote computer linked to the internet, as is the user were using a text-based terminal directly attached to that computer. Telnet is part of the TCP/IP suite of protocols.*

UDP—User Datagram Protocol. A connectionless protocol that converts data messages generated by an application into packets to be sent via IP but does not verify that message have been delivered correctly.*

VPN—Virtual Private Network. A set of nodes on a public network such as the internet that communicate among themselves using encryption technology so that their messages are as safe from being intercepted and understood by unauthorized users as if the nodes were connected by private lines.**

WEP—Wired Equivalent Privacy. A security protocol for wireless local area networks defined in the IEEE 802.11b standard. WEP is designed to provide the same level of security as that of a wired local area network.**

---

*Microsoft Press Computer Dictionary, Third Edition. 1997.
**Webopedia: On-line Dictionary for Computer and Internet Terms. www.webopedia.com
***Microsoft Developer Network, July 2000.

Section V

System Perspective
and
Enterprise Energy Systems

Chapter 19

Building Control Systems & the Enterprise

Toby Considine

ABSTRACT

Many current journal articles and most current building control system product literature make claims about building controls interacting with the enterprise. Far from being ready to interact with the enterprise, today's control systems have not yet matured into enterprise functions and so are not ready to interact with other enterprise functions. Merely using the current enterprise protocols such as XML and SOAP are not enough. It is only by developing the operation of building automation systems into an enterprise function that control systems can ready themselves for a role as an enterprise asset. This chapter describes a path to this mature function, and introduces the term building automation for the enterprise (BAE) to describe it.

INTRODUCTION

The controls industry today, is abuzz with talk of opening up control systems to the enterprise. Every marketing brochure, every position paper, every salesman talks of it, and has some verbiage about it on their PowerPoint presentations.

These messages are misleading, often to those who speak them as well as those that hear them. They have only the vaguest notion of the enterprise, or what value the enterprise might find in controls. All too often, it is implied that merely using a modern protocol like SOAP or web services is sufficient without thinking about what information is needed by the enterprise. The model for securing the information in these systems is rudimentary or is bolted on as an afterthought. Sometimes these new systems put good engineering at risk, threatening the reliability and safety so carefully built into those systems.

What is needed is a clear understanding of enterprise architectures, and the requirements of the protocols underneath them. These architectures require common standards to develop classifications (taxonomies) that deal with the big picture rather than the minute details. They must also have a refined (nuanced) concept of security that interacts well with the enterprise. In other words, these systems must be ready to be participants in an enterprise Service Oriented Architecture (SOA).

Today's discipline of Building Automation Systems (BAS) deals only with the building automation control systems (BACS). The language, and methods, and design are all best suited to isolated control systems installed by controls engineers and maintained by facilities personnel. The discipline is control-centric and BACS requires practitioners who are trained in the intricacies controls and the physical nature of the systems they control. There is another aspect of BAS, what I call building automation for the enterprise (BAE) which completes BAS, and makes interaction with the BACS safe for the enterprise programmer, one who is trained in business process and enterprise protocols rather than in control systems and building operations.

Today's enterprise systems are assembled from coherent modular systems. These systems have well defined interfaces that have abstracted their internal operations to suit the needs of the enterprise. This abstraction serves three strong purposes. (1) It exposes the internal functions of each system in ways that make sense to the enterprise. (2) It hides the inner complexity of the systems to protect the inner operations of the systems from inappropriate interference at the enterprise level. (3) It enables different systems to be competitively swapped out at as modules, allowing competition to drive innovation and competence within each system.

To support BAE, BAS must develop in ways that other business functions have already. It must develop an abstract lexicon, wherein intricate interactions necessary to BACS are described in terms of higher-level business functionality. It must embrace objects, which provide defined interfaces while hiding their inner complexity, both to simplify higher level programming and to protect complex operations from meddling. BAE must support the protocols of the enterprise, complete with standard taxonomies, or ways of talking about the operations, that persist across the control system life-cycle. When BAE has developed, the full value of control system operations will be made available to the enterprise. Then and only then will the BAS function be ready for the enterprise.

THE ENTERPRISE

The enterprise is, quite simply, the normal business of a corporation or agency. Management guru Peter Drucker once observed that no one should work in a role for a corporation from which one could never rise to be an executive of that corporation. Any employee in such a position would never find his work valued. Drucker said that such activities should be outsourced, and he further noted that this would only improve life for that employee who would spend his working life in an enterprise that valued his work, and whose core values supported him in that work. For example, John Aker, CEO of IBM in the 70s and 80s started his career on the IBM loading dock.) What this means to me is that the enterprise identifies all the core activities from which a business derives its value.

On the edge of the enterprise are what I call "hygiene processes." These processes are ones that are needed to keep the enterprise operating smoothly, such as janitorial services, security, or communications, but they are not part of the core mission of the enterprise. No one is hired for their personal hygiene, although people have been fired for neglecting it. A cottage in the country can get by with an outhouse; but a town that does not upgrade its hygiene requirements as it grows smells bad. Hygiene processes must simply be dealt with, and their costs minimized. We rarely describe people, systems, or companies in terms of their hygiene unless that hygiene is sub-standard. As an enterprise grows in sophistication, it requires that those hygiene processes be upgraded to meet the expanding needs of the enterprise.

Today, BAS is merely a hygiene process. No one likes to think about it. The only time that the enterprise thinks hard about BAS is when it fails. There are low expectations for strategic value of BAS. Rarely does the quality of service (QOS) of the BAS get discussed in the enterprise. Rarely does the enterprise think of building control systems at all except when the elevator jars the tenant, the too-hot/too-cold call comes in, or the security

system fails.

Systems that do not interact with others in the enterprise are often referred to as process silos. A silo has thick walls to prevent outside influences (insects, weather and livestock) silo from getting in. A process silo goes straight up and down, not interacting with any layers of the organization; a silo stands alone, a monolith on the farm, but somehow not part of the farm. Figure 19-1 shows business process silos in an organization prior to the integration of across silos. Note that there are no interactions between these silos.

At one time, each of the process silos shown in Figure 19-1 might have been considered a hygiene process. When the cost of carrying capital was less significant, the primary purpose of inventory control was to make sure that manufacturing didn't stop. Some companies treated workers as interchangeable parts with no strategic function. Salesmen kept their information to themselves as long as they made their quotas. These functions were what the company had to do, not what it "did."

These hygiene processes took direction only from the top of the corporation. Like the farm silo, they were tall and thin, with well-defined walls that preclude any interaction with other business processes. The narrative of the last 40 years of information technology is a tale of hygiene processes moving to the enterprise, by opening up interactions outside their silo of operations. Figure 19-2 shows how the communication between processes has integrated the enterprise.

BAS has yet to interoperate with the enterprise. It still remains isolated, with thick walls of process, or terminology, and often of networking infrastructure in place between BAS and the enterprise.

Today, BAS, with a few exceptions such as verified

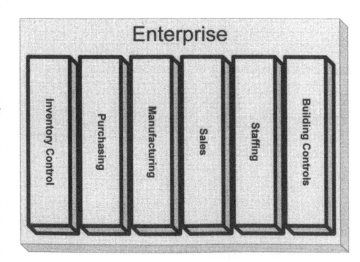

Figure 19-1. Isolated Process Silos

environment systems for pharmaceutical operations, are treated as little more than hygiene processes. As such, their operations are relegated to the dankest room in the back office, with little respect and outsourced if at all possible. BAS are noticed only when they fail (and perhaps by the odor). BAS are certainly not managed as a strategic enterprise function.

But what can control systems provide the enterprise?

The simplest, most elegant answer I know is by example. There is no control whose complete operation is simpler to understand than that of the black mat at the store to open the door for each customer when it is trod upon. One large retailer has enterprise-enabled its door openers to provide live foot-traffic information to its sales staffing operations. This information is also used by its advertising and marketing groups to analyze the direct effects of their activities. A simple convenience control with very little smarts is now a corporate asset to three separate enterprise activities.

A LITTLE HISTORY LESSON

When a business process is first computerized, little more is attempted than some efficiency in doing the same old thing. Paper time sheets and accounting systems were little different from the computerized payroll and accounting systems that followed them. In Boston, where I began my career, the downtown streets are laid out in the same confusion as the original settlement. This process is called "paving the cow paths."

For me, professional computing began with automated purchasing systems in the late 70s. The systems

I wrote logged purchase orders, printed invoices, and worked the factory stock room. Some paper handling was eliminated, but even more paper was printed since the printed receiving documents were now filed. Some efficiencies were gained, but no changes were made to the core business processes; and this core process did not interact with any other business process. We stayed in the silo and continued to pave the cow paths.

During the 80s, these purchasing systems were expanded, exploding the silo. Detailed work flows and manufacturing schedules dictated exactly what was purchased and when. Off-shore suppliers built materials to meet detailed sales projections and deliver them to the warehouse just as they were needed for shipment to customers. Material requirements planning (MRP) grew through several iterations, including more and more processes (MRP II), until we have today's enterprise requirements planning (ERP), encompassing almost all of the core needs and processes of the enterprise. Each activity—sales, manufacturing, distribution, personnel, and fulfillment—interacts with all the others.

To me, BAS remains, for the most part, in the first phase. A janitor used to turn on the furnace in the basement—and tenants on each floor could turn valves on their radiators. We have made that process more efficient, by installing control systems to replace those turncocks, and bringing the setpoints and decisions to a man in the control room. BACS is more efficient, perhaps more precise, but essentially unchanged. We have only paved the cow path. We have automated rather than innovated; the process is more efficient. But BAS has not brought any real change to the process; BAS remains BACS.

DOES THE ENTERPRISE NEED BAS?

Health and safety concerns, green buildings, and sustainability are beginning to make BACS a strategic asset. Qualities of services (QOS) requirements are increasingly being written into leases, not only for comfort control but also for power management, for business intelligence, and even for support of new areas of enterprise regulatory requirements.

New regulatory mandates bring new requirements to building automation systems. For example, medical facilities are being required to protect patient information as never before under the Health Insurance Portability and Accountability Act (HIPPA). This means that access to records must be carefully monitored. This can include limiting unsupervised access to records areas, say on weekends or evenings. This requirement naturally drives

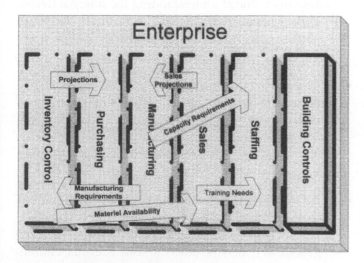

Figure 19-2. The integrated Enterprise

medical facilities from simple key access to automated access control, able to limit employee access outside of their work shift and to automatically log that access.

Providing access to user-friendly information enables people to make better decisions. At present, much of the information available about our building and utility systems is generated for a specific user group focused on a narrow area of interest. The data are provided at a level of detail, and in a terminology, that is not useful to enterprise audiences. This unnecessarily visible complexity actually reduces the utility of the system.

Putting in place a monitoring and information system that expands opportunities for access and analysis would improve communication and decision making across a range of operational and academic departments and users.

Current attempts at data gathering often create winners and losers. One silo is made more robust and credible while another is weakened. Providing a data gathering network that expands choices and possibilities for all who might be interested in putting the information to productive use is logical and elegant. Just as the internet has made quick access to ever more detailed and credible information available to anybody with interactive connectivity, an improved campus communication network is needed to better share information among diverse parties interested in improving their decision-making capabilities.

A building may have a card-access control system to control who enters each area. A simple relay may trigger the HVAC system from the access control system. There may be a machine-readable electric meter to track energy use (and thereby cost). If these functions were abstracted, and able to fully interact with each other as well as other enterprise functions, such as human resources and accounting, then the card swipe that opens the door would not only turn on the air handler but could also bill the energy use to the appropriate department, facilitate staff scheduling decisions, reduce peak demand, and improve occupant comfort and productivity. This would create interaction between silos (processes), and would build new highways rather than paving the old cow-paths.

ENTERPRISE-READY
SYSTEMS ARE ABSTRACT

Interaction with today's control systems requires too much knowledge of the inner workings of the control system. A control protocol like LON or BACnet may tell you that an actuator is open (these data are called tags); the enterprise only needs to know whether the system is turned on. Another tag may report temperature as milliamps currently going through a probe. Unless that data are converted to a temperature in degrees Celsius with a known accuracy, the information is useless to the enterprise.

While a mechanic or engineer might be interested to know the fan speed at a particular air handling unit, that data will not facilitate metering, billing, occupant scheduling, demand management, utility expansion plans, or fiscal accountability. The details of the internal operations must be hidden.

You would not want an employee to whom you had to give detailed instructions on every little task every time. Still less would you want to go for a walk with a friend if you had to instruct him on the proper clothes to wear, or worse, every muscle movement required to take each and every step. Instead, we "abstract" the activities of those with whom we interact. "Give me the monthly report on the first Monday of the month." "Meet me at the corner at 6:00 AM and we will go for a jog." We assume the person we are talking to will handle the details and we will see them at the appointed time. Many processes, particularly biological ones, are ones we do not wish to know about at all, unless, perhaps, if that friend is stricken ill on that run. We want those complex tasks encapsulated, hidden in a black box, with an abstract higher interface that signifies the activity desired.

Control system interfaces today are not abstract enough; and they do not hide enough. This lack of abstraction has at last five bad effects:

- Systems that are not abstract require too much of the integrator; no one can interoperate with today's systems without understanding the internal details of the system. Programmers at the enterprise level, however, do not routinely have engineering training and expertise in control systems.

- Systems that are not sufficiently abstract are able to pass data, but not information. In today's installed systems, you can find a number representing milliamps passing through an analog sensor named "temperature." Without calibration data, this number provides no information.

- An incompletely abstracted system provides information that is not fully qualified. A fully qualified temperature data point presents a number in degrees, *and* it also includes the scale (Celsius, Fahrenheit, Kelvin, etc.) used, with a known accuracy.

- Without abstraction, and the boxing of functionality that such abstraction enables, it is unsafe for a system to expose its processes to the enterprise. Such systems can only have a simple concept of security and cannot distinguish between managing setpoints and managing configuration. This is dangerous not only because it enables the enterprise programmer, not trained to understand the control system, to reconfigure the system, but also because it relieves the engineer of responsibility for the outcomes of the system. Without the ability to have levels of access, we cannot allow the occupant to use the network to adjust the thermostat to change the temperature of the room where she works. Abstraction enables a security systems with different levels of access based upon user role and prevents the diffusion of responsibility.

- Without abstraction, each system integration is unique; true plug and play interoperability is not achievable. If abstraction is somehow managed during the initial system installation, the feat is not likely to be repeated during sub-system swap-out.

System functions must be encapsulated and kept safe from the enterprise. Enterprise programmers and business managers have no understanding of system configuration and calibration. If they could modify a control system's sequence of operations, they could damage the system or even endanger the occupants of a facility. Control systems should display their operations to the enterprise as a car displays its operations on a dashboard to a driver: a few simple abstract measures of speed and engine health, a few simple controls to cause the car to speed up, slow down, or turn. All other functions are made available only to the trained mechanic.

In the Toronto airport today, gate lighting is brightened, and HVAC is increased when a plane is cleared to land by the control tower. This integration occurs because the gate's control system understands enterprise protocols and exposes its complex operations as simplified abstract interfaces. The system integrator does not need to know about the details of the control tower process, about the safety margins between planes, about transponders and post-9/11 communications requirements. The integrator merely wants to know that flight 1374 is coming in. In the same way, the integrator does not want to know about sensor calibration, or loop tuning, or variable speed fans. The system integrator wants to know what the system can do and how to ask it to do it.

ENTERPRISE SYSTEMS HAVE COMMON TAXONOMIES

With abstractions come taxonomies. When enterprise programmers speak of taxonomies, they mean hierarchically arranged classifications of information, in which meaning is abstracted in standard ways (and with standard nomenclatures) to allow interoperability and discovery. If we are going to abstract systems away from their atomic activities, we have to give things names that are well defined and based upon well-known standards. Such standards are developed using open processes, usually by consortia with representatives from all communities who need the standard.

The closest that most control systems get today to a common taxonomy is in tagging standards. For a given building, the point tags may be those used by the original owner, the contractor, the integrator, or the designer. A vigilant owner may have a detailed tagging standard in place for use in all control points. A good designer may diligently apply that standard in each and every drawing. A top-notch integrator may apply those standard tagging requirements. An especially vigilant owner will verify the application of this standard during a commissioning process for the building. To the extent that each of these activities has been consistently performed, a building may have a tagging schema that is similar to that of a biological taxonomy today.

A tag-based schema is not easily reusable for purposes other than those for which it was initially designed. A chilled water coil return tag on the third floor might look like CWR342, or 3CWR42 or CW3R42. Requesting access to all the flow meters on the third floor, or all chilled water tags in the building, or all return temperatures is difficult. This means that any interaction with those tags is arbitrary and labor intensive.

Control protocols such as LON and BACnet provide solid control-level taxa (names) for many control functions. Because the functions these taxa describe are not abstract enough for the enterprise, they do not provide the taxonomy needed by the enterprise.

ENTERPRISE PROTOCOL SOUP

Today, product sales literature for control systems and the sales forces that use them sprinkle IT acronyms across their product descriptions. While the enterprise practitioner expects a layered implementation wherein each of protocols named by the acronyms fits together, too often both the sales force and the product engineer, as well as the installer, implement only part of one of them, and claim that the entire standard has been delivered.

Real life implementations of standards-based protocols include many protocols, stacked one on top of another, to meet the need. Users may often simplify conversation by referring to only one of them. But if any of the rest fail, the whole stack falls.

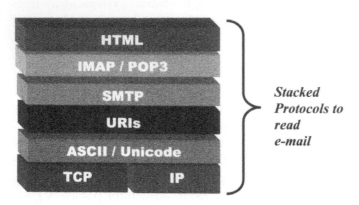

Stacked Protocols to read e-mail

Salesmen and controls practitioners alike often wonder if XML, web services, and SOAP are the same thing, or they claim all are in use when only one is. These standards are merely three faces of the same standard. XML (extended mark-up language) is a way to describe data in text. XML extends to data the techniques that are used to describe formatting in HTML (hypertext mark-up language), the language of the world wide web. SOAP (simple object access protocol) is the emerging open standard for requesting and receiving self-describing XML data from point-sources of information on an as-needed basis in the same way one can find and request documents from the world wide web on an as-needed basis. SOAP need not be on web servers, nor need it use the well-known web protocols HTTP (hypertext transfer protocol) and HTTPS (the secure form used for transfers such as credit purchases on the web). If it does, then it is known as a web service.

XML and web services specify only how the data look and how to interact with distributed systems across a network. For each vertical industry and business process, consortia are defining the taxonomy of services and objects within that process. Purchase orders, invoices, personnel actions, even inter-agency work flows for governments, each have developing standard taxonomies and XML schema. If the taxonomy is understood, then a networked system can contact a service point, query that service point about its capabilities and functions, and begin interacting. This negotiation is considered characteristic of web services.

This means that by their nature, web services are uniquely appropriate for integrating building automation systems. Rarely are two building control systems identical, as rarely are two buildings identical. Many interests in the enterprise would benefit from access to information from building control systems. When a common XML schema is used as the representation of a common taxonomy across building control systems, then these systems can be queries about the functionality they offer as well as their current operating state. SOAP also provides us with a high-level standard for interoperability between building control systems, now isolated in individual control silos.

In the discussion below, I often refer to Scalable Vector Graphics (SVG). Every CAD drawing is a database of linear equations. SVG is an XML schema for linear equations of the type used in CAD. SVG is a graphics standard developed by the World Wide Web consortium, the standards body for core web protocols (*www.w3.org*). Because W3 is vector based, it can be manipulated like a CAD drawing. Because SVG is XML-based, it can be delivered in SOAP, as can data that are more traditional. This offers the possibility of self-describing web services that provide schematics and drawings as well as data and methods.

DISINTEGRATION OF CONTROL SYSTEMS

As this process of abstracting and "black boxing" systems becomes applied to each of the systems in a building, both traditional and non-traditional, it is no longer necessary to tightly couple the control systems. Individual building functions can be isolated behind a clean well-defined interface for interoperating with all of the enterprise processes, whether they are enterprise business functions or other BACS systems.

This will break current large expensive systems into several smaller coordinated systems, each able to be upgraded and tuned based upon its own unique operational requirements. This will lead in time to better performance of each module with its own better defined requirements, a benefit that will more than outweigh any loss of tight coupling. New types of systems can be brought into a building, then, without disturbing existing systems; all can be loosely coupled at a more abstract level.

Instead of tightly integrating control systems, we can disintegrate them. When we do so, we open the door to loose integrations across new classes of control systems.

At the University of North Carolina, I have tenant refrigeration equipment used for research. From a maintenance standpoint, I want my refrigeration mechanics to have the same operational information available from this portable tenant equipment as they do from the embedded built-in systems that are part of the building. From a day-to-day monitoring and operations viewpoint, these systems are the responsibility of the tenant (in this case the researcher who needs to control how the temperature of the samples is maintained).

In a similar way, life-safety systems need to be focused on their mission, and perform it flawlessly, without interference from the energy management systems or the access control systems. Fume hoods, for example, despite moving a lot of air, have a life-safety mission rather than an HVAC mission.

This way of thinking leads to a confederated systems model. Each system is optimized to perform its own mission, and to defend its own business and system needs. Each system is able to provide information to and interact with any number of other systems. All information exchange is abstracted, so similar functions from dissimilar systems are presented to other systems, including enterprise systems, in similar ways. The resulting systems are neither monolithic, with a single architecture, mission, and purpose, nor stand-alone, with little information exchange; they present a united front along with their allies for any external functions, and remain independent in their internal operations.

If we achieve this, it will cause a great flowering of modular control systems. Owners will be more willing to upgrade modules within the enterprise controls than they are to upgrade any part of a monolithic system. We can imagine tactical upgrades, where a package system is added just for the summer, or during a renovation, and that system will be managed by the enterprise system.

Freed from the need to span large networks, controls system engineers will be able to focus better on the simpler, more contained control system modules. This will enable a more performance-oriented market, as engineers compete to produce the best-tuned control application, confident that it will fit as a module into the larger structure.

THE INTELLIGENT BUILDING TALKS TO THE ENTERPRISE

Traditional BACS systems provide silos of interaction, summarized at the top, and locked into homogenous actors. In the past, these silos interacted only through communications at the central level. Now, we begin to see buildings filled with autonomous networked control objects, each supervised by the old silo, but able to interact through abstract communications with the other control systems in the building. BACS systems interact with security systems, with access control systems, with elevators, with fume hoods. Each is able to interact with its peers by providing and requesting abstract information.

- Someone has entered the building.
- Someone is authorized, so disable the intrusion detection.
- Someone is allowed to go to the third floor.
- Someone has left the building, leaving the widows open, so summon security.

Sensors of interest to multiple systems can be placed on nodes, and thus be available through the network to each of those systems directly.

Non-traditional sources of data such as SOAP-enabled weather stations and green building sensors can provide additional information outside the traditional silo of control. For example, in our research program, special purpose sensors in laboratories, tracked with SOAP, will be provide information to tenant applications that are also able to draw information directly from the BAE interface to the BACS sensors in the BAS.

Maintenance personnel with proper authorization will receive just-in-time operations data delivered directly from the building systems. This information will include XML-based schematic information delivered in SVG directly from the system controllers. These same personnel will receive identical information from tenant-owned equipment if they are responsible for that maintenance.

BAS TO THE ENTERPRISE

Today, BAS is not ready for the enterprise. The component BACS have only achieved the first step of automating their pre-existing functions. Today's systems are not abstract enough, are not encapsulated enough, are not intelligent enough, and do not even have enough of a common taxonomy to support even the needs of the Operator, mush less the tenant or the owner. Without this, we will not get to the enterprise.

There are six discrete business processes in acquiring and operating a building: programming, design, construction, commissioning, operation, and condition assessment. Only rarely is there any relationship between performing one of these steps and performing any of the others. When we have integrated these steps, we will have created an enterprise function ready to interoperate with other enterprise activities.

Before a building is built, there are design goals, i.e., the intents for performance that should provide the basis for all steps of owning and operating a building. All too often, these goals are ignored after the initial design phase. If done properly, this information about the design goals would be codified in both an abstract machine and human-readable format that would provide the base description of the building and the structure under which all design and equipment decisions should be made.

The actual design of the systems usually pays brief lip service to the design intent, and then launches into a sequence of steps that is only rarely structured properly. Large CAD documents are produced with detailed schematics, circuit designs, sequences of operations, and tag lists. While these all appear on the same sheet of paper, they often have only the vaguest relationship to each other. The schematics and the circuit diagram are two different drawings, each made from scratch. The sequence of operations is prepared separately in a word processor and pasted into the CAD drawing. The tag list often comes from a separate spreadsheet maintained in a separate process. Close inspection of these tags and operations reveals that they were really merely copied, without correction, from the drawing for the floor above or below.

Given the quality of the design project, it is surely no surprise that construction does not match the design. I am reminded of the role of the medieval dissectionist. In medieval medical schools, the professor of anatomy would never touch a cadaver but would read from a classical text. As these texts were prepared by dissecting non-human animals, they often did not describe what was actually being found during the dissection. A dissectionist was successful to the extent that he could anticipate these problems and, at the right time, root around inside the cadaver, and pull out some other bit to match the flawed description. Some of the same skills are required of today's mechanical contractor and his controls subcontractor working from today's flawed design documents.

Following after these processes, commissioning can be a messy, expensive affair. The commissioning process begins without a clear catalog of the components and systems that should be in the building. Rarely are the tag standards of the design documents verified. This is due in equal parts to inaccuracies in the initial drawings and the lack of an automatable way to recapture this information. Even figuring out what to commission can be a daunting task.

When commissioning is performed, it is often limited to making sure that things really work, (e.g., that air is actually moving) rather than ensuring that the initial performance goals of the design intent were met. As described above, there are rarely clear descriptions remaining from the initial design intent of what performance standards are expected from each component system. Pre-construction performance models of the building systems rarely exist except for such government mandates as LEEDS or green buildings. Even if we can figure out what to commission, we do not know to what standard we should hold the systems.

The operation of the building from a control center (the process we usually talk about in building automation) is no better than it has to be. If we have a single building with a single contractor, it usually works well enough. If there is a marked deficiency in the initial design intent, or in the design, or in the construction, it will be fixed with a service order, as an operating cost. Perhaps that service order will be paid for under warranty. If the facility is part of a larger process with an integrated control system spanning multiple buildings, such as a campus or some type of complex, and even worse, multiple contractors, then a large task looms. These costs will probably be covered by operations.

Condition assessment is external to all the processes above. Underfunded colleges and universities strive to catalog unfunded maintenance deficiencies. Commercial real estate owners try to commoditize and monetize systems to rationalize their markets. But the tools are inadequate. Condition assessment may be based upon the number of "too hot, too cold" calls, or upon obvious dripping pans, or, rarely, complete failures. There are no measures of performance, whether optimal or degraded. There is little basis for analyzing systems that are drifting out of control.

It doesn't have to be this way. Design intent can be codified in machine readable form, and the performance goals for systems can be recorded. Modern CADD systems, if properly used, can produce models to predict performance of the systems under design. These models can generate all of the parts of the mechanical drawing, from schematic to sequence of operations to tag lists as views and reports of the same underlying model, intrinsically unifying the components of the design. With accurate design documents, it will be possible to build what was designed. This is the first part of the problem;

and it requires common data standards and common taxonomies across the processes.

This part, at least, is coming together, thanks to the world's largest landlord, the US General Services Agency (GSA). Beginning in fiscal 2006, the GSA will accept transmissions for the design, construction, and acquisition of buildings only in AECXML, an ISO standard accepted internationally. The taxonomy wars are over; we know what the data looks like.

AECXML is an XML-based taxonomy used to represent information in the architecture, engineering, and construction (AEC) industry. AECXML was developed under the coordination of the National Institute for Building Sciences (NIBS). AECXML is based upon the Industry Foundation Classes (IFCs) data model developed by NIBS to assemble a computer-readable model of a facility that contains all the information on the parts and their relationships to be shared among project participants. The intent is to provide a means of passing a complete, thorough and accurate building data model from the computer application used by one participant to another with no loss of information.

The IFCs (and AECXML) are used to create a building information model (BIM). BIMs create systematized, easily usable data storehouses for 3-D modeling to handle cost, schedule, fabrication, maintenance, energy, and other information across facility lifecycles. Current versions of the major CAD software packages can export AECXML directly from the design. Further information on the IFCs and AECXML can be found at the International Alliance for Interoperability—North America web site (*http://www.iai-na.org/*).

Closely related to AECXML is Green Building XML (GBXML), which is used to model the performance of designed systems. GBXML is a lighter-weight derivative of AECXML developed to enable interoperability between building design models and engineering analysis tools. As of 2004, GBXML generation from CAD is supported by the CADD systems from Autodesk, Graphisoft, and Bentley, eliminating the need to manually transfer information from drawing to model. The exported GBXML can be fed directly into system modeling software. Services that model building performance based upon submitted GBXML are even available over the internet today. Further information on GBXML can be found at *www.gbxml.org*.

With proper procedures, design intent can be stated in GBXML as performance requirements. The results can be described in the AECXML required for US federal contracts. With these inducements, there is no reason not to use those same models, and the tools

that create them to prepare construction documents that have higher quality and are automatically internally consistent. We can do this; we have the tools and the taxonomy.

The next part, where most BAS discussions begin, should leverage this work. Commissioning, building operation, and condition assessment can each leverage this work to address today's shortcomings. By adopting AECXML taxonomy across the board in the operation of buildings, these three processes become linked with each other. By using performance metrics compatible with GBXML performance modeling, we can verify our processes and systems.

Commissioning should be based upon an inventory that is derived directly from the AECXML inventory taken during the construction process. Because the data are in a standard format, the market place of tools to assist in commissioning will mature. But this is still only the first step, even if it is better than is often done today. Those same tools will be able to read the original design intents, moving commissioning to a higher level.

The first level is seeing if the actuators actually respond when asked, and in the way expected. The next level will be when we automate the answer to the more interesting question; do the systems perform as desired, in conformance with the original design intent? This is the type of abstract question, dealing with performance of function rather than mere movement of parts that is beginning to be more interesting to the enterprise.

The BAE Operations center, then, will run systems based upon the same data structures. The inventory of control points will come from those same data structures, with knowledge derived directly from the commissioning process. The sequence of operations will come from the original design models, as described in AECXML and GBXML. This defines the path to integration of campuses or building portfolios.

Because the modeled performance of the systems is known to and intrinsic to the operations center, tools will be developed to compare the current operations of a building to its original operation as documented during commissioning as well as the original design intent. The current condition of systems then becomes the known equipment (from the original inventory) operating within a known variance of its design specifications. This will lead to standard formulas for assessing the current state of building systems and to predictive systems to guide BACS replacement.

This condition assessment flows from the operations of the building, using known and standard data structures which can be modeled and remodeled. This leads directly to continuous commissioning, wherein the operational status of the building systems is verified and re-verified as a natural outgrowth of normal operations.

Standardizing the assessment of building systems, their performance, and their current state will lead to formulas for monetizing (or quantifying the benefits of) the building control systems. While monetized value is what drives commercial real estate, particularly the REIT (real estate investment trust) world, it is outside the scope of this discussion. Suffice it to say that once we can monetize the control systems, then control systems will be an enterprise asset.

When we accomplish this integration of the processes of the building systems, we will have created an enterprise function of building automation. This function will have described its internal processes and operations through a common taxonomy, hidden its internal complexity and heterogeneity through abstraction and boxing, and defined its performance metrics through the understanding built by modeling and tuned by performance. This new building automation function, now an enterprise activity, will be ready to talk to the enterprise.

EXAMPLES AND POTENTIAL USE CASES

The provost's office sets up class schedules and space assignments and the building automation systems automatically schedule the appropriate spaces for occupancy

1. A research building occupant can access real-time energy use data to take advantage of time-of-day energy rates when running major pieces of equipment

2. A secretary in a conference facility can have a desktop application that enables room scheduling and setpoint control for the entire conference facility without any interaction with BACS staff

3. A professor in the School of Business can assign undergraduate research that easily makes use of near real-time data from energy-producing and consuming equipment to develop business models

4. Load profile modeling can provide decision support for configuration and operation of energy-producing utilities such as cogeneration and chilled water distribution systems

5. Hourly electric rate structures analyzed in conjunction with weather models and load profiles enable optimization of combined cycle cogeneration plants and thermal storage systems

6. HVAC analytical programs that rely on historical data for trend analysis and comparison with design models support maintenance, renewal and replacement programs and allow HVAC system optimization using rules-based analysis

7. Construction documents generated using XML and SVG are able to become part of the real-time BACS operator interface after project completion

8. GIS utility maps make use of real-time metering data to identify possible utility delivery problems before customers are impacted

9. Utility billing systems are able to make use of multiple rate structures to allow customers to choose which best suits their operation without additional staff requirements since historical metering data can be automatically integrated into the bill

10. Indoor air quality information (both historical and real time) can be accessed directly by environmental health and safety personnel to support healthy buildings and identify unhealthy ones— without the support of BACS staff

12. A maintenance management system that interacts with a building control system can perform predictive maintenance and support run time maintenance models for complex systems. Such a system might track when a process is drifting out of control rather than merely noting a parameter outside its limits. If necessary, the system might generate work requests that are based on real time information. This same system will interact with enterprise functions for materials management to ensure parts and tools required for scheduled maintenance are available.

WHERE DO WE GO FROM HERE?

To become a vital part of the enterprise, to drive market growth and faster replacement cycles, and to provide full value to the owners, operators, and tenants, Building automation control systems need to follow the path of other business support systems and remodel themselves as enterprise functions. To become an enterprise function, BACS must not only accept the technology standards of the enterprise, but also the systems approaches that enable those functions to be full enterprise players. Building operators who fail to heed this call will find themselves increasingly irrelevant to the day-to-day operations of the enterprise.

Today's enterprise systems are assembled from modular coherent systems. BAS must do the same. BAS must have well-defined interfaces that have abstracted their internal operations to a level that suits the needs of the enterprise. This abstraction must expose the internal functions of each in ways that make sense to the enterprise. These systems must hide the inner complexity of the systems to protect the inner operations of the systems from inappropriate interference at the enterprise level. The abstraction must enable different systems to be competitively swapped out at as modules, allowing competition to drive innovation and competence within each system.

The architecture of this integration, today, is called service oriented architecture—it will have other names in the future. The protocol for integrating the enterprise today is the simple object access protocol (SOAP), which because it is message oriented, standards-based, and self describing, is uniquely positioned for spanning large heterogeneous systems touching every aspect of the enterprise. SOAP messages are encoded in XML. This architecture and protocol will drive the next generation of systems. Protocols and architectures are not enough; BACS must move beyond automation to innovation.

Building operation is just a small part of the overall function of facilities within an enterprise. Building Operation should leverage the taxonomies of design, construction, and contract to provide an open, self-managing asset to the enterprise throughout its life cycle. These taxonomies will provide the framework for abstracting building operations. This abstraction will enable automation of important enterprise functions such as building commissioning and condition assessment.

These common taxonomies and methods overlaid across the underlying technologies will free us up for better and easier integration across building systems. This will reduce the effort spent today on integrating disparate systems that are forced to be too intimate with each other's details. Building systems will become testable provable components working within the larger BACS realm. Such systems can be more easily replaced and upgraded without breaking the whole, thus shortening control system replacement cycles. The new systems will then compete more efficiently in performance.

By becoming abstract systems operating across the building life-cycle as part of an integrated enterprise function, BACS will intrinsically be ready to participate in overall enterprise functions. They will support bi-directional information sharing not only within the BACS realm, but also with non-traditional control and sensor systems as well as with diverse enterprise systems.

Systems providing this level of functionality will be worth upgrading to sooner than today's automated systems. The practitioner who understands this type of system and who can provide this level of functionality will be able to supply higher value to the enterprise than his competitor. This will drive innovation across building control systems to shorten life-cycles and expand market at a time when the underlying controls will increasingly become ever cheaper commodities. For the building system professional, this cannot be ignored.

Chapter 20

Disney's Approach to Energy Management System Design

Paul J. Allen
Chris Sandberg

INTRODUCTION

Walt Disney Parks and Resorts approach to designing, installing and operating an energy management system (EMS) bucks the multi-vendor "interoperability" trend by standardizing on one EMS system (i.e., one manufacturer). Maintaining a competitive procurement process with one EMS vendor might sound impossible, but Disney's EMS design approach makes this happen. The ability to design, install, program, create graphics, and continuously improve the EMS in-house is important to Disney. Costs for training, spare parts and EMS software licenses costs reduced by using one EMS. This methodology works well. There are no surprises and in the end, it all works, which is the important thing.

INDUSTRY TRENDS IN ENERGY MANAGEMENT SYSTEMS

Aside from the impact that IT will have on future EMS, there are some fundamental characteristics that owners have always desired and will continue to desire from a new EMS:

- Single-seat User Interface
- Compatible with Existing EMS
- Easy-to-use
- Easily expandable
- Competitive and Low-cost
- Owner Maintainable

There have been several changes made by the EMS industry to help satisfy some of these desires. The creation of open protocols such as LonWorks and BacNet has made field panel interoperability plausible. The development of overlay systems that communicate to multiple EMS vendor systems has made a single-seat operation possible. However, each has introduced their own levels of difficulties and additional cost. There is currently little commonality between different EMS vendors' low-level panel programming, and different service tool software is needed for each EMS vendors' system regardless of the "open" protocol.

Disney's EMS design approach is simple and straightforward. By standardizing on one manufacturer's EMS, Disney obtains all of the above-mentioned desired features. The key to the success is that the Disney energy management team is involved throughout the design, construction, startup and ongoing operation of the project to ensure the best EMS possible.

EMS DESIGN STRATEGIES

There are two strategies available for the design and specification of EMS:

1. Specify a multi-vendor interoperable EMS.
2. Standardize on one EMS manufacturers system.

Specifying a multi-vendor interoperable EMS is probably the most popular choice of the facility design community. The engineer's controls design is more schematic and the specifications more performance-based using this approach. The second approach is based on standardizing on one EMS manufacturer's system. To create competition and keep installation cost low, the engineer must create the EMS design as part of the design documents and prescriptively specify all components of the EMS. Let's take a closer look at each of these EMS design strategies in detail.

MULTI-VENDOR INTEROPERABLE EMS DESIGN APPROACH

Although this EMS design process has all the right goals and objectives, the results generally fall short of the owner's ultimate desires. This is the step-by-step process which most companies follow today:

A/E Design

1. The mechanical engineer completes the HVAC design—including chiller and boiler plant, pipe and pumping systems, air handlers, ductwork and temperature controls.
2. The control design is generally the last item to be completed because all of the other components must be designed first to know what is needed for controls.
3. The control drawings are generally schematic in nature and the specifications performance-based.
4. To allow for competitive bidding, several EMS vendors systems are called out as acceptable.
5. In an attempt to standardize the EMS for the owner, the specifications call out for the EMS panels to communicate with an open protocol. The thought is that the system selected could be compatible with the existing EMS installations in some form.

Construction

6. The project is bid and a mechanical contractor is awarded the contract, more often than not having used a rough order of magnitude estimate for the controls based on a percentage of the mechanical contract.
7. The mechanical contractor solicits bids from several controls contractors and generally selects the lowest bidder.
8. The controls contractors scope is all-inclusive—controllers, sensors, actuators, cable, conduit, etc.
9. The temperature controls contractor prepares a submittal which shows the actual EMS design and materials required for the job.
10. The mechanical engineer reviews the submittal to determine compliance with the plans and specifications.
11. The EMS is installed by the controls contractor, programmed to meet the desired sequence of operation, started up and commissioned.
12. The system is turned over to the owner.

Operation

13. The EMS vendor contacts the owner and requests a service contract to maintain the system.

Let's pick this approach apart and see where the weak links are. First, the engineer's controls design is often not much more than a set of "typical" schematics accompanied by a narrative specification which describes the desired results with little or no "how-to." In other words, the engineer delegates the responsibility for the detailed EMS design

to the controls contractor. This must be done, because the engineer does not yet know which EMS vendor will be selected. Even if a single EMS vendor was selected, it is very rare that a design engineer would be familiar enough with this system to produce a detailed design. Thus, the resulting EMS design is by nature somewhat vague and entirely performance-based.

Many times the engineer will specify that the EMS protocol be open to have the systems be interoperable. This sounds good, but there is currently little commonality between different EMS vendors' low-level panel programming, and different service tool software is needed for each EMS vendors' system regardless of the open protocol.

In the end, the owner gets a system which meets the original specification, but might not be the same as or even compatible with the existing facility EMS. The new EMS might require the owner to obtain an annual service contract because the system is so difficult to change, re-program, and repair. The owner operates this new system as just one of many disparate EMS systems in their EMS portfolio.

Several companies offer products which will interface to different EMS vendors systems to provide a common user interface and result in a "single seat." This approach comes with additional costs (both hardware and software) and results in an added level of complexity to the existing EMS. Furthermore, training, spare parts, and software licenses required for each EMS vendors system is still required.

STANDARDIZE ON ONE EMS MANUFACTURER'S SYSTEM

This approach is based on two fundamental principles: (1) to ensure competitive bidding the EMS hardware shall be owner furnished (2) the design documents and specifications must be prescriptively define the EMS design. Let's look at this EMS design process in more detail:

A/E Design

1. The mechanical engineer prepares all the HVAC design—including chiller and boiler plant, pipe and pumping systems, air handlers, ductwork, water and air flow diagrams and temperature controls.
2. The mechanical engineer fills in a temperature controls drawing showing the controls points and the sequence of operation and includes a controls specification based the owners EMS.
3. The design drawings also show all EMS panel locations with 120V power, locations of all space temperature, humidity and CO_2 sensors.
4. The electrical drawings detail all motor and lighting

control schematics including control relays. All underground conduits are also shown on the electrical drawings.

EMS Design

5. The energy management engineer prepares the EMS design.
6. An EMS design drawing is prepared that includes the unit control panel (UCP) wirelist showing point numbers and sensor types for all input/output cables. EMS communication block diagram and the UCP enclosure mounting details.
7. A specification for the EMS panel work describes the work to be completed inside the EMS panel (i.e., cable termination standards, UCP programming standards).

Construction

8. The project is bid and a mechanical contractor is awarded the contract.

Owner

9. All of the EMS modules, EMS panel and other components used inside the UCP are owner Furnished.

Controls Contractor—"Outside-UCP" Work

10. The mechanical contractor competitively bids the "outside-the-UCP" work which includes the wire/conduit/sensors/actuators as detailed EMS design drawings and specifications. A controls contractor is hired for this work.
11. The controls contractor prepares a submittal which show the equipment details for the job. This might be as simple as a set of material cut sheets, or they may produce control diagrams for each piece of equipment with the point number and cable information from the EMS design drawing.
12. The mechanical engineer reviews submittal to determine compliance with plans and specifications.
13. The controls contractor installs, labels and terminates all of the field devices and pulls cables back to the EMS panel leaving 10 feet of slack for final termination by owner or the EMS contractor.
14. The controls contractor completes a cable/device checklist to verify proper installation and labeling. A signed copy is submitted to the owner as notification for the EMS contractor to begin.

EMS Contractor—"Inside-the-UCP" Work

15. The owner or a separate EMS contractor familiar with the owners EMS is hired to perform the "Inside-the-UCP" work. This includes wire termination on EMS modules, programming, and start-up.
16. The temperature control contractor and the EMS contractor work together to commission the EMS and verify the operation of each control point.
17. The system is turned over to the owner after building acceptance.

Operation

18. The owner coordinates the connection of the EMS to the corporate Ethernet network and uploads all EMS panels to the central EMS server.
19. The owner prepares the EMS graphics and fine-tunes time and setpoint schedules.

With this EMS design approach, the EMS design is completed during the design phase and all details about the EMS are prescriptively defined in the construction documents. The owner knows exactly what they are getting and knows that it will seamlessly integrate into their existing EMS.

Competitive bidding is used to select the controls contractor that will comprise the largest portion of the EMS installation cost. The installation of all the wire/conduit/sensors and actuators can be competitively bid because it was included in the design documents and shows each input/output point number for the EMS.

The owner negotiates with the preferred EMS vendor on unit pricing for their EMS control modules. These modules are provided as owner furnished material to the project based on the EMS design. Keep in mind that the success of this process is predicated on the idea of direct parts sale from the preferred EMS vendor to the owner at deeply discounted prices. Some EMS vendors may not be willing to do this.

Once the controls contractor has all of the field wiring properly labeled and pulled into the EMS panel, the EMS contractor completes the final wire termination, programming and startup. The owner's energy management team coordinates the addition of the new panels to the central EMS server as well as prepares the EMS graphical interface.

CASE STUDY— CORONADO SPRINGS EXHIBIT HALL

Disney's Coronado Springs Resort celebrates the character and traditions of the American Southwest and northern Mexico. Here, palm-shaded courtyards and Spanish-style haciendas create the perfect climate for business and pleasure. The resort has over 1,900 guest rooms and

220,000 square feet of meeting and exhibit space. Disney's Coronado Springs Resort is the premier single-level hotel convention facility on the East Coast.

In June 2005, the 100,000-square-foot Veracruz Exhibit Hall was as added to Disney's Coronado Springs Resort Convention Facility. During the design phase of this project, the EMS control design was included as part of the design documents and specifications using the method described previously.

The mechanical system for this facility included a new chiller and boiler plant, 13 air handling units and 17 VAV boxes. The chiller plant design used a variable flow primary system with variable speed drive chillers. The energy management system controlled the HVAC operation and was designed using Disney's EMS design, procurement and operation process.

As with most new projects, there is a period of test and adjustment that results when going from the drawing board into actual operation. In consultation with the design engineer and the chiller manufacturer, several changes were made that resulted in reducing the chiller plant energy usage by approximately 30%.

Sub-metering of the chiller plant operation was part of the original EMS design for the convention center. Each chiller came equipped with an on-board electric meter, chilled water supply and return temperature sensors. A chilled water flow meter was added to each chiller to allow for the chilled water tons to be calculated. Additional electric sub-meters were included to measure the condenser and chilled water pump motors and the cooling tower fans. The energy management system collected the energy data, chiller operational parameters and the convention center space temperature and relative humidity values.

The chiller plant was designed with several energy saving features used to control the operation. The following changes to the original sequence of operation resulted in the 30% reduction in chiller plant energy usage:

- Chiller leaving water temperature setpoint was raised from 40°F to 44°F.

- Chilled water pumps maintained a differential pressure at the farthest air handler unit. Variable speed drives were used to vary the motor speed and the resulting pump flow.

- The minimum chilled water flow through each chiller was lowered from 600 gpm to 400 gpm, This resulted in an increased differential temperature at each chiller.

- The air handlers in the convention center exhibit space were set up as single-zone variable air volume systems. The discharge air temperature was fixed and variable speed drives on each unit were modulate the fan speed to maintain the space temperature set point.

- The minimum speed on for each air handler variable frequency drive was lowered from 75% to 20%.

- CO_2 sensors were used to measure the occupancy. If the CO_2 sensor was below the set point, the outside air dampers were kept at their minimum values. As the CO_2 levels increased, the outside air dampers were modulated open to maintain the CO_2 levels at set point.

- During unoccupied periods (11 pm to 6 am) the entire chiller plant was shutdown down. Each air handling unit and all exhaust fans were also simultaneously shut down. The temperature and humidity in the convention center were monitored and quickly recovered in the morning before any convention activity occurred.

- During the winter months, when the outside air temperature is cool and dry, the chiller plant is shutdown down completely and the economizer algorithm controlling the handlers outside air dampers is opened to cool the Exhibit Hall using outside air.

CONCLUSION

Disney's EMS design approach is both simple and straightforward. By designing their EMS into the construction documents, Disney takes control and responsibility over their EMS at design time. The resulting EMS design is well defined and all details specified prescriptively in the construction documents. There is no vagueness in the EMS design or the scope of work for the EMS installation contractors.

Disney is able to use this EMS design approach because it has an intimate understanding how their EMS works—both hardware and software. Disney's EMS design expertise comes from day-to-day EMS operational experience and from the development of EMS design standards. New EMS designs benefit from this EMS know-how and result in EMS installations that work well and integrate seamlessly into their existing EMS. Disney keep's their focus on the end result -facilities that use less energy and an EMS that is flexible enough to meet their changing needs.

Chapter 21

Using Custom Programs to Enhance Building Tune-up Efforts

Paul Allen
Rich Remke
David Green

ABSTRACT

Energy management system (EMS) control building heating, ventilation and air conditioning (HVAC) and is programmed to provide comfortable occupant temperatures/humidity levels and minimize energy savings. Keeping an EMS at optimal conditions is a difficult task. Changes in the building occupancy, equipment can result in changes in temperature/humidity and equipment schedules. As equipment ages, failures of sensors, actuators, dampers and valves can occur. Timely response to changes can keep an EMS operating at optimal conditions. There are several names for this process: Building tune-up, re-commissioning, and continuous commissioning [1]. Although difficult to predict for a single building, facility energy savings resulting from this effort can be expected to be between 5-15%. The building tune-up process is one of the most cost-effective energy conservation projects available for an energy manager.

This chapter describes the basics of the building tune-up process and showcases three custom programs that aid in that process: (1) the facility time schedule (FTS) Program and (2) the building tune-up system (BTUS) and (3) the Carrier Alarm Notification (CAN) program.

BUILDING TUNE-UP PROCESS

The building tune-up (BTU) process is one of the most cost-effective energy management projects available to an energy manager. The actions taken are typically low-cost or no-cost adjustments to an existing EMS and will not only minimize current operating costs but will also lower future maintenance costs. The BTU process does not necessarily involve the purchase and installation of new equipment or technology. Instead it requires an investigative-style approach to ensure that the EMS controls are working and controlling the HVAC and lighting systems optimally.

The first step in the BTU process is to review the HVAC and lighting systems time/setpoint schedules with the "building owners." This is best accomplished in a meeting with all the stakeholders that would be affected by changes in heat/air conditioning operating schedules and setpoints. The time schedules and setpoints for each building HVAC system are established until all of the building systems have been reviewed.

Alarms are programmed into the EMS to continuously monitor for equipment failures or overrides to the normal equipment schedules. Even with the best EMS, control equipment failures can and do occur over time. Once notified of the problem, the Maintenance Department can perform timely repairs to bring the equipment back into normal operation.

CUSTOM PROGRAMS TO ENHANCE BTU PROCESS

Three programs are used to enhance the BTU process. The facility time schedule (FTS) program manages the equipment time schedules and temperature setpoints using a server-side program and automatically resets time schedules and temperature setpoints on a daily basis. The building tune-up system (BTUS) is a web-based program that was developed to provide the "building owners" a view into the EMS control settings, without actually having access to the EMS. The Carrier Alarm Notification (CAN) program sends EMS alarms to the appropriate maintenance personnel via email/page. Each of these programs will be further described in the following sections.

The Facility Time Schedule (FTS) Program

The facility time schedule (FTS) program is a custom client/server program that interfaces with the Carrier

ComfortView energy management system. The purpose of the FTS program is to provide the energy manager a method to manage the time and setpoint schedules for a large campus facility in a master schedule database. Time schedules can be set up as "relative schedules" that incorporate the facility opening/closing times, dusk/dawn times and by day of week. Each day, the FTS program determines the appropriate open/close/dusk/dawn times and calculates actual time schedules that are broadcast to the Carrier EMS controllers on the Carrier ComfortView Network (CCN). Without this automatic reset feature that is created when the master schedules are downloaded each night, the time schedules and setpoints would eventually get changed from their optimal settings.

The FTS program can also handle special events that occur after the normal open/close time schedules by sending additional time schedules that effectively increase the HVAC/lighting equipment run-time to accommodate the special event. The FTS program provides some additional operational features:

- Users can make local EMS panel adjustments to both time and setpoint schedules to respond to building conditions without worrying that the changes would be permanent. All schedule changes will revert to the master schedule at the programmed download time the next day.

- Setpoint schedules can be grouped into common areas or types and can be programmed with a bias offset. This offset can be used during load-shed conditions to change the setpoint low and high values to a user adjustable level, reducing energy consumption. Schedule groups can also be used to pre-cool or pre-heat an area during special functions. The setpoint bias will be removed from the setpoints after the next automatic download.

- These master schedules are sent automatically to each Carrier EMS controller on a daily basis. For each schedule, the system administrator can choose whether and when to send either or both time and setpoint schedules.

- If the facility open/close times are changed during a given day, the new time schedules are re-calculated and downloaded again to the EMS controllers to reflect these changes.

The Building Tune-up System (BTUS)

The building tune-up system is a web-based program that provides information on each air conditioning system and shows the time and setpoint schedules that are in effect. The purpose of the BTUS is to give building owners a view into their buildings HVAC and lighting systems without accessing the EMS directly. The building owners provide the information on how their HVAC systems are controlled by establishing the time and setpoint schedules. These data are input in the BTUS database and provide a permanent record of the equipment time schedules and setpoints. Because it is web-based, it also allows a broader audience to access it via its web browser interface.

The BTUS shows the following information for each HVAC system:

- Description of area serviced. A color-coded floor plan can be displayed if available.

- Time and setpoint schedules. Includes both desired schedules and a link to look at the most recent schedules broadcast by the FTS program.

- Shows equipment in need of repair.

- HVAC temperature, humidity and status trends can be graphically displayed if available

The main control on the BTUS is a drop-down menu that allows the user to select the area desired. The user is then presented with a list of buildings from which to pick. Once the user selects an individual building, the detailed data for each HVAC system is displayed. Figure 21-2 shows a screen shot that shows a list of the HVAC systems in the building PAVILION1. Links are available to show the detailed information on each HVAC system.

The BTUS displays the HVAC system time schedules and setpoint schedules for each HVAC system. Clicking on the link for these schedules displays the latest actual schedules that were downloaded the previous night to the EMS controllers by the FTS program. Figures 21-3 and 21-4 show screen shots for these displays.

Another useful feature of the BTUS is to display a graphical floor plan that shows what each HVAC system covers. The floor plan is color coded to show the coverage areas for each HVAC system.

If an EMS trend report is available for a particular HVAC system, a link under the ID# will be highlighted. This makes an easy-to-use method to display trend data graphically.

The Carrier Alarm Notification (CAN) Program

Programming alarms into an EMS can notify the maintenance department when equipment problems occur. The alarms need to be prioritized to some extent, since

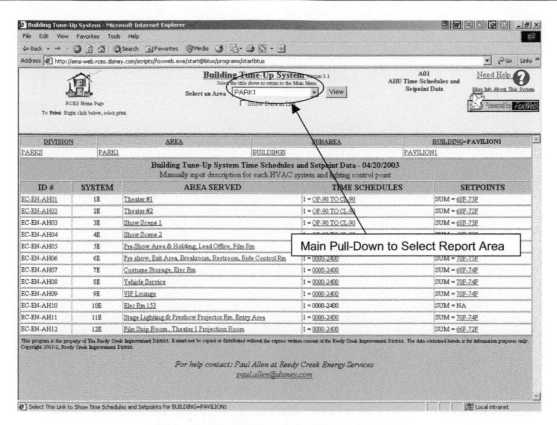

Figure 21- 1. BTUS Main Display Screen

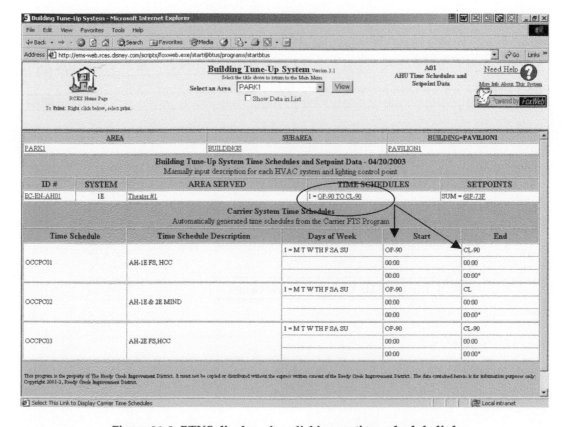

Figure 21-2. BTUS display after clicking on time schedule link

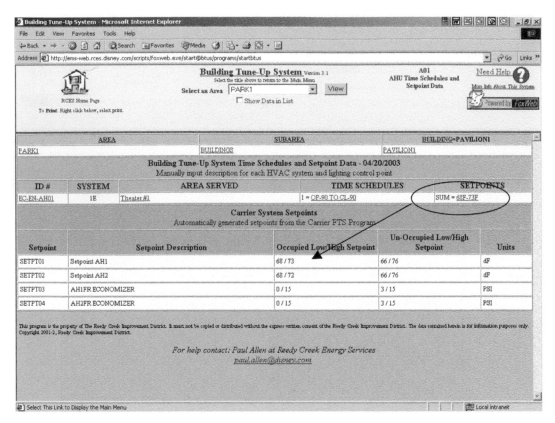

Figure 21-4. BTUS display showing EMS trend reports and floor plan links

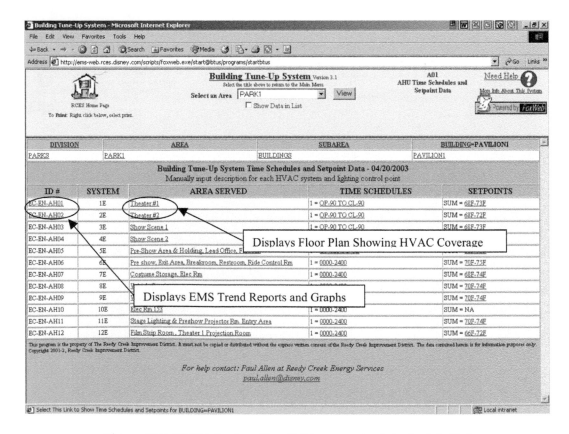

Figure 21-3. BTUS display after clicking on setpoint schedule link

the number of alarms could be overwhelming. The high priority alarms should be sent immediately by email/page to the appropriate maintenance personnel. An example of a high priority alarm would be a high temperature alarm for the chilled water system. This alarm would indicate something wrong with a chiller, chilled water pump or cooling tower that would require immediate attention.

There is another group of alarms that can notify the maintenance department when the HVAC/lighting systems are not operating properly. Typical EMS alarms are shown below:

- Chilled water valve closed, but display of cold supply-air temperature indicates that the valve is not closed.

- Fan that is commanded off, but the fan status shows that it remained on.

- Forcing of EMS controls to manual, rather than automatic control, which frequently suggests an operational or maintenance deficiency.

- EMS communication failures to EMS controllers.

Low priority alarms would not be emailed or paged immediately but would instead be logged into a database and included in a weekly summary report of all EMS alarms requiring attention. The weekly report is sent to the responsible maintenance department for follow-up corrective action.

The CAN program is a custom program that interfaces with the Carrier ComfortView energy management system. The CAN program uses two separate programs in its operation. The first program runs on a dedicated PC and is set up to run every 5 minutes to query the Carrier ComfortView SQL database alarms table. This table contains all of the alarms generated by the Carrier ComfortView program. These data are then compared to alarm definition table that contains the information about the alarms being monitored. If there is a match between the active alarms read from the Comfortview SQL table and the alarm definition table, the information is either emailed/paged out to the email/page distribution list assigned to the alarm or just logged to the CAN alarms history database.

The second program is a web-based system to display data in alarms history table along with data from other Carrier database tables containing manual overrides (Forces) and EMS Communication errors. This allows the User to quickly see all the EMS issues needing further at-

tention. Clicking on the ALARMS button will display all of the alarms received within the dates and area selected. A sample report is shown in Figure 5.

In Figure 21-5, AH-1 SUPPLY FAN STATUS description means that AH-1 had a problem with the fan status. Either AH-1 is staying ON when commanded OFF by EMS (most likely), or AH-1 is OFF when commanded ON by EMS (least likely). This alarm occurred 6 times in the dates selected. In order to look at the details of this alarm, clicking on the "6" in the right most column will generate the report shown in Figure 21-6.

The report shows that every night the same alarm occurs and shows "AH-1 SUPPLY FAN STATUS 1." The "1" indicates that the fan is staying ON when it should have been OFF. The most likely cause for this is that AH-1 is in MANUAL Control at the motor control center and needs to be put back into AUTO.

Figure 21-7 shows "AH-3 CHECK FOR DEFECTIVE CHW VALVE" which means that AH-3's chilled water valve was showing closed, but the supply air temperature was cold indicating the valve was probably not closing. This alarm occurred 11 times during the dates selected. This chilled water valve should be further tested by maintenance technicians and repaired or replaced to make sure it is operating properly.

Clicking on the FORCE button will display all of the EMS points that were overridden (or "FORCED") within the dates and area selected. The report shows who did the force, when the force was done, and to what point on the EMS was forced.

Clicking on the COMM button will display all of the EMS controllers that were off-line and in EMS communication failure within the dates and area selected. The EMS communication to these controllers needs to be investigated and brought back on-line. A common reason for controllers being off-line is lightning damage to EMS equipment. This report provides the maintenance technicians a list of EMS controllers to look at and get the EMS communications to the server working again.

CONCLUSION

The building tune-up process is a systematic approach to fine tuning an energy management system for optimal performance. This effort can be considered one of those proverbial "low-hanging fruit" energy projects that all energy managers should focus on.

The custom programs described in this chapter keep the energy management system settings at their optimal state. The FTS program automatically resets the energy management system time schedules and setpoints to their optimal valves. The BTUS program allows all users to view the time schedules and setpoints from their own PCs using web browser software. The CAN program notifies the maintenance department when equipment gets out of normal operation so that timely repairs can be performed.

The building tune-up process helps users understand how their HVAC system operates and provides a method to prevent EMS degradation by auto-resetting time and setpoint schedules on a daily basis. By automatically setting HVAC at established company standards and keeping EMS equipment working properly results in lower energy consumption and costs.

References

[1]   Continuous Commissioning SM in Energy Conservation Programs, W. Dan Turner, Ph.D., P.E., Energy Systems Lab, Texas A&M University, 409-862-8480, e-mail: dturner@esl.tamu.edu

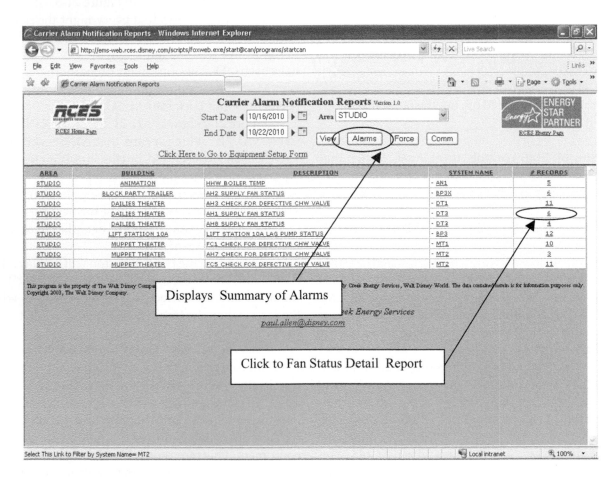

Figure 21-5. Alarms Report Display Page

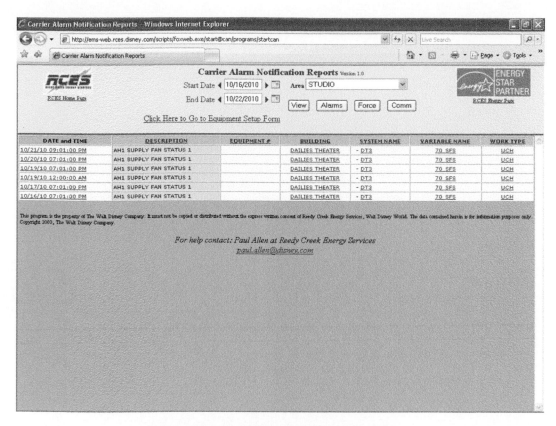

Figure 21-6. Fan Status Report Display Page

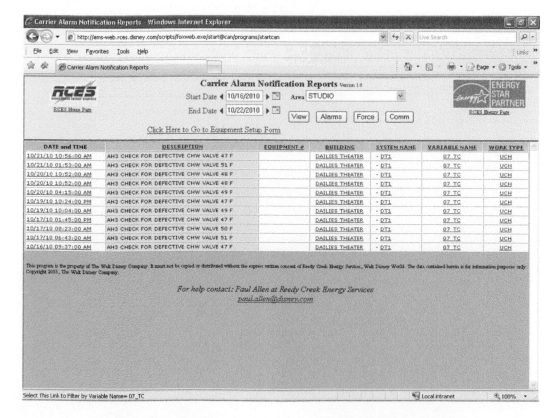

Figure 21-7. Defective Chilled Water Valve Report Display Page

Chapter 22

An IT Approach to Optimization and Diagnosing Complex System Interactions

Bill Gnerre, CEO
Greg Cmar, CTO
Kevin Fuller, EVP

ABSTRACT

Most facilities personnel do not know what is really happening with their building operations. They can't because they see only a tiny fraction of the operational data that would give them real insights. Today, optimization happens at a component or sub-system level with little or no information on how the parts work together. Diagnosing complex problems is often impossible because you cannot see how all the affected systems interact. Actual hourly operating cost data, and any ability to see how control changes, weather, occupancy, etc. influence costs is simply beyond most organizations.

This case study of a hospital in the southeastern U.S. illustrates how an IT approach, which makes all operational information available, quickly uncovers the interdependencies of the chilled water plant and building systems. The diagnosis explores VAV box, AHU, secondary pump, and chiller operations. Interval data provide an in-depth look at the unintended side effects of control engineering, operator decisions, and resultant energy waste caused by addressing the wrong issue—and the verified savings from addressing the correct issues.

Critical to identifying and solving challenges such as these is the availability of continuous, historical operational data from both the plant and buildings. This chapter describes how data are used to show cause-and-effect relationships between (and within) building and plant operations, and unnoticed system instability. It also demonstrates how continuously available data reduces the time demands for operations and maintenance, making such analysis both feasible and cost-effective.

INTRODUCTION

A recent news story told how UPS is saving millions of dollars by reducing gas consumption of its delivery vehicles. They use data to analyze dispatch plans and optimize delivery routes and times, saving UPS almost 14 million gallons of fuel annually. It's the data that make the improvements possible, and enable them to measure and verify results (such as reducing mileage by 100 million miles) as well. Mike Eskew, UPS's chairman and CEO commented, "At UPS, we're never satisfied with the status quo."

What does UPS's gas consumption have to do with your HVAC system? A lot. It's all about using the data to improve operations.

1. Both UPS and your facilities operations group are service businesses. Whether formally structured that way or not, facilities operations is a services business, and the more it is run like a business, the better it will perform.

2. Both strive to provide a high level of service (package delivery or building comfort) that consumes a substantial amount of energy to deliver.

3. Both have customers that take the service somewhat for granted, expect nearly perfect reliability, and give little thought to the costs incurred to make it happen.

4. Both employ technicians/mechanics to keep their equipment (delivery trucks or HVAC systems) running, and benefit if the personnel can work more efficiently and proactively leading to labor savings and extended equipment life.

5. Both can optimize performance, using data and information technology (IT) to calculate detailed operational costs. But, in all probability, UPS does a much better job of this than your facility.

Yes, there are some differences. UPS uses IT to leverage their operational data to great benefit. Most facilities operations groups don't. Without the data, you're guessing, and optimization is not possible on any meaningful scale. Without the data, you're accepting the status quo.

BACKGROUND

Before we get started, here's some background information to help give context to the diagnostic information presented later in the chapter.

The Facility

The hospital in this case study is a 2,000,000 square foot facility. The hospital campus includes a half dozen buildings with a total of 115 air handlers served by the physical plant. The plant, partially shown in Figure 22-1, has a total of 5,700 tons of chilling capacity, a bi-directional bypass, and a secondary loop with four 75hp pumps that service seven zones. The chiller plant also includes three cooling towers and a 400 ton absorber with its own cooling tower.

In Figure 22-2 you can see the standard configuration for an air handling unit. A VFD fan provides the air flow and two cooling coils chill the air. Dampers regulate return air, exhaust, and outside air. This AHU has 24 VAV boxes, each with a reheat coil.

Data Collection

A building automation system (BAS) will contain thousands of monitoring and control points (over

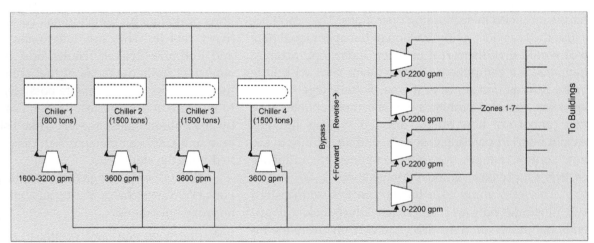

Figure 22-1. Chilled Water Plant—Chillers with Primary and Secondary loop.

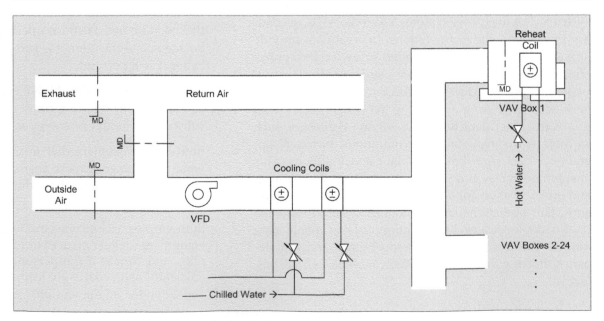

Figure 22-2. Air Handling Unit and VAV boxes.

Table 22-1. Sample of Interval Data Points from Hospital Systems

System	Partial Points List
Chillers & Absorber	tons output; CHW flow, setpoint, supply temp, return temp, ΔT^*; CW flow, supply temp, return temp, ΔT^*; % full load amps; efficiency*; kW/ton*; production balance*; pump brake horsepower*
Cooling Towers	CW supply temp, setpoint, low limit, high limit; CW return temp; CW flow, % full load amps; fan speed, load, kW*, kW-hour; fan motor speed, volts, amps, estimated make-up water*; estimated make-up water BTU*
Secondary Pumps	CHW supply, return, & differential pressure; zone differential pressures & setpoint; average valve % open; secondary CHW supply temp, return temp, flow; bypass flow & direction; kW*;
AHUs	return air humidity, humidity setpoint, temp; discharge air temp, setpoint; cooling valve positions; outside air CFM, CFM setpoint, damper position; supply fan static pressure, setpoint; average VAV heating valve position
VAVs	zone temp, heating limit, cooling limit; discharge air temp; heating valve position; CFM min, max, heating, calc & actual; damper position
Weather	outside air temp, humidity, wetbulb

*calculated point

100,000 for larger campuses). You need them all. The hospital is collecting data from about 18,000 points (see Table 22-1 for a partial listing). Ideally, you want data not just from the BAS (although it's a better place to start than meters), but from every data source:

- Building automation systems (all of them if you have more than one)
- Advanced meter systems
- Manually read meters
- Utility interval data
- Utility billing data (including all rate information)
- Weather (from local airport as BAS weather sensors are often faulty)

The hospital collects each point at standard 15-minute time intervals to enable synchronized views, for example, showing chilled water supply temperature, AHU discharge air temperature, and outside wetbulb at the same time. Saving the data indefinitely (for the life of the equipment, 20 years or so) allows the review of past operations and provides a historical record of all the equipment in the facility.

THE IMPACT OF A SINGLE PLANT CONTROL DECISION

There's a saying in this business that chiller plants are run by legend. Unfortunately there is a lot of truth to it. Instead of a disciplined engineering process, based on complete operational data, facilities rely on estimates based on a tiny data sample and educated guesswork.

Nowhere else are decisions with such a large financial impact based on so little information. Even small decisions get more thorough analysis. Imagine buying a computer only knowing the processor speed and hard drive size—it would be a crap shoot. You wouldn't make a $2,000 PC decision not knowing if the screen was large enough to be readable, if there was enough memory to handle your workload or just enough to boot the operating system, if it came with an operating system at all, or even if it were a laptop or desktop. But facilities groups make decisions with 100 times the financial consequences of buying the wrong PC every day as part of standard operating procedure.

Facilities business decisions don't have to be left to random chance. You can put the engineering back into your operations and make informed decisions based on a complete set of facts. You do it with data and an IT approach that turns the raw system data into actionable information.

Inside the Chilled Water Distribution Loop

During an investigation of the chiller plant, it was found that the secondary chilled water flow exceeds the primary flow quite frequently. When this occurs, the bypass flow reverses and the secondary pumps, of course, work harder creating the additional flow. This one situation has an impact across a wide assortment of air handling units—this chapter focuses on just one.

Experience has shown that every time you collect data and thoroughly examine them, you will find scores of problems. It has also shown that building HVAC systems are complex, interactive, and self-compensating, requiring a systemic view to really identify root causes of problems. Once you have a starting point, diagnostics is a stream-of-consciousness process that requires immediate interactive access to data or else the effort gets derailed. You'll see this in practice in the sections that follow.

Note: the starting point, in this case chilled water flows in the distribution loops, is not particularly important. It's just where this diagnosis started based on what was noticed first. The data exploration through a business-oriented IT system is the important part, and would have led to the same conclusions if the process had begun inside the air handler or with the VAV boxes.

Figure 22-3 illustrates the flow problems that are commonplace. You can see the primary chilled water flow drop suddenly at the same time the secondary flow jumps. At the same time, the bypass flow increases and reverses direction.

There is a cost associated with this behavior—that of the secondary pumps that are now working overtime. The pumps are maxing out as can be seen by the total pump kW. This isn't as costly as it might be, as this is all happening during off-peak hours when electric rates are lower, but the consumption is notably up.

What is causing the demand for greater flow? The air handlers start to tell the story.

Cooling Valves and Discharge Air

Looking at data for one of the hospital's rooftop units during the same time period shows more details. The cooling valve opens to 100% instead of the 45-60% range where one sees it at other times (Figure 22-5). This demands more chilled water, making the secondary

Figure 22-3. Coincident Flows in System—the Primary CHW Flow (dotted line) drops as a chiller is shut down, and at the same time, both the Secondary CHW Flow (gray line) and Bypass Flow (thin black line) jump. You can also see the Bypass direction reverse (1 is forward, 0 is reverse).

Figure 22-4. Pump Electrical Consumption Maxes Out—each time the Secondary CHW Flow (gray line) jumps to 7000 gpm, the total kW (dotted line) of the pumps reaches their maximum.

Figure 22-5. Cooling Valve Opening 100%--the Secondary CHW Flow (gray line) is peaking because the Cooling Valve (dotted line) is 100% open.

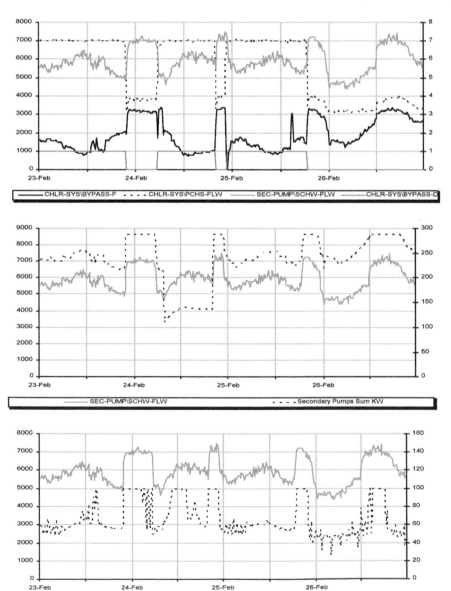

pumps work harder to deliver the increased flow. The correlation between this single valve and total secondary system water flow indicates that the issue is systemic.

The cooling valves need a reason to open up, and that reason is the discharge air temperature (Figure 22-6). It fails to meet its 55°F setpoint at the times in question, and calls for more cooling, causing the cooling valves to open to meet that request.

All of this activity is happening at night or on the weekend. The outside temperature and the building load are low most of these times. So why can't discharge air meet setpoint?

Chilled Water Temperature

In order to determine what was causing the discharge air to exceed setpoint the chilled water temperature was examined. Figure 22-7 shows the discharge air

temperature and setpoint and the secondary chilled water supply temperature. At the times in question, the secondary system is delivering 47-50°F water instead of around 42.5°F, where it stays most of the weekday hours.

Chiller Operations

February in Florida does require cooling, but the load is not high, especially nights and weekends. During those off hours, the hospital ran only one chiller, as it was assumed it would save money. Figure 22-8 shows that every time the system cuts back to one chiller, all the other issues start to arise.

Now look at the supply temperatures in Figure 22-9. The chiller 4 and primary loop supply temperatures hold tight, as they should, except in our problem areas. At those times the primary loop loses about 1.5°F. This is due to the drag flow through chillers 2 and 3 (Figure

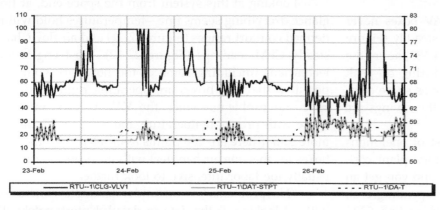

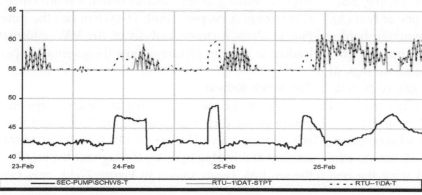

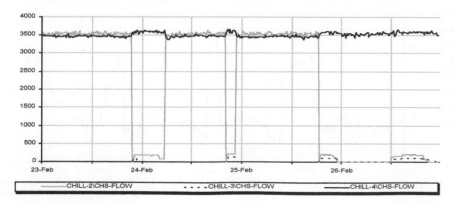

Figure 22-6. AHU Discharge Air Can't Make Setpoint—the Cooling Valve (thin black line) is at 100% open because the Discharge Air Temperature (dotted line) does not maintain Setpoint (gray line).

Figure 22-7. Secondary CHW Supply Temp Rises—the Discharge Air Temperature (dotted line) does not maintain Setpoint (gray line) when the Secondary CHW Supply Temperature jumps.

Figure 22-8. Chiller Flows—Chiller 4 (black line) stays on at a fairly constant 3,500 gpm flow, while Chiller 2 (gray line) is turned off during the times in question. Note the small flow amounts for Chiller 2 and Chiller 3 (dotted line) as there is some drag flow through those chillers.

22-8), where 200gpm of warm return water is mixing with the properly chilled output from chiller 4.

The additional 2°F jump between primary and secondary supply temperatures is the result of the reverse bypass flow. When the secondary loop flow is greater than the primary loop, the bypass reverses direction (Figure 22-2) mixing return water with supply water, increasing the temperature.

That circles us all the way back to Figure 22-3 where you can see, of course, that going from running two chillers to one cuts the primary flow in half since chillers 2, 3, and 4 are all fixed-capacity 3,600gpm units (Figure 22-1). When the chiller shutdown occurs, even though the space load is low, there is a chain reaction due to the interdependencies of the chiller plant, chilled water loop, and air handlers.

Did the Strategy Really Save Money?

Throughout this little ride, the VAV boxes never noticed anything going on, and operate steadily (Figure 22-12). One could argue that saving $34-$37/hour (the cost of running one of these chillers during off-peak hours) is worth a little instability as long as space comfort isn't affected. But the truth is that the savings are only a fraction of that.

Figure 22-10 shows the hourly cost of operations for various plant components and the total plant cost. While you do cut the cost of one chiller, and you get an extra $2/hour in reduced cooling tower operating cost, the remaining chiller is working harder, costing $20/hour more to run than when running as one of two (A). Then when the secondary pumps start working harder to compensate, they increase another $3/hour (a 25% jump). In total, there is a small operational savings of about $10/hour (B), but on a ton-hour basis, costs actually increased 7% (C). Factor in the instability introduced and the added wear on equipment—is it worth it?

This is great. The analysis has shown what occurred, how the plant and building systems interact, how to take the instability out of the operations, and what the cost

implications are of these choices. There's only one problem… none of this actually fixes anything.

"AS DESIGNED" VERSUS "AS NEEDED"

What do we mean that it didn't fix anything? This is a classic case of treating the symptoms instead of the real root cause. The one-chiller-or-two decision affects behavior as far out as the air handler. But the real problem with the discharge air not maintaining setpoint isn't too little chilled water; it's that the AHU's discharge air setpoint is too low in those circumstances. The issue isn't originating at the plant, but at the other end of the system… at the VAV boxes.

Space Comfort and Air Flow

Looking at this system from the space end, at first glance everything seems fine—temperatures hold within defined comfort ranges (Figure 22-11) and air flows have steady CFM readings (Figure 22-12). Most engineers/technicians (including some at the hospital) would report that the VAV boxes are operating right on design spec—no problems—since the VAV is supplying the space with proper temperature and air flow. Gathering more data (there are 288 points associated with the VAV boxes for this one air handler) without probable cause is usually too labor intensive to look further.

"No problems" is an understandable conclusion without looking at the data in detail. Unfortunately, it's also wrong. A properly built IT system has the information to allow a closer analysis of the VAV information, leading to a better picture of how things should operate.

Too Much Reheat

As can be seen by looking at the discharge air temperatures for each VAV box (Figure 22-13) and the VAV box heating valve (Figure 22-14), there is a substantial amount of reheat happening throughout the building. Of the ten VAV boxes shown in Figure 22-14, only two are

Figure 22-9. Chilled Water Supply Temps—there is an increase in chilled water temperature as it circulates through the system. The Chiller 4 Supply Temperature (gray line) is the lowest, with the Primary CHW Supply Temperature (dotted line) above it, and finally the Secondary CHW Supply Temperature has the highest temperature.

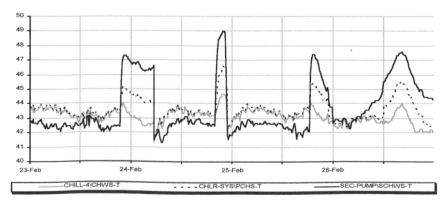

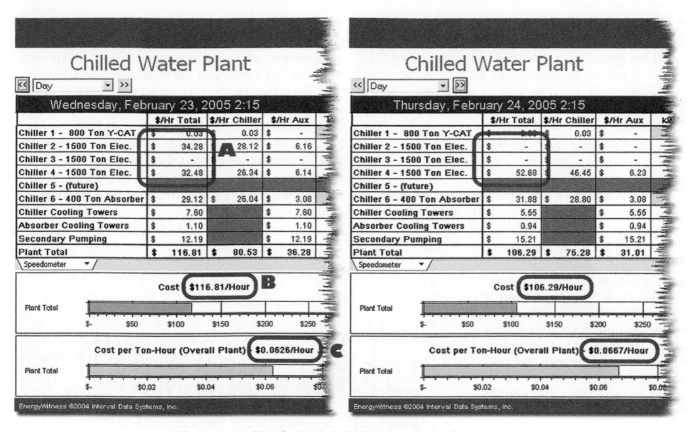

Figure 22-10. Hourly Cost of Chiller Plant Operations

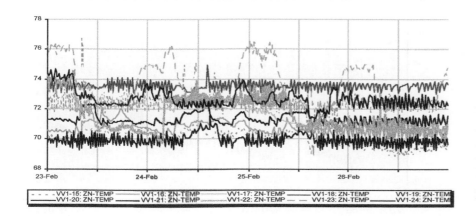

Figure 22-11. Space Temperatures at 10 VAV Boxes—the Zone Temperatures are all fairly constant and within a comfortable range of 69-74°F. The one temperature that jumps up to 76°F does so in response to occupants adjusting the thermostat.

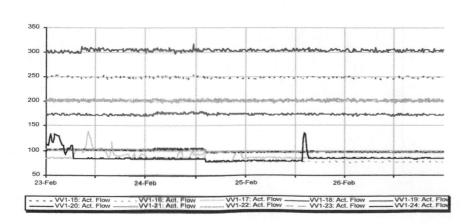

Figure 22-12. Air Flows (CFM) for 10 VAV Boxes—only two of the VAV boxes ever operated above their minimum CFM levels.

not supplying a significant amount of reheat.

The chiller plant is operating under the false load created by reheating the air that was just chilled. The hospital is paying to cool the air to 55°F in the AHU, then paying again to warm that air to 70-100+°F in the VAV boxes.

Heating valves are frequently opening to 100%, even during daytime hours for some VAV boxes.

VAV Influence on the Whole System

The VAV behavior, specifically the frequent reheat, is the real culprit in the system instability shown earlier.

There is a lot of information in Figure 22-15 that serves to pull the story together. From the AHU there is

the discharge air temperature and setpoint, and the cooling valve setting (there are two, operating in unison). In addition you can see the average reheat valve setting from the VAV boxes and the outside temperature.

Even though there is reheat occurring during the day, the outdoor temperature is low enough that the AHU discharge air setpoint stays at its 55°F minimum. The need for 55°F air is based on meeting summer daytime cooling demands, but these data are for February. Even with air flow at minimum CFM, discharge air at 55°F causes reheating to occur, creating a feedback loop demanding more cooling.

One chiller could not deliver enough cooling to keep the AHU discharge air at setpoint, causing the in-

Figure 22-13. Discharge Air Temperatures for 10 VAV Boxes—only a couple of the VAV boxes Discharge Air Temperatures track near the air handler Discharge Air Temperature (black line across bottom), the others are performing significant reheat, but tend to spike rather than provide steady temperatures.

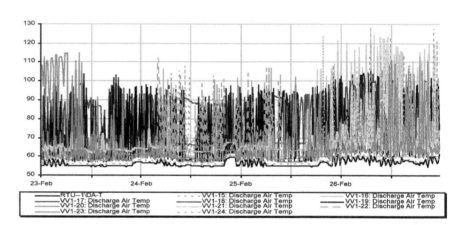

Figure 22-14. Reheat Valve Position on 10 VAV Boxes—many VAV box Reheat Valves are opening to 100% and cycle widely between 0-100%. The chart also shows Outside Air Temperature (upper black line) and the Average Heating Valve Position (lower thick black line).

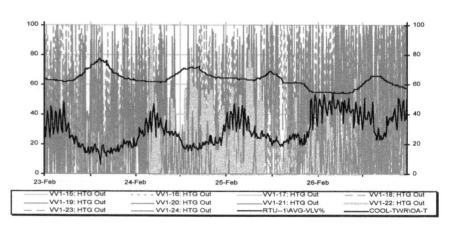

Figure 22-15. Reheat Causes Systemic Instability from VAV to AHU to Plant—Average Reheat Valve Position (bottom, dotted line); Discharge Air Setpoint (dashed line); Discharge Air Temperature (light gray line); Outside Air Temperature (medium gray line); and Cooling Valve Position (top thin black line).

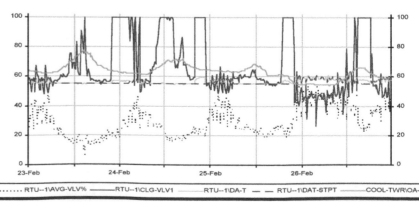

creased secondary loop flow and bypass reversal shown earlier. Eventually, the average reheat valve positions impact the discharge air setpoint, moving it up 3-5°F, although oscillating by about 3°F due to the cycling reheat valve positions. The discharge air temperature reaches setpoint again either when a second chiller comes online or when reheat pushes the setpoint high enough so that the cooling valves don't need to be 100% open and driving the demand on the chilled water loop.

Implementing the Solution and Measuring the Savings

The AHU and VAV boxes needed rebalancing based on actual needs for operating the space properly (as opposed to original design specs). Balancing is an iterative process—make changes, re-examine the data, make more changes as necessary, etc.

The hospital changed the control program on the AHU's discharge air setpoint to stop using the average reheat setting. They employed a simpler approach based on outside air temperature and a higher baseline setpoint. We monitored the affected spaces and also adjusted air flow settings for several of the VAV boxes.

It's worth noting that this whole scenario and strategy is based on winter conditions. The summer will bring a new set of circumstances and needs, which the data will make clear when it happens.

The results of the optimization didn't change space comfort (which was fine and never an issue). It did, however, have a significant impact on the operating costs, as seen in Table 22-2.

All of the adjustments made were within the buildings, to the air handler unit and the VAV boxes. Yet as you can see from the chart above, the majority of the savings are realized in the central plant, with approximately equal portions coming from cooling and heating savings. While much of the optimization work happens in the buildings, most of the dollar savings occurs in the plant.

Repeat the Process 115 More Times

That optimization did a lot to improve operations and lower costs for the air handler. In the big scheme of things, it is only a start at reducing the hospital load enough so that next winter they could indeed run with just one chiller. Another 115 air handlers need optimization as well.

The implementation of the changes to the rest of the AHUs is still underway as this chapter is being written. However, the hospital's facilities director and staff have a plan in place to prioritize and systematically address each one. The director suspected that issues existed for some time, but prior to this analysis didn't have the evidence and facts to put an action plan into place.

Using the data we were able to evaluate every AHU, determine if space comfort requirements were being met, analyze the operations (whether meeting comfort levels or not), and make recommendations where needed. The plan exists because the data exist.

Over one third of the 115 air handlers had issues. A partial list includes:

- AHU discharge air temperature is 50°F but cooling valve is closed.
- AHU discharge air temperature setpoint is 50°F. Cooling valve at 100% all the time
- Discharge air temperature resets to 49°F based on humidity, but humidity rises because space is overcooled.
- Unit provides 51°F discharge air temperature, but not making return air temp setpoint.
- Unit is not making discharge air temperature. Outside air pre-cooling has capacity.

Table 22-2. Cost Savings Calculation

Temp Range	Differential DA Temp	Hours at Temp	Delta Enthalpy	Avg Tons Saved	Cooling Savings	Heating Savings	Fan Savings	Total Savings
45-50°F	4.64°F	314	1.13	967	$70	$80	$17	$172
50-55°F	4.74°F	371	1.15	1,170	$90	$97	$20	$208
55-60°F	3.55°F	617	0.86	1,457	$113	$121	$33	$267
60-65°F	2.54°F	854	0.62	1,442	$112	$119	$46	$277
65-70°F	2.29°F	1,005	0.56	1,531	$118	$127	$54	$300
Totals		3,160		6,567	$ 508	$ 544	$171	$1,223

Total system operating cost: $6,612   Percent reduction: 18.5%

- Unit is not making pre-cooling temperature, but is making discharge air temp. Pre-cooling valve is 100% open.

- Unit is not making return air temperature, set at 72°F, despite cooling valve @100%.

- Cooling valve is locked at 20% open and space is below setpoint.

- Zone temperature is 68°F but cooling valve is closed.

Oh, did I forget to mention, one diagnostician did the entire analysis and report in less than three (count 'em, 3) days?

THE ANSWERS ARE IN THE DATA

The exploration of the hospital's operations shows two main points:

- First, it showed in detail how the system operates; how changes have a systemic impact since the overall system is self-compensating; and how "as designed" is not "as needed" regarding how the buildings should operate.

- Second, the exercise of looking at interval data, in detail, was the key to unlocking real insights about this facility in an extremely efficient manner.

It doesn't really matter what the questions are, the answers are somewhere in the data. This case—identifying where problems existed, tracking them throughout the facility, measuring the cost impact, finding the true root cause, determining a plan to fix the problem, and measuring the results—took only three hours of engineering/diagnostician time. Another four to five hours went into communicating with the hospital over the course of a week. And, it was done without ever stepping foot in the hospital—completely diagnosed from 1,000 miles away using the data.

The optimization discussion in this chapter is only the tip of the iceberg in terms of what you can accomplish with an IT approach to optimization and diagnostics. Effective IT systems are driven by data. Data collection is not a waste of time, but does represent a lot of time wasted. Let me explain. Collecting data and having

them at your fingertips is invaluable. However, in most facilities today, data collection consumes a huge amount of time by senior staff, a situation that is either not realized or accepted as "part of the job." Conservatively speaking, an engineer is likely to spend 5-8 times as long collecting data as doing real engineering analysis. Think of the value your organization would realize if the data collection to analysis ratio were reversed.

Today's HVAC systems are complex and the components are highly interdependent. The self-compensating control logic often masks problems so they go unnoticed or show up as symptoms in an entirely different part of the facility. A complete view of the data from the plant all the way to the terminal boxes, which can track and show the interdependencies, is the only way to effectively manage and diagnose these systems. The artificial walls that organizations create separating the plant from the buildings, or operations from utilities, are worse than counterproductive, they impede the facilities from operating at peak efficiency. Business-oriented IT systems that sit above the engineering/control systems break down those walls of inefficiency.

Operational and cost data are the cornerstones to getting a complete understanding of how your facilities operate. It provides you with the engineering facts to establish a systematic plan to make real progress. And once you look at the data, you will want to establish a new plan—any that you had before will be tossed once you actually see what is going on. Data create a different world for operations that completely changes what you can do and how long it takes.

Or, as Yoda might say, "The answers you seek, in the data they are."

References

Cmar, G.; Gnerre, W. "Defining the Next Generation Enterprise Energy Management System" In *Web-Based Energy Information and Control Systems: Case Studies and Applications*; Capehart, B., Capehart, L., Eds.; Fairmont Press: Atlanta, GA, 2005; Chapter 32.

Cmar, G.; Gnerre, W.; Rubin, L. "Diagnosing Complex System Interactions (from 1000 miles away)," Proceedings from the 13th National Conference on Building Commissioning. New York, 2005.

Cmar, G.; Gnerre, W. "Optimizing Building HVAC through Data and Costs," Proceedings from the 2005 World Energy Engineering Congress. Austin, TX, 2005.

Mills, E., Friedman, H., Powell, T., Bourassa, N., Claridge, D., Haasl, T., Piette, M., 2004, "The Cost-Effectiveness of Commercial-Buildings Commissioning: A Meta-Analysis of Energy and Non-Energy Impacts in Existing Buildings and New Construction in the United States."

Chapter 23

EModel: A New Energy Optimization Modeling Language

Timothy Middelkoop
Herbert Ingley

ABSTRACT

Energy and climate pressures in today's society have created a demand to use energy more efficiently and intelligently. In this chapter we present an energy modeling language for the optimization of web-based intelligent energy control systems for heating, ventilating, and air conditioning (HVAC). The framework is based on a uniform extensible markup language (XML) representation of the system and it's components. From this XML representation, an energy model of the system is built and transformed into an optimization program to maximize efficiency. By taking a systems perspective, the framework is capable of optimizing energy usage across the entire system. We present the model in the context of an example system based on a pilot site; however, the framework is designed to support wide-scale deployment for a wide variety of energy configurations and technologies.

INTRODUCTION

In this chapter we present a new energy modeling language for energy systems. This modeling language is part of a larger framework for clean energy optimization, control, and integration for heating, ventilating, and air conditioning (HVAC) systems. The framework is designed to manage all aspects of an energy system from design and operation to performance evaluation and service. In this chapter we present the energy modeling language in the context of finding an optimal operating policy using a linear programming (LP) formulation of the energy balance equations. The LP formulation is novel in the sense that it is free of any integer variables. The energy modeling language is designed to represent an energy system in a compact manner independent of the target optimization problem so that it can be used for a wide variety of optimization formulations and problems. In this way, advances in energy optimization can be easily applied to problems already encoded in the energy modeling language eliminating the need to recode them in the target formulation. This allows practitioners to take advantage of advances in energy optimization research and researchers to test new models on benchmark problems.

Mathematically, modeling energy systems is not new and there is a large number of academic papers on this topic (for an extensive review see [1]). However, most of these models are hand-constructed and implemented either directly as a mixed integer linear program (MILP), or similar formulations, using formats such as MPS [2] and OSiL [3] or using a modeling language such as AMPL [4]. The construction of such models is a time consuming and often overlooked step in the process. Currently, in order to generate these models automatically, either a modeling language for complex physical systems such as Modelica ([5]) must be used or the information must be constructed from data exchange formats such as gbXML [6] and ifcXML [7] (for an informative comparison of these see [8]). Modeling languages handle the complexities of the equations well but provide little support for representing the geometry of the system. Data exchange formats provide the details (and are often available) but can contain too much detail and provide little or no support for handling the model's equations.

The goal of this work is to provide an intermediary modeling language that allows energy systems to be easily representable and at the same time can be converted into an optimization program. Although this modeling language is bias towards a particular formulation, the goal is to provide an optimization neutral way of representing the energy system. Most formulations are very similar in the sense that choices must be made about when to operate equipment subject to operational constraints. It is these similarities that the energy modeling language capitalizes on. We now discuss a number of the common energy model formulation techniques.

One approach is to formulate the problem as a mixed integer linear program (MILP). Here the equipment is turned on and off with binary decision variables for discrete time periods and the energy balance equations are expressed as linear relationships. Due to the integer component of the formulation, the problem can quickly become computationally infeasible. Recent work of [9] uses such an approach but must resort to a linear relaxation to find near optimal solutions for scheduling the operation of equipment. Their approach only takes into consideration the energy required to operate the equipment in different configurations. Similarly, the work of [10] schedules energy usage but over a year. The MILP formulation uses 24 candidate days taking into account energy balance, heat loss, and performance characteristics. This model is used for planning capital expenditure and takes cost of capital plus energy costs as the objective function. The work of [11] takes a operational systems approach by integrating manufacturing planning with energy production in the form of a MILP. This work uses mass and energy balance equations and integrates performance information by using piecewise linear approximations. Results indicate energy savings but require substantial computation.

There are many other formulations of the energy management problem in the literature. Since many global optimization problems can be difficult to solve, there are also a number of heuristic solutions, many of which utilize genetic algorithm to find a solution (for examples see [12]). Both supply (in the case of many renewable energy sources) and demand (weather and building occupation) can also be considered stochastic and there are also a number of approaches that take this into consideration (for example see [13]). The comfort of the occupants can also be considered in the formulation, for example, in the review paper of [14] the authors present a number of techniques (multi-agent system, fuzzy logic, and neural networks) for developing intelligent controllers that take comfort into consideration. Although this is an important aspect, it is difficult to model user preferences and the results are hard to quantify due to the subjective nature of the variables. From the perspective of the model presented in this paper, we are only considering energy savings given pre-defined comfort settings. Changing comfort levels only changes the heating an cooling requirements of the solution.

In order to maximize the effectiveness of an energy control system, the optimization must be performed system-wide as demonstrated by [15]. Although this work only considers the operation of a traditional chilled water system, it focuses on the non-linear modeling and solution aspects of the problem and demonstrates the value of a systems approach. However, to maximize the savings the system must have the ability to substantially shift production and/or consumption of heat energy through storage. The work of [16] evaluates the effectiveness of both active and passive thermal storage systems providing evidence of the potential savings of such systems. Although the work is similar to that in this paper, it focuses on the non-linear model of energy storage. Storage is also important when integrating renewable energy sources into the system because most have high levels of variable and uncertain production.

Although there are many ways to formulate the energy optimization problem, the primary focus of the paper is the presentation of the energy modeling language. To reduce the computational complexity of the problem we have formulated the energy management problem as a pure linear program (LP), which can be solved computationally with relative ease. We first present an overview of the energy model and the example system. In subsequent sections we present the energy modeling language followed by the LP formulation. We close the paper with some numerical solutions and conclusions.

OVERVIEW

In this section we give a brief overview of the different components of the chapter. Since many of the concepts are best explained in terms of others, this section provides a brief overview to support the detailed presentation of the topics later in the chapter.

Energy Model

The energy model is based around the concept that a system is comprised of a number of energy loops that transfer energy from one system to another and these systems convert one form of energy to another. The loops, when in operation, operate in equilibrium in the short term. The equilibrium equations consist of the energy balance for the loops and the energy balance for the equipment that they connect. Equipment, such as chillers, have equations that convert one form of energy (electricity) to another (chilled water). As conditions change we assume that a new set of equilibrium equations need to be constructed resulting in a discrete time formulation. The change in conditions can be either operational (a building opens) or environmental (the temperature changes). By using this mechanism potentially nonlinear effects become constant during equilibrium maintaining the linear properties of the model.

Model Representation

The energy model, or EModel, is represented using the eXtensible Markup Language [17] commonly referred to as XML. XML is a structured text based document that is formed by a hierarchy of ordered *elements*. Each element contains zero or more *attributes* and optionally by *text* or child *elements*. An *element* consists of a starting tag and end tag, where tag is an identifier. If the element has no children or text the start and end tag may be combined. The start *tag* contains the optional attributes in the form of key="value" where key is another identifier. Listing 1 demonstrates the basic structure of an XML element. The structure of an XML document may be defined in terms of an optional schema [18], which defines the valid ways in which tags and attributes may be used in the document.

```
<element e:name="global . id" key="value">
    <tag>3.2</tag>
    <childless attribute="4"/>
</element>
```

Listing 1: Sample XML fragment

In this document we refer to specific elements by their tag name using a fixed-pitch font. For example, the top level element of the EModel document is referred to as the e:model element. Since XML tags are case sensitive we use the capitalization of the tag. We now present the major components of the model, and where appropriate refer to them by tag name.

System Components

A link is defined as a directed arc from one point to another, for example a chilled water pipe or an air duct. Each end of the link is connected to either the in port or the out port of a node. A node represents equipment in the energy model and consists of zero or more in (port), out (port), connection, internal, and external elements. A connection element represents the transfer of energy and material between an in port and out port, for example a cooling coil. Changes in internal state (for example energy storage) are represented by a internal element. Interactions external to the model (for example heat loss or electricity supply) are represented by external elements. Each element in the node element

has an e:name attribute that represents a variable in the energy balance system. In this way it is easy to represent the energy balance of a node.

Likewise a loop is used to represent the transfer of energy and mass from one node to another in a closed loop. A loop is constructed of multiple link elements and connection elements forming a valid operating configuration. Different connection elements can be used to represent different operating conditions (for example different configurations of valves) of a node. Since a loop is closed, the energy and mass must also be balanced. By combining loops and nodes the energy and mass balance equations can be formed for the entire system.

Links, nodes, and loops represent the physical configuration of equipment in the energy model. Valid operating configurations in the system are represented by the mode element. Modes combine loops and nodes that can operate in equilibrium (for example the operation of the heating system) in a collection of named loop elements and connection elements. Since modes also describe the way in which a collection of equipment operates (cooling or heating), it also contains the equations that represent the transformation of energy from one form to another (for example converting electricity into heat). As discussed earlier, these equations describe the system in equilibrium during a period and are expressed as a transform element. However, there are situations where the state of a system depends on it's previous state (for example energy storage, cooling of a room). For this case the update element is used to carry values from one period into the next.

Modes only express valid operating configurations in the system, not the entire system itself. The configurations element is used to express valid combinations of modes for the entire system. For example, it may be impossible, or undesirable, to run the cooling and heating system at the same time.

Energy Modeling Language

The energy modeling language (EModel) presented in this chapter by itself does not generate a solution. The modeling language is used by an intermediary program to produce an optimization problem formulation to be solved by a solver. The solver is responsible for computing the optimal solution from an intermediary representation of the problem. In our case we use the optimization services instance language (OSiL) [3] to represent the problem and use the COIN-OR CLP solver [19]. One of the motivations for developing an energy modeling language is due to the lack of any good open source modeling languages. The other motivation for a domain specific modeling language was to make it accessible to researchers and practitioners in the HVAC community.

The operation of the intermediary program responsible for generating and solving the optimization program must be completely defined by the EModel specification and the target energy optimization model. By taking this approach, it pushes all model design to the EModel and optimization problem representation to the solver leaving only implementation.

EXAMPLE SYSTEM

The example is taken from an existing restaurant in Tampa, Florida, that consists of a solar powered absorption chiller with heat medium storage and an electric chiller with ice storage for backup. The system is design to cool the space and supply hot water for the kitchen, restrooms and for the reheat section of the cooling system. We assume energy costs are $0.10/kWh during peak time and $0.05/kWh during off-peak hours with no demand charges. Currently the system is controlled using a legacy system and is in the process of being transitioned to a system that can run an EModel directly for control.

We now present the a brief overview of the equipment that is used in the example system. The components in the energy system are given short names that are used in combination in the EModel as unique identifiers. A list of these names can be found in Table 1 and how they are connected is shown in Figure 23-1.

ENERGY MODELING LANGUAGE

We now present the energy modeling language in the context of the example system. We present the language in the order in which it is defined in the XML document. Please refer to Table 23-1 for the names used in the examples. The complete schema is included in the electronic resources for this chapter.

XML Document and Schema Definition

The energy modeling language is an XML document defined by an XML schema definition (XSD) file. The XML document begins with the XML document and schema definition shown in Listing 2. The top level element of the document is the e:model element where e: is the namespace prefix. The namespace prefix is not ordinarily necessary except for globally defined attributes such as e:name and e:ref. The required use of the prefix for global attributes is an artifact of the XML schema definition.

Named Entities

Whenever an object (an entity with an e:name) in the EModel is defined, it is given a globally unique name. The e:name is an XML ID and, by convention, uses dot separated list of names from most significant to least significant. For example, the e:name of the example model is e2.model.linear, which indicates "example two" "models" "linear." The structure of the e:name allows the objects to be organized in a hierarchical manner starting with the site. It also makes it easy for an implementation to store and retrieve entities from a

Symbol	Equipment	Capacity	COP/Loss
ac	absorption chiller	75.7 kW	0.72
ahu	air handler unit		
ahu.cool	cooling load	73.8kW	
ahu.heat	reheat load	50.6kW	
ct	cooling tower	157kW	
cwt	chilled water tank		
hhwt	heat tank	5.678m^3	226 W/°C
dhw	hot water demand	123kW	
dhwt	domestic hot water	0.401m^3	20 W/°C
dhwt.electric	electric backup	20kW	
ec	electric chiller	17.2 kW	3.38 at 27.8°C
			2.92 at 35.0°C
			2.76 at 37.8°C
ice	ice storage	518GJ, 25kW	
sc	solar collector	352kW	

Table 23-1. Equipment

database based on the global e:name.

An e:name is often used to declare variables that are used later in the EModel. Most times a variable is paired with a state attribute, which specifically indicates what is being measured or tracked. One can think of variables as points of interest and the state, along with time, as different ways of looking at that point. State is also used to automate the process of generating the energy balance equations. Energy balance of a node or loop is the sum of all variables with the same state (Q) at a node or loop.

Links, Nodes, and Loops

As discussed earlier, links, nodes, and loops form the physical and logical configurations of the system. In this section we give a brief outline of the EModel representation of each as defined by the XML schema.

Nodes are connected to other nodes through links. Links are connected to nodes via in and out elements and each in/out has an e:name and medium attribute. For example (see Figure 1), a hot water tank (hhwt) connected to a solar collector (sc) via a heat exchanger (hx) would have an in element for the glycol (to prevent freezing) entering the heat exchanger node from the solar collector and an out element for the water going back to the solar collector as shown in Listing 3, which are named e2.hx.sc.s (supply) and e2.hx.sc.r (return) respectively.

The connection element is used to indicate a possible internal connection between an in and out element. A connection matches corresponding ins and outs in order to create an internal link between them, which is used later to form loops. Each connection has an e:name, medium, in, and out attribute. The in and

```
<?xml version="1.0" encoding="UTF-8"?>
<e:model version="1.0" e:name=" e2.model.linear"
      xmlns:e=" http://example .com/schema/EModel/1.0"
      xsi:schemaLocation=" http://example .com/schema/EModel/1.0
http://example .com/schema/EModel/EModel −1.0. xsd "
      xmlns:xsi="http:/ /www.w3. org/2001/XMLSchema−instance">
```

Listing 2: EModel XML document schema declaration

Nodes represent physical units within the model and are used to declare names (which in turn are transformed into optimization variables) that are used by other elements to construct the system. All energy coming into a node is positive while energy leaving is negative. Storage tanks, chillers and solar collectors are examples of nodes. In their current form, nodes are strictly not necessary; however, they are important structures for building user friendly tool that manage the construction of EModels.

out attribute refer to the e:name of the in and out elements in the node element. In the hot water tank/solar collector example shown in Listing 3 the exchanger has connections for the water and glycol sides of the exchanger.

Energy transfer external to the system and change in internal state are represented by the external and

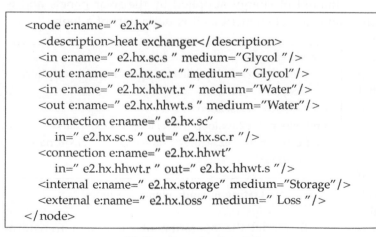

```
<node e:name=" e2.hx">
  <description>heat exchanger</description>
  <in e:name=" e2.hx.sc.s " medium="Glycol "/>
  <out e:name=" e2.hx.sc.r " medium=" Glycol"/>
  <in e:name=" e2.hx.hhwt.r " medium="Water"/>
  <out e:name=" e2.hx.hhwt.s " medium="Water"/>
  <connection e:name=" e2.hx.sc"
      in=" e2.hx.sc.s " out=" e2.hx.sc.r "/>
  <connection e:name=" e2.hx.hhwt"
      in=" e2.hx.hhwt.r " out=" e2.hx.hhwt.s "/>
  <internal e:name=" e2.hx.storage" medium="Storage"/>
  <external e:name=" e2.hx.loss" medium=" Loss "/>
</node>
```

Listing 3: node element

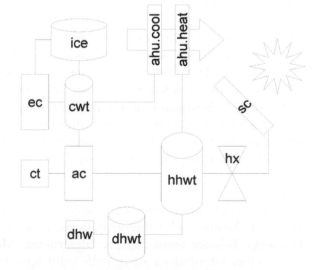

Figure 23-1. Layout of the example energy system

internal elements respectively. The external element is used for representing such things as heat loss and external energy sources such as electrical power and solar power. In the case of the hot water tank/solar collector example, heat loss is a external element and is used to model the heat loss due to the temperature differential between the tank and the ambient air. The internal element is similar but tracks changes in internal state (such as storage or temperature) for a given node. Both the internal and external elements have e:name and medium attributes.

Links connect nodes together and are directed. In a more physical sense they are the pipes of the system. They each have an e:name, medium, src, and dest attributes. The src and dest attributes refer to the e:name of the out and in elements of the source and destination of the link respectively. The medium is used to ensure that loops only carry a single medium (from source to destination) as well as providing a means to automatically connect them to nodes based on the medium. Listing 4 illustrates the use of a link connecting the chilled water supply (s) of the absorption chiller (ac) to the chilled water tank (cwt). Note the construction of the e:name from the properties of the Link. This is only done for convince; however, it also could have been named by some other means such as from construction drawings.

```
<link e:name=" e2.link.ac.cwt.s " medium="Water"
    src=" e2.ac.cwt.s " dest=" e2.cwt.ac.s " />
```

Listing 4: link element

Loops are what transfer energy to and from nodes and are formed by grouping links and connections. When the system is in equilibrium the net energy change in the loop will be zero. The connections element contains a collection of ref elements for which the e:connection attribute points to a connection element declared in a node with the same e:name. The links element is similarly used to declare what links are in the loop. Listing 5 shows an example declaration of a loop between a heat exchanger and solar collector. Now that nodes, links and loops are defined we can now present modes that use these elements (Listing 5).

Modes

The nodes, links, and loops form a network of energy transfers and transformations from which the energy balance equations are constructed. Modes combine this information along with valid operational configurations and modeling equations. A mode de-

```
<loop e:name=" e2.sc.loop">
    <links>
        <ref e:link=" e2.link.hx.sc.r "/>
        <ref e:link=" e2.link.sc.hx.s "/>
    </links>
    <connections>
        <ref e:connection=" e2.sc.hx"/>
        <ref e:connection=" e2.hx.sc"/>
    </connections>
</loop>
```

Listing 5: loop element

scribes a single equilibrium condition that a piece of, or a collection of, equipment can operate in. Many equipment/nodes will have multiple modes that describe either independent operation or different configurations. For example, the electric chiller (ec) can either produce chilled water for the cooling system or for ice generation. For independent operations, the air handling unit (ahu) can run the cooling and heating coils independently. More specifically, a mode is a valid configuration of loops that can be turned on or off. For the solar collector and hot water tank operation recall that a heat exchanger is utilized. In this case, a mode is formed from both loops since it does not make sense to run one loop without the other. There are also modes that manage internal state such as storage and heat loss.

Modes consist of the following elements: loops, externals, internals, transforms, and updates. As discussed earlier, the mode element defines a single valid operating condition and the energy equations associated with this operating condition. The mode element contains references to loops declared in the previous section. An example of a simple mode can be found in Listing 6. In this example the variable e2.sc.IDN is the amount of energy supplied by the solar panels and is an external element as it is not a part of a loop.

```
<mode e:name=" e2.sc.mode">
    <description>Solar collector operation</description>
    <loops>
<ref e:loop=" e2.sc.loop "/>
<ref e:loop=" e2.hx.loop"/> </loops> <externals>
<ref e:external=" e2.sc.IDN" state="Q"/> </externals>
    </mode>
```

Listing 6: mode element – loops

To model the operation of equipment transform elements are used. The transform equations are used to model the conversion of one variable (heat energy) to another. For example, the operation of the electric chiller (ec) converts electrical energy to chilled water with a coefficient of performance (COP) as shown in Listing 7. Each term element is a term in a linear expression that will be inserted as a constraint into the optimization program. The var attribute refers to a e:name declared in the loop or node elements. The term must also have a state attribute, which relates to the different type of information/parameter being affected (Q, J, or D). When the constraint is constructed the variable in the term element is multiplied by the product of any additional attributes in the term element. The constant attribute refers to constants declared in the constant section of the EModel (see Listing 11 for an example). The param attribute is similar to a constant but the value can change over time. Similarly, if a dt attribute exists it is multiplied the length of the period. In this way it is easy to integrate values over time. Listing 8 shows the use of the dt attribute. Simply put, a term in a constraint is constructed by multiplying the all attributes contained within a term element.

```
<transform name=" e2.ec.cw.electric.
    transform">
  <term var=" e2.ec.electric" state="Q"
      param=" e2.p.ec.cop "/>
  <term var=" e2.ec.cwt" state=
      "Q" const="-1"/>
</transform>
```

Listing 7: mode element – transforms

Storage is represented similarly as an update element with the addition of a var and state attribute. These two attributes indicate the term in the linear expression that is in the next period. Updates transfers values to the next time period and are generally used for storage in water tanks. An example of a complete node along with transform and update elements can be found in Listing 8. Note that the state *D* indicates the temperature difference between the tank and the environment (Listing 8).

Operation

The operation element is used to define the operation of the system. It indicates which modes are valid

```
<mode e:name=" e2.dhwt.mode">
  <description>DHWT operation</description>
  <externals>
    <ref e:external=" e2.dhwt. loss" state="Q"/>
  </externals>
  <internals>
    <ref e:internal=" e2.dhwt.storage" state="Q"/>
  </internals>
  <transforms>
    <linear>
      <transform>
        <term var=" e2.dhwt. loss" state="Q"/>
        <term var=" e2.dhwt. loss" state="D
              const=" e2.c.dhwt. lossfactor"/>
      </transform>
    </linear>
  </transforms>
  <updates>
    <linear>
      <update var=" e2.dhwt.storage" state="J">
        <term var=" e2.dhwt.storage" state="J"/>
        <term var=" e2.dhwt.storage" state="Q" dt="-1"/>
      </update>
      <update var=" e2.dhwt. loss" state="D">
        <term var=" e2.dhwt. loss" state="D"/>
        <term var=" e2.dhwt.storage"
              state="Q" const=" e2.c.dhwt. differential" dt="-1"/>
      </update>
    </linear>
  </updates>
</mode>
```

Listing 8: mode element – loops

in what combinations, places bounds on variables, sets the default values of parameters, and indicate for which state (Q in this example) nodes and loops need to be balanced. The children of this element are detailed in the remainder of this subsection.

The configurations element is used to combine modes in meaningful ways and are either required, parallel, or exclusive. Modes that must always be operational are referenced in a required element and are typically modes that have storage and losses involved. Modes that can be run in parallel are referenced in a parallel element and modes that are mutually exclusive are referenced in a exclusive element. Mutually exclusive modes must contain a variable for which to construct a limiting constraint. The constraint is constructed by limiting the fractional sum of the exclusive mode variables to be less than one. The fraction is computed as the variable over the maximum value (bound) of the variable during that period. Listing 9 limits the electric chiller (ec) to either run in the ice mode (e2.ice.charge.mode) to charge the ice storage or in cool mode (e2.ec.mode) by using the fractional output going to either the chilled water tank (ec.cwt) or the ice storage (ec.ise).

```
<exclusive>
   <modes e:name=" e2.ec.exclusive">
      <ref e:mode=" e2.ec.mode" var=" e2.ec.cwt"
         state="Q"/>
      <ref e:mode=" e2.ice.charge.mode" var=" e2.ec.ice"
         state="Q"/>
   </modes>
</exclusive>
```

Listing 9: exclusive modes configuration

The bound element is used to place bounds variables to ensure that they have the correct sign and to limit their values. The ub and lb attributes set the upper bound and lower bound respectively for the state of a variable. For example, Listing 10 places a bound on the charge and discharge rate of the ice storage to 25 kW and limits the charging capacity of the electric chiller (ec.ice) to 17.2 kW. It is important to ensure variables have the correct sign otherwise the optimization program may attempt to construct a solution that is thermodynamically impossible.

```
<bounds>
<bound var=" e2 . ice . storage" state="Q" lb="−25e3 "
   ub="25e3 "/>
<bound var=" e2.ec.ice" state="Q" lb="0" ub="17.2e3 "/>
</bounds>
```

Listing 10: bound element

The balance element is used to indicate which states need to be balanced for nodes and loops. For the example system in this paper heat energy (Q) is balanced for nodes and loops.

The parameter element declares the default values of parameters, which are constants that change over time. This value is later overwritten by values declared in the analysis part of the EModel. An example of a parameter is the COP of a chiller, which changes with respect to the outside temperature.

Analysis

The analysis section of the EModel is used to set the duration and number of periods, to construct the

objective function, and to alter the bounds of variables over time. The construction is straightforward and for the sake of brevity is discussed below. For more information see the complete example and the XML schema definition (XSD) available from the authors.

ENERGY MODEL

As mentioned previously, the goal of the EModel is to represent a HVAC system using a wide variety of modeling techniques. This paper presents the first modeling technique, a linear discrete time formulation, in order to present both the energy model as well as the energy modeling language simultaneously. The model is presented in the remainder of this section.

Notation

The primary purpose of HVAC systems is to transfer or transform energy and in this paper, we are primarily concerned with the associated energy balance equations for the system. The energy decision variables are presented in Table 23-2 and the optimization model notation summarized in Table 23-3.

Table 23-2. Decision variables

Variable	Unit	Description
Q	W	Heat transfer
W	W	External energy
J	J	Energy storage (heat/ice)
D	°C	Temperature differential

Each energy parameter is indexed by node, time, and an additional index. The maximum value of the last index is dependent on the type and number of decision variables present at a node forming a ragged data structure. Superscripts indicate time (t) and/or qualify a constant based on a decision variable, for example b^Q, b^W, b^J or b^D are constants for Q, W, J and D decision variables respectively. The linear discrete time formulation for an EModel is presented in the next subsection.

LP Formulation

The EModel presented in the previous section is used to generate a discrete time energy balance linear programming formulation of the energy optimization problem. The energy balance equations are generated by collecting all the variables utilized by nodes and loops and by transversing the energy network generated by the modes. Equations to transform and store energy are extracted from the transform and update elements and inserted directly into the model taking into consideration the discrete time formulation. Recall that a transform is a linear equality constraint within a node for a single period and an update is similar except one term is in the next period.

In words, the optimization problem is the minimization of the cost of running the system (Equation 1) over multiple periods subject to the energy balance of nodes and loops for each period considering the storage and transformation of energy. Formally we have the following minimization problem:

$$min \sum_{t=0}^{P} \sum_{i=1}^{N} \sum_{w=1}^{m_i^W} c_{i,w}^t W_{i,w}^t \qquad (1)$$

subject to:

$$\sum_{k}^{n_i} Q_{i,k}^t = 0 \qquad \begin{array}{l} i = 1 \dots N \\ t = 0 \dots P \end{array} \qquad (2)$$

All of the $Q_{i,k}^t$, $W_{i,w}^t$, $J_{i,j}^t$, $D_{i,d}^t$ decision variables belong to $\Re$ and bounded by $= a_{.,.}^{St}$, where S is replaced by the appropriate Q, W, J, or D and the dot notation

Table 23-3. Notation

Symbol	Description
N	number of nodes
L	number of loops
P	number of periods
i	index of node
l	index of loop
w	index of transform
j	index of storage update
d	index of temperature update
e	index of exclusive mode constraint
k	index of a Q variable at a node
v	index of some variable at a node
t	period
n_i	number of Q state variables at node i
m_i^Q	number of Q transform constraints at node i
m_i^W	number of W transform constraints at node i
m_i^J	number of storage (J) update constraints for node i
m_i^D	number of temperature (D) update constraints for node i
m_i^E	number of exclusive constraints for node i
L_l	set of connections for loop l
$Q_{i,k}^t$	a Q decision variable k at node i for period t
$W_{i,w}^t$	a W decision variable w at node i for period t
$J_{i,j}^t$	a J decision variable j at node i for period t
$D_{i,d}^t$	a D decision variable d at node i for period t
$c_{i,k}^t$	cost of energy for $W_{i,w}^t$
$b_{i,k,q}^{Qt}$	constant for $Q_{i,k}^t$ in transform $Q_{i,q}^t$
$b_{i,k,w}^{Wt}$	constant for $Q_{i,k}^t$ in transform $W_{i,w}^t$
$b_{i,k,j}^{Jt}$	constant for $Q_{i,k}^t$ in update $J_{i,j}^{t+1}$
$b_{i,k,d}^{Dt}$	constant for $Q_{i,k}^t$ in update $D_{i,d}^{t+1}$
$b_{i,k,e}^{Et}$	constant for $Q_{i,k}^t$ for the set of exclusive modes e at node i, period p
$\underline{a}_{i,v}^{St}, \overline{a}_{i,v}^{St}$	respective lower and upper bounds for decision variable $S_{i,v}$ at time t

indicates the decision variable index.

Equation 23-2 forms the energy balance constraints for nodes and is constructed from the EModel by taking all connection, external, and internal variables from active modes. Active modes are any mode listed in the configurations section of the operation element. All the variables with state="Q" from each node are collected and form a single Equation 23-2 constraint for each node and for each period. Equation 23-8 represents the additional constraint required by a group of exclusive modes to ensure that the use of a mode is not oversubscribed. The constant $b_{i,k,e}^{Et}$ is set to $1/\underline{a}_{i,k}^{Qt}$ or $1/\bar{a}_{i,k}^{Qt}$ (based on the sign of Q) if $Q_{i,k}^{t}$ is in the set of exclusive modes (e) otherwise it is set to 0. Equation 23-3 forms the loop energy balance constraints. These constraints are constructed from the EModel by collecting all variables with state="Q" from a single loop. A constraint is formed for each loop at each period.

Equation 23-4 forms the energy transform constraints that are constructed from the EModel using transform elements. These constraints model the operation of equipment, such as absorption chillers, that use heat energy to run a heat pump/chiller. There are m_i^Q equations at each node i. The constant $b_{i,k,q}^{Qt}$ is used to manage the sign differences as well as the coefficient of performance (COP) of the equipment. For example, for the absorption chiller, the EModel index style equation would be $Q_{ac.cwt} = b_{ac.COP} \, Q_{ac.hhwt}$.

Transform constraints for equipment with electrical energy sources are represented by Equation 23-5. Typically, $W_{i,w}^{t}$ will be an external variable since electrical power is external to the EModel system and does not need to be balanced. In the example system this equation represents the electric chiller, which converts electrical energy to chilled water. In this case the constant will be 1/COP, where COP is the coefficient of performance. The electric chiller uses a parameter in the EModel since the COP will change based on the temperature of the outside air. Since the electric chiller has both cooling and ice generation modes there will be two sets of equations, one for each mode.

Equation 23-6 represents the storage constraints at a node. Storage is modeled as the current storage level ($J_{i,j}^{t}$) plus or minus any transfers in or out of storage. The result is equal to the next periods i,j storage value ($J_{i,j}^{t+1}$). In most cases the $b_{i,k,j}^{Jt}$ constants will be either 0, 1 or −1. There will be an update constraint for each storage device at every time period. Similarly, Equation 23-7 represents update constraints for tracking temperature. Additionally, the binary $b_{i,0,d}^{Dt}$ constant is used (when set to one) to allow Equation 23-7 to represent temperature transforms as well.

The temperature differential between a tank and the environment is used to both model heat loss and energy storage. Additional energy can be stored in a tank by raising the temperature and is computed by taking the product of the additional heat energy (Q) with a constant and the duration of the period. This equation equates net heat gain with a rise in temperature. Using the EModel index style this is $D_{hhwt}^{t+1} = D_{hhwt}^{t} + b_{differential}^{Dt} Q_{loss}^{t}$, where $b_{differential}^{Dt}$ is the inverse of the product of the C_p (taken to be 4.186 kJ/(kgK) for the example) of the storage medium, the total mass of the storage medium (5678 kg for the example) and the length of the period. Listing 11 shows the EModel representation of this update for the example and the associated constant. Note that the length of period is represented by dt="-1," which also changes the sign of the term.

The large number of indexes and constants illustrates the need for an intermediary modeling language.

```
<constant e:name=" e2.c.hhwt.differential" value="42.1e−9"/>
<update var=" e2.hhwt. loss" state="D">
   <term var=" e2.hhwt.loss" state="D"/>
   <term var=" e2.hhwt.storage" state="Q"
      const=" e2.c.hhwt.differential" dt="−1"/>
</update>
```

Listing 11: EModel representation of loss

For a typical analysis of the example presented in this paper (48 one hour periods) there are 2,021 decision variables, 192 objective function variables, and 1,632 constraints with 4,224 terms. Fortunately the LP formulation allows this to be solved in less than a second on a typical machine manufactured in 2007. We present the transformation of the EModel into a linear program to be solved in the next section.

MODEL TRANSFORMATION

In order for the EModel to be useful it must be transformed into the optimization model presented in the previous section. The EModel is transformed into a LP in the OSiL format using an intermediary program. The choice of language for this program is somewhat arbitrary ([20] was used for web-based reporting in this case), but mainly requires advanced data structures (associative arrays) and the ability to easily manipulate XML documents. The program constructs the objective function and constraints, as defined by the LP, by transversing the nodes, loops, and modes. This process generates the energy balance equations and inserts update and transform equations into the constraint set. In addition, each decision variable is given an default upper and lower bound as defined in the bounds section of the operation element. This bound can be changed for any period in the analysis element of the EModel. Using this mechanism the operating conditions of the system can be modeled. A portion of the analysis section of the EModel for the example used in this paper is shown in Listing 12.

Example Computation

The restaurant presented earlier in the paper was used for the verification of the optimization model and transformation software. The EModel was constructed based on Figure 23-1 and equipment described in Table 23-1 for a 48-hour period during the summer months using one hour periods. The system starts without any ice storage and with the minimum allowed temperature differential of 8°C (the maximum is 16°C) for heat storage. To simplify the presentation we assume that 1 liter of storage medium (water) has a constant mass of 1 kg and a constant heat capacity (C_p) of 4.186 kJ/(kgK). The demand for hot water and cooling (with reheat) is from 08:00 to 17:00. The demand for domestic hot water is half (61.5 kW) during off-peak hours, which is from 08:00 to 09:00 in the morning and from 15:00 to 16:00 in the after-

```
<analysis>
   <time periods="48 " duration=" 3600 "/>
   <costs>
      <fixed>
         <cost var=" e2.ec.electric" state="Q"
            value=" 27. 7 e−9" dt="1"/>
      </fixed>
      <variable>
         <cost var=" e2.ec.electric.ice" state="Q" dt="1">
            <value period="0" value="18.5e−9"/>
         </cost>
      </variable>
   </costs>
   <external>
      <bounds>
         <bound var=" e2 .dhw.Demand" state="Q">
            <value period="8" lb="−615e2 " ub="−615e2 "/>
         </bound>
      </bounds>
   </external>
   <parameters>
      <set param=" e2.p.ec.cop ">
         <value period="0" const=" e2.c.ec.82.cop "/>
      </set>
   </parameters>
</analysis>
```

Listing 12: analysis element

noon. The outside air temperature, for the consideration of the electric chiller COP, is 35.0°C from 06:00 until 20:00 and 27.8°C during the evening hours. The model of solar energy collection is simplified to be 25% at 07:00 and 17:00, 50% at 08:00 and 16:00, 75% at 09:00 and 15:00, and 0% at night and 100% during the day.

The example system is encoded, along with the time-based bounds, into an EModel, which is then transformed into the LP presented earlier in the paper. The result of the transformation is a LP in OSiL format which is then solved by the CBC solver [19]. The results of the optimization are shown in Figure 23-2. This figure shows the heat flow in the system over time (refer to Table 23-1 for notation). The solid lines indicate supply and demand and the dashed lines indicate storage. Storage is shown as negative for charging and cooling (cool, ice, etc.) is shown as positive. The results show that ice is generated at night when the COP and rates are better and used to back up the absorption chiller during the day. The hot water storage tank (hhwt) temperature is raised for storage at 09:00 in anticipation of increased demand in the next

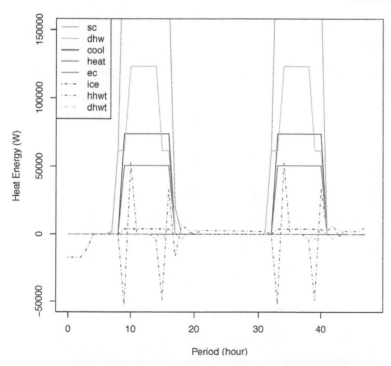

Figure 23-2. Heat transfer for the nominal case

period. Likewise when the demand is reduced at 15:00 the excess capacity is shifted to the next period when the energy from the solar collectors (sc) is lower.

For the second example, the solar collector performance is reduced to 200 kW between 10:00 and 14:00 (a cloudy afternoon). In this case, the electric chiller (ec) must be used to cool the space as shown by Figure 23-3. The electric chiller in cooling mode alone is unable to meet the demand; therefore, ice is generated at night in anticipation of the need and is used during the day.

Overall, the result show the ability of the EModel to represent an energy system that can be transformed into a useful optimization program to manage diverse energy sources over time in an economic fashion. The LP model takes into consideration differential billing, chiller performance, hot water tank storage and loss, and ice storage and use. The LP formulation allows the solution to the optimization to be found little computational effort making it suitable for control.

CONCLUSIONS

In this chapter we present a new energy modeling language to represent a clean energy system. This modeling language is used to generate a novel linear programming formulation to represent the energy balance in the system taking into consideration ice and heat medium storage, heat loss, and equipment that can operate in different configurations. The resulting linear program produced a feasible schedule minimizing the use of electrical power by scheduling the use of storage. By using a systems approach to controlling the HVAC system savings can be realized by coordinating the use of storage devices in the

This work is a part of a larger effort to develop an integrated energy modeling and control environment for HVAC systems utilizing clean energy technology and is the first step in developing high fidelity performance models of equipment in energy systems. Advanced performance models are required for the formulation of advanced non-linear optimization models for system control. By using high fidelity models and non-linear optimization more opportunities for energy savings can be realized by operating equipment not when the equipment is most efficient but when the system is most efficient.

The new energy modeling language presented in this paper provides a way to manage the complexity and multiple dimensions of an energy system in an

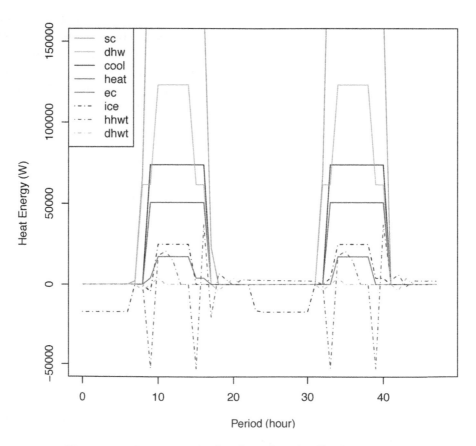

Figure 23-3. Heat transfer for the reduced collection system.

easy to understand and programmatically manipulative format. The modeling language is a balance between expressing a HVAC system in a pure mathematical programming format and developing sophisticated software to generate the equations from detailed building information. Importantly, the isolation between the mathematical model and the system allows existing EModels to utilize additional energy model formulations as they are developed. The machine friendly format also allows graphical tools to be developed to handle the management and creation of these models by end users while at the same time making them usable by researchers. It is our hope that this modeling language will evolve into a system that researchers and practitioners alike can use to advance research and adoption of system wide operation and optimization of HVAC systems.

Acknowledgements

This work was supported in part by a grant from Solarsa Inc. and the Florida Energy Office (SO424). The authors would also like acknowledge the work of the following students: Scott Moody and Xun Jai.

References

[1] S. Jebaraja and S. Iniyanb, "A review of energy models," *Renewable and Sustainable Energy Reviews*, vol. 10, no. 4, pp. 281–311, 2006.

[2] "MPS file format." http://lpsolve.sourceforgge.net/5.5/mps-format.htm, 2010.

[3] R. Fourer, J. Ma, and K. Martin, "OSiL: An instance language for optimization," *Computational Optimization and Applications*, vol. 45, no. 1, pp. 181–203, 2010.

[4] R. Fourer, D.M. Gay, and B.W. Kernighan, "A modeling language for mathematical programming," *Manage. Sci.*, vol. 36, no. 5, pp. 519–554, 1990.

[5] S.E. Mattsson, H. Elmqvist, and M. Otter, "Physical system modeling with Modelica," *Control Engineering Practice*, vol. 6, pp. 501–510, 1998.

[6] "Green building XML." http://www.gbxml.org/, 2010.

[7] "Industry foundation classes XML." http://www.iai-tech.org/products/ifc_specification/ifcxml-releases, 2010.

[8] B. Dong, K. Lam, Y. Huang, and G. Dobbs, "A comparative study of the IFC and gbXML informational infrastructures for data exchange in computational design support environments," in *Proceedings: Building Simulation 2007*, pp. 1530–1537, 2007.

[9] S. Gustafsson, M. Rönnqvist, and M. Claesson, "Optimization models and solution methods for load management," *International Journal of Energy Research*, vol. 28, pp. 299–317, 2004.

[10] M. Casisi, P. Pinamonti, and M. Reini, "Optimal lay-out and operation of combined heat & power (chp) distributed generation systems," *Energy*, 2009.

[11] M.H. Agha, R. Thery, G. Hetreux, A. Hait, and J.M.L. Lann, "Integrated production and utility system approach for optimizing industrial unit operations," *Energy*, vol. 35, pp. 611–627, 2010.

[12] A. Kusiak and M. Li, "Cooling output optimization of an air handling unit," Applied Energy, vol. 87, no. 3, pp. 901–090, 2010.

[13] D. Livengood and R. Larson, "The energy box: Locally automated optimal control of residential electricity usage," *Service Science*, vol. 1, no. 1, pp. 1–16, 2009.

[14] A.I. Dounis and C. Caraiscos, "Advanced control systems engineering for energy and comfort management in a building environment—a review," *Renewable and Sustainable Energy Reviews*, vol. 13, pp. 1246–1261, 2009.

[15] K. Fong, V. Hanby, and T. Chow, "System optimization for hvac energy management using the robust evolutionary algorithm," *Applied Thermal Engineering*, vol. 29, pp. 2327–2334, 2009.

[16] G.P. Henze, C. Felsmann, and G. Knabe, "Evaluation of optimal control for active and passive building thermal storage," *International Journal of Thermal Sciences*, vol. 43, pp. 173–183, 2004.

[17] "Extensible markup language."http://www.w3.org/XML/, 2010.

[18] "XML schema definition." http://www.w3.org/XML/Schema/, 2010.

[19] "Computational infrastructure for operations research."http://coin-or.org, 2010.

[20] "PHP: Hypertext preprocessor." http://php.net, 2010.

Section VI

Energy Information Systems Development

Chapter 24

Developing an Energy Information System: Rapid Requirements Analysis

David C. Green

ABSTRACT

This chapter describes a simple method of evaluating the importance of specific Energy Information System (EIS) requirements with respect to Information Technology (IT). This method helps organizations that have identified a need to develop an EIS using IT resources but have difficulty defining the complex requirements in a timely manner. Energy management teams typically know a lot about the need for each requirement but may know little about the cost or effort to produce it. On the other hand, IT development teams know more about costs and effort then the relevant importance of each of the requirements. This method breaks the requirements definition down into manageable tasks, evaluates the best balance between need and cost for each of the requirement tasks and then produces a priority of work document to guide the development team.

INTRODUCTION

With any EIS the purpose of requirements analysis is to translate the needs and goals of the energy management team into a document that the IT development team understands and can use to produce a cost estimate for the project. It seems requirements for an EIS may be more complex than many other software development projects. EIS requirements involve data collection, sometimes from remote locations, conversion formulas, input forms and reports for a wide variety of data. [1] The utility data has unique characteristics. Meters produce the data in a wide variety of units of measurement and at many different intervals. The energy management team understands these complexities very well. However, the IT development teams often times don't understand the utility data well. The roles of the energy management team and the IT development team are quite different. But, they need to work together to take advantage of both their strengths in order to work through the complex requirements of any EIS.

EIS requirements fall into several categories. There are collection tasks, input forms, reporting tasks, graphing, security issues and configuration requirements. The EIS must collect data from meters automatically, by importing data or through input forms. The characteristics of the data vary from meter to meter so each meter must have a unique configuration description in the EIS. The EIS uses this configuration to collect data and report data for each meter. This configuration description might include the interval at which the data is collected, minute, hour, day or month. Varying collection intervals sometimes present a problem to EIS processing tasks so they need to be part of the system configuration as well. Input forms have to validate data and make sure only certain users can change certain data at certain times. Security features may link users to input screens based on which data is changed or on the time period. The system will likely prevent users from changing previous month or year values. This is necessary to insure the ability to reproduce reports from the past exactly as they appeared at the time. There are conversion formulas to present utility data in units other then those used to measure the utility at the meters.

Utility data have unique characteristics that make the requirements for an EIS complex. First, the EIS usually collects utility data on an hourly basis, if not even more often. This creates a huge amount of data. Systems collect data from many different locations and report it on different levels. The data is usually date specific and requires trending over time to be useful. The data come from many types of utilities and are collected in a variety of units of measure. They may need to be broken down into groups for reporting purposes. Looking at the utility usage for different "groups" is a reasonable expectation. The electric consumption of a building means just as much as for a whole complex of buildings. It might also be necessary to combine data from several meters and report it as one data point or fractionalize data from one meter and report it as more than one data point. Systems record all data with a time stamp. The data are only meaningful when presented in the context of time. Hourly data for specific days and daily data spread over the period of a month provide the most meaningful reports and graphs. Comparisons to other data, such as outside temperature, are also helpful. Sometimes averaging data or summing for each utility is required. Systems can average or sum data together over periods of time or by combining meters at various locations. These conditions obviously require the energy management team to be deeply involved in the EIS development. [2]

The energy management team understands the processes that currently aid them in doing business. So, they know the priorities, formulas, problems, plans for improvements and potential benefits of an EIS. They know which tasks must absolutely be done and they know how to collect, input and store the data in such a way to produce the reports they need. They work around the problems that come up from day to day such as missing data or changes in prices. They maintain formulas to calculate the desired results. They keep track of adjustments when needed. The energy management team needs to express these requirements to the IT development team in a clear and concise manner. They also need to give them specific tasks that are measurable and easily determined as being complete or not complete.

The IT development team has unique capabilities that the energy management team does not possess. They probably have developed applications similar in nature to the EIS desired. They have the ability to store, analyze and present data in a user friendly manner. They understand the maintenance requirements and all the hardware and software required for the system. They understand the staffing requirements, strengths and weaknesses of the available personnel and effort required of them to do the job. Even though it is not necessary to put an exact cost estimate on each task, it should be easy for the IT development team to assign a *relative* cost to each individual task. They will still need to collaborate in some manner with the energy management team to complete the relative cost assessment.

Bringing teams together to work on IT projects has been a problem for some time. Even well defined systems development life cycle methods have failed to bridge the gap. Statistics show that more than half of large IT projects have significant cost overruns and require constant maintenance. Also, more then half the errors result from poor requirements gathering and poor functional specifications. Surveys show that most of the time users and developers fail to cooperate and coordinate with each other. Efforts to teach people to work together better may have had some limited short-term effect. However, it seems these problems persist even today in some organizations. Dealing with them is not an affordable option. Time is more important in systems development today then ever. Yet, organizations continue to attack the problem with more and more conversation and analysis rather then stepping back and trying alternative approaches. [3]

A case study concerning conversations among analysts and clients during requirements gathering shows results demonstrating that more conversation may not be the solution. The study showed that brief comments during the discussions were well informed and pertinent to the topics. However, there was a clear failure to put them all together toward common overall goals. This could be because of frequent topic changes and backtracking during the conversation. Interestingly, the study also shows that most participants gained satisfaction from the conversation based on how well the social interaction went rather than how many goals the group accomplished. [4]

It's clear that bringing two diverse teams together to deal with such complex data could be a challenge. It's important that the two teams meet to discuss the overall purpose of the project and what resources are available. The energy management team understands the data much better and can clearly see the priority of each task. As long as the energy management team can organize the requirements in such a way as to keep the IT development team focused on one task at a time then the whole process becomes a stepwise effort. The IT development team clearly sees the amount of effort required for each task and will want to complete the easy tasks first. Merging the strengths of both teams together to produce a development plan in a reasonable amount of time is the goal of rapid requirements analysis.

GOALS OF RAPID REQUIREMENTS ANALYSIS

Rapid requirements analysis provides a way to independently evaluate the tasks required to complete an EIS yet combine the results of those independent evaluations into a balanced requirements definition and work plan. And, it seeks to do it in a short period of time so that those involved maintain interest and budgeting is easier. It provides a way for the energy management team to present a clear picture of how important one task is relative to others by establishing a prioritized list of required tasks based on *need*. In other words, for each task required, the team asks the question: "how much is this needed?" relative to other tasks. Ultimately, this tells the IT development team which tasks are absolutely required and which tasks are not.

The IT development team then presents a clear picture of how *costly* each task is relative to others. They rank the requirements in some logical order based on cost and effort required to complete the task. The *less costly* tasks get a higher priority just as the *most needed* tasks get a higher priority. This encourages everyone involved to examine individual tasks in-depth independently of each other. A facilitator can answer questions concerning the details of the tasks or act as a liaison between the teams to clarify the task requirements. The project manager or a consultant may fill this role. This is a more efficient manner of detailed analysis then conducting meetings with everyone involved since many times the details of any particular task may only involve a few people. Later, a meeting with everyone would be a good way to combine the results of their investigations together.

The final list of tasks is a combined evaluation of the requirements tasks with the most needed and least costly tasks near the top of the list and the least needed and most costly tasks near the bottom of the list. This list, prioritized by a combined need and cost ranking, gives the teams a preliminary look at which tasks may be too costly to include in the project. The final list is the beginning of a detailed requirements document and priority of work that allows the teams to work together more efficiently.

Now that the teams have evaluated the requirements list in this manner they can stay focused on one task at a time and be sure they are working in the correct priority. Evaluating the tasks independently prevents "cross contamination" of results due to the effects of one criterion on the other. For instance, if the teams meet together to discuss one particularly important task and they place a cost estimate on the task at that time, the effect of knowing the cost estimate might be to lower the tasks importance in order to save time and money. In the same manner tasks that seem less costly might tend to get a rating of higher importance just because they appear to be easy.

So, this method of independently evaluating tasks using the strengths of each team is very important and often overlooked. Later combining the results of the evaluations into a comprehensive list of requirements helps to organize thoughts and processes while the project proceeds. Rapid requirements analysis encourages a well organized and accountable project lifecycle.

Rapid requirements analysis has been successfully used in both large and small projects. But, its design was founded as a result of experiences with large organizations. If initial meetings concerning a project involve ten or more people from diverse parts of an organization then this method is well suited. These are the conditions that

cause the meetings to drag on for what seems an endless cycle of "what if" and "I'll look into it" contributions. Contributions that are enlightening but don't really get any closer to defining or solving the problem at hand. In fact, sometimes they make the problem seem insurmountable when actually it isn't. So, conducting a rapid requirements analysis will help the principles stay focused within their area of expertise initially then provide a comprehensive guide for mutual discussion later. The following section explains how it's done.

CONDUCTING A RAPID REQUIREMENTS ANALYSIS

The following steps illustrate how to conduct a rapid requirements analysis. The project manager may wish to add other tasks to suit an organization's system design rules. However, this framework is a proven method of requirements gathering. It has worked under the toughest of conditions and shows improvement over conventional methods in concept-to-prototype cycle times.

Step 1

Identify members. The energy management team should identify who is involved. List everyone involved in the project and what their roles are to be. A contact list with phone numbers, email addresses, office locations and short description of their responsibility is helpful.

Step 2

Identify tasks. The energy management team should list all primary and secondary tasks that the EIS must perform in order to completely replace the existing processes and accomplish the desired result. The IT development team will have an opportunity to add supporting tasks in a subsequent step. A mission statement can help to qualify which tasks the project actually needs to accomplish its goals. The team should list the tasks as clear and concise descriptions but not so detailed as to make them cumbersome to read. Detailed analysis of the task at a later time can clarify any questions that arise. See Figure 24-1 for an example of how to list tasks.

Step 3

Indentify criteria. The energy management team and the IT development team need to agree on criteria for evaluating the requirement tasks and a rating scheme for the criteria. It's likely the criteria will be "need" and "cost." Either team can add other criteria as desired. However, the more criteria the more difficult and time consuming the analysis becomes. Our examples will assume *need* and *cost* are the two criteria as shown in Figure 24-2. The rating scheme determines how tasks compare to each other. For instance, the most common method is to assign the most needed tasks a lower number and the least needed tasks a higher number on a linear scale. Or, in other words, "lowest is best." The same method applies to cost evaluations. The tasks with the lowest cost should have a low value assignment and tasks with a high cost should have a high cost value assignment.

Step 4

Independently evaluate. The energy management team independently evaluates the tasks based on *need*. They assign a value to the "NEED" column associated with each task that portrays its *need* relative to the other tasks. The team should assign values in large increments, so that other tasks will fit between them later. At the same time, the IT development team independently evaluates the same tasks based on *cost* in the same manner. They should only evaluate the tasks according to *relative cost* and not try to estimate an exact cost for each task. They can add new tasks that are required to complete other tasks but only if their *need* value is inherited from the tasks that require the new task. These additional required tasks should be few in number. Any other tasks required to complete listed tasks should be part of the cost and effort involved in completing that task. Figures 24-3 and 24-4 show independent evaluations of tasks based on need and cost.

Step 5

Merge the evaluations. The project manager independently ranks the *need* values and the *cost* values then merges the rankings together by adding the *need ranking* to the *cost ranking* as a "TOTAL RANKING" column. The project manager then sorts the list by "TOTAL RANKING" and calls a

meeting to discuss the results. It could be that some tasks are not in the appropriate position in the list. Adjusting the *need* and *cost* values in the list produces a new prioritized list. This "adjusting" is repeated until the teams agree on a final requirements task list. Figure 5 shows the "NEED" and "COST" columns with the rankings (in parentheses) summed together as "TOTAL RANKING" and then sorted by "TOTAL RANKING."

CONCLUSION

EIS project requirements analysis is never going to be easy. As more and more technological innovations work their way into the energy management arena it will only become more complex. As our responsibilities grow and time becomes more and more scarce it will be more difficult to get large groups of people to work together on such complex projects. I suppose that is one reason why outsourcing is so popular these days. This may also explain why energy information Systems have been slow to prevail. Yet the advantages of EIS projects continue to multiply rapidly. Political, environmental and weather trends among others effect our energy consumption and cost. For the most part these factors are uncontrollable.

RAPID REQUIREMENTS ANALYSIS

EIS Project
Task
Mission: To automatically collect energy data on a daily basis and display user-friendly reports to any interested parties via the Web.
Maintain List of Utility account numbers
Maintain Electric Billing Days
A security plan describing all security issues and procedures.
Degree days
Generator gasoline
By Facility
Distribution list
Data will be reported at various levels: by year, month, day, hour.
Specific input screens will require security login procedures.
Graphs will be an image allowing them to be copied and pasted to other applications.
Data will be stored at various levels: by year, month, day, hour.
Data collected automatically will be available for reporting on an daily basis.
Email notification of change to persons on distribution list.
Maintain list of Name, email address, Report URL, Frequency of delivery
Email link to reports according to email subscriptions on a daily basis.
Graphs will allow for any specified data set as the auxiliary data on the right axis.
Graphs will allow for multiple data sets, combined line and bar graphs and auxiliary data on the right axis.

Figure 24-1. Task List

Monitoring consumption regularly is the most productive way to cut costs. This justifies an effort to streamline the requirements gathering process and produce energy information systems specific to the needs of an organization.

This rapid requirements analysis method attempts to bridge the gap between two diverse professional teams in an effort to take advantage of each of their respective strengths toward developing an EIS. It avoids the well known pitfalls of large lengthy meetings hammering out details of an EIS project but without losing the detailed analysis needed to get the job done. It prioritizes tasks so that their need and costs compare with other tasks easily. It brings everything together in the end to a well organized document that shows the goals of both development teams.

As with any other idea in this fast-paced world, this method is open to adaptation. Categories may be helpful to show tasks in phases or areas of responsibility. Also, the project manager may want to add a reference column to link to other information for each task. The project manager should be cautious to avoid over-complicating the process since simplicity is one of its virtues. Follow the progress of this rapid requirements analysis methodology at http://www.utilityreporting.com/rraonline and take advantage of the free online tool.

References

[1] Burns, Kathleen; "Energy Information Systems: Knowledge About Power"; *Energy User News*, August 2000, http://energy-usernews.com.

[2] Capehart, Barney Ph.D., C.E.M.; "Utility Data Web Page Design: An Introduction"; *Information Technology for Energy Managers*; Chapter 23; Fairmont Press, Inc., Lilburn, Ga. 2004.

[3] Jennerich, Bill; "Joint Application Design"; Internet page, http://www.bee.net/bluebird/jaddoc.htm, accessed 8/2/2004; Bluebird Enterprises Inc., Berwyn, PA

[4] Urquhart, Cathy Ph.D.; "Strategies for Conversation and Systems Analysis in Requirements Gathering: A Qualitative View of Analyst-Client Communication"; *The Qualitative Report*, Volume 4, Number ½, January, 2000; (http://www.nova.edu/ssss/QR/QR4-1/urquhart.html).

RAPID REQUIREMENTS ANALYSIS

EIS Project			
Task	Need	Cost	Total Ranking
Mission: To automatically collect energy data on a daily basis and display user-friendly reports to any interested parties via the Web.	0 (1)	0 (1)	2
Maintain List of Utility account numbers	100 (2)	100 (2)	4
Maintain Electric Billing Days	110 (3)	150 (7)	10
A security plan describing all security issues and procedures.	130 (5)	140 (6)	11
Degree days	150 (9)	110 (3)	12
Generator gasoline	120 (4)	200 (8)	12
By Facility	150 (8)	130 (5)	13
Distribution list	160 (10)	120 (4)	14
Data will be reported at various levels: by year, month, day, hour.	140 (6)	400 (10)	16
Specific input screens will require security login procedures.	150 (7)	410 (11)	18
Graphs will be an image allowing them to be copied and pasted to other applications.	350 (13)	300 (9)	22
Data will be stored at various levels: by year, month, day, hour.	200 (11)	420 (12)	23
Data collected automatically will be available for reporting on an daily basis.	210 (12)	430 (13)	25
Email notification of change to persons on distribution list.	400 (14)	500 (14)	28
Maintain list of Name, email address, Report URL, Frequency of delivery	600 (15)	510 (15)	30
Email link to reports according to email subscriptions on a daily basis.	610 (16)	520 (16)	32
Graphs will allow for any specified data set as the auxiliary data on the right axis.	630 (18)	600 (17)	35
Graphs will allow for multiple data sets, combined line and bar graphs and auxiliary data on the right axis.	620 (17)	610 (18)	35

Figure 24-2. Criteria

RAPID REQUIREMENTS ANALYSIS

EIS Project	
Task	**Need**
Mission: To automatically collect energy data on a daily basis and display user-friendly reports to any interested parties via the Web.	0
Maintain List of Utility account numbers	100
Maintain Electric Billing Days	110
Generator gasoline	120
A security plan describing all security issues and procedures.	130
Data will be reported at various levels: by year, month, day, hour.	140
Specific input screens will require security login procedures.	150
By Facility	150
Degree days	150
Distribution list	160
Data will be stored at various levels: by year, month, day, hour.	200
Data collected automatically will be available for reporting on an daily basis.	210
Graphs will be an image allowing them to be copied and pasted to other applications.	350
Email notification of change to persons on distribution list.	400
Maintain list of Name, email address, Report URL, Frequency of delivery	600
Email link to reports according to email subscriptions on a daily basis.	610
Graphs will allow for multiple data sets, combined line and bar graphs and auxiliary data on the right axis.	620
Graphs will allow for any specified data set as the auxiliary data on the right axis.	630

Figure 24-3. Independent Evaluation by Need

RAPID REQUIREMENTS ANALYSIS

EIS Project	
Task	**Cost**
Mission: To automatically collect energy data on a daily basis and display user-friendly reports to any interested parties via the Web.	0
Maintain List of Utility account numbers	100
Degree days	110
Distribution list	120
By Facility	130
A security plan describing all security issues and procedures.	140
Maintain Electric Billing Days	150
Generator gasoline	200
Graphs will be an image allowing them to be copied and pasted to other applications.	300
Data will be reported at various levels: by year, month, day, hour.	400
Specific input screens will require security login procedures.	410
Data will be stored at various levels: by year, month, day, hour.	420
Data collected automatically will be available for reporting on an daily basis.	430
Email notification of change to persons on distribution list.	500
Maintain list of Name, email address, Report URL, Frequency of delivery	510
Email link to reports according to email subscriptions on a daily basis.	520
Graphs will allow for any specified data set as the auxiliary data on the right axis.	600
Graphs will allow for multiple data sets, combined line and bar graphs and auxiliary data on the right axis.	610

Figure 24-4. Independent Evaluation by Cost

RAPID REQUIREMENTS ANALYSIS

EIS Project			
Task	Need	Cost	Total Ranking
Mission: To automatically collect energy data on a daily basis and display user-friendly reports to any interested parties via the Web.	0 (1)	0 (1)	2
Maintain List of Utility account numbers	100 (2)	100 (2)	4
Maintain Electric Billing Days	110 (3)	150 (7)	10
A security plan describing all security issues and procedures.	130 (5)	140 (6)	11
Degree days	150 (9)	110 (3)	12
Generator gasoline	120 (4)	200 (8)	12
By Facility	150 (8)	130 (5)	13
Distribution list	160 (10)	120 (4)	14
Data will be reported at various levels: by year, month, day, hour.	140 (6)	400 (10)	16
Specific input screens will require security login procedures.	150 (7)	410 (11)	18
Graphs will be an image allowing them to be copied and pasted to other applications.	350 (13)	300 (9)	22
Data will be stored at various levels: by year, month, day, hour.	200 (11)	420 (12)	23
Data collected automatically will be available for reporting on an daily basis.	210 (12)	430 (13)	25
Email notification of change to persons on distribution list.	400 (14)	500 (14)	28
Maintain list of Name, email address, Report URL, Frequency of delivery	600 (15)	510 (15)	30
Email link to reports according to email subscriptions on a daily basis.	610 (16)	520 (16)	32
Graphs will allow for any specified data set as the auxiliary data on the right axis.	630 (18)	600 (17)	35
Graphs will allow for multiple data sets, combined line and bar graphs and auxiliary data on the right axis.	620 (17)	610 (18)	35

Figure 24-5. Merge Evaluations

Chapter 25

Developing an Energy Information System: Custom Design vs. Off-the-shelf Software

David C. Green

ABSTRACT

A custom designed and developed energy information system (EIS) has distinct advantages over off-the-shelf software. This chapter examines those advantages in terms of project goals, data acquisition, presentation, support, integration and cost. Commercial software development has grown significantly over the last 20 years. But, has it provided the best comprehensive solutions we need or simply acceptable substitutes? This chapter presents the argument that custom designed software provides more opportunity for innovation and ultimately lower cost.

INTRODUCTION

When an organization decides to implement an energy information system (EIS) an enormous number of questions arise. But one of the first (and most difficult) is whether to develop custom software in-house (with employee or contract programmers) or buy one of the many off-the-shelf (OTS) products. There are certain criteria for the decision which show up time and time again. One of those criteria is "how well does the software meet our *goals*, is it focused on our goals?" Another obvious criterion is *cost*. That is initial cost and ongoing cost. As a part of the cost factor one might want to include considering return on investment. How about *support*? Which method is the easiest to support? Are we prepared to support it in-house? Is the OTS vendor prepared to support it? What exactly are the support requirements? Are there end user support requirements as well as administrative ones? Let's try to find a way to organize all these criteria in a way that makes it easier to decide whether to develop the software on our own or buy it.

One can use a decision matrix to help track the relative strength of these *criteria* for each *choice*. The example table below shows the two *choices* across top and each of the pertinent *criteria* along the left.

The numbers shown are *rankings*. Decision makers rank the two choices in terms of each criteria, "2" being the best solution. This, of course, is the difficult part. It requires a lot of painstaking evaluation for each of the criteria. But ultimately it will be worth it. This is only an example the mission statement and rankings. This example shown does not apply to every situation. Remember in evaluating each of the criteria that the choice that is the best solution is the one that best supports the mission statement for the project. After the rankings are complete they are summed. The highest total ranking wins and is the best suited for your situation, *theoretically*. [1]

This is not "rocket science" and one might be tempted

Project Mission Statement: To create an Energy Information System to automatically collect consumption data from individual meters for all services and provide a way for many users to analyze the data across multiple levels in order to make decisions relevant to conservation programs.		
Criteria	**Custom Developed**	**Off-the-Shelf**
Goals	2	1
Data Acquisition	2	1
Presentation	1	2
Support	2	1
Integration	2	1
Cost	1.5	1.5
Total	10.5	7.5

to think that it is too simplistic for tough decisions such as this. However, the process might bring to light some issues that hadn't been considered before. At the very least it is a simple way to keep everyone's thoughts organized over time. The results of the decision matrix are not necessarily the best answer either. In fact, the results may be a tie. But, the decision matrix approach does two things. One, it gets people thinking about the project in a detailed manner. It provides a way to divide the work up among different groups of people that are experts with regard to a certain criteria. And two, it keeps everyone's work organized. Each choice becomes accountable for its values with regard to each of the criteria. For example let's consider *goals*. Which choice has goals that are similar to our goals?

GOALS

It's not that difficult to determine which type of software choice has goals that match those of the project. I'm sure the number one goal of any commercial software company is to *sell software* and a lot of it. Virtually all of its business systems are oriented toward that goal of winning *market share*. Otherwise, they may not stay in business. Another secondary goal is to lower their operating costs because they have a great amount of overhead to produce so many versions of their product and support them all. So, of course, they *outsource* as much of the labor as possible.[2] I suppose this may not hold true for all software companies. However, it's a trend that we've all noticed for some time now. Do they have the goal to create quality EIS software? Sure, they do. However, it's a goal to create quality software for many organizations, a "one size fits all" approach that leaves gaps in any detailed analysis goals an organization needs to have.

How about the custom developer's goals? For the most part they should match the goals of the project exactly. After all, they were tasked or hired to perform this particular project. They may have distractions and time restraints. It may take them a while to develop the software. But, their goal is to develop quality software in a reasonable amount of time at the lowest cost with *specific* regard to the project requirements. In fact, one of their primary goals is to understand the project requirements in detail. Their secondary goal is to take those requirements and using the nearly unlimited number of tools available to them translate them into results that support the project mission statement. For instance, contrary to commercial software developers, their goal is to provide a conduit for all data into the EIS from any of the many disparate data sources that exist in order to provide the in-depth look at the consumption patterns.

DATA ACQUISITION

Data acquisition is the most critical part of an EIS. It's the process of getting the data into the database. This is where custom developers really have an edge over OTS software. Energy data, in fact most kinds of data, arise from many different types networks, operating systems, devices and software. It is stored in many different file formats and the data is organized in an unlimited number of ways. It's safe to say it will be impossible for commercial software developers to account for all of those possibilities without some amount of customization. Even though efforts are in place to standardize the file formats and data items it's a long way in coming that all energy data will be accounted for by these standardized formats.[3] Initially, EIS software developers surmounted this issue by providing manual input forms to get the data into the database. This approach is tedious and costly not to mention the errors that may be produced. So, unless a software company is willing to customize the data collection process it may very well be that it could not provide the "automated data collection" requirement mentioned in the example project mission statement above. Even if provided, it would likely be very expensive.

On the other hand, in-house developers can customize the data collection process to fit nearly any format that the energy data is found in. They have the ability to customize the transfer, storage, archiving and translation of the data for each disparate data source if need be. For instance, an EIS using utility company data may need to collect data from a municipal water company that can only provide the data in spreadsheets. The arrangement of the data in these spreadsheets could exist in any number of ways. And, it may change over time. In-house developers simply create a small customized process to transfer and import the data in the specified format and if the format changes they create a new process. An important aspect of the customized data collection process is that developers can include quality control checks which insure the presentation of the data is clear and accurate.

PRESENTATION

At first, one might think that the presentation aspect of the project would be the same for both OTS and custom developed software. We've all pretty much resigned to the fact that web based presentation is the best. A popular analysis scenario is comparing current consumption to prior year or baseline data in order to quantify a reduction in consumption over time. Also, benchmarking buildings is popular. Trend analysis is useful and both choices seem to handle it well. [4] However, let's get into the details of presenting the data.

First of all, there are a virtually unlimited number of values that can be derived from the raw consumption data that might be useful in analyzing energy use. Commercial software doesn't provide this "unlimited" number of values. It may provide for summarizing or combining some values such at kWh per square foot. But the possibilities are endless and commercial software just can't know all values the users are going to want to look at. Conversely, the custom developer can create calculated values for any number of summations or combinations as the project requires. This is the same for reporting formats.

Legacy software or spreadsheets may be producing reports that are useful and need to be replicated in a new system. Commercial software design provides a limited number of report designs and those are typically static reports with no interaction capability. [5] The custom developers can build reports exactly as desired and provide links to interact with other reports or information. The overall appearance of a custom designed EIS is likely much less complex than an OTS software version since a custom designed EIS will only include configuration items that are required. An OTS system has to account for many different configuration possibilities of the many different organizations that will be using the software. This makes the configuration and perhaps even the presentation of the reports more difficult. And, of course, it makes the software more complicated and difficult to use and support.

SUPPORT

Commercial software vendors provide support for hundreds maybe even thousands of customers. This is why many of those vendors have chosen to outsource those tasks at a cheaper rate. That's not to say that those support personnel are not qualified. Outsourcing is the nature of the beast these days and may be necessary to meet the needs of commercial software support in general. However, it is a "weaker" type of support in that requests for changes to the way the software functions can not be supported in most cases. It's just not practical to constantly change commercial software to fit the user's whims. Changes are made however. Unfortunately, sometimes those changes (marketed as *improvements*) simply create more unnecessary complexity and require more training.

Custom developed software on the other hand is readily adaptable. And in fact, the successive revision technique [6] of developing features for custom software is widely accepted. Support personnel for custom developed software are guaranteed to be highly qualified and likely may even be the same people who developed the software initially. Changes can usually be made quickly within a matter of hours or days by a dedicated support team. Any requests for changes or improvements are likely made as a consensus of all or most of the users. The majority of these requests are likely to be for some sort of integration capability.

INTEGRATION

Integration with other email, word processing and presentation software is demanding for commercial software developers because of the need to accommodate so many different types of such software. Because of this it's likely that the integration requirements for any one organization won't be met completely. This is where the limitations of the commercial software really begin to have a negative effect on the organization. It's not uncommon for existing processes and software in an organization (that were working perfectly well) to change simply to integrate well with new EIS software.

In-house developers need only integrate with the software already available for their particular organization. This is still a challenge no doubt. But, the ability to automate data collection, emailing of reports, exporting to other formats and even copying and pasting in and out of the application is obviously easier for in-house developers. Their best advantage is the ability to integrate with other custom software developed in-house at a reasonable cost.

COST

It might *seem* that OTS software is always going to be more economical than paying to develop software. However, there are a lot of hidden costs involved in OTS software such as training and support. The often overwhelming complexity of OTS software, made to accommodate the needs of a wide range of customers, requires that users spend a good deal of time learning the system. [7] There may also be a requirement to re-train after a major upgrade to the OTS software. These upgrades require a reinstall of the software causing downtime. OTS software sometimes requires special networking protocols not already available on the network that have to be installed and supported. This may require paying for a support contract in addition to the cost of the software itself. Then if you add in the routine costs of inputting data (that can't be automatically collected), analyzing and distributing the reports it becomes a quite expensive endeavor.

Custom developed software on the other hand holds its costs down by only developing what features are included in the scope of work for the project. These features are the minimum features required to accomplish the project mission statement. So the complexity and required training are held to a minimum as well. A cost analysis will show that the labor costs involved are similar to those labor costs already effecting the organization. Since the costs of support, upgrades and training are done by in-house personnel or contractors it fits better into the existing budget. The support costs actually go down over time because the support personnel become more and more familiar with the application and the changes take less time. As the users become more familiar with the application the need for additional training drops off. On the contrary, OTS software becomes more complicated to accommodate the needs of the multitude of customers using the system. Some of those added components may not be of use to some users at all.

CONCLUSION

So, it really comes down to a matter of efficiency. Is it more efficient to buy software off the shelf that is designed for a multitude of organizations and supported by people who know nearly nothing about the particular energy program that the software is supposed to serve? Or, does it make sense to design and develop a system using the experience of the engineers and managers who understand the organization's energy program requirements first hand. I know it's tempting to just buy the software, conduct the training and hope that some usefulness comes out of it. However, is that our goal, to get "some usefulness" out of our energy conservation programs?

I think more likely than not our goals should be to get *as much* out of our energy conservation programs as we possibly can. Opportunities to conserve will continue to present themselves in a variety of ways. We need to be prepared to confront them with detailed analysis specific to our situation in order to arrive at the best solution *for our organization*. [8]

I'm sure this dilemma will not be resolved anytime soon. There will continue to be OTS software developed and sold and some organizations will choose to develop their own. Use the decision matrix approach (mentioned in the introduction) for your next project and see what results you get. Carefully consider the hidden costs of OTS software and the capabilities of your in-house or contract software developers. But, more important than anything, try to consider the long range effects with regard to each criterion. After all, we have a long way to go to be *completely energy efficient*.

References

[1]   "How to use a decision matrix to streamline your decision making process"; Internet page, http://www.time-management-guide.com/decision-matrix.html; Time-Management-Guide.com 2005

[2]   "Statistics Related to Offshore Outsourcing"; Internet page, http://www.rttsweb.com/outsourcing/statistics/; Real-Time Technology Solutions, March 24, 2010

[3]   VanAusdall, Steve; "Standard Energy Usage Information"; NIST Smart Grid Project, November 24, 2009

[4]   Granderson, Jessica M.A.; Piette, Ghatikar and P. Price. 2009; Building Energy Information Systems: State of the Technology and User Case Studies, Lawrence Berkley National Laboratory, LBNL-2899E, p. 21.

[5]   Granderson, Jessica M.A.; Piette, Ghatikar and P. Price. 2009; Building Energy Information Systems: State of the Technology and User Case Studies, Lawrence Berkley National Laboratory, LBNL-2899E, p. 47.

[6]   Kan, Stephen H.; "The Iterative Development Process Model," *Metrics and Models in Software Quality Engineering*, Addison-Wesley Publishing Company, Reading, MA 1995, p. 25.

[7]   Granderson, Jessica M.A.; Piette, Ghatikar and P. Price. 2009; Building Energy Information Systems: State of the Technology and User Case Studies, Lawrence Berkley National Laboratory, LBNL-2899E, p. 8.

[8]   Granderson, Jessica M.A.; Piette, Ghatikar and P. Price. 2009; Building Energy Information Systems: State of the Technology and User Case Studies, Lawrence Berkley National Laboratory, LBNL-2899E, p. 49.

Chapter 26

Developing an Energy Information System: A New Look

David C. Green

ABSTRACT

Energy information system (EIS) reporting tools take on a variety of "looks" these days. Many of them produce a standard set of reports allowing the users to substitute different meter or building data into them. Some go as far as to allow for "drag and drop" operations. Almost all require some sort of option selection to get the desired data set displayed. This chapter describes a unique approach that gives the users an almost infinite number of reports while providing instant results from their mouse clicks. The users click their way through designing reports and graphs that summarize data at multiple levels and show pertinent trends in the form of colorful graphs. This EIS design provides the novice user with a very short "learning curve" and the expert user can take advantage of the design flexibility. It provides an ideal solution for energy managers that need simple, easy-to-use tools that everyone can use to analyze their energy consumption data in just a few minutes per day. Energy managers can use these tenants to develop their own EIS, evaluate the effectiveness of the EIS they already have or use them to help decide which EIS to purchase.

INTRODUCTION

Energy "dashboards" are providing a long awaited link between the energy meter and the computer desktop. Smart meters and data collection devices move the data nearly seamlessly (sometimes wirelessly) from the point of consumption to a computer software program. However, these software programs still lack the full featured functionality needed to effectively analyze energy consumption quickly. Set aside the fact that installing the data collection devices sometimes involves an electrician, the software used to analyze energy data has always suffered with respect to either functionality or web usability standards. This makes it difficult for *everyone* to take advantage of the progress made in collecting the data. The user-friendly, simple and consistent software interface that energy analysis software lacks today omits a whole group of potential energy conservation "hawks." They could be making a difference in our energy appetite if given the correct tools to make maximum use of their limited time and money. One bright spot, the release of multitouch-touch tablets, will bring a new revolution in computing we have not seen since the 1980's when PCs dramatically lit up our desktops.

Without a doubt, our goal is to someday bring energy consumption data directly from the meter to a user's touch screen computer and provide simple yet comprehensive analysis tools. This software should provide scalable trend analysis graphs of not only consumption values but any number of useful calculated values. The web based software should be flexible allowing users to create customized innovative reports "on the fly." It must allow users to take full advantage of the touch screen technology to instantly drill down through the data's "time and space" interactively reporting on targeted values then saving, copying or emailing them as needed. The vital link between comprehensive energy data analysis and any energy conservation program is the ability to collaborate among all those involved. This software must not give up the things that most web based software often does with respect to integration but rather embrace the features of the browser and the Internet protocols that are available today. It is essential that printing, copying/pasting, exporting and emailing be part of the software.

In this chapter we describe an EIS design that may someday allow individual homeowners, property managers, facility engineers, energy managers and directors the same maximum opportunity to be innovative in their approach to energy conservation. Not only will it provide the same consistent user-friendly interface to a wide audience but also a wide range of data. Reporting on energy data for a single meter in real time is now as easy as reporting the monthly consumption of whole complexes. Its simple design will allow for cloning the presentation into a catalog of reports and templates to suite the users' particular needs. Uploading of data from a variety of sources will remove the mystery of "how to get the data into it" and scheduled alerts can provide notification of critical conditions. Energy "dashboards," touch screen technology and this progressively designed software will provide the catalyst to solidly bridge the gap between meter and desktop.

DASHBOARDS

Energy dashboards are becoming a very popular method of monitoring home electric consumption. They are providing the valuable link between the meter and the desktop for energy data that can be displayed and analyzed. Most of them connect to a circuit panel and send real-time consumption data to a desktop device. That device either displays the data or stores it so that a computer, hooked to the device, can display it. The software used to display the data is somewhat limited in its ability to perform any in depth analysis. However, this new innovation has sparked a new interest in energy monitoring that hopefully will spread to the workplace. Those who are more aware of their energy consumption are more likely to conserve. The more options we have to become aware of our consumption then the more likely we are to implement them and thus lead to a behavioral change at home *and* at work.

At the lower and less expensive end of the range of dashboard options are small devices that plug into a wall outlet and act as a receptacle but also track the energy that passes through it. The "Kill-A-Watt" plug-in device (http://www.p3international.com/products/special/P4400/P4400-CE.html) displays a cumulative usage over time but there is no way to move that data off of the device for further analysis. The "PicoWatt" device (http://tenrehte.com/products/) is a similar plug-in module but it has a wireless transmitter that sends the data to a web site for viewing and analysis. The obvious shortfalls to these devices are that they only monitor one appliance at time so monitoring the whole house would require several plug-in devices. A new concept is the "Pinch-A-Watt" method (http://thornproducts.com/devices.html) where small inexpensive devices are attached to the switch or wall sockets to collect that data and send it to a central receiver unit. The unit then stores or forwards the data on to other devices or web sites to view or analyze.

The most widely available of all choices are monitoring devices that have circuit transformers (CT) that connect adjacent to the meter or circuit panel and then send the data to a desktop device by wire or wirelessly. Some of these use scanners on the meter face instead of the CTs.

"The Energy Detective" (TED, http://www.theenergydetective.com/) is a popular example of one of these. The included software installed on the computer displays real-time and historic data but is still rather limited in its depth of analysis features. An additional advantage to TED is that it can send the data to Google Power Meter (http://www.google.com/powermeter/about/about.html) a free online tool for analyzing energy data. And this is where energy dashboards have really excelled. The integration with the Google Power Meter Application Programming Interface (API) allows dashboards to send energy data from the meter to *any desktop* throughout the world.

Smart meters, typically installed by the utility company, require little or no human intervention to collect real-time energy data but only a small percentage of electric meters are of the "smart" variety. They are also quite expensive. The advantage of "smart meters" is that they can also send consumption information back to the utility company to help them institute a load management program. Without a doubt someday all electric meters will be smart meters. However, that day is likely far in the future. Until then we need to bridge that gap that exists in delivering energy data to the computer desktop for meaningful analysis.

All of these devices, and there are many more then mentioned here, help to bridge that gap and deliver meaningful data to the consumer. However, the real contributors will be the ones that provide the best software to display and analyze the data. Small plug-in devices provide little analytical capabilities. Even those that integrate into the Google Power Meter via its API lack the customization

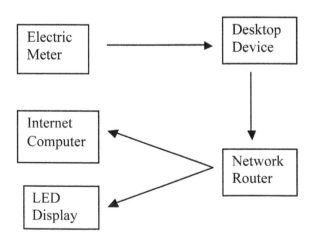

Figure 26-1. Energy Monitor Flow Chart

opportunities required to do meaningful in-depth analysis of the data. It would be helpful if the device manufacturers provided access to the raw data for custom analysis. But, none of the ones reviewed advertise that capability. Access to the raw data would allow developers to create customized displays suited to a select group of users. It might also open up a flood of "open source" type contributors to help develop a whole catalog of reports and graphs to choose from. Computer users still represent a wide variety of experience levels. They need an interface that is simple and employs good web usability standards in order for them to enjoy the benefits of all of this new found energy data.

WEB USABILITY STANDARDS

We need simple user-friendly pages that are consistent in design and produce a fast response. Web usability standards are transforming toward a model more suited to the "web surfer" than a web application user. Terms like "white space" and "oversized font" remind us of the traits that made print media so compelling to us for such a long time. Those traits are becoming more and more important on the web now that our time is more limited and our patience is short. The challenge is to keep all of those characteristics and at the same time provide an online application that is robust and flexible enough to provide everyone from home owners to energy managers the tools they need to analyze energy consumption patterns effectively. A simple interface design can go a long way toward helping a web application user navigate the site efficiently. Figure 26-2 is an example of an EIS report that acts both as the report and the means to navigate the site (through its links) in order to design other reports.

An application doesn't have to be packed with features to provide an effective energy management tool. On the other hand, the number of possible features can grow quickly as users develop a hunger for reports, graphing and collaboration. Perhaps the goal is to "hide" the features from novice users and let expert users discover them in time. This decreases the learning curve in the beginning yet still provides a feature packed application. It's still important to limit the features to only those that support the focus of the application. The more features there are the more frustrated the user becomes and the more that can go wrong. [1] Keeping the interface simple and consistent reduces the chances of frustration.

Consistency takes advantage of the user's spatial memory, the ability to recall where objects are related to other objects. This is how people are able to navigate through their homes in the dark without bumping into

Institution = Energy Information Institute
Campus = Greenville

Year	kWh	Prior Year kWh	Prior Year kWh Variance	% Variance Prior Year kWh	Base Year kWh	Base Year kWh Variance	% Variance Base Year kWh
				ELECTRIC			
2005	22,588,895.756	23,066,283.739	-477,387.98	-2.07 %	23,066,283.739	-477,387.98	-2.07 %
2006	22,097,313.199	22,588,895.756	-491,582.56	-2.18 %	23,066,283.739	-968,970.54	-4.20 %
2007	21,700,062.256	22,097,313.199	-397,250.94	-1.80 %	23,066,283.739	-1,366,221.48	-5.92 %
2008	21,390,057.575	21,700,062.256	-310,004.68	-1.43 %	23,066,283.739	-1,676,226.16	-7.27 %
2009	20,819,364.411	21,390,057.575	-570,693.16	-2.67 %	23,066,283.739	-2,246,919.33	-9.74 %
2010	17,174,436.609	20,819,364.411	-3,644,927.80	-17.51 %	23,066,283.739	-5,891,847.13	-25.54 %

Figure 26-2. Efficient Report Design

things. They remember how many steps it is to the doorway, which way to turn to the kitchen, etc. In the same way, computer users remember mouse clicks and the position of the mouse when clicked relative to other items on the screen. As long as the pages are designed with consistent alignment and links the user becomes more and more comfortable in navigating the site. This produces the "good feeling" in them that they are succeeding in their task. [2] A similar "good feeling" results from clicking on a link or button and getting an instant response.

This instant response leads one to believe they are on the right track. If the resulting page moves them closer to their end result it satisfies their curiosity. It doesn't really matter how long or how many clicks it takes to get there as long as progress is being made. Conversely, web applications in the past have taken on the characteristics of the older desktop applications in which many items may need to be selected before a result can be seen. Pull down menus, radio buttons, checkboxes and form fields all contribute to this maze of questions that makes the user think too hard before getting any results. [3] Making the user think too hard my not lead to frustration but it may leave them with an uneasy feeling, like they are wasting time.

Instead the application should be "polite" in its presentation. It should provide an easy pathway to what the user is trying to do without any interruptions. Vague error messages that pop up on the screen and require user intervention are some of the worst examples of an impolite or "rude" application. Omit the errors or steer the user around them with little effort on their part. Applications that require the user to read instructions before accomplishing their tasks are rude. The user should be able to immediately experiment with the application clicking their way toward the desired result as easily as surfing the web for new shoes. [4] This brings that energy dashboard data not only from the meter to the desktop but almost effortlessly into the user's realm of thought.

Web usability standards will be important criteria for deciding which systems most effectively bring energy data from the meter to the desktop and make it available for effective analysis. As the number of people with energy data available to them increases the number of novice computer users looking at that data increases and the need for simple but effective web applications to analyze the data will expand. The touch screen computer places even more stringent criteria on the web interface to provide the functionality needed to navigate through a sea of energy data.

TOUCH SCREEN

Touch screen technology has been around for a long time but is just now starting to become widely used. You may have noticed the news reporters now using these devices to illustrate their stories on television. Conference rooms use the technology for more effective presentations. Will this trend continue all the way to my desktop? Well, if you are an Apple iPad user the answer is yes. As for the rest of us, I'm not sure, but I think the possibility is real.

The touch screen does provide the simple user interface that many novice computer users require and some of the experts *desire*. But, it leaves out some actions that users have been taking advantage of for some time. The rollover actions used to scroll down the page with the roller wheel of the mouse is not available on touch screens. Also, the mouse over actions will not work unless someone comes up with a new idea. The touch screens lack the fine grain control of the mouse leading to bigger fonts and larger buttons. So redesigning web sites to work effectively with an iPad may soon be a big issue. However, the richness of the touch screen's availability and simplicity will lead to greater innovation in the usefulness of this tool. [5]

It's already becoming clear that the touch screen is meant for much more than the iPhone. Using a large touch screen display allows multiple users at one time standing next to a wall size screen to interact at the same time. [6] The touch screen lends itself nicely to the drill-down operations of an energy data display. Changing between buildings, months, years and hours can be accomplished with one click (or touch) of the screen. Icons can still be used to pull up alternative features as needed. The drag zoom feature of touch screens could be very helpful to examine large energy reports. A one-touch graphing link would make meaningful trend analysis easy on touch screen displays.

TREND ANALYSIS

The primary goal of the EIS of the future is to take advantage of more widely available energy data to provide meaningful trend analysis quickly and easily to a wide variety of users. Raw energy data does not contain all the values necessary to effectively analyze the consumption trends. For instance, electric consumption for a building over time is effected by weather conditions. Comparing kWh month to month is meaningless without removing the effects of weather. The best way to do this is by dividing the consumption values by degree days creating a calculated kWh/degree day value that can then be compared month to month or year to year without the effects of weather. Similarly, buildings can be compared to each other without the effects of floor area by dividing the consumption by floor area. There are many calculated values the can enhance the analytical power of energy data. Once variances are determined in these calculated values a percent of variance can be used to show improvements or lack thereof over time. Figure 26-3 shows a simple trend analysis of kWh by month.

There is a nearly unlimited number of combinations of these calculated values trending over time to analyze. Deciding which ones to use must be left to the *users* of the software not the *developers* of the software. Given the tools the users will innovate as needed to do the best job of analyzing the data.

INNOVATION

Simplicity and consistency in web application design feeds the potential for the user to be innovative. Options, settings and choices all get in the way of innovation. A reporting tool that is interactive in nature with one-click opportunities to change the data or the format of the report gives the user that "good feeling" and inspires innovation. [7] The links in Figure 26-4 provide the flexibility to modify this report one click at a time.

The ability to instantly change the scale of the data presented gives the user more power to compare and perceive anomalies in the data that may be evidence of a mechanical problem wasting valuable energy dollars. A simple web page design allows copying and pasting of information freely from the application to any other. An EIS such as this, available free on online, could spur a flurry of "open source" type contributions in terms of calculated values, report design, trend analysis and integration.

INTEGRATION

A good EIS should integrate well with other applications in order to help the users collaborate with others on energy data analysis. Even though printing of reports is not nearly as popular as in the past it is important that EIS reports are printable. Printing has been replaced mostly by email. So, emailing the reports, either automatically or manually is a requirement. Copying and pasting reports and graphs into emails is helpful as well. Also, the ability to copy and paste into MS Word or MS Excel is a helpful feature. Believe it or not integration with the browser itself is not always a given these days. Some web applications do not allow saving pages as a favorite and retrieving them later. It's important that the EIS web application retain the scope of the user's session within the URLs so that any report can be saved as a favorite and retrieved later. This is a simple and valuable organizational feature since it is difficult to determine exactly how wide the scope of the user or the data is during EIS development.

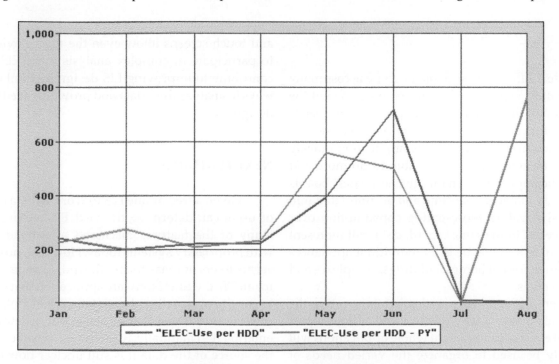

Figure 26-3. Trend Analysis Chart

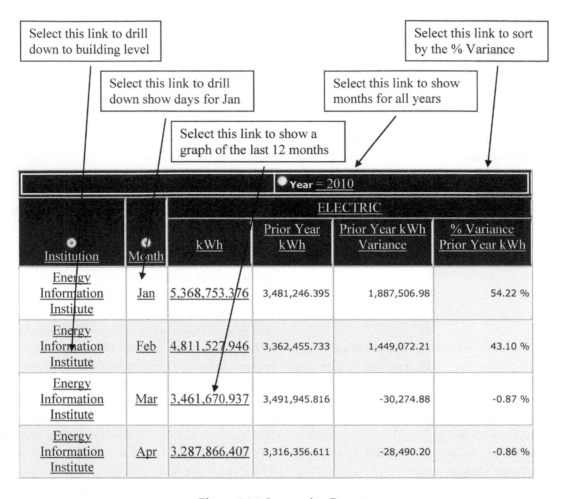

Figure 26-4. Interactive Report

SCOPE

The ideal EIS will be scalable in scope across many different realms, expertise of the users, time span of the data, location of the data, report designs and number of calculated values derived from the data. A novice user will be able to pick up the basics of the application quickly and as they become an expert be able to do more and more with the application. Thanks to energy dashboards data will be available in real-time and accumulated into comprehensive values representing consumption over months or years. In the same regard, data will represent individual meters or maybe even individual appliances and roll up to represent homes, buildings, complexes and maybe even cities.

As users become more and more innovative the number of calculated values will grow and various report designs will evolve. Templates of whole applications will undoubtedly be used to organize the varied needs of the users. Now that the entire scope of data is available

and touch screens allow even the most novice of users to participate in complex analysis more EIS users can contribute to improving EIS design by thinking of new ways to analyze their data and providing feedback to EIS designers.

NEXT FEATURES

Given a free online EIS to work with, a community of users can determine the next EIS features. Granted many of the features discussed are already available with proprietary systems today. However, no one system seems to encompass them all. And, change comes slow for an Off-the-Shelf software application that is expensive to purchase, install and maintain. A free online EIS will provide an optional method of uploading data rather than entering it in manually. Even though dashboards may be the source of the data it is still unclear how easy it will be to upload it into a customized EIS. Other sources of data such as utility companies and building automation systems must be accounted for as well. The EPA's Energy Star Portfolio Manager (http://www.energystar.gov/index.cfm?c=evaluate_performance.bus_portfoliomanager) may be a potential source of data. The growing number of reports and calculated values will spur the need for some sort of catalog from which to share and choose those possibilities. A convenient way to integrate reports into emails is to provide a *subscription* process allowing users to design a report and then subscribe to have it sent to them on a regular basis.

The same is true for alert notifications concerning data that fall outside a specified range. Emails can notify users of conditions in real-time and since many users carry smart phones those emails can be retrieved immediately no matter where the user is. Report subscriptions and alerts are valuable additions that characterize the emerging transformation of data from the energy meter to the desktop.

CONCLUSION

The energy information system is undergoing a transformation in "look and feel" due in part to the contribution of energy dashboards in making data more available and helping to provide a wider range of user interest. Proprietary energy reporting systems have slowed the advancement of energy reporting by holding their secrets close as close as their market share. Google Power Meter breaks the mold but still falls short of providing

the comprehensive set of customization required to allow users to design analysis processes to fit their energy management program. Progression of the web user persona and advancements in touch screen technology call for a much quicker and efficient method of navigating web applications. A project to provide a free online EIS for a community of energy conservation enthusiasts can help speed up the innovation of data translation from meter to desktop and beyond. Follow the progress of this project at http://energyinformationsystem.com.

References

[1] Hoekman, Jr., Robert; "Build Only What is Absolutely Necessary," Designing the Obvius: A Common Sense Approach to Web Application Design, New Riders, Berkley, CA 2007, p. 54.

[2] Hoekman, Jr., Robert; "Design for Uniformity, Consistency and Meaning," Designing the Obvius: A Common Sense Approach to Web Application Design, New Riders, Berkley, CA 2007, p. 189.

[3] Krug, Steve; "Design for Uniformity, Consistency and Meaning," Don't Make Me Think: A Common Sense Approach to Web Usability, Second Edition, New Riders, Berkley, CA 2006, p. 41.

[4] Hoekman, Jr., Robert; "Don't Innovate When You Can Elevate," Designing the Obvius: A Common Sense Approach to Web Application Design, New Riders, Berkley, CA 2007, p. 237.

[5] "Redesigning the web for touch screens"; Internet page, http://www.technologyreview.com/computing/25236/; Technology review 2010

[6] "Redesigning the web for touch screens"; Internet page, http://www.technologyreview.com/Infotech/18079/; Technology review 2010

[7] Hoekman, Jr., Robert; "Turn Beginners into Intermediates, Immediately," Designing the Obvius: A Common Sense Approach to Web Application Design, New Riders, Berkley, CA 2007, p. 133.

Figure 26-5. Report Subscription Form

Chapter 27

Interoperability of Manufacturing Control and Web Based Facility Management Systems: Trends, Technologies, & Case Studies

John Weber

ABSTRACT

For years, facilities management and energy management systems have been primarily systems offered by major industry players seeking to provide the entire solution for the end user's application. There are excellent systems available; however, they all tend to share one thing in common, a general lack of openness. When users want to integrate or interface these systems with other plant systems, other vendor's hardware, or plant control systems, the cost of integration grows dramatically or users are told they just cannot do it.

This same challenge faced the manufacturing automation industry in the 1980s and early 1990s. Since then, open standards and open technologies have taken great leaps in decreasing the cost of integration. Technologies from the commercial information technology market, such as Ethernet, have made a significant impact on connectivity on the plant floor. There still is a long road to travel to reach the perhaps mythical "plug-and-play" control system. It may never be reached because it's not always in the best interest of vendors of control hardware and software who prefer to provide the customer everything under their brand. In the manufacturing automation industry, the drive for decreased automation system integration costs through open systems and standards continues as manufacturers are forced by the changing global economic landscape to become more competitive.

The first half of this chapter will explore two trends in the manufacturing automation industry already having an impact on facilities and energy information systems: open connectivity standards and wireless connectivity. The second half of this chapter is dedicated to coverage of case studies where integration, interoperability, or utilization of technologies from the commercial information technology world or manufacturing automation world have made a significant impact on a facilities management system.

INTRODUCTION

Nearly every manufacturing operation involves some type of control system. Whether a single machine, a work cell, or an entire line, there is an electrical control system involved. That control system may be simple or complex, it may operate the entire machine without a human operator, or it may perform basic functions only upon the command of the operator. For purposes of this discussion, the term "manufacturing control" will be used to refer to techniques, technologies, and systems used in controlling the actual manufacturing process.

Of course, all manufacturing operations are housed in some form of a facility, and there are systems that manage the facility's energy usage, temperature, humidity, air quality, electrical system, and more. The needs of the facility's control systems are typically driven by the specific manufacturing and safety requirements of the actual manufacturing process, with consideration of any regulatory requirements included. For purposes of this discussion, the term "facilities management" will be used to refer to techniques, technologies, and systems used in managing the facility.

Without the proper facility and infrastructure, the manufacturing process cannot operate. Without a manufacturing process inside to generate products that people want to buy in sufficient volume to be profitable, there is no need for a facility or an energy management system. With the growing need for efficient use of energy resources, facilities operation data is required as a crucial component of the overall data needed for general facilities management as well as good manufacturing decision-making

Despite this obvious co-dependency, for years there has been a divide between manufacturing control and facilities management systems. Sometimes the systems are interconnected, often times they are not. Technologies used in each type of system sometimes overlap in use of hardware components or software, but often times they do not. In some cases, the lack of sharing of common technologies is justified as the specific needs of facilities management and manufacturing control in a specific business may not overlap. In other cases, there is great overlap.

Examples
- If you break down each system to core components, they each are looking at analog or digital signals.
- Based on those signals, decisions about what to do are made by the system.
- The values of the signals are communicated to machine operators, facility managers, and other interested parties.
- Information communicated to interested parties should be delivered to the type of device in the format and technology that suits their working style: Devices such as cell phone, PDA, tablet computer, laptop or desktop computer. Browser, thin application, and desktop gadgets that work like Windows news, weather, and stock gadgets are becoming increasingly popular commercial technologies used in facilities & manufacturing automation & control.

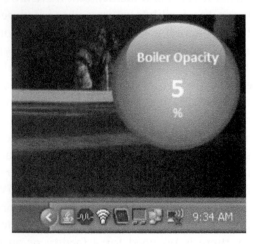

Figure 27-1. Example of a boiler health indicator displayed directly on a user's desktop using Windows gadget technology, courtesy of Factorywidgets.com

This gap between the systems results in missed opportunities for better coordination between manufacturing and facilities management and missed profitability improvement opportunities. Examples of where better coordination within the business can occur, but are often missed, include:

- Use of open, standards based systems from manufacturing control to provide more flexible and open systems in facilities management.
- Use of similar hardware components allows for common spare parts, and common maintenance training
- Use of the same software systems allows for close and low-cost integration of facilities data into process operations displays and vice-versa
- Integrated systems allow for quicker and easier reporting and analysis to aid in drawing valuable correlations that can improve profitability, such as:
  — Product quality with facility conditions (air, humidity, temperature)
  — Energy usage with product production mix and time of day
  — Energy usage with machine operating conditions—i.e. machines that need maintenance may use more energy and thus provide an early indicator of machine maintenance needs.
  — And many others—take a moment and think about questions you might ask about your business and could get answers to, if only you could quickly get the data needed to answer the question.

As more energy management systems turn to the use of a group of technologies loosely grouped as "web-based" or "internet" or "intranet," the availability of information is increasing for the facilities or energy manager. Manufacturing control systems have been following a similar trend for over 10 years.

This explores the state of the use of open systems and web-based technologies in manufacturing control systems. It will show how opportunities that can benefit the entire business exist for sharing of technologies, ideas, and techniques from the world of manufacturing control with facilities management systems.

TRENDS IN MANUFACTURING CONTROL SYSTEMS

There are two trends in the manufacturing control systems domain to consider and discuss how they relate to the use of web-based systems in energy and facility management: Open systems and the use of web based technologies.

Open Systems

Since about 1995, there has been a general trend towards openness in manufacturing control systems. By openness, we mean the ability for the owner of the control system hardware or software to easily interface the product to that of another vendor, or easily modify the operation of the system to suit the user's particular needs.

You could in theory take the idea of open to a logical extreme where everything in the system was open for modification, as in open-source computer software, where you can obtain the source code and modify the system to suit your specific needs. There are even some examples of open-source software in the manufacturing control systems market; however there are not signs of wide spread adoption of these systems.

What most users of manufacturing control systems seek are secure, standardized interfaces at all physical and logical connection points to the particular piece of control system hardware. They also seek easily implemented ways to customize the software or hardware within reason. Some examples will help illustrate these points.

Example 1: In a Programmable Logic Controller (PLC), it is common today to have a TCP/IP Ethernet interface available either as an option module or built into the controller. Although the protocol used at the application layer of the TCP/IP packet is unique to that vendor (i.e. AB, GE, Modicon), the hardware required to integrate the controller into plant and business networks is standard off-the-shelf commercial Ethernet hardware such as switches and routers.

Example 2: There a variety of types of networks available to connect a PLC to field sensors, instruments and devices. These networks are typically called "field busses." There are a number of field bus standards in the manufacturing control world, each suited to be strong in specific application types. Representative names include Profibus, Profinet, Foundation Fieldbus, Hart, Devicenet, CAN, ASI, FIP, Interbus and more. Users purchasing a PLC expect to be able to obtain from the PLC manufacturer or a third-party a module that will allow them to be able to connect the field bus(es) of their choice to that PLC and easily integrate sensors and field devices from many vendors to that PLC

Example 3: Human Machine Interface (HMI) software vendors began to adopt the OPC standard for connectivity to communications drivers back in the late 1990's.

OPC is a software interface standard that was created by automation vendors and managed by an independent organization. OPC interfaces enable vendors to share information in a standardize format between systems. Today it is a basic checklist requirement to offer OPC connectivity if a vendor plans to try and sell HMI software. By using OPC for device and inter-system connectivity, the software vendor allows their user to choose from a wide range of vendors for the connectivity software required to connect to both manufacturing control and facilities management systems.

Example 4: Evolution of the PLC into the PAC. As the cost of computing power has dropped, and power increased, the PLC has evolved in the scope of what it can do and manufacturing control system vendors now refer to their systems as Programmable Automation Controllers or PACs. The state of the art PAC today often offers multiple programming languages, built-in machine user interface applications, and even sometimes includes a web server in the PAC along with the ability to build web page user interfaces in the PAC. After the marginally successful attempts at "PC-based control" in the 1990s, the PAC has become to resemble a PC more and more in it's functionality, but without the moving parts and operating system risks that held back adoption of PC-based control in the 1990's.

Example 5: Entry of the OPC standard into the facilities automation space. Today, there are off-the-shelf commercial software applications, known as OPC servers, that have the ability to communicate to traditionally closed networks of facilities automation systems. The BACNet Ethernet standard in the facilities automation industry has provided a place for software vendors to write OPC software that can speak the BACNet Ethernet protocol to a variety of control system hardware on one side, and standardized OPC to a wide range of software applications on the other side.

Example 6: Using open standards such as OPC, and the commonplace availability of real-time power information from independent power operators over the internet, it is now possible to deliver real-time price information for electricity into control systems that can make decisions in real-time to curtail loads to avoid or shift cost.

Example 7: Invensys Operations Management, now offers a commercial energy management application that is entirely built upon their Wonderware product line software applications designed and proven in their 20+ year history in industrial automation, fulfilling a trend that started in the 1990s when ingenious system integrators would pull together PLCs, HMIs and other software from automation and control and implement custom systems based on open hardware and software.

In manufacturing control systems, users have found that often, by demanding the flexibility of open systems for connectivity to other systems and devices, they are able to choose best of breed components and integrate them to provide a system that meets their exact needs. There is a cost to such integration. The user should carefully weigh their integration costs versus the benefits they gain from choosing particular components from multiple vendors. They should also look carefully at any interoperability information available from a vendor before embarking down the path of a system based on multiple vendors, tied together with open system interfaces.

They key is the trend towards openness has resulted in greater freedom of choice for users. The trend towards more openness has reduced, but not eliminated, the possibility of getting locked into a single vendor's offering.

Web Based Technologies

From around 1999 to 2010, Ethernet became the largest of the web or internet technologies in use in manufacturing. As the internet boom and bust made commercial Ethernet hardware for PCs and setting up networks low cost and ubiquitous, providers of manufacturing control systems have seized upon the opportunity to leverage the low cost, high volume production of these components. Nearly every control system now offers an Ethernet interface. Those that can't widely support the use of low cost Ethernet to Serial converters to bridge their hardware to use Ethernet. Many control systems offer secure wireless Ethernet interfaces as options. Providers of industrial grade network switches, wireless access points, and even cable components are commonplace in the manufacturing control systems Market.

Ethernet has not completely replaced the many proprietary control networks that were used prior to Ethernets rise in popularity; however, their influence and usage have shifted to applications that are able to leverage their unique strengths as compared to off-the-shelf Ethernet. The reasons for this are multiple.

First is the existing installed base of networks from a variety of vendors. Users have significant investments in cabling, support tools, and training. If the networks are working, doing their jobs, and can accept additional capacity, then there is no incentive to change.

Second is the typical life cycle in control systems, which is about 10 years. That is the average. It is not uncommon to see 20+ year old control systems in use as companies seek to maximize their return on the capital invested in control systems. Over time, as older systems have been replaced, the installation rate of Ethernet has increased dramatically.

The third reason is there are applications where vendor proprietary control networks are needed. One example involves what is known as "deterministic response" or "determinism." Control systems engineers use the term deterministic response to mean that when a request is sent from one device to another on the control network, the time it takes for the transaction to complete is predictable, and within an application specific tolerance, repeatable from one trial to another. Different applications can accept different levels of variance in the two variables discussed that comprise a deterministic response. It has been shown in published studies that using 100baseT switched Ethernet, a level of determinism can be achieved that is the same as some popular proprietary control networks. That said, there will be cases though where the customer's definition of what constitutes acceptable "repeatability" and "predictability" will not be able to be met by even the best switched Ethernet networks. Thus, there will likely always be some need for vendor proprietary control networks to meet specific functionality required by specific systems and applications, either in security, capacity, level of determinism or other functional reasons. As the use of open systems has proliferated, concerns about security have actually helped foster a renewed interest in some proprietary systems. Debate is widespread on whether an open system that can be tested, tried, and rapidly improved will yield a more secure system, or whether one that is closed to a single vendor's developers is more secure. The debate is similar to the open-source Linux vs. Windows debate regarding security, and is far from being settled and may never be settled.

Any serious provider of Human-Machine Interface software offers web-browser based user interface options to their systems with the objective of providing information from the manufacturing control system anywhere. With the improvements in web browser performance, graphics, and the proliferation of browsers on everything from cell phones, to PDAs, to tablet computers, the options available to "see it in a browser" have grown exponentially since about 2008. Technologies such as HTML5, Flash, and Silverlight are capable of delivering the same type of rich, interactive user experience to a browser that used to require a full computer and software application to be installed.

Cloud computing is the most current trend that has arrived in the manufacturing automation business and is available to any facilities automation application today. Cloud computing is based on the idea that computing resources are becoming like electric and water utilities. The concept depends on universal access to the internet, as reasonable speeds. The idea is that rather than running applications on your own computer hardware, and having to maintain, secure, support, and manage the hardware yourself, you run your applications on a remote system, only paying for resources as needed and elastically being able to get more resources on demand in exchange for more money. Private Cloud Computing uses Virtual Private

Servers (VPS) which are simply fully dedicated instances of virtual servers running on someone else's hardware, that you access over the internet or "in the cloud," providing all the same benefits of a dedicate computer without the headaches.

Vendors in the automation industry are offering applications in either public cloud based applications for securely viewing industrial data in a browser, even developing your screens in a browser without any other software located on your PC, or private cloud based solutions that can run in your virtual private server, delivering a rich, Silverlight graphics experience to the user, along with full screen development capabilities without installing any software on the user or developer's PC.

Regardless of whether you choose a public or private cloud based solution, facilities automation users have the option to deliver a rich, secure, experience to users anywhere, without having to worry about buying, managing, and supporting the infrastructure on their own.

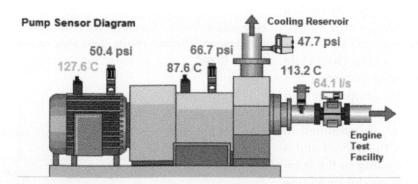

Figure 27-2. Example of cloud based, browser delivered machine interface, courtesy of IndustrialFalcon.com

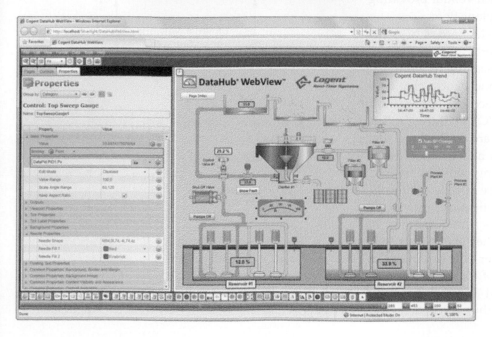

Figure 27-3. Example of a rich, Silverlight, browser based user interface delivered using a private cloud based application, the Cogent DataHub® WebView™.

CONNECTIVITY SOFTWARE AND WIRELESS TECHNOLOGIES IN MANUFACTURING CONTROL AND FACILITIES AUTOMATION

Connectivity Software for Manufacturing Controls and Networks

OPC Standards

Software for connectivity in the manufacturing control world can be handled in several ways, but the most popular way is using software that adheres to the OPC standard. OPC today stands for "Openness, Productivity and Connectivity." The OPC standard is an open standard for software-to-software connectivity that was first created in 1996 by a group of hardware and software vendors serving the manufacturing control business who came together to form the OPC Foundation. The OPC standard is managed by the independent OPC Foundation (www.opcfoundation.org). The OPC Foundation's primary mission is to foster open interoperability in automation software.

Today, the OPC standards comprise a wide range of specifications for exchange of different types of information between automation/manufacturing control software. When the OPC Foundation was first started in 1996, the first standard created was the OPC Data Access (OPC DA) standard. Prior to the OPC Data Access standard, every company that provided applications that needed data from the shop floor had to write their own communications software for the myriad of device types. (Figure 27-4). This resulted in a great deal of excess work done in the market and a wide variance of functionality and quality in the communications software offered by each vendor providing software that interfaced with the plant floor.

The purpose of the OPC DA standard is to create a way in which communications software ("OPC Servers") can be created that communicate with wide and varied types of control networks and expose that data through standard software interfaces to applications that need to consume the data ("OPC Clients"). The primary objective is to allow software applications that support the OPC Client interfaces and need access to read/write data over manufacturing control networks to be agnostic relative to the device specific protocols. (Figure 27-5) The OPC client applications only have to be aware of tag names and have a standard way of reading and writing the data to the OPC server software applications. The OPC Server software applications handle all the requirements of the device specific network topologies, hardware, and addressing.

The result of this for users is they can obtain their connectivity software from any vendor that is able provide the solution that meets the user's requirements. For developers of OPC client applications, they can exit the business of writing connectivity software and let companies solely focused on connectivity provide the OPC server's software. For users, they are no longer limited to the connectivity offered by the OPC client application developer.

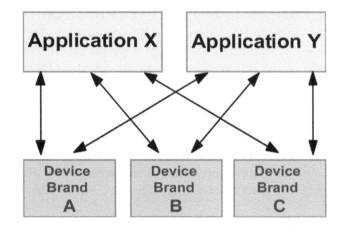

Figure 27-4. Device Connectivity Without OPC requires each Application Vendor to write their own drivers to talk to each device brand

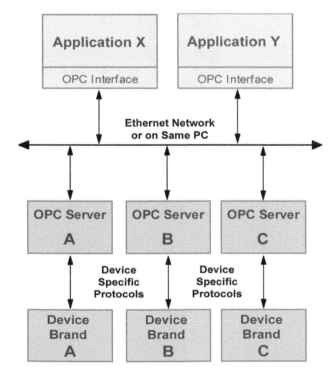

Figure 27-5. Utilization of OPC for Device Communications Separates Communications details from the Application

When talking about OPC compliant applications, the terms "server" and "client" are often heard. These are not hardware terms. These terms speak to the nature of the data handled by the OPC compliant software. In the world of OPC a "server" application provides data (Figure 27-7). It speaks a vendor specific protocol through one interface, and it serves the data to applications that need the data through it's OPC standard software interface.

An OPC client application (Figure 27-4) is one that wants data from a device or system. The OPC client gets its data from a server application. The same logic applies when writing data. An OPC client wants to write data to a device and it tells the server to perform the write on its behalf. It is the OPC server's job to perform the write of data to the device or system and report back a success or failure.

After OPC Data Access, the Foundation moved on to address the exchange of complex data, historical data, batch process data, and alarms and events data between software applications

Today there are over 300 member companies in the OPC Foundation offering OPC compliant products. Member companies are able to obtain OPC certification for their products through a series of tests and interoperability workshops that strive to give users assurances that products carrying the OPC-certified logo can deliver on the promises of interoperability in automation. The OPC Foundation maintains a catalog of OPC-certified products on their website at www.opcfoundation.org.

The current state-of-the art of OPC has fully embraced the use of internet and web technologies in the OPC Unified Architecture (OPC UA) standard. OPC UA took all of the various first generation standards for Data Access, Historical Data Access, Alarm & Events, etc and has unified them into a common standard. Redundant code and specifications were eliminated, and rich information models are now possible using OPC UA.

OPC UA also moved to a completely web-services oriented architecture. Rather than relying on Microsoft COM and DCOM, they now ride on a standard TCP/IP Ethernet stack and are able to leverage a variety of application layer transports on top of the Ethernet connection for inter-software communication. By removing the single operating system dependency, OPC UA is now able to run on multiple operating systems and in lightweight embedded devices. OPC UA is starting to even appear at the device level in some situations. Because OPC UA uses web services, it can run on any TCP/IP port number, and communications can be secured using standard SSL certificate encryption. For geographically distributed business, information interchange within the business is simplified because OPC UA can run over a single TCP/IP port, which was not possible without third-party software applications when using the original OPC standards.

For users of existing OPC software, migration to OPC UA has been simplified in three ways. First, the basic concepts of reading, writing, and subscribing

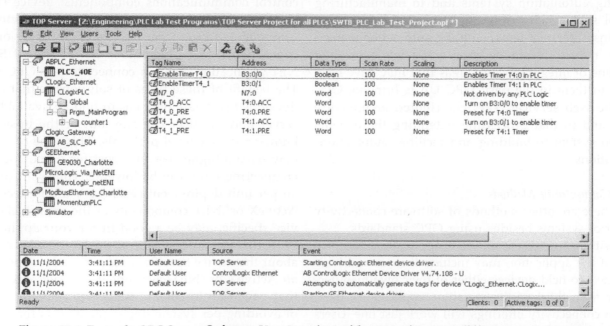

Figure 27-6. Example OPC Server Software User Interface with connections to 6 different control system types plus a simulation interface for testing.

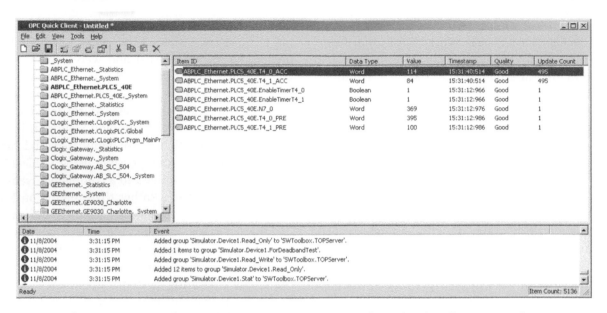

Figure 27-7. Example OPC Client Software User Interface Showing live process data

to information remain the same. Second, a variety of toolkits are available to speed developers adding OPC UA to their products. Third, and most importantly, a variety of OPC UA to DA "convertor" software applications are available that make it point-and-click to wrap OPC UA around an existing OPC DA product without having any access to the existing product source code.

For the building automation system manager, the relevance of OPC is that they can leverage the open connectivity options offered by the over 300 companies offering OPC solutions to provide connections to their building automation systems and to manufacturing control systems. By utilizing OPC standards based solutions in their systems, building automation managers can reduce their integration cost while expanding the scope of integration into areas that may have been closed to them in the past. OPC UA is bringing OPC into the web services, service-oriented architecture world and in that processing extending the benefits of open system to building and facilities automations applications.

Other Connectivity Methods

There are other methods of software connectivity to control systems besides using OPC standards.

Within any given software application, the provider of the application may include their own drivers to connect to field devices either of their own manufacture or from third parties. In some applications, these built-in connectivity offerings will work just fine. Users should of course investigate these offerings with their own requirements in mind though to insure that they

can get to the data that the user needs, at the data rates required, and in the volume of data required. There are some cases where connectivity software that is provided with an application is done to meet a basic set of requirements, but advanced functionality is not supported.

For the user looking to develop their own applications using tools such as Visual Basic (VB), Visual Basic.NET or Visual Studio, there are tools available in the form of ActiveX control software or .NET components that can handle communications to control system devices. Generally there are two classes of these control communications components: device specific or OPC connectors. The latter, OPC connectors are components in the form of ActiveX software or .NET components that plug into a VB or Visual Studio. NET application and can connect to any OPC server. The benefit of these types of solutions is that custom developed software only has to find an available OPC server and the user can stay out of the business of knowing device level protocols. This type of solution may have a higher per unit cost for deployment, but engineering costs can be lower. If you want to avoid all per unit deployment costs (i.e. runtime fees), then ActiveX or .NET components of the second class, device specific, may be a good fit for your applications. With these components you do have to know more about the device specific protocols, and you will need an ActiveX or .NET component for each device type that you wish to communicate with. For the savings of avoiding per system deployment costs for software licenses, you assume more responsibility for device level communications.

Web Connectivity

Connectivity to building automation systems and manufacturing control systems using web/browser based technologies has taken many leaps forwarding in the last few years. Until recently, displaying process data in a web browser could be done, but had its issues.

The most common method of displaying live data in a web browser was to have the web server read the data from process control systems and deliver it to the browser as a static value. To update the data, the user would have to refresh the entire web-page. Other solutions could deliver data to the browser and have the data update without a page refresh, but required you to install software on the client PC or the use of ActiveX controls or Java Applets in the web browser. Current security measures in companies could often prevent these small pieces of software from loading in the browser, thus preventing these types of systems from working.

In the last couple of years, the use of XML and web services has changed the way data can be delivered to a web browser. A web service is an intelligent application that runs on a web server and the web browser can load a web page that will automatically pull new data from the web service without refreshing the entire page. As long as you have Internet Explorer 5.5 or higher on a Windows PC, you can consume a web service from a web server.

There are now software products available that will run on a web server, connect to OPC server software on the web server, and allow you to publish process data in a way that users can see the live process data in a remote web browser, updating, without having to refresh the entire web page or load software on the client PC. One example is the OPC Web Client product (www.opcwebclient.com and Figure 27-5).

The OPC Web Client software is loaded on a web server, talks to local or remote OPC servers, which in turn talk to building automation systems or manufacturing control systems. The system implementer builds web pages that access the OPC Web Services provided by the OPC Web Client to display the live data in the users web browser. The user simply points their browser to web pages on the server and they are able to read and write data from their browser. (Figure 27-6)

Wireless Technologies in Manufacturing Control Systems

There has been a growing trend towards the use of wireless technologies in manufacturing control systems. There are two areas where wireless technologies are typically used in manufacturing control systems:

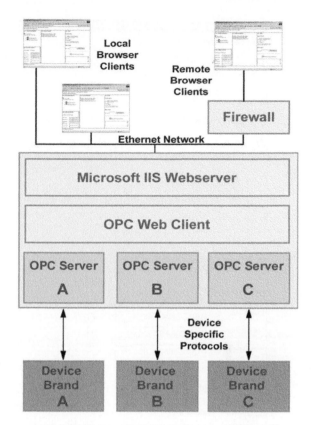

Figure 27-8. OPC Web Client System Architecture Overview

in-plant and remote site communications.

In-Plant Applications

With in-plant applications of wireless, the objective is typically to avoid the costs and time delays involved in running network cable to reach a remote device. In some cases, it may not even be practical to run additional cable. Using off-the-shelf wireless technologies used in commercial information technology applications, plants can quickly create Ethernet connections from one part of a plant to another.

Typical ranges with in-plant wireless technologies can vary greatly depending on the physical site. Wireless routers and access points are available that advertise 108 Mbps or higher speeds at up to 400 ft and in some cases longer; however, the range can drop dramatically as signals have to pass through walls and other obstructions. The more obstructions that exist from the device to the nearest wireless access point, the weaker the signal and thus full speed may not be reliably achieved. Some manufacturers offer the ability to include repeaters or additional wireless access points to help with physical obstructions. Users planning to deploy wireless should carefully test signal strength by placing routers at proposed locations and then us-

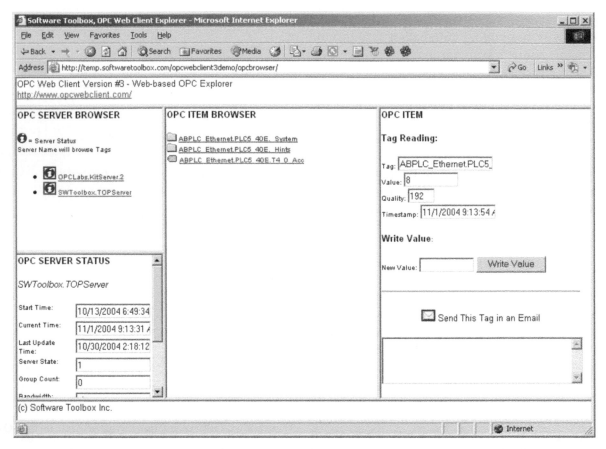

Figure 27-9. Web Browser displaying data from OPC servers using OPC Web Client

ing a portable laptop computer with a wireless access adapter to identify the practical bounds of their wireless network.

When implementing wireless networks, one must also be careful to not assume infinite throughput. It is easy to think "oh, this is 108 Mbps; I'll never use that much bandwidth." As previously discussed, the rated speed of wireless hardware is the maximum speed under proper conditions. Obstructions and interference can easily drop the practical speed on the wireless network. Also, wireless networks are not necessarily switched networks—which means in some cases with multiple devices talking concurrently, collisions can occur, just like on non-switched Ethernet, which means the more traffic on the network, the greater the risk of degradation of performance. Wireless can perform well if applied carefully and not used as if it is a connection of infinite bandwidth.

Security is a very important consideration in implementing commercial wireless technologies. Users should NEVER connect a wireless router or access point to their plant network without first at a minimum changing the default password for access to the router.

Commercial wireless hardware typically ships in a non-secure state, with a default password. If a wireless router or access point is installed without changing the default password, an intruder can easily find the device on your network and access the login screen. Nearly all wireless hardware "advertises" the model and make on the login screen. (Figure 27-10) An intruder can go to the internet, download manuals from the manufacturer, obtain the default password, and take over control of the wireless network. By changing the default password, you prevent this.

An additional security measure that should be taken is to change the default wireless network name in the router and tell the router to not broadcast it's network name for discovery. This can help thwart an intruder finding the network. Next, you should consider only buying wireless hardware that supports encryption and use the highest level of encryption supported by your hardware. When configuring the encryption setup in the hardware, never use pass phrases that are included in manuals or tutorials in the hardware documentation. Create your own pass phrases that are long and alphanumeric in nature, just like you would

Figure 27-10. A NetGear Router Exposing It's Model Number through the login screen

create a secure password for your login to a network. Never write passkey codes or other such information on the hardware as a means of convenience under the assumption that intruders won't actually see the physical network hardware. Intruders many times can come from within the facility, not just random outsiders.

Manufacturers of wireless hardware include Netgear, D-Link, Belkin, and Linksys. There are also numerous providers of industrial grade wireless hardware including Black Box, B&B Electronics, Digi, Atop Technologies, and others.

Remote Site Wireless

For remote site applications, traditional IT technologies discussed previously are not a good fit because of their distance limitations. There are a number of emerging technologies for inter-building or inter-campus wireless connections that promise up to 30 miles line of site connectivity. Users needing remote site connectivity should stay abreast of what new technologies are being adopted in widespread fashion in the commercial IT world that may be applied to facilities or manufacturing automation applications.

Presently, wireless technologies used for remote site applications are either private radio systems or public cellular based systems. Only recently have public cellular based technologies reached levels of performance and cost to make them practical alternatives to private radio networks.

Private radio networks have the advantage of being under total control of the user and can provide very high levels of security since everything is in the user's control. However, private networks can be costly to set up because towers must be built, licenses potentially obtained, and the entire network must be managed and maintained by the user. The water treatment industry has typically taken this route because they had large

numbers of remote sites to manage and the economics of managing the network made sense for them.

Presently, the CDMA and GSM cellular technologies can offer up to 1 Mbps or higher download speeds, increasing it seems with each jump in the generation of cell phone technologies. The advantage of public cellular networks are that someone else bears the capital costs of setting up the network and the ongoing maintenance costs. You simply pay a monthly fee for each device that has access to the network. Plans vary in cost based on whether you are using metered amounts of data or want unlimited data volume. Depending on your application, the unlimited volume plans can be the most economical solution as most plans are less than $50 per month and discounts can be obtained for multiple device contracts. Obviously local pricing will vary and it is important to work with local carriers to negotiate a solution that meets your needs. The downside of a public cellular network is that you do not control the network. This means that you are sharing the network capacity with others and there are greater security risks. The security risks can be mitigated with proper use of firewalling software and hardware. You should discuss capacity and security concerns with local carriers based on your requirements.

Regardless of the wireless technology chosen, it is important to insure that the communications software you choose can work well with the wireless technologies you choose. You should integrate the selection of communication software with your wireless technology search as the software provider may have vendors of wireless hardware that they have found work better than others. The communications software must also offer the flexibility to deal with the additional delays in transmission time that wireless technologies for remote locations can impose. If you are using public cellular radios, your radios will likely have network IP addresses that can change. The radios must be able to report their new IP address to the host computer automatically and the communications software (i.e. OPC server software for example) must be able to receive those reports and handle the situation appropriately.

Now let's look at some case studies that utilize the various topics that have been discussed in this chapter.

CASE STUDIES

The case studies presented in this chapter provide examples of how manufacturing control systems and technologies have been integrated with and interoperate with energy or facilities management systems. Some of the systems are web based, others are not. Regardless, the value to the reader is in generating creative thought about the many possibilities that may exist in their own business for integration and technology sharing.

All case studies are based on real-world user applications that are currently in operation. For privacy and company confidentiality reasons, the exact names of the subjects in some of the case studies are omitted.

Case Study: Gardening supplies manufacturer

Technologies Used:

- OPC Server Software
- Off-the-shelf web server
- Programmable Logic Controllers (PLCs)

A major provider of supplies to plant nurseries in the United States has facilities in several locations around the country. Their maintenance manager needed a way to access operating conditions and predictive maintenance information from the sites quickly and effectively over the internet. The systems they wished to connect to were all controlled by GE Fanuc Series 90 Programmable Controllers (PLCs). Connection options for each PLC were serial network or Ethernet.

A key design objective in building the system was the ability to access the system information anytime, anywhere, without having to load any software on a client PC. The system also needed to be cost effective and easy to implement.

The customer evaluated a number of off-the-shelf human machine interface solutions that could also offer internet connectivity to view the operating data. Solution costs ranged from $10,000 to $20,000 total, with purchased software license costs of around $10,000 included in the total. The customer decided to also evaluate solutions that would require some engineering work on the part of the customer and utilize open standards and technologies to allow the user to build their monitoring application on their own.

The customer chose to utilize a solution using OPC Server software to connect to the GE Fanuc PLCs over an Ethernet network and an ActiveX component that would interoperate with Microsoft Internet Information Server (IIS) to display the operating data from the OPC server in custom web pages designed by the user. Using Microsoft Front Page as a web page editor, and utilizing example web pages provided with the products chosen, the maintenance manager was able to build their application in approximately 1 week. The total out-of-pocket software cost was $1500.

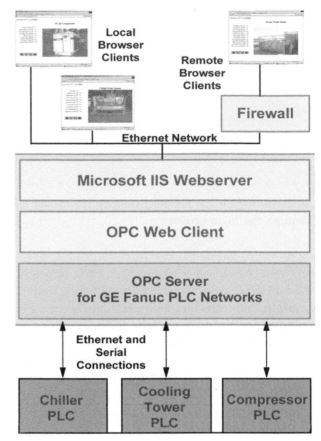

Figure 27-11. Web Based Facilities Monitoring Application Architecture

The system implemented monitors control systems on critical facilities equipment such as ventilation, chillers, air compressors, tower cooling units, pneumatic air dryers, and air reservoirs. As a result of installing this system, the customer has seen efficiency gains of nearly 15% and their annual maintenance budget is at an all time low. The customer attributes the improvements to "Knowing what's going on in our critical facilities systems at anytime from anywhere. With this system we literally feel like we can be in two places at one time. The knowledge we have obtained from our application allows us to make intelligent, informed decisions about what systems need maintenance and when. We are better serving our customers now with improved on-time delivery of products, reliability, and product quality."

The customer chose the OPC Web Client because it provided data in a browser that would update without refreshing the page, did not require any software to be loaded on client PCs, and was easy to implement. The OPC Web Client has allowed the customer to access equipment with the click of a button anywhere on site, corporate wide, or even from home. He no longer

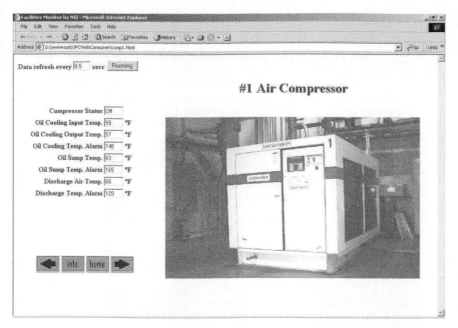

Figure 27-12. Example Screen Monitoring Air Compressor Status

has to wait on reports from facilities because he can see what's going on with critical equipment by clicking on our maintenance web site.

Products used in this application:
- TOP Server OPC Server—www.toolboxopc.com
- OPC Web Client—www.opcwebclient.com
- Case Study: Supermarket Facilities Management and Haccp Compliance

Technologies used:
- I/O Server Software
- Industrial Human Machine Interface Software
- Visual Basic
- Internet remote access

In this application, Ukrops, a major supermarket operator in central Virginia sought to implement a monitoring and management system in their stores that would provide a standardized user interface from one store to another, provide local and remote monitoring, and easily integrate data from all facilities infrastructure in the store. The new system needed to be intuitive, utilize open systems, and easily integrate with other facilities systems without 100% dependence on a single vendor.

The types of systems monitored in each store include cold storage systems in the stock room, generators, HVAC units, refrigerator and freezer cabinets on the store floor, lighting, and facility temperatures. Each

type of system was often provided by a different company with their own means of connectivity or lack thereof. In some cases, the manufacturer of the equipment, for example refrigerator cases, would provide their own proprietary software applications for the monitoring of their equipment and their equipment only. Or they would provide a closed system that could only be expanded to integrate other systems by purchasing additional hardware and consulting services from that manufacturer.

Also, Ukrops needed to implement an automated means of tracking compliance with government Haccp regulations governing the tracking of handling of food products. Haccp stands for Hazard Analysis and Critical Control Point, and is pronounced "hassip."[1] Haccp regulations require grocers to track when conditions exist that could cause hazards in food, log corrective actions taken, and the amount of time the hazard existed.

Over the years prior to this project, Steve Little of Ukrops had been working with Edward Stafford of Electronic Technologies Corporation (ETC) of Street, Maryland to implement a number of systems for management of the facilities. The existing systems utilized the latest technology at the time, and served their purpose, but over the years, the team had identified shortcomings they wished to overcome in a new system.

The existing systems were text-based systems that required training for use and operation. When a problem occurred, only a trained maintenance person or operator could get details on the alarm condition and take action. The result of this was ongoing training costs as personnel moved from store to store, unknown impact on stock spoilage due to lack of response to alarms, and lack of management visibility into operating conditions. If a store manager or regional manager wanted to find out what current conditions were in a store, they had to go to a trained maintenance person who could gather the data from the monitoring system.

With these limitations in mind, a solution was developed that would meet their requirements for their next generation monitoring system. The new system is based on an off-the-shelf Human Machine Interface software package named InTouch, provided by Inven-

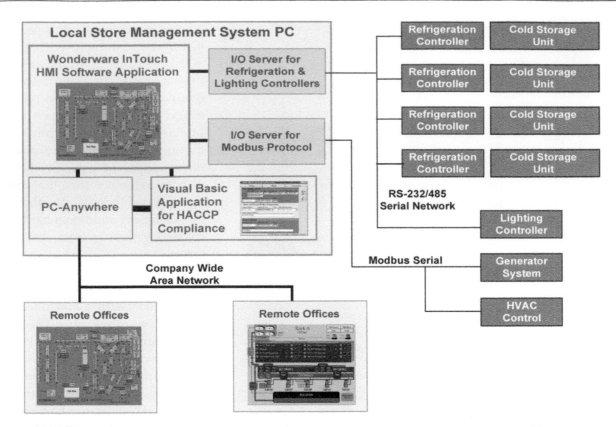

Figure 27-13. Supermarket Facilities Management & HACCP Compliance System Architecture

sys Wonderware and their local representative InSource Software Solutions. The InTouch software package provided ETC with the ability to build user friendly graphical screens to display operating data and open,

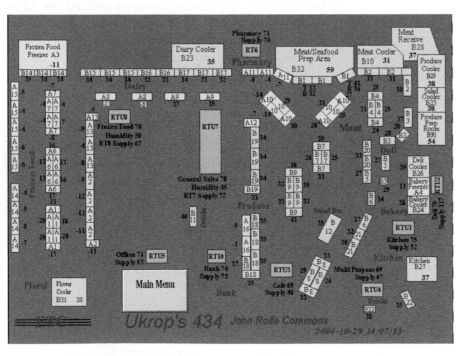

Figure 27-14. Store Overview Graphic in the InTouch HMI

standards based methods of gathering data from the various facilities systems. (Figures 27-14 and 27-15) InTouch provides an open OPC client interface for connecting to OPC servers for data sources. InTouch also provides an open Modbus Serial and Ethernet connection that is used to connect to generators and some of the other systems in the store.

Part of the integration challenge in this application was connecting to the refrigeration controllers. The controllers in each refrigeration case on the store floor were connected over an RS-485 network using a protocol developed by the refrigeration controller manufacturer. The refrigeration controller manufacturer offered their own user interface software package for gathering data from their systems and to some degree, integration with other systems. There were no "open" communications software drivers available for connecting to the refrigeration controllers.

ETC worked with Software Toolbox Inc. of Charlotte, NC to de-

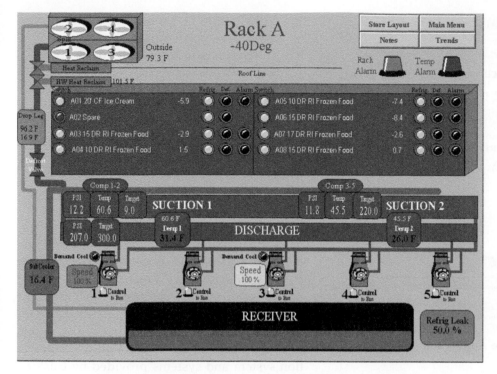

Figure 27-15. Refrigeration Rack Detail Screen in InTouch

velop an I/O server software application that would communicate to the refrigeration controllers. ETC was able to procure from the refrigeration controller provider the necessary protocol information for the controllers and test hardware for development of the I/O server. Software Toolbox provided a driver for their TOP Server OPC and I/O Server application that implemented the necessary protocol for communicating to the refrigeration controllers.

In this application, the system takes advantage of a native interface offered by the InTouch Industrial HMI system called Suitelink to connect to the I/O server instead of OPC because the Suitelink connection in InTouch is easier to configure than InTouch's OPC interface. However, by developing the communications software so that it could also connect to other applications using the OPC standard, ETC and Ukrops could be assured they could change the Human Machine Interface software application later or connect other applications to the OPC server software later without writing custom code. Essentially, Software Toolbox provided a means to take data from the closed system and expose it to any software application using an open standard.

Development of the connectivity to the refrigeration controllers was not without its challenges. Close coordination with the hardware manufacturer was required because during the design process, it was determined that the refrigeration controller's communi-

cations protocol did not provide all the functionality needed to meet the supermarket operator's requirements. The refrigeration controller did not allow digital points in the controller to be turned on and off from a remote connection through the device's serial port unless you were using the vendor's proprietary software. Ukrops and ETC worked with the supplier of the hardware to persuade them to add the functionality to the communications protocol available for external connections to the controller. By upgrading the firmware in the refrigeration controller units, Ukrops and ETC were then able to perform the desired writes to the system from the InTouch HMI and I/O Server software. This application illustrates the importance of working with hardware companies who will be responsive and cooperative when the users of their hardware wish to interface to their systems with other applications. Without the protocol documentation and cooperation of the hardware vendor that was secured by ETC, Software Toolbox could not have provided the software required by the application.

To implement the Haccp tracking requirements, ETC developed an application in Visual Basic (VB). (Figure 27-13) The application connects to InTouch through an open DDE interface on the InTouch application and monitors for new alarms. When an alarm occurs, such as a high temperature in a freezer cabinet, InTouch displays the alarm on screen, and the VB application automatically creates a database record with the date/time/location of the alarm and current condition, and the currently logged in operator. When the operator acknowledges the alarm, he is required to input the corrective action taken, any temperature measurements taken from the food, when the problem was resolved, and if the food was moved to temporary cold storage to prevent spoilage. Operators cannot dismiss an alarm until the necessary Haccp information is entered into the system.

Remote access was implemented using PC-Anywhere operating over the supermarket operator's Local Area Network (LAN) in store and across their Wide Area Network (WAN) through the region. Management

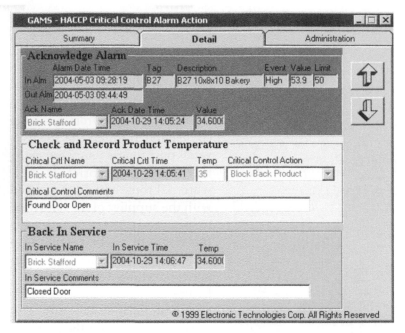

Figure 27-16. HAACP Compliance Application User Interface

at the central office can connect to the systems in each store across the network using PC-Anywhere and see the operational screens the same way as if they were physically present in the store.

With the rich graphical user interface provided by Intouch and the connectivity provided by the TOP Server I/O server, Ukrops now has a system where anyone in the store who has had basic training can view alarms, react to them quickly to take corrective action, and notify maintenance. Store managers can go to any store in the regional chain and find the same system in place and know how to use it. Since store managers move about once a year from one store to another, this consistency of operation is important to the overall effectiveness of the managers. By reducing reaction times to problems, spoilage has been reduced, increasing the profitability of the stores.

Because the system is an open system, stores can easily add additional equipment to the displays and make changes to the displays. By using a device connectivity solution that offers both native interfaces to their current HMI and also an open standards based interface using OPC, stores are not locked into a single software application for the display of the data. This application is presently in use in 25 stores in central Virginia.

Products used in this application:
- TOP Server OPC and I/O Server—www.tool-boxopc.com

- OPC and I/O Server development services—www.softwaretoolbox.com
- Invensys Wonderware HMI Software—www.wonderware.com
- Integration and Development by Electronic Technologies Corp, Street, MD, Edward Stafford.

Case Study: Integration of Chilled and Hot Water Production and Delivery Systems

Custom-Flo Incorporated of Cincinnati, Ohio is a maker of skid-mounted pumping systems used in chilled and hot water production and delivery. Custom-Flo is often responsible for custody and delivery of chilled water and/or hot water in their customer's facilities. The facilities automation manager is responsible for the final end use of the resources. As a result, tight integration between the building automation system and systems provided by Custom-Flo is a standard requirement.

Custom-Flo has standardized their controls package for their skid-mounted systems on a PC-Based Control platform for over 4 years. Custom-Flo chose this platform because their customer ends up with a non-proprietary hardware and software solution. If a customer in the future changes direction or needs, he doesn't have to be tied to Custom Flo, he can go forth on his own. The customer can also upgrade to the next generation of PC technology and port the application up to the next revision without having to rewrite the application. The totally open solution offered by Custom-Flo also has saved their customers significant amounts of money whenever their requirements changed or new systems were added.

This case study covers two areas:
1. Pumping system/balance of utility plant control system
2. Integration with building automation systems using manufacturing control technologies.

Pumping System Control System

The control system provided by Custom-Flo on their skid mounted systems consists of several parts.

Human Machine Interface (HMI)—Custom Flo uses Advantech Web Studio, an off-the-shelf, user configurable human machine interface application. With this application, Custom-Flo provides a rich graphical user interface in a package that is easily modified later without custom programming. All projects installed by

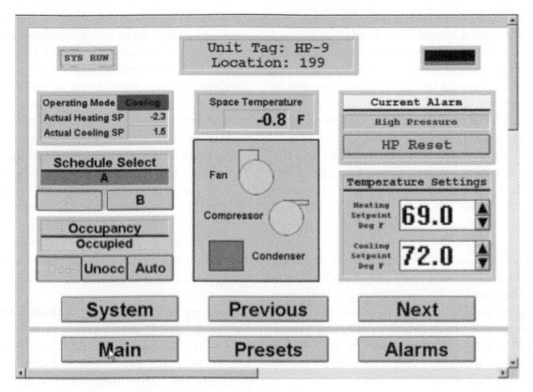

Figure 27-17. Operator Screen from the HMI System on the Skid

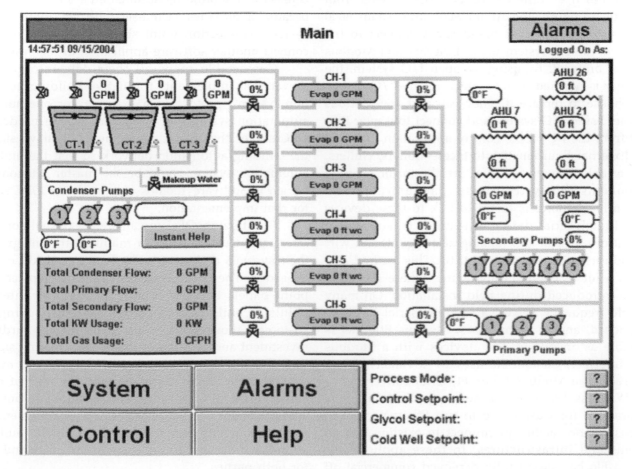

Figure 27-18. Overview Screen

Custom-Flo are web enabled for remote access. The remote access allows Custom Flo to connect through a secure port on the company's network or dialup to support their systems and users, and it allows the users a means for remote access to the systems provided by Custom Flo. In most installations the systems provided by Custom-Flo will be outside the company firewall but have their own router to secure them.

PC-Based Control—Custom Flo uses an IEC 1131 based PC control package, MultiProg Embedded Soft Logic Control, provided by Advantech. This system operates on Advantech TPC-1260 Industrial PCs with display built in so they can run the HMI and SoftLogic on the same PC. The PC-Based control package uses a real-time kernel for the control logic so the Windows operating system can crash, but control will continue to function. Custom Flo believes strongly that users implementing PC based control should not cut corners on the hardware. They have found that by implementing robust, industrial grade PCs, they may pay more up front, but they see a return quickly in overall system uptime and reliability.

Ethernet Based I/O—For remote and local I/O, Custom-Flo uses Ethernet based I/O systems from Advantech, mainly the Adam 6000 and Adam 5000 systems. These Ethernet based racks connect to the PC based control system using Ethernet and Modbus Ethernet protocol. By using an open I/O system, the customer is not tied to Advantech for the I/O if they ever choose to replace it. They can add any Modbus Ethernet based I/O system and connect to it over Ethernet from the PC-Based Control application.

External Equipment Interfaces—Boilers, chillers, and variable frequency drives are the major external components the system must communicate with. Custom-Flo's choice is to try to talk to these devices via a digital/serial data link because it reduces the amount of physical I/O (analog and digital) and wiring and increases the system MTBF. This method also avoids a lot of conduits and wires where problems can be introduced during field installation. On many variable frequency drives, a Modbus serial interface is standard, and in many cases, Modbus Ethernet is available as an option. For devices with a Modbus serial interface, Custom Flo uses an Advantech Adam 5000 rack as a Modbus Ethernet/serial converter.

Wireless Ethernet—in some applications, it is not practical for Custom-Flo to run Ethernet cable in the customer's facility to connect to existing control systems and infrastructures. In these applications, Custom-Flo has successfully deployed commercial off the shelf IEEE 802.11b and 802.11g wireless Ethernet networks. They have used wireless Ethernet to connect to their own Ethernet based I/O systems and to other operator interface nodes in their HMI systems. Wireless has also been used to facilitate an Ethernet linkage to existing building automation systems. When implementing wireless networks, Custom-Flo takes advantage of all available security measures in the wireless routers and works closely with the customer's IT department to insure a secure network is implemented.

By using an open system for the control system, Custom Flo provides their customer with investment protection and flexibility. Where other vendors charge significant amounts for system changes or adding additional monitoring points, Custom-Flo is able to make changes quickly, easily, and at low cost for their users.

Interfacing to the Building Automation Systems—No Hardwired Point Panels

When Custom-Flo connects to building automation systems at their customer locations, they first try to utilize the native network interface offered by the building automation system. Custom Flo prefers to connect to the building automation system using OPC because it provides their user with yet another open systems connection point should they ever need to connect another software application to the system.

Custom Flo maps all their data into an OPC Server that implements the native building automation system's protocol. For example, for one building automation system network, Custom-Flo needed an OPC server that would implement the slave side of the building automation network so that the building automation system could be the master. Custom Flo turned to Software Toolbox to provide the OPC server for that implementation.

By utilizing OPC for all their external software interfaces, Custom Flo can maintain the integrity of their systems and cope with changes quickly on site if needed. The OPC Server software has replaced the point panels that are commonly found at interface points in building automation systems, saving the customers thousands of dollars in up front hardware investment and installation costs, and creating significant flexibility for future system changes and expansion. Custom-Flo works to standardize a set of point names in the OPC server for each building automation system vendor, so that once they have implemented a few systems with an interface to a particular building automation system, future systems are quick and easy for both parties.

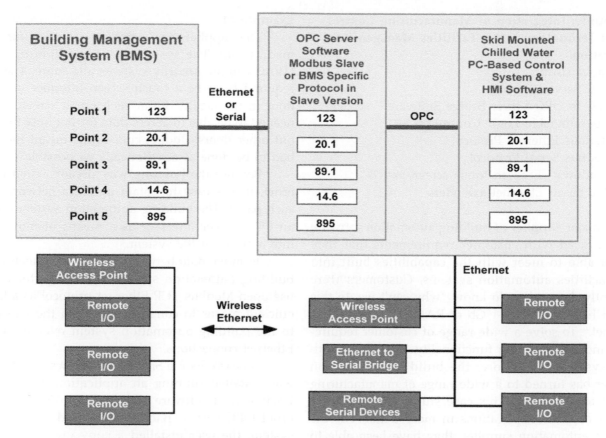

Figure 27-19. Building Management System to Skid Mounted System Interface Overview

Example—Airport Facilities Management:

In a commercial airport, Custom Flo supplied 6 hot water systems in 6 concourses and interfaced them to a building automation system using an OPC server supplied by Software Toolbox Inc. After the systems had been running for a few years, the operator of the airport decided to replace their building automation system with one from another vendor. The change in the interface for Custom-Flo simply involved changing the OPC servers from one using the native building automation system protocol to Modbus Slave OPC servers from Software Toolbox. The changeover of the interface was done in one day. There were no costs associated with replacing or rewiring point panels, or hardware involved in the interface change with the Custom Flo systems.

General advice to users implementing open systems:

During the research for this case study, Custom-Flo felt it important to emphasize some key points to users considering implementing open systems in facilities automation:

- Do not cut corners on PC hardware cost
- Anyone can purchase the pieces needed to imple-

ment an open system; however, it takes a competent, experienced controls engineer or integrator to implement the system and get the best return.

- The best way to evaluate the competency of an integrator is through references.
- If you choose to implement wireless Ethernet, you must realize that it too has a capacity limit, and that you should evaluate your data update rates and network utilization needs when implementing. 11 Mb/s or 54 Mb/s may sound like a lot of capacity, but if you take the "oh read everything that's there as fast as you can" approach, you can easily overload a wireless network.

Products used in this application:
- TOP Server OPC Server—www.toolboxopc.com
- OPC Server development services—www.softwaretoolbox.com
- Advantech Industrial Computers, Human Machine Interface Software, and PC Based • Control Software—www.advantech.com
- Sample applications online—http://www.custom-flo.com

Case Study: Integration of Manufacturing Control System Technology into a Facilities Management Application

Technologies Used:
- OPC Server Software
- OPC to OPC Server Bridge Software
- Programmable Logic Controllers
- Modbus Ethernet Protocol
- Modbus Serial Protocol
- Windows CE based touch screen panels
- DSL based Ethernet extenders

A major supplier of building automation software is often faced with customer requirements that they are not able to meet with the capabilities built into their facilities automation systems. Customers were frequently demanding to know "why can't we do this when I have a 3 GHz, 1 Gb of RAM PC on my desk at home!." To solve a wide range of customer requirements including adding functionality, integration with other systems, and more, the building automation supplier has turned to a wide range of manufacturing control technologies. This case study will explore a number of these applications in more detail. For the building automation supplier, they have been able to meet customer requirements their competitors could not meet, save their customers money, and provide their customers with more flexibility than they otherwise would have obtained from a traditional building automation system.

Example #1

The supplier was installing a system for a large medical lab. The system managed lighting control, HVAC control, security systems and more. The project required there be a touch screen interface on each of floor in the facility. From each touch screen, the user needed to be able to access data for not only their floor, but other floors. Budget constraints meant the system had to be done as economically as possible.

Because the building was already wired for Ethernet, it was easy to obtain an extra network jack at each panel. The building automation system provided an OPC server interface as a means of moving data into and out of the system.

To move data between the remote panels and the building automation system, the integrator chose to use open Modbus TCP Ethernet protocol as a fast and efficient means to communicate from the panels back to the building automation system over the existing Ethernet connections.

On each floor, a Siemens Windows CE Multipanel was installed running an application using Siemens ProTool HMI software and an INGEAR Modbus Ethernet OPC server. Back at the building automation system, the user installed a copy of the TOP Server Modbus Ethernet Slave driver and the Linkmaster OPC to OPC bridge. Linkmaster maps all necessary data from the building automation system to the Modbus Ethernet Slave. Each remote panel reads/writes data to/from the Modbus slave using the INGEAR Modbus

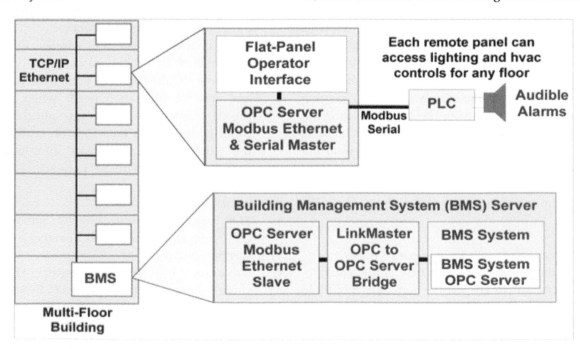

Figure 27-20. Medical Lab BMS to Remote Panel Interface System Architecture

Ethernet OPC server.

So how does it work? Let's look at some examples. Remember, all these steps are happening in sub-second time over switched Ethernet.

Writing Data from the Touch Screen to the Building Automation System

When the user wants to write a value to the building automation system, the HMI software writes the data to the INGEAR Modbus Ethernet OPC server, which in turn writes the data to the TOP Server Modbus Ethernet Slave. The Modbus Ethernet Slave automatically notifies Linkmaster of the new value, and Linkmaster in turn writes the data into the building automation system's OPC server.

Reading Data from the Building Automation System and Displaying it on the Touch Screen & Sounding Alarms

When values change in the building automation system, the building automation system OPC server automatically notifies Linkmaster of the new value. Linkmaster notified the TOP Server Modbus Ethernet Slave of the new value. The INGEAR Modbus OPC server on the Touch screen is periodically polling the slave for new data—as soon as the new data is received, the INGEAR OPC Server notifies the HMI application which updates the users screen.

In addition to the interface to the building automation system, the application also had a Siemens micro S7 connected to the touch screen via its RS-232 port. The system uses the PLC to drive a local annunciator alarm/whistle for the customer. The Ingear Modbus

OPC server on the panel is used to interface to the S7 PLC through an available Modbus interface on the PLC. The application sets PLC alarm tags from both Vbscript in the panel, as well as tags up in the building management system OPC interface. This solution was chosen because it overcomes the lack of an alarm device on the particular touch screen chosen. Through use of the open Modbus protocol and a low cost, off-the-shelf PLC, the solution for an audible alarm was implemented for a fraction of the cost of a custom solution.

Example #2

The supplier needed to integrate several small, very low cost touch panels into a building automation system. The touch panels offered a Modbus Serial protocol master communications interface. To make the connection to the building automation system, the supplier used the TOP Server Modbus Serial slave driver and LinkMaster OPC bridge.

On a PC, the supplier installed a multi-port serial port board, and connected all of the touch panels to these ports using RS-422 connections. On the same PC, the TOP Server Modbus Serial slave OPC server was installed and configured to listen for data requests from the touch panels connected to each serial port.

The building automation system exposes an OPC server interface. To bridge from the TOP Server Modbus Slave OPC Server to the building automation system the supplier used the Linkmaster OPC to OPC bridge. Data points were mapped in software through a point-and-click interface between the modbus slave connections to each touch panel and the points in the

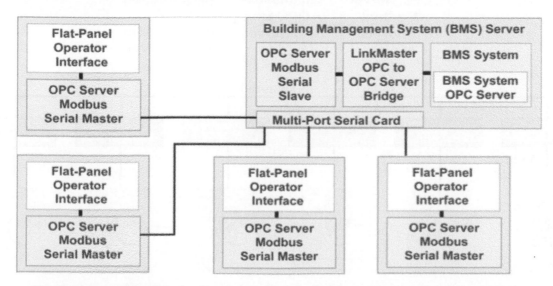

Figure 27-21. Application for Connecting Remote Panels to BMS via Serial Connections

building automation system. Once configured, data is able to flow bi-directionally between the touch panels and the building automation system. Changes are easily made through a point-and-click user interface.

Example #3

The supplier needed to integrate a large number of remote energy meters into the building automation system. Each energy meter exposes a modbus serial interface. The supplier used the TOP Server Modbus Serial Master OPC server along with Ethernet to serial convertors to connect to the remote meters over Ethernet. The connection from the TOP Server OPC server to the Building Automation system was made using the LinkMaster OPC to OPC bridge in the same fashion as other applications.

Products used in these applications:
- TOP Server OPC Server – www.toolboxopc.com
- Linkmaster OPC Bridge – www.toolboxopc.com/linkmaster
- Siemens MultiPanels and Siemens S7-200 PLCs – www.sea.siemens.com
- Patton Electronics DSL based Ethernet Extenders – www.patton.com

Case Study: DCS to Building Automation Application
Technologies Used:
- OPC Server Software
- Modbus Ethernet Protocol

A major petroleum producer needed to interface an Emerson DeltaV Distributed Control System (DCS) to a facilities automation system based on the Invensys Wonderware Intouch off-the-shelf industrial human machine interface software.

Design standards for the DCS system required that in all interfaces to external systems, the DCS system be the master. This means the DCS system would initiate all communications. The InTouch HMI system was designed to operate the same way, as it is typically the master when connecting to its data sources. The customer utilized OPC server software solutions from Software Toolbox to bridge the gap between these systems.

They installed the TOP Server Modbus Ethernet Slave OPC server on the same computer as the InTouch HMI based building management system (BMS). In the InTouch HMI system, they mapped all data points that were to be shared with the DeltaV DCS system (read or write) into points in the OPC server.

In the DCS system, they configured the Modbus Ethernet Master interface to connect to the OPC Server over Ethernet, using the open Modbus TCP protocol. When the DCS wants to read data from the InTouch HMI system, it reads the Modbus addresses in the OPC server that correspond to the data points mapped from InTouch into the OPC server. When the DCS wants to write data, it writes to the modbus addresses in the OPC server. The OPC Server automatically updates the InTouch HMI with the new values.

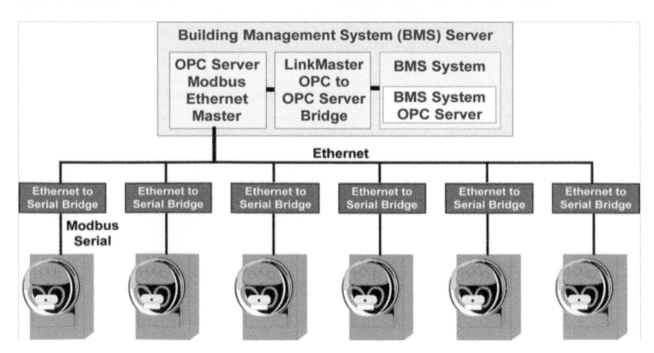

Figure 27-22. Connecting remote meters using modbus, serial to ethernet bridges, and OPC

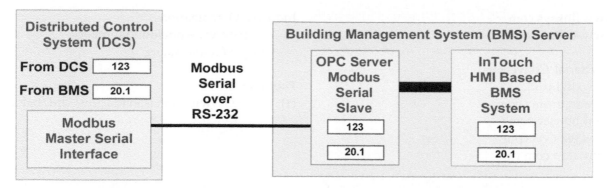

Figure 27-23. Distributed Control System (DCS) to Building Management System (BMS) Integration

The use of an open system based on OPC and Modbus generated significant savings in integration time and engineering for the user of the application. In addition to these savings, the open system made system checkout possible. The prime contractor for the job was in Europe, with an integrator in the Middle East, a planned installation in the Middle East, and a system checkout planned in the United States. Through the use of open systems, all interconnections were planned ahead of time via email correspondence between the parties. At checkout, a technician skilled in the use of OPC and Modbus was contracted in the United States to attend the system checkout on behalf of the integrator from the Middle East. In one day, all interface points where tested and checked, saving the integrator the expense and time of traveling to the United States for system checkout.

Products used in this application:
- TOP Server OPC Server – www.toolboxopc.com
- Consulting Services – www.softwaretoolbox.com
- Emerson DeltaV – www.easydeltav.com
- Invensys Operations Management Wonderware InTouch – www.wonderware.com

CONCLUSION

The use of open systems and web based technologies from the commercial information technology world and from the manufacturing automation industry have become a major influence in the facilities automation market in the last 10 years, and will continue to accelerate their influence. Implementers and managers of facilities automation systems no longer accept the answers "well you just can't do that" or "we can do that but it will take a customization that costs $X." To be sure there will be cases where both answers will still

apply. However, it will become more common to see technologies such as OPC, OPC UA, Bacnet, Human Machine Interface (HMI) software, Modbus, Modbus Ethernet, and Wireless Ethernet used to cleanly integrate Building Management systems into the overall enterprise. Cloud based computing and the rapid move of browser based computing to highly portable devices will further pull open the doors for open systems in facilities automation and lower the total cost of implementation and ownership of systems. As the first 10 years were a time of change of opening up systems, the next 10 years will also a time of change, taking users to new possibilities never before imagined. For those who embrace the change and the newfound openness possible, the returns to their businesses, whether they be suppliers, implementers of systems, or users of systems, will be significant.

Reference Websites:
Products Used in Case Studies
> www.softwaretoolbox.com
> www.softwaretoolbox.com/cogentdatahub
> www.paneldrivers.com
> www.toolboxopc.com
> www.factorywidgets.com
> www.toolboxopc.com/linkmaster
> www.industrialfalcon.com
> www.opcwebclient.com
> www.ingearopc.com
> www.siemens.com
> www.advantech.com
> www.patton.com

Other Hardware and Software Resources
Wireless networking hardware
> www.netgear.com
> www.dlink.com
> www.belkin.com

www.linksys.com
www.airlink.com

Ethernet/Serial Convertors
www.digi.com
www.lantronix.com
www.bb-elec.com
www.sealevel.com
www.atop.com.tw

Industry Organizations
http://www.opcfoundation.org
http://www.isa.org

Footnote
{1}   http://www.cfsan.fsan.fda.gov/~lrd/bghaccp.html

Section VII

Building Commissioning, Maintenance and Modeling

Chapter 28

Electric Usage Monitoring Per Socket/Switch: Economical Alternative

Dave Thorn
Paul Greenis

ABSTRACT

ThornProducts' patent pending technology, Pinch-A-Watt™ allows electricity usage monitoring to be imbedded in electric sockets, switches, etc with simple, reliable communication to the building receiver-gateway. This significantly reduces the cost of individually monitoring all electrical usage in the home/business for providing actionable feedback to the user for conservation and cost savings results of 10% or more. The system sends compiled data to electric utility company, Google Power Meter,[1] Microsoft Hohm™ or other web or database portal for report distribution or consumer access.

The technology and products can also be directly applied to flex tenant metering (or sub-metering) applications in commercial environments. This provides appropriate electricity cost distribution to tenants by the landlord or building manager, easy re-apportioning of distribution points, alerting of any overcurrent points and of course encouragement w/info to tenant for conservation.

INTRODUCTION

Smart meters and similar implementations will provide only whole house/building electricity energy usage to the consumer, but per appliance and per electrical socket usage would provide the detailed incentive for consumers to take effective and long term energy conservation actions.

Studies have shown that more detailed electricity information to users can provide 10% and more conservation even over the long term. One Ontario study actually tracked this over one and a half years.

Smart Meters and several existing products provide whole house usage only which does not get down to the detail required for more significant conservation improvements.

Present products and implementations providing usage information are:

- Single whole house measurements
- Limited number (e.g. 3) of large appliance or section measurements often via current clamps
- Smart Appliances connected to receiver/gateway unit, (coming in 2011 or later)
- Spot application of plug-load modules
- Breaker panel level only monitoring

Utilizing Figure 28-1 and the analysis below it, we can see that without significant coverage of individual electricity usage points we fail to cover from 40% to 47% or more of the typical home electricity usage.

The Pinch-A-Watt collected data can be used stand-alone with graphical reports or can complement existing management systems that presently only monitor smaller portions of usage. These would include systems that work with smart appliances and smart meters.

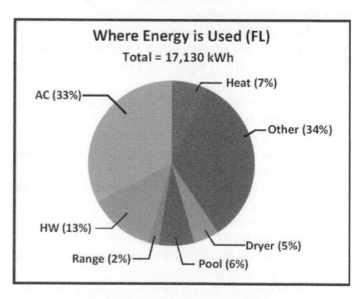

Figure 28-1. Energy Usage in Florida

This chart is based on some information from the report "Energy Efficiency—The First Priority in Solving Energy Issues" by the Florida Solar Energy Center whose work in energy conservation is greatly appreciated.

This chart is data for Florida USA where HVAC, specifically air conditioning, is the most significant part of the usage.

Other sections of USA would either exchange the Air Conditioning section for heating or actually have reduced HVAC and thus an even larger percent of "other" un-monitored detailed usage.

USAGE CHART, one implementation scenario: In an all-electric home, current clamp and other systems could cover the AC, Heat and Hot Water portions leaving 47% of the usage unmonitored and thus unavailable for direct use/by the user for conservation improvements.

Smart appliances & associated reporting systems when available would cover a portion of this 47% but only for some of the appliances and only when they become available and when existing appliance replacement is justifiable.

Plug-load module systems could cover a portion of this 47% (not lights, fans, hard wired appliances), but at a significant cost & physical management if many devices were to be deployed. Pinch-A-Watt built-in technology would cover all of this 47% as well as all the other AC, heat and hot water areas.

BACKGROUND/PROBLEMS

The existing methods are large plug-in ("plug-load"), modules and current sensor clamps utilizing a variety of data transmission protocols.

The problems with these methods are:
- Expensive per appliance and per socket devices
  - Results in only equipping a few locations instead of the whole house.
- Requires consumer set up and maintenance including databases & location changes
  - Results in inconsistent data credibility.
  - Data unlikely to be accepted by utilities companies, etc.
  - May result in security issues when used by electric utilities.
- No easy provision for light switches & need current clamps for HVAC, water heater, etc.
  - Results in non-full house/building coverage due to complex installation.
- Use of complex data transmission protocols within the home/building.

- Different from brand to brand, reduces reliability and increases size and cost.
- Does not allow an easy mass deployment solution
  - Thus used in too few homes/buildings to produce significant conservation results.

SOLUTION

A simple method that's easily and economically deployed throughout the whole house/building.

- A monitoring method built into sockets, light switches, HVAC switches, etc.
- A small snap or screw add-on device to the sockets and switches as an alternative.

Thus simple replacement with inexpensive "monitoring" sockets, light switches, HVAC cut-off switches, etc. can be performed for the whole house/building at one time. Even if installation is performed by the consumer, reliable data and long equipment life will be provided.

Employ a simple and reliable transmission method to send collected data to a main receiver unit. This unit can then send compiled data to electric utility company, Google Power Meter, Microsoft Hohm™ [1] or other web or database portal for report distribution or consumer access.

Data delivery would also be compatible with "Smart Meter" and other standard protocols. Coupling these sender units with the main receiver and data distribution unit provides a solution for the problems described above.

EXAMPLES OF EXISTING METHODS AND PRODUCTS

- Kill-A-Watt™: large, no network connection, consumer installation & maintenance.
- PicoWatt™: large, consumer installation and maintenance.
- TED ® The Energy Detective™: Senders; limited # of monitor points, clamp on sensors.
- Alert Me Energy: – Only whole house (electric meter) monitoring.

These existing methods are primarily designed for purchase, installation and use by the consumer instead of data acquisition, control and reporting use by the electric utility or by building management.

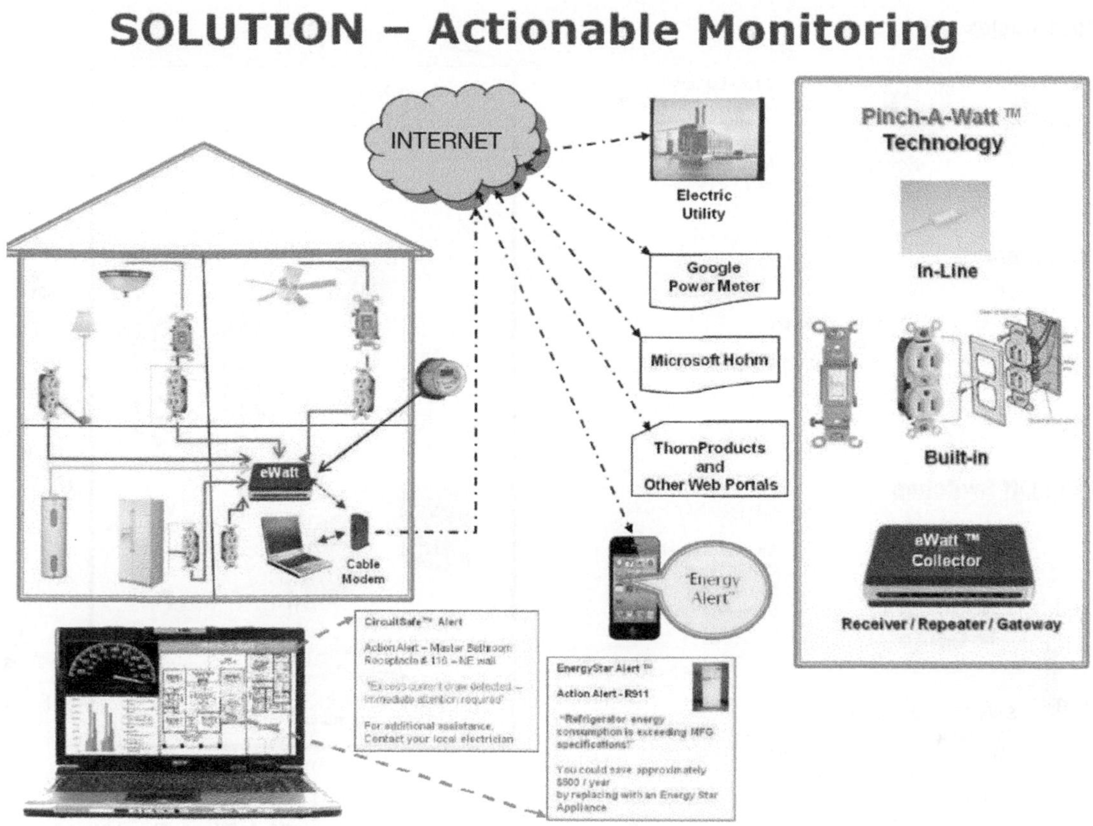

Figure 28-2. Solution—Actionable Monitoring

CONTROL

In addition, the compiled data can be used to send commands to third-party control devices which would be installed only in those places where control and demand-response is needed and practical.

The compiled data can also be sent to third-party energy management systems to compliment them with usage data from all devices in the home/building.

EXAMPLES OF SYSTEMS COMPLEMENTED BY THE SYSTEM

• GE Nucleus
• Tendril Insight or Vantage.
• Other energy management systems

CONCLUSION

These devices can go a long way to provide unprecedented visibility for the homeowner of their electric consumption. Detailed analysis of each and every appliance is now much closer than the day when all of our appliances are "smart" appliances. Couple this technology with a truly comprehensive analysis software application and the homeowner can learn a lot about benchmarking and trend analysis. Hopefully they will take what they can learn from this energy monitoring system to the workplace where similar efforts there can really make a difference in our energy consumption appetite and carbon footprint. You can follow the progress of this project at http://thornproducts.com.

The application of the technology and products to Flex Tenant Metering (or sub-metering) applications in commercial environments provide a major increase in both the accuracy and the flexibility of electricity cost apportioning. The fixed (in the recepticle, switch, junction box) characteristics also provides a level of required tamper resistance not provided by existing products for outlet point monitoring.

References
[1] Note that these references may be trademarked or otherwise owned by their respective owners: Google Power Meter (Google Inc.); Microsoft Hohm (Microsoft), PicoWatt (Tenrehte Technologies); TED & The Energy Detective (Energy Inc.); Alert Me Energy (AlertMe); Kill-A-Watt5 (P3 International); GE Nucleus (GE); Tendril, Kill-A-Watt, etc.

Sockets and Receptacles

Table Lamps
Entertainment Electronics
Washer, Dryer
Refrigerator
Microwave Oven
Any / all plug in devices

Switches and Dimmers

Ceiling Lamps
Fans
Switched Sockets

Breakers and Cut Off Switches

Air Conditioners
Water Heater
Dishwasher
Stove/Range

In-Line wired in

Built-In to appliances, or
Alternative for any of the above devices
Including alternative to breaker & cut off switch
for Hard wired devices like:
    Intercoms
    Some Stoves
    A/C units
    Ceiling Lamp fixtures
    Garbage Disposal

External Plug-in and Clamp-on Monitors temporary needs or demonstration

© Copyright March, 2010, Thornl

Figure 28-3. ThornProducts, LLC
Sockets, Switches & Breakers with Usage Monitoring Built In

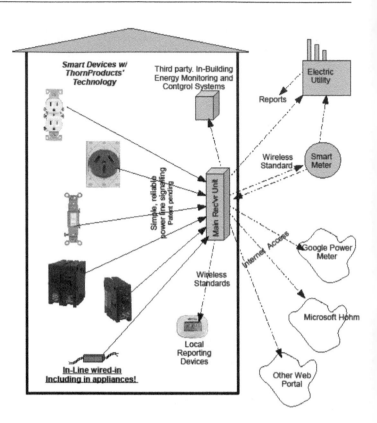

Figure 28-4. ThornProducts, LLC
Sockets, Switches & Breakers with Usage Monitoring Built In
Interface to other systems from our Electric Components

Chapter 29

Computerized Maintenance Management Systems (CMMS): The Evolution of a Maintenance Management Program

Carla Fair-Wright

ABSTRACT

There are two approaches to acquiring a computerized maintenance management system (CMMS) system: build or buy. Cost and time are key considerations for any project but are crucial factors to assess when selecting a CMMS solution. Managers faced with evaluating any CMMS system must make the build vs. buy decision early in the process. While many businesses choose commercial off-the-shelf solutions, there are distinct advantages to building an application in-house.

This chapter will discuss the software development process and address some of the primary issues managers face when considering an in-house CMMS solution. It will illustrate the key principles for successfully upgrading a CMMS application from a monolithic application model to n-tier architecture capable of supporting http protocol. Both the technical and business aspects of the project will be addressed.

The lessons learned in this case study include: (1) The need to understand that to build smart, flexible, and integrated software there must be reliable and robust tools available, (2) the need to get the cooperation of all interested parties in a large complex project, (3) the importance of project management, and (4) the obstacles faced when incongruent systems are required to communicate.

INTRODUCTION

Inadequate maintenance of energy-using systems is a major cause of energy waste in both the federal government and the private sector. Energy losses from steam, water and air leaks, uninsulated lines, maladjusted or inoperable controls, and other losses from poor maintenance are often considerable. Good maintenance practices can generate substantial energy savings and should be considered a resource. Moreover, improvements to facility maintenance programs can often be accomplished immediately and at a relatively low cost.[1]

Good maintenance practices are wrapped in a blanket woven from a few simple concepts. First, the equipment in operation must be suitable for the job. Second, there should be an adequate supply of spare parts and a skilled labor force to install them. Finally and most importantly there should be a preventive maintenance program in place which utilizes a good CMMS.

Commercially available software packages do not inherently integrate with existing energy management and control systems (EMCS) as well as property management systems. For an organization running heterogeneous software systems, finding a compatible product can be very costly and time consuming. Customization of some software packages can almost equal the price of the program itself. One way to handle the problem is to develop the software in-house.

In the next few pages the particulars of CMMS software planning and design will be discussed. The goal in writing this chapter is to demystify the process of software construction and provide the reader with a better understanding of CMMS systems and how they work. The goal is not to teach programming, but there are various software topics that are specific to the development process which will be discussed. Also, there are many acronyms in this chapter. They are defined before being used.

This chapter is divided into four sections: a historical view of equipment maintenance at Cooper Cameron Corporation, a general description of CMMS applications, a technical examination of the application itself—internal structures and data specifications, and testing and deployment strategies.

EQUIPMENT MAINTENANCE AT COOPER CAMERON—HISTORICAL VIEW

Facility Background

Cooper Cameron Corporation is a leading international manufacturer of oil and gas pressure control equipment, including valves, wellheads, controls, chokes, blowout preventers and assembled systems for oil and gas drilling production and transmission used in onshore, offshore and subsea applications, and provides oil and gas separation equipment. Cooper Cameron is also a leading manufacturer of centrifugal air compressors, integral and separable gas compressors and turbochargers. With annual sales of approximately $1.2 billion, the company is divided into three divisions: Cameron, Cooper Cameron Valves, and Cooper Compression. Cooper Cameron operates in over 115 countries around the world with headquarters in Houston, Texas.

The Maintenance Technology Services (MTS) department is part of the services organization of Cooper Compression. MTS provides customers with value-added, integrated equipment operating and maintenance service solutions, generally in the form of fixed price, risk assumptive long term operating and/ or service agreements with varying performance incentive structures.

Service agreements, typically in the form of operating and/or maintenance contracts, can cover total (i.e., scheduled and unscheduled) maintenance, full or partial scheduled maintenance, operations, or maintenance management assistance to customer organizations. These agreements typically cover power and compression equipment with coverage options for related support systems as well as the facilities within which such equipment is installed.

Service agreements offer multiple benefits to Cooper Compression's customers:

- Maintenance-related equipment downtime is minimized so that equipment availability and productivity is maximized
- Maintenance expenditures are defined and controlled throughout the life of the agreement

- Customer fleet and asset management is optimized
- Risk, as well as direct and indirect operating and maintenance costs, is transferred to a qualified, knowledgeable service provider
- The service provider absorbs the peaks and valleys inherent in maintenance resource deployment

The goal of the MTS department is to help the client improve his competitiveness and productivity with effective maintenance management technology. A properly designed service agreement provides quality parts, experienced service technicians, machine shop services and the accumulation of years of experience in predictive maintenance, planning and scheduling.

MTS develops client-specific maintenance plans for both Cooper and non-Cooper manufactured equipment. Every service maintenance planner knows from experience that optimized maintenance planning and scheduling are the foundation of a cost-effective maintenance program. With a good predictive maintenance program in place, there are major improvements in equipment availability, reliability, safety and performance.

Maintenance Management System

Rusty Creekmore, Director, Maintenance Technology Services, Cooper Compression has said:

> For Cooper Compression to be successful in providing operating and maintenance services for our customers' power and compression equipment, we must be able to maximize equipment availability, reliability, productivity and safety at the lowest possible cost. We rely on predictive maintenance technology to accomplish these goals, and our proprietary Computerized Maintenance Management System was developed and is being enhanced specifically to provide Cooper Compression with that competitive advantage.

In the mid 1980s the predecessors to today's maintenance technology services group were unable to find a suitable CMMS software package that met the unique requirements of power and compression equipment. Therefore, they developed an in-house system called simply *maintenance management system* or "MMS." Over the years the software proved to be a valuable asset supporting Cooper's field personnel on numerous service agreements. With subsequent enhancements, MMS became the primary tool to process, analyze, and interpret equipment operating data supplied to it from hand-held data collectors and from equipment control systems based on operating systems such as Invensys' Wonderware. In the process of several reorganizations in the early 1990s, Cooper's IT department assumed control of the MMS software, but a shortage of computer personnel skilled in relational database management systems (RDBMS) programming led to limited development and maintenance support. Ultimately, the software languished and by 2001 was in critical need of a major upgrade. Some of the underlying drivers for the upgrade program were:

- The software ran on Microsoft's Windows 98 operating system, which is considered to be obsolete in today's computer technology environment.
- Much of the interface was carried over from the original DOS version with only minor Windows-based enhancements.
- The system architecture was top-down structured and not object-oriented programming (OOP), which is the current industry standard.
- The software could not communicate with the newly installed plan maintenance (PM) module of the company's SAP R/3 enterprise management software.
- The software contained irregularities and inconsistencies that were a constant source of discontent.

Unfortunately, not all of Cooper Compression's senior managers understood the software application's critical role in Cooper's operating and maintenance business segment and this presented a challenge in obtaining the resources required to update or replace MMS. Ultimately, however, with the support of Cooper's executive-level services management, a small project team was tasked to find or develop an appropriate system that would meet the operating and maintenance information requirements for the future.

The MTS service maintenance planners assigned to this team found the CMMS systems that were commercially available at that time were limited in functionality. Their conclusion was supported by several authorities. According to Labib, the new generations of CMMSs are complicated and lack user friendliness. These systems are difficult for either production operators or maintenance engineers to handle. The author describes these systems as more accounting and/or IT

oriented than engineering based.[2]

Based on their investigation, the service maintenance planners decided to upgrade the current software which had been written in Microsoft FoxPro version 2.6a. The program had a number of analysis and reporting capabilities as well as equipment operating data collection, filtering, analysis, trending and reporting functions far superior to anything the planners had found in the current marketplace. Few, if any, of its competitors had attained anything approaching this level of sophistication. The main reason for using FoxPro was its ability to handle the enormous volumes of operating and maintenance data used in the predictive analytical modules. The upgrade was written using the object-oriented relational database management system. Microsoft Visual FoxPro version 8.0.

OVERVIEW OF CMMS

Before we discuss the architecture and functions of a CMMS, we should have an understanding of what we are attempting to achieve in the software design process. What is exactly is a CMMS? What are the benefits of using it?

Prior to 1970, facilities seldom focused on identifying and managing the maintenance function. The primary focus was on direct manufacturing, operations and materials; generally, maintenance was looked upon as a cost of doing business. During the period from 1970 to 1980, there was more focus on documenting maintenance as it relates to preventive maintenance and equipment uptime. Machine maintenance is viewed as a business process that can be augmented by computer software. The role of the maintenance planner and scheduler expanded as it is became necessary to organize maintenance functions into quantitative tasks.. As we moved into the mid 1980's there was an increasing focus on preventive maintenance as well as predictive maintenance. The concept and practice of total productive maintenance (TPM) began to receive attention and become implemented which involved the self directed work force [3]. Traditional maintenance regimes are normally reactive and produce high levels of unplanned expenditures. Technology-driven regimes are more proactive and will typically increase planned maintenance cost but will reduce unplanned and overall maintenance cost while providing increased equipment productivity and profitability.

Significant savings can be achieved by identifying and preventing expensive equipment breakdowns before they occur; an added benefit is that the efficiency of routine maintenance tasks improves. In fact, many insurance companies have recognized that proper use of a CMMS reduces the chance of costly insurance claims and will reflect this through reduction of premiums. Depending upon the client, unplanned equipment downtime can cost a plant $1 million per day or more.

Studies have shown that optimized reliability and maintenance programs can increase operational availability up to 10%, decrease maintenance costs by 10-20%, and reduce both the number and severity of unplanned process interruptions. [4]

A technology-driven predictive maintenance program can not only decrease maintenance costs but, more importantly, improve operational availability and efficiency, thereby potentially enabling a client to achieve significant financial improvements to the bottom line. An added bonus is that the plant or facility becomes safer because required maintenance is performed on schedule. But the greatest potential utility of CMMS is its data storage capability related to all equipment operating and maintenance functions. For example, optimal use of a CMMS can alert an organization to under-performing or high-operating-cost equipment. It can highlight labor inefficiencies and can provide essential data for root cause failure analysis to eliminate chronic equipment failures. With this type of information at hand, plant and facility management can make better, more accurate data-driven decisions.

What Does a Computerized Maintenance Management System (CMMS) Do?

CMMSs are often perceived to be no more than a means of scheduling maintenance work. While preventive maintenance scheduling is normally part of a computerized system, most of them are capable of much more. Virtually all aspects of a maintenance department's work can be managed by the modern, integrated software packages. [5]
The following is a list of basic functions all CMMS applications are designed to perform:

1.   Managing Assets
2.   Scheduling Planned Maintenance
3.   Recording Unplanned Maintenance
4.   Creating and Validating Work Orders
5.   Allocating Labor Resources
6.   Monitoring Conditions
7.   Reporting Statistics

Managing Assets

This is the entry point for starting a CMMS program. Gathering and storing asset information is an important component in building the maintenance plan. Most CMMS applications will require data on functional category, location, serial number, model number, make, and description of equipment as a minimum to set up the plan. Optional information can range from identifying information (such as an assigned asset barcode number), engineering drawings, or digitized pictures or diagrams of internal systems and components.

Scheduling Planned Maintenance

Tactics win battles, but strategy wins the war. Tactics in the maintenance planning business are the day-to-day operations. The strategy is the maintenance plan, a list of short and long-term objectives. The maintenance plan is simply a listing of all the preventive maintenance (PM) activities for each piece of monitored equipment. This list tells the plant manager or reliability engineer when and how often a specific set of predefined tasks should be performed.

The sample report from MMS shown in Figure 29-1 tells the maintenance planner what maintenance type or level is due for this particular type of piece of equipment and the specific hour meter reading at which the service should be performed. The report's header information consist of a description of the equipment, the period of the scheduled maintenance (1680 hours or 70 days), and the amount of time we will allow for overlap. Just like the odometer of a car, the number of hours a unit has been in operation provides an interval to base maintenance on. The term overlap refers to the time between scheduled maintenance events.

Activities are set by intervals that are either calendar time-based, operating time-based or operating cycle-based. Just as an automobile manufacturer might suggest that engine sparkplugs be changed every 100,000 miles, a machinery maintenance plan may specify that industrial engine sparkplugs be changed every 4,000 operating hours. The maintenance planner may import these maintenance task descriptions from an external application or enter them manually. The maintenance plan is the heart of a maintenance program. But as you will see later, monitoring equipment condition is its soul.

Recording Unplanned Maintenance

The history of a machine is used to analyze and improve performance and streamline expense. There-

REPORT DATE: 10/11/04

Cooper Compression - Customer Maintenance Services
MAINTENANCE MANAGEMENT SYSTEM
MAINTENANCE FORECAST

EQUIPMENT CODE: CEMFSACTCPAAAAAAAAAA
EQUIPMENT DESC: SUPERIOR 16SGTB/W74 COMPRESSION UNIT
MAINTENANCE NOW DUE OR DUE WITHIN: 1680 HOURS, 70 DAYS FROM TODAY'S DATE
OVERLAP PERCENTAGE: 80% HOURS, 80% DAYS

LEVELS INCLUDED: ABCD

	|<------- HOURS ------->|			|<------- DAYS ------->|		
LEVEL	LAST	NEXT	DUE IN	LAST	NEXT	DUE IN
A	N/A	N/A	N/A	05/10/04	10/11/04	0
A	117056.90	117392.90	0.00	N/A	N/A	N/A
A	N/A	N/A	N/A	05/10/04	10/25/04	14
A	117056.90	117728.90	336.00	N/A	N/A	N/A
A	N/A	N/A	N/A	05/10/04	11/08/04	28
A	117056.90	118064.90	672.00	N/A	N/A	N/A
A	N/A	N/A	N/A	05/10/04	11/22/04	42
B	N/A	N/A	N/A	04/28/04	11/22/04	42
A	117056.90	118400.90	1008.00	N/A	N/A	N/A
A	N/A	N/A	N/A	05/10/04	12/06/04	56
A	117056.90	118736.90	1344.00	N/A	N/A	N/A
B	116783.90	118783.90	1391.00	N/A	N/A	N/A
A	N/A	N/A	N/A	05/10/04	12/20/04	70
A	117056.90	119072.90	1680.00	N/A	N/A	N/A

Oil is checked every 14 days or 336 hours of operation

Calibrations and alignments are check every 42 days or 1000 hours of operation

Figure 29-1. Scheduled Maintenance Report from MMS

fore, a CMMS must be able to record unplanned work (repairs and/or breakdowns). What, when, and why a failure has occurred is important information that can be used to re-evaluate present maintenance practices or identify possible operating, process or engineering problems. Analysts can review the repair information and fashion questions to further refine the maintenance plan. For example, a planner may find by reviewing unscheduled work orders that the cost of changing a component during the preventive maintenance (PM) activity to prevent a certain type of failure is less costly than making unscheduled repairs and incurring unscheduled process downtime and loss of equipment production.

Creating and Validating Work Orders

The work order will detail estimated and capture actual labor hours for service or repairs, the labor skills required, type of work (repair or PM), instructions for performing the work, required parts, tools, expendable materials, and associated accounting costs. Some CMMS applications can generate formal work orders and some can not. But all CMMS programs must be able to validate the work orders for any asset within the database. Validation is necessary because parts and

labor are charged to the work order. This information is a critical part of preventive maintenance; it determines the allocation of man, material, and time resources.

The information in a work order can be essential in diagnosing the health and life expectancy of the asset. Even if the application is not used to create the work order, it should save the critical actual "as performed" elements of the work order into a maintenance history file. This can be done via an interface to the company enterprise system or by importing data pulled from it.

Allocating Labor Resources

Keeping track of maintenance and repair labor costs is another benefit of a CMMS. Labor time and cost data can be captured via a variety of technologies including: badge-based, biometric, touch-screen kiosk, mobile/wireless, telephone/Interactive Voice Response (IVR), and PC/Web-based devices. Once collected, this information can be ported to the CMMS or keyed in manually.

Monitoring Conditions

Is it possible to predict potential failure and prevent costly repairs? Yes, it is possible—although only

31% of companies operating in the United States are doing so.[6] Studies done as recently as late 2000 still show reactive maintenance as the predominant mode of maintenance in the United States. Reactive maintenance, or the "run it till it breaks" maintenance mode, is the easiest maintenance model to implement. But, it usually results in the greatest overall equipment costs over time.

Condition monitoring is the beloved companion of predictive or condition-based maintenance (CBM) techniques. It refers to the use of advanced technologies to determine the condition of a mechanized area of a plant or equipment. It is based on the idea that identifying certain changes in the condition of a machine will indicate that some potential failure may be developing.

Once the physical characteristics are pinpointed that identify the normal condition of a machine, they can then be measured, analyzed, and recorded to reveal certain trends. Sensors may be used in sampling operational parameters. These sensors can detect flow rate, motion, weight, quantity, phases of electrical power and much more. The sensors usually come with switch contacts in an opened or closed state. A change in state alerts the operator to possible problems in the plant or facility.

Another method used to gather status information includes the use of embedded equipment control software and equipment monitoring software. At set intervals a reading is taken from the monitoring points and transmitted to a database with the origin, date, and time. Of course, not all points can be monitored by software. Figure 29-2 shows a snapshot of the upload screen from MMS. The software is flexible enough to retrieve information from a variety of sources.

If the signal exceeds or falls below a pre-determined range, or the rate of change of the measured parameter exceeds a pre-determined rate, then an alarm status is returned. But, this simple alarm system is not the only technology applied, as seen in Table 29-1.

Statistical Reporting

A CMMS is only valuable if you can access the information in it. It should be able to provide immediate answers in a repair situation or intricate details for a long-term equipment utilization, availability and reliability, cost or condition analysis. In fact, one of the most significant functions of a CMMS is the report generator.

Reports enable the analyst to make efficient use of the extensive information contained within a CMMS for fault analysis, costing and work statistics. Most systems supply a number of pre-defined report queries

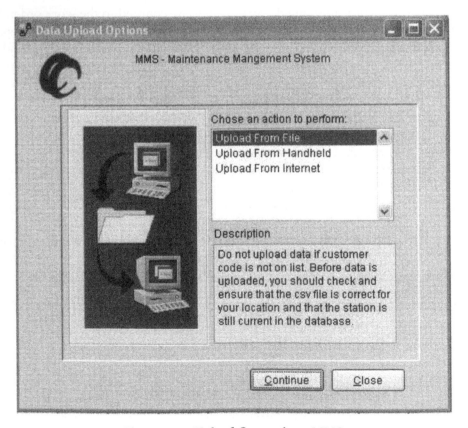

Figure 29-2. Upload Screen from MMS

Table 29-1. Condition Monitoring—adapted from Rockwell Automation[7]

Condition Monitoring Tools	
Vibration Analysis	The Vibration Collection and Analysis process monitors the response of the equipment to internal and external forces being applied. The response is measured by a general-purpose transducer at the pump and motor bearings and is passed to an analyzer for interpretation. The Analysis can provide early indications of problems such as machinery imbalance, misalignment, bearing wear, worn gears, etc. The results can be downloaded to the CMMS or asset management software for analysis and recommendations.
Temperature Analysis	There are a variety of devices used to measure temperature. These devices will be designed to measure either surface or gaseous temperature. Normally, thermocouples or templugs are used to determine surface temperature from inside an engine
	Fine wire resistance thermometer, Infrared, Spectroscopy, Hot Wire Anemometry (HWA), or 3-Colour pyrometry are used for gaseous temperatures. The concept of a thermometer and infrared are familiar to most and work as implied. Spectroscopy measures the intensity of light emitted at wavelengths in the infra red, ultra violet or visible light spectrum by a gas. Hot wire anemometry is a technique using a thin wire stretched between the tips of two prongs to measure time resolved flow in the cylinder. Three color pyrometry looks at the relationship between the temperature of an object and the intensity of the radiation that it emits. Identifying an abnormal rise in the temperature of machinery that could result from problems such as bearing wear, lack of lubrication, poor electrical connections, etc.
Oil Analysis	The oil sample is taken and sent to a lab for analysis where it is subjected to physical testing, spectroanalysis, and particle testing. The physical testing looks at viscosity and water content greater than 1%. The oil is heated and run through the viscosity bath. The results are then compared to the new oil specification. This test is valuable in determining the condition of the oil and an indicator of water contamination and oxidation. Spectroanalysis examines metal content and additive package. This test checks around 19 elements and reports them in parts per million. The particle count is a critical part used to measure the efficiency of the system filtration. The particle count measures all particulate in the oil larger than 5 microns. Particulate include: dirt, carbon, metals, fiber, bug parts, etc. Using either laser or optical methods does this. The laser method reports the quantity, size and distribution of particulate, but not what they are. The optical method gives a quantity, size, distribution and identification. Controlling viscosity and other chemical impurities is the key to preventing premature failure of mechanical systems such as bearings, gears, etc.[1]
Motor Current Signature Analysis	Analyzing the unique signature patterns in the motor supply current created by fluctuations in load, high resistance joints in rotor bars and end rings or uneven air gap between the rotor and stator.

Table 29-1. (*Continued*)

Operating State Dynamic Analysis	Correlation of machinery operating parameters such as load, pressure, speed, etc. to the dynamic characteristics of the machinery such as vibration and motor current signature.
Ultrasonic Leak Detection	Use of ultrasonic measurements to identify changes in sound patterns caused by high-pressure gas/steam leaks, bearing wear or other deterioration.
Balancing	The leading cause of imbalance in rotating machinery is vibration. The imbalance may vary with time and can be hard to correct. Without detection and adjustment, machine problems can occur through mechanical stress. In the case of power tool operators, long-term physical harm to humans called "White Finger" can occur. This is a condition in which the blood supply is interrupted to the hand.
	Vibration monitoring equipment can enable maintenance staff to detect, diagnose and correct misalignment and imbalance in rotating machinery.

[1]*http://www.oillab.com/oil.html* web site contains one of the best descriptions of the process.

with customization capabilities. A good system will provide powerful ad-hoc querying with multiple report output formats, including Word, Excel, HTML and PDF. A difficult challenge for any complex field of endeavor such as the facility maintenance function is sharing meaningful information with executive management. The reports generated in a first class CMMS are readable and easy to interpret.

One of the advantages of using a web-based system is that these reports can be shared independent of the platform. The data can be reviewed in real time via any internet-connected computer that has a browser. Previously, the system users created complex accounting reports manually by exporting data from MMS into Microsoft Excel, performing various calculations and then formatting as needed. The report was then ready to be distributed by email. Now they can simply press a button to get the same information in all the right formats—from basic reports to reports for the executives to government-mandated reports.

THE APPLICATION INTERIOR

Software Lifecycle Planning

A lifecycle is the procedure by which a piece of software is created, implemented, tested, and finally maintained. There are a variety of lifecycle systems in use today and the numbers continue to grow as internet programming derails the static models built in the last decade.

Every software-development effort goes through a "lifecycle," which consists of all the activities between the time that version 1.0 of a system begins life as a gleam in someone's eye and the time it is retired from production. A lifecycle model is a prescriptive model of what should happen between first glimmer and last breath. [8]

The waterfall method developed in the 1970s is one of the oldest lifecycle models still in use. In the waterfall method a project moves sequentially through non-overlapping phases. These transitions from one phase to another, often called "throwing it over the wall," reflect the assembly-line mentality of the method. Each phase is fully documented and must be completed before the beginning of the next phase. The waterfall method is considered too hard and too expensive to implement in its pure form. To capture all the system requirements and furnish a complete analysis before the design process can begin is extraordinarily costly in terms of time and manpower.

The Sashimi model, a modification of the waterfall method, allows overlapping of the stages. This normally results in a cleaner product definition, but it makes project management difficult to chart and can lead to ambiguous milestones. The sashimi model is best suited for software projects that are well understood, but intricate to design. This is because the complexity can be tackled early on in the project lifecycle during the requirements gathering phase. CMMS programs are enormously complex, so for our project, we used the sashimi lifecycle model.

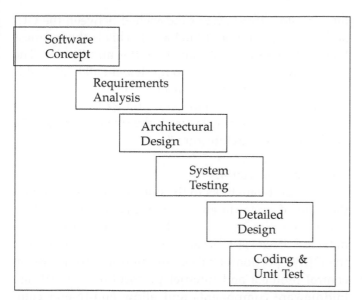

Figure 29-3. Sashimi Model—First implemented by a Japanese hardware manufacturer. Sashimi is the Japanese method of presenting fish sliced raw with the sides overlapping.[8]

Software Concept

Our MMS system had been operational since 1980 and was reflective of the command and control hierarchies of 20 years ago. Much of the work in the conceptual phase of the update was refinement and expansion of the original system. The application had to be enhanced to work with internet protocols and employ standard security and encryption technology. This revision recognized the reality that applications are beginning to migrate from a standalone executable sitting on a user's hard drive to a distributed application delivered by a web server across the internet. MMS needed to be redesigned to utilize extensible markup language (XML) based documents. The extensible markup language is the latest offering in the world of data access. XML is quickly becoming the universal protocol for transferring information from site to site via HTTP.

Requirements Analysis

In this phase of our project, we gather and itemize the high-level requirements of the system and identify those basic functions that the software must perform. These requirements are called **business rules**. According to the Object Management Group, business rules can be thought of as conditions that *must* be satisfied for business operations to be correct. [9] The business rule basically tells us what a system should do and what it should not. For example, a business rule would

state that every piece of operating equipment must have a serial number or some identifying tag.

In capturing the business rules we used Ivar Jacobson's [10] approach to documenting system requirements. This object oriented approach called "use case modeling" has become the de facto standard in modern development methodology. Use cases define the system in short narratives. They help keep requirements gathering simple and understandable for all involved. Here is a short example of a use case:

Entering Update Information:

A service technician receives a work order for scheduled maintenance on compressor unit #4. He performs the 2000 hour maintenance. He completes the work order and submits it to the office. The data entry person logs into the system and enters the maintenance details, which the system validates and records. The system updates the resource data (labor and parts) and the schedule. The data entry person runs a new schedule for the upcoming week.

A formal use case is more involved than this, but the basic concept is the same: discover and write functional requirements by creating a narrative of the steps. The process begins with the domain expert. This is a person with extensive knowledge of the equipment or system. We were fortunate enough to have access to several domain experts at our facility, who could provide a rich and detailed accounting of the system.

ARCHITECTURAL DESIGN

A major part of any software project is the architectural design. Architectural design is the organization of data and other program components necessary to build a computer-based system. It encompasses the identification of all interfaces, relationships, and database layouts. It can also require the use of notation systems like the unified modeling language (UML), which is a graphical shorthand for constructing and documenting software applications. The unified modeling language (UML) is the successor to the wave of object-oriented analysis and design (OOA&D) methods that appeared in the late '80s and early '90s. [11]

From Monolithic to N-tier

The original DOS-based MMS was created in the monolithic configuration model. In computer programming, the term monolithic is used to describe applications where all the logic and any resources it may need to function are contained within the application. Monolithic software is confined to a single computer and cannot support multiple users on a networked drive. Some operating systems fall into this category. Monolithic software is simple, but it is stunted by the inflexible structure that defines it.

As the need for applications that could be shared across a network grew, a change in the way applications communicated with the data source was needed. The term client-server was first coined in the 1980s to describe the personal computer (PC) on a network. The client was the application on the user's PC. The server was a powerful computer with processes committed to running the drives (file server), printers (print server), or network traffic (network server).

The client-server concept was extended to mean breaking logic and data apart into modules and deploying the data across a network. Initially there were two layers: the application and a database server. The application consisted of a graphical user interface (GUI) and the code to control the data retrieval and updating. The GUI is exposed to a user; it is the screen or web page, a place to enter or read information. The design of the GUI determines a system's usability or user-friendliness. The data access layer is simply the database or a collection of files where the data is stored. Web applications introduced the three-tier model by default: the browser is the client tier, the database the back-end tier, and the web server and its extensions became the middle tier. [12]

We used a three-tier architecture for MMS. Popularized by Rational Software and Microsoft, three-tier architecture breaks the GUI and logic apart. The logic is

moved into a third layer sandwiched between the GUI and persistent storage (databases). This is the business logic ("business rules") layer or the middle tier. The BusinessRules layer serves to implement all logic that is outlined in the system requirements. It validates data, performs calculations, and reads and writes data. When the middle tier itself is layered, the overall architecture is called "n-tier architecture."

Middleware Components

The software talks to the data source (server) with the help of open database connectivity (ODBC) drivers, which *translate* the application's requests to the database. Because it resides between the two entities, the ODBC portion is called middleware. Transmission control protocol and internet protocol or TCP/IP[1] are middleware components and allow end-to-end communication from one point (computer) across the network to another point or host computer. Without TCP/IP, dubbed "the cockroach of protocols" by Microsoft guru Don Box, the internet could not exist.

Detailed Design

Object-oriented programming (OOP) is a programming language model structured around "objects" rather than step-by-step routines. In top-down

---

[1]The higher layer, TCP, manages the assembling and reassembling of a message or file into smaller packets for transmission over the Internet. The lower layer, Internet Protocol, handles the address part of each packet so that it gets to the right destination.

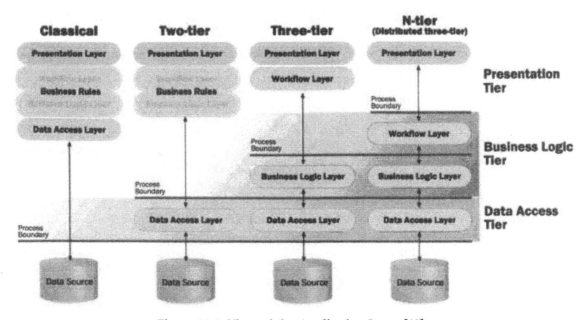

Figure 29-4. View of the Application Layers[13]

structured programming, the problem is broken into natural pieces and each piece is solved independently in sequence. Logic flows from the top to the bottom. Data are input, processed, and output.

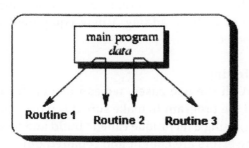

Figure 29-5. Program Constructs

In object-oriented programming we build islands of code and call them objects. These objects perform specific functions. They communicate with each other by sending and receiving messages. Each message contains a request for a particular set of instructions to be executed by the receiving object. We refer to a block of instructions as a method. An object can have as many methods as it needs to perform the tasks assigned during the design phase.. The term "class" refers to the static form of object. If you think of an object as a house, then the class would be the blueprint.

Data modeling is the first step in OOP. The purpose is to find and identify all the procedures that are used in a business and determine the relationships among these procedures. The next step is to refine these procedures into classes that have properties and methods. Properties define the kind of data the object

contains and methods are the logic sequences that can manipulate it. The class is a nothing more than a blueprint.

In our system, we began with several base classes and used them to define the objects in our system. For example, we have a print service class to handle all external outputs from the system. There is also a data service class that talks directly to the tables and other files. Then there are a plethora of functional classes, each with its own specific identity and function. The customer class knows what tasks and what data to use when we need a customer process. The maintenance plan class understands maintenance planning functions. When these "classes are in use" they become objects. An object is an instance of a class. They talk among themselves by passing messages and share tasks and information. As in a team of players, each object is as a distinct "person" with a common goal.

Relational Database Management Systems (RDBMS)

To support our n-tier strategy we needed a relational database on the back-end. A relational database is a series of tables that represent entities related to one another. We began by mapping all fields within each table and describing each field's function in the program. After each field was defined and categorized, we determined if the field existed in more than one table. If so, it was moved to a single table. The field's destination was based upon the dictates of the business rules and the relational model.

For example, our business rules for the customer told us that a customer can have many addresses. There is a shipping address, billing address, and the

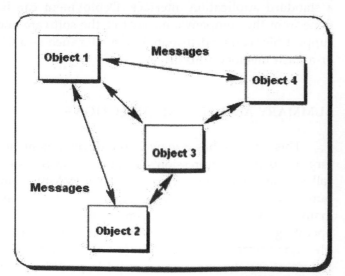

Figure 29-6 shows a simple interaction of several objects.

Object 1 has sent messages to objects 2, 3, and 4 telling them to execute one of their methods.

Object 2 and Object 4 receive the message execute the code and return a success message to Object 1.

To process the method on Object 3 additional information from Object 2 and Object 4 is needed. Object 3 sends and receives a message to the other objects. Once the information is received Object 3 continues processing the original request and returns a message back to Object 1.

Figure 29-6. Object Constructs

personal contact address. The address portion of the CUSTOMER table was removed and placed in a separate ADDRESS table. Why? Mathematics is at the heart of relational database theory. Based on predicate logic and set theory, the relational model tells that we must remove any unnecessary redundancy to be efficient. To store multiple customer addresses in the CUSTOMER table meant repeating the customer name and other details for each address. By moving the address to its own table and placing an identifying key on each record to tell us which customer the record belongs to, we saved space and increased our ability to maintain the integrity of the various relationships between tables.

To design the database, we used xCase from F1 Technologies and Visio. The xCase software allowed us to conceptualize the table structures. It also permitted us to relationally view and edit the physical data with integrity. For instance, if we decide to delete a customer from the database, it is not enough just to remove the customer from the CUSTOMER table, for this would leave dangling references to that customer in the ADDRESS table, among others.

Data flow diagramming shows business processes and the data that flows between them. Think of a process model as a formal way of representing how a business operates. To capture these details we used Visio, a Microsoft application for business and technical drawing and diagramming.

Coding & Unit Testing

Coding is the construction phase of the software. It is often the longest and most difficult part of the process. Earlier we defined a method as a set of coding instructions contained within an object. A good design will have many small methods in separate classes, and at times it can be hard to understand the over-all sequence of behavior. Unified modeling language or UML was conceived out of a need for a standard notation for modeling system architectures and behaviors. In the coding the new system we used UML models to sequence the movement of data through the system. These sequence diagrams were produced from a modeling software package called Visual UML made by Visual Object Modelers, Inc.

Sequence Diagrams

UML sequence diagrams model the flow of logic within the system in a visual manner. They enable the software designer to document and validate the program logic, and are commonly used for both analysis and design purposes. Sequence diagrams are a favorite

tool for dynamic modeling.

Sequence diagrams are about deciding and modeling. It details how the system will accomplish what the use case models have defined in the requirements gathering phase. Here is where we decide and describe how the system will create and store maintenance schedules or how adding a new equipment item will be handled.

Unit Testing

With the use cases, we can test each functional area of the program to make sure it is working. By daily testing, using a technique called "build and smoke" testing, the errors are caught and repaired quickly and the process moves forward. Once unit testing is complete, the software is assembled and made ready for system testing.

System Testing

This is the final phase before deployment of the software. The goal of system testing is to verify that the code satisfies all requirements. The objectives of system testing are as follows:

1. To measure the stability of the system and determine operating system specifications.
2. To measure the system's performance. (Issues such as speed and accuracy of the data retrieval process)
3. To measure the overall usability of the system.

Once the system has passed system testing it can be deployed. Our application has two user interfaces: a browser interface for our sites with web access and a standard application interface. Deployment can be done from the company's intranet or the software and support files can be distributed on media such as a CD or a flash memory USB drive.

SUMMARY AND FUTURE DIRECTIONS

This chapter has reviewed the development history of Cooper Compression's in-house CMMS system called maintenance management system (MMS). Computerized maintenance management system software maintains a schedule for preventive maintenance on operating equipment. The software can generate complex analytical reports and record the maintenance history of any machinery. Compared to periodic maintenance, predictive maintenance is what is known as

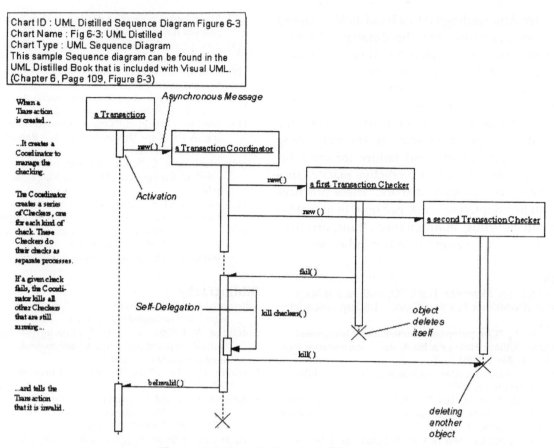

Figure 29-7. Sample Sequence Diagram

condition-based maintenance. It records trend values, by measuring and analyzing data and uses a surveillance system, designed to monitor conditions via an on-line system.

A good CMMS will increase the effectiveness and overall efficiency of any company's maintenance program. MMS is helping the Maintenance Technology Services department control costs, save administrative time, and increase customer service and satisfaction. And, because MMS was designed to be a scaleable

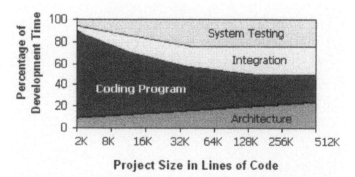

Figure 29-8. Determining Development Time by Software Size[8]

system, it supports the client's needs today and will continue to do so well into the future.

As a rule, web-based systems can be rapidly deployed. To implement a standard CMMS system can be complex and normally requires the coordination of several departments. One of the chief benefits of using web-based software is that the software does not have to be installed on the computer, all support and maintenance is handled on the back end at the server. This means there are no computers to configure or costly hardware upgrades. Having the software run from a single source also simplifies troubleshooting and error detection efforts.

Another advantage of using web-based software is mobility. Web-based programs can be designed to operate in the occasionally connected computing (OCC) environment. In the past, maintaining an accurate history of equipment in remote locations was an involved process for Cooper Compression field service personnel. Large volumes of data entered into the system from remote areas had to be manually uploaded to the LAN for planners to review and evaluate.

By enabling the program to work off-line, technicians can enter data readings into a hand-held or laptop for later synchronization into the database. Because the upload process is automated, the time to populate a data source is streamlined. Computer resources are used efficiently and "manual" steps in the data synchronization process are reduced.

Today, with competition in industry at an all time high, cost effective practices may be the only thing that stands between growth and failure for some organizations. Technology-driven predictive maintenance programs have been proven to work. These practices and tools can be adapted to work not only in industrial plants, but in building maintenance, transportation, construction, and in a variety of other industries.

References

[1] Sullivan, Pugh, Melendez, Hunt, "Operations and Maintenance (O&M) Best Practices Guide," US Department of Energy, 2002

[2] Labib, A. W., "Computerized Maintenance Management Systems—A Black Hole or a Black Box," Maintenance Journal, Vol. 17 No 1, 2004

[3] Oberg, P., "Managing Maintenance as a Business," EPAC Systems Handbook, 2004

[4] Universal Oil Products (UOP), "Reliability, Availability, Inspection, and Equipment Support Services," *http://www.uop.com/services/2060.html*.

[5] Weir, "An Impartial View of CMMS Functions, Selection and Implementation," The Plant Maintenance Resource Center, 2002

[6] US Department of Energy, *http://www.eere.energy.gov*

[7] *http://domino.automation.rockwell.com/applications/gs/region/gtswebst.nsf/pages/Integrated_Condition_Monitoring*

[8] McConnell, S., "Rapid Development," Redmond, Wash, Microsoft Press. 1996.

[9] Ulrich, W., "Legacy Business Rule Capture: Last Piece of the Transformation Puzzle," *http://www.systemtransformation.com/IT_Arch_Transformation_Articles/arch_legacybrc.htm*, 2004

[10] Jacobson, Ivar et al, "Object-Oriented Software Engineering: A Use-Case Driven Approach," Reading, MA, Addison-Wesley, 1992

[11] Fair-Wright, C. "UML Distilled: Interview with Martin Fowler," Component Developer Magazine, Issue 3. 2001. p34

[12] Komatineni, S., "Qualities of a Good Middle-Tier Architecture," *http://www.radford.edu/~wkovarik/design/ch1.html*, 2003

[13] "Microsoft Architecture Decisions for Dynamic Web Applications: Performance, Scalability, and Reliability," *http://msdn.microsoft.com/library/en-us/dnnile/html/docu2kbench.asp*

Bibliography

Automated Buildings website, *http://www.automatedbuildings.com*

Coleridge, R. (1999). "An Introduction to the Duwamish Books Sample Application," *http://www.aspxnet.de/XML_RK/visualstudio/duwamish/dw1intro.htm*

Edelstein, H. (May 1994). "Unraveling Client/Server Architecture," DBMS 7, 5: 34(7).

Federspiel, C.C. and L. Villafana. (2003). "A Tenant Interface for Energy and Maintenance Systems." CHI 2003, Florida, USA. ACM 1-58113-630-7/03/0004.

Levi, M., D. McBride, S. May, M.A. Piette, and S. Kinney. (2002). "GEMnet Status and Accomplishments: GSA's Energy and Maintenance Network." Proceedings of 2002 ACEEE Summer Study on Energy Efficiency in Buildings. LBNL-50733.

Norman, D. (1998). The Invisible Computer. Cambridge, MA: The MIT Press, 113-115.

Chapter 30

Automated Commissioning for Lower-cost, Widely Deployed Building Commissioning of the Future*

Michael R. Brambley
Srinivas Katipamula

How is commissioning likely to be performed 20 or 30 years from now? Will technology advance to the state where building systems and equipment are automatically commissioned on an on-going basis as conditions demand? The authors of this chapter believe it will and that technologies emerging from research and development today—wireless communication, automated fault detection and diagnostics, advanced controls, and inexpensive smart sensors—are the seeds for the technology that will provide these capabilities in the future.

This chapter takes a brief look at the benefits of commissioning and describes a vision of the future where most of the objectives of commissioning will be accomplished automatically by capabilities built into the building systems themselves. Commissioning will become an activity that is performed continuously rather than periodically, and only repairs requiring replacement or overhaul of equipment will require manual intervention. This chapter then identifies some of the technologies that will be needed to realize this vision and ends with a call for all involved in the enterprise of building commissioning and automation to embrace and dedicate themselves to a future of automated commissioning.

THE VALUE OF COMMISSIONING

Retro-commissioning of existing buildings reduces energy use by anywhere from a few percent to over 60% of pre-retro-commissioning consumption, with most savings falling into the range of 10% to 30%. [3]–[6] These savings result from detecting and correcting faults such as:

- incorrectly installed equipment (e.g., backwards fans) and sensors,

- incorrectly implemented control algorithms (e.g., economizing cycles),

- inefficient set points,

- equipment and lighting operating during unscheduled times,

- equipment operating at degraded performance (e.g., low refrigerant charges in vapor-compression equipment) and

- missing, failed and uncalibrated sensors.

Energy savings are more difficult to establish for new construction, but by using building simulation, the savings associated with commissioning can be estimated. A meta-analysis found median cost savings on energy of 15% and payback periods of 0.7 years for retro-commissioning of existing buildings, while for commissioning of new buildings, the study found a median payback period of 4.8 years. [7]

While these savings are impressive, penetration of commissioning* in the buildings market remains small; in the U.S., less than 5% of new buildings and less than 0.03% of existing buildings have been commissioned. [8] A 2003 report on commissioning in public buildings states "While the concept of commissioning is increasingly accepted, there are still barriers—particularly with regard to cost—to implementation of the kind of thorough, independent third-party commissioning that

---

*This is an update and expansion of earlier versions of this chapter published as References [1] and [2].

*Throughout this chapter, the authors use the term "commissioning" to represent generically all forms of the process applied to both new buildings during design and construction and existing buildings already in operation. The context establishes which use is intended.

is necessary for the full benefits of commissioning to be realized." [9] Even when performed, pressures exist to keep costs down, which in some cases limits the thoroughness to which the commissioning is performed.

The two factors that likely play the most significant roles in limiting the widespread adoption of commissioning are: 1) lack of knowledge of the benefits of commissioning by key decision makers and 2) the cost of commissioning. This chapter focuses on the second of these by using automation of commissioning processes to reduce labor requirements and, as a result, decrease the cost of commissioning substantially. Fortunately, reducing cost also reduces barriers in decision making, making acceptance by decision makers easier even in the absence of greater information and knowledge. Changes to how commissioning is executed that reduce its cost and make it a routine part of how building systems are designed, installed, and operated could in the long run better promote the objectives of commissioning than keeping it a distinct practice, if by doing so activities producing equivalent results to commissioning penetrate the buildings segment more rapidly and more completely. Key to this is reducing the labor intensity of commissioning by automating as many of the processes involved as possible. Compared to the cost of labor, automation technology is inexpensive.

HYPOTHESIS—THERE'S A BETTER WAY

Commissioning provides important benefits to both new and existing buildings, but there may be a better way to achieve these benefits. In the long-term future (say 20 or 30 years from now), most (but not all) of the objectives of commissioning could be provided using automated processes, reducing barriers that exist today for commissioning and impacting a much larger portion of the building stock. Furthermore, doing so could increase consistency in the commissioning process, improve the reliability of building systems, and make the process of assessing performance, which is a critical part of retro-commissioning, a truly continuous process.

How might this be done? First, identify processes that produce the desired outcomes that could be done automatically. Then develop building equipment, systems, control systems, and tools that implement these processes automatically. Gaps between the automated processes could be filled by manual procedures, ensuring that the activities that absolutely require human intervention and tie the automated processes together are executed efficiently and cost effectively.

In the next two sections, we first provide a vision

for commissioning in the distant future after the automated processes are fully developed and implemented, then take a critical look at commissioning to identify the processes that require direct human involvement and processes that do not and could be automated. We then identify technologies that are key to realizing the automated processes and conclude by identifying research and development that will be essential to accomplishing our vision for commissioning of the future.

In doing this, we are not suggesting that commissioning and efforts to promote it be suddenly terminated today. On the contrary, commissioning as conducted today is important as a transition process. It provides important benefits, but as with most things, it will benefit from improvement over time. Our vision is for the long term. It will not be realized tomorrow, but the vision can help guide research and development (R&D) and product development decisions so they lead us to a future where the benefits of commissioning permeate the entire building enterprise. Likewise, as this technology and tools embodying it emerge, commissioning should adapt and change to assimilate these new capabilities, making the commissioning process faster, less expensive, more thorough, more consistent, more reliable, more cost effective, more continuous, and appealing to more of the market.

COMMISSIONING FOR THE FUTURE— HIGHLY AUTOMATED AND EFFICIENT

When operation of a new building or new piece of equipment or system in a building is started in the future, with the push of a "start" button, equipment and systems should all test themselves, identify any installation or configuration problems, automatically fix problems amenable to "soft" solutions, and report the need for "hard" solutions requiring replacement or installation of hardware. A report on the performance of all building systems and equipment should be automatically generated, delivered to key recipients, and stored electronically for future reference and updating. During initial operation (e.g., for the first year of a building's life) and continuing over its lifetime, the system should optimize itself, integrating its behavior with external constraints, such as occupancy levels, occupant behavior and feedback, energy prices, energy demand charges, and weather. Although most of the optimization should take place during this initial period of say a year, the building systems should continue to optimize themselves as prices change, spaces are converted to different uses, tenants change, and building equipment itself ages and wears. Using prognostics, systems should automatically inform building staff regarding expected lives and recommended service times for equipment plus the costs of waiting to service the equipment until some time in the future. Automated diagnostics should detect degradation of system performance and failure, make "soft" fixes when possible (alerting building staff to changes made and electronically documenting changes automatically made for future reference), and alert staff to performance degradation, impending failures and required maintenance to prevent them.

Realization of this vision will present unique technological challenges, but methods under development (some examples of which are provided in the next section) are addressing these. Technology will evolve over time and change the practice of commissioning, bringing benefits to building owners and occupants. Ours is a vision for buildings of the future that automatically will perform many of the actions required to meet the objectives of commissioning, where technology will provide the cornerstone for achieving this future.

WHAT PARTS OF COMMISSIONING COULD BE AUTOMATED?

The major tasks composing commissioning of new buildings and retro-commissioning of existing buildings are shown in Table 30-1 by phase of the project. Some processes are critical to accomplishing the objectives of commissioning. Other processes exist specifically to support commissioning as it is performed today. If the overall process of commissioning were changed, some of these processes might become unnecessary or would be modified considerably. The same overall objectives might be achieved without an identical set of processes. Table 30-1 includes designations of whether activities would be manual (M) or automated (A) in a highly automated future for commissioning. Commissioning objectives would be established largely by standards and specialized to specific projects by detailed objectives being automatically inherited by associating objectives with generic types of commissioning projects. For retro-commissioning, most activities would become a routine part of building operation and maintenance and might not require explicit development of objectives. Similarly, objectives associated with equipment startup would be unneeded because all equipment would be automatically started up and tested with standard automated start-up routines satisfying standard objectives.

Commissioning plans similarly would be developed automatically based on information obtained from design documents, and only limited information would be input manually (e.g., special constraints on schedules). Design information would be automatically stored as developed and shared throughout the life of the building. This would include the objectives and the intent behind the design. This sort of information is beginning to be captured and shared today with building information models (BIM).[10] Automatic storage and universal data sharing protocols (using BIM) would eliminate the need to manually take off information from drawings or re-input information developed in earlier phases, which is labor intensive today. Ultimately, even existing buildings that were designed before automated data storage was routinely used, will possess systems that will automatically detect all aspects of the building, systems and equipment installed, generate the equivalent as-built documents, and evaluate the design.

Designs will be automatically evaluated with respect to meeting design intents as well as energy and other standards. For many years, researchers have studied the design process and developed methods for automating both design generation and evaluation. A sampling of issues and advancements in design automation can be found in References [10]-[15]. Research in these fields will provide the basis for automation of the review and the revision of designs performed as part of commissioning.

To the extent that commissioning specifications are still required in bid documents far in the future, most of the required language might be generated automatically, reviewed manually, and revised manually in special cases where required. Eventually, though, all documents will be reviewed automatically. Computer-based tools will parse text, "interpret" the meaning, and evaluate the design with respect to needs and design criteria. Given information about the equipment and systems in a building or specified by the design, checklists (to the extent they are still needed) could be generated automatically. With some exceptions, most checklists would be eliminated because checks would be performed automatically. Lists might remain only as an organizational form for presenting results to human users. Just-in-time facility documentation [16] may become the basis for operation and even parts of commissioning and retro-commissioning. Even proper installation of equipment (e.g., whether any fans are installed backwards) could initially be checked automatically. Some problems might require visual inspection after initial automatic detection, but the labor for this would be highly targeted to problem situations, limiting labor requirements and costs.

Table 30-1. Major Commissioning and Retro-commissioning Activities

New Construction Commissioning	*Retro-commissioning Existing Buildings*

1. Conceptual or pre-design phase
  a. Develop commissioning objectives (A)
  b. Hire commissioning provider (M)
  c. Develop design phase commissioning requirements (A)
  d. Choose the design team (M)
	1. Planning phase
	  a. Develop commissioning objectives (A)
	  b. Hire commissioning provider (M)
	  c. Review available documentation and obtain historical utility data (A)
	  d. Develop retro-commissioning plan (A)

2. Design phase
  a. Commissioning review of design intent (A)
  b. Write commissioning specifications for documents (A)
  c. Award job to contractor (M)
	(No design phase activities)

3. Construction/installation phase
  a. Gather and review documentation (A)
  b. Hold commissioning scoping meeting and finalize plan (M)
  c. Develop pre-test checklists (A)
  d. Start up equipment or perform pre-test checklists to ensure readiness for functional testing during acceptance (A)
	2. Investigation phase
	  a. Perform site assessment (M/A)
	  b. Obtain or develop missing documentation (A)
	  c. Develop and execute diagnostic monitoring and test plans (A)
	  d. Develop and execute functional test plans (A)
	  e. Analyze results (A)
	  f. Develop master list of deficiencies and improvements (A)
	  g. Recommend most cost-effective improvements for implementation (A)

4. Acceptance phase
  a. Execute functional tests and diagnostics (A)
  b. Fix deficiencies (M)
  c. Re-test and monitor as needed (A)
  d. Verify operator training (A)
  e. Review O&M manuals (A)
  f. Building/retrofit accepted by owner (M)
	3. Implementation phase
	  a. Implement repairs and improvements (M)
	  b. Re-test and re-monitor for results (A)
	  c. Fine-tune improvements if needed (A)
	  d. Revise estimated energy savings calculations (A)

5. Post-acceptance phase
  a. Prepare and submit final report (M/A)
  b. Perform deferred tests (if needed) (A)
  c. Develop re-commissioning plan/schedule (A)
	4. Project hand-off and integration phase
	  a. Prepare and submit final report (M/A)
	  b. Perform deferred tests (if needed) (A)
	  c. Develop re-commissioning plan/schedule (A)

Source of original table without M and A designations: Haasl and Sharp 1999 [4]

All testing, data collection, analysis, and interpretation of results would be performed automatically. Examples of how some tests could be executed automatically today are given by references [17]-[19]. These capabilities are based on research and development in the fields of automated fault detection, diagnostics and prognostics. (See references [20] and [21] for a comprehensive review of fault detection, diagnostic, and prognostic methods and reference [22] for a review of early fault detection and diagnostic tools.) "Fixing deficiencies" and "implementing repairs and improvements" are designated in Table 30-1 as being done manually; however, only repairs and improvements requiring

physical repair, replacement, or reinstallation require human intervention. As shown in references [17] and [18], some repairs, such as revising control code, changing set points, and recalibrating sensors, might be done automatically with no human intervention except to read a short report from the computerized system regarding actions it took. Algorithms used to automatically detect, diagnose, characterize and cor¬rect selected sensor, damper and control faults in heating, ventilating and air-conditioning systems are given in references [24] and [25]. Automatically re-tuning of control algorithms is also possible today for some applications, and most tuning will be done automatically in the long-term future.

As indicated in Table 30-1, most commissioning activities will be done automatically at some time in the future. People will still need to coordinate the processes and ensure that reporting to owners and management is appropriate, but many of the commissioning activities executed manually today will become automatic. This transformation will reduce the labor, time, and cost of commissioning and help overcome some of the key barriers that widespread application of commissioning faces today. Reaching that future, however, will require advances in key enabling technologies and then application of them to building systems. Table 30-2 provides a list of key technologies needed to achieve this future and the capabilities for commissioning that each might provide.

Wireless data communication will eliminate many of the wires required today to collect data or transmit control signals to device actuators. Wires can represent a significant fraction of the cost of a sensor or control point. As a result, wireless communication for sensors and controls will enable more ubiquitous use of sensing, increasing information on the operating state of systems and equipment available at any point in time and enabling better control and maintenance [23]. Plug and play controls and equipment will enable quicker installation and set up of physical systems and controls. Controls will ultimately become self-writing, given some input on the performance objectives for the building and equipment characteristics. Small, embedded, networked processors will distribute control to a greater degree than today's control systems, leading to better, higher resolution, system response while coordinating through networking with other subsystems and components to achieve building-level objectives.

Automated fault detection and diagnostics

Table 30-2. Technologies Needed for Highly Automated Commissioning

Technology	Potential Applications
Wireless sensing, data acquisition, and control	Cost effective sensing and data collection Condition monitoring
Plug and play building equipment and controls	Self-identifying equipment and automatic system design recognition Rapid automatic self-configuration of controls Automatic control algorithm selection and application
Embedded networked sensing and processing	Highly distributed processing of information with local control capabilities coordinated to meet system and building level objectives
Automated fault detection, diagnostics, and prognostics	Automatic detection and diagnosis of operation, equipment, and control faults Automatic detection and diagnosis of designs and hardware installations Anticipation of system and equipment degradation based on condition monitoring Automatic generation of maintenance plans Condition-based maintenance
Automated proactive testing	Automated start up and functional tests, analysis of data, and interpretation of results Continual automated monitoring and testing
Automatic records management and data exchange protocols	Automatic generation of plans and reports Automatic storage of data Automated asset tracking Automatic project management assistance

will lead to greater awareness of system conditions throughout buildings on a continuous basis (see Figure 30-1). Corrective actions will be enacted automatically by "aware" agents capable of correcting faults in some cases (e.g., correcting a control schedule or fixing an incorrect set point). In cases where automatic fault correction is not possible, notifications will be provided to building staff and management regarding faults and their costs. No longer will faults go unrecognized or will an engineer need to study data patterns to detect them. The operating state of building systems will be known, along with the performance and cost impacts of problems, so priorities for operation and maintenance can be made with complete information. Prognostic techniques will automatically predict the remaining serviceable life of equipment and suggest condition-based maintenance actions. Automated proactive testing (as in reference [24]) will be the basis for short-term functional testing. These tests allow a wide range of conditions to be simulated over a relatively short period of time so that problems can be detected faster than if only passive observation of routine operation is used. Proactive testing will enable consistent performance of functional tests automatically during initial commissioning and then at regular periods or when needed throughout the life of the building.

Fault detection and diagnostic methods will have applications in design review in addition to use on physical components. Diagnosis of design is similar to diagnosis of a physical device. First a problem or fault is detected with the design. Evaluation of the design indicates that it does not satisfy some design criterion (requirement). This is analogous to fault detection. Then the reason for the fault (its cause) is identified or isolated, which is analogous to fault diagnosis or isolation. The design then needs to be revised to correct the deficiency, which parallels fault correction. When this entire process is automated, it will provide continuous review and evaluation of designs as they evolve. This will likely be done by automated agents (software processes whose purpose is to execute part of the design review and report the results), each of which is responsible for evaluation with respect to a small subdomain. Some of these agents will specifically handle evaluations from the perspectives of commissioning.

Data exchange protocols will provide the basis for sharing data among automated agents as well as commissioning professionals, operating staff, and facility management. Radio frequency identification tags will also play a role in tracking assets as well as enabling easy, automatic identification of each piece of equipment and component, enabling automatic checking for consistency with specifications as equipment arrives on the construction site, and assessing its installation. Tags may also provide physical and performance characteristics from manufacturer tests, which then will become available to processes that evaluate the correctness of installation, develop control algorithms, evaluate functional test results, and monitor performance. Geographic information systems (GIS) as well as localization algorithms using wireless communications information (see, for example, reference [24]) will be used to determine locations of equipment. Together these technologies will enable realization of highly automated commissioning and operation.

THE PATH TO THIS FUTURE

The impediments to realizing a future where building commissioning and retro-commissioning are largely automated are technological, social and institutional. Without the technology, however, the vision is not possible. With it, automated capabilities for executing all but the repair, replacement, and some management activities of retro-commissioning could be delivered as parts of equipment packages and control systems. With the addition of design tools, this could be extended to commissioning of new buildings. Efforts already underway are beginning to develop tools that automate parts of the commissioning process or provide assistance with parts of commissioning. [17]-[19],[25]-[27] These are initial attempts at using automation to improve commissioning. To realize the full benefits of automated commissioning, advances in each of the technologies identified in Table 30-2 will be needed.

Because the buildings industries are highly fragmented, public R&D organizations will need to provide leadership to produce this technology. Even then, a market demand will need to develop to drive the creation of new equipment and control systems with automated commissioning capabilities. The building commissioning industry will evolve, gaining market share over time as energy and electric power prices increase and more burden for management of the electric power grid is pushed to end users (see, for example, http://gridwise.pnl.gov/ for a vision of the future electric power grid in which "customers" play an active role). Penetrating the market will require improved cost effectiveness for commissioning, as well as education of building owners and operators regarding the benefits of commissioning. Commissioning will need to change in ways that reduce cost while preserving or

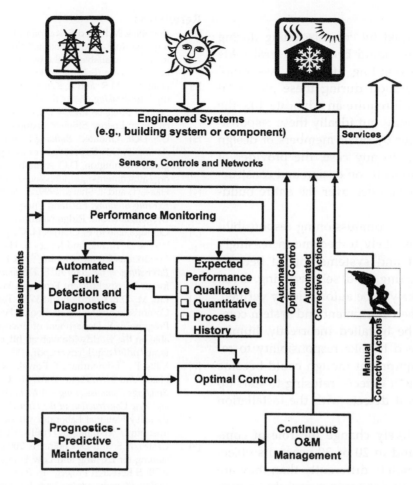

Figure 30-1. Advanced Automated Retro-Commissioning Process

even enhancing the returns on it. The practice of commissioning will likely change gradually over time with the introduction of new tools that automate parts of the process. Enabling this, however, will require investment in research and development of new automated capabilities.

Market transformation programs at the federal, regional, state and local levels can help spread the application of commissioning for the public good. Research, development, deployment and market transformation programs will be needed to accelerate the introduction of automated capabilities and the spread of commissioning, improving the performance of the building stock and bringing energy and environmental benefits. Still, the willingness of the commissioning profession to accept and embrace these technologies will be critical to determining their rate of penetration. Resistance won't stop the introduction of the technologies, only delay their application, but earlier acceptance will help accelerate capture of the benefits associated with high-quality, widespread commissioning even if the mechanism of delivery changes.

CONCLUSIONS—A CHALLENGE FOR THE COMMISSIONING COMMUNITY

Automation could change the nature of the commissioning process for both new and existing buildings. The services performed today as part of commissioning of existing buildings should become part of routine operation and maintenance with automated monitoring, testing, and diagnosis continually performed by the building systems and equipment themselves, taking much of the responsibility off humans.

Still, repairs and replacement of hardware will continue to require human intervention. Deteriorating bearings in pumps, failed windings in fan motors, and leaking valves will need humans to repair or replace them. Automation can only prompt repair technicians to take action to make repairs. Likewise, equipment found to be installed incorrectly during construction (e.g., a fan installed backwards) would require human involvement to remove and reinstall it properly. For the commissioning cycle to be complete, even in the long term, humans will still need to respond to information provided automatically.

The services provided by commissioning during design and construction should become integral parts of those phases of the building life cycle. Assurance of their proper consideration during these phases of building projects may still require an advocate, like the commissioning agent today, but ideally these responsibilities will be taken over by other members of design and construction teams. In any case, the provider of these services is likely to focus on design and construction, rather than responsibilities over the entire building life cycle.

System start up, like commissioning responsibilities during operation, is likely to become increasingly automated. Equipment and systems should become self-configuring, self-testing, and self-verifying. Even proper installation is likely to be automatically verified. Once again, though, when equipment and system components are found to be installed incorrectly, human technicians will still need to take responsibility to repair the installation. Equipment though could become intolerant of some faults it detects, refusing to start up until all such problems it detects with the installation are corrected.

Automation will likely change the role of commissioning over time and in 20 or 30 years, its objectives may be met completely differently than they are today. These changes will not occur overnight or even in a few years, but rather over many years, but they should lead to more cost effective delivery of the outcomes promoted by commissioning to a much broader segment of the commercial buildings market. Change is inevitable and will bring benefits. As with use of automation in design [28], detractors will find objections to greater use of automation in commissioning; proponents will grasp increased automation as an opportunity. The authors recommend that the building commissioning community embrace the opportunities posed by new technology and employ them to deliver better services.

Research and development will be required to achieve the benefits of greater automation in commissioning but so will adoption by the various players in the commissioning and broader buildings communities. Researchers and providers of services alike have the opportunity to transform the delivery of commissioning's objectives by working in concert to pursue a vision in which those objectives are delivered faster, less expensively, more thoroughly, more consistently, more reliably, more cost effectively, and more continuously, to a broader market through automation.

References

[1]   Brambley, M.R. and S. Katipamula. 2005. "Beyond Commissioning: The Role of Automation." Published by Automatedbuildings.com. Available on the worldwide web at www.automatedbuildings.com (February 2005).

[2]   Brambley, M.R. and S. Katipamula. 2004. "Beyond Commissioning." In *Breaking Out of the Box, Proceedings of the 2004 ACEEE Summer Study on Energy Efficiency in Buildings.* American Council for an Energy Efficient Economy, Washington, DC.

[3]   U.S. DOE. Undated. *Building Commissioning—The Key to Quality Assurance.* U.S. Department of Energy, Rebuild America Program, Washington, D.C. Available online: http://www.rebuild.org/attachments/guidebooks/commissioningguide.pdf.

[4]   Haasl, T. and T. Sharp. 1999. *A Practical Guide for Commissioning Existing Building.* Portland Energy Conservation Inc., Portland, Oregon, and Oak Ridge National Laboratory, Oak Ridge, TN.

[5]   Claridge, David E, Charles H. Culp, Mengsheng Liu, S. Deng, Wayne D. Turner, and Jeffery S. Haberl. 2000. "Campus-Wide Continuous Commissioning[SM] of University Buildings." In *Proceedings of the 2000 ACEEE Summer Study.* American Council for an Energy Efficient Economy, Washington, D.C.

[6]   Liu, M., D.E. Claridge, and W.D. Turner. 2002. "Continuous Commissioning[SM] Guide Book." Federal Energy Management Program, U.S. Department of Energy, Washington, D.C. Available on the worldwide web at: http://www.eere.energy.gov/femp/pdfs/ccg01_covers.pdf.

[7]   Mills, E., H. Friedman, T. Powell, N. Bourassa, D. Claridge, T. Haasl, and M.A. Piette. 2004. *The Cost-Effectiveness of Commercial Buildings Commissioning—A Meta-Analysis of Existing Buildings and New Construction in the United States.* LBNL-56637. Lawrence Berkeley National Laboratory, Berkeley, California. Available online: http://eetd.lbl.gov/emills/PUBS/Cx-Costs-Benefits.html.

[8]   Castro, N.S. and D. Choineire. 2006. "Cost Effective Commissioning for Existing and Low Energy Buildings—A New IEA ECBCS Research Project." In *Proceedings: National Conference on Building Commissioning 2006.* Portland Energy Conservation Inc., Portland, Oregon.

[9]   Quantum Consulting. 2003. *Market Progress Evaluation Report—Commissioning In Public Buildings Project, No. 3.* Report #E03-107. Northwest Energy Efficiency Alliance, Portland, Oregon. Available online: http://www.nwalliance.org/resources/reports/107.pdf.

[10] Eastman, C., P. Teicholz, R. Sacks and K. Liston. 2008. BIM Handbook: A Guide to Building Information Modeling for Owners, Managers, Designers, Engineers and Contractors, Wiley, Hoboken, New Jersey.

[11] Gero, J.S. 2000. "Developments in Computer-Aided Design." In *INCITE 2000,* H. Li, Q. Shen, D. Scott and P. Love (eds.), pp. 16-24. HKPU Press, Hung Hom, Kowloon, Hong Kong.

[12] Caldas, L.G. and L.K. Norford. 2002. "A Design Optimization Tool Based on a Genetic Algorithm." *Automation in Construction* 11(2):173-184.

[13] Iliescu, S., P. Fazio and K. Gowri. 2000. "Similarity Assessment in a Case-Based Reasoning Framework for Building Envelope Design." In *INCITE 2000,* H. Li, Q. Shen, D. Scott and P. Love (eds.), pp. 697-713. HKPU Press, Hung Hom, Kowloon, Hong Kong.

[14] Fleming, U. and R. Waterbury. 1995. "Software Environment to Support Early Phases in Building Design (SEED): Overview." *Journal of Architectural Engineering* 1(4):147-152.

[15] Fleming, U. and Z. Aygen. 2001. "A Hybrid Representation of Architectural Precedents." *Automation in Construction* 10(6):687-699.

[16] Song, Y., M.J. Clayton, and R.E. Johnson. 2002. "Anticipating Reuse: Documenting Buildings for Operations Using Web Technology." *Automation in Construction* 11(2):185-197.

[17] Katipamula, S., M.R. Brambley, and L. Luskay. 2003. "Auto-

mated Proactive Techniques for Commissioning Air-Handling Units." *ASME Journal of Solar Energy Engineering, Transactions of the ASME*, Special Issue on Emerging Trends in Building Design, Diagnosis and Operation 125(1):282-291.

[18]   Portland Energy Conservation, Inc. (PECI) and Battelle Northwest Division. 2003. *Methods for Automated and Continuous Commissioning of Building Systems.* Final Report. ARTI-21CR/610-30040-01. Air- Conditioning & Refrigeration Technology Institute, Washington, D.C. Available online: www.arti-21cr. org/research/completed/finalreports/30040-final.pdf.

[19]   Brambley, M.R. and S. Katipamula. 2003. "Automating Commissioning Activities: Update with Examples." *In Proceedings of the 11th National Conference on Building Commissioning*, May 20-22, 2003. Portland Energy Conservation Inc., Portland, Oregon.

[20]   Katipamula, S. and M.R. Brambley. 2005. "Methods for Fault Detection, Diagnostics and Prognostics for Building Systems—A Review Part 1." *International Journal of Heating, Ventilating, Air-Conditioning and Refrigerating Research* 11(1):3-25.

[21]   Katipamula, S. and M.R. Brambley. 2005. "Methods for Fault Detection, Diagnostics and Prognostics for Building Systems—A Review Part 2." *International Journal of Heating, Ventilating, Air-Conditioning and Refrigerating Research* 11(2):169-187.

[22]   Friedman, H. and M.A. Piette. 2001. *Comparative Guide to Emerging Diagnostic Tools for Large Commercial HVAC Systems.* LBNL 48629. Lawrence Berkeley National Laboratory, Berkeley, CA. Available online: http://www.peci.org/library/PECI_DxTools-Guide1_1002.pdf.

[23]   Brambley M.R., M. Kintner-Meyer, S. Katipamula, and P. O'Neill. 2005. "Wireless Sensor Applications for Building Operation and Management." Chapter 27 in *Information Technology for Energy Managers, Volume II—Web Based Energy Information and Control Systems Case Studies and Applications*, B.L. Capehart and L.C. Capehart (eds.), pp. 341-367. Fairmont Press/CRC Press, Lilburn, GA.

[24]   Fernandez, N., M.R. Brambley and S. Katipamula. 2009. *Self-Correcting HVAC Controls: Algorithms for Sensors and Dampers in Air-Handling Units*, PNNL-19104. Pacific Northwest National Laboratory, Richland, WA.

[25]   Fernandez, N., M.R. Brambley, S. Katipamula, H. Cho, J. Goddard and L. Dinh. 2009. *Self-Correcting HVAC Controls Project Final Report*, PNNL-19074. Pacific Northwest National Laboratory, Richland, WA.

[26]   Castro, N.S. and H. Vaezi-Nejad. 2005. "CITE-AHU: An Automated Commissioning Tool for Air Handling Units." In *Proceedings: National Conference on Building Commissioning 2005*, Portland Energy Conservation Inc., Portland, Oregon.

[27]   Salsbury, T.I. and R.C. Diamond. 1999. "Automated Testing of HVAC Systems for Commissioning." In *Proceedings of the National Conference on Building Commissioning 1999*, Portland Energy Conservation Inc., Portland, Oregon.

[28]   Chastain, T., Y.E. Kalay and C. Peri. 2002. "Square Peg in a Round Hole or Horseless Carriage? Reflections on the Use of Computing in Architecture." Automation in Construction 11(2):237-248.

APPENDIX: PROCESS FOR AUTOMATED RETRO-COMMISSIONING OF SENSORS [17]

Isolation of Outdoor-, Return- and Mixed-Air Temperature Sensor Problems

The process described in this appendix follows identification of a problem with the outdoor-, return-,

*See references [17] and [18] for detailed information on the passive fault detection and diagnosis process.

or mixed-air temperature sensor in an air-handling unit (AHU) by routine passive (observational) automated monitoring and fault detection during operation.* One of these sensors is faulty but which specific one is not known from the passive process. The active process described here would be executed automatically immediately following detection of the problem or at some later time (e.g., overnight while the building is unoccupied) to isolate the cause of the fault and to automatically implement a (possibly temporary) corrective action.

In an AHU, the return- and outdoor-air streams are mixed and the resulting air stream is called the mixed-air stream. Therefore, the fundamental equations for sensible energy balance along with positioning of the return-air and the outdoor-air dampers can be used to isolate the fault. Placing the dampers at specific positions in this case provides analytical redundancy, which provides additional information.

As shown in Figure 30-2, the first step in the proactive diagnostic process is to close the outdoor-air damper completely and wait for the conditions to reach steady-state, which usually occurs within a few minutes. While keeping the outdoor-air damper fully closed, the return-air and mixed-air temperatures are sampled for a few minutes. With 100% of the return-air recirculated, the average mixed-air temperature should nearly equal the return-air temperature. If this is found, then the return-air and mixed-air temperature sensors are consistent with one another and, because one of the three sensors has failed, the outdoor-air temperature sensor must be faulty.

If the return-air and the mixed-air temperatures are not approximately equal, command the outdoor-air dampers to open fully and wait until steady-state conditions are achieved. When the outdoor-air damper is fully open, no return air recirculates and the average mixed-air temperature should approximately equal the average outdoor-air temperature during the sampling period. If this condition is found, then the outdoor-air and mixed-air temperature sensors are consistent with one another, and the return-air temperature sensor is faulty. If the measured mixed-air temperature does not equal the measured outdoor-air temperature, then the mixed-air temperature sensor is faulty (because earlier the return-air temperature sensor was found fault-free).

After isolating the faulty sensor, further diagnosis can identify the underlying cause or nature of the problem. In contrast to relative humidity, air flow, fluid flow, and pressure sensors, temperature sensors are more reliable, but they do exhibit erratic behavior occasionally. In addition to random noise, temperature sensors commonly drift and acquire bias over time. A process for detecting and estimating bias in temperature measurements is described in the next subsection. The ability to detect the drift over time does not require proactive testing; it can be detected using passive methods (see PECI and Battelle [18]).

Some notes of caution are appropriate for users of the process described here because tolerances of mechanical components can vary widely and change over time. All dampers possess seals to prevent leakage when they are fully closed. Some leakage, however, occurs around the seals, and as the AHU ages, the seals deteriorate, increasing the leakage. Under these conditions, when the return-air dampers are closed, the mixed air consists mostly of outdoor air but mixed with some leaked return air. As a result, the mixed-air temperature may not equal the outdoor-air tempera-

ture precisely. Therefore, in addition to allowing for measurement inaccuracies of the sensors, the equality tests in Figure 30-2 should also account for damper leakage. Compensation for these sources of uncertainty can be accomplished by relaxing (i.e., increasing) the tolerances on the equality tests. This may sometimes lead to incorrect identification of a faulty mixed-air sensor even when the outdoor-air or the return-air temperature sensor is slightly biased (because it is the least resistive path on the flow chart in Figure 30-2). These sorts of trade-offs between sensitivity of diagnosis to detect problems and the potential for false alarms or false diagnoses are best determined through field tests and experience.

Stratification of air in the mixing box leads to another potential source of error. The measured mixed-air temperature may vary significantly across the duct cross-section. As a result, the mixed-air temperature measured at a single point may differ significantly from

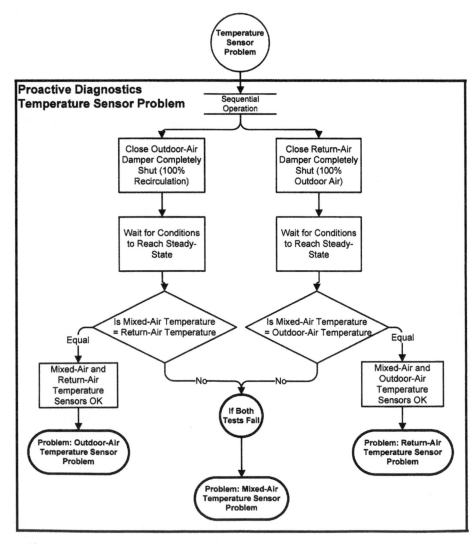

Figure 30-2. Decision Tree of Process to Isolate the Faulty Temperature Sensor

the average mixed-air temperature and lead to misleading diagnoses. To prevent this, the mixed-air temperature should always be measured across the duct and averaged using an averaging sensor.

Determining the Bias in an Outdoor-Air Temperature Sensor

In this section, we present an approach for classifying the nature of the fault found in an outdoor-air temperature sensor as an example of a method that can be applied to air-temperature sensors (see reference [18] for detailed schemes for other sensors).

Once a specific temperature sensor has been identified as faulty, further classification of the fault is possible. This section and the next describe a process for estimating the bias and reconfiguring the controls to compensate for it.

The first step in this proactive diagnostic process (see Figure 30-3) is to fully open the outdoor-air damper and wait for conditions to reach steady-state. In this case, values of the mixed-air temperature can be used to identify when steady-state conditions are attained, because at this point in the diagnostic process, we know that the mixed-air temperature sensor is good. One form of steady-state filter is based on the rate of change of the mixed-air temperature. If the rate of change is zero or below a predefined threshold, steady-state conditions have been achieved. After steady-state conditions are achieved, compute the difference between the outdoor-air and the mixed-air temperatures and store the result for further analysis.

The frequency of sampling and the duration of the proactive test depend on field conditions. A sampling rate of a minute or less and total test duration of 15 minutes should be sufficient in most cases. In some cases, the test may have to be performed at different times of the day to ensure that the bias is consistent at all hours of the day. In some cases, something as simple as positioning of the sensor may affect its readings. For example, an outdoor-air temperature sensor positioned so it is exposed to sunlight part of the day may read a few degrees high for those hours of the day, the amount depending on the position of the sun, but may otherwise read normal. This type of bias or problem is difficult to detect, unless the proactive test is repeated several times at different hours of the day and then correlated with other observations, such as solar position. An outdoor-air temperature sensor showing bias during certain hours of the day each day for many days in a row (but not at other hours) would indicate such a problem. As with uncertainty mentioned earlier, field tests are required to better understand these issues.

After the difference between the outdoor-air and the mixed-air temperatures is computed for the duration of the test at a desired sampling rate, the next step is the analysis of the stored data to confirm whether the difference is nearly constant over the entire test period. Commonly-used statistical tests such as the mean and the standard deviation of the sample are recommended. The mean provides the central tendency of the sampled data (the estimate of the bias), while the standard deviation provides the dispersion (how tightly the data are clustered around the mean).

In order for the test to be true (i.e., the difference nearly equal over the test period), the mean must be greater than the tolerance or the accuracy of the tem-

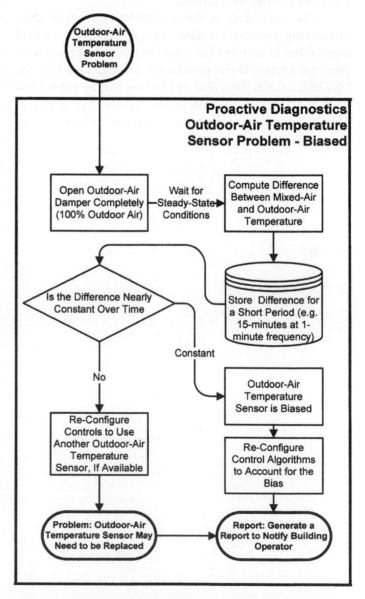

Figure 30-3. Decision Tree for the Process to Check Whether the Outdoor-Air Temperature Sensor is Biased and Implement a Temporary Correction.

perature sensors and the standard deviation should be reasonable. Another statistical metric called the coefficient of variation can be used to check whether the standard deviation is reasonable compared to the sample mean and the sensor tolerance. The coefficient of variation measures the relative scatter in data with respect to the mean; it is computed as the ratio of the standard deviation to the mean. A threshold for the coefficient of variation must be selected. Below this threshold, the standard deviation would be acceptable and the bias considered constant. Previous studies that used field data to develop empirical models have concluded that a coefficient of variation of about 15% is reasonable.

Reconfiguration of Controls

The final step in the automated proactive commissioning process involves reconfiguring the control algorithms to account for a bias in the outdoor-air temperature sensor. If the previous test concludes that the bias is constant, the value of the bias can be subtracted from the measured value of the outdoor-air temperature to obtain a correct value. If the test concludes that the temperature difference (bias) is not constant, then the controls can be reconfigured to use another properly functioning outdoor-air temperature sensor. Buildings often have several outdoor-air temperature sensors, and substitution of measurements from another outdoor-air temperature sensor should provide a reasonable value for control of the air handler with the faulty sensor. This obviously would not be possible for sensors measuring the return-air or mixed-air temperature, so when a variable bias is detected with one of them, the sensor must be replaced.

Any time controls have been reconfigured as the result of a proactive test, a report should be generated to notify the building manager or the building operator of this change. Then, when the sensor is repaired or replaced, this report will alert the manager or operator that the outdoor-air sensor used for control should be re-configured, removing any software corrections to measured values.

Proactive procedures similar to the one presented in this section can be developed for return-air, mixed-air, and supply-air temperatures (see reference [18]).

Section VIII

Energy Information Systems Case Studies

Chapter 31

The Utility Report Cards: An Energy Information System for Orange County Public Schools

Paul Allen, David Green, Safvat Kalaghchy
Bill Kivler, and Blanche Sheinkopf

ABSTRACT

The Utility Report Cards (URC) program is a web-based energy information system that reports and graphs monthly utility data for schools. The program was developed and prototyped using Orange County Public Schools (OCPS) utility data. Each month, a web-based report is automatically generated and emailed to school principals and staff to examine their school's electricity usage (energy efficiency) and to identify schools with high-energy consumption for further investigation. The easy-to-use web-style report includes hyperlinks to (1) drill-down into further meter details, (2) show graphs for a 12-month comparison to prior-year data, (3) filter the data to show selected schools, and (4) re-sort the data to rank schools based on the data selected. The URC is also intended for teachers and students to use as an instructional tool to learn about school energy use as a complement to the energy education materials available through the U.S. Department of Energy's EnergySmart Schools program (ESS). To run the URC, go to *www.utilityreportcards.com* and click on "URC Live."

WHY UTILITY REPORT CARDS?

The URC was created to help OCPS staff understand and therefore manage their utility consumption and associated costs. The URC program was designed as a web-based system to take full advantage of email and web-browser technologies. The URC allows each school principal to become aware of how his/her school is performing relative to a projected benchmark and to other schools of similar design and capacity. Giving awards to schools that improve performance from prior-year levels could create a spirit of competi-

tion with the opportunity to recognize success. Those schools identified as high-energy users become the focus of attention to determine the reasons for their consumption level and ultimately to decrease the energy used. All of this is done by using the monthly utility data that is provided electronically at minimal or no–cost by the utilities

The URC public/private partnership includes:

Orange County Public Schools, FL (OCPS)

The Orange County public school system is the 14th largest district out of more than 16,000 in the nation and is the 5th largest in Florida. The OCPS operates 152 schools with over 160,000 students. [1]

Florida Solar Energy Center (FSEC), Cocoa, FL

A nationally recognized energy research institute of the University of Central Florida, FSEC developed the URC program and provides monthly utility data translation and maintenance. This approach provides a standard report format and consistency between the utilities and the school district. [2]

Orlando Utilities Commission and Progress Energy, Orlando, FL

These two local utility companies provide electric utility service to OCPS and electronically send the utility billing data to FSEC on a monthly basis. The utilities benefit from the URC by the positive public relations and outreach associated with helping schools manage their energy usage. [3]

State of Florida Energy Office, Tallahassee, FL

This agency provided the initial funding to FSEC for development and implementation of the URC. [4]

U.S. Department of Energy's EnergySmart Schools (ESS), Washington, DC

The ESS is a program that works with schools to reduce utility consumption/costs throughout the United States. The URC provides ESS with an energy information system tool for school districts that will complement energy education programs already offered through the EnergySmart Schools. Additionally, through the existing national network of ESS partnership districts, the URC can be easily replicated and offered to schools nationwide. [5]

Walt Disney World Company, Lake Buena Vista, FL

Walt Disney World assisted OCPS in streamlining their facilities management processes and reducing utility costs. Walt Disney World has developed web-based energy information systems and provided technical support for the development of the URC. [6]

URC DATA COLLECTION

Data Sources

The utility data used in the URC program are based on the monthly electric billing data from Orlando Utilities Commission and Progress Energy. The OCPS superintendent authorized the utilities to release the OCPS utility data to FSEC, and each utility company was asked to provide OCPS utility data from the last two years. Both utilities agreed to provide the OCPS data electronically to FSEC at no-cost based on the positive public relations generated.

The URC also uses information from the Common Core of Data (CCD), a comprehensive national statistical database of information about all public schools and school districts. CCD is compiled annually by the U.S. Department of Education's National Center for Education Statistics. One component of data that was missing was the school square footage. A supplemental table was made for the square footage of each school. This was needed to calculate the consumption per square foot utility efficiency benchmarks for each school. Additionally, since the school's square footage could change as a result of school construction projects, the square footage table is updated monthly to account for school expansions.

Data Transfer

FSEC established an account on its server where the utility data files were transferred automatically using file transfer protocol (FTP). Although each utility had the OCPS utility data available electronically in its utility billing system in an ASCII delimited file format, there was no consistency in the data output formats provided to FSEC. Therefore, FSEC created a custom program called URC_DPP (URC Data Processing Program) that processed each utility's data file separately and loaded the data into a common database. A future enhancement for the URC would be to develop (specific?) standards for the format and method of delivery for electronic monthly utility data.

On a monthly basis, the utilities electronically transmit to FSEC the OCPS utility data, which then adds it to the URC relational database. FSEC sends an email to a designated email address at OCPS with a copy of the current month URC embedded in the email message (the URC format for February 2004 is shown in Figure 31-2). OCPS then forwards this email to their internal email distribution list for principals and staff. This makes it easy for users, since all they need to do is click on the hyperlinks in the URC email to produce graphs and detailed reports.

At this point, only electric utility data is being tracked. In future enhancements, the URC could include additional utilities, such as water and natural gas.

Relational Database

The Oracle relational database management system was used for the URC, primarily due to FSEC extensive use and experience with Oracle databases. The database tables used to store the URC information include:

- School Information: constructed from the CCD database

- District Information: constructed from the CCD database

- School Building Information: square footage and other building related data

- School Contact Information: contact information for school officials

- Utility Contact Information: contact information for utility personnel

- Service Type: types of services offered by the utilities

- Meter Information: school where located, meter ID, meter description

- Meter Readings: monthly reading from the meters

- Weather Information: cooling degree day and heating degree day

URC WEB-PROGRAM FUNCTIONALITY

The URC program is designed to be informative, intuitive and flexible for all users. It takes advantage of extensive use of hyperlinks that create graphs and detailed reports from an overall summary report listing all schools. Users are able to view graphs and see the electric consumption patterns by simply clicking on hyperlinks in the URC web page. They can also modify the reports by using hyperlinks to sort and filter the data. Hyperlinks provide the users with instant results, making the application exciting to work with and easy to understand.

The URC home page (*http://www.utilityreportcards.com*) serves two functions: (1) contains hyperlinks that provide information about the URC and its development partners and (2) it provides a launching point to run the URC program. Figure 31-1 shows the URC home page.

When the User selects the "URC Live" hyperlink, a summary report (see example in Figure 31-2) shows totals for each school type in the entire school district—primary (elementary) schools, middle schools, high schools, etc.—as the rows in the report. The data

presented in the columns include the electric consumption (kWh), the cost (dollars) and the efficiency (Btu/sqft). To provide meaningful comparisons to the same time in the prior year, the URC program divides the data by the number of days in the billing period to produce per day values. The URC also allows the user to change the values shown to per month figures which is just the per day values multiplied by 30 days. Figure 31-2 shows the overall school district summary report for February, 2004.

The URC program interface makes the program easy to use considering the enormous amount of data available. The top down approach lets users view different levels of data from the overall district-wide summary to individual meters in a single school. For example, the user can click on a school type to display the details for each school. Clicking on the "On" link in the legend will toggle the color flags on or off. Figure 31-3 shows the result of clicking on the hyperlink "Regular High School": a report on all high schools in the database for February 2004. The report is sorted based on the percent change in kWh usage from the prior year levels. The schools that changed the most are at the top of the list. Re-sorting the schools is accomplished by simply clicking on the column title. To produce the report shown in Figure 31-3, the original report sorted by Efficiency Percent Change was re-sorted by clicking on the Consumption Percent Change column.

Graphing is accomplished by clicking on any current period value in the report. To display a 12-month graph for Apopka Senior High School, clicking on the number 22,399 would produce the graph shown in Figure 31-4. Note that the consumption levels are significantly higher than prior year levels for this school. Focusing in on the reasons for this increase should be the next step for the OCPS facility personnel. Once it is determined the increase is not due to other factors such as increased enrollment or extreme weather, personnel can consider making adjustments to the energy management system controls to turn the trend around. Clicking

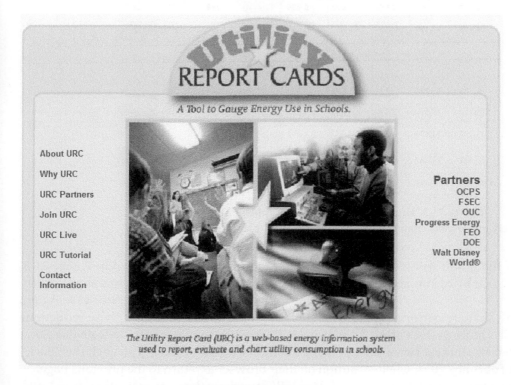

Figure 31-1. URC Home Page

UTILITY REPORT CARDS

Electric Per Day   ORANGE COUNTY PUBLIC SCHOOLS   << February ▾ 2004 ▾ >>

School Type	Consumption (kWh/day)			Cost ($/day)			Efficiency (Btu/sq ft/day)		
	Current Period	Previous Period	Percent Change	Current Period	Previous Period	Percent Change	Current Period	Previous Period	Percent Change
Regular Primary School	178,073	181,162	-2 %	$ 13,923	$ 12,012	16 %	146	149	-2 %
Regular Middle School	121,735	124,550	-2 %	$ 9,119	$ 7,984	14 %	139	142	-2 %
Regular High School	168,196	167,136	1 %	$ 12,135	$ 10,121	20 %	161	160	1 %
Special Other	6,351	6,092	4 %	$ 479	$ 404	19 %	143	137	4 %
Vocational High School	5,910	6,572	-10 %	$ 469	$ 448	5 %	106	118	-10 %
Grand Total ORANGE COUNTY PUBLIC SCHOOLS	480,266	485,511	-1 %	$ 36,125	$ 30,969	17 %	148	150	-1 %

On   Denotes Increase From Previous Year
On   Denotes Decrease From Previous Year by ◀ 0 % ▶ or more

Figure 31-2. Overall School District Summary Page

UTILITY REPORT CARDS

Electric Per Day   ORANGE COUNTY PUBLIC SCHOOLS   << February ▾ 2004 ▾ >>

SCHOOL	Consumption (kWh/day)			Cost ($/day)			Efficiency (Btu/sq ft/day)		
Regular High School	Current Period	Previous Period	Percent Change	Current Period	Previous Period	Percent Change	Current Period	Previous Period	Percent Change
APOPKA SENIOR HIGH SCHOOL	22,399	20,360	10 %	$ 1,777	$ 1,375	29 %	211	192	10 %
CYPRESS CREEK SENIOR HIGH SCHOOL	16,932	19,245	-12 %	$ 1,176	$ 1,108	6 %	153	174	-12 %
EVANS HIGH SCHOOL	17,108	16,526	4 %	$ 1,229	$ 1,022	20 %	156	151	4 %
OAK RIDGE HIGH SCHOOL	12,940	14,669	-12 %	$ 908	$ 855	6 %	148	168	-12 %
OLYMPIA HIGH SCHOOL (FORMERLY DR. PHILL	13,441	13,119	2 %	$ 980	$ 805	22 %	118	116	2 %
ROBERT HUNGERFORD PREPARATORY HIGH SCHOOL (FORMERL	5,525	5,727	-4 %	$ 424	$ 393	8 %	192	199	-4 %
TIMBER CREEK HIGH SCHOOL	15,133	15,379	-2 %	$ 1,081	$ 924	17 %	134	136	-2 %
UNIVERSITY HIGH SCHOOL	20,713	19,056	9 %	$ 1,482	$ 1,131	31 %	163	150	9 %
WEST ORANGE HIGH SCHOOL	21,375	21,898	-2 %	$ 1,473	$ 1,261	17 %	190	194	-2 %
WINTER PARK HIGH SCHOOL	22,631	21,157	7 %	$ 1,604	$ 1,247	29 %	168	157	7 %
Regular High School	168,196	167,136	1 %	$ 12,135	$ 10,121	20 %	161	160	1 %

**Figure 31-3. High School Summary Report
sorted by percent change in consumption from prior year.**

on the "Off" link in the legend of the graph will toggle on and off a display (above the graph) of the actual data for each month.

Users can also look at schools that are performing better than prior year kWh levels. In Figure 31-3, Cypress Creek Senior High School showed the largest percent change from prior year levels in February 2004. Clicking on the number 16,934 next to Cypress Creek Senior High School produced the graph shown in Figure 31-5. This graph shows that Cypress Creek has significantly lowered its kWh consumption from the prior year. Again, focusing on the changes that produced these reductions will help OCPS facilities management replicate these best practices in other schools. The user can click the "Back" button on their browser to return to the previous page.

The URC program has several other functions.

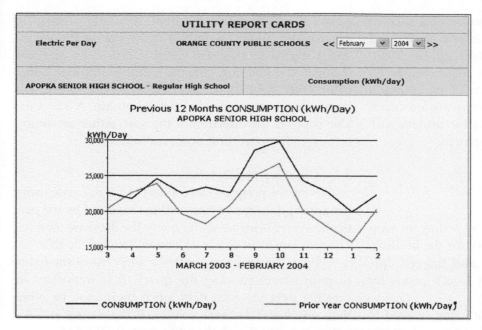

Figure 31-4. Apopka Senior High School 12-Month kWh graph.

Clicking on the school name produces a report that shows data for each electric meter at the school. Clicking on the efficiency column of the report in Figure 31-3 re-sorts the report and shows which schools are more efficient on a per square foot basis. In the case of OCPS, the newer schools have proven to be more efficient than the older schools. Readers are encouraged to try out the URC functionality themselves by visiting the URC website at *http://www.utilityreportcards.com/*. The driving forces behind all of this functionality are dynamic web programs and relational database tables.

URC WEB-PROGRAM DETAILS

Two web programs make up the application interface. One is for reporting and one is for graphing. The programs are written in PHP (Hypertext Preprocessor). "PHP is a widely used Open Source general-purpose scripting language that is especially suited for Web development and can be embedded into HTML" according to PHPBuilder.com. This means that the PHP application itself is not for sale. It is developed and supported solely by volunteers. What makes it an attractive choice is that the developers' code is part of the HTML page itself. PHP is ideal for connecting to databases and running SQL queries, to return dynamic content to a web page. [8]

Using PHP, the reports are designed to accommodate both the novice and the expert user. Hyperlinks in the reports pass values back into the same two programs in a recursive manner using the CGI query string. These values are processed by a set of predetermined rules built into the program by the developer to incrementally change the appearance of the reports to suit the user. The hyperlink construction includes messages using the *onmouseover* event to explain the action of the hyperlink. Total and sub-total lines provide summary information. Data that show an increase or decrease from the previous year are flagged with a different background cell color. The user can define *percent criteria* for marking decreases from the prior year because this is helpful

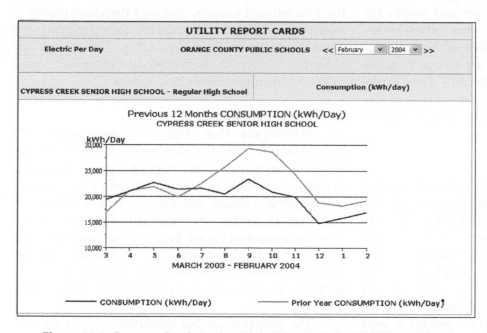

Figure 31-5. Cypress Creek Senior High School 12-Month kWh graph.

in tracking progress toward a particular goal such as 5% decrease from previous year. Graphs are created using KavaCharts, a collection of Java applets available from Visual Engineering at *http://www.ve.com*. [9] The following section describes some of the technical programming details in the order they are executed. We hope these details will be useful for readers with some experience in program development.

Reporting Program

1. Retrieving Values

The program algorithm begins by setting up variables to define the background colors for the heading, sorted heading column, data cells and flagged data cells. The HTML table width is set to 670 pixels for portrait printing (see below).

Next, the program retrieves the values passed in from the various hyperlinks using the CGI query string or sets them to default values. Values for the time period of the report are first. The default month is set to last month since that is likely the most recent full month of data. The program retrieves values associated with the design of the reports such as the column to sort by and the direction of sort (ascending or descending). There are also values associated with turning the flags on and off and setting the percent criteria. Another value controls whether or not the report is normalized to an *average per day* basis or simply reports the totals per month. Data reported on an *average per day* basis is more meaningful since each meter and month has a unique number of billing days. Other values are associated with what data to show such as consumption, cost, efficiency or all three. There are values to define filtering of the data. The user can *drill down* in the report to look at data for only one school type or one school. Retrieving CGI values in PHP is done as at the bottom of this page.

2. Using the Values

The retrieved values help to design the report. If the flag1 value is "off," then the program sets the background color for flagged cells to the normal background color. It does the same for the flag2 value. If a school *type and level* hyperlink was selected, a name value pair such as "typelevel=11" is sent back into the program. So a $filterby variable is set to "typelevel" and a $filter variable is set to "11." This makes it easy to add filters to the SQL query. The same process is used if a school hyperlink is selected. The $filterby variable is set to "school" and the $filter variable is set to the

school ID of the selected school. Similarly, sorting by a particular column is passed in by one of the values and assigned to a $sortby variable. Then the program changes the heading color for that column to some color other then the normal color so that users will know which column is the sorted column. A $dir variable controls the direction of the sort, either ascending or descending (see first program below).

3. Querying the Database

Now the program can connect to the database using the appropriate connection settings required by the particular server being used and query the database for data. The program must construct a query string to do this.

First, the program creates a *select list* of the fields to return from running the query. It is important to *sum* the data for each field since there may be more than one meter for each school. If the *average per day* value is set to "yes" then the sum is divided by the number of billing days in the month. Next, the query joins the tables together using an SQL join or where clauses. Tables are typically together using key fields such as *school ID, meter ID*, etc. Then conditions are used to filter the data for the time period, school type and level, or school selected using SQL *where* clauses. The *group by* clause is used to sum the correct values. If a *total row* was selected then *group by* [school type and level]. If a school was selected then *group by* [meter], otherwise, *group by* [school]. You can use the *order by* clause to sort by school type and level, then apply the selected sort column and sort direction passed in as described above. The program runs the query and stores those results in a data array. The second program below gives a simple version of an SQL query to show consumption for one school.

4. Displaying the Report

The program then displays the main heading. On the report title, a hyperlink is created to reset the sort and filter and display the default report. Also, a hyperlink is available to toggle back and forth between *average per day* values and *total per month* values. The hyperlink to switch from *average per day* to *total per month* is created as shown below. Other hyperlinks are created in a similar manner. Notice that all values required to produce the desired report design are passed back into the program in each and every hyperlink (at bottom).

The program creates variables to hold the previous month and next month values. If the current month is January, then it sets the previous month to

December of the previous year. If the current month is December then it sets the next month to January of the next year. Then it creates a hyperlink to change the month to the previous month as well as a hyperlink to change the month to the next month. The current month and year are displayed between the two hyperlinks (<< and >>).

5. *Looping through the Records*

Before looping through the data records returned from the query, the program initializes a variable to hold the value of the current school *type and level*. Variables for totals and the grand totals are initialized as well. While looping through the data records the program checks for conditions to display sub-headings and total rows at the appropriate place.

```
$cellcolorheading = "#CCCC99";        //Color of heading background.
$cellcolorsorted = "#CCCCCC";         //Color of sorted column heading background.
$cellbgcolor = "#FFFFFF";             //Color of data background.
$cellcolorflag1 = "#FF9933";          //Color of data background flagged for increase from previous year.
$cellcolorflag2 = "#99FF00";          //Color of data background flagged for decrease from previous year.
$tablewidth = 670;                    //Set the table width to 670 pixels for portrait printing.
```

```
//Get the date variables or set them to default values.
if(!$month) {                         //Check for a $month variable, if there is none,
    $month = $_POST['month'];         //then get the value from the POST query string.
}
if(!$month) {                         //Check for a $month variable, if there is none,
    $month = $_GET['month'];          //then get the value from the GET query string.
}
if(!$month) {                         //Check for a $month variable, if there is none,
    $month = date("m") - 1;           //set the current month to last month.
    if ($month == 12) {               //If last month is December,
        $year = date("Y") - 1; }      //then set the $year variable to last year.
```

```
if ($flag1 <> "yes") {$cellcolorflag1 = $cellbgcolor;}    //If flag1 is turned off, change cell color to normal background color.
if ($flag2 <> "yes") {$cellcolorflag2 = $cellbgcolor;}
                                                          //If flag2 is turned off, change cell color to normal background color.

if ($sortby == "consumption") {       //If sorting by consumption,
$cellcolorkwh = $cellcolorsorted;     //set the consumption heading background color different then the others.
} else {                              //Otherwise,
$cellcolorkwh = $cellcolorheading;    //set the consumption heading background color to the normal heading background
                                      color
```

```
Select    sum(c.consumption) as consumption,
          a.School_name, b.meter_id, a.school_id, a.type_code, a.level_code
From school_info a, meter_info b, reading_info c
Where c.school_id = $filter
Group by b.meter_id, a.school_name, a.school_id, a.type_code, a.level_code
Order by $sortby $dir
```

```
//Create to a hyperlink to switch between average per day and total per month
echo "<a href=query.php?school=$school&sortby=$sortby&dir=$dir&$filterby=";
echo urlencode($filter)&content=$content&district=urlencode($district);
echo "&month=$prevmonth&year=$prevyear&flag1=$flag1&flag2=$flag2";
echo "&dpercent=$dpercent&perday=no ";
echo "onmouseover=\"window.status='Select This Link to Show Total Values for the Month'; return true\" >";
    //Add a message to explain the hyperlink.
echo "Day"; }
```

It creates string descriptions of the school type code and school level code. If this record is a new school *type and level* the program adds a total line (if not the first record) then it displays the school *type and level* as well as the sub-headings.

The total line has a hyperlink to show only the total lines when selected. The data values for total consumption, cost and efficiency for each school type and level have hyperlinks which link to a program that generates a graph of the values for the last 12 months. Other values for prior year and percent difference are displayed without hyperlinks.

Next, the sub-heading row is added with hyperlinks to generate a sort by column; school (or meter), consumption, previous year consumption, percent difference, etc. The hyperlinks act as toggle switches changing the sorting back and forth between ascending and descending.

Now the program displays the detailed data lines. The first cell is the school name as well as a hyperlink to display the data for just the school. The next cell shows the consumption and is a hyperlink to graph consumption for the last 12 months. The prior year consumption and percent difference are displayed without hyperlinks. The same is done for cost and efficiency.

After displaying the data line, the current school *type and level* is assigned to the variable created for that purpose. Then the totals and grand totals are updated. The program then goes to the next record.

After looping through all of the data, a total line is added in the same manner as before for the last school *type and level*. Then the grand total line with hyperlinks to show only the grand total and to graph the grand total values for the last 12 months is added. Below the report are toggle hyperlinks for turning flags on or off and adjusting the percent decrease criteria by 5% in either direction. The final report is shown in Figure 31-8.

Graphing Program

1. Querying the data

The graphing program algorithm is nearly the same as the reporting program algorithm. The first exception is that data is queried for the last 12 months including the current month rather than the current month alone.

//Filter for all records less than or equal to the current month and greater than the same month last year.
$query = $query. " and ((a.bill_month <= $month and a.bill_year = $year) or (a.bill_month > $month and a.bill_year = $pyear)) ";

Secondly, all data are sorted by year and month to produce a data display and trend graph of the last 12 months. Figure 31-9 shows the graphic that is produced.

$query = $query. " order by a.bill_year, a.bill_month ";   //Sort by year and month

2. Displaying the data

While looping through the data, aside from displaying the values as in reporting, text strings are created to assign to the Java applet parameters that produce the graph. The KavaChart applet parameter requires a comma-separated format of values.

<param name=dataset0yValues value='2602, 2966.66, 3276.36, 2562, 2715.625, 3271.03, 3651.33, 3145.625, 2992.41, 2726, 2541.76, 3626.20'>

A toggle hyperlink is added to turn the data display on or off.

UTILITY REPORT CARDS		
Electric Per Day	ORANGE COUNTY PUBLIC SCHOOLS	<< February ▾ 2004 ▾ >>

Figure 31-6. URC Title Rows

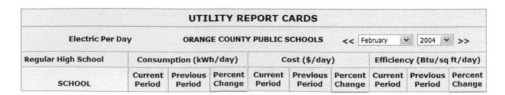

UTILITY REPORT CARDS									
Electric Per Day	ORANGE COUNTY PUBLIC SCHOOLS					<< February ▾ 2004 ▾ >>			
Regular High School	Consumption (kWh/day)			Cost ($/day)			Efficiency (Btu/sq ft/day)		
SCHOOL	Current Period	Previous Period	Percent Change	Current Period	Previous Period	Percent Change	Current Period	Previous Period	Percent Change

Figure 31-7. URC Column Row

UTILITY REPORT CARDS

Electric Per Day	ORANGE COUNTY PUBLIC SCHOOLS		<< February ∨ 2004 ∨ >>						
Regular High School	**Consumption (kWh/day)**			**Cost ($/day)**			**Efficiency (Btu/sq ft/day)**		
SCHOOL	Current Period	Previous Period	Percent Change	Current Period	Previous Period	Percent Change	Current Period	Previous Period	Percent Change
APOPKA SENIOR HIGH SCHOOL	22,399	20,360	10 %	$ 1,777	$ 1,375	29 %	211	192	10 %
CYPRESS CREEK SENIOR HIGH SCHOOL	16,932	19,245	-12 %	$ 1,176	$ 1,108	6 %	153	174	-12 %
EVANS HIGH SCHOOL	17,108	16,526	4 %	$ 1,229	$ 1,022	20 %	156	151	4 %
OAK RIDGE HIGH SCHOOL	12,940	14,669	-12 %	$ 908	$ 855	6 %	148	168	-12 %
OLYMPIA HIGH SCHOOL (FORMERLY DR. PHILL	13,441	13,119	2 %	$ 980	$ 805	22 %	118	116	2 %
ROBERT HUNGERFORD PREPARATORY HIGH SCHOOL (FORMERL	5,525	5,727	-4 %	$ 424	$ 393	8 %	192	199	-4 %
TIMBER CREEK HIGH SCHOOL	15,133	15,379	-2 %	$ 1,081	$ 924	17 %	134	136	-2 %
UNIVERSITY HIGH SCHOOL	20,713	19,056	9 %	$ 1,482	$ 1,131	31 %	163	150	9 %
WEST ORANGE HIGH SCHOOL	21,375	21,898	-2 %	$ 1,473	$ 1,261	17 %	190	194	-2 %
WINTER PARK HIGH SCHOOL	22,631	21,157	7 %	$ 1,604	$ 1,247	29 %	168	157	7 %
Regular High School	**168,196**	**167,136**	**1 %**	**$ 12,135**	**$ 10,121**	**20 %**	**161**	**160**	**1 %**

On	Denotes Increase From Previous Year
On	Denotes Decrease From Previous Year by ◀ 0 % ▶ or more

Figure 31-8. Completed URC Report

URC MEDIA EVENT

On April 5, 2004, the URC was unveiled to OCPS and the nation in a media event held at Citrus Elementary School in Ocoee, Florida. U.S. Secretary of Energy, Spencer Abraham, along with dignitaries from the URC partners and a group of fifth grade students attended the event. The media covered the event through television, radio and newspaper reports to inform the public about the URC.

Secretary Abraham said the report cards would allow schools to save money while also teaching children to be responsible energy consumers. The goal, Abraham said, is education. "If we have more money to spend on students, that means more teachers and more equipment," he said. "If we can save on the energy side and spend it on the student side, in our opinion, that's great."[7]

Schools can use this as just one tool in their arsenal to have better visibility of their energy costs," said Bill Kivler, director of engineering services for the Walt Disney World Co. To run the URC, go to *www.utilityreportcards.com* and click on "URC Live."

CONCLUSION

The U.S. Department of Energy and others have identified the need to help our nations' schools lower their energy consumption costs. The URC provides a way for schools to save money on energy consumption. Some day students may use the URC to learn to become efficient consumers themselves. The URC is a challenging and cooperative effort to collect utility data from various sources and put it all together in meaningful reports and graphs. In addition, the URC is designed as a comparison and analysis tool providing many different "views" into the data quickly and easily using its interactive web reporting features. The URC was developed using some of the most current database and open-source programming tools available. Hopefully, this approach will allow the URC to continue to help reduce energy costs far into the future.

At the time of this writing, the URC was just unveiled so it is too early to measure the impact the URC will have on reducing utility consumption for OCPS. Based on the initial positive comments, we think the URC will help OCPS focus attention on school energy consumption. One of the main benefits of the URC is that the school principals will know their own school's energy usage pattern. This knowledge allows the principals to take appropriate actions to focus on the reasons for the increases and ultimately return their schools to their normal consumption levels.

The URC reporting format is also not limited to schools. A facility manager could use the URC to list any similar group of facilities that are separately metered. Buildings on a university campus, resorts hotels in a city, supermarkets in the same geographic area are a few examples that come to mind.

The future for the URC looks bright. Future URC enhancements include (1) reporting other school districts (2) adding other utilities in addition to electricity (3) adding weather information (4) developing a standard for utility data transfer and (5) integrating the EPA's Portfolio Manager Energy benchmarks for each school.

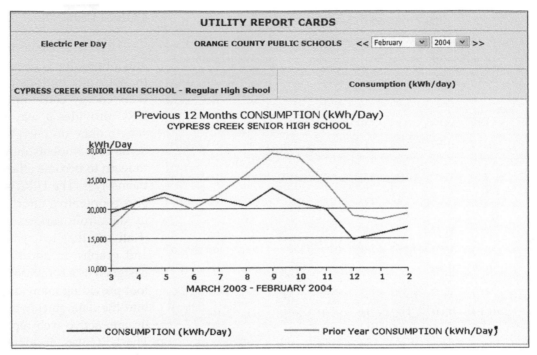

Figure 31-9. Graphic showing 12-Month kWh Usage

Figure 31-10. Left to right: U.S. Secretary of Energy Spencer Abraham, and Kym Murphy, Walt Disney Company Senior Vice President of Corporate Environmental Policy, at OCPS media event April 5, 2004.

School districts that are interested in establishing their own URC are encouraged to find out more by visiting the URC website at *http://www.utilityreportcards.com*.

References

[1]   Orange County Public Schools, Orlando, Florida; *http://www.ocps.k12.fl.us/*, Internet page accessed April, 2004.

[2]   Florida Solar Energy Center; "A Research Institute of the University of Central Florida"; *http://www.fsec.ucf.edu/*, Internet page accessed April, 2004.

[3]   Orlando Utilities Commission, "OUC The Reliable One"; *http://www.ouc.com*, Progress Energy, People, Performance, Excellence"; *http://www.progress-energy. com*, Internet page accessed April, 2004.

[4]   State of Florida Energy Office; *http://dlis.dos.state.fl.us/ fgils/agencies/energy*.html, Internet page accessed April, 2004.

[5]   U.S. Department of Energy's EnergySmart Schools; "Rebuild America, Helping Schools Make Smart Choices About Energy"; *http://www.rebuild.org/sectors/ ess/index*.asp, Internet page accessed April, 2004.

[6]   Walt Disney World Company; "Disney's Environmentality"; *http://www.disneysenvironmentality.com*, Internet page accessed October, 2004.

[7]   "Report Cards Track Schools' Utilities"; Orlando Sentinel; 4/6/2004.

[8]   The PHP Group; "PHP Manual, Preface"; *http://www. phpbuilder.com/manual/preface.php*; Internet page accessed April, 2004; last updated, 10/26/2002.

[9]   Visual Engineering; "KavaChart, The Complete Solution for Java-Based Charting"; *http://www.ve.com/*; Internet page accessed April, 2004.

ACKNOWLEDGMENTS

I would like to thank Robert J. Lewis for assuring me that it is ok to use recursive hyperlinks in web applications. The technique is not applicable to all projects, but it works well for the URC. Robert is a talented and accomplished programmer whose innovations inspire me greatly. —*David Green*

Chapter 32

Machine to Machine (M2M) Technology in Demand Responsive Commercial Buildings

David S. Watson
Mary Ann Piette
Osman Sezgen
Naoya Motegi

ABSTRACT

Machine-to-Machine (M2M) is a term used to describe the technologies that enable computers, embedded processors, smart sensors, actuators and mobile devices to communicate with one another, take measurements and make decisions—often without human intervention.

M2M technology was applied to five commercial buildings in a test. The goal was to reduce electric demand when a remote price signal rose above a predetermined price. In this system, a variable price signal was generated from a single source on the internet and distributed using the meta-language, XML (extensible markup language). Each of five commercial building sites monitored the common price signal and automatically shed site-specific electric loads when the price increased above predetermined thresholds. Other than price signal scheduling, which was set up in advance by the project researchers, the system was designed to operate without human intervention during the two-week test period.

Although the buildings responded to the same price signal, the communication infrastructures used at each building were substantially different. This study provides an overview of the technologies used to enable automated demand response functionality at each building site, the price server and each link in between. Network architecture, security, data visualization and site-specific system features are characterized.

The results of the test are discussed, including system architecture and characteristics of each site. These findings are used to define attributes of state-of-the-art automated demand response systems.

INTRODUCTION

This chapter provides a summary of the control and communications systems evaluated and reported on as part of a larger research report. [1] The objective of the study was to evaluate the technological performance of automated demand response hardware and software systems in large facilities. The concept in the evaluation was to conduct a test using a fictitious electricity price to trigger demand-response events over the *internet*. Two related papers describe the measurement of the electric demand shedding and the decision making issues with the site energy managers. [2,3]

The two main drivers for widespread demand responsiveness are the prevention of future electricity crises and the reduction of average electricity prices. Demand response has been identified as an important element of the State of California's Energy Action Plan, which was developed by the California Energy Commission (CEC), California Public Utilities Commission (CPUC), and Consumer Power and Conservation Financing Authority (CPA). The CEC's 2003 Integrated Energy Policy Report also advocates Demand Response.

A demand responsive building responds to a remote signal to reduce electric demand. This is usually done by altering the behavior of building equipment such as heating ventilating and air conditioning (HVAC) systems and/or lighting systems so as to operate at reduced electrical loads. This reduction is known as "shedding" electric loads. Demand responsiveness and shedding can be accomplished by building operators manually turning off equipment in response to a phone call or other type of alert.

In this paper, the term *"automated demand response"* or *"auto-DR"* is used to describe "fully automated" demand response where electric loads are shed automatically based on a remote internet based price signal. Although the facility operating staff can choose to manually override the auto-DR system if desired, these systems normally operate without human intervention.

Previous Research

The California Energy Commission (CEC) and the New York State Energy Research and Development Agency (NYSERDA) have been leaders in the demonstration of demand response programs utilizing enabling technologies. Several studies associated with the California and New York efforts investigated the effectiveness of demand responsive technologies. In California, Nexant was charged with evaluating CEC's Peak Load Reduction Program. The Nexant reports document the performance of all the California funded technology projects including the magnitude of the response and the cost associated with it.[4,5]

In addition to research concerning utility programs, controls, and communications systems, several research studies have examined various topics concerning DR in commercial buildings, including how to operate buildings to maximize demand response and minimize loss of services. Kinney *et al.* reported on weather sensitivity of peak load shedding and power savings from increasing the setpoint of temperatures in buildings to reduce cooling loads. [6] This research project also builds on previous LBNL work concerning the features and characteristics of web-based energy information systems (EIS) for energy efficiency and demand response (DR). [7]

PROJECT DESCRIPTION

The automated DR research project took approximately two years, beginning with a planning activity in summer, 2002, successful pilot tests in November 2003 and final reporting in March 14, 2004 (Piette, et al 2004). The building sites, including their use, floor area, and equipment loads shed during the auto DR tests are listed in Table 32-1.

System Geography

Although all of the auto-DR pilot sites were in California, the supporting communications infrastructure and several of the developers were distributed throughout North America (see Figure 32-1).

AUTOMATED DEMAND RESPONSE SYSTEM DESCRIPTION

The automated demand response system published a fictional price for electricity on a single *server* that was accessible over the internet (Figure 32-2). Each of five commercial building sites had *client* software that frequently checked the common price signal and automatically shed site-specific electric loads when the price increased beyond predetermined thresholds. Other than price signal scheduling, which was set up in advance by the project researchers, the system was designed to operate without human intervention during two one-week pilot periods. The test process followed these steps:

1. LBNL defined the price vs. time schedule and sent it to the *price server*.

Table 32-1. Summary of Sites

	Albertsons	BofA	GSA	Roche	UCSB
Location	Oakland	Concord	Oakland	Palo Alto	Santa Barbara
Use	Supermarket	Office	Office	Pharmaceutical laboratory (Office & Cafeteria)	Library
Floor Area (ft²)	50,000	211,000	978,000	192,000	289,000
Equipment loads shed during test	50% of overhead lighting, Anti-sweat heaters	Supply fan duct static pressure setpoint	Global zone setpoint setup and setback	Constant volume fan shut off	Fan speed reduction, Chilled water valves closed

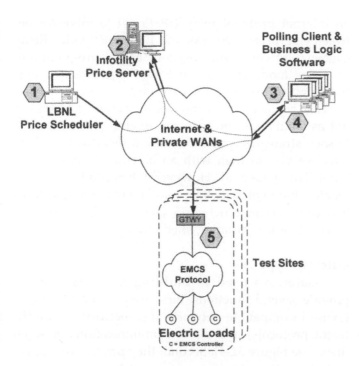

Figure 32-2. Auto-DR Network Communications Sequence

2. The current price was published on the server.
3. Clients requested the latest price from the server every few minutes.
4. *Business logic* determined actions based on price.
5. *Energy Management Control System* (*EMCS*) carried out shed commands based on logic.

Web Services/XML

The infrastructure of the auto-DR system is based on a set of technologies known as *web services*. Web services have emerged as an important new type of application used in creating distributed computing solutions over the internet. Properly designed web services are completely independent of computer platform (i.e., Microsoft, Linux, Unix, Mac, etc.). The following analogy helps to describe web services: web pages are for people to view information on the internet, web services are for computers to share information on the internet. Since human intervention is not required, this technology is sometimes referred to as "*machine-to-machine*" or "*M2M*." M2M is a superset of technologies that includes some XML/web services-based systems.

XML is a "meta-language" (for describing other languages) that allows design of customized markup languages for different types of documents on the web. [8] It allows designers to create their own customized tags, enabling the definition, transmission, validation,

and interpretation of data between applications and between organizations. [9] Standard communication protocols (*TCP/IP*, *HTTP* and *SOAP*) are used on the internet and *LAN/WAN*s (local area network/wide area network) to transfer XML messages across the network.

Price Scheduling Software

Researchers at the LBNL used a software application to set-up the price vs. time profile published in the price server. The price profile could be set up hours, days or weeks in advance.

Price Server

The central infotility server published the current price for electricity ($/kWh). Although the price used in the test was fictitious, it was designed to represent a price signal that could be used by utilities or independent system operators (*ISO*) in future programs that could be offered to ratepayers.

Web Services Clients

The *polling client* is the software application that checks (polls) the web services server to get the latest price data. The polling client resides on a computer managed by the building operators (or their representatives) for each site. In the pilot test, each client polled the server at a user-defined frequency of once every 1 to 5 minutes. The building operators were not given any prior knowledge of upcoming price increases planned by researchers. By checking their automatic price polling clients, operators could only see the current, most recently published price.

Polling-Client Price Verification

The price server included a feature that verified that each client received correct pricing information. This feature was implemented by requiring that each time the client requested the latest price from the server, it included its current price (from the client's perspective) and a *time stamp*. All pricing data were stored in a database. Although the intent of this feature was to verify client receipt of the latest pricing, there was another unforeseen benefit as well. When pre-testing began, researchers could see which sites were polling the server as each came on-line. After all systems were on-line, there were several cases where clients would stop polling for known or unknown reasons. When program managers observed these problems, they were able to manually make phone calls to the site system administrators, who restored proper communications.

Controls and Communications Upgrades

In order to add auto-DR functionality to each pilot site, some upgrades and modification to the controls and communications systems were required. The upgrades were built to work in conjunction with the existing EMCS and energy information system (EIS) remote monitoring and control infrastructure in place at each site. For this project, custom software was written for each site, including: price polling client, business logic, and site-specific EMCS modifications.

Electric Price Signal and Test Description

Figure 32-3 shows the fictitious price signal that was in effect on the afternoon of November 19, 2003. During the rest of that day, the price remained at $0.10/kWh.

Auto-DR System Architecture Overview

Some auto-DR facilities hosted the polling client software on-site and others hosted it at remote *co-location* sites (see Table 32-2). The geographic location of the computer that hosts the polling client is less important than the type of environment where it is hosted. Professional co-location hosting services, or *"co-los"* offer highly secure environments for hosting computers and servers. Co-los generally provide battery and generator backed electrical systems, controlled temperature and humidity, seismic upgrades and 24/7 guarded access control. For companies that don't have similarly equipped data centers, co-los fill an important need. For computer applications where high system *availability* is important, co-location facilities are often used.

Systems with a high level of integration between *enterprise* networks and EMCS networks tend to allow direct access to any or all control points in the EMCS without a need for excessive *point mapping*. Direct remote control of EMCS points from enterprise networks allows the business logic computer to send commands over the network(s) directly to the EMCS *I/O controller* to shed HVAC or lighting equipment. In a highly integrated system, the EMCS becomes an extension of the enterprise. In these types of integrated systems, a *gateway* device is used to translate between the different protocols used in enterprise networks and EMCS networks.

Alternately, some systems used

an internet protocol relay (*IP Relay*) to interface between enterprise networks and EMCS networks. Relay contacts are commonly used in EMCS programming to define mode changes in HVAC equipment operation (e.g., smoke detector contacts). However, the use of relay contacts as an interface between networks is not as flexible as the gateway devices. Modifications to shed strategies would be more difficult with a relay interface system than with an integrated system with a translating gateway. However, when properly implemented, both gateway-based and relay-based interfaces between enterprise networks and EMCS networks can be effective for initiating shed strategies.

Gateway Type

Gateways used in building *telemetry* systems provide several functions. First, they connect two otherwise incompatible networks (i.e., networks with different protocols) and allow communications between them (see Figure 32-4). Second, they provide *translation* and *abstraction* of messages passed between two networks. Third, they often provide other features such as *data logging*, and control and monitoring of I/O points.

Of the five auto-DR sites, two used *embedded* two-way communicating gateways to connect each site's EMCS networks to its enterprise networks (Table 32-3). *Embedded devices* are generally preferred over PC-based gateway solutions for scaleable, ongoing system deployments. Embedded devices have the following advantages:

- More physically robust. There are no hard drives or other moving parts.
- Less susceptible to viruses and other types of hacker attacks due to custom-designed operating systems and applications.

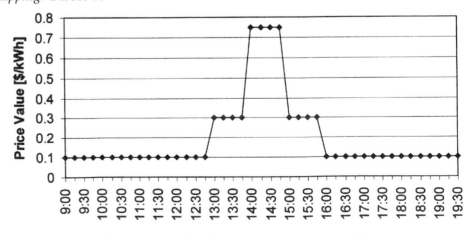

Figure 32-3. Price Signal on November 19, 2003

Table 32-2. Characteristics of Auto-DR Systems—Architecture

	Albertsons	B of A	GSA	Roche	UCSB
Client hosted at co-lo	Yes	Yes	No	No	No
Remote polling client	Yes	Yes	Yes	No	No
Remote control via Internet Gateway	No	Yes	No	Yes	No

- Less susceptible to human error. Once they are set up to function, there is no reason for site personnel to interact with the device. Since they are not "general purpose" computers, there is no risk of memory overloads due to computer games, screen savers and other applications that may be inadvertently loaded onto them.

- Better form factor. Embedded devices are usually smaller than PCs and are designed to be mounted in secure server rooms with other IT equipment.

- Lower cost. Although volume dependent, application-specific embedded devices can be produced in volume for lower cost than PCs.

At Albertsons, an embedded *IP I/O device* (Engage EPIM™) was used for power monitoring and shed mode control. The EPIM provided power monitoring by directly counting *pulses* from power meters. The EPIM set various shed modes into operation by opening and closing *onboard* relay contacts. Although the EPIM IP I/O device effectively provides the interface between the EMCS and enterprise networks, it does not fit the most basic definition of a gateway because it does not connect the protocols of the two networks.

At UCSB, gateway functionality for monitoring was provided by software running on a PC. A previous version of gateway software also provided remote control functionality, but this feature was unavailable at the time of the auto-DR test due to incompatibility issues that occurred after a software upgrade. To meet the remote control requirements of the auto-DR test, an embedded *IP relay* device was added. This device had onboard relay contacts similar to the EPIM, but direct measurement of I/O points (such as power meters) was not required.

The common source of electricity price and the communications protocol translations between the business logic and the final control element (relays, valves etc.) controllers that actually shed the electric loads is shown in Figure 32-4, "Network architecture overview of five combined auto-DR sites." Gateways or other devices are used to transfer necessary communications between dissimilar network protocols.

Integration

For purposes of this study, integration between EMCS and EIS can be characterized by asking two key questions. First, can data from the EMCS and EIS be viewed and analyzed with one *human machine interface (HMI)*? Second, do the EMCS devices such as energy meters reside on the same network as the EMCS devices? Table 32-4 summarizes the answers to these questions for each of the sites. Albertsons, B of A, and GSA either don't have EIS or else they are not integrated with the EMCSs at those sites. At Roche, the Tridium system integrates most of the EMCS points and a small percentage of the electric meters into a comprehensive HMI for viewing, archiving and analysis. UCSB has extensive monitoring of most of the electric meters and sub-meters throughout the campus. Data from these meters are available for visual representation, archiving and analysis through the SiE (Itron/Silicon Energy) server along with relevant EMCS points that have been mapped over to it.

One distinguishing characteristic of the auto-DR sites was whether they leveraged the existing corporate or campus enterprise network to transmit EMCS and/or EIS data. Use of the existing enterprise network for this purpose has many advantages. System installation costs can be much lower if existing enterprise networks are used for communications instead of installing new, separate networks solely for EMCSs and EISs. In addition, the information technology department that manages the enterprise is often better equipped to assure network reliability and security than the facilities group that traditionally maintains the EMCS and EIS.

Each facility has different functional requirements and organizational structures that dictate how the enterprise, EMCS and EIS networks are designed, installed and maintained. Of the five sites in the auto-DR test, three of them shared mission critical enterprise networks with EMCS/EIS/auto-DR systems. Although *bandwidth* requirements for EMCS/EIS/auto-DR systems are low, other organizational impediments may prevent the sharing of enterprise networks for non-standard purposes.

At GSA, a completely separate enterprise network was created for the GEMnet EMCS/EIS/auto-DR sys-

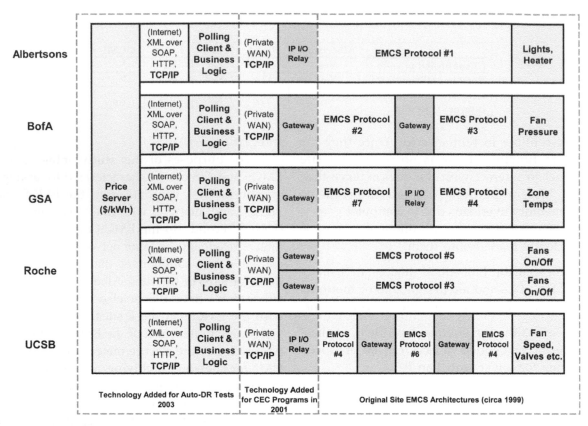

Figure 32-4. Network architecture overview of five combined auto-DR sites

tem. This was the logical choice for this facility because it was not practical to share the existing enterprise networks with other tenants at the site: the Government Services Administration (GSA) and the Federal Bureau of Investigation (FBI). In such circumstances, it is easier to create a new enterprise network for local and remote access to EMCS and EIS data than it is to resolve complex security and maintenance issues associated sharing an enterprise network with another department or organization.

Shed Control Characteristics

Each auto-DR site used different shed strategies. The control characteristics of these strategies also varied substantially. This section describes the characteristics of each shed strategy (Table 32-5). The number of shed control points that were adjusted or altered to invoke the shed strategy at each site is one characteristic of a given

Table 32-3. Characteristics of Auto-DR Systems—Gateways

	Gateway type	Interface Device Description for Remote Shed Control	Remote Monitoring Description
Albertsons	Embedded IP I/O device	IP Relay (2 contacts)	Meter pulses monitored via EPIM™
B of A	Embedded Gateway	Control of EMCS via Internet gateway	Monitoring of EMCS via gateway
GSA	Embedded IP I/O device	IP Relay (2 contacts)	None – Local trending only
Roche	Embedded Gateway	Control of EMCS via Internet gateway	Monitoring of EMCS via gateway
UCSB	PC based Gateway for monitoring, Embedded IP relay for control	IP Relay (3 contacts)	Selected EMCS points monitored via gateway

Table 32-4. Characteristics of Auto-DR Systems—Integration

	Albertsons	B of A	GSA	Roche	UCSB
Integrated EMCS & EIS	No	No	No	Partial	Partial
Primary enterprise network shared with EMCS/EIS/Auto-DR systems	Yes	Yes	No	Yes	Partial

auto-DR implementation. Shed control points include hardware control points (for example, valve position) and software points (for example, *setpoints*) that were altered during the shed. Software points other than setpoints were not included. Control granularity and closed loop shed control are additional characteristics that influence the likelihood and degree to which some occupants may be negatively affected by a given shed strategy.

Albertsons had only two control points (overhead lights and anti-sweat heaters). Because the size of the store is comparatively small, this was rated as "medium" control granularity. Switching off half of the overhead lights is an *open loop* type of control (i.e., there is no feedback to the system). The anti-sweat heater remained in closed loop control during the shed by operating with a reduced setpoint.

B of A had just one control point (duct static pressure setpoint) for the entire 211,000 ft² building, hence the "coarse" control granularity rating. The shed strategy of resetting the duct static pressure setpoint while maintaining zone temperature is a type of closed loop control, but the dearth of temperature *sensors* in the zones reduced the closed loop rating to "partial."

In stark contrast to the rest of the sites, the GSA building used a fine granularity, closed-loop shed control strategy. The zone temperature setpoints for each of 1,050 VAV terminal boxes (1,400 including reheat side of dual duct boxes) were "relaxed" during the shed.

In other words, the cooling setpoints were raised and the heating setpoints were lowered. This approach had an energy saving effect on the central HVAC systems while assuring a reasonable level of service modification to the occupants.

The Roche site used a coarse open loop shed strategy of shutting off fans during the shed.

UCSB used a variety of shed strategies of medium granularity. The shed strategies (including closing cooling valves, and reducing duct static pressure) were all open loop. The outside air dampers were opened to 100%, a strategy that could backfire in extremely hot conditions. The temperate climate in Santa Barbara made this scenario unlikely.

Open Standards

In the EMCS and EIS fields, protocols refer to the low-level communication languages that devices use to "talk" to one another on the network. Of course, one device can only talk to another if they are speaking the same language. Traditionally, each control system manufacturer built controllers and other devices that only spoke their own unpublished *proprietary protocol* (Table 32-6). Once a system is built using a proprietary protocol, the original manufacturer or their representatives are the only parties that can make substantial additions or changes to the system. Some control companies use proprietary protocols as a "lock" on their customers' systems so as to ensure future business and

Table 32-5. Characteristics of Auto-DR Systems—Shed Control

	Albertsons	B of A	GSA	Roche	UCSB
Number of Shed Control Points	2	1	~ 1,400	7	42
Shed Control Points per 10,000 ft.²	0.4	0.05	14	0.4	1.4
Control Granularity	Medium	Coarse	Very Fine	Medium	Fine
Closed loop shed Control	Partial	Partial	Yes	No	No

Table 32-6. Characteristics of Auto-DR Systems—Open Standards

	Albertsons	B of A	GSA	Roche	UCSB
Open Protocol EMCS	No	Yes	No	Partial	No
Open Protocol EIS	Yes	Partial	NA	Partial	No
Open Protocol Auto-DR	Yes	Yes	Yes	Yes	Yes
Open Standards Auto-DR	No	No	No	No	No
Data Archiving in Open Database	NA	NA	Yes	NA	NA

NA = Not Available

high profit margins.

Over the past fifteen years or so, there has been a movement toward "open" protocols in the EMCS and EIS industries. *Open protocols* are based on published standards open to the public. Interested companies can build products that communicate using open standards. In a truly open, interoperable system, products from a variety of open product vendors could be added at any time by skilled installers from independent companies. Several sites in the auto-DR test use open EMCS and/or EIS products that include the *BACnet*, *LonTalk* (EIA-709) and *Modbus* open protocols.

Even with considerable interest from building owners few, if any, new or existing building EMCS or EIS systems are truly open and interoperable. Even when open protocols are used, they are often installed as part of a system that requires use of proprietary software or components at the higher levels of the system architecture. Another way "openness" is reduced is by designing products and systems that require proprietary software tools for installation.

In the IT marketplace, open protocols (e.g., TCP/IP), open database interface standards (e.g., *ODBC*) and open hardware standards (e.g., *SIMM*) have helped the industry thrive. This has allowed products from a wide variety of vendors to communicate with one another on internal LANs, WANs and the internet. A service industry of independent *Systems Integrators* has grown to fill the need of integrating multiple vendor networks into cohesive systems.

Another important trend in the IT industry is the use of a new set of open standards, protocols and languages collectively known as XML/web services. The use of XML/web services in the building controls industry is increasing. This trend will help increase the ability to easily distribute, share and use data from disparate EMCS, EIS and other business systems. This will create opportunities for new products and services that will improve comfort and efficiency in buildings.

In the auto-DR test, the use of XML/web services over the internet provided an overarching open-standards platform by which all of the proprietary and partially open EMCS and EIS systems could communicate. Although the number of commands transmitted between the systems in the 2003 test was minimal (e.g., price, shed mode, etc.), the implications of XML based "add-on" interoperability are very powerful.

RESULTS

Aggregated Whole Building Power and Savings

Figure 32-5 shows the aggregated whole building power and associated savings for all five sites during the shed. The shed period was from 1:00 pm until 4:00 pm on November 19, 2003.

The average savings (load shed) is shown in Figure 32-6. Each bar represents the average savings over

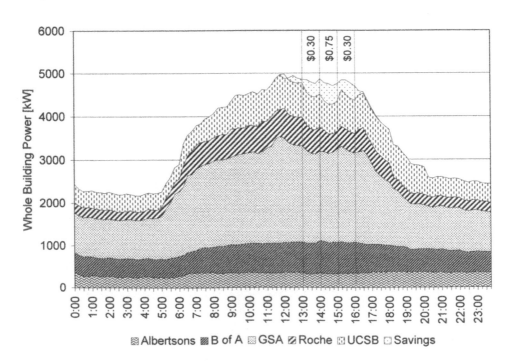

Figure 32-5. Aggregated Power and Savings of All Sites

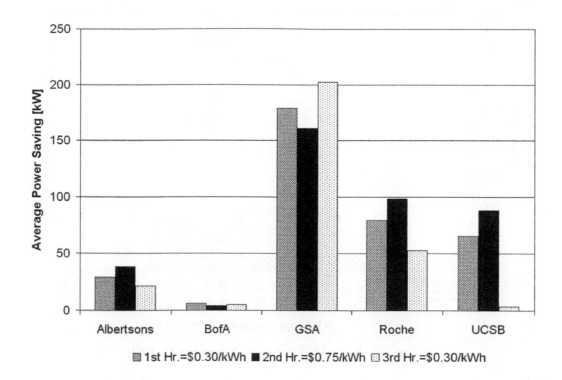

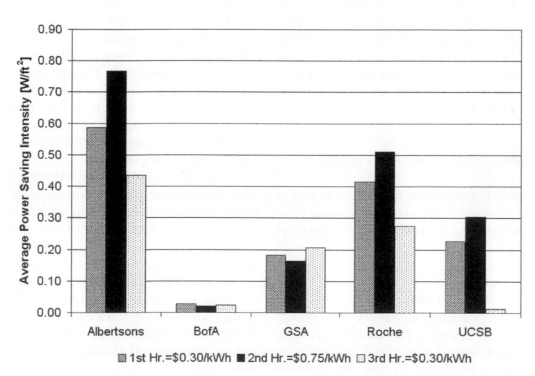

Figure 32-6. Average shed (kW and W/ft²) during the 3-hour test on November 19, 2003

one hour of the 3-hour elevated price test. The electricity price during the first, second and third hours was $0.30/kWh, $0.75/kWh and $0.30/kWh, respectively. [2] The left graph shows the total average power savings per site, while the right graph shows the savings normalized by floor area. This view presents a comparison of the aggressiveness and/or effectiveness of each shed strategy on an area normalized basis.

As an example, the results from an individual site (Roche) are shown in Figure 32-7. Savings are determined by comparing actual metered power on the day of the shed with a calculated normal (non-shed)

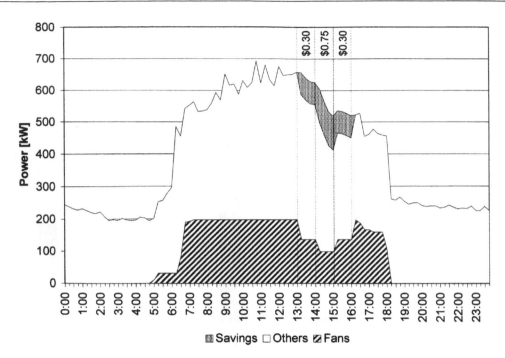

Figure 32-7. Roche Electricity Use, November 19, 2003

baseline. [2] The vertical lines show boundaries of the price range. The fan load component (cross-hatched) is superimposed on the profile of the whole building electric load (white). The savings due to the shed are shown both in the gray section above the whole building load profile and the inverted hat shape missing from the fan load profile.

SYSTEM CHARACTERISTICS OF EACH SITE

This section identifies the unique attributes of each participating auto-DR facility. Controls and communications infrastructures and shed strategies are discussed for each system.

Albertsons (Supermarket)
System Overview (Figure 32-8)

The Albertsons building telemetry data system is split between two systems. The EMCS (excluding electric power monitoring) is accessed via a dial-up modem. EIS data is available via any web browser through the EIS web site. The segregated nature of the EMCS and EIS make it a burdensome task for the facility operator to change a temperature setpoint or lighting schedule and then observe the effect on electric usage.

However, the integration between the enterprise networks and *control networks* is rather tight. The corporate WAN (wide area network) is used to communi-

cate between the business logic/polling client and the on-site internet protocol input/output (IP I/O) relay device. The enterprise network is also used for mission critical *point of sale* data communications within the nationwide organization. The fact that the energy data are shared and communicated over the mission critical enterprise network indicates a high level of collaboration and trust between the Albertsons energy managers and other department managers involved with the core business of the organization.

The shed strategy was not objectionable to the store managers or patrons. Although the transition between 100% overhead lighting to 50% was noticeable, there were no complaints. The reduction of overhead lighting appeared to make the other light sources in the store, such as case lights, seem more intense. There is no evidence that the freezer doors fogged up during the shed, even though the setpoint of the anti-sweat heaters was reduced. If the transition of overhead lights to 50% were gradual (e.g., through use of dimmable ballasts) the entire shed would probably not be noticeable.

BofA (Bank Office) System
Overview (Figure 32-9)

Integration between enterprise networks and control networks at this site is tight. The BofA corporate WAN is used to communicate across the country to the on-site gateway. This network is also used for mission critical financial data communications within

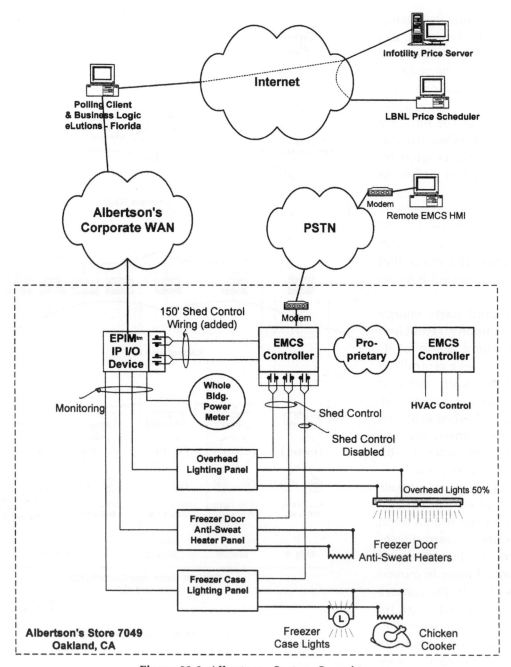

Figure 32-8. Albertsons System Overview

the BofA organization. Like Albertsons, the fact that energy and HVAC (heating, ventilation, and air conditioning) data are shared over the mission critical enterprise network indicates a high level of collaboration and trust between BofA's energy managers, IT (information technology) security managers and other department managers involved with the core business of the organization. The use of highly secure and reliable hardware *VPN* (virtual private network) routers and the use of a co-location site to host the polling client and business logic computers are indications that

system availability and security are high priorities.

With regard to the shed strategy employed at this site, there is no evidence that a modest reduction in duct static pressure for short durations caused any negative comfort effects to the occupants during the test. However, as shown in the measured data, the extent of the electric demand shed is negligible. If this strategy were extended so as to produce significant sized electric sheds, the method may pose some fundamental drawbacks. When the duct static pressure is reduced below the minimum required by the terminal boxes in

VAV (variable air volume) systems, airflow is reduced in the zones. But the reduction is not shared evenly between all the boxes. The zones of greatest demand are the ones that are starved for air most immediately and most severely. In the building used in the auto-DR pilot, the potential problem is exacerbated by the lack of sensors. Fan airflow is not measured and only nine "representative" zone temperature sensors are available for the entire 211,000 ft² building. There were not enough sensors to estimate the effect that reductions in airflow would have on occupants.

When the third party energy management company (WebGen) takes action to reduce energy at its connected sites, it uses a centralized control paradigm. While demand response systems are inherently centralized (signals to shed loads are generated in a one-to-many relationship), centralized control for day-to-day operation is less common.

In most control system markets (commercial buildings, industrial controls, etc.) there has been a trend for several decades toward decentralized control. In decentralized control, the control logic is moved (physically) as close to the sensors and final control elements (e.g., relays, valves, etc.) as possible. Decentralized control systems have traditionally been less costly, more flexible and more robust. However, in the IT community, there has been a movement in certain areas toward hosted solutions, application service providers and other centralized solutions. Ubiquitous internet connectivity and other IT technology advances make these systems less costly, more flexible and more robust for certain applications.

The WebGen system alternates between centralized and decentralized paradigms on cycles as short as twenty minutes. At the end of one cycle, a fan system maintains a setpoint entered by on-site building operators. In the next minute, a neural network algorithm may define the setpoint from over 3,000 miles away.

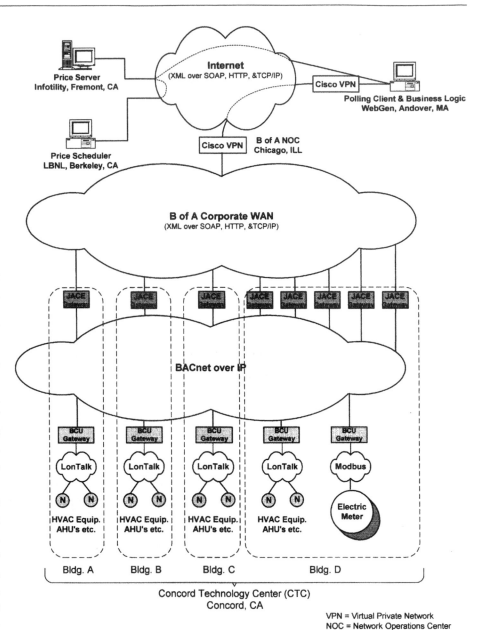

Figure 32-9. BofA System Overview

GSA (Government Office) System Overview (Figure 32-10)

The enterprise and EMCS infrastructures used to enable auto-DR at this site are linked together in a long series of serial components and communication links. The prototype system was assembled at low cost using spare parts. With so many links, it is not surprising that there were communication failures due to an unexplained equipment lock-up during the first test. To make the system more robust, a review of the components and architecture should be conducted.

The second test was quite successful, as communications were functional from end-to-end. The shed strategy produced an electric shed about as large as the other four sites combined. Because the temperature setpoint reset was at the zone level, comfort for each occupant could be maintained within the revised, relaxed constraints (Table 32-7). To implement this strategy, it was necessary to revise the software parameters and some logic in each of the 1,050 VAV terminal box controllers. For most EMCS systems, the labor required to make these revisions would be substantial (1-3 weeks). In this building, the process had been somewhat automated by previous system upgrades. This allowed EMCS reprogramming for auto-DR to be conducted in about three hours.

Roche (Offices and Cafeteria)
System Overview (Figure 32-11)

A third-party software framework (Itron/Silicon Energy) ties together three different EMCS protocols at Roche in a seamless fashion. The web interface provides operators with compete monitoring and control capability from anywhere on the campus. It was relatively straightforward to interface the auto-DR polling client and associated business logic to the system. The most challenging part of the project was setting up the "extra" computer outside of the Roche *firewall* and establishing communications to devices inside of the secure corporate network.

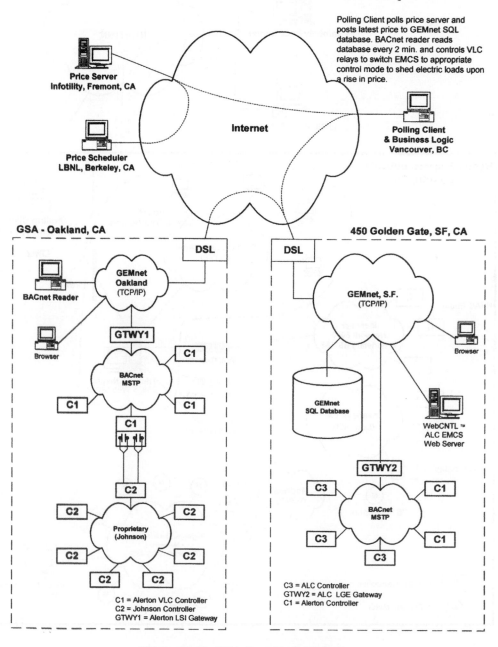

Figure 32-10. GSA System Overview

Table 32-7. Oakland GSA Zone Temperature Setpoints—Normal and Shed Modes

Shed Mode	Zone Heating Setpoint	Zone Cooling Setpoint
Normal ($0.10/kWh)	70°F	72°F
Level 1 ($0.30/kWh)	68°F	76°F
Level 2 ($0.75/kWh)	66°F	78°F

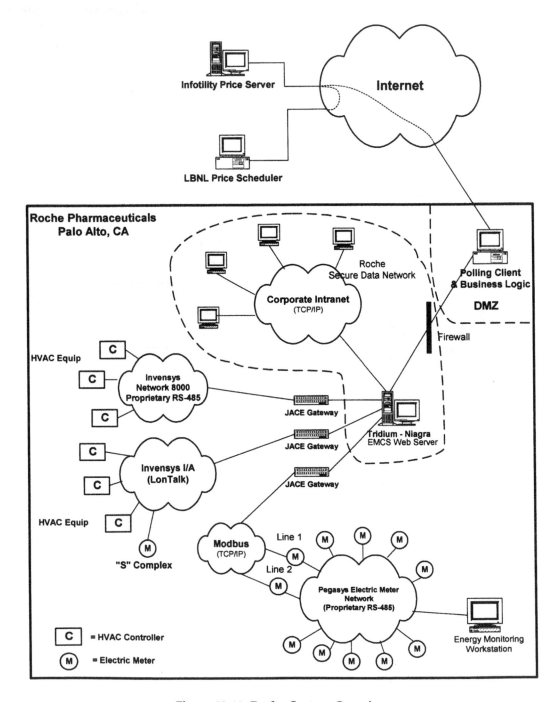

Figure 32-11. Roche System Overview

trol and monitoring elements (e.g., sensors), high security, and high system availability.

Leveraging Trends in Technology

The lower the installed cost of state-of-the-art auto-DR systems, the sooner they will find their way into mainstream use. One of the most important ways to keep costs low is to leverage existing trends in technology. For example, existing IT technology should be used in auto-DR systems wherever possible. The public internet and private corporate LAN/WANs are ideal platforms for auto-DR controls and communications due to their ubiquity, especially in large commercial buildings. In addition to the availability of networks, the performance of IT equipment (e.g., routers, firewalls, etc.) continues to improve and prices for this equipment continue to drop.

Enterprise, EMCS & EIS Integration

Another way to obtain high system performance and keep the system costs low is through increased integration within the building. Since energy data from EISs is simply another type of measured data, EISs and EMCSs should share the same networks so as to maximize system performance and functionality and minimize cost. In addition to eliminating a redundant EIS network, other aspects of the system are also unified through this approach. Use of an integrated EMCS/EIS database and associated archiving and visualization tools increases user functionality while reducing cost. The ability to change setpoints for HVAC equipment and observe and analyze the effect on electric consumption from the same Human Machine Interface (HMI) is an important enhancement to both the EMCS and EIS.

State-of-the-art auto-DR systems should also have tight integration between the EMCS/EISs network and enterprise networks within buildings. Once the integrated controls and communications infrastructures are in place, many applications in addition to auto-DR are enabled. Some other telemetry applications include: energy management, aggregation, equipment maintenance, access control and regulatory record keeping.

The network architecture of a state-of-the-art auto-DR system normally tends to be flatter than most of the sites in the November 2003 pilot. A flat architecture is one in which there are a minimum number of layers of control networks and protocols between the HMI and the final control and monitoring elements. The most robust and least costly systems should have

UCSB (University Library) System Overview (Figure 32-12)

Remote monitoring and control of the EMCS and EIS was available over the internet prior to the auto-DR pilot. However, at the time of the test, remote control of the EMCS was not available. The software gateway between the enterprise network and the EMCS network lost remote control functionality during an "upgrade" of Itron/Silicon Energy's third-party server software. To meet the test schedule of the auto-DR pilot, an IP I/O relay was added to allow the auto-DR business logic to initiate the control functions such as initiating sheds. The shed strategy proved to be very effective. The books and other thermal mass in the library buildings acted as a thermal "flywheel" to help keep the space comfortable during the shed periods. In addition, the shed strategy reduced airflow without shutting off fans completely. The coastal climate of the site helped provide a temperate airflow even when the cooling and heating valves were closed.

STATE OF THE ART IN AUTOMATED DEMAND RESPONSIVE SYSTEMS

By evaluating the systems demonstrated in the November 2003 auto-DR test, along with other existing technologies found in the EMCSs, EIS and the IT Industries, state-of-the-art auto-DR systems can be envisioned. The five participating sites all successfully met the functionality criteria of the pilot (under tight schedules and limited budgets). However, a truly state-of-the-art system would use the "best of the best" components, systems and strategies from end to end. Such a system would be designed from scratch to meet a very specific set of requirements. The "best" system would meet or exceed the requirements at the lowest installed cost. State-of-the-art auto-DR systems should have the following characteristics.

Flexible Designs for the Future

Today's state-of-the-art auto-DR technology could be applied in many different ways, depending on the scenarios and applications that they are designed to satisfy. As the scenarios, applications, and driving forces behind auto-DR become better defined, systems will be designed and deployed accordingly. Since these design criteria are likely to remain in flux, auto-DR system flexibility and future-proofing have a very high priority.

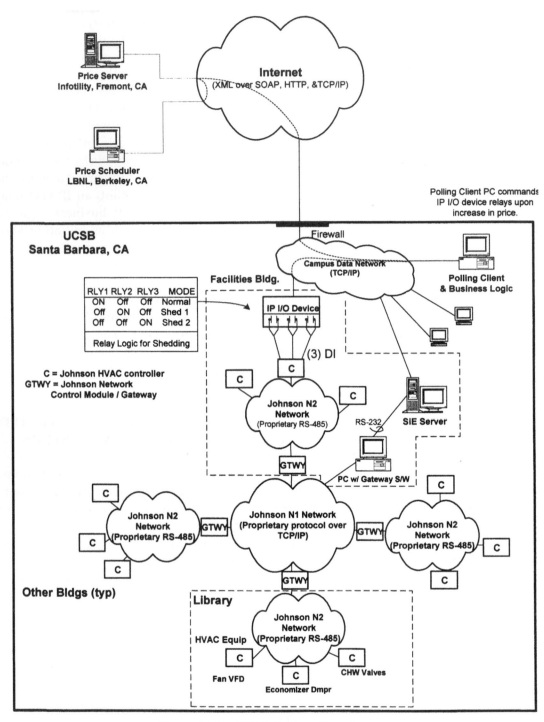

Figure 32-12. UCSB System Overview

Features

Customers should have numerous options about how they can participate in auto-DR programs. For any given motivating force that drives customers to consider auto-DR (i.e., price), each will have different circumstances under which they will want to participate. Any state-of-the-art auto-DR system must have sufficient flexibility to meet the needs of a variety of customers. They should have the ability to use custom business logic that is applicable to their own operations. Some may choose to allow remote *real-time* control (for its extra value) while others may want some advanced warning (via pagers, cell phones, etc.) and the ability to opt out, if desired. Other important features in state-of-the-art auto-DR systems include real-time two-way communications to the final con-no more than one enterprise network protocol and one control network protocol.

Open Standards

For flexibility and future-proofing as well as the option to choose "best of breed" products, state-of-the-art auto-DR systems should use open standards wherever possible. Unlike proprietary systems, truly open systems are interoperable. In other words, a device from one company will easily and naturally reside on a network with products from other companies. Most products in enterprise networks are interoperable. They communicate using the TCP/IP protocol and can be set up and managed using common network management tools. TCP/IP is clearly the worldwide protocol of choice for LAN/WAN, internet and enterprise networks.

There are several open standards control networks including BACnet (ASHRAE Standard 135-2001) and LonTalk (ANSI/EIA/CEA 709.1). Several database formats have become de facto open standards as well. Although the use of the meta-language XML is becoming a standard framework for communicating over enterprise networks and the internet, XML alone does not define data formats that could be used to convey measured building or energy data. Standards of this type are being developed by OPC XML, *oBIX*, BACnet XML and other organizations.

With only two network protocols in the state-of-the-art auto-DR system, only one type of gateway is required for translation/abstraction between them. An embedded gateway device that conforms with IT industry standards for reliability, security, and network management should be used.

Shed Strategies

State-of-the-art shed strategies should be designed to minimized discomfort, inconvenience and loss of revenue to the participating sites. Shed strategies should be devised by customers to meet their needs. In general, shed strategies that use fine granularity closed loop control are less likely to negatively impact building occupants. Ideally, sheds would vary, commensurate with a variable shed signal. Transitions would be fast enough to be effective, but slow enough to minimize attracting the attention of building occupants.

In addition to HVAC control strategies, lighting and switch-able plug loads should be considered for sheds as well. By increasing the controlled load to the point where it approaches the whole building load,

each load type (HVAC, lighting, etc.) would need to shed a smaller amount in order to achieve a given shed target for the whole building.

Future Directions

In the industrial controls marketplace it is becoming more common for the TCP/IP based industrial Ethernet to be used all the way down to the device level. In these systems, traditional open control protocols such as BACnet and LonTalk are eliminated all together. TCP/IP could be used in an end-to-end integrated enterprise, EMCS/EIS system. This trend is likely to gain momentum once the next *generation* of internet protocol, IPv6 is implemented. For greater flexibility, increased control granularity and lower costs, increased use of wireless devices in auto-DR, EMCS and EISs is likely to occur.

ACKNOWLEDGMENTS

The authors are grateful for the extensive support from Ron Hofmann, a consultant to the California Energy Commission, Laurie ten Hope (CEC), Gaymond Yee (CIEE) and Karen Herter (LBNL), Joe Desmond and Nicolas Kardas (Infotility), and numerous individuals from the participating sites. This project was supported by the California Energy Commission's Public Interest Energy Research Program and by the Assistant Secretary for Energy Efficiency and Renewable Energy, Office of Building Technology, State and Community Programs of the U.S. Department of Energy under Contract No. DE-AC03-76SF00098.

References

[1] Piette, Mary Ann, David S. Watson, Osman Sezgen, Naoya Motegi, Christine Shockman and Ron Hofmann. Forthcoming in 2004. "Development and Evaluation of Fully Automated Demand Response in Large Facilities." LBNL Report #55085.

[2] Motegi, Naoya, Mary Ann Piette, David S. Watson, and Osman Sezgen Lawrence Berkeley National Laboratory, "Measurement and Evaluation Techniques for Automated Demand Response Demonstration," 2004 ACEEE Summer Study on Energy Efficiency in Buildings. LBNL Report #55086.

[3] Shockman, Christine. 2004. "Market Transformation Lessons Learned from an Automated Demand Response Test in the Summer and Fall of 2003." Proceedings of the ACEEE 2004 Summer Study on Energy Efficiency in Buildings. American Council for an Energy-Efficiency Economy.

[4] Nexant. 2001. AB 979, AB 29X and SB 5X Peak Load Reduction Programs. December 2001 Annual Report. Submitted to California Energy Commission and the California State Legislature.

[5] Nexant. 2002. AB 29X and SB 5X Program Evaluation. First Quarter Report (January 1 to March 1,2002). Submitted to California Energy Commission. Contract #400-00-070.

[6]   Kinney, S., M.A. Piette, L. Gu, and P. Haves. 2001. Demand Relief and Weather Sensitivity in Large California Commercial Office Buildings, Proceedings of the 2001 International Conference for Enhanced Building Operations, LBNL Report 48285

[7]   Motegi, Naoya, Mary Ann Piette, Satkerter Kinny and Karen Herter. 2003. "Web-based Energy Information Systems for Energy Management and Demand Response in Commercial Buildings." High Performance Commercial Building Systems, PIER Program. LBNL#-52510. *http://buildings.lbl.gov/hpcbs/Pubs.html*

[8]   Flynn, P. (Editor). 2003. "The XML FAQ" Maintained on behalf of the World Wide Web Consortium's XML Special Interest Group v. 3.01 (2003-01-14).

[9]   Webopedia. 2004. *http://www.webopedia.com/TERM/X/XML.html* Viewed on Mar. 15, 2004.

Chapter 33

Participation through Automation: Fully Automated Critical Peak Pricing in Commercial Buildings*

Mary Ann Piette, David Watson,
Naoya Motegi, Sila Kiliccote, and
Eric Linkugel

ABSTRACT

California electric utilities have been exploring the use of dynamic critical peak prices (CPP) and other demand response programs to help reduce peaks in customer electric loads. CPP is a tariff design to promote demand response (DR). Levels of automation in DR can be defined as follows. Manual demand response involves a potentially labor-intensive approach such as manually turning off or changing comfort set points at each equipment switch or controller. Semi-automated demand response involves a pre-programmed demand response strategy initiated by a person via centralized control system. Fully automated demand response does not involve human intervention, but is initiated at a home, building, or facility through receipt of an external communications signal. The receipt of the external signal initiates pre-programmed demand response strategies. We refer to this as auto-DR.

This chapter describes the development, testing, and results from automated CPP (Auto-CPP) as part of a utility project in California. The chapter presents the project description and test methodology. This is followed by a discussion of auto-DR strategies used in the field test buildings. We present a sample auto-CPP load shape case study, and a selection of the auto-CPP response data from September 29, 2005. If all twelve sites reached their maximum saving simultaneously, a total of approximately 2 MW of DR is available from these twelve sites that represent about two million ft². The average DR was about half that value, at about 1 MW. These savings translate to about 0.5 to 1.0 W/ft² of de-

mand reduction. We are continuing field demonstrations and economic evaluations to pursue increasing penetrations of automated DR that has demonstrated ability to provide a valuable DR resource for California.

BACKGROUND

California electric utilities have been exploring the use of critical peak prices (CPP) and other demand response programs to help reduce peak demands from customer electric loads. CPP is a form of price-responsive demand response. Recent evaluations have shown that customers have limited knowledge of how to operate their facilities to reduce their electricity costs under CPP (Quantum Consulting and Summit Blue, 2004). While lack of knowledge of how to develop and implement DR control strategies is a barrier to participation in DR programs like CPP, another barrier is the lack of automation in DR systems. Most DR activities are manual and require people to first receive emails, phone calls, and pager signals, and second, for people to act on these signals to execute DR strategies.

Levels of automation in DR can be defined as follows. **Manual demand response** involves a labor-intensive approach such as manually turning off or changing comfort set points at each equipment switch or controller. **Semi-automated demand response** involves a pre-programmed demand response strategy initiated by a person via centralized control system. Fully automated demand response does not involve human intervention, but is initiated at a home, building, or facility through receipt of an external communications signal. The receipt of the external signal initiates pre-programmed demand response strategies. We refer to this as auto-DR. One im-

*This chapter was previously published in the Proceedings of the American Council for an Energy Efficient Economy's 2006 Summer Study on Energy Efficiency in Buildings.

portant concept in auto-DR is that a homeowner or facility manager should be able to "opt out" or "override" a DR event if the event comes at a time when the reduction in end-use services is not desirable. Participation of more then 30 large facilities in the last three years of demonstrations has shown that the automation can be provided with minimal resistance from facility operators.

The PIER Demand Response Research Center conducted a series of tests during the summers of 2003, 2004, and 2005. The objectives of these tests were two fold. First, we sought to develop and evaluate communications technology to send DR signals to commercial buildings. This was necessary because buildings use controls with diverse protocols and communication capabilities. Second, we sought to understand and evaluate the type of control strategies facility owners and managers would be willing to test in their buildings. During these past three years we have evaluated auto-DR in 28 facilities; the average demand reductions were about 8% over the three to six hour DR events. Many electricity customers have suggested that automation will help them institutionalize and "harden" their electric demand savings, improving overall response and repeatability. The evaluation of the California's 2004 DR programs found that ten to fifteen of the sites that participated in their study could not participate in the DR event because the person in charge of the demand reduction was not in the facility on the day of the event (Quantum Consulting and Summit Blue, 2004).

Table 33-1 shows the number of sites that participated in each year's field tests along with the average and maximum peak demand savings. The electricity savings data are based on weather sensitive baseline models developed for each building that predicts how much electricity each site would have used without the DR strategies. Further details about this research are available in previous reports (Piette et al., 2005a and b). One key distinction between the 2005 and the previous tests is that the 2005 test sites were actually on a CPP tariff, while the 2003 and 2004 tests used fictitious prices

and there was no actual economic incentive for the sites. The "fictitious" test consisted of an actual shed based on fictitious prices. There were no DR economics incentives. The sites were willing to conduct the DR to understand their DR capability and automation infrastructure.

The focus of the rest of this chapter is the design and results from the 2005 auto-CPP field tests, with some additional comments about the previous years' tests. The next section describes the auto-CPP project description and test methodology. This is followed by a discussion of auto-DR strategies used in the field test buildings. We then present a sample auto-CPP load shape case study, and a selection of the auto-CPP DR data from September 29, 2005. The summary section provides an overview of key findings. Since the buildings only participated within the program during the later DR events of the summer 2005, we do not have detailed economics on the impact of CPP. Each site, however, saved money.

Automated CPP Project Description

PG&E's critical peak pricing (CPP) program is a voluntary alternative to traditional time-of-use rates. The CPP program only operates during the summer months (May 1 through October 31). Under the program, PG&E charges program participants' higher prices for power on up to 12 hot afternoons between May 1 and October 31. Manual CPP customers are notified by email and phone by 3 p.m. the previous day that the following day is a CPP day. The customer sees lower electricity costs on non-CPP days. The price of electricity rises on a maximum of 12 hot days, with the DR event triggered by temperature. The additional energy charges for customers on this tariff on CPP operating days are as follows (Figure 33-1):

*   **CPP Moderate-Price Period Usage**: The electricity charge for usage during the CPP moderate-price period was three times the customer's summer part-peak energy rate under their otherwise-applicable rate schedule multiplied by the actual

Table 33-1. Average and Maximum Peak Demand Savings during Automated DR Tests.

Results by Year	# of sites	Duration of Event (Hours)	Average Savings During (%)	Highest Max Hourly Savings (%)
2003	5	3	8	28
2004	18	3	7	56
2005	12*	6	9	38

*Some of the sites recruited were not successful during the 2005 CPP events because of delays with advanced meters and control work, but are expected to be ready for the 2006 tests.

energy usage. The CPP moderate-price period was from 12:00 Noon to 3:00 p.m. on the CPP operating days.

- **CPP High-Price Period Usage**: The total electricity charge for usage during the CPP high-price period was five times the customer's summer on-peak energy rate under their otherwise-applicable rate schedule multiplied by the actual energy usage. The CPP High-Price period was from 3:00 p.m. to 6:00 p.m. on the CPP operating days.

The 2005 auto-DR project design was a collaboration between LBNL, the DRRC, and PG&E. PG&E had offered voluntary critical peak pricing in 2004, with over 250 sites participating. We recruited 15 PG&E customer facilities to participate in fully automated response critical peak pricing. There were three categories of recruits. First, five of the sites had participated in the 2004 auto-DR tests and were willing to move from the fictitious tests to the actual tariff. Second, we worked with the PG&E customer account representatives to recruit two sites that had been on CPP to include them in the auto-CPP tests. Third, eight sites were recruited for the 2005 tests that had not been on CPP or had not participated in the previous auto-DR tests.

Demand Response Automation Server

PG&E sent the critical peak price signals to each participating facility using the demand response automation server developed by LBNL and Akuacom. The automation server communicated via XML with PG&E DR communications system, Interact II. Qualified sites were configured to respond to automated price signals transmitted over the internet using relays and gateways that send standardized signals to the energy management control system (EMCS). A few sites used the day-ahead automation notification for their pre-cooling strategies. Most of the sites used the signal in real time that alerted them at noon on the CPP day that the event was triggered. During the 2005 summer test period, as the electricity price increases during a CPP event, pre-selected electric loads were automatically curtailed based on each facility's control strategy. The automation server uses the public internet and private corporate and government intranets to communicate CPP event signals that initiate reductions in electric load in commercial buildings. The researchers worked with the facility managers to evaluate the control strategies programmed in the energy management and control systems (EMCS), which executed pre-determined demand response strategies at the appropriate times.

Connectivity was provided by either an internet gateway or internet relay (as shown in Figure 33-2). The internet gateways typically connect the internet communication protocol (TCP/IP) to the protocol of a given EMCS. This means that a different internet gateway type is usually required to communicate with each different EMCS brand or product line. Gateways provide a variety of functions further described in Piette et al. (2005). An internet relay is a device with relay contacts that can be actuated remotely over a LAN, WAN or the internet using internet protocols (IP). The internet is based on a standard protocol (TCP/IP) and all EMCS can sense the state of relay contact closures (regardless of their particular EMCS protocol). Because of this, internet relays can be used on virtually any commercial building that has a standard connection to the internet. Internet connectivity directly to the EMCS is not required.

The four elements of the diagram are as follows:

1. PG&E uses their standard InterAct II system to notify the automation server of an upcoming CPP event (notification occurs day-ahead).

2. The automation server posts two pieces of information on its Web services server:

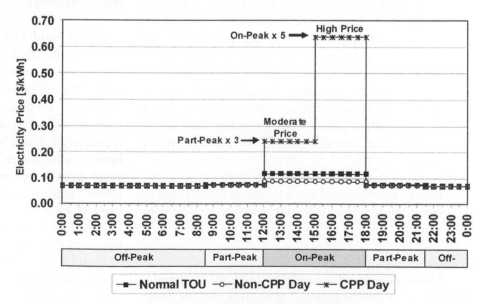

Figure 33-1. Critical Peak Pricing Tariff

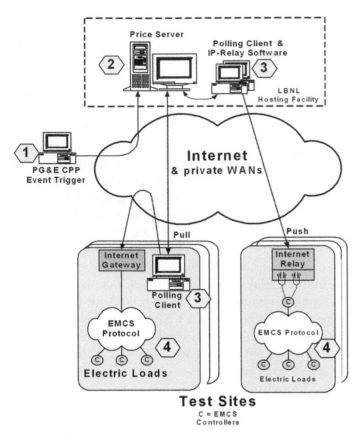

Figure 33-2. Building type, size, year in Auto-DR, and DR control strategy used.

— There is a pending event. This is posted immediately upon receipt from PG&E at approximately 3:00 p.m. the day ahead.

— There is an active event of a given level. Moderate-level demand response events are posted between 12:00-3:00 p.m. High-level demand response events are posted between 3:00 p.m.-6:00 p.m. on the day of the event.

3.   Polling clients request information each minute. Logic software determines actions based upon latest information polled from the automation server. Actions are initiated based on predetermined logic.

4.   Energy management control system (EMCS) carries out predetermined demand response control commands.

Evaluation Methodology

LBNL developed an electric load shape baseline model to estimate the demand shed from the DR strategies for each building. First we collected the electric

consumption data from Interact for each site. We subtracted the actual metered electric consumption from the baseline-modeled consumption to derive an estimate of demand savings for each 15-minute period. The model is described in previous papers (Piette et al, 2005). PG&E uses a baseline for the CPP evaluation. The demand response strategy was considered effective if in either or both of the moderate price and the high price periods, the average power savings over the 3-hour period was larger than the average of the standard error in the baseline model. For each building we derived the hourly electric load savings, percent savings in whole-building load, and power density reduction (W/ft^2). Sample results for the auto-CPP events are shown below.

The CPP baseline used by PG&E does not include weather data, but is based on the average hourly load shape of 3 highest consumption days in the last 10 working days (excluding holidays). The baseline algorithm considers the site electric consumption from the period of noon to 6 p.m. to choose the highest 3 days. CPP event days are excluded from the reference days. The CPP baseline estimate may be lower than the actual demand if the site's demand is weather-sensitive, since a CPP day typically occurs on a higher temperature day. If the ten previous working days were cooler than the CPP day, the baseline will be lower than weather normalized baseline.

There are a few other features about the project that we do not have space to review in this chapter. The evaluation included post-event surveys to determine how well each strategy performed and if there were any outstanding issues in the DR control strategies. The evaluation also examined the cost to program control strategies in the EMCS and to connect the internet gateways and relays.

Auto-DR Field Test Results

In 2003, 2004, and 2005 we conducted automated DR tests in 28 buildings listed in Table 33-2. Table 33-3 shows the entire list of sites and which years they participated. The tests included numerous building types such as office buildings, a high school, a museum, laboratories, a cafeteria, data centers, a postal facility, a library, retail chains, and a supermarket. The buildings range from large campuses, to small research and laboratory facilities. The table lists the DR control strategies used at each building. The full reports from the auto-DR field tests describe these strategies in greater details, and they are also discussed in Watson et al. (2006). The global zone temperature adjustment was the most commonly used strategy, though 16 other strategies are listed. Nearly all of these strategies were based on direct connections to the EMCS. Further details on pre-

Table 33-2. Building Type, Size, Year in Auto-DR, and DR Control Strategy Used.

				Participation			HVAC														Light, Misc.				
	Building use	Total conditioned area	# of bldg	2003	2004	2005	Global temp. adjustment	Fan-coil unit off	SAT reset	Fan VFD limit	Duct static pres. reset	Fan quantity reduction	Electric humidifier off	CHW temp. reset	CHW current limit	Chiller demand limit	Boiler lockout	Pre-cooling	Extended shed period	Slow recovery	Common area light dim	Office area light dim	Anti-sweat heater shed	Fountain pump off	Transfer pump off
300 CapMall	Office	383,000	1		X		X			X		X		X										X	
ACWD	Office, lab	51,200	1		X	X	X		X		X			X	X		X		X						
Albertsons	Supermarket	50,000	1	X																		X	X		
B of A	Office, data center	708,000	4	X	X	X				X	X	X		X	X							X	X			
Chabot	Museum	86,000	2		X		X											X							
Cal EPA	Office	950,000	1		X						X									X	X			
CETC	Research facility	18,000	1		X								X	X											
Cisco	Office, tech lab	4,466,000	24		X				X	X								X			X	X			
2530 Arnold	Office	131,000	1	X	X	X													X						
50 Douglas	Office	90,000	1	X	X	X													X						
Echelon	Corporate Headquarter	75,000	1	X	X	X			X			X	X								X	X			
GSA 450 GG	Federal office	1,424,000	1		X	X																		
GSA NARA	Archive storage	202,000	1		X	X																		
GSA Oakland	Federal office	978,000	1	X	X	X																		
Gilead 300	Office	83,000	1		X				X															
Gilead 342	Office, Lab	32,000	1		X	X			X															
Gilead 357	Office, Lab	33,000	1		X	X			X															
Irvington	Highschool	N/A	1		X	X										X							
IKEA	Retail	300,000	1		X	X																		
Kadent	Material process	-	1		X																		X
LBNL OSF	Data center, Office	70,000	1		X	X										X							
Monterey	Office	170,000	1		X														X				
Oracle	Office	100,000	2		X	X					X													
OSIsoft	Office	60,000	1		X		X																	
Roche	Cafeteria, auditorium	192,000	3	X	X							X											
Target	Retail	130,000	1		X							X							X				
UCSB Library	Library	289,000	3	X	X					X	X				X								
USPS	Postal service	390,000	1		X	X										X		X					

cooling research, which may prove to be an important DR control strategy, are presented in Peng et al, (2004 and 2005).

Example of Demand Response from an Office Building

This section provides an example of the DR electric load shape data for the 130,000 ft² Contra Costa County office building. The graph shows the electric load shape during an actual auto-CPP event on September 29, 2005. The baseline power peaks around 400 kW, with the weather sensitive LBNL baseline and the PG&E CPP baseline also shown. The vertical line at each baseline power datum point is the standard error of the regression estimate. The vertical lines at noon, 3 p.m., and 6 p.m. indicate price signal changes. The building shed about 20% of the electric loads for six hours by setting up the zone temperatures from 74 to 76 during the first three hours and 76 to 78 F during the second three hours. This strategy reduced the whole-building power density by an average of 0.8 W/ft² during the six hours.

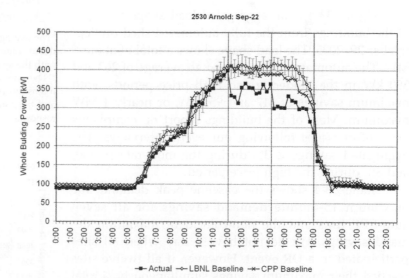

Figure 33-3. Baseline and Office Building Electric Load Shape during Auto-DR Event

Aggregated Automated Demand Response

The auto-CPP tests consisted of seven events that took place from August through November 2005. Configuring many of the sites to participate in the auto-CPP was time consuming because of complications related to control programming and meter installation delays. Thus, several of the CPP events took place before our sites were configured. To account for this delay, we developed a series of fully automated mock-CPP tests that took place in October and November 2005. These days were not as warm as actual CPP days and the DR events show lower electric savings than we observed on warmer days.

Table 33-3 shows sample results from eight buildings that participated in auto-CPP on an actual CPP day. The table lists the average and maximum peak demand savings, whole building percentage savings, and power density savings during the two three-hour price periods: Moderate and High. The average reductions per building ranged from 2 to 184 kW, with maximum savings of 31 to 291 kW. The table shows the total DR (Shed kW), whole-building power reduction (WBP %), and power density reduction (W/ft^2). The columns list both the average and maximum savings for the moderate and high priced CPP periods. The maximum is the fifteen-minute max demand response in the six hour monitoring period. Average percentage reductions ranged from zero (negligible) to 28% savings, and maximum percentage reductions from 3 to 37%. The average power density reductions ranged from 0.02 to 1.95 W ft^2, with maximum demand reductions 0.21 to 4.68 W/ft^2. The Bank of America site dominates the aggregated demand response.

Figure 33-4 is an aggregated load shape for eight of the buildings from the fully automated shed on September 29, 2005. The load shape shows a total of about 8 MW. The automated DR provided an average of 263 and 590 kW in the moderate and high price periods, with maximum savings of 617 and 952 kW, or nearly 1 MW maximum. Most of the buildings report no complaints or comfort issues following our event interviews. The aggregated savings is 3% during the moderate period, and 8% during the high price period.

Table 33-4 shows the baseline peak demand, the maximum 15-minute demand savings for all seven auto-CPP tests and the non-coincident maximum demand savings. We do not have a day when all sites participated in a DR event. However, if all twelve sites reached their maximum savings simultaneously, a total of approximately 2 MW of demand response is available from these twelve sites that represent about two million ft^2. Using the sum of the average demand response for each of the twelve sites shows the average demand response was about 1 MW. These results indicate that 1 to 2 MW of demand response can be expected for two million ft^2 of buildings (0.5 to 1.0 W/ft^2 of demand saving) with this type of automation. As mentioned, fol-

Table 33-3. Average Demand Response by Price Period, September 29th

Unit	Site Name	Average		Max	
		Moderate	High	Moderate	High
Shed kW	ACWD	67	57	101	72
	B of A	22	184	132	291
	Chabot	2	32	31	88
	2530 Arnold	34	58	90	89
	Echelon	32	109	42	143
	Gilead 342	45	55	73	75
	Gilead 357	48	62	94	150
	Target	14	33	53	44
Total: Σ(ΔP)		263	590	617	952
WBP %	ACWD	24%	19%	38%	23%
	B of A	0%	4%	3%	6%
	Chabot	0%	3%	10%	28%
	2530 Arnold	8%	14%	21%	21%
	Echelon	9%	28%	12%	37%
	Gilead 342	13%	15%	19%	20%
	Gilead 357	9%	11%	16%	25%
	Target	4%	9%	15%	12%
Total: Σ(ΔP)/Σ(BP)*		3%	8%	8%	12%
Average: Σ(ΔP/BP)/N		9%	13%	17%	21%
W/ft^2	ACWD	1.53	1.29	2.30	1.63
	B of A	0.04	0.30	0.21	0.47
	Chabot	0.02	0.37	0.35	1.02
	2530 Arnold	0.26	0.44	0.69	0.68
	Echelon	0.43	1.45	0.56	1.91
	Gilead 342	1.39	1.72	2.30	2.36
	Gilead 357	1.50	1.95	2.95	4.68
	Target	0.13	0.30	0.48	0.40
Total: Σ(ΔP)/Σ(A)**		0.23	0.52	0.55	0.85
Average: Σ(ΔP/A)/N		0.66	0.98	1.23	1.64

*The average of the individual average whole building response and the average of the maximum individual DR results are shown, along with the aggregated shed compared to the total baseline power.

**The power densities are also shown for the average of the demand intensities (sum all building densities and divide by the sample size) and the sum of the total area and the total aggregated total demand response.

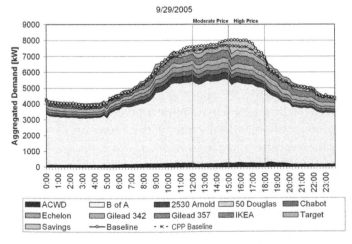

Figure 33-4. Automated CPP Aggregated Demand Saving Results, September 29th

lowing each event LBNL interviewed building managers to evaluate if any problems occurred. There were some minor complaints in a few cases. Overall the sites were able to provide good demand response with minimal disruptions. We have begun to explore the costs required to configure the auto-DR communication systems and program DR control strategies within an EMCS. Initial research suggests we can configure auto-CPP systems with the existing financial incentives available as part of California utility DR technical assistance funds. Ideally auto-DR systems would be installed as part of retro-commissioning programs. With their knowledge and skills, today's retro-commissioning engineers may be key players in providing building control tune-ups and developing custom DR strategies during field work (Piette et al, 2006). Installation and configuration of auto-DR systems require a good understanding of HVAC, lighting, and control strategies.

SUMMARY AND FUTURE DIRECTIONS

The auto-CPP tests in 2005 have demonstrated the technical feasibility of fully automated DR. While there are considerable challenges in auto-DR in general and auto-CPP specifically, the research demonstrates that this can be done with reasonable levels of effort with today's technology. New knowledge on what strategies are available for different types of buildings has been obtained and is the subject of another ACEEE paper (Watson et al, 2006).

During 2006 we will be pursuing a larger number of tests throughout California. The research may also move beyond CPP into other DR programs such as demand bidding. The primary objective of this new research will be to better understand the economics of installing and configuring automated systems, exploring connectivity and control strategies in more building types, including industrial facilities, and evaluating the peak demand reduction levels for different weather. We are also interested in "heat storm" performance that moves beyond single day DR participation, to several hot days in a row.

In the long term this research aims at transform communications in commercial and industrial facilities to explore literally "connecting" the demand and supply side systems with the technologies and approaches explored in this project. Our goal is to understand how to configure buildings to be "DR ready" in a low cost way, developing requirements for new buildings through future codes and embedding such communications directly into future EMCS. Additional research is also needed

Table 33-4. Maximum Demand Response for all Seven Event Days

		Aug-08	Sep-22	Sep-29	Oct-06	Oct-13	Oct-25	Nov-10	2004	Max
ACWD	Baseline Peak kW			330	253	290	238			330
	Max Shed kW			101	74	83	77			101
B of A	Baseline Peak kW			5311		5163	5053			5053
	Max Shed kW			291		219	552			552
Chabot	Baseline Peak kW		225	308	244	270				308
	Max Shed kW		19	88	36	42				88
2530 Arnold	Baseline Peak kW	505	419	431	404	406	345			505
	Max Shed kW	176	119	90	63	89	40			176
50 Douglas	Baseline Peak kW	381					259			381
	Max Shed kW	95					78			95
Echelon	Baseline Peak kW		334	403	363	359	304			403
	Max Shed kW		115	143	132	117	84			143
Gilead 342	Baseline Peak kW		288	384	289	340	278			288
	Max Shed kW		94	75	45	55	80			94
Gilead 357	Baseline Peak kW			607		455	443			607
	Max Shed kW			150		119	145			150
IKEA	Baseline Peak kW					1982	1803			1982
	Max Shed kW					321	223			321
Oracle	Baseline Peak kW							507		507
	Max Shed kW							65		65
Target	Baseline Peak kW		314	364	328	341	296			341
	Max Shed kW		52	53	60	64	49			64
USPS*	Baseline Peak kW								1483	1483
	Max Shed kW								333	333
Total	Baseline Peak kW	886	1579	8138	1881	9608	9020	507	1483	12189
	Max Shed kW	272	399	992	410	1108	1329	65	333	2182

* 2004 data (Oct-13) is used for USPS because USPS failed to conduct demand shed in 2005.

to integrate price and reliability DR signals, which we believe can co-exist on similar communications systems. Finally, there is a need to better understand advanced controls for simultaneous use applications of energy efficiency and demand response. We need to define explicit "low power" building operating modes for DR events. Daily advanced energy efficient operations with granular controls provide the best starting point for DR capability. New technologies such as dimmable ballasts and wireless HVAC control are likely to provide such new levels of granularity that can be optimized to provide both daily and enable advanced DR strategies. Along with such new technology is the need for improved energy management and financial feedback systems. As the DR economics mature, better real-time economic feedback is needed if energy managers and facility operators are going to understand the value of participating in DR events.

Acknowledgements

The authors are grateful for the extensive support from numerous individuals who assisted in this project. Many thanks to the engineers and staff at each building site. Special thanks to Ron Hofmann for his conceptualization of this project and ongoing technical support. Thanks also to Laurie ten Hope, Mark Rawson, and Dave Michel at the California Energy Commission. Thanks to the Pacific Gas and Electric Company who funded the automated CPP research. This work described in this report was coordinated by the Demand Response Research Center and funded by the California Energy Commission, Public Interest Energy Research Program, under Work for Others Contract No.150-99-003, Am #1 and by the U.S. Department of Energy under Contract No. DE-AC03-76SF00098.

References

Piette, Mary Ann, David S. Watson, Naoya Motegi, Norman Bourassa and Christine Shockman. 2005a. "Findings from the 2004 Fully Automated Demand Response Tests in Large Facilities" September. CEC-500-03-026. LBNL-58178. Available at http://drrc.lbl.gov/drrc-pubs1.html

Piette, Mary Ann, Osman Sezgen, David S. Watson, Naoya Motegi, and Christine Shockman. 2005b. "Development and Evaluation of Fully Automated Demand Response in Large Facilities," January. CEC-500-2005-013. LBNL-55085. Available at http://drrc.lbl.gov/drrc-pubs1.html

Piette, Mary Ann., David S. Watson, Naoya Motegi, Sila Kiliccote. 2006. Automated Critical Peak Pricing Field Tests: Program Description and Results, LBNL Report 59351. March.

Piette, Mary Ann., David S. Watson, Naoya Motegi, Sila Kiliccote, and Eric Linkugel. 2006. "Automated Demand Response Strategies and Commissioning Commercial Building Controls, 2006 National Conference on Building Commissioning. April.

Quantum Consulting Inc., and Summit Blue LCC. 2004. Working Group 2 Demand Response Program Evaluation—Program Year 2004, Prepared for the Working Group 2 Measurement and Evaluation Committee, December.

Watson, David S., Naoya Motegi, Mary Ann Piette, Sila Kiliccote. 2006. Automated Demand Response Control Strategies in Commercial Buildings, Forthcoming Proceedings of 2006 ACEEE Summer Study on Energy Efficiency in Buildings. Pacific Grove, CA. Forthcoming.

Xu, Peng, Philip Haves. 2005. Case Study of Demand Shifting With Thermal Mass in Two Large Commercial Buildings. ASHRAE Transactions. LBNL-58649.

Xu, Peng, Philip Haves, and Mary Ann Piette, and James Braun. 2004. Peak Demand Reduction from Pre-cooling with Zone Temperature Reset of HVAC in an Office. Proceedings of 2004 ACEEE Summer Study on Energy Efficiency in Buildings. Pacific Grove, CA. LBNL-55800.

Chapter 34

Web Based Wireless Controls for Commercial Building Energy Management

Clifford Federspiel, Ph.D., PE

ACKNOWLEDGEMENTS

The project described in this chapter was funded by direct and in-kind support from the California Energy Commission's Public Interest Energy Research (PIER) Program and the Iowa Energy Center Energy Resource Station. Martha Brook and John House provided project oversight. John House and Curt Klaassen provided input on the design of the experiments. Joe Zhou provided valuable technical support throughout the project. Kurt Federspiel was the software development engineer for this project.

INTRODUCTION

This case study describes a web-based, wireless, supervisory control system for commercial-building heating, ventilating, and air-conditioning (HVAC) systems that has been designed to convert constant air volume (CAV) HVAC systems to variable air volume (VAV) operation in a way that is non-intrusive (no terminal retrofits or static pressure controls), inexpensive, and easily maintainable. The system design avoids the need for asbestos abatement, which is commonly required for buildings that have CAV HVAC systems.

There are three common types of CAV systems that serve multiple zones: single-duct systems with terminal reheat, dual-duct systems, and multi-zone systems. Single-duct systems deliver cooled air to each zone then re-heat it as needed to keep the zone space temperature close to a setpoint. Dual-duct systems supply heated and cooled air to each zone, then mix the two to maintain the space temperature. Multizone systems are a special case of a dual-duct system where the mixing dampers are part of the air-handling unit rather than located at each zone.

CAV systems of the types described above are inefficient. In states with strict energy codes, such as California, they are prohibited in new construction. They are also prohibited by ASHRAE Standard 90.1. For HVAC systems that serve multiple zones, it is now common to use variable-air-volume (VAV) systems.

VAV systems have variable-speed fans and terminal dampers that are controlled so that the amount of simultaneous heating and cooling or re-heating is significantly reduced. There are two common kinds of VAV systems: single-duct and dual-duct. Single-duct VAV systems supply cooled air to each zone terminal unit, where it is metered with a control damper when cooling is required or re-heated when heating is required. When heating, the amount of cooled air is reduced to a low level by the terminal controls, so there is much less wasted re-heat energy than a single-duct CAV system. Dual-duct systems deliver heated air and cooled air all the way to each zone terminal unit with separate air ducts. Dual-duct VAV terminal units have independent dampers that modulate the hot airflow rate to heat a zone and modulate the cold airflow rate to cool a zone. Unlike the dual-duct CAV system, the dual-duct VAV system does very little mixing. Most of the time it supplies a variable amount of hot air when heating and a variable amount of cooled air when cooling. It only mixes air when the zone load is small so that adequate ventilation air is provided.

Although CAV systems are less common in new construction, there is still a large installed base. According to the Energy Information Agency (Boedecker, 2005), CAV systems that serve multiple zones condition 3.5 billion square feet of commercial building floor space in the U.S. Since they are inefficient, retrofit strategies have been developed to modify their design and operation in order to make them more efficient. These strategies require mechanical modifications to the HVAC system.

Mechanical modifications are disruptive to the commercial operations in the building and they will always require asbestos abatement if asbestos is present.

Existing Solutions

For single-duct CAV systems, Liu et al. (2002) recommend adding a VFD to the fan to reduce the fan speed during after-hours operation. During occupied hours the fan is operated at full speed. This strategy does not save energy for systems that are shut off after hours. Even when there is after-hours operation, this method is not cost effective unless the system is large because the energy savings are limited.

For dual-duct CAV systems, Liu and Claridge (1999) describe a means for improving energy performance without retrofitting terminal units. They add a damper to the hot duct and use it to control the pressure in the hot duct. This strategy still requires a mechanical modification, which is intrusive and requires that the system be shut down. It also requires the installation of pressure sensors in the hot air duct and cold air duct.

For multi-zone CAV systems, Liu et al. (2002) describe a means for improving the energy performance by adding a VFD to the supply fan and controlling the supply fan speed so that the most-open mixing damper is 95% open to the hot deck in the heating season. In the cooling season their strategy controls the fan speed so that the most-open mixing damper is 95% open to the cold deck. They do not describe how the strategy works in swing seasons when the unit could be heating some zones while cooling others. The command to the VFD comes from a Proportional-Integral-Derivative (PID) controller that takes the most-open damper position as input. This strategy requires that position sensors be added to the mixing dampers. Position sensors are expensive and difficult to install. Resistive position sensors are prone to vibration-induced pre-mature failure. This strategy cannot be applied to single-duct CAV systems because they do not have mixing dampers.

Johnson (1984) describes a case where a single-duct re-heat system with cooling-only operation was modified for VAV operation by eliminating the zoning, regulating the discharge air temperature, and modulating the average zone temperature by adjusting the fan speed. This strategy yielded large annual energy savings (46.5% reduction in HVAC energy for one unit and 53.9% for another), but could not have worked in a system with heating and most likely had a negative impact on thermal comfort since the zoning was eliminated (i.e., the system was operated as a large single-zone system after the retrofit).

Existing products that are used to convert CAV systems include VAV retrofit terminals and VAV diffusers. VAV retrofit terminals are VAV boxes installed in existing supply ducts where CAV terminals (re-heat units or mixing boxes) are located in a CAV system. VAV retrofit terminals are sold by many manufacturers. Examples of VAV retrofit terminal units are shown in Figure 34-1. When CAV systems are retrofit with VAV retrofit terminal units, the system must be shut down and workers near the terminal location must move because the retrofit requires significant mechanical and electrical modifications to the existing HVAC system.

VAV diffusers are variable-area diffusers that have an actuation mechanism combined with temperature feedback so that the open area of the diffuser is modulated to maintain the local temperature. Modern VAV diffusers can accommodate switchover from heating to cooling mode, and some come with embedded DDC controls. VAV diffusers are sold by Acutherm, Price, Titus, and others. Figure 34-2 shows examples of VAV diffusers. Retrofitting a CAV system to VAV with VAV diffusers is expensive because every diffuser must be retrofit, and most zones have air supplied by several diffusers.

APPROACH

The project involved building and testing a new solution for converting CAV systems to VAV operation.

Figure 34-1. VAV retrofit terminal units.

We designed a web-enabled, wireless control system that included a new control application called discharge air regulation technique (DART) to achieve the following goals:

1. Short installation time
2. Minimal disruption of occupants
3. Flexible configuration
4. No need for asbestos abatement, should asbestos be present
5. Standalone operation if necessary
6. Long battery life
7. Remote monitoring and alarming
8. Browser-based human-machine interface (HMI)
9. Modular design so that it can be used to deploy other applications

Goals 1-5, when combined with the large energy savings from CAV to VAV retrofits, result in a short payback period. Goals 6-8 yield a system that is easy to maintain. Goals 1-5 are facilitated by the use of low-power wireless sensing and control modules that utilize self-healing, mesh networking. The particular wireless technology that we selected uses a time-synchronized mesh network that enables extended battery life (several years on AA lithium batteries) and mesh networking capability for every node in the network. Both the gateway for the wireless network and the supervisory control system are web-enabled devices. Both of these devices communicate using XML-RPC, both run a web server, and both are configurable via a browser. Additionally, the supervisory controller can send alarms via the internet to an email address or a pager, and it can display time series data. The design of both the hardware and software is modular. We could use a different wireless network, yet still use the FSC and the software running on it. This modular design allows us to use this same platform to deliver other supervisory energy management applications in addition to DART.

DART Application

DART works by reducing the supply fan and return fan speeds at part-load conditions. When there is no load, the discharge air temperature is equal to the zone temperature, so reducing the fan speeds has no impact on the room temperature. At part-load conditions, the discharge air temperatures are somewhat higher or lower than the zone temperature, depending on whether the zone is being heated or cooled. Reducing the fan speeds at part-load conditions causes the discharge air temperatures to increase or decrease, depending on whether the zone is being heated or cooled, so that the heat transfer rate to the zone doesn't change. DART maintains the highest discharge air temperature close to a high-temperature setpoint or the lowest discharge air temperature close to a low-temperature setpoint, which has the effect of keeping the fan speed low when the load is low but causing it to increase as the load increases. Lowering the fan speed not only reduces fan energy consumption, but it also reduces the amount of mechanical cooling, and reduces the amount of (re-) heating.

Control System Architecture

The control system architecture consists of the components shown in Figure 34-3. A supervisory control computer, called an FSC, is connected via a LAN/WAN to at least one wireless network gateway, which is called a manager. The FSC and manager do not need to be co-located. The FSC could be located in a data center in one city while the manager is located in a building in another city. Additionally, systems can be configured with a remote FSC and multiple managers located in different parts of a single building or located in different buildings. The FSC is an embedded, fanless computer with a x86 architecture and a 600 MHz clock. The FSC runs a web server for its HMI. The HMI supports configuration tables, dynamic data tables, and time series graphics. The manager is a web-enabled, embedded, fanless computer with ethernet and serial ports. The manager soft-

Figure 34-2. VAV diffusers.

ware includes a web server that can be used to remotely configure the wireless network. Having the manager and the FSC both web-enabled makes the system highly flexible and easy to manage and service remotely. Sensor inputs and control command outputs are provided by wireless modules, each of which is equipped with analog and digital I/O. To enable the modules to measure temperature with an external probe, a thermistor circuit was added to the HD-15 connector on the wireless modules. To allow the modules to produce 0-10 VDC control commands, we designed a digital-to-analog converter (DAC) circuit that connected to the HD-15 connector of the wireless modules.

The wireless modules and the wireless network manager automatically form a mesh network such as the one depicted by Figure 34-4. The wireless networking hardware we selected operates in the 902-928 MHz ISM band. The modules can use any one of 50 channels (frequencies) in this band. Individual modules in a single network may use different channels. The modules change their channel and their routing parents dynamically in search of a clear channel to avoid interference. These features help improve the reliability of the

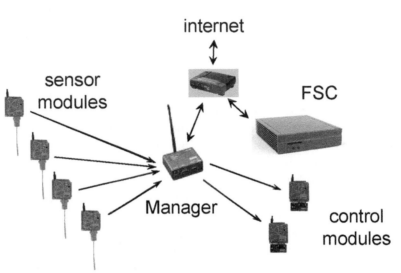

Figure 34-3. Control system components.

wireless communications.

For the DART application, we used wireless temperature sensors to measure zone temperatures, discharge air temperatures (Figure 34-5), supply air temperature, and heating hot water temperature. The wireless modules have an internal temperature sensor that was used for zone temperature measurement instead of the external probe shown in Figure 34-5. The zone temperature sensors were used primarily for monitoring purposes. The supply air temperature sensor and the heating hot water temperature sensor were used to allow the low-temperature and high-temperature setpoints of the DART application to follow the resets on the supply air temperature and heating hot water temperature. Using these two sensors this way eliminated the need to program these reset schedules into DART. By installing the discharge air temperatures in the ceiling plane as shown in Figure 34-5, we could avoid asbestos that might be used to insulate supply ducts or structural components of an older building above the ceiling plane.

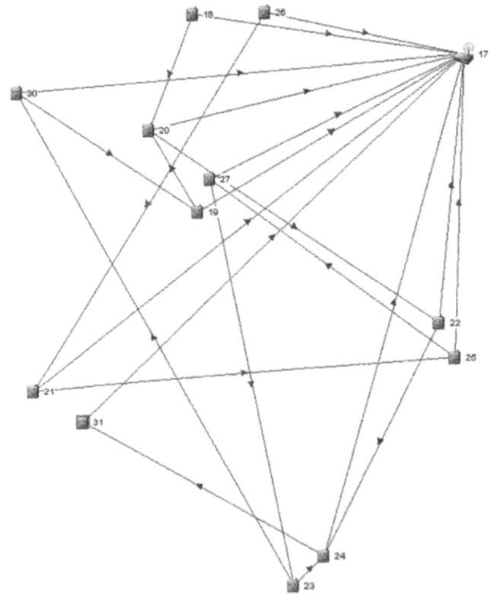

The wireless sensors operate on two AA lithium batteries. The wireless network uses a time-synchronized mesh protocol that allows the sensor to be in a deep sleep mode that uses very little power most of the time. Battery life is dependent on the network configuration and how much routing a module performs. Battery life of up to eight years is achievable for modules that do not re-route.

Figure 34-6 shows the wireless control modules. The control modules are line powered. For this application, they were powered from the 24 VDC power supply provided by the variable frequency drive (VFD). They returned a 0-10 VDC speed control signal back to the analog input of the VFD.

The FSC and the manager communicate using XML-RPC. XML (eXtensible markup language) is a standard for creating markup languages that describe the structure of data. It is not a fixed set of elements like HTML, but is a meta-language, or a language for describing languages. XML enables authors to define

Figure 34-4. Mesh network at the ERS

Figure 34-5. Discharge air temperature sensor.

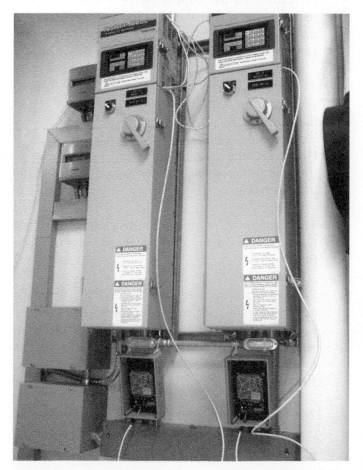

Figure 34-6. Control modules installed beneath the VFDs.

their own tags. XML is a formal specification of the World Wide Web Consortium. XML-RPC is a protocol that allows software running on disparate operating systems, in different environments to make procedure calls over the internet. It is remote procedure call using HTTP as the transport and XML as the encoding. XML-RPC is designed to be as simple as possible, while allowing complex data structures to be transmitted, processed and returned. The FSC uses XML-RPC to configure wireless network settings, to retrieve certain kinds of network data such as battery life and module status (connected or unreachable), and to set the output voltages of the control modules. Sensor data (e.g., voltages measured by the sensor modules that correspond to temperature) are pushed by the manager to the FSC asynchronously. The FSC parses the incoming XML document and places the sensor data in a database.

The FSC has five basic software components. They include the XML-RPC driver for the manager, application control software, a database, a web server, and HMI software (web interface). The XML-RPC driver and the application control software are written in C++, which results in fast, compact code that is beneficial for running on an embedded platform. The driver and application control software are object-oriented, which simplifies software maintenance and upgrades. The application software includes web-based alarms in addition to the DART application software. The system is programmed to send a message to an email address or pager when certain events such as loss of communication between the FSC and the manager or an unreachable control module occur. This feature proved useful during the first week of testing, when a network glitch caused the system to stop. We received the alarm in California, contacted the ERS staff and had them reset the system, so that the controls were quickly operational again. We selected MySQL for the database, Apache for the web server, and PHP for the HMI software because they are all open-source software components, but are also widely used and highly reliable. For this case study, the FSC used the Windows XP operating system, but all of the five software components could run under the Linux operating system.

Energy Resource Station

The wireless control system was installed and tested at the Iowa Energy Center's Energy Resource Station (ERS). The ERS has been designed for side-by-side tests of competing HVAC control technologies. The building has two nominally identical HVAC systems that each serve four zones in the building (three perimeter zones and one internal zone). The building is oriented with the long axis north-south so that the zones of each system nominally have the same solar exposure. Figure 34-7 shows a floor plan of the ERS. The supply and return ducts for the test rooms are color-coded. The numbered circles show the locations of the wireless control devices during the first week of testing. Each number is that device's network ID. Node 17 is the wireless network gateway. It and the supervisory control computer were located in a telecom closet. Both were connected to the internet via a switch, and each used a static IP address provided by the ERS.

The ERS has submeters on all energy consuming HVAC loads including all pumps motors, fan motors, chillers, and the heating hot water boiler. The ERS has approximately 800 monitoring points that are trended every minute for temperatures, humidities, flows, pres-

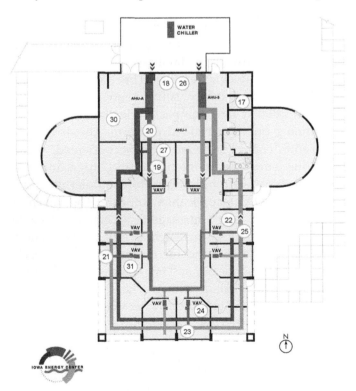

Figure 34-7. Floor plan of the ERS showing mote locations during the first week of testing.

sures, lighting conditions, and weather conditions. The sensors are calibrated regularly. The ubiquitous and accurate sensing at the ERS enhance the ability to accurately measure the energy performance difference between competing control strategies and technologies.

Additional details about the ERS can be found at http://www.energy.iastate.edu/ers/.

Test Conditions

During the first week of testing, the following configuration was used:

- The A system was controlled with DART, while the B system was run as a CAV reheat system.

- The design flow to the perimeter rooms was 650 CFM (2.4 CFM/sf), while the design flow to the interior room was 300 CFM (1.1 CFM/sf).

- The VAV box dampers were fixed to deliver the design flow to each room at 100% (60 Hz) fan speed.

- We calibrated the supply fan – return fan speed relationship so that the supply flow was approximately equal to the return flow and configured DART to use this relationship.

- The return fan speed of the CAV system (B) was set so that the supply flow was equal to the return flow.

- Discharge air was delivered with one diffuser in the perimeter rooms, but with two diffusers in the interior room.

- The blinds in all perimeter rooms were lowered but the slats were maintained in the horizontal position.

- Low-temperature and high-temperature setpoints of DART were reset based on the supply air temperature and the heating hot water temperature, respectively. These setpoints determine the maximum absolute difference between the zone temperatures and the discharge air temperatures.

- Zone temperature controls of the Metasys system used proportional plus integral (PI) control to modulate reheat coil valves.

- The supply air temperature setpoint was manually reset each morning based on the high-temperature outdoor air temperature forecast for that day. The schedule was 55 to 60 degF as the maximum outdoor air temperature forecast ranged from 50 to 20 degF, respectively.

- The systems were operated 24/7.

- The test ran from February 11, 2006 through February 19, 2006

- The minimum supply fan speed was 40%.

Figure 34-8 shows the false load per room at the beginning of the test. The loads were reduced by 100 Watts (base load turned off for the remainder of the test) at 6 p.m. on February 13 because the peak loads significantly exceeded the capacity of the system. Beginning February 14, the supply air temperature reset schedule was changed to the 52-57 degF range so that the cooling capacity would match the peak loads better. The peak load density (3.7 W/sf at the beginning of the test and 3.4 W/sf after these changes were made) is significantly higher than average. Wilkins and McGaffin (1994) reported load densities from office equipment ranging from 0.48 W/sf to 1.08 W/sf, with an average value of 0.81 W/sf. With the same diversity factor for occupants that was observed for the office equipment, the occupant load density would be 0.53 W/sf, which would give an average load density of 1.34 W/sf. The peak load density in this test was, therefore, 2.5 times higher than average.

Figure 34-9 shows the lighting schedule for each room each day. The peak lighting load was 2.2 W/sf, which is higher than the 1.5 W/sf limit set by modern codes and standards.

During the second week of testing, the following changes were made to the test configuration:

- The B system was controlled with DART, while the A system was run as a CAV reheat system.

- We re-calibrated the supply fan – return fan speed relationship so that the supply flow of AHU-B was approximately equal to the return flow and configured DART to use this relationship. The relationship for this week was not the same as the first week (A and B unit fan characteristics were not the same).

- Low-temperature and high-temperature setpoints of DART were fixed at 60 degF and 90 degF, respectively.

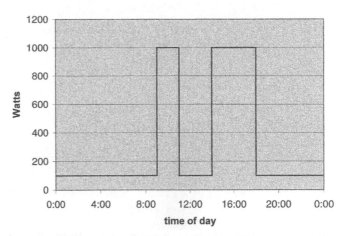

Figure 34-8. False load in each room (zone) during the first week of testing.

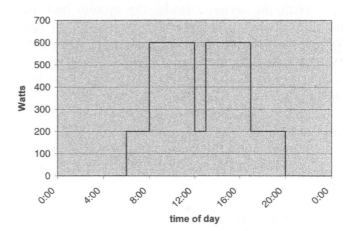

Figure 34-9. Lighting load in each room during the tests.

- Zone temperature controls of the Metasys system used proportional-only control with a 3 degF proportional band to emulate the operation of pneumatic controls.

- The supply air temperature setpoint was fixed at 55 degF at all times.

- The test ran from March 4, 2006 through March 12, 2006

Figure 34-10 shows the false load per room during the second week of testing. The loads were lowered to 1.12 W/sf so that they would more closely represent actual loads in buildings. The base load was provided by a desktop computer because it is increasingly common to keep computers on 24/7 for after-hours maintenance.

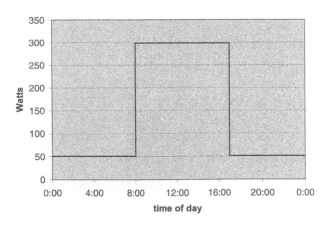

Figure 34-10. False load in each room (zone) during the second week of testing.

With the lowered loads, the system had many daytime operating hours at the minimum fan speed. To exercise the system, the loads were increased by 900 Watts per room on two days (March 8 and 9) from 9:30 a.m. to 11 a.m. and again from 1:30 p.m. to 3 p.m. in just the internal rooms. We picked the internal rooms for the increased loads to emulate the use of an internal conference room.

The lighting loads during the second week of testing were the same as during the first week of testing.

PROJECT RESULTS

Energy Performance Comparison

Table 34-1 summarizes the average energy performance from 6 a.m. to 6 p.m. each day for the first week.

Table 34-1. Energy consumption and savings during the first week of testing (from 6 a.m. to 6 p.m.).

	DART	CAV	% saved
Supply fan kWh/day	15.9	23.5	32.1
Return fan kWh/day	4.31	4.92	12.3
Therms/day	3.16	3.86	18.2

Based on the 12-hour schedule (6 a.m. to 6 p.m.) and the average weekly operating hours reported in the Energy Information Agency's commercial building energy consumption survey (CBECS), which is 61 hours/week, we estimate that DART would save 2.0 kWh/sf/yr and 0.17 therms/sf/yr under this load condition.

Table 34-2 summarizes the average energy performance from 6 a.m. to 6 p.m. for the second week.

Table 34-2. Energy consumption and savings during the second week of testing (from 6 a.m. to 6 p.m.).

	DART	CAV	% saved
Supply fan kWh/day	6.82	23.5	71.0
Return fan kWh/day	2.15	6.24	65.5
Therms/day	2.04	4.07	49.8
Chiller kWh/day	1.44	2.3	37.1

Based on the 12-hour schedule (6 a.m. to 6 p.m.) and the average weekly operating hours reported in the CBECS (61 hours/week), we estimate that DART would save 5.2 kWh/sf/yr and 0.49 therms/sf/yr under this load condition.

If we estimate typical energy savings by averaging the savings from the two weeks, then we get an estimated annual energy savings of 3.7 kWh/sf/yr and 0.34 therms/sf/yr. Using average energy costs reported by EIA for California ($0.1198/kWh and $1.08/therm for 2005), this equates to an energy cost savings of $0.81/sf/yr in California. For Iowa, where the utility costs are $0.0697/kWh and $1.066/therm, the annual energy cost savings should be $0.62/sf/yr. For the U.S. as a whole, where average utility rates are $0.0867/kWh and $1.157/therm, the annual energy cost savings should be $0.71/sf/yr. At these saving rates, the payback period of the entire system, including the VFDs, should be less than two years.

The savings figures are higher than anticipated based on published savings from conventional CAV to VAV retrofits. The high electrical energy savings are probably due to the fact that conventional retrofits don't normally use static pressure reset for the supply fan control, so the supply fan energy isn't reduced as much as possible with a conventional CAV to VAV retrofit. DART operates the supply fan as low as possible while ensuring that the zones are still in control, which yields fan energy savings that should be comparable to VAV operation with static pressure reset. The thermal energy savings may be higher than anticipated because the base case during the second week didn't use supply air temperature reset, and the reset during the first week was modest. Supply air temperature reset is a way for CAV systems with DDC controls on the air-handling unit (AHU) to reduce reheat and mechanical cooling. Since supply air temperature reset doesn't affect fan energy consumption, and since the mechanical cooling savings were small in this demonstration (due to the cool outdoor air temperatures allowing the system to cool entirely with an economizer most of the time), not accounting for supply air temperature reset cannot inflate the potential electrical energy savings. We expect that thermal energy savings for systems that already use a large supply air temperature reset will be about half of the levels observed in these tests.

Temperature Control Performance Comparison

During the first week of testing, the zone temperature control performance of the DART system was more oscillatory than that of the CAV system. The oscillatory behavior was caused by at least the following two factors: 1) the discharge air temperatures were often not stable, 2) the economizer was often not stable. These instabilities were present in both the DART system and the CAV system, but they remain localized in the CAV system, whereas they can become distributed by the DART system because the variable fan speed affects all parts of the system. Additionally, the lower discharge velocities and higher discharge air temperatures under heating conditions may have resulted in a stratification layer that could also result in oscillatory temperature readings at the zone thermostats.

After the first week of testing we made changes to the DART software to make it less sensitive to discharge air and supply air instabilities, switched from Proportional-Integral (PI) to Proportional-Only control for the zone temperatures, and used fixed setpoints for DART. We anticipated that these changes would result in less zone temperature variability. The Proportional-Only zone temperature control emulates the behavior of pneumatic controls, which are still commonly used with legacy CAV systems.

Figure 34-11 shows the maximum absolute deviation of the zone temperatures from the average zone temperature by zone from 8 a.m. to 6 p.m. during the second week. In three of the four zones, the maximum excursion occurred in the system controlled by DART, but the largest excursion occurred in the West zone of the CAV system.

Figure 34-12 shows the average absolute deviation of the zone temperatures from the average zone temperature for the same period on the same scale as Figure 34-11. The average zone temperature variability was highest with the DART system, but it was less than 0.5 degF, so we do not anticipate that DART will result in thermal discomfort due to temperature variability.

We also installed a vertical temperature sensing tree in the East zones to measure stratification. The trees were located half-way between the supply diffuser and the return grill in each room. Figure 34-13 shows the vertical temperature profile in East-A (CAV) and East-B (DART) at 6 a.m. on the coldest morning of the test (27.8 degF outdoor air temperature). At this point in time, the

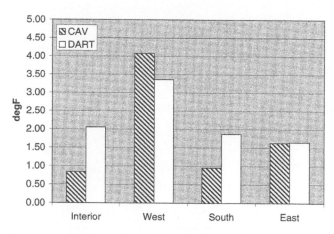

Figure 34-11. Maximum zone temperature excursions during the second week.

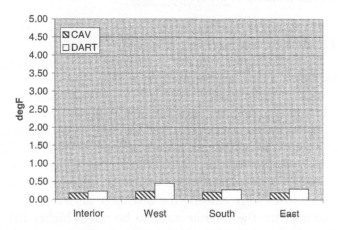

Figure 34-12. Average zone temperature variability during the second week.

discharge air temperature of the East-A (CAV) zone was 71.4 degF, while the discharge air temperature for the East-B (DART) zone was 88 degF with a fan speed of 40%. The CAV system has better mixing, and a nearly uniform vertical temperature profile. The DART system has a clearly increasing vertical temperature profile. The profile is less steep in the occupied zone, and the vertical temperature difference from the ankles to the head (0.5 feet to 5.5 feet) is 5.1 degF, which is less than the 5.4 degF requirement of ASHRAE Standard 55-2004. We conclude that stratification should not be a problem either for thermal comfort or temperature control stability, at least not under load conditions similar to these. In practice there will probably be less temperature stratification because movement of people, mixing between rooms at open doorways, cross-flow in open plan offices (none of which was allowed to occur in these tests) and additional office equipment will mix the air.

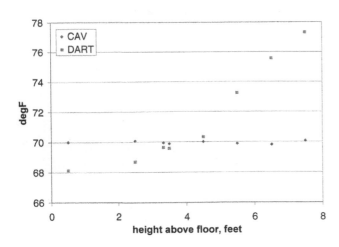

Figure 34-13. Vertical temperature profile at 6 a.m. on the coldest morning of the second week.

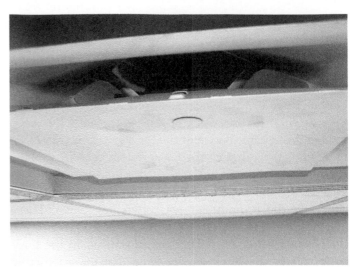

Figure 34-14. Installation of the interior zone discharge air temperature sensor.

Wireless Sensor Network Evaluation

We installed the wireless sensors and controls in locations and orientations that were convenient, and not necessarily ideal for radio transmission. Figure 34-5 shows the installation of one of the discharge air temperature modules. Three of the four zones had this configuration, but in the interior zone we placed the sensor inside the diffuser, as shown in Figure 34-14, because the lower discharge velocities and cross-flow at the diffuser were causing the discharge air temperature readings for the interior zone to be much higher than the other zones when all reheat valves were closed. The diffuser and the connecting ductwork are all constructed of sheet metal, so this module was essentially shrouded in metal. Even in this semi-shielded location, the radio performance was good. The discharge air temperature module installed inside the diffuser (Figure 34-14) provided just 1% fewer packets than the other discharge air temperature modules. We never had an instance where a module was unreachable or where the data were more than three minutes old, which was a criterion that we used for communication failure.

The control modules were installed close to the VFDs (Figure 34-6), which are a potential source of interference. We found that the control modules took longer than the other kinds of modules to join the network, presumably because they had to search longer than the other modules to find a clear channel. However, once they had joined the network, the communications between the control modules and the wireless network manager never failed.

The benefits of wireless communications and mesh networking became quite clear after the installation. The entire network, which consisted of 10 sensor modules, 2 control modules, one manager, and one FSC were installed and operational in less than two hours. By the time we had completed the installation of the sensor modules and control modules and returned to the manager and FSC, the mesh network had configured itself. We did not need to add repeaters to the network or relocate modules to improve the reliability of the wireless communications. Packet loss during the two weeks of testing was negligible.

CONCLUSIONS

We conclude the following from this project:

1. The energy savings potential from DART is high, even higher than anticipated based on published savings from conventional CAV to VAV retrofits.

2. The web-enabled manager and FSC, combined with wireless mesh networking make the system easy to install, commission, and maintain.

3. DART increases the zone temperature variability and heating mode stratification, but the increase in the temperature variability is small and the heating mode stratification is within the bounds of ASHRAE Standard 55-2004.

4. The reliability of the wireless sensor network technology used for this demonstration was good.

5. The wireless control platform designed for this project could easily be used to provide other supervisory energy management applications.

References

Boedecker, E., 2005, Statistician, Energy Information Agency, U.S. Department of Energy, personal communication.

Johnson, G.A., 1984, "Retrofit of a Constant Volume Air System for Variable Speed Fan Control," *ASHRAE Transactions*, 90(2B), 201-212.

Liu, M. and D.E. Claridge, 1999, "Converting Dual-Duct Constant-Volume Systems to Variable-Volume Systems without Retrofitting the Terminal Boxes," *ASHRAE Transactions*, 105(1), 66-70.

Liu, M., D.E. Claridge, and W.D. Turner, 2002, "Chapter 4: CC Measures for AHU Systems," *Continuous Commissioning Guidebook for Federal Energy Managers*, Federal Energy Management Program, U.S. Department of Energy.

Wilkins, C.K. and N. McGaffin, 1994, "Measuring computer equipment loads in office buildings," *ASHRAE Journal*, 36(8), 21-24.

GLOSSARY

AHU: air-handling unit
ASHRAE: American Society of Heating, Refrigerating, and Air-conditioning Engineers
CAV: constant air volume
CBECS: commercial building energy consumption survey
CFM: cubic feet per minute
DAC: digital to analog converter
DART: discharge air regulation technique
FSC: Federspiel supervisory controller
HMI: human-machine interface
HTML: hyper-text markup language
HVAC: heating, ventilating, and air-conditioning
ISM: industrial, scientific, medical
LAN: local area network
PHP: Hypertext preprocessor (originally called Personal Home Page)
PI: proportional-integral
PID: proportional-integral-derivative
RPC: remote procedure call
SQL: structured query language
VAV: variable air volume
VDC: volts direct current
VFD: variable frequency drive
WAN: wide area network
XML: extensible markup language

Chapter 35

Wireless Sensor Applications for Building Operation and Management

Michael R. Brambley, Michael Kintner-Meyer, and
Srinivas Katipamula, Patrick J. O'Neil

ABSTRACT

The emerging technology of wireless sensing shows promise for changing the way sensors are used in buildings. Lower cost, easier to install, sensing devices that require no connections by wires will potentially usher in an age in which ubiquitous sensors will provide the data required to cost-effectively operate, manage, and maintain commercial buildings at peak performance. This chapter provides an introduction to wireless sensing technology, its potential applications in buildings, three practical examples of tests in real buildings, estimates of impacts on energy consumption, discussions of costs and practical issues in implementation, and some ideas on applications likely in the near future.

INTRODUCTION

Wireless communication has been with us since the invention of the radio by Marconi around 1895. We have benefited from the broadcast of information for purposes of informing and entertaining. Radio technology has also enabled point-to-point communication, for example, for emergency response by police and fire protection, dispatch of various service providers, military communications, communication to remote parts of the world, and even communication into space.

We commonly think of communication between people by voice when thinking of radio frequency (RF) communication technology but need to look no further than a television set to realize that other forms of information, such as video, can also be transmitted. In fact, RF technology can be used to transfer data in a wide variety of forms between machines and people and even among machines without human intervention. This more generic wireless RF transfer of data and its application to operating and maintaining buildings is the focus of this chapter.

Wireless communication of data via WiFi (or IEEE 802.11 standards) is now routine in many homes, offices and even airports.[1,2] Rather than ripping walls open or fishing networking cable through them to install computer networks in existing homes and commercial buildings, many users opt to use wireless technology. These standards use license-free frequency bands and relatively low power to provide connections up to several hundred feet (although additional parts of IEEE 802.11 are currently under development for much longer ranges of up to 20 miles and higher data transfer rates). These standards are generally for relatively high bandwidth so that large files can be transported over reasonable time periods.

In contrast to the data rates required for general computer networking and communication, most sensor data collection can get by with much slower rates with as little as a few bits every second, every minute, 10 minutes, or even less frequently. Sensing generally imposes (or loosens) other constraints as well. For example, if the value of a single sensor point is low, its total installed cost must be very low as well. Furthermore, if power for sensing and communication is not conveniently available where sensor measurements are needed, an on-board power source may be needed. In general, we'd like to put sensors in place and then forget about them, so they should have long lives and require little attention. If a sensor requires frequent maintenance, the cost for its use increases rapidly, so power sources, like batteries, with lives of 10 years or more would be ideal. These requirements for sensors and sensor networks are leading to the evolution of wireless sensor network technology and standards that provide specifically for convenient, widespread use of large numbers of sensors from which data are collected wirelessly.

The ideal wireless sensor would have very low installed cost, which would require that its hardware cost be very low and that it be installed quickly and easily using limited labor. One concept calls for wireless sensors that you "peel, stick and forget." The radio frequency identification (RFID) tag industry debatably has reached a cost as low as about $0.20 per tag and seeks to reach $0.05 per tag with a production of 30 billion tags per year for inventory tracking purposes. [3] Wireless sensors for active property measurements like those suitable for use in building operations still cost on average two to four orders of magnitude more than this.

To achieve easy and low-cost installation, wireless sensor networks, which provide the means for moving data from the collection points to where it can be used, will probably need to be *self-configuring*. This means that the sensors would assign themselves identifications, recognize their neighboring sensors, and establish communication paths to places where their data are used (e.g., on a personal workstation or a receiver connected to a building automation system). A self-configuring wireless sensor network would only require placing the sensors where the measurements need to be made and possibly providing a connection to a user interface or computer network.

To reduce the cost of maintenance, the sensors and sensor network would need to be *self-maintaining* and *self-healing*. For example, if a metal cabinet were moved into the communication path between two sensors, blocking communication between them, the network would automatically reroute the signal by another path with no human intervention. In addition, the sensors would need to maintain their own calibration reliably over their lifetimes (be *self-calibrating*), actively ensuring that they are within calibration periodically. These capabilities are critical to ensuring low cost and reliable sensor networks. If each sensor has to be maintained by technicians periodically during its life, the cost will be too high to justify its use in all but the most critical and high-value applications. To increase sensor use, lower life-cycle costs are essential.

Some wireless sensors may have access to hard-wired power, but for many applications the sensor and its radio must be *self-powered*, using a battery that lasts for many years or harvesting power from the ambient environment. [4] In 2004, some manufacturers of wireless sensors claim battery lives as long as 7 years for some applications. Wireless sensors that use environmental vibrations as a source of power have also been developed for a limited set of applications, [5] but most ambient power harvesting schemes are still under development. Complementary developments are underway for a wide range of applications that reduce the power requirements of electronic circuits. Examples for sensor networks include: intelligent management of on-board power use by sensor radios to limit power requirements (and battery drainage), using sleep modes, transmitting only as frequently as absolutely required, and minimizing message size. Power requirements are also tied directly to the distance over which signals must be transmitted. By decreasing this distance and using multiple hops to span a long distance, power can be conserved. The mesh networking schemes described later in this chapter have the potential to significantly reduce the power requirements for wireless sensors.

These are some of the capabilities of the ideal wireless sensor. In the sections that follow, an introduction to wireless sensor technology is provided, potential applications for wireless sensors in buildings are described, potential benefits are discussed, a few real-world cases are presented, and the current state of wireless sensing and likely future developments are described. Three primary concerns are frequently raised in discussion of wireless sensing for building operation: cost, reliability, and security. This chapter addresses each of these, providing references for the reader interested in more detail. Some practical guidance for using wireless sensors in buildings today is also provided.

Why use Wireless Sensing in Buildings?

The cost of wiring for sensors and controls varies widely from about 20% to as much as 80% of the cost of a sensor or control point. The precise costs depend on the specific circumstances, e.g., whether the installation is in new construction or is a retrofit in an existing building, the type of construction, and the length of the wiring run. For situations where wiring costs are high, eliminating the wires may produce significant cost reductions.

Too often today operators are not able to effectively monitor the condition of the vast array of equipment in a large commercial building. Field studies and retro-commissioning of commercial buildings show that dirty filters, clogged coils, inoperable dampers, and incorrectly-implemented controls are all too common. [6, 7] Pressures to reduce operation and maintenance costs only exacerbate this problem. The problem can be even worse in small commercial buildings, which frequently don't even have an operator on site. Keeping apprised of the condition of equipment and systems in these buildings is nearly impossible for an off-site operator. If an equipment problem does not directly affect the occupants of a building (and this is quite common when the systems compensate by running harder and using more energy), it will usually continue undetected and uncorrected until conditions deteriorate and the occupants complain. This is often long after the problem started wasting energy and costing the bill payers money. Annual or semi-annual service visits by maintenance technicians, often catch only the most obvious problems. Incorrectly-implemented controls can go undetected for years unless major retro-commissioning of the building is undertaken.

More sensors to monitor the condition of equipment and systems, as well as conditions in the building, are needed along with software tools that automatically sort through data as it arrives and alert building operations and maintenance staff (or service providers) to problems. Building owners, however, often cite the need to keep costs down as the reason for not installing these sensors. By doing this, they are trading lower initial costs for higher expenditures on energy and lost revenue from tenant churn caused by poor environmental conditions in the building. This might be addressed by education and more evidence of the net value of good operation and maintenance over the building ownership life cycle, but lowering the cost of collecting data and obtaining useful results from it may be a more direct approach. This chapter focuses on the data collection issue by presenting information on wireless sensing; the need for tools that automatically process the data is a companion problem that is just as critical, but that is the subject of Chapter 18 in this book.

Better sensing in commercial buildings would lead to greater awareness of the condition of buildings and their systems. Operation and maintenance (O&M) staff would have the information to recognize degradation and faults in building equipment and systems and prioritize problems based on cost and other impacts. Today, most building staffs do not have this information. With it, the most costly and impactful problems could be identified, even those that are not usually recognized today.

The benefits of more data and tools that provide useful information from that data would be: lower energy and operating costs, longer-equipment lives, and better, more consistent conditions provided to building occupants. The value of these should all well exceed the cost of collecting and processing the information. With new, lower cost means such as wireless sensing for gathering data, first costs should also decrease making the financial decision to make this investment easier for building owners.

There are also some advantages directly attributable to the unique characteristics of wireless sensing beyond lower cost. Wireless sensors having their own power sources are mobile. Such a sensor can be readily moved from one location to another to investigate a problem. If a particular office, for example, were chronically reported as too hot, a wireless air-temperature sensor might be moved to that office or an additional one added to the wireless sensor network for that office to verify that the temperature was indeed unacceptably hot, then used to verify whether the corrective actions were successful. New sensors could be added to equipment for similar purposes without installing additional wiring. For example, if a pump motor were thought to be intermittently running hot, a wireless sensor might be installed on it to monitor its temperature and verify the need for repairs. If not wired, these sensors could be placed temporarily and then used at different locations as needed; no wiring costs would be necessary. One of the benefits of a wireless sensor network is that once it is in place in the building, sensors can be added or moved easily without installing new cables. As a result, wireless sensors have unique value for diagnostics.

Wireless Sensor Networks
Primary Components

Each wireless radio frequency (RF) sensor requires three critical components to sense a condition and communicate it to a point at which it can be used (whether by a human or directly by another machine): 1) a *sensor* that responds to a condition and converts it to a signal (usually electrical) that can be related to the value of the condition sensed, 2) a *radio transmitter* that transmits the signal, and 3) a *radio receiver* that receives the RF signal and converts it to a form (e.g., protocol) that can be recognized by another communication system, another device, or computer hardware/software. This is the simplest communication configuration for wireless sensing (see Figure 35-1).

At the sensor the device usually consists of signal processes circuitry as well as the sensor probe itself. This circuitry may transform the signal with filtering, analog to digital conversion, or amplification. The transmitter, in addition to modulating and sending a signal, may encode it using a protocol shared with the receiver. At the receiver, electronic circuits will perform similar operations, such as filtering, amplification, digital to analog conversion, embedding in another communication protocol (e.g. Ethernet or RS-232 serial), and transmission as output.

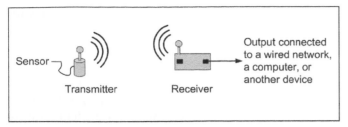

Figure 35-1.

Many wireless networks replace the transmitter and receiver with radio transceivers (which have combined transmitting and receiving abilities). This permits 2-way communication so that the radio at the receiving point can send requests for data transmissions (poll the sensor transmitter) and send messages acknowledging receipt of both data and messages transmitted from the sensor's radio. The sensor's transceiver can receive requests and acknowledgments from the transceiver at the receiving point, as well as send the sensor data. In addition to these functions, both radios formulate packets of data that precede and follow the main data or messages sent that are specified as part of the protocol the radios use for communication purposes.

All of these components require electric power to operate and, therefore, a power supply, which is usually either wired power or a battery. The power supply then converts the source power to the form (e.g., direct current, DC) and voltage required by the device. Battery operated devices generally have sophisticated power management schemes implemented to conserve the battery's energy by powering the electronics down between transmissions. Another source of power for distributed devices under development is power-scavenging technology, which can extend battery lifetime or even fully substitute for a battery. Power-scavenging devices convert ambient energy forms such as vibrations, light, kinetic energy inflows, and temperature differentials into electric energy.

Networks of sensor nodes (the combination of a radio, other electronic circuitry, and the sensor) can be formed from the basic principle illustrated in Figure 35-1, but many sensor nodes transmit data to points of reception. Wireless sensor networks can have tens, hundreds, even thousands of nodes in the network, providing measurements from different kinds of sensors that might be located at many different positions. For example, a wireless network might measure many temperatures, humidities, and pressures throughout many HVAC systems, the electric power use of all major equipment, as well as the temperature and occupancy of rooms throughout a building, all reported

to one receiver that sends the data to a computer for processing or display.

Network Topology

Wireless sensor networks have different requirements than computer networks and, thus, different network topologies and communication protocols have evolved for them. The simplest is the *point-to-point topology* (see Figure 35-2) in which two nodes communicate directly with each other. The *point-to-multipoint* or *star topology* is an extension of the point-to-point configuration in which many nodes communicate with a central receiving or gateway node. In the star and point-to-point network topologies, sensor nodes might have pure transmitters, which provide one-way communication only, or transceivers, which enable two-way communication and verification of the receipt of messages. Gateways provide a means to convert and pass data between one protocol and another (e.g., from a wireless sensor network protocol to the wired Ethernet protocol).

The communication range of the point-to-point and star topologies is limited by the maximum communication range between the sensor node at which the measured data originate and the receiver (or gateway) node. This range can be extended by using repeaters, which receive transmissions from sensor nodes and then re-transmit them, usually at higher power than the original transmissions from the sensor nodes. By employing repeaters, several "stars" can communicate data to one central gateway node, thus expanding the coverage of star networks.

In the *mesh network topology* each sensor node includes a transceiver that can communicate directly with any other node within its communication range. These networks connect many devices to many other devices, thus, forming a mesh of nodes in which signals are transmitted between distant points via multiple hops. This approach decreases the distance over which each node must communicate and reduces the power use of each node substantially, making them more compatible with on-board power sources such as batteries. In addition to these basic topologies, hybrid network structures can be formed using a combination of the basic topologies. For example, a mesh network of star networks or star network of mesh networks could be used (see Figure 35-3).

Point-to-Point

In a point-to-point network configuration each single device (or sensor node) connects wirelessly to a receiver or gateway. An example would be a remote

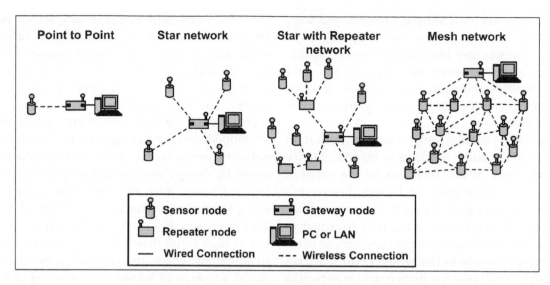

Figure 35-2.

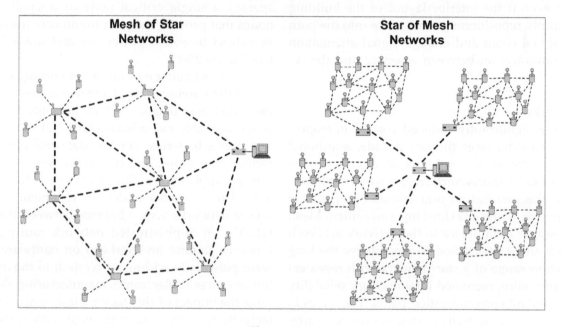

Figure 35-3.

control for a TV, a garage door opener, or a wireless PLC (programmable logic controller) to turn on/off a remote pump or light. The communication can be kept simple with identification schemes that are either set up in the hardware with dip switches or by software during the initial configuration. Point-to-point wireless architectures apply a simple master/slave communication protocol whereby the master station issues a command for a single dedicated slave.

Star Networks

The star network is an extension to the point-to-point configuration. One central node broadcasts to many end nodes in the network (i.e. point to multipoint). Alternatively, the communication can originate from the end nodes, communicating to one single central point (i.e. multipoint to point). The latter is a typical architecture for currently available in-home and building security products. Remote sensors on doors and windows, when triggered, communicate to one central station, which then issues an alarm and performs other pre-programmed procedures such as dialing the police or fire department. A star topology can be used in building operation for monitoring zone-air temperatures with wireless sensors as described in References 8, 9 and 10.

The star network is a simple network topology to support many sensors. Before standard integrated-circuit (IC) manufacturing technologies were capable of making high performance RF chipsets, the only cost-effective wireless network was the star network because the sensor nodes often had only transmitters and not transceivers.

This topology provides only one communication path for each sensor node, so there is no redundancy in the network. As a result, each link in the network infrastructure is a single point of failure. Ensuring a reliable communication path for each sensor is critical, and a thorough RF site survey must be performed to determine the need and locations for repeaters to carry each sensor signal reliably to the receiver. Sufficient resilience should be built into the design of star networks so that reliable communications of all sensors can be maintained even if the interior layout of the building changes. Simply repositioning a bookcase into the path of a weak signal could add enough signal attenuation to stop communication between a sensor and the receiver.

Mesh Networks

With the significantly reduced cost of microprocessors and memory over the last decade, additional computational power at the device level can now be used to operate a more complex network that simplifies both the installation and commissioning of a sensor network while maximizing reliability. Mesh networks—where each device in the network acts both as a repeater and a sensor node—can achieve the long communication range of a star network with repeaters while also providing increased total network reliability through redundant communication paths. The nodes in a mesh network automatically determine which nearby neighbors can communicate effectively and route data through the network accordingly, changing the routing dynamically as conditions change. Having multiple links in a network provides built-in redundancy so data can be effectively routed around blocked links. This means that there are few single points of failure in the system, so the overall network is extremely reliable even if individual wireless links are not. Mesh networks also pass data from one node to another in the network, making the placement of additional sensors or controllers in the network akin to building out additional infrastructure. As additional devices are placed in a mesh network, the number of communication paths increases, thereby improving network reliability.

The most-used nodes in any sensor network use the most energy. So if the routing is static, even in a mesh network (when the "best" communication routes don't change with time), the energy demands will vary among nodes with those used most expending the most energy. For battery-powered nodes, this demand can rapidly drain the battery. Network protocols are being developed that are "energy aware." To help maximize network performance time, these protocols even account for energy use along each potential communication path and check the remaining charge of batteries along the paths in selecting the preferred route. [11, 12] This approach, however, works best where node density is high throughout the area covered by a network. In situations where node density is not high (as during initial adoption of wireless monitoring in buildings or other cases where sensor node deployment may be sparse), a single critical node or a small number of nodes that provide the path for all communication will be subject to excess power use and lower battery life (see Figure 35-4).

A disadvantage of mesh networking could be the use of the wireless data channels for network management and maintenance, which not only takes up part of the available radio bandwidth, but also uses power and drains batteries. For low-data-rate applications in facility monitoring and control as well as many other sensing applications, this limitation is likely manageable. The protocols under development for wireless sensor networks seek a balance between these factors. [11, 12, 13] Sophisticated network routing schemes, however, impose an overhead on hardware and firmware potentially adding a premium to the overall cost, but advances in electronics manufacturing should minimize the impact of this factor. Mesh sensor networking technology is in a nascent stage with early products just beginning to enter the building automation and monitoring market.

Frequency Bands

To minimize interference and provide adequately for the many uses of radio frequency communication, frequency bands are allocated internationally and by most countries. The International Telecommunication Union (ITU) is the organization within which governments coordinate global telecommunication networks and services. The United States is a member of the ITU through the Federal Communications Commission (FCC). The ITU maintains a Table of Frequency Allocation that specifies regionally and by country the allocations of radio spectrum. [14] The ISM (indus-

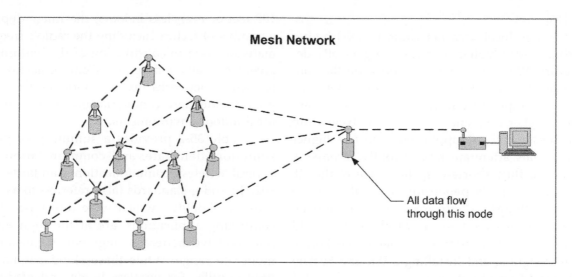

Figure 35-4.

Table 35-1. ISM Frequency Band Allocations and Applications. [13]

Frequency band	Center Frequency	Band-width	Applications
6,765-6,795 kHz	6,780 kHz	30 kHz	Personal radios
13,553-13,567 kHz	13,560 kHz	14 kHz	
26,957-27,283 kHz	27,120 kHz	326 kHz	
40.66-40.70 MHz	40.68 MHz	40 kHz	Mobile radios
902-928 MHz	915 MHz	26 MHz	In the US, applications includes Railcar and Toll road applications. The band has been divided into narrow band sources and wide band (spread spectrum type) sources. Europe uses this band for cellular telephony services (GSM)
2,400-2,500 MHz	2,450 MHz	100 MHz	A recognized ISM band in most parts of the world. IEEE 802.11, Bluetooth recognizes this band as acceptable for RF communications and both spread spectrum and narrow band systems are in use.
Cordless phones 5,725-5,875 MHz	5,800 MHz	150 MHz	Cordless phones. The FCC have been requested to provide a spectrum allocation of 75 MHz in the 5.85-5.925 GHz band for Intelligent Transportation Services use.
24-24.25 GHz	24.125 GHz	250 MHz	Allocated for future use
61-61.5 GHz	61.25 GHz	500 MHz	
122-123 GHz	122.5 GHz	1 GHz	
244-246 GHz	245 GHz	2 GHz	

trial, scientific, medical) bands provide frequencies for license-free radio communications given a set of power output constraints. The ISM frequencies and common applications are shown in Table 35-1.

Consumer products ranging from cordless telephones to wireless local area networks use the 2.4 GHz band. The trend for selecting higher frequencies is primarily driven by the need for higher data rates. As can be seen in Table 35-1, the bandwidth is greater at higher frequencies. Bandwidth is defined as the width of a particular frequency band. For instance, the 900 MHz band has a bandwidth of 26 MHz (928 MHz—902 MHz, see Table 35-1). Data rates and bandwidth of a frequency band are related. According to Nyquist, the maximum data rate in bits per second (bps) that can be achieved in a noiseless transmission system of bandwidth B is $2B$. [15]Using the Nyquist theorem for the example of a bandwidth of 26 MHz, we would obtain a theoretical data rate limit of 52 Mbs. In practical applications where we encounter signal noise, the signal-to-noise ratio limits the actually achievable data rate to a value less than that determined by the Nyquist theorem. [16]

For wireless local area networks (LANs) higher bandwidth provides higher data rates, a generally desirable feature. Wireless sensor networks, on the contrary, are generally low-data-rate applications sending, for instance, a temperature measurement every 5 minutes. Hence, higher frequencies provide no bandwidth benefit for sensor network applications. In fact, higher frequency signals attenuate more rapidly in passing through media, thus shortening the range of the RF transmission as signals penetrate materials, e.g., in walls and furnishings. [17] To maximize transmission range, a low transmission frequency technology should be selected (see the discussion on signal attenuation in the section Designing and Installing a Wireless System Today: Practical Considerations).

Communication Protocols

There are a large number of wireless technologies on the market today, and "wireless networks" as a technology span applications from cellular phone networks to wireless temperature sensors. In building automation applications where line power is not available, power consumption is of critical importance. For example, battery-powered "peel-and-stick" temperature sensors will only be practical if they and their network use power at a very low rate. In general, a 3- to 5-year battery lifetime is believed to be a reasonable minimum. Although power is generally available in commercial buildings, it is often not conveniently available at the precise location at which a sensor is needed. Thus, for many wireless sensors, some kind of onboard power, such as a battery is necessary to keep the installed cost low. To maximize battery life, communication protocols for wireless sensor networks must minimize energy use.

Beyond power requirements, communication range is important. A radio that has a maximum line-of-sight range of 500 feet outdoors may be limited to 100 feet or even less indoors, the range depending on a number of factors including the radio's frequency, the materials used in construction of the building, and the layout of walls and spaces. Communication protocols for sensor networks installed indoors, therefore, must provide adequate communication ranges in less than ideal indoor environments.

Table 35-2 provides a summary of power consumption, data rate, and communication range for several wireless communication standards. The IEEE 802.11b and g standards (also referred to as "WiFi" for Wireless Fidelity), which were developed for mobile computing applications, are at the high end of data rate and have moderately high power consumption and moderate range. While these standards have proven very popular for wireless home and office networking and mobile web browsing, they are not suitable for most building sensor applications because of their high power consumption. Furthermore, in the long run, 802.11b and g are likely to see quite limited use for sensor networking because of their limits on the number of devices in a network and the cost and complexity of their radio chipsets, compared to simpler, ultimately lower cost, wireless sensor networking standards.

Bluetooth, another wireless communications standard, was developed for personal area networks (PANs) and has proven popular for wireless headsets, printers, and other computer peripherals. [18] The data rate and power consumption of Bluetooth radios are both lower than for WiFi, which puts them closer to the needs of the building automation applications, but the battery life of a Bluetooth-enabled temperature sensor is still only in the range of weeks to months, not the 3 to 5-years minimum requirement for building applications, and the communication range is limited to about 30 feet (100 feet in an extended form of Bluetooth). The number of devices in a Bluetooth network is also severely limited, making the technology applicable for only the smallest in-building deployments.

Table 35-2. Basic characteristics of some wireless networking standards.

Network Name/Standard	Power Use (Watts)	Data Rate (kb/sec)	Line-of-site Range (meters)
Mobile telecommunications			
GSM/GPRS/3G	1 to 10	5 to >100	>1000
Wi-Fi IEEE 802.11b	0.5 to 1	1000 to 11,000	1 to 100
Wi-Fi IEEE 802.11g	0.03 to 0.7	1000 to 54,000	>100
Bluetooth IEEE 802.15.1	0.05 to 0.1	100 to 1000	1 to 10
ZigBee with IEEE 802.15.4	0.01 to 0.03	20 to 250	1 to >100

The IEEE 802.15.4 standard [19, 20] for the hardware layers together with the Zigbee standard [21] for the software layers provides a new standards-based solution for wireless sensor networks. IEEE 802.15.4, which was approved in 2003, is designed specifically for low data-rate, low power consumption applications including building automation as well as devices ranging from toys, wireless keyboards and mouses to industrial monitoring and control [19, 20]. For battery-powered devices, this technology is built to specifically address applications where a "trickle" of data is coming back from sensors or being sent out to actuators. The standard defines star and meshed network topologies, as well as a "hybrid" known as a cluster-tree network. The communication range of 802.15.4 radio devices is 100 to 300 feet for typical buildings, which, when coupled with an effective network architecture, should provide excellent functionality for typical building automation applications.

The industry group ZigBee Alliance developed the ZigBee specification that is built upon the physical radio specification of the IEEE 802.15.4 Standard [21]. ZigBee adds logical network, security, application interfaces, and application layers on top the IEEE 802.15.4 standard. It was created to address the market need for a cost-effective, standards-based wireless networking solution that supports low data rates, low power consumption, security, and reliability. ZigBee uses both star and meshed network topologies, and provides a variety of data security features and interoperable application profiles.

Non-standardized radios operating with proprietary communication protocols make up the majority of today's commercially available wireless sensors. They usually offer improved power consumption with optimized features for building automation applications. These radios operate in the unlicensed ISM frequency bands and offer a range of advanced features which depend on their target applications.

Technical Issues in Buildings

The primary issues of applying wireless sensor technologies in buildings are associated with 1) interference caused by signals from other radio transmitters (such as wireless LANs) and microwave ovens that leak electromagnetic energy, 2) attenuation as the RF signal travels from the transmitter through walls, furnishings, and even air to reach the receiver, and 3) security.

Interference generally stems from electromagnetic noise originating from other wireless devices or random thermal noise that may impact or overshadow a sensor signal. Spread spectrum techniques are used to increase immunity to interference from a single- frequency source by spreading the signal over a defined spectrum. Spread spectrum techniques utilize the available bandwidth such that many transmitters can operate in a common frequency band without interfering with one another. Spread spectrum, however, is not guaranteed to be completely immune to interference, particularly if the frequency band is heavily loaded, say with hundreds of wireless devices sending messages. Early technology demonstration projects with 30 to 100 wireless sensors in buildings have not revealed any problems with crosstalk or loss of data in the transmission; however, it remains unclear whether reliable communications can be maintained as the frequency band becomes crowded with hundreds or thousands of wireless devices. Experiences with the technology over time will reveal how wireless technology will perform under these conditions.

Signal attenuation is a weakening of the RF signal. It is a function of distance and the properties of the material through which the signal travels. Signal attenuation can be compensated by using repeaters that receive signals, amplify them, and then retransmit them to increase the transmission range.

With steadily increasing threats from hackers to the networking infrastructure, the *security* needs of modern facility automation systems have grown. The vulnerability of wireless networks is of particular concern because no direct "hard" physical link is required to connect. Data encryption techniques have been successfully applied to wireless LAN systems to combat intrusion and provide security. These techniques encode data in a format that is not readable except by someone with the "key" to decode the data. Encryption, however, requires additional computational power on each wireless device, which runs counter to the general attempt to simplify technology in order to reduce cost. These challenges are currently being addressed by researchers, technology vendors and standards committees to provide technology solutions with the necessary technical performance that the market demands.

Costs

Costs of commercially available sensor network components in 2004 are shown in Table 35-3. Excluded from the table are single point-to-single point systems based on RF modems. The table shows that costs vary widely, and as with many technologies, costs are expected to decrease with time.

According to a recent market assessment of the wireless sensor networks, the cost of the radio frequency (RF) modules for sensors is projected to drop below $12 per unit in 2005 and to $4 per unit by 2010. [22] While these costs reflect only one portion of a wireless sensor device, the cost of the sensor element itself is also expected to decrease with technology advancements. For instance, digital integrated humidity and temperature sensors at high volumes are currently commercially available for less then $3 per sensor probe.* The general trend toward greater use of solid state technology in sensors is likely to lead to lower cost sensors for mass markets.

To date, end users are caught between the enthusiastic reports of the benefits that wireless sensing and control can provide and skepticism regarding whether the technology will operate reliably compared to the wired solution. While advancements in wireless local area networks (LAN) have paved the road for wireless technology market adoption, it also has made end users aware of the inherent reliability challenges of wireless transmission in buildings and facilities.

Types of Wireless Sensing Applications for Buildings

Applications of wireless sensing in buildings can be placed into two broad categories that significantly affect requirements on the underlying wireless technology and its performance: 1) applications for which at least some (and often most) of the devices must be self-powered (e.g., with an on-board battery) and 2) applications for which line power is available for each device. In this section, we describe experiences in field testing both types of applications. The first (Building Condition Monitoring) is illustrated with wireless sensors used to measure the air temperature in build-

*Quote by SenSolution, Newberry Park, CA, February 2004.

ings at a much higher resolution than possible with the wired thermostats usually installed. In the second (Equipment Condition Monitoring), data for continually monitoring the performance of rooftop packaged HVAC units is collected using a wireless sensor data acquisition system.

Building Condition Monitoring

As discussed above, eliminating the need for wiring makes wireless sensor technology particularly appealing and well suited for monitoring space and equipment conditions in buildings of all sizes. Without the wires though, some additional care must be exercised in engineering and installing the wireless network to ensure sufficient robustness of communication.

Starting in 2002, Pacific Northwest National Laboratory (PNNL) conducted some of the first demonstrations to assess the performance of commercially available wireless sensor technology in real buildings and to compare the cost of the wireless solution with that of a conventional wired system. The first demonstration building was an office building with 70,000 square feet of open office floor space on three floors and a mechanical room in the basement. The building is a heavy steel-concrete structure constructed in the early 1960s. The second demonstration building represents a more modern and structurally lighter building style with individual offices totaling 200,000 square feet of floor space in a laboratory building completed in 1997.

Demonstration 1:
In-Building Central Plant Retrofit Application

The building is located in Richland, Washington. The HVAC system consists of a central chiller, boiler, and air distribution system with 100 variable-air-volume (VAV) boxes with reheat distributed in

Table 35-3. Cost ranges of commercially available wireless sensor network components in 2004.

Network Component	Cost Range ($)
Sensor transmitter unit	$50 - $270
Repeaters	$250 - $1050
Receivers	$200 - $900
BAS Integration units	$450*

*Only one is currently commercially available in 2004 specifically for connecting a wireless sensor network to a building automation system.

the ceiling throughout the building. A central energy management and control system (EMCS) controls the central plant and the lighting system. Zone temperature control is provided by means of stand-alone and non-programmable thermostats controlling individual VAV boxes. The centralized control system receives no zone temperature information and cannot control the VAV boxes. The long-term goal of PNNL facility management is to network the 100 VAV boxes into the central control infrastructure to improve controllability of the indoor environment. As an intermediate step toward this, a wireless temperature sensor network with 30 temperature sensors was installed to provide zone air temperature information to the EMCS. The wireless sensor network consists of a series of Inovonics wireless products including an integration module that interfaces the sensor network to a Johnson Controls N2 network bus.* The zone air temperatures are then used as input for a chilled-water reset algorithm designed to improve the energy efficiency of the centrifugal chiller under part-load conditions and reduce the building's peak demand.

---

*N2 bus is the Johnson Controls network protocol.

The Wireless Temperature Sensor Network

The wireless network consists of a commercially available wireless temperature sensor system from Inovonics Wireless Corporation. It encompasses 30 temperature transmitters, 3 repeaters, 1 receiver, and an integration module to interface the sensor network to a Johnson Controls EMCS N2 network. The layout of the wireless temperature network is shown in Figure 35-5.

The operating frequency of the wireless network is 902 to 928 MHz, which requires no license per FCC Part 15 Certification [23]. The technology employs spread spectrum frequency hopping techniques to enhance the robustness and reliability of the transmission. The transmitter has an open field range of 2500 feet and is battery-powered with a standard 123 size 3-volt $LiMnO_2$ battery with a nominal capacity of 1400 mAh. The battery life depends on the rate of transmission, which can be specified in the transmitter. The manufacturer estimates a battery life of up to 5 years with a 10-minute time between transmissions. The transmitter has an automatic battery test procedure with a 'low-battery' notification via the wireless network. This feature will alert the facility operator through the EMCS that the useful life of the battery in a specific transmitter is approaching its end. The repeaters are

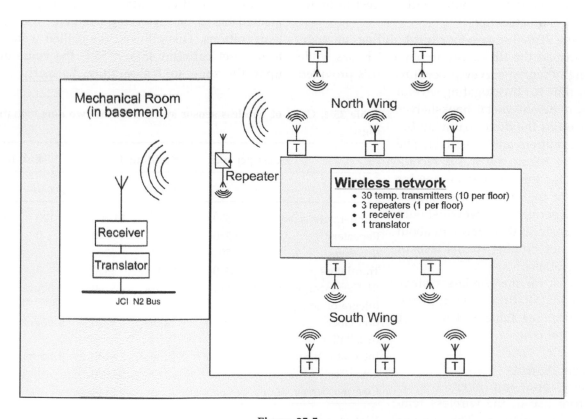

Figure 35-5.

powered from ordinary 120 volts alternating current (VAC) wall outlets and have a battery backup. Three repeaters were installed, one on each floor. Because the repeaters are line powered, the repeater operates at high power and provides up to 4 miles of open field range. The receiver and the translator are installed in the mechanical room in the basement. The translator connects the receiver with the Johnson EMCS system.

Design and Installation Considerations:

Installation of the wireless network requires a radio frequency (RF) survey to determine the proper locations for the repeaters to ensure that the received signal strength is sufficient for robust operation of the wireless network. RF surveying is an essential engineering task in the design of the wireless network topology. The signal attenuation in metal-rich indoor environments caused by metal bookshelves, filing cabinets, or structural elements such as metal studs or bundles of electric or communication wiring placed in the walls can pose a significant challenge to achieving robust wireless communication. Background RF noise emitted from cordless phones and other sources can also impair the transmission such that the receiver cannot distinguish noise from the real signal. There is no practical substitute for RF surveying a building because each building is unique with respect to its RF attenuation characteristics.

For the 70,000-square-foot test building, an engineer performed the RF survey in about 4 hours while instructing others in survey procedures. This provided sufficient time for investigating several scenarios, whereby metal bookshelves were placed in the direct pathway between transmitters and a receiver. The result of the RF survey was a recommendation for three repeaters, one for each floor of the building (see Figure 35-5). An experienced surveyor should be able to perform this survey in about 2 hours, if not running special tests or instructing others.

The cost for the wireless system, including installation, was approximately $4000. See Table 35-4 for more details on the cost.

Operational Benefits

Operational improvements resulted from use of the wireless temperature sensor network. The wireless sensors enabled facility staff to respond to 'hot' and 'cold' complaints much more effectively. Because sensors can be easily moved and new ones readily introduced into the network, a spare sensor can be easily taped directly into a localized problem area for monitoring air temperature over a few hours or days. The much higher spatial resolution provided by the 30 zone air-temperature sensors enabled facility staff to identify individual VAV boxes that were causing uneven supply air. These malfunctioning boxes spread the range of air temperatures through the building. After repairing the faulty VAV boxes, the facility staff was able to raise the supply-air temperature by 2°F, alleviating the need for overcooling some zones in order to deliver enough cooling capacity through the faulty VAV boxes. Repair of VAV boxes improved the thermal comfort of occupants and eliminated the occasional use of space heaters during the early morning hours in both summer and winter months.

Energy Efficiency Benefits

The energy savings resulted directly from repairing several VAV box controllers, resetting the supply air temperature by 2°F during cooling periods, and reducing the use of small space heaters by occupants who were previously uncomfortably cool at times. In addition, a chilled-water reset strategy was implemented based on an average value of the 30 zone air temperatures. This allowed the chilled water set point to be reset between 45 and 55°F, the value depending upon the zone air temperature. Formerly, the chilled

Table 35-4. Costs of wireless sensor systems in the two demonstration buildings.

	Cost per unit	Building 1		Building 2	
		Quantity	*Total*	*Quantity*	*Total*
Temperature sensors	$50	30	$1,500	120	$6,000
Repeaters	$250	3	$750	0	$0
Receivers	$200	1	$200	3	$600
Translators	$450	1	$450	3	$1,350
RF Surveying Labor	$80/hour	2 hours*	$160	2 hours	$160
Integrator configuration labor	$80/hour	4 hours	$320	8 hours	$640
Installation of Integrator labor	$80/hour	8 hours	$640	8 hours	$640
Total Cost			$4,020		$9,390
Cost per Sensor			$134		$78

*For an experienced surveyor.

water temperature was fixed at 45°F. The average zone air temperature was used as an indicator for meeting the cooling loads. As a result the average coefficient of performance (COP) increased by about 7% due to the higher chilled water temperatures. The fan power for any given cooling load increased some but not nearly enough to offset the savings. The net result was an estimated cost savings of about $3500 over the cooling season (May through September). Additional energy savings were achieved by avoiding the use of space heaters and resetting the supply air temperature for a total estimated annual cost savings of about $6000. Based on the costs and estimated savings, the simple payback period for this wireless system was about 7 months.

Demonstration 2: Laboratory/Office Building

The second building, opened for occupancy in 1997, houses laboratories and offices. The gross floor space is about 200,000 square feet with three protruding office wings of about 49,000 square feet each. Only the office area was used for the demonstration. Each office wing has a separate air-handling unit and a variable-air-volume (VAV) ventilation system. Each VAV box supplies air to two offices controlled by a thermostat located in one of the two offices. The construction of the office area consists of metal studs with gypsum wall. The offices contain metal book shelves, and at a minimum, two computers with large screen monitors. The office space is relatively metal-rich, posing a challenge for wireless transmission from the sensors to the receivers.

Facility staff explored night setback options for the ventilation of the office space that would turn off the air-handling unit during the night hours after 6 p.m. The decision to implement such a strategy was suspended out of concern that those offices without a thermostat might be occupied during late hours and if so, that the air temperature in those offices could exceed the thermal comfort limits. Because of this concern, the ventilation system operated on a 7-day per week, 24-hour per day schedule. It was believed that if each office were equipped with one zone temperature sensor, the night setback could be implemented and then overridden if the zone temperature exceeded an upper threshold of 78 °F. A cursory cost estimate from a controls vendor for installing wired temperature sensors in the offices without thermostats yielded an installed cost per sensor of about $500, which exceeded acceptable costs.

After the initial positive experiences with wire-less sensors in the other building, facility staff re-examined the viability of the ventilation night setback using a wireless solution and implemented wireless temperature sensors in early 2004. The same wireless temperature sensor network technology as deployed in Building 1 was used. Familiarity with the technology and experience gained from the first wireless demonstration greatly reduced the level of effort for a RF survey of the building and the wireless network setup.

The Wireless Temperature Sensor Network

Each office not previously equipped received a wireless temperature sensor. Forty wireless temperature sensors were deployed in each of the three office wings of Building 2, bringing the total to 120 sensors (see Figure 35-6). The temperature signals were read by three receivers, each located where the office wing meets the main hallway and connected via an integrator to the Johnson Controls network control module. The wireless network consisted of a total of 120 sensors, three receivers and three integrators. Facility staff tested the need for repeaters and found that with the use of one receiver for each wing, the communication was sufficiently robust. An alternative wireless network design was considered that would use one receiver in the middle wing and repeaters in each of the side wings to assure communication from the most distant transmitters in the exterior wings to the receiver. The integrator has a limit of 100 transmitters. Since this alternative used only one integrator, it could not support enough sensors for all the offices, and it therefore was rejected.

The temperature sensors are programmed to transmit a temperature measurement every 10 minutes. A sensor will transmit early when a temperature change is sensed that exceeds a pre-set limit. This is to enable detection of rapid temperature changes as quickly as possible.

Installation and Setup of Wireless Network

The installation costs for the wireless sensor network were minimal. They included a 2-hour RF survey, an initial setup of the integrator device to specify the number and ID numbers of the sensors, and the physical connection of the integrator and the Johnson Controls network control module. Configuration of the integrators was done in stages (each wing at a time) and the total time for setup of all 120 sensors was conservatively estimated afterward to be 8 hours. The integrator installation involves physically connecting the 24 VAC power supply provided in the Johnson Control

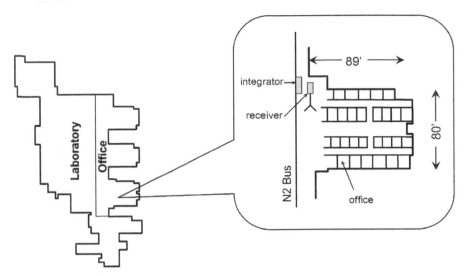

Figure 35-6.

network module and connecting the Johnson Controls N2 bus to the integrator using a 3-wire shielded cable. A short 4-wire cable connects the integrator and the receiver providing power supply and communication between the two devices. This work was performed by an instrument technician. The sensors were then attached to the office walls using double-sided tape.

Table 35-4 presents the cost components for the two demonstration buildings. The capital costs for the hardware represent the costs to PNNL and are representative of costs for a wholesaler. List prices would commonly be 75% to 100% higher than those shown.

Energy Savings

The supervisory control program was augmented to schedule night setback starting at 6 p.m. and suspending it if an office zone temperature exceeded a threshold temperature of 78°F during the cooling season or dropped below 55°F during the heating season, instead of maintaining the temperature continuously at a set point of 72°F. Initial estimates concluded that energy savings are largely attributable to the shut down of the supply and return fans and, to a lesser degree, to reduced thermal loss during the night as the temperature is allowed to float (rise in the cooling season and drop in the heating season). Trend-logs of run time using the new night setback strategy were used to estimate the electric energy savings. Preliminary estimates suggest that the night setback will achieve savings of approximately $5,000 annually. Verification of the savings is planned after one full year of night setback operation is completed. We attribute the cost savings to the wireless sensors because they enabled implementation of the ventila-

tion night setback, something the facility operations staff was unwilling to do without the additional information provided by these sensors. Based on these energy savings, the wireless sensor system (which had an installed cost $9390) has a simple payback period of less than 2 years (22.5 months).

Other Impacts

Building operators also implemented a temperature averaging scheme for controlling the distribution system VAV boxes based on the average of the office temperatures in the zone served by each box. Although no energy savings resulted from this change, the building operators report that the number of occupant complaints about temperature has decreased significantly, saving building staff time and enabling them to devote that time to other improvements in operation.

Discussion on Costs for Demonstration Projects

Cost for the sensor and controls technology is a critical factor for the viability of any retrofit project or even in new construction. The wireless sensor solution was slightly more cost effective compared to an equivalent wired solution for Building 1. [9] For Building 2, the wireless sensor cost ($78/sensor) was significantly less than the estimated cost for the wired sensor retrofit (~$500/sensor). These examples tend to show that wireless sensor networks can compete with wired sensing on the basis of cost for retrofit projects. In both demonstration buildings, the wireless network infrastructure is sufficient to accommodate many more sensors at the cost of sensors alone. No additional infrastructure (repeaters, receivers, or translators) is needed to accommodate additional sensors. This enables facility staff to add sensors at the cost of the sensor itself plus a minimal setup time (a few minutes) for configuring the integrator.

Figure 35-7 shows cost curves for both demonstration buildings as a function of number of sensors installed. These curves are nearly identical. For 30 sensors, for example, the difference in cost is $22/sensor ($160-$138), and for 120 sensors, the difference is $6/sensor ($78-$72). This suggests that the cost of the wireless system per sensor might be nearly independent of the building itself but highly dependent on the number of sensors installed. The curves are actually dependent on the costs of the wireless components. The two curves shown are for the same brand and models of hardware. Average costs per sensor for systems built from components with substantially different costs will lie on other curves. Unless signal attenuation differs so significantly among buildings that it affects the number of sensors that can be served by each repeater or receiver, the curves for different buildings using the same wireless components should lie very close to one another. This observation proves useful in simplifying estimation of costs for wireless sensor systems.

The second insight from Figure 35-7 is that at high quantities of the sensors, the system cost on a per-sensor basis asymptotically approaches the cost of a sensor (in this case, $50/sensor). Therefore, for densely deployed sensors (high numbers of sensors per unit of building area), further cost reductions for wireless sensor networks must come from re-

ducing the cost of the sensor modules (sensors plus transmitting radio) rather than decreasing the cost of infrastructure components—the receiver, repeaters and translators. In the short-term, however, while wireless sensing technology is just beginning to be deployed, sensor densities are likely to be relatively low, and as a result, all components will have a significant impact on cost. Users should realize, though, that once a wireless sensor network is installed in a building, additional sensors generally can be added to the network in the area covered by the network at the incremental cost of the additional sensors. The more uses the building staff can find for the wireless sensor network, the more cost-effective its installation becomes.

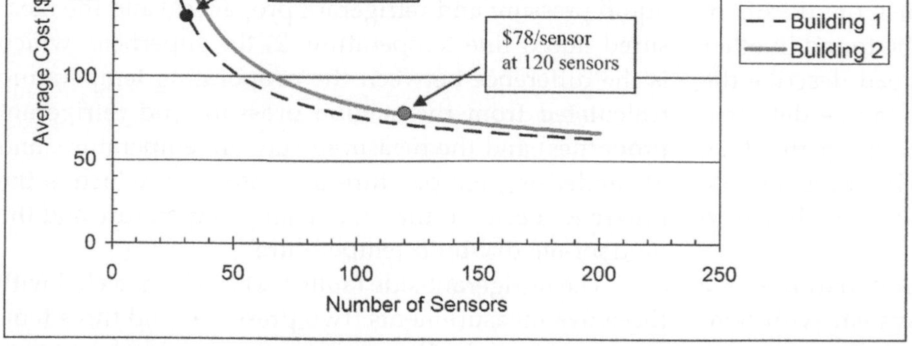

Wireless Monitoring of Equipment Conditions

Heating, ventilating, and air-conditioning equipment is often run until it completely fails ("hard" faults), for example from a failed compressor, failed condenser fan, failed supply fan, or significant loss of refrigerant. Upon complete failure, the owner, operator, or building occupant calls a service company to repair the unit. Complete failure, though, is often preventable. Avoiding failures by properly maintaining the equipment would reduce repair costs, increase operating efficiency, extend equipment life, and ensure comfortable conditions, but this would require awareness of equipment condition and when the equipment needs servicing. Furthermore, several studies have noted that building systems operate under degraded conditions caused by insufficient refrigerant charge, broken dampers, stuck dampers, mis-calibrated and failed sensors, improperly implemented controls (e.g., incorrect schedules), electrical problems, and clogged heat exchangers [6, 24-29]. Many of these faults do not result in occupant discomfort because the system compensates by working harder (and expending more energy), and therefore, these faults are not reported nor are they corrected. Some of the faults require a service technician to correct, but many can be fixed with minor adjustments to controls or schedules; these faults are referred to as "soft" faults in this chapter.

Figure 35-7.

With increasing pressure to reduce operation and maintenance (O&M) costs and with reduced staff in today's facilities, regular visual inspection by staff is out of the question. For small buildings without on-site operators, this was never a possibility. Service contracts providing scheduled but infrequent inspection and servicing alone are not likely the solution to this problem. Without a lower cost solution, package units are likely to continue to be maintained poorly and operated inefficiently.

Automated continuous condition monitoring provides a potential solution, but its cost is generally perceived as too high. Even installation of adequate sensors alone is usually viewed as too costly. Studies have shown, however, that automated monitoring and diagnostics implemented with wireless sensing and data acquisition can provide a cost effective solution [6, 8, 30]. In this section, we describe a wireless system for monitoring the condition and performance of packaged air conditioners and heat pumps, which are widely used on small commercial buildings.

*Wireless System for Automated
Fault Detection and Diagnostics*

Functionally, packaged rooftop units can be divided into two primary systems: 1) air side and 2) refrigerant side. The air-side system consists of the indoor fan, the air side of the indoor coil, and the ventilation damper system (including its use for air-side economizing), while the refrigerant-side components include the compressor, the refrigerant side of indoor and outdoor heat exchangers, the condenser fan, the expansion valve, and the reversing valve (for heat pumps).

The choice of the fault detection and diagnostic (FDD) approach depends on the type of faults to be identified and the sensor measurements available. Many researchers have developed FDD algorithms to detect and diagnose faults in air-conditioning equipment. In this chapter we do not discuss the details of the diagnostic approaches, which can be found in other references [e.g., 31, 32, 33, 34] but instead describe the measurements needed, the faults that can be detected, and the system for collecting and processing the data. This system, which can be applied to both the air side and the refrigerant side of a heat pump is shown in Figure 35-8.

The minimum set of information required for monitoring the state of the air-side system with temperature-based economizer controls or no economizing includes: 1) outdoor-air dry-bulb temperature, 2)

return-air dry-bulb temperature, 3) mixed-air dry-bulb temperature, 4) outdoor-air damper-position signal, 5) supply-fan status, and 6) heating/cooling mode. To identify whether the system is actually in heating or cooling mode, the status of the compressor (and the reversing valve for heat pumps) is required. If these measurements are available, economizer operations and ventilation requirements can be monitored and evaluated to verify their correct performance. If an enthalpy-based economizer control is used, then the outdoor-air relative humidity (or dew-point temperature) and return-air relative humidity (if differential enthalpy controls are used) are required in addition to the 6 measurements needed to monitor the performance of systems with temperature-based economizer controls. If supply-air temperature is also measured, additional faults relating to control of supply-air temperature can be detected and diagnosed. Details of the approach for detecting and diagnosing air-side faults are given in References 32 and 33.

Faults detected on the air side can be grouped into four categories: 1) inadequate ventilation, 2) energy waste, 3) temperature sensor and other miscellaneous problems including control problems, and 4) missing or out-of-range inputs. For more details on the faults that can be detected on the air-side, see References 6 and 32.

The minimum set of measurements required to monitor refrigerant-side performance include: 1) outdoor-air dry-bulb temperature, 2) liquid-line temperature (refrigerant temperature as it leaves the condenser), 3) liquid line pressure (as it leaves the condenser), 4) suction line temperature (refrigerant temperature at the compressor inlet), and 5) suction line pressure (refrigerant pressure at the compressor inlet). In addition to the five measured quantities, several derived quantities are used in monitoring the refrigerant-side performance: 1) liquid sub-cooling, which is estimated as a difference between the condensing temperature (calculated from liquid pressure and refrigerant properties) and the measured liquid line temperature, 2) the superheat, which is the difference between the evaporating temperature (calculated from the suction pressure and refrigerant properties) and the measured suction temperature, and 3) condensing temperature over ambient, which is the difference between the condensing temperature and the outdoor-air dry-bulb temperature.

The refrigerant-side faults that can be detected with these five measurements (two pressures and three temperatures) include: 1) evaporator (indoor coil) heat transfer problems, 2) compressor valve leakage (compressor

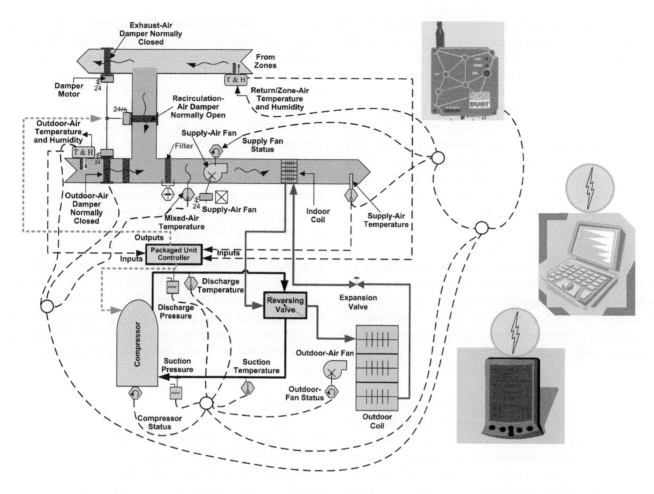

Figure 35-8.

fault), 3) condenser (outdoor coil) heat transfer problems, 4) improper supply-fan speed, 5) expansion device fault, 6) improper charge (too little or too much refrigerant), and 7) non-condensable substances in the refrigerant, such as air in the system. Details of diagnostics for the refrigerant side can be found in References 29 and 34.

Additional measurements that improve diagnostic capability and also increase the number of faults that can be detected include: 1) supply-air dry-bulb temperature, 2) mixed-air dry-bulb temperature, 3) mixed-air relative humidity (or dew point), 4) surface temperature of the condenser, 5) surface temperature of the evaporator, and 6) compressor power consumption. These measurements enable refinement of the diagnostics provided by the minimum set of sensors. In addition, cooling/heating capacity and efficiency degradation can be computed and tracked with these additional measurements. Although having pressure measurements makes diagnosis of the faults more reliable, pressure sensors are expensive compared to temperature and humidity sensors. The pressure sensors

can be replaced with surface temperature sensors at the evaporator and condenser [31], and the temperature measurements can then be used as indicators of saturation temperature in the evaporator and condenser. Although the use of temperatures to estimate superheat and subcooling may lead to some error, their use will reduce the system cost and should still provide adequate diagnostics.

A wireless system providing data collection and diagnostics for only the air side of package HVAC units had a total installed cost per sensor of approximately half that of a wired system providing the same capabilities ($78 per point compared to $147-$193 per point for the wired system). [30] This wireless system uses one radio on each packaged unit, sending measurements from 4 thermocouples and a current switch used to measure the on/off status of the supply fan of the unit. Six units are monitored using one receiver unit, distributing its cost over the 30 sensors it serves. Power is tapped off the power supply for the packaged HVAC unit, so no batteries are used. Both the cost and benefits of a wireless condition monitoring system depend on several parameters, such as number of roof top units to be monitored, the size of the units, the size of building, the local climate, and potential savings from use of the monitoring and diagnostic tool. For a typical application on an 18,000 square foot 2-story building with six 7.5-ton units, the simple payback will be less than 3 years for most U.S. climates (assuming energy savings of 15% are achieved through better operation and maintenance) [30]. Paybacks will be shorter for larger units in more severe climates and longer for smaller units or units in milder climates.

Deploying Wireless Condition Monitoring

There are several ways to deploy wireless condition monitoring: 1) centralized data collection and processing at each building, 2) distributed or on-demand diagnostics and 3) centralized data collection and processing at a remote server—an application service provider model.

Method 1

The first approach is a conventional approach where all data from wireless monitors are collected by a wireless receiver that is directly connected to a computer. The data are continuously or periodically processed using automated software and results provided to the user through a simple and user-friendly graphical user interface. The authors have tested a prototype wireless monitoring and diagnostic system described in the previous section using this approach. Although the prototype system was capable of monitoring both the air- and refrigerant-side performance, only air-side diagnostics were tested. In this approach, data from packaged roof top units are automatically obtained at a user-specified sub-hourly frequency and averaged to create hourly values that are stored in a database. As new hourly values become available in the database, the diagnostic module automatically processes the data and produces diagnostic results that are also placed in the database. The user can then open the user interface at any time to see the latest diagnostic results, and can also browse historical results.

Method 2

Detailed diagnosis often requires historical data to isolate the primary cause of a fault or performance degradation; however, some faults can be detected with instantaneous or short-term measurements. The second deployment uses wireless data collected while servicing units along with simple rules-of-thumb to

determine the condition of equipment. For example, data from rooftop packaged units might be accessed wirelessly by a technician visiting the site using a Personal Digital Assistant (PDA) with compatible wireless communication capabilities. This method can be effective in identifying incorrect refrigerant charge, blocked heat exchangers, and blocked refrigerant lines. The technician could get a report on each unit without even opening the units. Time at the site could then be devoted mostly to the units with faults or degraded performance. The authors have not yet demonstrated this approach, but a wired system with these sorts of diagnostic capabilities is available commercially. [35] The wired system requires physically connecting to previously installed sensors on each unit or connecting the instrument's sensors before use. Once the sensor system has been installed, the wireless approach is likely to save time and enable service technicians to identify units requiring the most attention immediately upon arriving at a site, improving the quality of service while decreasing cost.

Method 3

The third approach is similar to the first approach but all data are collected and sent to a central server possibly hosted by a third party—an application service provider (ASP). Ideally, the data are received at a central location at each building or site and then transferred to the central server. The transfer of data can be by phone line (wired or wireless) or through an existing wide area network (wired or wireless). The ASP provides access to software and data via subscriptions. For payment of a monthly subscription fee, users obtain access to software on the world wide web using nothing more than a web browser to access it. The software needs to be installed on only one computer, the web server, rather than on the individual work station of every user. To provide reliability, usually the software is installed by the ASP on several redundant servers to provide backup in case a computer fails. Many users are then able to access a small number of installed copies of the software. User files are also maintained on the ASP's servers and backed up in a similar manner. The wireless monitoring equipment can be purchased by the owner or can be leased from the ASP for a subscription fee. This type of approach is still in its infancy. The authors will soon be testing this delivery approach.

The three approaches may also be combined to provide information on equipment condition more flexibly. For example, once the wireless sensing and data acquisition infrastructure is installed on the equipment at a building, it can be connected for remote monitoring by building operations staff/management or at a service provider's office and also be accessed by service technicians when they visit the site. Availability of information on equipment condition and performance would provide the basis for a conditioned-based maintenance program that would help ensure that equipment gets serviced and repaired when needed rather than more frequently than needed or less frequently (which is all too common, especially for package equipment).

Long-distance Data Transmission

So far, this chapter has focused on short-range wireless data acquisition at a building for monitoring indoor conditions and equipment conditions and performance. Although not widely used yet, wireless communications have also proven effective in transmitting data between individual building sites and central monitoring systems. Deployment of this model by an ASP was discussed briefly in the preceding section. Central monitoring using wireless communication of data, however, can be implemented by any organization having geographically distributed facilities and the willingness to maintain the computer infrastructure necessary to implement and maintain such as system. This requires appropriate security and backup to ensure the system meets the necessary performance and reliability demands.

An example system is shown in Figure 35-9. Data collected from electric meters and sensors on equipment are transmitted by a wireless pager network to the operations center of a wireless carrier. Data are then sent through the Internet to the operations center of the ASP providing the service. There, the data are stored securely in databases and processed by the tools provided by the ASP. Customers can then securely access the processed results from their buildings from any computer with a web browser. The monitoring equipment for collecting and transmitting the data is provided by the ASP.

**Designing and Installing a Wireless
System Today: Practical Considerations**

Laying out a wireless network indoors is probably as much art as it is science. Every building is unique, if not in its construction and floor plan, at least in the type and layout of its furnishings. Predicting wireless signal strength throughout a building would require characterizing the structure, its layout, and the furnishings and equipment in it and using that information to model RF signal propagation. No tools are available today for accurately doing this. Furthermore, when space use changes or furnishings are moved or change over time, radio signals encounter new obstacles in new positions. Despite these difficulties, there are several practical considerations for the design of a wireless network that are helpful for generating bills of materials and budget estimates and laying out wireless sensing networks.

Determining the Receiver Location

The decision with perhaps the most impact on the design of a wireless sensor network for in-building monitoring is determining the number and locations of the receivers. A stand-alone wireless network (not connected to a wired control network) may have some flexibility in choosing the location of the receiver. The best location from a communications perspective is one that is open and provides the best line-of-sight pathways between the most wireless sensors and the receiver. Convenient connection to a computer where data will be processed and viewed is another important consideration. These factors must be balanced. If the design requires integration of the wireless sensor network with an existing building automation system (BAS) infrastructure, then receivers must be located near points of connection to the BAS. Locations are constrained somewhat in this case, but there are typically still many options. Frequently, a convenient integration point is a control panel that provides easy access to the communication cables as well as electricity to power the receiver and integration devices. In commercial buildings, the BAS network wires are often laid in cabling conduits (open or closed) above the ceiling panel and are relatively easily accessible. Often the lack of electric power in the ceiling space, however, renders this location less convenient than a control panel.

Signal Attenuation and Range of Transmitters

Estimating the range of the transmitting devices is important from a cost point of view. If the transmission range from a transmitting device to the ultimate end-node cannot be accomplished with a single transmission path, additional hardware is required for signal amplification adding to the total cost of the installation. The discussion below is designed to provide a general overview of this topic that may lead to generating some rough estimates of how many repeater or amplification devices an installation may need. It does not replace a thorough RF survey of a facility to determine the exact number and locations of receivers, repeaters, or interme-

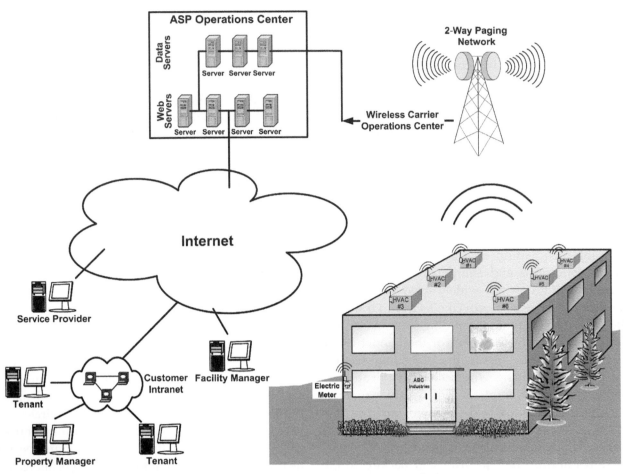

Figure 35-9. Example of a long-distance data transmission system

diate nodes necessary to assure robust communication.

The range of a transmitter depends on the three key variables: 1) attenuation because of distance between wireless devices, 2) attenuation caused by the signals traveling through construction material along the signal pathways, and 3) overall electromagnetic noise levels in the facility.

The attenuation of the signal strength due to distance between the transmitter and receiver (free path loss) is governed by the relation of the electromagnetic energy per unit area of the transmitter to the distance of the receiving surface (see Figure 35-10). The energy per unit area at a distance d from the transmitter decreases proportionately to $1/d^2$. Therefore, for every doubling of the distance d, the energy density or signal strength received decreases to one-fourth of its previous strength. This relationship accounts only for the dispersion of the signal across a larger area with distance from the source. In practice, other factors affect the strength of the signal received, even for an unobstructed path, including absorption by moisture in the air, absorption by the ground, partial signal can-

cellation by waves reflected by the ground, and other reflections. In general, this causes the signal strength at a distance d from the transmitter to decrease in practice in proportion to $1/d^m$, where $2 < m < 4$. [11]

The following example illustrates signal attenuation with distance from the transmitter in free air for a 900 MHz transmitter. This example shows how simple relations can be used to obtain an estimate of potential transmission range.

For this example, assume that the signal strength of a small transmitter has been measured to be 100 mW/cm^2 at a distance of 5 cm from the transmitter's antenna. The transmission path efficiency or transmission loss is customarily expressed in decibels, a logarithmic measure of a power ratio. It is defined as

$$dB = 10 \log_{10} (p_1/p_0),$$

where p_1 is the power density in W/cm^2 and p_0 is a reference power density (i.e., the power density at a reference point) in W/cm^2.

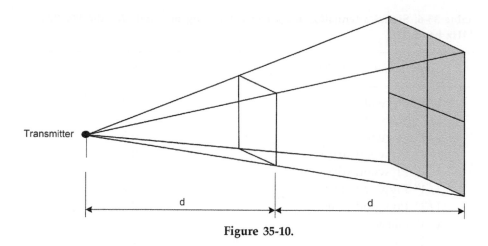

Figure 35-10.

We choose the power density measured at 5 cm distance from the transmitter's antenna as the reference power density p_0. Table 35-5 shows the attenuation of the emitted signal as a function of distance from the transmitter for a signal traveling through air only. For every doubling of the distance, the signal strength decreases by 6 dB or, stated alternatively, the attenuation increases by 6 dB.

Further, assume that the ambient noise is measured to be -75 dB. For a signal to be detectable above the surrounding noise level, the strength of the signal should be at least 10 dB above the noise level (i.e., signal margin of 10 dB or greater is recommended) [36]. Using the results of Table 35-5, we can determine the transmission range of the wireless system in our example that meets the 10 dB signal margin requirements to be 80 meters, since -75 dB +10 dB = -65 dB, which is less than -64 dB at 80 meters.

Next, we extend this example to consider attenuation inside buildings. Suppose that the receiver is placed in a mechanical room of a building and that the signal from the furthest transmitter must go through two brick walls and two layers of drywall. Using signal attenuation estimates from Table 35-6, the combined attenuation of the brick and drywall is 14.6 dB [2 × 0.3 (for the 1/2" drywall) + 2 × 7 (for 10.5" brick wall) = 14.6], for practical purposes say 15 dB. Adding the ma-

terial-related attenuation of 15 dB to the -65 dB signal strength requirement yields -50 dB as the new indoor signal strength requirement for the free air transmission segment. Using Table 35-5, we conclude that the transmission range is between 10 and 20 meters, only 1/8 to 1/4 of the range in open air. This example illustrates how significantly radio signals can be attenuated indoors compared to outdoors simply by the structure itself. Furniture further adds to attenuation and complicates prediction of the signal strength as a function of location in buildings. Therefore, to characterize indoor environments with respect to RF signal propagation, empirical surveying is recommended.

RF Surveying

The purpose of an RF facility survey is to determine the actual attenuation of RF signal strength throughout the facility. This information, together with knowledge of the locations at which sensors will be positioned, is used to lay out the wireless network. The layout will include the number of repeaters and receivers in the network and their locations. For instance, for a multi-story facility there may be good reasons for placing one receiver on each floor, provided the data are needed only on each floor (e.g., one user per floor for that floor) or there is another means to communicate the data between floors (such as a BAS connection on each floor). If the

Table 35-5. Attenuation of an RF signal in free air as a function of distance.

Distance in m	0.05	1	2.5	5	10	20	40	80
(ft)	(0.2)	(3)	8)	(16)	(33)	(66)	(131)	(262)
Signal strength in dB	0	-26	-34	-40	-46	-52	-58	-64
Attenuation along line-of-sight in dB	0	26	34	40	46	52	58	64

Table 35-6. Signal attenuation for selected building materials for the 902-928 MHz band. [38]

Construction Material	Attenuation (dB)
1/4" Drywall	0.2
1/2" Drywall	0.3
3/4" Drywall	0.5
1/4" Plywood (dry)	0.5
1/2" Plywood (dry)	0.6
1/4" Plywood (wet)	1.7
1/2" Plywood (wet)	2
1/4" Glass	0.8
1/2" Glass	2
3/4" Glass	3
1.5" Lumber	3
3" Lumber	3
6.75" Lumber	6
3.5" Brick	4
10.5" Brick	7
8" Reinforced concrete with 1% ReBar mesh	27

data are needed at a computer located on a specific floor (such as a control room in the basement), a repeater might be used on each floor to transmit signals to the location of a central receiver located close to where the data are needed. If communication between receivers on different floors is not sufficient, there may be opportunities to route signals inside an elevator shaft, stair case, or on the exterior of the building. The most cost-effective solution is in most cases determined by the difference in cost between repeaters and receivers and the cost of interfacing the receivers to pre-existing wired networks. The layout with the lowest total cost that provides sufficient (reliable) communication is generally optimal.

Most vendors of wireless sensor networks offer RF survey kits that are specific for the vendors' technologies. These kits consist of a transmitter and a receiver. The transmitter is often a modified sensor transmitter that is programmed to transmit at frequent time intervals. The receiver generally is connected to (or part of) an indicator of signal strength, together making a wireless signal-strength meter. These meters may simply give an indication whether the signal strength is adequate or provide numerical values of signal strength and background noise levels from which the adequacy of signal strength can be determined.

Before the RF facility survey is performed, potential receiver and sensor locations need to be known. The survey is then performed by placing the trans-

mitter in anticipated locations for the receivers, then moving the signal-strength meter to locations where sensors will be positioned and taking measurements. By taking measurements throughout the facility, the limits of transmission range where the signal can no longer be detected (or is not of sufficient strength) can be identified. Repeaters will then need to be located in the layout within the transmission range to extend the range further.

The RF surveying is generally done by the wireless technology vendor or installer. Depending on the diversity of noise level in the facility and the complexity of its interior layout, an RF survey can be performed for office buildings with a floor space of 100,000 square feet in 2 to 4 hours.

Although RF surveys are critical for successfully designing and installing a wireless network that uses a star topology, systems using a mesh network topology with sufficient sensor density will ultimately not require RF surveys for installation. With sufficient densities of sensors (i.e., relatively short distances between sensors and multiple neighboring sensors within the communication range of each node), these networks will be self-configuring with the multiple potential transmission paths ensuring reliable, consistent communications. In the near term, care should be exercised in assuming that mesh networks will perform reliably for every application, especially in cases where high sensor density is not anticipated. For low sensor density installations, communication over long distances may require a higher-power repeater to connect a local mesh network to the point where the data are needed or a daisy-chain of nodes to communicate. In these cases, the advantages of mesh networking are lost in the region where individual devices carry all data communicated and those devices become potential single points of failure for the entire mesh that they connect to the point of data use.

Other Practical Considerations

Several other factors should be considered in deciding to use wireless sensing in buildings. Peter Stein [38] provides a nice summary of practical considerations for monitoring with wireless sensor networks. In addition to communication range, some of the key considerations that need to be assessed when selecting a wireless sensing network are:

- component prices
- availability of support
- compatibility with different types of sensors with different outputs
- battery backup for line powered devices
- low-battery indicators for battery-powered devices
- on-board memory
- proper packaging and technical specifications for the environment where devices will be located
- battery life and factors that affect it
- frequency of data collection and its relationship to battery life (where applicable)
- need for and availability of integration boxes or gateways to connect wireless sensor networks to BASs, other local area networks, or the Internet
- availability of software for viewing or processing the data for the intended purpose
- compatibility among products from different vendors—this is rare today but will improve with manufacturer adoption of new standards [e.g. IEEE 802.15.4 [19, 20] with Zigbee [21]]
- tools for configuring, commissioning, repairing, and adding nodes to the sensor network
- software to monitor network performance

Most important is ensuring the selected wireless network meets the requirements of the intended application. All factors need to be considered and assessed with respect to satisfying the requirements of the application and the specific facility. Each installation is unique.

The Future of Wireless Sensing in Buildings

The steadily growing number of technology companies offering products and services for monitoring and control applications fuels the expectation that the sub-$10 wireless sensor is likely to be available in the near future [22]. When we reach that point of technological advancement, the cost of the battery may then be the single largest cost item of a wireless module. Even the battery may be replaceable by ambient power scavenging devices that obviate the need for a battery as a power source. A self-powered sensor device creates fundamentally new measurement applications, unthinkable with battery- or line-powered technology. For instance, sensors could be fully embedded in building materials, such as structural members or wall components. They can measure properties in the host material that currently cannot be accessed easily or continuously by external measurement probes. In the energy efficiency domain, new diagnostic methods could be envisioned that use embedded sensors for early fault detection and diagnostics to prevent equipment failure and degradation of energy efficiency. Researchers are exploring different ambient sources for the extraction of electric power. Mechanical vibration emanating from rotary energy conversion equipment, such as internal combustion engines, pumps, compressors, and fans can be converted into electric power by induction driving a magnetic element inside a coil. Alternatively, piezoelectric materials can generate an electric potential when mechanically strained. Present research and technology development focuses on maximizing the energy extraction of mechanical energy by adaptive techniques that sense and adjust to a given vibration frequency and amplitude to maximize power extraction. [39] Thermoelectrical power generators utilize the Seabeck Effect, commonly used in thermocouple probes for temperature measurements. A temperature differential of a few degrees Celsius can, in cleverly designed probes, generate power in the micro-Watt range. [40] The small power generation from ambient power devices can be used to recharge a battery or stored in a super-capacitor to operate the wireless sensors when communication is required. Recent prototypes of ambient energy scavenging devices that generate sufficient electric power to operate a wireless sensor show promise for these revolutionary technologies to soon be commercially available. [41]

With an optimistic outlook on cost projections of wireless sensors and revolutionary self-powering devices, what are the likely impacts and opportunities of this technology for the building sector in general, and for energy efficiency improvement opportunities in buildings in particular? While the scenario of ubiquitous sensing by miniaturizing sensors to the size of paint pigments that can be painted on a wall may be in the realm of science fiction, there are real near-term opportunities for low-cost wireless devices providing value in the building sector now. Some of the applications where wireless sensing should have impact soon include:

HVAC fault detection, diagnostics, and control
- Higher spatial resolution of measurements of zone temperature and humidity to help assure better thermal comfort. Causes of localized hot and cold conditions can be detected and diagnosed. Each office or cubicle would be equipped with one or more temperature/humidity sensors.

- Expand terminal box control from a common single thermostat control point to multiple sensors located throughout the zone served. An average temperature that is more representative of the thermal needs could be used to control terminal boxes.

- Retrofit of terminal boxes for condition and performance monitoring. Because there are hundreds, sometimes thousands, of VAV boxes in commercial buildings, they receive very little inspection or maintenance except when suspected of causing a comfort problem. Wireless sensors placed on these units could be used to measure airflow rates, temperatures, and equipment status to enable central monitoring, performance-based alarms, and diagnostics that would support condition-based maintenance of this largely neglected equipment.

- Additional outdoor-air temperature sensors for improved economizer control. Ideally, place one or more air-temperature sensors near air intakes to air-handlers to minimize bias from radiative heat transfer and sensor failure.

- Equip packaged rooftop HVAC systems with sensors to continuously and automatically monitor performance.

Lighting control and monitoring
- In open-space office buildings, retrofit lighting controls for individual and localized control from the occupants' desks.

- Retrofit reconfigurable lighting systems with indi-

vidually addressable dimmable ballasts.

- Retrofit light sensors at the work task location to turn off or dim lighting fixtures where daylight is adequate.

- Retrofit wireless occupancy sensors and control points on lighting panels to turn off lights during unoccupied periods.

Security and access control
- Motion sensors and door sensors for physical security systems.

- Environmental monitoring and physical security for IT systems and server rooms.

- Access control systems for retrofits and new construction.

Demand Responsiveness
- Retrofit wireless power meters for electricity end-use metering

- Retrofit wireless power meters and control on major loads to modulate or switch off power during grid emergencies or during periods of high power prices.

- Retrofit large appliances with wireless devices for receiving price signals or load control instructions from the power grid to respond to stress on the power grid.

CONCLUSION

Application of wireless communication for monitoring the conditions inside buildings and the performance of building equipment is feasible today. For retrofits, wireless sensing can be installed in many situations at lower cost than an equivalent wired system. Savings on energy, extended equipment life, lower total maintenance cost over equipment lifetimes, and maintenance of better conditions for occupants can even justify sensors using wireless communication where wired sensing has not been used previously.

Very few wireless products for building monitoring are available on the market today, but the technology is poised for rapid introduction soon. Generic hardware is available that can be adapted to building applications. Care should be exercised by those considering wireless technology for these purposes to ensure that wireless communication best matches the application requirements and that the specific system selected is the one best meeting needs. Every application is unique, and wireless technologies should be evaluated with respect to each project's unique requirements. Furthermore, special steps such as RF surveys of facilities in which wireless sensing is planned should be used to plan the proper layout of equipment to ensure reliable communication over the system life.

Data on the condition and performance of equipment can be used to implement condition-based maintenance for building equipment that may previously have been largely run until failure. Information collected from wireless sensor systems installed where no sensing previously existed can be used to improve control by adjusting set points, using sets of measurements throughout a zone rather than measurements at a single point in a zone as inputs for control, and diagnosing hot and cold spots. Control directly from wireless sensors is also possible but less developed and tested than monitoring applications, but today's wireless networks are not suitable for control requiring rapid response on the order of seconds or less. The network and its adaptation must be matched to the needs of the application.

Although wireless sensing can bring benefits not previously possible with wired systems, it is not a panacea for all monitoring and control applications in buildings. As pointed out recently by an author from a major building controls company:

> Part of the answer, at least for the near term, is that wireless networks can provide tangible benefits to engineers, consultants and clients alike. However, as we have witnessed with so many other fast-growing technologies coming of age, only time will tell if the technology will become an accepted and vital part of the HVAC industry. For now, all-wireless control of a facility is neither sensible nor realistic. Conversely, wireless technology cannot be ignored. Although every facility is unique with its own specific requirements, the most sensible building control solution could well be a balanced blend of wired and wireless devices that are strategically integrated for optimum performance and cost savings. [42]

Wireless technology for monitoring and control in buildings is emerging and can be used cost effectively today with care. In the next few years, new technology and products will make application of wireless easier and more reliable. Experience will build widespread support for this technology. Applications of sensors in buildings not fathomable yesterday will emerge based on wireless communication, bringing cost, comfort, safety, health, and productivity benefits.

ACKNOWLEDGMENTS

Work reported in this chapter was supported in part by the U.S Department of Energy Building Technologies Program of the Office of Energy Efficiency and Renewable Energy.

References

[1] *IEEE 802.11-1997.* Standard for Information Technology, Telecommunications and Information Exchange Between Systems, Local and Metropolitan Area Networks, Specific Requirements, Part 11: Wireless LAN Medium Access Control (MAC) and Physical Layer (PHY) Specifications: Higher Speed Physical Layer Extension in the 2.4 GHz Band. *Institute of Electrical and Electronic Engineers, New York.*

[2] *IEEE 802.11-1999.* Supplement to Standard for Information Technology—Telecommunications and Information Exchange Between Systems—Local and Metropolitan Area Networks, Specific Requirements—Part 11: Wireless LAN Medium Access Control (MAC) and Physical Layer (PHY) Specifications: Higher Speed Physical Layer Extension in the 2.4 GHz Band. *Institute of Electrical and Electronic Engineers, New York.*

[3] *RFiD Journal.* 2003. "The 5c RFID Tag." RFiD Journal 1(1):30-34 (January 2004).

[4] S. Roundy, P.K. Wright and J.M. Rabaey. 2003. *Energy Scavenging for Wireless Sensor Networks with Special Focus on Vibrations.* Kluwer Academic Publishers, Boston.

[5] Ferro Solutions. 2004. Energy Harvesters and Sensors (brochure). Ferro Solutions, Cambridge, Massachusetts. Available on the World Wide Web at *http://www.ferrosi.com/files/FS_product_sheet_wint04.pdf.*

[16] Katipamula, S., M.R. Brambley, N.N. Bauman, and R.G. Pratt. 2003. "Enhancing Building Operations through Automated Diagnostics: Field Test Results." In *Proceedings of the Third International Conference for Enhanced Building Operations.* Texas A&M University, College Station, Texas.

[7] Jacobs, P. 2003. *Small HVAC Problems and Potential Savings Reports.* Technical Report P500-03-082-A-25. California Energy Commission, Sacramento, California.

[8] Kintner-Meyer M., M.R. Brambley, T.A. Carlon, and N.N. Bauman. 2002. "Wireless Sensors: Technology and Cost-Savings for Commercial Buildings." In *Teaming for Efficiency: Proceedings, 2002 ACEEE Summer Study on Energy Efficiency in Buildings: Aug. 18-23, 2002, Vol. 7; Information and Electronic Technologies; Promises and Pitfalls,* pp. 7.121-7.134. American Council for Energy Efficient Economy, Washington, D.C.

[9] Kintner-Meyer M., and M.R. Brambley. 2002. "Pros & Cons of Wireless." *ASHRAE Journal* 44(11):54-61.

[10] Kintner-Meyer, M. and R. Conant. 2004. "Opportunities of Wireless Sensors and Controls for Building Operation." 2004 ACEEE Summer Study on Energy Efficiency in Buildings. American Council for an Energy-Efficient Economy.

Washington, D.C. 2004.

[11] Su, W., O.B. Akun, and E. Cayirici. 2004. "Communication Protocols for Sensor Networks." In *Wireless Sensor Networks*, eds. C. S. Raghavendra, K.M. Sivalingam and T. Znati, pp. 21-50. Kluwer Academic Publishers, Boston, Massachusetts.

[12] Raghunathan, V., C. Schurgers, S. Park, and M.B. Srivastava. 2004. "Energy Efficient Design of Wireless Sensor Nodes." In *Wireless Sensor Networks*, eds. C.S. Raghavendra, K.M. Sivalingam and T. Znati, pp. 51-69. Kluwer Academic Publishers, Boston, Massachusetts.

[13] Ye, W. and J. Heideman. 2004. "Medium Access Control in Wireless Sensor Networks." In *Wireless Sensor Networks*, eds. C.S. Raghavendra, K.M. Sivalingam and T. Znati, pp. 73-91. Kluwer Academic Publishers, Boston, Massachusetts.

[14] FCC. 2004. The FCC's On-Line Table of Frequency Allocations. 47 C.F.R. § 2.106. Revised August 1, 2004. Federal Communications Commission. Office of Engineering and Technology Policy and Rules Division, Washington, D.C. Available on the world wide web at *http://www.fcc.gov/oet/spectrum/table/fcctable.pdf*.

[15] Nyquist, H. 1928. "Certain topics in telegraph transmission theory," *Trans. AIEE*, vol. 47, pp. 617-644, April 1928.

[16] Shannon, C.E. 1949. "Communication in the presence of noise," *Proc. Institute of Radio Engineers*, vol. 37, no. 1, pp. 10-21, January 1949.

[17] Pozar, D. 1997. *Microwave Engineering*, 2nd edition, Wiley, New York.

[18] Bluetooth SIG Inc. 2001. *Specification of the Bluetooth System-Core*. Version 1.1, February 22, 2001.

[19] IEEE 802.15.4. 2003. *Part 15.4: Wireless Medium Access Control (MAC) and Physical Layer (PHY) Specifications for Low-Rate Wireless Personal Area Networks (LR-WPANs)*. The Institute of Electrical and Electronics Engineers, Inc., New York.

[20] José A. Gutierrez, Ed Callaway and Raymond Barrett, eds. 2003. *Low-Rate Wireless Personal Area Networks. Enabling Wireless Sensors with IEEE 802.15.4*. ISBN 0-7381-3557-7; Product No.: SP1131-TBR. The Institute of Electrical and Electronics Engineers, Inc., New York.

[21] Kinney, P. 2003. "ZigBee Technology: Wireless Control that Simply Works." ZigBee Alliance, Inc. Available on the world wide web at *http://www.zigbee.org/resources/documents/ZigBee_Technology_Sept2003.doc*.

[22] Chi, C. and M. Hatler. 2004. "Wireless Sensor Network. Mass Market Opportunities." ON World, San Diego, California. Available on-line at: *http://www.onworld.com*. February 2004.

[23] FCC Part 15, 1998. *Part 15 Radio Frequency Devices. Code of Federal Regulation 47 CFR Ch. I (10–1–98 Edition)*, Federal Communications Commission, Washington, D.C.

[24] Ardehali, M.M. and T.F. Smith. 2002. *Literature Review to Identify Existing Case Studies of Controls-Related Energy-Inefficiencies in Buildings*. Technical Report: ME-TFS-01-007. Department of Mechanical and Industrial Engineering, The University of Iowa, Iowa City, Iowa.

[25] Ardehali, M.M., T.F. Smith, J.M. House, and C.J. Klaassen. 2003. "Building Energy Use and Control Problems: An Assessment of Case Studies." *ASHRAE Transactions*, Vol. 109, Pt. 2.

[26] Lunneberg, T. 1999. "When Good Economizers Go Bad." E Source Report ER-99-14, E Source, Boulder, Colorado.

[27] Portland Energy Conservation Inc. (PECI). 1997. *Commissioning for Better Buildings in Oregon*. Oregon Office of Energy, Salem, Oregon.

[28] Stouppe, D.E., and Y.S., Lau. 1989. "Air Conditioning and Refrigeration Equipment Failures." *National Engineer* 93(9): 14-17.

[29] Breuker, M.S., and J.E. Braun. 1998. "Common faults and their impacts for rooftop air conditioners." *International Journal of Heating, Ventilating, Air Conditioning and Refrigerating Research* 4(3): 303-318.

[30] Katipamula, S., and M.R. Brambley. 2004. "Wireless Condition Monitoring and Maintenance for Rooftop Packaged Heating, Ventilating and Air-Conditioning." *Proceedings, 2004 ACEEE Summer Study on Energy Efficiency in Buildings: Aug. 22-27, 2004*. American Council for Energy Efficient Economy, Washington, D.C.

[31] Breuker, M.S. and J.E. Braun. 1998. "Evaluating the Performance of a Fault Detection and Diagnostic System for Vapor Compression Equipment." International Journal of Heating, Ventilating, Air Conditioning and Refrigerating Research 4(4):401-425.

[32] Katipamula S., M.R. Brambley, and L. Luskay. 2003. "Automated Proactive Commissioning of Air-Handling Units." Report PNWD-3272, Battelle Pacific Northwest Division, Richland, WA. Also published by the Air-Conditioning & Refrigeration Technology Institute, Washington, DC. Available on the world wide web at *www.arti-21cr.org/research/completed/finalreports/30040-final.pdf*.

[33] Katipamula S., M.R. Brambley, and L. Luskay. 2003b. "Automated Proactive Techniques for Commissioning Air-Handling Units." Journal of Solar Energy Engineering—Transactions of the ASME 125(3):282-291.

[34] Rossi, T.M. and J.E. Braun. 1997. "A Statistical, Rule-Based Fault Detection and Diagnostic Method for Vapor Compression Air Conditioners." *International Journal of Heating, Ventilation, Air Conditioning and Refrigeration Research* 3(1):19-37.

[35] Honeywell. 2003. "The HVAC Service Assistant." Honeywell Home and Building Controls, Golden Valley, Minnesota. Available on the world wide web at *http://customer.honeywell.com/buildings/CBWPServiceAssistant.asp*

[36] Inovonics. 1997. *FA116 Executive Programmer, User Manual for FA416, FA426 and FA464 Frequency AgileTM Receivers*. Inovonics Corporation, Louisville, Colorado.

[37] Stone, William. 1997. *Electromagnetic Signal Attenuation in Construction Materials*. NIST Construction Automation Program Report No. 3. NISTIR 6055. Building and Fire Research Laboratory. National Institute of Standards and Technology, Gaithersburg, Maryland.

[38] Stein, Peter. 2004. "Practical Considerations for Environmental Monitoring with Wireless Sensor Networks." *Remote Site & Equipment Management*, June/July 2004 (*www.remotemagazine.com*).

[39] Roundy, S., P.K. Wright, and J.M. Rabaey. 2004. *Energy Scavenging for Wireless Sensor Networks with Special Focus on Vibrations*. Kluwer Academic Publishers, Norwell, Massachusetts.

[40] DeSteese, J.G., D.J. Hammerstrom, and L.A. Schienbein. 2000. *Electric Power from Ambient Energy Sources*. PNNL-13336. Pacific Northwest National Laboratory, Richland, Washington.

[41] Ferro Solutions. 2004. "Energy Harvesters and Sensors." Ferro Solutions. Roslindale, Massachusetts. Available at *http://www.ferrosi.com/files/FS_product_sheet_wint04.pdf*.

[42] Wills, Jeff. 2004. "Will HVAC Control Go Wireless?" *ASHRAE Journal* 46(7): 46-52 (July 2004).

Section IX

Enterprise Energy Systems Case Studies

Chapter 36

How Disney Saves Energy
(Hint: It's Not *Magic*)

Paul J. Allen, P.E.

ABSTRACT

The Walt Disney World Resort near Orlando, Fla., is among the most highly visited destinations on earth. Its "campus" consists of 47 square miles containing hundreds of buildings that include world-class hotel and conference centers, theme parks and exotic ride adventures, and precisely controlled spaces for horticulture and animal care.

In addition to a Wall-Street eye on the bottom line, Walt Disney himself encoded the company's DNA with an ethic toward conserving natural resources and the environment that remains to this day as a program called Environmentality. Environmentality is a way of thinking, acting, and doing business in an environmentally conscientious way—from saving energy and water to reducing waste and other environmental impacts.

At Disney, energy management is a key to success. Air-conditioning, refrigeration, compressed air, and water-moving systems for buildings, rides, and transportation all run primarily on electricity and natural gas. To maximize energy conservation and efficiency while minimizing costs and environmental concerns, the Walt Disney World Resort has implemented a state-of-the-art energy management program (EMP) that can, and has served as a role model to owners and administrators of public and private facilities.

This chapter describes the energy management program at the Walt Disney World Resort near Orlando and discusses its results in terms of energy and cost savings. Perhaps in doing so, other facility owners worldwide will develop their own energy management programs and cultivate the economic, energy, and environmental benefits enjoyed by Disney [1,4,6,9,10].

THE ENERGY STAR FOUNDATION

The cornerstone of the Disney EMP is its strong relationship with the U.S. Environmental Protection Agency (EPA) through the EPA Energy Star Buildings program, which has five main components:

- Building tune-up (recommissioning).
- Energy-efficient lighting (Green Lights).
- Load reductions.
- Fan-system upgrades.
- Heating-and-cooling-system upgrades.

The relationship between the Walt Disney World Resort and the Energy Star Buildings program was established in 1996, when Disney implemented the EPA Green Lights program across 17 million sq ft of facilities. This was completed in 1998 and resulted in annual electrical savings of 46 million kWh. Also in 1998, the Walt Disney World Resort began the implementation of numerous other cost-effective energy-saving projects.

Disney's projects included:

- Optimizing compressed-air-system controls
- Upgrading hot-water-boiler controls
- Retrofitting variable-speed-drives into air, pumping, and chilled-water systems
- Retrofitting demand-controlled ventilation into convention-center spaces
- Upgrading and integrating energy-management-systems (EMS), including networking one EMS vendor's stand-alone EMS to centralized network-based servers.
- Installing utility-submetering systems in areas operated by non-Disney companies working in Disney facilities for utility cost recovery purposes.

In aggregate, the efforts Disney has undertaken since 1996 have resulted in a 53-percent internal rate of return (IRR) and metered annual reductions of approximately 100 million kWh of electricity and 1 million therms of natural gas.

THE DISNEY EMP FRAMEWORK

Disney's multifaceted EMP has three main components: the energy management systems (EMS) that are installed in each building or facility; the energy information system (EIS), which is a suite of information technologies that works with the EMS to provide data and information to energy managers and other stakeholders; and Disney staff (called "cast members"), who collectively participate in the EMP. It's the combination of technology and people that makes Disney's EMP successful and sustainable.

ENERGY MANAGEMENT SYSTEM

Overview

The energy management systems (EMS) used at the Walt Disney World Resort are used to control energy consuming equipment—primarily for heating, ventilating and air conditioning (HVAC) equipment and lighting control. The parameters of greatest interest are temperature and humidity setpoints and equipment operating time schedules.

Over the years, Disney has installed a variety of energy management systems from different vendors, which it continues to operate. One vendor's system controls more than 80 percent of the installed EMS base. This system was upgraded to a centralized server-based system connected to the corporate Ethernet-based intranet. This upgrade provided Disney's EMS with a standard and stable hardware and software platform along with the other benefits shown below:

- Review of EMS field panel programming and real-time operation can be made "globally" through any desktop PC on the corporate network.

- The EMS program and data is stored on network servers that are maintained by Disney's Information Services Team. Backups are made daily.

- Automatic reset of equipment time and setpoint schedules are made daily from a server-side control program.

- Data collection for both EMS point trends and utility meter data can be collected and used by the Energy Information System for quick and easy display.

- Maintenance and training of the EMS is simplified.

- Services contracts are minimized or eliminated.

- EMS spare parts inventory is minimized.

ENERGY INFORMATION SYSTEM

Utility Reporting System

The philosophy, "If you can measure it, you can manage it," is critical to a sustainable EMP. Measurement for management is the job of the EIS. The EIS is a suite of programs and computers that take data from the EMS and other data collection sources and churn it into actionable information for use by operators and managers. The EIS measures energy at the facility level and tracks the resulting energy conservation efforts over time.

Continuous feedback on utility performance pinpoints problems in the EMS that need attention. Such feedback also drives Disney's incentive program, which keeps people actively seeking to reduce consumption and expenses without creating new problems.

Disney created their own web-based EIS that uses an off-the shelf database management system to store the vast amount of energy data they collect [5,7,8]. The custom program, called the Utility Reporting System (URS), resides on a network web server. The URS gathers, stores, and processes monthly utility bill data and hourly meter data from a variety of data collection sources. The URS's reports are created in Web-accessible (HTML) formats and can be reached via the Disney intranet.

One popular feature of the URS is a "report-card" format for publishing utility data and historical information. The report card is distributed via e-mail on a monthly basis, with each message containing high-level (summary) information and hyperlinks allowing "point-and-click" access to greater detail. Some links are to graphs that compare current data to data from up to 12 previous months. Also, data can be filtered to compare one Disney area against others. For example, how is Epcot performing relative to Animal Kingdom? Such comparisons foster a healthy spirit of competition among area managers.

Specialized reports are used to monitor and report utility usage in areas operated by non-Disney companies working in Disney facilities, which helps to keep them aware of their usage rates. By measuring actual energy consumption instead of a square foot allocation, operating participants are motivated to manage their energy usage to keep their utility expenses low.

Disney Goes to School

The Walt Disney World Resort participated in a public/private effort to develop an energy information system, called utility report cards (URC), to help Orange County Public Schools (OCPS) better manage energy costs. The URC program was based on the energy information system methods and techniques developed at the Walt

Disney World Resort. The URC is a Web-based energy-information system that reports and graphs monthly utility data for schools.

Each month, a web-based report is automatically generated and e-mailed to school principals and staff as encouragement to examine their school's electricity usage (energy efficiency) and to identify schools with high-energy consumption needing further investigation. The URC also is intended for teachers and students to use as an instructional tool to learn about school energy use as a complement to the energy-education materials available through the U.S. Department of Energy's EnergySmart Schools program (ESS). To see how the URC operates, go to http://www.utilityreportcards.com and click on "URC Live."

The URC was created to help OCPS staff understand and, therefore, manage their utility consumption and associated costs. The URC allows school principals to become aware of how their school is performing relative to a projected benchmark and to other schools of similar design and capacity. Giving recognition to schools that improve performance from prior-year levels could create a spirit of competition with the opportunity to recognize success. Those schools identified as high-energy users become the focus of attention to determine the reasons for their consumption level and ultimately to decrease the energy used. All of this is done by using the monthly utility data that is provided electronically at minimal or no cost to the schools by the utilities.

PEOPLE ARE THE REAL ENERGY STARS

Organization

Conservation has always been one of Disney's core values. In a public-service announcement recorded while he was the honorary chairman of National Wildlife Week, Walt Disney defined "conservation" and thereby set a tone for Walt Disney's Environmentality ethic:

"You've probably heard people talk about conservation. Well, conservation isn't just the business of a few people. It's a matter that concerns all of us. It's a science whose principles are written in the oldest code in the world, the laws of nature. The natural resources of our vast continent are not inexhaustible. But if we will use our riches wisely, if we will protect our wildlife and preserve our lakes and streams, these things will last us for generations to come."

Disney's Environmentality Program provides the framework behind Disney's resource conservation efforts. Everyone has a role to play. There is a dedicated staff of energy conservation engineers and technicians who orchestrate the energy conservation efforts and keep the program moving forward by refining the EMS, EIS and other program components.

Management supports the EMP by promoting and encouraging energy savings efforts and authorizing budgets sufficient to get meaningful work done. New projects are considered based on their expected internal rate of return (IRR). There may also be other non-financial benefits that weigh in to the energy project funding decision.

Disney also recognizes that cast members need to be involved in the EMP to establish a facility-wide sense of ownership and accountability for energy usage. Through Disney's Environmental Circles of Excellence, Disney cultivates Environmentality instead of dictating it. These local teams meet monthly and work on various resource conservation projects in their respective park, resort or support area. Using the Environmentality motto, "Every little bit makes a BIG difference" lets cast members participate in identifying energy waste no matter how small the detail.

Energy Star Tool Bag

How Disney saves energy is not "magic." The Energy Star Tool Bag was created as a guide to help cast members look for energy waste.

Overall Building

1. Heating, Ventilating and Air Conditioning (HVAC).
   - Turn off units during unoccupied hours.
   - Adjust temperature and humidity setpoints to minimize unnecessary heating and cooling.

2. Turn off interior and exterior lighting when not required.

3. Perform walk-through's—look for energy waste
   - Any exterior lighting on during the day?
   - Note "too cold" or "too hot" areas.
   - Note any areas that are "too humid"
   - Close open doors during hot or cold weather
   - Is all non-essential lighting turned off/dimmed down?
   - Are there any PCs left on?
   - Are there any decorative fountains on?
   - Can building facade or other decorative lighting be turned off?

4. Review utility metering reports and look for energy waste.

In the Office
1. Turn your lights off when you leave your office or conference room.
2. Program your PC monitor, printer and copier to "go to sleep" during extended periods of non-activity
3. Turn your computer off completely when you leave to go home.

In the Kitchen
1. Minimize Kitchen equipment pre-heat times.
2. Turn cooking equipment down or off during slow periods of the day
3. Eliminate water waste, report leaking faucets.
4. Turn off kitchen hoods after closing.
5. Turn off or reduce lighting levels in dining areas & kitchen after closing.
6. Keep refrigerator/freezer doors closed. Install plastic strip doors on refrigerator/cooler doors.

In Convention Areas
1. Turn off lighting and HVAC equipment during unoccupied hours.

Swimming Pools
1. Adjust pool water heating temperatures to minimize natural gas consumption during winter months.

In Guest Rooms
1. Setback Guest Room Thermostat to low cool.
2. Close Drapes in Guest Rooms.
3. Keep sliding doors closed.
4. Turn off lights in Guest Rooms.

PROJECT SUPPORT

New projects and renovations provide a great opportunity to incorporate energy saving products into the design. Even though the incremental cost to install an energy saving project would most likely be lowest if installed during a scheduled facility downtime or as part of new construction, the project budget might not be able to support the increased incremental cost. Estimating the potential internal rate of return (IIR) resulting from the expected annual cost savings helps justify increased project budget. A business case can be prepared that details the scope, shows alternatives, describes potential risks and rewards of the project. A strong business case and a high IRR will certainly help sell the project to management.

Replacement of heating, ventilating and air conditioning (HVAC) equipment is an opportune time to incorporate more efficient equipment and new energy management system controls. For example, with the Walt Disney World Resort's hot humid climate, the addition of a heat pipe wrapped around the cooling coil in a 100% makeup air unit provides an efficient method to control humidity while minimizing cooling and reheat energy costs [11].

CONCLUSION

Environmentality is part of the way of life at the Walt Disney World Resort. Energy management programs are good for the environment and make good business sense.

The Disney EMP began by working with and learning from the well-established Energy Star Buildings program, which is available to everyone at http://www.energystar.gov.

Disney adopted Energy Star and then tailored it by integrating commercial energy management systems with a custom energy information system. This technology-based solution is used throughout the Disney World Resort organization by administrative managers, engineering, operations and maintenance staff, and cast members. This combination of people and technology has resulted in a sustainable energy management program at Walt Disney World. As the Walt Disney World Resort continues to expand, these programs will continue to play an important role in reducing energy costs in both new and existing facilities.

References

[1] How Disney Saves Energy and Operating Costs, Paul J. Allen, P.E., Heating/Piping/Air Conditioning (HPAC) Engineering, January 2005.
[2] Continuous Commissioning^SM in Energy Conservation Programs, W. Dan Turner, Ph.D., P.E., Energy Systems Lab, Texas A&M University, Downloaded from http://esl.tamu.edu/cc on Dec. 20, 2004.
[3] ENERGY STAR Buildings Upgrade Manual—Stage 2 Building Tune-Up, US EPA Office of Air and Radiation, 6202J EPA 430-B-97-024B, May 1998
[4] Disney's "Environmentality Program," Paul J. Allen, Brett Rohring, Proceedings of the Energy 2003 Workshop and Exposition, August 17-20, 2003
[5] Information Technology Basics for Energy Managers—How a Web-Based Energy Information System Works, Barney Capehart, Paul J. Allen, Klaus Pawlik, David Green, Proceedings of the 25th World Energy Engineering Congress October 9-11, 2002
[6] Sustainable Energy Management—Walt Disney World's Approach, Paul J. Allen, Proceedings of the Energy 2002 Workshop and Exposition, June 2-5, 2002
[7] Managing Energy Data Using an Intranet—Walt Disney World's Approach, Paul J. Allen, David C. Green, Proceedings of the Business Energy Solutions Expo November 28-28, 2001
[8] Measuring Utility Performance Through an Intranet-Based Utility Monitoring System, Paul J. Allen, Ed Godwin, Proceedings of the 23rd World Energy Engineering Congress October 25-27, 2000

[9] Walt Disney World's Environmentality Program, Paul J. Allen, Bob Colburn, Proceedings of the Business Energy Solutions Expo December 1-2, 1999

[10] Walt Disney World's Energy Management Program, Paul J. Allen, Ed Godwin, Proceedings of the Business Energy Solutions Expo December 9-10, 1998

[11] Applications of Heat Pipes for HVAC Dehumidification at Walt Disney World, Paul J. Allen, Khanh Dinh, Proceedings of the 15th World Energy Engineering Congress October 27-30, 1992

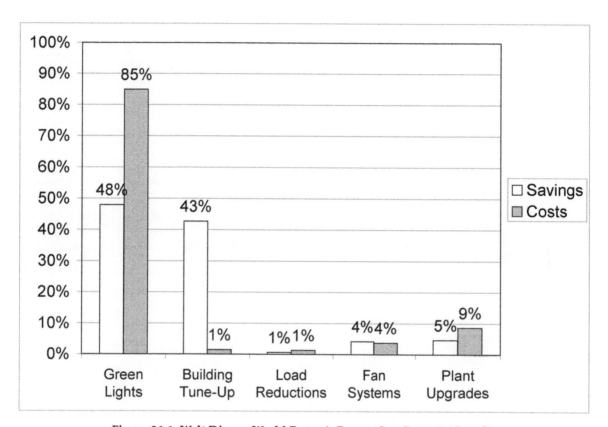

Figure 36-1. Walt Disney World Resort's Energy Star Program Results

Chapter 37

The Business Value of Enterprise Energy Management at DFW Airport

Rusty T. Hodapp, P.E., CEM, CEP, LEED™ AP

ABSTRACT

The Dallas/Fort Worth International Airport has a long track record of success in conventional energy management. For 20 years, this technically oriented program existed principally as an initiative of the airport's maintenance department and flourished in a stable environment characterized by plentiful resources and little competitive pressure. Although successful in producing technical accomplishments and cost reductions, the program never achieved broad corporate impact.

In the mid-1990s, under the leadership of a new CEO, DFW adopted a business-oriented posture focusing on service quality and competitiveness. Although slow to adapt to the changing internal and external environments, by 1999 the maintenance department succeeded in reinventing itself by radically changing its business model and adapting its structure and processes to the new competitive landscape. New department leadership leveraged existing core competencies to recreate the energy management program with an enterprise orientation. They were subsequently able to demonstrate to executive management how the new model supported strategic business objectives and directly contributed to DFW's competitive advantage. Enterprise energy management was represented as a core business function that supported internal objectives (business growth, customer satisfaction, asset renewal) and addressed external factors (electric industry deregulation, environmental issues) by virtue of its positive impact on cost effectiveness, asset productivity and performance, resource utilization, and regional public policy.

Having established credibility and demonstrated the business value of enterprise energy management, the department received unparalleled support from the DFW executive team and board of directors. Corporate policies were enacted to mandate energy efficiency, commissioning, clean fuel vehicles, and energy efficient building codes. New business strategies were developed including energy master planning, evaluation of large-scale onsite power generation, adoption of sustainable practices in investment evaluation, design, construction, operation, and procurement, and development of an integrated/interoperable technology approach to enterprise energy and asset management. Substantial financial and human resources were committed to support program objectives, and the maintenance department was renamed to Energy & Asset Management to signal its new stature and enterprise orientation. These outcomes reflected DFW Airport's renewed, top-down commitment to enterprise energy management as a source of competitive advantage, and their persistence over time has confirmed the validity of the basic value proposition.

BACKGROUND

The DFW International Airport, which first opened to traffic a few minutes past midnight on January 13, 1974, is jointly owned by the cities of Dallas and Fort Worth and governed by the DFW airport board. Today, DFW is the world's third busiest airport serving over 59 million passengers a year. The airport maintenance department manages the airport board's multi-billion dollar facility and infrastructure asset portfolio and provides a variety of services including energy management, thermal energy production and distribution, potable water and sewer system operation, transit system operation, facility management, fleet management, and infrastructure repair and renewal. The original airport maintenance department organizational structure was purely functional in design, and its business processes were dominated by a task orientation.

An energy management program was initiated at DFW soon after the airport opened. This program was managed as an airport maintenance department function, and it achieved many notable technological successes. The program's objectives consisted solely of sound operating and maintenance practices, retrofits of existing systems, and incorporation of energy efficient technology in new construction. Significant economic benefits and reductions in energy consumption were produced; however, energy management remained a department-level initiative with modest recognition of its value at the corporate level.

Little change occurred within the department over the next 20 years as competitive pressures were virtually non-existent and resources relatively plentiful. The energy management program remained focused on applications of technology and achieved relatively little visibility at the corporate level.

MOTIVATION AND OPPORTUNITY TO CHANGE

The early 1990's saw the entire commercial aviation industry experience severe financial losses resulting in strong pressure on airports to reduce operating costs. In 1993, a new chief executive officer assumed leadership of the airport board. He established a vision of "running the airport like a business." The next several years produced corporate-level reorganizations creating various business development oriented units, an aggressive program of diversifying airport revenues, and a continuation of the cost containment focus.

The airport maintenance department was slow to respond to these changes in the internal and external environment. Consequently, the department came to be viewed by senior management as resource intensive, inflexible, dominated by an internal perspective (e.g., maintenance as an end in itself rather than a means to achieve a corporate goal), and out of alignment with the evolving corporate culture and business objectives. The tenure of a new Vice President selected in 1994 produced limited improvement in airport maintenance, and the author; a 12-year member of the department's management team, replaced him in 1997. The CEO personally communicated to the author the need to change the department and bring it into alignment with the new corporate model.

Under this new leadership and with a mandate to change, a comprehensive performance improvement program was designed and implemented, resulting in a near total revision of the department's structure and business processes.

Reinvention

The result was a comprehensive program designed to reinvent the airport maintenance department at its most fundamental levels including:

- Role and direction
- Strategies and processes
- Structure and image
- Culture

A major objective was to establish a clear direction and shared set of values. A comprehensive situation analysis was performed and from it goals, strategies, and objectives developed and implemented. One major outcome involved changing the department's core business model to one employing a total asset management approach. The total asset management model incorporates the full lifecycle of an asset from acquisition through decommissioning and thus requires an enterprise orientation. From this broader perspective, the department began to evaluate its core business functions and competencies.

Assessing the Existing Energy Management Program

From his background in energy engineering, the author was strongly committed to energy management as a core business function and a core competency of the department. Viewed from a total asset management (i.e., lifecycle) perspective, he considered energy management to be a potential source of strategic value to a capital asset- intense enterprise such as a large commercial airport.

In evaluating the department's energy management program a basic SWOT analysis was conducted. It concluded that major strengths consisted of expertise in technical energy management, best-in-class district energy system operation, and willingness to change in order to improve. Major weaknesses included the basic lack of alignment with corporate organization and strategic business objectives and the resulting lack of internal credibility.

The mandate to change established by the CEO introduced a major opportunity to recreate the existing energy management program as an enterprise business function. A second was identified in the impending deregulation of the electric utility industry in Texas and its foreseeable impact on electricity consumers. Ironically, deregulation also created a certain amount of threat as business entities evolving or materializing to operate in the future deregulated markets began approaching key decision makers with various alternatives. The result was a series of mixed messages relative to the viability of energy management as a core airport business function.

SELLING ENTERPRISE ENERGY MANAGEMENT

Upon concluding that energy management, if applied with an enterprise orientation, would contribute fundamentally to the airport's core business success, it also became apparent that selling this concept could provide a critical means of demonstrating the maintenance department's value added contributions and corporate alignment.

The basic strategy developed to sell enterprise energy management involved leveraging a currently successful business operation against a new opportunity. This strategy would establish credibility internally and then explicitly link energy management outcomes to key business objectives. In this case, the maintenance department's district energy (DE) system operation (thermal energy production & distribution business process) was leveraged against the opportunity presented by deregulation of Texas' electric markets. The airport DE system's full cost of service had been benchmarked at best in class levels for years demonstrating efficient operation and capable management. Numerous business entities positioning themselves for the post-deregulation market place were approaching the airport about selling or outsourcing the DE system. The physical facilities and operating/financial records were opened to all business entities desiring to make a proposal to purchase or contract for O&M of the DE system. The low number and limited nature of responses demonstrated forcefully the success of the airport's DE system operation in the competitive marketplace. Consequently, the department now had a platform for initiating a dialog with the CEO regarding energy management and airport business objectives.

Gaining an Audience

The opportunity to sell the business value of enterprise energy management to the CEO came during a presentation on the results of the DE system acquisition and/or outsourcing proposal process. The publicity associated with impending deregulation in Texas and the associated business offers being conveyed directly or indirectly to the CEO stimulated his direct personal interest in the internal analysis. His interest in evaluating the changes being implemented in the maintenance department also predisposed him to participate actively in the dialog.

As the department Vice President, the author along with the utility business unit manager delivered the presentation and key messages to the CEO and the Executive Vice President of Development.

The results of the process established several critical assurances to the executive team:

1. The department management team's willingness to evaluate and embrace change.
2. Their comprehension of corporate business objectives and ability to manage in alignment with them.
3. Energy management as a core competency.

The circumstances also provided an opportunity to extend the discussion and present potential energy management outcomes in the specific context of the airport's $2.5 billion expansion program, then in its initial programming stages. Energy management strategies supporting four vital strategic objectives were proposed:

1. Airport development (expansion and redevelopment of the DE system)
2. Infrastructure renewal (renew 30 year old assets)
3. Electric utility deregulation (position the airport to operate cost effectively in competitive energy markets)
4. Air emission reductions (reduce point source emissions to comply with regulatory mandates)

Two principal challenges had to be overcome in order to convince the CEO of the business value of maintaining ownership and management of the DE system as well as significantly increasing the capital invested in that particular enterprise. Countering the differing viewpoints of other influencers, primarily outside firms, and demonstrating that risks (real and perceived) associated with the proposed changes were manageable proved to be essential in selling the results of the analysis and the proposed changes.

Key Messages

The airport strategic plan developed in 1999 identified two key elements of success for DFW: the capacity to grow by developing its facilities and infrastructure; and a low operating cost structure. To communicate to the CEO how enterprise energy management would contribute to key business objectives, program outcomes were linked to these basic success factors.

An inherent factor in a large commercial airport's ability to grow is the need to attain necessary environmental approvals. Virtually all major commercial airports are located in urban areas with moderate to severe air quality issues. Thus, the emission reductions that would be created by decreasing energy consumption, including those originating on the airport and those resulting from regional power generation, constituted a key message linking energy management explicitly to enterprise business objectives. Similarly, reducing the demand on the airport's energy production and delivery infrastructure would result in improved asset utilization, thus enabling additional development from the existing fixed asset base and deferring capital expansion.

The contribution of an effective energy management program to a low overall operating cost structure was fairly easy to demonstrate. The existing program's track record of demonstrated success in reducing cost and consumption and the additional benefits that would result from broader application formed another key message.

Finally, a number of other strategic business objectives were eventually shown to benefit from enterprise energy management outcomes, including:

Objective—Customer Friendly Facilities
Benefit—Improved asset performance and occupant comfort/satisfaction

Objective —Industry Leading Environmental Programs and Practices
Benefit — Emission reductions, reduced natural resource use, energy efficient building code, purchasing policies

Objective —Revenue Growth
Benefit —Tenant energy supply chain management, expanded thermal energy services

Objective —Total Asset Management
Benefit —Commissioning, lifecycle cost analysis

Objective —Superior Management
Benefit —Industry leading energy and environmental programs

In some cases, the success metrics proposed to evaluate the effectiveness of enterprise energy management were qualitative in nature and described as enhancements along the airport's value chain (i.e., industry leadership, reduced resource use). In others, explicit quantitative measures were offered (i.e., percent reductions in lifecycle cost due to commissioning, energy use reductions resulting from efficient code and purchasing policy, emissions reduced, etc.).

Outcomes

In general, the presentation of the results of the DE system acquisition/outsourcing proposal process produced four explicit outcomes:

1. CEO and Executive agreement with the proposal to continue internal O&M of the airport's DE system.

2. Approval of the proposed alternative reconfiguration of the DE system expansion programmed in support of the airport capital development program to also address renewal of the aging energy infrastructure, flexibility required to operate effectively in deregulated energy markets and reduce regional air emissions. This resulted in an increased investment in the project of approximately $88 million.

3. Direction to pursue detailed engineering and economic analysis of incorporating combined heat and power to supply 100% of the airport's total electric power needs.

4. The opportunity to provide a full briefing to the airport board of directors on energy management accomplishments and proposals.

More important, an ongoing dialog with the CEO and Executive team on energy issues was initiated, as was their appreciation of, and commitment to, the business value of energy management. This commitment was initially signaled to the organization through the approval of the DE system expansion and reconfiguration recommended. In approving a four-fold increase in the capital investment originally programmed and conferring control over the project's design intent and operating business plan to the maintenance department, a powerful and unmistakable message of support was delivered.

The CEO's continued commitment was further communicated in both formal and informal ways. Formal corporate statements of policy were adopted establishing principles in support of energy efficiency, commissioning of all new airport construction, industry collaboration to expand the availability of efficient technologies, and support for legislative action creating incentives to stimulate deployment of clean and efficient technologies. Revisions to a number of existing business processes and adoption of new ones were also authorized, including:

•   Reorganization to create an Energy & Facility Services business unit.

•   Participation in a retail electric competition pilot program.

•   Strategy for energy procurement and management of this function with internal energy staff.

•   Adoption of the International Energy Conservation Code

•   Development of a strategic energy management plan.

In addition, it became easier to secure approval for resources and participation in initiatives to elevate the visibility and influence of the airport's energy management accomplishments. Examples include:

- Addition of energy engineer and energy analyst positions to augment the energy manager's staff.
- A full time staff position to function as the airport's commissioning authority.
- Annual investments in energy efficiency projects identified in the airport's 10-year capital program.
- Application for, and acceptance of, numerous grants for energy audits and demonstration or acquisition of clean and efficient technologies.
- Participation in federal, state, and NGO initiatives to study airport energy use, sustainable practices, etc.
- Memberships in high profile organizations and initiatives including Energy Star, Rebuild America, U.S. Green Building Council, Texas Energy Partnership, etc.

Informal means included continued visibility of energy management objectives, accomplishments, and plans at the CEO and board of director levels through regular briefings.

The airport's reconstituted energy management program has since been recognized by industry and governmental organizations, the trade press, and with regional, State, and international awards. In a telling measure of the CEO's continuing commitment, in 2003, he directed the maintenance department be renamed to Energy & Asset Management to more accurately reflect its enterprise orientation.

DE SYSTEM EXPANSION PROJECT

The approved reconfiguration of the DE system expansion project addressed each of the four strategic objectives noted previously by installing new chilled water, thermal energy storage, heating, preconditioned air (PCA) and controls systems.

- Chilled Water—(6) 5,500 ton chillers and 90,000 ton-hour stratified chilled water storage
- Steam—(4) 40 MMBH and (1) 100 MBH medium pressure boilers with ultra-low NO$_x$ (9ppm) burners
- PCA—(5) 1,350 ton chillers and (6) 1,130 ton heat exchangers for precooling or heating
- Cooling Towers—35% capacity increase using ex-

isting structure with optimized fill, water flow & distribution, and increased airflow
- Controls—new industrial distributed controls

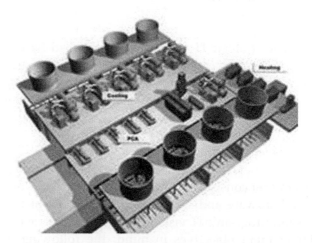

Figure 37-1. DFW District Energy Plant Layout

Thermal energy service was provided with the new system to existing loads in 2004 and substantial completion of the project coincided with the opening of the new international passenger terminal in July 2005.

DE System Controls and Automation

To coordinate, monitor, control and troubleshoot the complex thermal systems a robust and sophisticated automation system was required. An industrial grade distributed control system (DCS) was selected to provide the multi-level redundancy necessary for a business critical operation.

Information Technology and Airport Expansion

As noted previously, DFW was engaged at this time in a multi-billion dollar expansion program which was the principal driver for the DE system expansion project. The program centered on a new, 2.1 million square foot international passenger terminal and a new automated people mover system. Both project elements would involve significant information technology (IT) components.

The scale and complexity of IT applications coupled with the advent of entirely new business processes (for DFW) dictated a strategic approach to technology planning for the program. The program's technology (or technology enabled) goals included:

- Common use equipment
- High levels of customer service and amenities
- Ability to effectively manage assets and resources
- Integration with campus IT infrastructure
- Integration with enterprise business applications

- Enhanced information availability and access
- Enhanced situational awareness
- Operation efficiency
- Cost effectiveness
- Flexibility

The strategic IT vision thus established for the expansion program created another opportunity to demonstrate alignment of enterprise energy management and to leverage project specific IT applications to achieve strategic business objectives.

DE Project Automation Goals

The new DCS replaced an existing 18 year old system that consisted entirely of vendor specific, proprietary hardware and software. In the context of the expansion program's IT vision and with enterprise energy management objectives in mind, the following goals were established for the DE project's automation system.

- Replace the existing obsolete DCS
- Maintain use of industrial grade automation system for the DE production and distribution processes
- Specify a system based on open architecture
- Substantially improve the overall automation of DE plant equipment and unit operations
- Leverage the centralized monitoring and operator interface facilities and capabilities
- Interoperability with campus building management and process control systems
- Integrate with enterprise business applications to achieve a high degree of information sharing
- Provide DE customers with access to billing and usage information

While certain of these goals were specific and limited to the DE project, (DCS system replacement, automation enhancement), the integration and access goals were developed with a view towards improving energy management capabilities at the enterprise level. The nature of the expansion program itself—huge scale involving hundreds of design firms, contractors and dozens of project elements, coupled with the public procurement environment that DFW is obligated to operate in prevented a single technology or technology company solution to all IT applications, or in some cases (building automation systems for example) to similar applications in different construction elements. With that being the case, the energy management team saw the DE project's DCS as means to achieve integration of disparate automation systems (new and existing) as well as certain

enterprise business applications such as asset management, service requests and dispatch, procurement, performance measurement and reporting, and emergency operations management.

With this direction the project team developed functional block diagrams showing the DE plant DCS as the platform for integration of building management systems, other process control systems and high value enterprise business applications.

Similarly, network architecture diagrams established connectivity criteria for the project which included use of DCS communication bus features within the DE plant, corporate network for external connectivity with building management and other systems/applications and internet access for DE customers.

Finally, with no building automation system integration standard (e.g., BACNet or LONWorks) having been established for the expansion program as a whole (nor was a de facto one established through existing airport systems), the DCS specifications were developed around open, interoperable standards-based technologies to facilitate connections across the enterprise.

DCS Solutions

To implement the project automation goals the specifications provided for a single bid for the DCS product and integration services. The selected vendor was required to provide the hardware and integration services as one turn-key solution.

The system/service procurement was by bid under the in-place construction contract and resulted in the selection of Emerson's Delta V automation system for the hardware solution with integration services being provided by a major player in the integration services market.

DCS Based Enterprise Integration

The Delta V system makes extensive use of commercial off-the-shelf (COTS) technologies which facilitates easy connections across the enterprise and reduces the dependency on proprietary hardware and software. Embedded Foundation fieldbus and other digital busses provided relatively simple plug-and-play integration solutions within the DE plant and other process automation applications, particularly where programmable logic controllers (PLCs) were used. Open, interoperable standards including Extensible Markup Language (XML) and Object Linking and Embedding for Process Control (OPC) provided easy connections with Microsoft Office applications and enterprise business systems, and accessibility over the internet.

Numerous PLCs were utilized for process automa-

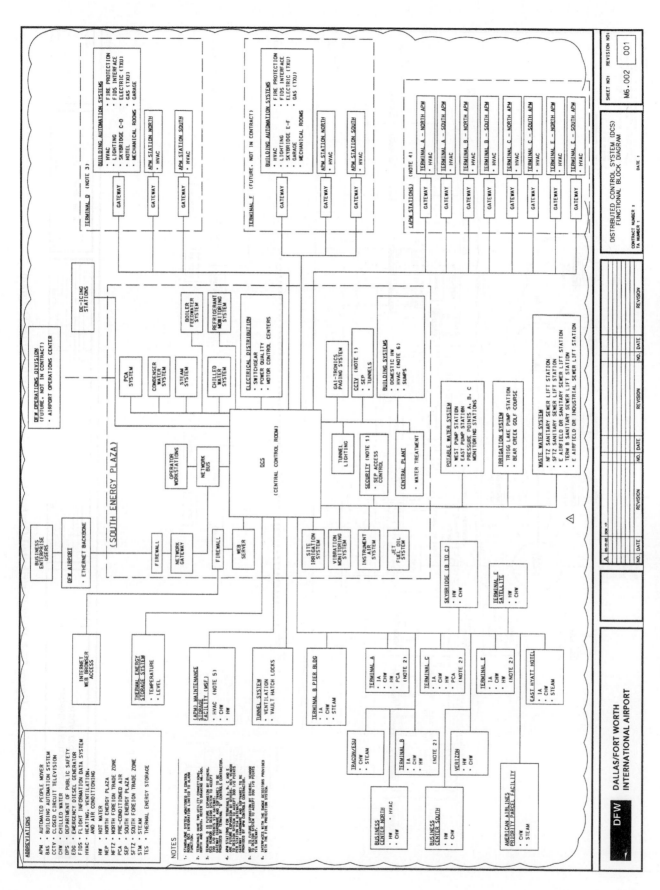

Figure 37-2. DFW Distributed Control System Functional Block Diagram

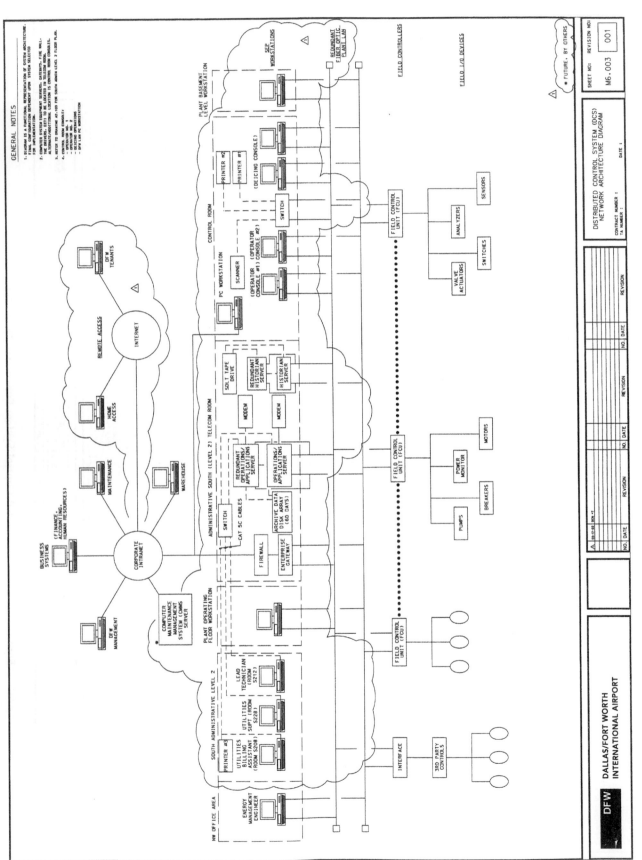

Figure 37-3. DFW Distributed Control System Network Architecture Diagram

tion throughout the DE plant and other utility processes on the airport including water distribution, wastewater collection, and collection, storage and treatment of spent aircraft deicing fluid. These systems were easily integrated with the Delta V using open bus standards (Modbus or Data Highway).

Integration of building automation systems with the Delta V DCS was more challenging. Building automation systems had been installed as part of the construction of ten new facilities totaling over 2.7 million square feet. In conjunction with building automation systems in existing airport facilities, products from a wide variety of vendors were represented. Given the diversity of products installed, neither BAC-Net nor LONWorks would provide a single integration solution. However all systems targeted for integration were OPC compliant. Consequently, OPC provided the interoperability solution necessary to integrate both new and existing building automation systems via the Delta V DCS.

A redundant Delta V network provides connectivity within the DE plant while an OPC network services BAS applications. Remote systems utilize the DFW LAN for access to the OPC network. Remote users have access via the internet.

OPC and XML provided the primary means of integrating the DCS (and other integrated automation systems) with enterprise and local user business applications.

Enterprise Asset Management

The integrated system developed through the DE project delivers significant value through its information sharing capabilities in addition to the efficiencies of automating energy and utility process operations. Experienced DE plant operators now have process control and operating information from systems across the airport and enterprise—thermal energy, water, wastewater, irrigation, HVAC, weather data, status of maintenance actions—at their fingertips and all through a common user interface. Summary reports of production, distribution, performance, exceptions, history, etc. are available for management and engineering—and may be shared via the network. The airport's asset management system (DataStream 7i) interfaces directly with systems and facilities through the integrated DCS allowing for automated tracking of critical asset information, generation of work requests, emergency response, etc. In addition, the asset management system itself accesses information from throughout the enterprise and interfaces with numerous other enterprise business applications supporting a wide array of users and critical business processes.

DE service invoices are automated and electronically transferred to accounts receivable and customers have web access to their real time and historical consumption and cost data. The strategies, technologies and techniques employed have effectively created an enterprise asset management system that improves efficiency through automation and interoperability and delivers value by connecting islands of information.

DE Project Results

The DE project has delivered impressive results through a combination of sound design principles, efficient equipment technology and a sophisticated automation strategy, all driven by enterprise energy management objectives. Nitrogen oxides (NO_x) emissions from combustion operations were reduced by 86% exceeded the regulatory mandate (70%). While adding 2.9 million square feet of service area, the DE plant energy consumption per square foot served has decreased by 47% ($5 million annually at current energy prices). The thermal energy storage system and associated operating strategy have demonstrated the ability to shift over 15MW of electric load off-peak. The combination of off-peak commodity pricing and reduced transmission and distribution charges has reduced annual operating costs by as much as $750,000.

Perhaps as important as the directly quantifiable economic returns accruing from the project is the creation of an enterprise enabled platform through which DFW will achieve multiple benefits in managing its energy, physical and information assets, resources and environmental footprint. The open standards-based approach enables a centralized and remote approach to monitoring, maintenance, control and management of the operating environment. The use of COTS and interoperable technologies provides a pathway for continuous integration of existing and new facilities, systems and business applications at the enterprise level.

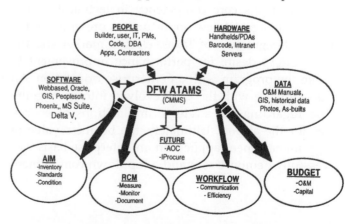

Figure 37-4. DFW Automated Asset Management System (ATAMS)

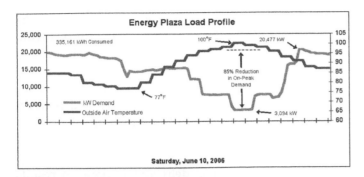

Figure 37-5. DFW District Energy Plant Load Profile with TES

Capital and operational expenditures will be reduced by limiting the need for multiple proprietary networks and enhancing the ability to manage large amounts of data while ensuring persistence of energy management measures. The productivity of assets and other resources will be increased through more flexible work process and scalable tools that enable improved collaboration and connectivity. The connections created between uses and users of formerly disparate islands of information will greatly enhance situational awareness and facilitate effective management of key business processes.

CONCLUSION

For 20 years, the DFW Airport's conventional energy management program produced technical achievements but as a maintenance department initiative, remained limited in application, visibility, and impact. As the commercial aviation industry was subjected to severe competitive pressure, the lack of an enterprise orientation resulted in the department's being out of alignment with the evolving corporate culture, creating a necessity to change. New leadership leveraged the department's core competency in technical energy management against new opportunities presented in the pending deregulation of Texas' electric market and the airport CEO's desire for performance improvement to recreate the program with an enterprise orientation. An initial audience with the CEO evolved into an ongoing dialog regarding the strategic implications of energy issues for the airport. Ultimately, the business value of energy management applied on an enterprise basis was successfully demonstrated and a top-down commitment

to it as an important contributor to the airport's competitiveness realized.

The application of open/interoperable standards based information and commercial off-the-shelf technologies provided a cost-effective and efficient means of developing an enterprise energy and asset management system. The technologies and strategies deployed produced significant efficiency gains, improved productivity and operational flexibility and have provided a pathway for continuous development of integrated enterprise solutions for management of key business processes.

The process of reinventing DFW's energy management program and gaining the CEO's commitment to it may offer a few lessons for consideration by other similarly situated organizations:

- Critically evaluate (situation assessment) existing energy management programs and practices.
- Seek alignment of energy management program and practices with corporate strategies and business objectives.
- Identify energy management contributions to the corporate value chain.
- Link energy management outcomes to key business objectives (key messages).
- Demonstrate (or establish) and then leverage credibility (personal and/or organizational) for access and/or to reinforce the key messages.
- Know your audience (CEO's perspective, critical issues, success metrics).
- Understand associated risks (real and perceived) and show they are manageable.
- Understand the value of information as a core asset of the enterprise—and how that value is enhanced when its uses are connected.
- Consider the value potential inherent in the convergence of IT, energy management systems, building automation systems and the web to manage effectively across the enterprise.

Acknowledgement

An earlier version of the material in this chapter appeared in the article, "The Business Value of Enterprise Energy Management at DFW Airport," by Rusty T. Hodapp, *Strategic Planning for Energy and the Environment*, Summer 2005.

Chapter 38

Cisco Connected Real Estate

David Clute and Wolfgang Wagener

ABSTRACT

This chapter was developed in response to a number of business drivers that are taking place in the industry. Cisco's own experience in managing a global portfolio of approximately 20 million square feet and almost 400 buildings has provided opportunities to streamline its own web-based facility management and energy management systems. Cisco's Workplace Resources Organization (WPR) is responsible for complete life-cycle management of the Cisco global real estate portfolio including strategic planning, real estate transactions and lease administration, design and construction, building operations and maintenance and for the safety and security of Cisco employees world-wide.

The Cisco Connected Real Estate (CCRE) program was initiated in the WPR organization and with the support of many cross-functional stakeholders in various business units, particularly Cisco's IT organization, this program has gained tremendous visibility across the company and now throughout the real estate industry. The successful deployment of the technologies discussed in this chapter depend heavily on the interaction of people, process and tools across a complex "eco-system" of employees, partners, vendors and suppliers. The methodologies presented here are changing the way that real estate is developed, used and managed, and is shifting the basis of the real estate business model from one based solely on space to the provision of service.

INTRODUCTION

Responsiveness. Innovation. Agility. Adaptability. All qualities that organizations must possess in order to thrive in today's highly competitive global economy. Until recently, however, these qualities have not been readily associated with the real estate that organizations use. That is changing. The real estate sector is in a state of transformation, driven largely by customers from a broad range of industries, demanding more from their assets.

These demands and needs are converging to create a shift in the way that buildings are both conceived and used. The changing real estate business climate is being driven by:

- Customers searching for ways to achieve visibility, transparency, and control over their entire real estate portfolio.

- A drive for innovation and sustainable capabilities to reduce capital and operational expenditure.

- Key stakeholders searching for opportunities to optimize value in the real estate life cycle.

- Industry searching for means to improve competitiveness and differentiation of its offering.

- Saving on energy consumption and achieving environmental sustainability.

- Technology adoption accelerating transformation

- Cisco Connected Real Estate drives value by transforming the way real estate stakeholders—for example developers, landlords, tenants, and others—design, build, operate, and use real estate.

Telephony, IT technology, and Building Controls Converge

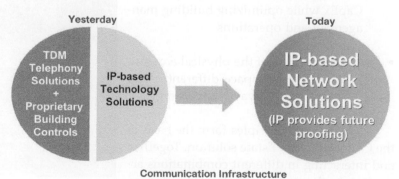

Figure 38-1. Building and Technology Solutions are Merging

511

- In the workplace, Connected Real Estate simplifies the business of providing real estate and allows landlords and owners to deliver effective work environments that drive workforce productivity. This relates directly to the ability of IP communications and innovative technology to drive higher productivity and greater cost savings.

- In a building, Connected Real Estate enables the delivery of powerful services or revenue-generating opportunities that drive business growth by combining real estate and IT. This is possible due to IP being installed as the fourth utility over which these services can be delivered, in-house or as managed services.

- In most parts of the world, energy is expensive and excessive consumption is becoming socially unacceptable; Connected Real Estate enables energy efficiency through planning, visibility, monitoring, and control.

- Last but not least, Connected Real Estate improves safety and security by transforming how building operators and owners can protect their people and assets.

Cisco Connected Real Estate does all of this by harnessing and integrating the power of IP networks. Connected Real Estate is predicated on the three fundamental principles relating to real estate and network interactions:

- Creating a "building information network" or flexible and scalable network foundation as the facility's fourth utility

- The convergence and integration of ICT and building systems onto a common IP network, reducing OpEx and CapEx while optimizing building management and operations.

- Transformation of the physical environment; delivering space differently, and introducing new ways of working.

These three principles form the basis of the Connected Real Estate solution. Together, and interacting in different combinations according to the various needs of developers, owners and occupiers, the principles are

driving the next wave of transformation in the construction and use of real estate.

This chapter explores those principles in detail and demonstrates how Connected Real Estate is delivering huge financial and operational advantages not only to the construction, real estate and property services industries but also to their customers such as hotel operators, multiplexed retail outlets, and corporate tenants in sectors as diverse as hospitality, healthcare, education and retail finance. It is changing the way that real estate is developed, used and managed, and is shifting the basis of the real estate business model from one based solely on offering space to service provision.

This new approach looks along the entire building life cycle, from concept, design and construction through to maintenance and operation. The network and the transformational capabilities it delivers is at the heart of this new approach. And to use it successfully means doing things differently, literally from the drawing board.

DELIVERING BUILDING INFORMATION NETWORKS: THE NETWORK AS THE FOURTH UTILITY

"After four years of market research it was time to renovate our business model. Our strategy was to differentiate this building and all our assets in an already saturated market. One of the methods to achieve this goal was the creation of a unique communications network that would connect Adgar tower with all our buildings in Canada and Europe. The Cisco Connected Real Estate initiative matched our aims."

Roy Gadish, CEO
Adgar Investment and Developments
Tel Aviv, Israel.

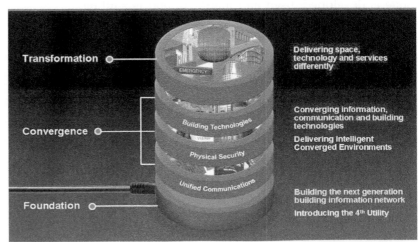

Figure 38-2. Developing Building Infrastructure

Power, water and heating are taken for granted in the construction of a building. To date however, the provision of communications and information networks has been left to tenants. This means that in multi-occupancy buildings, a number of parallel networks are likely to be installed on a piecemeal basis, with each tenant responsible for meeting its own requirements.

Today, however, the demand for connectivity creates a new business model for landlords and developers. The network becomes part of the fabric of the building, supplied to tenants just as water, light and heating are today. By providing the network infrastructure as part of the building's platform, developers provide a point of competitive differentiation to attract prospective tenants. This applies to all forms of real estate whether commercial office space, retail developments, hotels or even residential developments. Connected Real Estate provides landlords and owners with the ability to provide services that respond to the needs of their users, shifting the business model from space to service.

Traditional commercial buildings offer landlords limited opportunities for generating additional revenue from their tenants. And generally the only way that they can compete for prospective tenants is through location and lease rates. In short, landlords sell space. The integration of the network in the fabric of the building, becoming the building information network as the facility's fourth utility, removes those barriers and provides numerous new revenue generation opportunities. Rather than simply selling space, the network allows landlords to move into a business model based on service provision.

As an example, tenants in One America Plaza in San Diego have instant, secure access to communications and data networks providing them with the connectivity that they would otherwise have to acquire themselves.

For the building owners, the provision of the network as a utility means that they have a real source of competitive differentiation in the property market, attested to by the occupancy rates of 95 percent compared to an 88 percent average in the San Diego region. One America Plaza tenants are able to connect their operations to the network almost instantaneously (compared to the average 30-day turnaround when ordering from an ISP) and at far lower cost than sourcing provision independently. One America Plaza has a wireless network, meaning that literally the whole building is connected, allowing tenants the mobility and flexibility of working styles that characterize business today.

It's not simply commercial office space that benefits from the integrated offering of IP networks in the fabric of the building. One of the largest real estate developments in the Arabian Gulf, the $1.3 billion Greenfield, mixed-use development, Amwaj Island, will provide residential, commercial, and hospitality tenants immediate access to a range of communications and data services through the creation of a single IP infrastructure backbone that will provide connectivity to every home, business and hotel room on the island. This means that Amwaj Island will be able to provide added security and extra services to tenants such as video on demand for apartments and hotels and digital signage in retail and hospitality facilities over a single network, enhancing the revenue potential for landlords and providing enhanced services at lower costs for tenants.

Having an IP network at the heart of a building does more than deliver service to tenants. It enables significant reduction of move-in and retrofit costs and increases the speed at which they can set up their businesses. Such a network:

- Reduces tenant operating costs
  - Creates more productive work environments
  - Improves flexibility and start-up
  - Enhances user/landlord responsiveness
  - Enables 24-hour availability using Contact Center
  - Provides the ability to optimize building and tenant management (such as track work-order status)

A network that is integrated in the fabric of the building also provides landlords and users with greater control and security of the building's operations. For all building owners integrating communications, security, and building systems into one IP network

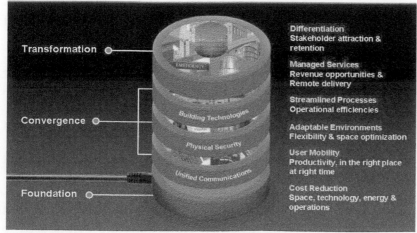

Figure 38-3. The Connected Real Estate Value Foundation

creates significant financial and operational advantages.

Boston Properties is a real estate investment company (REIT) that owns and operates 120 properties in major centers in the US, covering more then 40 million square feet. Boston Properties uses an integrated network approach to property management that connects all of its buildings systems to a single converged network infrastructure. This approach provides Boston Properties with the ability to monitor systems including energy management, security, ventilation, and access control around the clock from a single control center. The tenants of the buildings in Boston Properties' portfolio also benefit from this integrated approach. The system allows them to make requests for service over the web, greatly enhancing the efficiency with which their requests are handled and allowing Boston Properties to manage its workflow far more effectively. More than 60 percent of all requests from tenants are now received through the dedicated website. The benefits of centralized building management for Boston Properties became even more apparent when they added a new building to their existing portfolio. They achieved a payback period for "connecting" this new property into the portfolio of less than five months by adding the building's operation to the centralized management infrastructure.

As well as transforming the way that buildings are managed, the network also allows landlords to create services that respond to tenants' needs. For example, in a retail development digital signage can be provided that carries promotional material for a particular store, and content can be tailored to fit the precise requirements of each retailer. Landlords are able to provide instant access to a range of communications services—such as connectivity, internet access, and IP telephony. The landlord, in effect, becomes a service provider responding to the needs of its customers as they develop over time and opening new sources of revenue in addition to rent.

CONVERGENCE AND INTEGRATION:
IT AND BUILDING SYSTEMS
ON A COMMON IP NETWORK

"The move towards increasing enterprise integration enhances the need for advanced BAS solutions. Companies across all vertical building markets are striving to increase integration across the entire enterprise to improve information management and optimize the strategic decision-making process. As BAS increasingly adopt IT standards, they are increasingly converging with traditional IT infrastructures. Adoption of IT standards in the BAS industry, and the inherent cost savings regarding BAS integration, is causing many *building owners to rethink the value proposition of integrated BAS"*

(Building Automation Systems Worldwide Outlook, Market Analysis and Forecast Through 2009)

A key element of the business case for the Cisco Connected Real Estate framework is based upon the convergence of information technology and communication systems, security, and building systems onto a single IP network. This next wave of convergence creates opportunities for key stakeholders in the building value chain.

Most buildings and campuses today are constructed with multiple proprietary networks to run systems such as heating, ventilation and air conditioning (HVAC), security and access controls, lighting, and fire and safety as well as separate voice, video, and data telecommunications networks. As a result we see buildings that are complex to operate, with high installation, integration and ongoing maintenance costs, and sub-optimal automation functionality. Typically these generate constraints and inefficiencies such as:

- High CapEx for design, engineering, and installation
- Building performance not optimized, limited functionality
- Expensive maintenance (OpEx)
- High integration cost when linking different devices
- Reduced management capabilities—limited reporting and monitoring options (isolated views)
- Less flexibility with closed systems, proprietary networks, custom processes, vendor dependency

The Cisco Connected Real Estate solution unites the disparate—and often proprietary—networks and systems over a single IP network that allows all communication, security, and building systems to be monitored and managed centrally.

This so-called "building and IT convergence" creates new opportunities to reduce a building's total cost of ownership (TCO), enhance the building's performance, and deliver new building services to tenants.

The Connected Real Estate approach applies not simply to the network within one building. Cisco Connected Real Estate allows identical levels of oversight and control to be exercised across a geographically dispersed portfolio of facilities, or unrelated remote properties when offering building management as a managed service. The same network architecture that is used to allow an organization to communicate and share in-

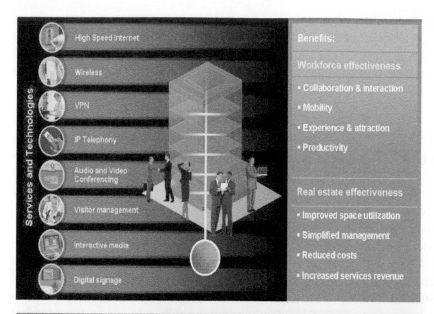

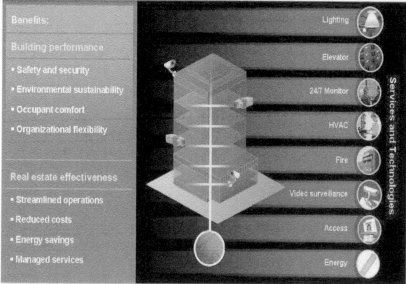

Figure 38-4. Moving from Multiple to a Single IP Network for Communications and Building Systems

Security, set out to measure the advantages of converged systems against those built on the traditional, separate model. By creating a model of an eight-storey office building suitable for 1,500 people, they were able to measure the capital and operating expenditure levels that both approaches would generate. The study found that the converged approach generated CapEx savings of 24 percent in the construction phase and reduced operating expenses (OpEx) by 30 percent over the economic life of the building.

Creating centralized capabilities for building management has a direct impact on the ongoing costs of building operation and maintenance. Centralization introduces economies of scale by which fewer staff can monitor and control far more properties in one or more real estate portfolios. For specific services, such as energy management, an intelligent network can provide constant visibility and monitoring of temperature and energy consumption and ensure that the system is adjusted to suit the demand. Because the system is constantly monitored, reaction times to unexpected developments such as energy surges and loss of power are significantly improved. In addition, centralized building management for energy issues eases the increasing regulatory pressure to comply with stringent environmental standards. An intelligent networked approach ensures that the risks of non-compliance can be managed effectively with a preventative rather than reactive approach.

Real estate assets spread across a wide geographical area benefit from this integrated approach, and in much the same way, existing campuses of many buildings in one physical location can also derive considerable operational and financial benefits from adopting this novel approach to building control.

Pharmaceutical company Pfizer operates 4 million square feet of R&D and manufacturing facilities on its site in the UK. More than 70 buildings are in active use and were operating with multiple building systems that created inconsistent and unwieldy control. Pfizer decided to link its building control network to its corporate IT, and the business now has consistent control from over 70,000 data points over one IP network and the internet. Pfizer has achieved considerable savings on an-

formation is also used to distribute information about activity within a building over the internet to any point on the network where it is needed. Security, for example, can be monitored and controlled across a broad campus or group of buildings from a single, central point. Furthermore, security personnel in a firm do not need to be worried about laying in their infrastructure since their cameras will be using IP. It is also easier and more cost-effective to add, replace and move IP cameras between locations.

A study commissioned by the Converged Buildings Technology Group (CBTG), a consortium of building system manufacturers including Tour Andover Controls, Molex Premises Networks and ADT Fire &

nual energy costs (5 percent or euro 8.6 million) as well as a 15 percent reduction in annual maintenance costs. Business continuity in mission critical facilities has been enhanced with facilities subject to far less downtime, and overall control of all real estate has been considerably enhanced.

CONVERGING SAFETY, SECURITY AND IT

The convergence of IT networks and buildings systems has a significant impact on the ability to create safer environments for building owners, operators, and users. Interoperability of devices and networks ensures that critical real-time data can be acted upon. Video surveillance, access control and asset management over the IP network can be used to drive more sophisticated and comprehensive physical security strategies.

Cisco Connected Real Estate provides a secure platform for consistent, real-time communication of emergency status and instructions through data, voice and video formats to multiple devices including PCs, IP telephones, and even information display and public address systems. This allows rapid communication of emergency information to tenants, visitors, and employees. For example, IP telephony applications allow security personnel rapidly to inform building occupants of security breaches that may require building evacuation.

Cisco Connected Real Estate enables state-of-the-art access control to buildings and car parks using a variety of recognition technologies. This not only allows close control of people and physical assets (such as workforce, car parking, equipment) but also lets building owners regulate access to buildings or areas, for example excluding personnel who are listed as on holiday or on sick leave (and may have been victims of identity theft).

IP-enabled closed circuit television (CCTV) has been proven to reduce vandalism and other forms of lawlessness. One example comes from the UK, where the local education authority in Newport, Wales installed IP-enabled CCTV to monitor school premises that were suffering from a high incidence of vandalism. By using CCTV installed over an IP network, the education authority has been able drastically to reduce the incidence of vandalism and reports considerable savings of both cash and teachers' time. The end result is a safer, happier and more productive school environment in which teachers can focus on teaching, rather than dealing with the

aftermath of vandalism. The flexibility of an IP network enables clear images to be monitored at any distance and instant action to be initiated. Digital storage of IP CCTV enables archived images to be instantly recalled without laboriously searching through videotapes.

Cisco Systems has implemented a worldwide security system that uses its IP network to provide video surveillance, security monitoring and access control in all of its 388 sites around the world. Each site is centrally monitored from one location in either the UK, United States, or Australia. Not only does CCTV over IP create an instantaneous reaction time so that facilities can be monitored in real time, but also other elements of the building operations (such at temperature control and leak detection) are visible on the network so that a potentially damaging change in the physical environment of a sensitive area can be spotted and controlled before long-term damage is done. Cisco saves on the physical presence required to control security without diminishing the level of security available to all of its sites around the world. The return on investment that Cisco generates from this integrated approach is $10 million (euro 8 million) each year.

TRANSFORMING HOW WE USE OUR ENVIRONMENT: DELIVERING SPACE DIFFERENTLY, INTEGRATING SPACE, TECHNOLOGY, AND SERVICES

"The great agent of change which makes new ways of working inevitable is, of course, information technology, the power, reliability, and robustness of which are already evident in their impact not only on work processes within the office but on every train, in every airport lounge, at every street corner, in every classroom, library and café. Office work, no

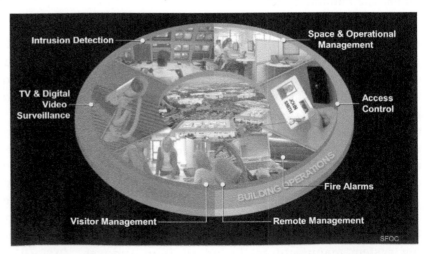

Figure 38-5. Convergence of Safety, Security and IT

longer confined to office buildings, is everywhere."
                                     Frank Duffy, Reinventing the Workplace

The nature of work is changing. Knowledge workers are mobile. They collaborate. They are no longer desk-bound by the processes they execute but are brought together by the projects they develop. This fundamental shift has not been widely reflected in the way that the space these workers occupy is designed and arranged. For many organizations today, much of the space they maintain is largely underutilized. An increasingly mobile workforce may mean that many of a building's intended occupants are absent for much of the time. Dedicated offices and cubicle spaces are empty, and spaces for meetings or other forms of interaction lie dormant. Yet, all these assets have to be maintained as if they were being occupied to maximum capacity. Improving the efficiency of the design and use of physical spaces will limit the need for space and result in cost reduction (rent and operations), while enhancing the productivity of the workforce.

Transforming the Workplace

Cisco's converged network for voice, video and data is enabling transformation of the workplace by helping organizations create flexible real estate portfolios, and supporting new workplace designs—at the same time improving organization-wide productivity, collaboration and mobility. A converged network:

- Improves employee mobility and remote working
- Delivers flexible and efficient workspace
- Enables new working practices and better collaboration
- Improves space optimization and use
- Reduces real estate costs

Robust, scalable and secure networks for voice, video and data improve employee productivity. Unified communications, wireless access and VPNs create flexible work environments, employee mobility and remote working initiatives. A converged network allows secure synchronous and asynchronous collaboration, email, voice mail, conference calls, video conferencing, knowledge management initiatives, intranets and instant messaging. This enables new working practices while reducing overall real estate requirements through, for example, hot-desking and VPN-based remote working.

IT and real estate executives can take advantage of a fully converged intelligent information network to create virtual workspaces that provide more flexible and efficient work environments. This converged intelligent information network provides wireless work areas that connect people to important corporate assets and building amenities to provide greater mobility, productivity and communication capabilities. This allows owners to achieve greater operational gains from the much more flexible and efficient use of their existing space and to use the network to derive more from their assets.

Cisco Connected Workplace in Action

Cisco Systems has put these principles into action with the redevelopment of office space at its headquarters in San Jose and around the world (Bangkok, Taipei, Charlotte, New York, and growing). Cisco's employees, like many in other organizations work differently than they did even as recently as a few years ago. An increasingly global work-force and customer base makes it more likely that employees need to work at nontraditional hours, leaving their offices vacant at other times. More complex business and technology issues increase the need for collaboration with team members in the same building or at various sites worldwide. Employees are often away from their desks, in meetings or workgroup discussions. Even Cisco employees who work on site are likely to be mobile within the building.

The Cisco Connected Workplace approach is built on the use of space rather than its allocation in accordance with head count. The design of the offices reflects the way that people work, i.e. collaborative and mobile both within and beyond the building and different types of space have been created that reflect the work modes that an individual employee may cycle through in the course of a working day. These range from quiet spaces for working alone, to areas designed for collaborative and more social ways of working. And critically, employees are supported at all times with access to data and communications through an IP network that provides various forms of connectivity. In Cisco's Building 14 Proof of Concept at headquarters, the result is that a space which under traditional configuration would have accommodated 88 workers, is now used by 200. Every aspect of space usage has been investigated and the approach modified to take account of needs as they develop. As employees have responded to the flexibility, so the workplace has evolved. For example, wireless access points have been installed in greater numbers in areas of the building where people tend to congregate, ensuring that no single access point is overloaded and thereby impairing employees' productivity.

The success of the Connected Workplace solution comes in two forms: reduced costs and improved employee satisfaction and productivity across a range of indicators. The costs of cabling and IT infrastructure have both been halved compared to a traditional office, as has spending on furniture. The savings generated by having more people productively using space more effectively means that rent and construction costs are also cut considerably (by 37 percent and 42 percent respectively). A more cost-effective solution is accompanied by higher levels of employee satisfaction with the new arrangements. Nearly 80 percent of employees say that they prefer the new environment, citing factors such as the greatly increased ease of finding a meeting space. Nearly two-thirds say that they enjoy coming to work more. In short, the connected workplace offers better and more productive use resources at lower cost; a proposition that few organizations can afford to ignore.

One example of an organization that has put this approach to building usage to work is Hillingdon Borough Council in the UK. There, the housing department identified that some 70 percent of its staff could work remotely provided that they had the necessary levels of connectivity eliminating the need to maintain the same level of office space. Staff now have access to the information, services and applications they need whenever, and wherever they are—whether working from home or on the road. The council has saved more than euro 4.3 million in annual office costs.

Another example is the UK's largest telecommunications service providers, British Telecom (BT). It has recognized the significant savings and productivity improvements available from rethinking the space it uses. Today, about 9,500 BT staff are contractually employed to work primarily from home and more than 63,000 others are able to work independently from any location at any time. BT now saves some euro 6.5 million annually on property costs, absenteeism has fallen by 63 percent and staff retention has increased dramatically

CISCO CONNECTED REAL ESTATE AND THE NEED TO DO THINGS DIFFERENTLY

New thinking about the use of the network in the deployment of real estate assets creates a wide range of exciting possibilities for the development of both the business models for the property value chain and the way that owners, operators and the building users will be able to use the buildings they inhabit.

But for these new models to become reality the design of the IT network will require a new place in the elaborate property development process. Though the design and consequences of water, electricity and gas infrastructures (the first three utilities of a building) are included in the very early stages of a property's con-

ception, network and communications needs are rarely given the same early attention. The network needs to become an up-front consideration from the planning stages and drawing board onwards.

Understanding the Building Life Cycle

A building life cycle comprises four phases: conceptualize, design, construct, maintain and operate.

- Conceptualize: The phase in which the building is scoped and financed, conceptualization, consumes about 2 percent of the total costs of the building life cycle and marks the beginning and end of each building life cycle.

- Design: During the design phase, architects and engineers plan the detailed layout, structure, and execution of the building.

- Construct: In the construct phase the building is erected to its design specifications. Together, the design and construct phases account for some 23 percent of the total costs of the building life cycle.

- Maintain and Operate: The maintenance and operation phase represents the time during which the building is used, typically 25 to 30 years in today's fast-moving environment—marked by its economic of functional life. It accounts for 75 percent or more of the total costs of the building life cycle.

With more than three-quarters of the total expense of a building arising during the maintain and operate period, rather than as initial capital expenditure, decisions taken in the design and construct phases can have far-reaching financial and operational effects. Therefore during those phases key stakeholders should carefully consider a building's network; especially because the Connected Real Estate framework will positively affect the functional use and design of the building and thus support the transformation of space and businesses. Decisions made during the early stages can effectively create the infrastructure that reduces ongoing operations costs over the life cycle of the building, and improves the opportunity to create revenue streams in the appropriate markets.

The inclusion of an IP network in the building design process, and its installation as early as possible in the construction process, provides immediate gains for building owners. The single IP network reduces capital costs during the construction process, because infrastructure can be laid more easily (rather than being retrofitted with consequent cost and disruption) and the single open standards cabling infrastructure reduces the requirement for multiple closed proprietary networks and the associated costs of installing them. Secondly, by installing networks early, building owners can extract value from the network over a longer period of time, increasing overall return on investment.

Cisco Connected Real Estate also helps lower operating expenses over the building's life-cycle. An open standards based building infrastructure encourages a centralized (and/or remote) approach to monitoring, maintenance and control of the building environment.

Higher levels of connectivity between building systems provides an array of benefits through access to and sharing of real time data including:

- Optimized remote control, monitoring and reporting of building systems including centralized management of a distributed property portfolio.

- Intelligent heating and lighting and cooling systems that reduce costs through increased energy efficiency.

- Improved staff productivity (maintenance, facilities and security personnel) and enhanced health and safety compliance.

- Improved asset management and tracking together with automated work scheduling, billing and help

desks linked to existing enterprise resource planning (ERP) systems.

The Cisco Connected Real Estate IP framework features embedded technologies that guarantee quality of service and high levels of security and resilience further reducing maintenance and repair costs. Furthermore, all components of the network are built entirely on open standards. Hardware, software, and services are designed using roadmaps that anticipate and support constantly changing business requirements.

Acknowledgements

There are many people that have been involved in the development and deployment of the CCRE program at Cisco, too many to mention here. There are several people however, that must be mentioned as key contributors to the success of this program.

Mark Golan—VP, WW Real Estate and Workplace Resources
Wolfgang Wagener—Manager, Workplace Resources
Andrew Thomson—Business Development Manager
Rick Huijbregts—WPR Program Manager
Oscar Thomas—Marketing Manager
Cori Caldwell—Marketing Manager
Agnieszka Jank—Integrated Marketing Communications Manager
Ray Rapuano—Strategic Account Manager

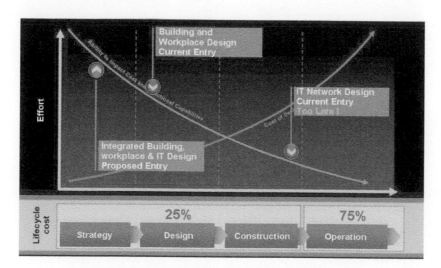

Figure 38-6. Understanding Building Lifecycle Costs

Chapter 39

Building Automation and Control Systems Case Study
Johnson Controls Headquarters, Glendale Wisconsin

Terry Hoffmann

INTRODUCTION

When the Director of Facilities and his staff arrive at the headquarters of Johnson Controls on a typical Monday morning they can be confident of two things. First, they have operational electrical and mechanical systems in place to provide a comfortable, safe and sustainable environment for all resources on the campus. Second, that they have a powerful tool in their Metasys® building automation and control system that provides them with the information they need to optimize energy usage, extend the life of capital equipment and minimize the impact of their facilities on the environment. This confidence does not come standard in buildings, even those designated as smart or green. It is the result of planning, process and execution by a talented and dedicated staff that knows how to apply all of the tools that have been made available to them. They specialize in using them to identify and focus on the things that matter.

PROJECT OVERVIEW

The Johnson Controls headquarters campus sits on a 33-acre site in Glendale, a near-in suburb of Milwaukee, Wisconsin. The six major buildings that make up the complex are the result of:
- Gutting and total renovation of two buildings with a combined 134,000 square feet used for corporate and divisional offices.
- New construction of a 95,000-square-foot divisional headquarters building for the Johnson Controls Power Solutions business unit.
- New construction of a 30,000-square-foot Amenities Building including a cafeteria, meeting rooms and a fitness center.

- Recent retrofit of the corporate headquarters, a 1960s design with architectural significance reflecting organizational heritage.
- Upgrade of a technology center that includes laboratories and test facilities.
- Construction of a four-level parking structure for 400 vehicles with dedicated hookups for a number of "plug and drive" electric vehicles.

Some of the sustainable construction highlights include;
- Locally harvested and manufactured materials account for more than 25 percent of project materials, including drywall, concrete, steel and stone.
- New construction waste recycling at 89% vs. an original 75 percent target.
- Participation by women- and minority-owned contractors at the goal of 20 percent.
- Sustainable technologies and strategies employed: Solar heat capture and generation to reduce the use of electricity and natural gas while reducing greenhouse gas emissions by over a million pounds/year.
- Solar thermal array on the roof of the new building.
- Ground-mounted photovoltaic solar array.
- Laminated solar on roof.

A closed-loop geothermal heat exchanger system using 100 percent groundwater designed to reduce the cost of heating and cooling.
- Well field located beneath the ground mounted solar array.
- Heat pumps (replacing old natural gas boilers) to reduce winter heating costs geothermal field. used for removing condenser heat in summer to reduce chiller operating costs.

Other features:

- More than 13 acres of native prairie vegetation, with a diverse mix of native plants to support local wildlife; wetlands restored; non-native species removed.
- Skylights and increased window space reduce use of energy for lighting.
- Control of lighting and sun blinds.
- "Green" (growing) section of the amenities building roof
- 30,000-gallon cistern to capture rainwater from all new roof surfaces for reuse; reducing potable water consumption for new bathroom fixtures by 77 percent, a savings of 595,000 gallons annually.
- 3-acre parking lot surfaced with permeable pavers, allowing rain and snowmelt to filter through pavers and then gravel base and soil before moving via groundwater to detention ponds. This replaces a blacktop lot that sent runoff contaminated with motor oils and other pollutants directly into sewer system and waterways.
- Underfloor air delivery system for greater efficiency in delivering the macro-environment.
- Personal Environments® modules for delivery of the micro-environment

As a result, the overall campus space has almost doubled, with a reduction of energy costs of 20 percent to date due to savings from geothermal heat pumps, solar generation, solar water heating and more efficient design and construction.

SYSTEMS PHILOSOPHY

In order to meet the needs of the updated corporate goals for energy and greenhouse gas reduction, the Metasys building management system for monitoring and controlling the campus was upgraded from an earlier version of the system and expanded to include numerous interfaces to additional equipment. The system was re-commissioned with two basic design tenants in mind.

First, the system is truly integrated to provide a single-seat interface to all of the mechanical and electrical systems in the facility. This includes not only heating, ventilation and air conditioning, but lighting, electrical monitoring, fire management, access control, emergency generators, and several specialty systems that are not normally monitored (to be detailed later). In part, this is because of the desire for the system to act as a showcase for the technology of the company to display all capabilities to visitors. More than a thousand people per year

take formal tours that include exposure to the building management system as well as the other systems and architectural features that make the facilities unique.

Second, the system greatly expands the number of end devices that are monitored in order to provide a basis for optimization of energy management strategies and the ability to include the maximum set of equipment in maintenance calculations. This includes over 100 meters for electrical systems as well as dozens of others for water and natural gas. This supports the more general accepted principle that you can't manage what you don't measure. Michael Porter has said that for any strategy to be valid the process should be measured, the results monitored and the information used as a basis for management.

In practice, Johnson Controls facilities management and staff have chosen to focus on several key performance indicators (KPIs) related to energy consumption and have implemented the system to provide maximum visibility to them. The most basic of these is electrical usage in Watts per sq. ft. This KPI is used to benchmark the buildings against each other as well as against owner occupied offices around the globe. Figure 1 shows the values in a graphic representation that is not only used by the facilities staff but also is available to all employees at an energy and sustainability kiosk. The display is located in a crossroads area so that the greatest number of people can self-monitor their own success in keeping consumption at a minimum and creating a good-natured competitive spirit among them as members of their respective buildings.

SYSTEM DESIGN

State of the Art

The current situation with computerized building automation systems (BAS) as they have evolved over nearly four decades is their transition from computerized data systems that were pasted on to mechanical systems and controls, to the sophisticated Internet Protocol (IP) controllers and servers of today. The original stated purpose of the BAS for saving energy and increasing productivity is greatly expanded with the ability to touch many other systems and data sources across networks that literally span the globe. This ability to communicate vital systems information to the highest levels of the enterprise is the foundation for managing facilities of any size. The latest BAS promise is best summarized as follows:

Building automation systems use current technology to provide safety for both occupants and assets. They contribute to the productivity of the enterprise by conserving energy and optimizing the efficiency of equipment throughout facilities and the people who are responsible for operating and maintaining them. They provide a foundation for sustainable programs and projects by providing the accurate and secure data required for decision making and verification.

This is a near-perfect match for the headquarters of a *Fortune* 100 manufacturer such as Johnson Controls. Three technologies appear to be having the greatest impact on these matters:

- Harmonized standards
- Wireless technology
- IP-based control and communication

Harmonized standards for hardware, software and communication—including standard protocols—provide the foundation for open, interoperable systems that allow facility management to select best-of-breed hardware and software for use in their buildings. They enable future systems expansion with minimal additional outlays for network infrastructure and provide a strategic foundation for future systems planning. Harmonized standards such as BACnet and Modbus make significant contributions to the Johnson Controls headquarters systems.

Wireless technology provides a number of important benefits, such as reduced infrastructure, maintenance and move/change costs. It also helps eliminate limitations, including distance, location or mobility. Wireless connections to controllers from sensors, as well as wireless connections from controllers to supervisory engines, provided reduced construction costs for many of the mechanical and electrical system applications throughout the Johnson Controls campus. Additionally, the maintenance costs of those systems are lower when the cost of troubleshooting poorly connected or severed wiring is taken into consideration. Even more importantly, the use of wireless technology eliminates the tether between system and operators that limits the use of mobile technologies for important user interface activities such as alarm notification and acknowledgement. Put simply, the wireless networks provide the ultimate freedom to connect.

IP-based control enables so many of the applications that it demands its own place on the list of important tools. Enterprise applications depend upon IP-connected web services for the flow of critical information. Mobile workstations take advantage of the wireless IP infrastructure. Maintenance and operation of critical system elements such as electrical switchgear and chilled water systems are dependent on the IP knowledge of the Information Technology staff. The reliability and availability of IP networks enable the interoperation of systems on a machine-to-machine basis. Finally, the standard IP-based web browsing capability on nearly every computing device provides ubiquitous access to the systems anywhere and at any time.

Application and Data Server, Extended

The application and data servers (ADS) are the components of the control system that manage the collection and presentation of large amounts of trend data, event messages, operator transactions and system configuration data. As site director, the ADS provides secure communication to the network of network automation engines and network interface engines. The user interface (UI) of the ADS provides a

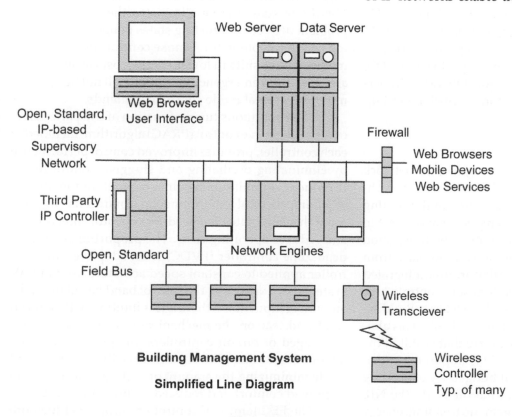

Building Management System

Simplified Line Diagram

Figure 39-1

flexible system navigation, user graphics, comprehensive alarm management, trend analysis, and summary reporting capabilities. Via a standard web browser, users can efficiently manage occupant comfort and energy usage, quickly respond to critical events, and optimize control strategies. The Metasys Advanced Reporting System (MARS) is available in conjunction with the Microsoft® SQL Server™ software components resident on the data server and offers a separate UI on which users can run reports on system configuration and performance.

Network Automation Engines

The network automation engine (NAE) uses web-based network technology to enhance the control system. On this system, the NAE uses the primary communication technology of the building automation industry, BACnet. This allows the user to monitor and supervise HVAC equipment, lighting, security and fire management systems. These engines support a wide variety of applications on the campus, such as alarm and event management, trending, energy management, data exchange and archiving, scheduling and communication at an area level. Users access data via the advanced user interface with a desktop, laptop or other computer using a standard web browser. The NAE allows users to access system data remotely over a Virtual Private Network. The NAE uses information technology (IT) standards and formats including internet protocol (IP), hypertext transfer protocol (HTTP), simple network time protocol (SNTP), simple mail transfer protocol (SMTP), simple network management protocol (SNMP), hypertext markup language (HTML) and extensible markup language (XML).

Network Interface Engines

Like the network automation engines, the network integrations engines (NIE) use web-based network technology as well as communication across an IP building network. Unlike the automation engines that connect to field gear that is open and expandable, the Integration engines talk to legacy equipment already in place from prior projects or existing buildings that are to be integrated into the control network. At the Johnson Controls HQ, this includes earlier versions of the Metasys Building Management System dating back to the 1980s. This ability to connect to operational hardware that is still in the heart of its usable product lifecycle is important in that it leverages the capital and fiscal budgets by expanding the options regarding updates. Like the NAE, the NIE provides alarm and event management, trending, energy management, data exchange and archiving, scheduling

and communication at an area level. Users can access data via the advanced user interface with a desktop, laptop or other computer using a standard web browser. The NIE allows users to access system data remotely over a virtual private network and uses the same standards and formats as NAEs.

Field Equipment Controllers

The field equipment controllers (FEC) are a complete family of BACnet-compatible field controllers and accessories designed with the flexibility to meet a wide range of HVAC control applications. Built on the American Society of Heating, Refrigerating and Air Conditioning Engineers (ASHRAE) standard for building automation system control and communication, these controllers support open communication standards and a wide variety of control options for users.

The FEC family includes the 10-point FEC1600 and the 17-point FEC2600, as well as I/O expandability and VAV application specific controllers, all integrated with the building management system via network engine compatibility. FEC controllers are installed with LCD display for local control and annunciation at the equipment. The FEC family is designed on solid, advanced technology platform. This starts with a finite state control engine that serves as the foundation for all controllers. This finite state engine eliminates the classic cycling problem between conflicting states of operation including heating, cooling, fire/smoke control and ventilation modes. This results in lower energy costs as well as the assurance that emergency operation will not be compromised by normal mode control commands.

The continuous tuning function of the pattern recognition adaptive control (PRAC) algorithm, included in each controller, provides improved control performance by eliminating oscillating and sluggish control loops. It reduces commissioning time as well, automatically adjusting to local environmental conditions. In most cases this eliminates the need for operator intervention and reduces operating costs. The proportional variable dead-zone controller (PVDC) is an adaptive flow controller applied to constant speed actuators used in VAV systems. By expanding the control band based upon the amount of input noise, there is less hunting with reduced wear and tear on the mechanical system components. For staged or on/off controllers, the pulse modulated adaptive controller (PMAC) controls within a given band while minimizing the amount of cycling. This results in improved comfort and reduced maintenance costs.

The FEC family of controllers supports Bluetooth wireless communications for commissioning and troubleshooting. This can eliminate the need to be tethered to a device located out of reach or squeezed into a tight location. The FEC family includes configurable and programmable controllers, expanded capacity through input/output modules and an application specific variable air volume controller. All of the devices in the FEC family communicate using the standard BACnet protocol and integrate seamlessly into the control system. The FECs are tested extensively for reliability and will be mounted in cabinets to meet shock and seismic requirements unique to combat vessels. Additional details on this device are included in the appendix.

Terminal Equipment Controllers

Terminal equipment controllers (TEC) are BACnet® master-slave/token-passing (MS/TP) networked devices that are used on the Johnson Controls campus to provide control of rooftop units (with or without economizers), heat pumps, and single- and multi-stage heating/cooling equipment including the underfloor air boxes used throughout the new facilities. Some of the TECs are installed with occupancy sensing capability built into the device. These devices provide energy savings in high-energy usage light spaces where occupancy is limited. The devices also maximize energy savings by using additional setpoint strategies during occupied times. They feature a building automation system BACnet MS/TP communication capability that enables remote monitoring and programming for efficient space control.

They feature an intuitive user interface with backlit display that makes setup and operation quick and easy. The thermostats also employ a unique, proportional-integral (PI) time-proportioning algorithm that reduces temperature offset associated with differential-based thermostats.

Enterprise Dashboards

In addition to the standard Metasys user interface and associated advanced reporting system, the facilities staff at Johnson Controls also relies on dashboards provided by the Metasys sustainability manager software. These dashboards consolidate, parse and rank information from all of the campus buildings. That information is then presented in ways that help management compare energy use, greenhouse gas emissions and areas where automated fault detection programs have been implemented. This precludes long hours of staff work generating and evaluating spreadsheets or configuring complex databases from scratch. This time can rather be spent on optimizing systems operation and further reducing energy use.

APPLICATIONS THAT KEEP PEOPLE SAFE AND COMFORTABLE IN AN EFFICIENT, SUSTAINABLE ENVIRONMENT

Solar

The Johnson Controls headquarters campus uses three separate solar energy features to reduce energy usage and lower the environmental impact of greenhouse gases. First, the new Power Solutions business unit headquarters includes a 1,330 square feet of solar thermal panels on the roof of the facility. These panels produce a majority of the hot water needs for two large buildings saving over 2,800 therms annually along with the resultant CO_2 impact that results.

A ground-mounted solar photovoltaic array is position on an open field area at the north end of the campus. It measures 31,115 square feet and is capable of producing 250 kVA of electrical power.

A roof-mounted, thin-film, laminated photovoltaic array on the roof of the Corporate North Building measures 14,355 square feet and is capable of producing up to 135 kVA. Together they made up the largest commercial solar photovoltaic installation in the state of Wisconsin when the facility was opened. They provide most of the power for one large building, reducing greenhouse gas emissions by 1.1 million pounds per year versus an equivalent amount of power from the electrical grid.

The BAS manages all electrical usage in the buildings with more than 100 meters tracking usage by end use. In most areas this includes, ventilation, heating, cooling, lighting and three different types of plug loads depending on whether they are fully backed up, standard or shedable loads. They are all trended with information stored in a central database and accessed via computers or hand-held devices. The output of the roof-mounted and field-mounted photovoltaic arrays are similarly trended, stored and accessed.

Geothermal Heat Pumps and Chiller Condenser Water

Beneath the ground-mounted solar array on the site are 272 wells, each 300 feet deep, which connect in a closed loop to a heat exchanger in the mechanical equipment room. They take advantage of the nearly constant subsurface temperatures for cooling the water circulated in the chillers and in conjunction with two 175-ton heat pumps using screw compressors.

The heat pumps reduce winter heating costs by nearly 30 percent. Operating costs are more than 50 percent lower than natural gas boilers when all factors are considered. Using the geothermal field for removing condenser heat reduces chiller operating costs by about 23 percent.

Over the first two summers of operation the geothermal solution has saved more than a million gallons of makeup water and hundreds of pounds of chemicals that would be required for water treatment. The BAS displays these complex systems in graphical format and trends important information for verification of savings and greenhouse gas reduction.

Lighting Control, Skylights and Sun Shades

All new and completely remodeled areas of the Johnson Controls headquarters feature advanced digital lighting systems that communicate over the open and interoperable BACnet protocol. The panels, manufactured by Lutron, are capable of dimming the T5 florescent tube lighting to match the needs of the occupants as programmed and reset by lighting sensors in the office spaces and hallways.

Automated sunshades work in concert with the overhead lighting to increase the amount of outdoor light that is available while reducing sun glare and unwanted solar radiation on warm days. The shades connect to the building weather station and use an extensive algorithm that takes into account the direction that the glazing faces, the time of day, date of the year, latitude and longitude of the building and the solar intensity. Shades on lower levels are lowered after sunset to reduce the "fishbowl effect" that occurs when buildings are lit inside but darkness masks people on the outside who may be watching.

The Corporate North and Amenities Buildings, which have two large conference rooms, workout facilities and a cafeteria that serves the campus, take advantage of skylights for maximum use of natural lighting as well as minimum electrical usage. They use a system of clouds that hang below them to reflect the light into the occupied space while eliminating unwanted glare from computer screens and desktops.

A 400-car parking garage feature hookups for hybrid plug-in electric vehicles and illumination provided by LED fixtures. LED lighting also is used for some outdoor security illumination and much of the architectural lighting on campus that highlights the award winning architecture.

The BMS provides complete control and override capability for all lighting and shades. Campus lighting averages less than .5 watts of energy per square ft. This compares to a national average of three times that amount.

Macro-micro Environmental HVAC Control

New and completely remodeled areas of the buildings feature underfloor delivery of conditioned air that uses very low static pressures for operation when com-pared to ceiling mounted VAV systems. The system uses a philosophy of "nose to toes" comfort that results from allowing air to plume from the floor as opposed to being pushed down from above. Boxes in the raised floor operate in a similar fashion to traditional VAV boxes but without the need for high pressures and velocity. This also leaves the ceiling space open for natural lighting. By cutting the delivery pressure in half, the fan energy can be cut by almost three quarters.

In the hallways, other open spaces and closed offices, the underfloor system and macro control of the space is sufficient for comfort, but in open office areas another element is added. Each desktop has a Johnson Controls Personal Environments® module, which places the control of critical environmental conditions at the fingertips of each individual. An easy-to-use desktop control unit gives each person the flexibility to adjust temperature, lighting, air flow and acoustic characteristics as often as necessary to maintain personal comfort levels, and an occupancy sensor helps reduce energy.

Other Features

The BAS is used in many novel ways to supplement other stand-alone systems to provide an extra layer of control and protection. For example, all of the defibrillator cabinets on campus are alarmed back to the system which sends messages to qualified first responders whenever a cabinet door is opened.

The unique design of the headquarters building places it as an island in the center of an ornamental pond. This means that it is important to measure the pond's water level in case the drainage is compromised and the level is close to finished floor. The system also uses wind speed indication to adjust the level of the fountain sprays for the safety of the people on walkways.

BAS Technologies of the Future

As progressive as the Johnson Controls headquarters campus is, with cutting-edge technologies, it is clear it's just the beginning. Many technology leaders in the BAS industry believe that the next big step in deriving benefits from installed systems will be through the use of new enterprise applications for functions such as enterprise energy optimization, asset allocation and tracking, resource planning, sustainability validation, dashboards, and smart grid applications.

In order to provide the greatest benefit to end users, these applications will require new network infrastructures that take advantage of concepts such as software as a service, a systems-oriented architecture, multicore and hybrid servers, virtualization and cloud computing.

Additional systems capabilities and applications that are under consideration include:

- **Semantic technologies** that will allow systems to take commands from operators in their natural and spoken language, providing feedback to acknowledge the intended operation before proceeding.

- **Augmented reality**, which may seem more useful in a video gaming environment, but will provide innovative applications for operators who need to envision the results of an intended command.

- **Context-aware control to** allow the controller itself to judge the surroundings and conditions before proceeding with a particular action, aided by state-based control. For example, consider two very different control actions for an air handling unit—one on a Sunday April evening in San Diego and the other on a Tuesday August afternoon in the Dubai.

- **Ubiquitous access, or** the ability of any device, on virtually any global network, to authenticate and operate with an appropriate level of security and safety dependent on the type of device in the hands of the requesting party. Refrigeration or HVAC experts in York, Pennsylvania, for example, could be patched in to view and correct conditions in a building anywhere in the world.

- **User experience**, which expands the concept of system usability beyond ease of use and the availability of context-sensitive help. Implemented correctly, it means that system users will actually delight in the experience of interacting with the BAS. Satisfaction of operators with their tasks could be a key to retention and a better return on the investment made in training them.

- **Mashups,** using information from different systems and devices provided by programmers to demonstrate the real value of systems.' Thus, if a correlation is determined, the historical control patterns of a chiller in Genoa, Italy, might be compared with data from the National Weather Service and information on the current atmospheric conditions to predict operation of an HVAC system on a roof in Las Colinas.

- **Enterprise social software,** similar to YouTube and Facebook might be of assistance to someone trying to locate maintainers or other experts to quickly evaluate the degree of criticality for a mechanical equipment problem.

At the time of this writing, the smart grid is undoubtedly the most talked about sustainability and energy efficiency issue in the eyes of facilities professionals. Two way communications between smart, automated buildings and the electrical power grid will provide for a much greater degree of customer interoperability and energy savings opportunities as this delivery network develops over the next decade. Pilot projects in place now such as Gridwise in the Pacific Northwest and standards being developed for grid communication including Open ADR point to a very bright future for distribution, generation and consumption of electricity.

CONCLUSION

While all of the systems integration and software capabilities implemented at the headquarters of Johnson Controls contribute to successful outcomes today, the systems architecture that has been put in place will support new features and functions as they become available in the future. The implementation of a scalable, flexible and mobile platform makes this possible. The convergence of the latest information technology, the latest mechanical and electrical systems and the ability to visualize all information on a single screen makes working in this building more comfortable, safe and efficient. It is truly a building efficiency showcase.

ABOUT JOHNSON CONTROLS

Johnson Controls is a global diversified technology and industrial leader serving customers in more than 150 countries. The company's 142,000 employees create quality products, services and solutions to optimize energy and operational efficiencies of buildings; lead-acid automotive batteries and advanced batteries for hybrid and electric vehicles; and interior systems for automobiles. Johnson Controls' commitment to sustainability dates back to its roots in 1885, with the invention of the first electric room thermostat. Through its growth strategies and by increasing market share, Johnson Controls is committed to delivering value to shareholders and making its customers successful.

Chapter 40

Bringing Building Automation Systems Under Control

James Lee

EXECUTIVE SUMMARY

Enterprise-wide thinking has been applied to many aspects of business: supply-chain management, manufacturing, inventory control, quality control, human resources, etc. Now it's time to apply this model to enterprise-wide facilities manage=ment. Instead of thinking of building automation as merely a series of controls, it is time to think of it as part of the overall IT infrastructure.

Significant structural problems exist today in the operating management of institutional real estate. These issues result in overspending on building maintenance and energy, undercutting asset profitability. Most real estate owners do not realize the economic impact of these inherent structural problems.

In general, energy costs for buildings are one of the largest variable components of the cost structure and are becoming a larger component as energy prices increase. Over 80% of buildings are never commissioned and the potential energy efficiencies never fully understood. Building management is sometimes assessed based on energy utilization, however tenant comfort is often achieved without a real understanding of energy cost impact.

Additionally, the management and operation of most buildings (even newer and highly sophisticated "smart buildings") is performed by facility personnel or is outsourced completely. While these personnel may be knowledgeable about operating basic mechanical equipment and making simple repairs, they often lack the training and the capacity to actively diagnose the performance of a building and to understand its implications for profitability. This lack of understanding of performance also results in reactive rather than proactive repair and maintenance operations, ultimately a more costly approach for the building owner.

Building owners are beginning to understand the inherent inefficiencies that exist in the operations of their real estate assets. They are proactively looking for ways to improve their return profile. Cimetrics provides independent data analysis and portfolio-wide consolidated reporting to the facilities and energy departments of building owners. These reports analyze and improve energy efficiency and reduce energy, operations, maintenance and regulatory costs. Cimetrics links into a facility's mechanical equipment (Heating, Ventilation, Air Conditioning, Lighting, etc.) through the building's automation system and utility meters, acquiring an ongoing flow of operational data. In the past, these raw data have been discarded by facility departments due to their volume and complexity. Now, however, Cimetrics collects this information portfolio-wide, and transmits it to a centralized database where Cimetrics' engineers use a set of proprietary algorithms to analyze and mine it for value.

The results are a wealth of ongoing unbiased recommendations and management information that enable firms to reduce energy and operational costs, identify potential equipment problems in order to avoid downtime, and enjoy the benefits of a continuously commissioned facility. This process of data acquisition, analysis, and reporting is called Infometrics. The Infometrics service provides a means to understand and manage building operations and their implications for cost structure and economic returns. By requiring little capital up front, Infometrics can create an immediate and lasting reduction in the operating cost structure of a building.

Infometrics has numerous potential benefits to building owners, including the following:

- Reduced energy consumption and energy cost
- Prioritization of equipment maintenance
- Reduced downtime caused by mechanical equipment failure
- Improved facility operations
- Ongoing commissioning of mechanical systems and control systems
- Reduced risk of indoor air quality problems
- Identification of profitable mechanical retrofit opportunities

- Improved occupant comfort
- Knowledge of facility energy consumption patterns and trends

Building controls companies, equipment and systems manufacturers, energy providers, utilities and design engineers will face increasing pressure to improve performance and reduce costs. These pressures drive the development, adoption and use of Infometrics.

CIMETRICS INC. AND INFOMETRICS

Cimetrics Inc. provides high-value energy and facilities management services to owners and occupants of commercial, institutional and industrial buildings. Through its Infometrics suite of products and services, the company collects real-time data from a customer's building automation systems (i.e., HVAC, lighting, fire control, etc.), integrates information from multiple facilities, applies proprietary algorithms, and generates regular and highly detailed reports for the building owner's facilities and energy departments. These reports identify opportunities to reduce energy, maintenance, operational and regulatory costs; uncover potential equipment problems; point to profitable retrofit projects; improve occupant comfort and enhance facility operations and uptime.

Cimetrics links into a facility's mechanical equipment through the building's automation system and utility meters, acquiring an ongoing flow of operational data. In the past, these raw data have been discarded by building operations staff due to their volume and complexity. Now, however, Cimetrics collects this information portfolio–wide (across multiple buildings) and transmits it to a centralized database where Cimetrics' engineers use a set of proprietary algorithms to analyze and mine it for value. The results are a wealth of ongoing unbiased recommendations and management information that enables firms to reduce energy and operational costs, identify potential equipment problems in order to avoid downtime, and enjoy the benefits of a continuously commissioned facility. This process of data acquisition, analysis, and reporting is called Infometrics.

The ability to create "smart buildings" is taking shape rapidly due to the proliferation of new technologies and the internet revolution, and Cimetrics has developed the technology and services to make the "smart building" concept a reality for the property owner/manager. To reduce its clients' energy costs and improve productivity, Cimetrics implements its Infometrics

solutions by: (i) connecting to clients' building automation systems; (ii) analyzing the data produced (through proprietary algorithms and other software technology); and (iii) assessing system performance in order to better manage utility and facilities costs. Infometrics allows building owners to integrate building automation systems and energy equipment at the building level with information systems at the corporate level.

Infometrics provides commercial, institutional and industrial building owners the ability to improve substantially the operating control, costs and efficiency of their buildings through greater communication and efficiency of building systems. Cimetrics was instrumental in developing and implementing the Building Automation and Control network ("BACnet®"), the dominant open standard (ISO 16464-5) in building automation communications world-wide. Cimetrics is the world leader in the development of BACnet® communication software, network analyzers and routers which, along with their analysis and recommendations, enable the Infometrics solution.

Infometrics offers the only complete remote monitoring solution in the marketplace, leveraging Cimetrics' depth of knowledge in system connectivity, proprietary analysis algorithms, engineering and high-touch consulting. A dedicated analyst is assigned to each client, providing unparalleled access and responsive service on a range of issues from reviewing periodic reports to maintaining communications and problem solving.

Cimetrics' professional team of engineers, project managers and analysts offers expertise in energy management and building operations, and in all aspects of optimizing facilities for energy and operational efficiency, maximizing clients' potential for significant savings. Cimetrics provides services ranging from energy cost savings analysis to long-term monitoring, analysis and reporting of building data. Cimetrics provides a complete, unbiased solution for a facility's needs by working solely for the building owner.

THE INFOMETRICS PROCESS

Collecting Data from Building Systems

The Infometrics system links into a facility's mechanical equipment through the building's automation system and utility meters, acquiring an ongoing flow of operational data. In the past, these raw data have been discarded by facility departments due to their volume and complexity.

Now, however, Cimetrics collects this information

Collect   **Analyze**   **Report**   **Save**

Figure 40-1

facility-wide from multiple disparate systems through the BACnet® protocol, and transmits it securely via the internet to their analysts. This scalable data processing technology is capable of collecting and analyzing information from thousands of buildings worldwide over long periods of time.

Data Analysis by Software and Engineers

After connecting to a building automation system, data relevant to the analysis are transmitted over a secure internet connection to the Infometrics data center. Cimetrics energy, electrical and mechanical engineers use proven algorithms and software to analyze building efficiency. Analysis algorithms use static data (equipment specs, system topology) and dynamic data (weather information and operational data collected on each piece of HVAC equipment as well as the entire building's mechanical systems).

Infometrics algorithms have been designed based on standard industry techniques and academic research. The analysis team has 100+ years of experience in energy engineering, controls, communications, and software development. Infometrics provides an independent measurement of building systems' efficiencies. The depth of the team's expertise enables Cimetrics to mine value from data for building owners. The Infometrics analytical approach, examining both static and dynamic data, focuses exclusively on adding value to customers' portfolios.

Complete Report Delivery

Periodically, Cimetrics analysts create a report on facility performance based on data which have been collected and processed by the Infometrics system. This report includes management information on energy consumption and mechanical system performance as well as specific prioritized recommendations. Target energy consumption and operational characteristics are identified and variances from predicted results are analyzed, problems identified and appropriate measures for remediation recommended to the owner.

The results are a wealth of ongoing unbiased recommendations and management information that enables firms to reduce energy and operational costs, identify potential equipment problems in order to avoid downtime, and enjoy the benefits of a continuously commissioned facility. Infometrics engineers work with facility staff and owners' contractors to deliver maximum value to the building owner.

Implementation of Recommendations and Savings

Infometrics' prioritized recommendations uncover hidden maintenance issues, providing a road map to immediate savings. Building staff can now act effectively to create value by optimizing equipment performance, reducing costly downtime and improving comfort.

Figure 40-2

Building Data

The value that Infometrics delivers is primarily derived from data collected from sensors and actuators that are connected to building control systems and building mechanical equipment. When combined with system and equipment set points, the data can tell a great deal about how well the building control system is performing and where the problems are, including issues that can't be detected through simple equipment observation.

There are vast amounts of data available in a large building—if you were to read just one sensor every

15 minutes, you would have 35,000 data samples per year from that one sensor alone. Some buildings have thousands of sensors and actuators. Special tools and expertise are needed to collect, manage and analyze all of this information. It is not surprising that most facility maintenance departments do very little effective analysis of building control system data.

Cimetrics has invested many person-years of effort into understanding how to extract valuable information from building data. Infometrics analysts review each facility's points list and building system documentation to determine what data should be collected.

Infometrics Relies on Data Collected from Customers' Building Control Systems

In order to maximize the potential value of the building systems data collected, they must be gathered at the right time and at the right frequency. Cimetrics has developed a special data collection device, called the Infometrics Cache, which gives our analysts excellent control over how information is collected. The Infometrics Cache is connected to the building control system at each customer facility and transmits critical information to Cimetrics via a secure, firewall-friendly internet connection. In most cases, the collection of data consumes a small fraction of the available network bandwidth.

Cimetrics engineers have considerable experience in connecting Infometrics-enabling equipment to different building control systems. BACnet® is the preferred communication protocol—if the control system uses BACnet® as its native network protocol, then the connection of the Infometrics Cache is very simple. If the building control system does not use BACnet®, or if BACnet® is not the primary protocol in use, then Cimetrics may need to arrange for the installation of a hardware or software gateway that will translate the necessary data into a format supported by our device.

The Infometrics Cache temporarily stores the data collected from the facility's building control system, then periodically transmits the information to the Cimetrics Data Center. In most cases, the best way to send the data to Cimetrics is to use an existing internet connection at the customer's facility. The Infometrics Cache transmits information using a secure industry-standard network protocol that is compatible with firewalls.

The Infometrics system has been designed to ensure that the security of customers' systems and data are maintained. Cimetrics I.T. experts are prepared to work with security-conscious customers to ensure that the Infometrics system meets their particular requirements.

Turning Building Data into Actionable Information

Facility managers need concise and accurate information to help them make decisions about how to maximize the performance of building systems. Infometrics was developed specifically to address this need.

Cimetrics delivers actionable information to Infometrics customers using a combination of state-of-the-art technology and analysis by experienced engineers. Infometrics reports include specific prioritized recommendations, most of which can be implemented at low cost, along with estimates of the resulting annual savings.

Infometrics analysis and recommendations are based on industry research in the areas of fault detection, fault diagnosis, building system optimization and commissioning. Cimetrics has used this research to develop algorithms and software tools that allow Infometrics staff to quickly and efficiently analyze the data that are continuously collected from customers' building systems.

An Infometrics analyst is assigned to every Infometrics customer. The analyst is responsible for creating a data collection strategy, analyzing the data, producing the Infometrics reports, reviewing recommendations with the customer, and being available to each customer for consultation when questions arise. Each of the Infometrics analysts has years of experience as an energy, electrical or mechanical engineer. Cimetrics analysts can also consult with staff engineers who are experts on building control systems, HVAC equipment and data analysis.

Cimetrics believes that the Infometrics approach—skilled analysts using state-of-the-art technology to analyze building data—is the best way to deliver actionable information to customers that want to maximize the performance of their building systems.

The following is a brief description of the Infometrics project development process:

Connectivity

The Infometrics system links into a facility's mechanical equipment (heating, ventilation, air conditioning, etc.) through the building's automation system and utility meters, acquiring an ongoing flow of operational data.

Cimetrics collects this information facility-wide from multiple disparate systems through the BACnet® protocol, and transmits it securely via the internet to the Data Center for analysis. This scalable data processing technology is capable of collecting and analyzing information from thousands of buildings worldwide over long periods of time.

Analysis

Cimetrics' energy, electrical and mechanical engineers use proprietary algorithms and software to analyze building efficiency. Infometrics algorithms have been designed based on standard industry techniques and academic research. The Cimetrics team has 100+ years of experience in energy engineering, controls, communications, and software development. Infometrics provides an independent measurement of building system efficiencies. The depth of Cimetrics' expertise enables Infometrics analysts to mine value from data for building owners. Infometrics' analytical approach, examining both static and dynamic data, focuses exclusively on adding value to customers' portfolios.

Reporting

Periodically, Infometrics analysts create a report on facility performance based on data which have been collected and processed by the Infometrics system. This report includes management information on energy consumption and mechanical system performance as well as specific recommendations. Target energy consumption and operational characteristics are identified and variances from predicted results are analyzed, problems identified and appropriate measures for remediation recommended to the owner.

The reports contain a wealth of ongoing unbiased recommendations and management information that enables firms to reduce energy and operational costs, identify potential equipment problems in order to avoid downtime, and enjoy the benefits of a continuously commissioned facility. Cimetrics' engineers work with facility staff and owners' contractors to deliver maximum value to the building owner.

Infometrics Project Timeline

Step 1: Facility Assessment

A facility assessment is intended to investigate the potential for Infometrics to provide energy and operational cost savings, as well as improved performance of facility systems. The assessment typically includes an analysis of two years of the facility's utility bills and a points/equipment list collected from the facility's building automation system. Operating personnel are interviewed to obtain general information about the facility (square footage, operating schedules, utility metering systems, energy conservation strategies already implemented, known operational problems, planned system changes, etc.). Based on the results of the facility assessment, Cimetrics develops a proposal for Infometrics services.

Step 2: Infometrics Connectivity

Once Cimetrics and the facility owner have reached an agreement on the scope of Infometrics services to be provided, Cimetrics establishes connectivity with the building systems that are to be monitored. A special Cimetrics device, the Infometrics Cache, connects to the facility's building systems via the BACnet® protocol, and internet connectivity is established. Every building automation point that is needed for Infometrics analysis is entered into the Infometrics database, and each point is assigned a standard name based on its function within the system. Often an Infometrics customer is a building owner with several facilities and varied facility control systems spread over a wide geographic area. Cimetrics' Infometrics solution enables connection and monitoring of all of these locations to produce the raw data that will ultimately be analyzed to produce recommendations.

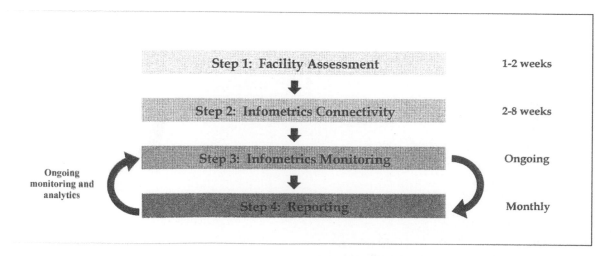

Figure 40-3. Infometrics Project Timeline

For example, at American University, Infometrics currently monitors three buildings, totaling 242,000 SF (see case study, page 100).

Step 3: Infometrics Monitoring

Cimetrics monitors the facility owner's electrical and mechanical systems 24 hours a day, and stores the data collected on average every 15 minutes from those systems in a database system located at Cimetrics' secure data center. The data are then trended and studied by Infometrics analysts using Cimetrics' proprietary analysis tools.

For example, in a technique called MicroTrend, a short-duration, high-frequency query is done on all pertinent data regarding a specific system component. For instance, a variable air volume box (computer controller damper) may have up to 16 variables that represent its real time performance. Characteristics that are measured may include: control loop constants, actuator travel, and/or energy consumption per unit of handled air. This technique enables analysts to uncover rapid fluctuations that would not otherwise be apparent.

Step 4: Reporting

Cimetrics provides clients with periodic reports which detail the facilities' targeted energy consumption and operational characteristics in a very usable format for clients, their energy departments, building managers,

maintenance personnel, etc. Variances from predicted results are analyzed. Problems are identified, prioritized and reported so that appropriate measures can be taken. An Infometrics Analyst reviews each report with the client during a scheduled conference call.

Technology Overview

The Infometrics system links into a facility's mechanical equipment through the building's automation system and to its utility meters using the firewall-friendly Infometrics Cache device, acquiring a secure and ongoing flow of operational data.

A simplified Infometrics system architecture is shown below. Major hardware components include a data collection device that is connected to each customer's building automation system, the database system, and the network operations center. A brief description of the components follows:

- **Data Collection Device ("Infometrics Cache").** Acquires data from the customer's building automation system, temporarily stores the data, and periodically transmits the data to the Infometrics database system over the internet using a secure, "firewall friendly" network protocol.

- **The Database System.** Stores configuration information, raw data collected from customer systems,

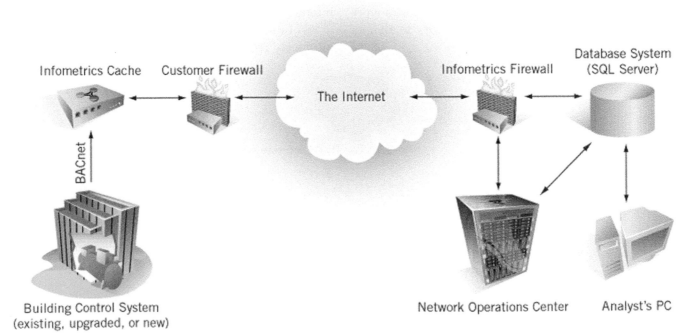

Figure 40-4. Infometrics System Architecture

processed data, and issue tracking data. The system is designed to easily scale out as needed.

- **The Network Operations Center ("NOC").** Monitors the performance of all components of the Infometrics system and critical devices in the customer's building automation system.

Cimetrics has developed a considerable amount of software to support Infometrics, and has also made use of several products that were developed for our BACnet® products business. Infometrics analysts use both commercial packaged software and proprietary software tools.

The Infometrics system collects data from a customer's building automation system by directly connecting to the system's network. If the network does not use the industry-standard BACnet® protocol, then Cimetrics and the customer arrange for the installation and configuration of a BACnet® gateway. BACnet® gateways are available for most popular building automation systems developed within the last ten to fifteen years. Cimetrics has developed considerable expertise on how to interface to popular building automation systems.

Point naming is of considerable importance in enabling the Infometrics service, and Cimetrics encourages the adoption of standard point names. By using a standardized point naming convention, several automated configuration tools can be used which reduce configuration costs.

Infometrics is fundamentally different from traditional commissioning services in that its value is derived primarily from the analysis of large amounts of building operation data collected over a long duration. The huge volume of data collected from every building each month (1,000 points per building average * data sampling frequencies of 15 minutes or less * 30 or so days per month = over 2 million data points per building per month; multiplied by a growing number of total buildings monitored = hundreds of millions of data points collected per month) must be converted into a high-value decision support tool for the building owner (i.e., a periodic report). Over the past five years, Cimetrics has developed scalable data processing technology capable of collecting and analyzing data from thousands of buildings. The Infometrics system takes advantage of data accessible through the building control system installed in each facility. Building automation systems are connected and points of data relevant to the analysis process are transmitted over the internet to the Cimetrics data center where they are processed using Infometrics algorithms. Cimetrics energy and mechanical engineers

work with facility staff and contractors to deliver value to the building owner.

Because of its deep industry experience, Cimetrics is able to provide a complete Infometrics solution from surveying buildings to the full deployment of data connectivity, analysis and service delivery. Because the company helped most manufacturers develop and test their BACnet® products, we have considerable knowledge of the communication capabilities of their products. We also leverage our years of work designing/installing routers and working as the liaison between building automation and IT to transmit data reliably and securely across the internet.

Cimetrics' Infometrics program provides powerful algorithm-based analysis of continuously collected building automation data. Valuable periodic reports enable owners to optimize efficiency and comfort, lower maintenance costs, and effectively manage facility staff and contractors across their portfolio of properties.

CUSTOMER BENEFITS

The benefits of ongoing commissioning have been demonstrated and are becoming better known. Several independent assessments have yielded estimates of cost

Table 40-1

Commissioning and Optimization Cost Savings

Study/Agency	Finding
Federal Energy Management Program (FEMPstudy)	20% on average (based on 130 facilities [Texas A&M])[1]
TIAX Report for US Department of Energy (DOE)	5-20% (guidelines)[2]
California Commissioning Market Characterization Study (CCMCS)	15% (existing facilities) 9% (new construction)[3]
National Institute of Standards and Technology (NIST)	US$0.16 per square foot (energy alone)[4]

[1]Liu, Minsheng, David Claridge and Dan Turner. "Continuous Commissioning Guidebook," October, 2002, page v.
[2]TIAX report for the DOE, "Energy Impact of Commercial Building Controls and Performance Diagnostics: Market Characterization, Energy Impact of Building Faults and Energy Savings Potential," November, 2005, page 9-137.
[3]Haasl, Tudi and Rafael Friedmann. "California Commissioning Market Characterization Study," Proceedings of the 9th National Conference on Building Commissioning, Cherry Hill, NJ, May 9-11, 2001.
[4]Chapman, Robert. "The Benefits and Costs of Research: A Case Study of Cybernetic Building Systems," NISTIR 6303, March 1999, page 80.

savings due to ongoing commissioning and optimization (see table below).

Infometrics' success has been driven by four main factors:

1. The growth of internet and broadband wide-area networks, which provide the communications infrastructure necessary for real-time remote monitoring and control;

2. Increasing adoption of BACnet® and other open communication standards for building automation and control, which reduce the complexity and cost of advanced building management systems;

3. The emergence of an enterprise-wide management paradigm; and

4. Rising energy prices, creating increased interest in energy efficiency and cost savings.

All of these factors have come into place very quickly over the past five years:

*   *The Growth of Internet and Broadband Wide Area Networks*: Until recently, communication between buildings typically required the use of telephone lines, but now low cost wide area network communications systems are available. This enables the cost-effective movement of large amounts of information across multiple locations.

*   *The Development of Open Systems*: Just as TCP/IP and other standard protocols were crucial to the development of the internet, open systems are critical to the development and deployment of new technologies in the building industries. The building industry has now adopted BACnet® as the open standard of choice for data communication. BACnet® makes the integration of building systems for various manufacturers and across buildings significantly more straightforward.

*   *The Emergence of an Enterprise-wide Management Paradigm*: Major corporations have embraced enterprise-wide management systems in supply chain management, human resources, finance, and manufacturing. Enterprise-wide management systems have now been expanded to building management, enabling building owners and managers to view their energy operations, maintenance and regulatory issues as part of the whole picture and

their building automation system as part of the IT infrastructure.

*   *The Recent Rise in Energy Prices*: With natural gas prices quadrupling since 2000 and electricity prices rising by as much as twenty percent, building owners are faced with a significant increase in their largest operating cost: energy. The energy problem has become a hot button issue with real estate owners.

The total savings afforded to Infometrics customers are comprised of multiple components, described below:

Summary of Infometrics Benefits

—   Reduced energy consumption and energy cost
—   Prioritization of equipment maintenance
—   Reduced downtime caused by mechanical equipment failure
—   Improved facility operations
—   Ongoing commissioning of mechanical systems and control systems
—   Reduced risk of indoor air quality problems
—   Identification of profitable mechanical retrofit opportunities
—   Improved occupant comfort
—   Knowledge of facility energy consumption patterns and trends
—   Integration with information technology systems

*   *Reduced energy consumption and energy cost.*—Access to and analysis of energy usage data delivers cost savings resulting from better control system performance, improved energy load management, and smarter energy purchasing.

*   *Prioritization of equipment maintenance.*—The information provided in Infometrics reports allows maintenance managers to prioritize maintenance activities, more effectively utilizing both in-house staff and outside service contractors.

*   *Reduced downtime caused by mechanical equipment failure.*—The use of Infometrics allows systems and equipment to operate under near-optimal conditions for extended periods of time. In addition, equipment and component malfunctions are diagnosed and remedied before catastrophic failure occurs. As a result, equipment life is extended, fewer replacements are required, and replacement costs decline. Furthermore, better diagnostic information enables support staff to more quickly and effectively repair equipment and components.

- *Improved facility operations.*—Prioritized recommendations allow facility managers to develop proactive maintenance plans. Catastrophic downtime is avoided with ongoing equipment assessment and proper maintenance, and life cycle costs are reduced.

- *Ongoing commissioning of mechanical systems and control systems.*—Systems operate at near-peak efficiency with proper maintenance, minimizing energy waste and extending equipment life. Purchasing and upgrades can be planned well before equipment failures are likely to occur.

- *Reduced risk of indoor air quality problems.*—HVAC system maintenance ensures proper air flow and the correct ratio of outdoor to recirculated air. Occupant health is maintained with maximized air quality and minimized temperature variation.

- *Identification of profitable mechanical retrofit opportunities.*—Equipment performance is continually assessed, highlighting those components which are operating at suboptimal levels and predicting cost savings to be gained by replacing them. Building owners are able to assess ROI for informed decision making on retrofits and replacements.

- *Improved occupant comfort.* Infometrics reports enable improved occupant comfort from enhanced operating performance of HVAC systems. Infometrics gives building operations staff the information needed to provide a more consistent environment which has a significant impact on worker productivity and tenant loyalty.

- *Knowledge of facility energy consumption patterns and trends.* Infometrics reports may enable building owners to aggregate energy buying and predict needs. Energy use trends are shown which can facilitate predictive buying and maintenance opportunities.

- *Integration with information technology systems.*—Infometrics enterprise-wide facility data can easily and cost-effectively integrate into existing information technology systems, providing more centralized control for operational, purchasing, and financial management.

With a low installation cost, Infometrics has achieved a track record of delivering immediate and lasting reductions in a building's operating cost structure. The Infometrics service also gives the customer necessary information for regulatory compliance, occupant comfort/safety and mission-critical monitoring. A sample of actual potential annual savings identified to date at current customer sites is provided in the table below.

Note that these figures only consist of measured energy cost savings and exclude maintenance, operational and regulatory benefits.

Short-term, Infometrics customers have identifiable and tangible savings in energy consumption, repair & maintenance and labor resources. Additionally, the Infometrics service gives the customer necessary information for regulatory compliance matters, occupant comfort/safety and mission-critical monitoring.

In the longer term, building owners and managers as well as utilities and energy providers realize intangible cumulative benefits critical to organizational effectiveness and key to taking advantage of energy deregulation.

CASE STUDY: REMOTE INTELLIGENCE CAPABILITIES HELP AMERICAN UNIVERSITY REDUCE COSTS, IMPROVE EFFICIENCY

Washington, D.C.—Universities large and small are under intense pressure to improve their educational environments and reduce costs to better accommodate faculty and students. American University faces unique challenges for its 84-acre campus in prestigious north-

Table 40-2

Sample Potential Annual Savings Identified To Date

Customer	Buildings Monitored	Points Monitored	Est. Per Point Savings	Annual Savings Identified
Customer 1	2	8,500	$28.50	$242,000
Customer 2	3	3,000	$22.00	$66,400
Customer 3	3	1,300	$40.00	$52,000
Customer 4	1	10,000	$72.90	$729,000

west Washington, D.C., as it transforms itself into "an academically distinctive, intensely engaged and student-centered community," one that provides the ideal balance between financial responsibility and educational priorities. Like most institutions, the university must maximize efficiency without compromising effectiveness.

Client Objectives

The university formulated a 15-point strategic plan that would transform the institution. The plan identified the reduction of costs and increased operational efficiency over three years as keys to the plan's success. In searching for ways to achieve these objectives, the remote analysis and optimization technology that powers Infometrics caught the attention of the university's physical plant management. Infometrics is a comprehensive, ongoing process, performed by industry specialists offsite, that helps institutions resolve operating problems, improve comfort, optimize energy use and identify retrofits for existing buildings and central plant facilities.

Solution Overview

A remote analysis and optimization program was created for three of the university's buildings, comprising three components: facility data acquisition, remote expert analysis and reporting. The aim was to give building operations personnel unbiased recommendations and management information so they could reduce costs and enjoy the many benefits of continuously commissioned buildings.

A site survey was conducted with a needs analysis and assessment of all facilities, including gathering relevant, existing performance data and site histories. Cimetrics' Infometrics team also ensured that the university's existing building automation system had the tools and capabilities to facilitate the collection and transmission of large amounts of real-time data.

Non-intrusive, secure BACnet® connectivity was established between the existing building automation system and the remote database to mine and transmit continuous, real time facility data. The Infometrics team installed and configured a firewall-friendly BACnet® routing device as the communications interface for 3,000 points.

The analysis and optimization program provides timely report recommendations designed to assist systems engineers with fault detection, troubleshooting and problem solving while prioritizing maintenance issues and reducing downtime.

With the intelligence provided in the analytical reports, the university is given unbiased energy, maintenance and operational recommendations that offer opportunities to reduce costs and optimize equipment for reliable operation.

Client Results

The university does not have the metering to verify building-by-building savings, but for the three monitored buildings, resolutions to the types of problems found could lead to annual savings in the range of $125,000 in energy alone.

Physical plant operations had an initial increase in the number of repair work orders in the monitored buildings related to the problems identified through the monitoring process. Most of the problems identified had the potential for, and some were actually having, a direct impact on occupant comfort. The increased ability to find and fix these problems before they resulted in an occupant comfort call is in line with the physical plant's strategic direction and is leading to fewer occupant comfort calls.

The remote analysis uncovers faults and produces value for the university—through the eyes of expert, unbiased professionals—that cannot otherwise be reasonably detected or uncovered from a one-shot survey of the buildings or addressed with an off-the-shelf software product.

"We're so busy handling day-to-day symptoms that we don't have time to dig deeper into the root causes of the problems. The remote analysis and optimization service gives us the big picture of how our facilities behave and what we should be doing to address the larger issues."

—Willy Suter, Director of Physican Plant Operations

Chapter 41

Ford Compressor Management System Case Study

Bill Allemon
Rick Avery
Sam Prud'homme

INTRODUCTION

Managing the production, distribution and use of compressed air is a frequently misunderstood process and one of the most expensive "products" created at manufacturing facilities. So much so, that compressed air is often referred to as the "Fourth Utility" because of its cost and wide spread use throughout industry.

Compressing air into a usable form can be an energy inefficient process, with up to 90% of the energy consumed by an electric air compressor lost as waste heat, due to mechanical friction and electric motor losses[1]. Some estimate that approximately seven times more energy is required to mechanically compress air, remove moisture, distribute it to the point of use, and convert it back into mechanical energy, than to directly use an electric motor to perform the same task. In response, some industries are converting from compressed air driven tools to direct-drive electric tools when the applicable uses are equivalent. However, there will always remain a market for compressed air driven tools, due to their inherent safety and convenience. The challenge remains: how to generate compressed air efficiently, while meeting the dynamic needs of a manufacturing environment.

In 2003, Ford Motor Company's North American Vehicle Operations Division embarked on a widespread project to significantly reduce compressed air production costs. With the help of the tier one supplier, Bay Controls, a comprehensive energy management system was installed, saving nearly one million dollars in electrical energy consumption within the first year. Vehicle Operations is Ford Motor Company's manufacturing division that includes stamping, body welding, painting and final assembly functions.

This study examines the integrated web-based compressor monitoring system installed as part of an overall control hardware upgrade project, which included Bay's ProTech microprocessor based compressor control devices, local networking and management systems, and BayWatch, a web based monitoring and analysis management system.

More specifically, this case study will focus on the Ford Compressor Management System (FCMS), the Ford application of the BayWatch product. This study will include the following sections: an overview of the capital project scope; a review of the preexisting state of compressed air control in the Ford facilities; the overriding concerns and needs that prompted the installation; the rationale behind selecting the Bay system; a description of how the system functions (both technically and from an end user standpoint); and finally, a survey of the primary economical and operational benefits that the system provides.

EXISTING CONDITIONS

The compressor controls capital project installed new control hardware at 19 manufacturing facilities across North American Vehicle Operations. A total of 132 compressors were modified, consisting of both positive displacement (reciprocating, rotary screw) and dynamic (centrifugal) types at each plant. Mixing compressor types and sizes is a common practice to meet the dynamic air volume needs of manufacturing operations.

Compressed air distribution infrastructures also varied from plant to plant. The majority of plants had centralized air generation with a separate oil-free system in the vehicle painting department. Some plants divided air generation into two systems, while others used a decentralized, point of use strategy. Each plant used common headers to distribute compressed air, with various cross-connections, valves, and back-feed loops for maintenance and redundancy.

Regarding compressor management, there was a

mix of prior Bay installations and various competitive systems, the later often based on PLC hardware. A number of plants had no modern control systems at all.

Overall, there lacked a centralized, automated information system to monitor, compare and analyze air generation across the Vehicle Operations division.

The diverse initial conditions between plants meant that compressed air systems were operating with varying levels of efficiency and an assortment of operating strategies. The number and complexity of Ford plants required a solution that was both flexible enough for installation at each location and capable enough to improve energy efficiency in variable conditions.

GOALS AND OBJECTIVES

The primary goal was to reduce the energy consumed in the production of compressed air and provide an application to manage the enterprise at both the plant and divisional levels. Since existing conditions at each plant were unique, the plant-specific deliverables also varied. This required many compressors to be upgraded with newer, more advanced instrumentation. Some compressors needed extensive upgrades to their monitoring and control equipment, while others needed little or no modification. Compressors vary in their complexity, so the number of monitor and control items ranged up to 64 points, and 15 control outputs for each machine. Typical metering points on a compressor include output air pressure and flow, motor (or primary mover) power consumption, and stage pressure, temperature, and vibration.

While most of the energy savings resulted from the new or upgraded ProTech compressor controls, the need for an enterprise energy management system was also identified. This system, which became FCMS, needed to address the following concerns:

- A desire to efficiently and automatically meter and verify energy consumption and savings resulting from the controller installation project.

- How to best extract useful long-term operating data from the built-in monitoring capabilities of the ProTech control systems.

- How to use these data to maintain and improve the operating performance of the compressed air systems.

- Use these data to address compressor system problems before they become serious and affect vehicle production.

- A desire to centrally monitor and benchmark compressed air systems across the Vehicle Operations Division.

PROJECT SOURCING AND FUNDING

Several factors were considered during the selection of the control system vendor for this project: the performance and capabilities of the resulting compressor control and management system; the pros and cons of externally hosting FCMS; and cost effectiveness.

Vendor Selection

Some key aspects in vendor selection included the proven success of Bay Control products at existing plants and Ford's understanding of the product's technical features. At the start of the controller replacement project, 45% of Ford Vehicle Operations plants were operating an earlier version of the Bay Controls system. These controllers would simply require an upgrade of their control hardware; the remaining plants would receive new controllers. Existing long-term installations had proven the reliability of ProTech as a product and Bay as a supplier. They also had proven the product's ability to reduce energy consumption and operate compressors reliably and safely.

System Selection

The FCMS system is based on the Bay Controls BayWatch product, a web based monitoring, data recording and performance reporting software, hardware and engineering service package. Ford Motor Company chose the BayWatch system for several reasons.

BayWatch is designed to integrate seamlessly with the ProTech control system, which made BayWatch easy to install and required minimal custom engineering work. BayWatch is also designed specifically to work with compressed air systems and to manage the complex metering and data recording that is required.

Additionally, the BayWatch system provided these comprehensive enterprise-wide compressed air management features:

- Real time monitoring of plant air compressor systems, as well as the ability to monitor the operating parameters of each individual compressor.

- Comprehensive data recording, reporting and analysis features.

- Straightforward, easy to use web based interface enables anywhere and anytime system management and access.

- Ability to handle multiple facilities and numerous compressors provides centralized monitoring of the entire North American Vehicle Operations compressor system.

- Includes an ongoing engineering service contract, wherein expert operators from Bay Controls perform daily analysis on the connected Ford compressor systems (the scope of this service is detailed later under FCMS Performance Enhancement Service). This, in turn, identifies opportunities for continuous efficiency improvements and preventive maintenance and diagnostics.

These features, combined with the excessive cost that a custom engineered solution would have required, made BayWatch a logical choice for FCMS.

System Hosting

A key decision involved whether to host the FCMS system internally or externally. Ford initially attempted to have the Bay system approved for connection to their internal LAN network, but ran into a number of IT related inhibitors and delays, common in large corporations.

In order to gain approval for internal hosting and connection to the Ford network, the Bay system would have required a lengthy testing and certification process. Corporate IT departments typically provide less support to facilities related systems as compared to systems that support product engineering and manufacturing. Thus, an unacceptable delay would have occurred prior to receiving approval to begin system installation and realize energy savings.

The BayWatch engineering service requires a real-time connection between personnel at the BayWatch Center, located in Maumee, Ohio, and the data hosting system. A connection of this type to internal Ford networks would have required multiple levels of management approval and system tests prior to being fully operational. After launch, management of this connection to comply with dynamic internal standards would have been costly, time consuming, and put the reliability of the data collection system at risk.

Finally, benchmarking verified that external hosting is becoming commonplace for the reasons mentioned above. After final review of all issues, it was decided that externally hosting FCMS would be a less complicated, more economical and generally superior solution.

Project Funding

The FCMS aspect of the project required a separate three year expense contract, justified through energy savings incremental to the overall compressor controls project. A historical incremental return of 20% was used to calculate a savings value and justify a three-year contract. Due to the law of diminishing returns, extending the contract beyond three years could not be justified solely using energy savings. The value of the BayWatch service would be reevaluated prior to contract expiration and either extended across the North American Vehicle Operations plants, continued at select plants, or discontinued due to changing business conditions.

TECHNICAL WORKINGS AND SYSTEM DESCRIPTION

Technical System

BayWatch, the underlying technology behind FCMS, relies on several key integrated hardware and software elements. The ProTech compressor controller is used as the individual compressor monitoring and control device, with a Bay Virtual Gateway connecting groups of controllers at each plant to a remote central server.

Monitoring Capabilities

The functionality of the web based management system is directly related to the monitoring capabilities of the ProTech unit. The controller is an advanced microprocessor based unit designed to work with all makes and models of compressors. Each standard configuration ProTech can support a combination of 64 analog and digital monitoring inputs, which are used to read such values as temperature, vibration, pressure and air flow. In addition, the ProTech controller includes built-in networking functionality, using RS-485 communication hardware to run Bay's proprietary C-Link networking protocol. The networking abilities are used to enable data communication between multiple ProTech controllers, and between a ProTech network and a remote data monitoring system.

In this fashion, the ProTech controller acts as the foundation for a plant wide or enterprise wide compressor management system.

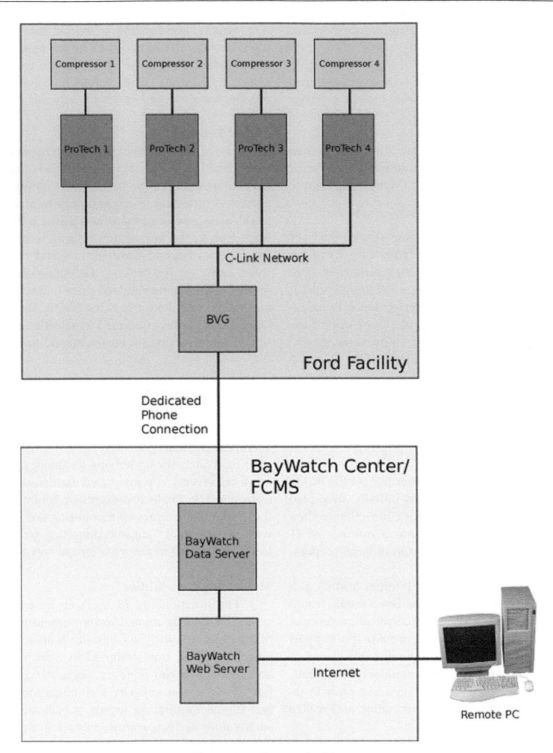

Figure 41-1. System Architecture

Data Connection

Each plant compressor network is connected to the BayWatch Center via a dedicated phone line or broadband connection. The BayWatch Center is a staffed central monitoring station at the Bay Controls offices (see FCMS Performance Enhancement Service for more information regarding the functions of this team.). Ford approved controls are in place to ensure the security of the internet connection and to prevent unauthorized use by plant personnel. A virtual gateway device is used to interface between the compressor network and the remote connection. In order to maintain continuity of data, each gateway contains internal memory to act as a storage buffer. For security purposes, the virtual gateway establishes the link with the BayWatch Center; it does not accept incoming connections.

BayWatch Servers

Remote data collection services are performed by a dedicated server running a linear, flat file based database. Information is collected for every monitoring and control point from each compressor. The state of the ProTech internal registers are also recorded, which give an additional snapshot of the controller's status. The data are collected continuously throughout the day, approximately every 15 seconds, and are transmitted to the BayWatch Center every few minutes. The database is designed to ensure varied analyses, future expansion, and modification of features.

An additional server supports security functions, maintaining account and log on information. Back up servers duplicate the databases to prevent loss of recorded data. All BayWatch servers use secure shell (SSH) data encryption to ensure the security of transmitted information.

Web-Based User Interface

A web portal provides access to an internet server which displays real-time and historical performance data and supports generation of reports. This server facilitates remote logon to the FCMS system via the public internet, making the system accessible both within and outside

of the Ford corporate firewall. Access to this system from any location with internet connectivity has greatly benefited Ford. As time constraints, travel requirements, and workloads increase, Ford requires easy access to information from any location. Ford personnel use the web-based interface to address issues at one plant while being located at another. The system has also been used to address emergency situations during holidays and weekends, without requiring travel to the site.

BAYWATCH USAGE AND INTERFACE

In this section, we will examine how the FCMS system works from an end user standpoint. FCMS is designed to provide several different categories of information, each of which is displayed on a separate screen.

Access to FCMS is password protected. A standard password screen prevents unauthorized users from proceeding to the main site.

System Overview and Facility List

After logging on to FCMS, the user is shown a list of authorized sites to chose from. Selecting a plant displays the system overview screen for that facility.

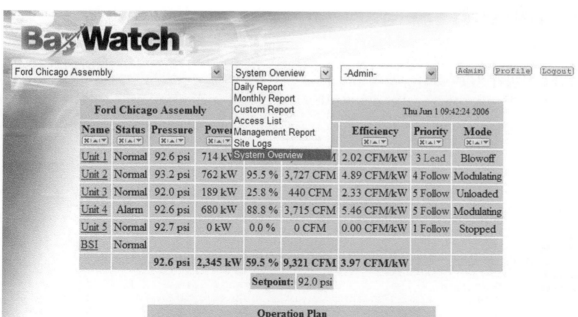

Figure 41-2. System overview shows a summary of the compressor system current operation.

The main system overview screen shows the current compressed air system status of one facility. Displayed on this screen is a tabled list of the air compressors in the plant, along with basic operations data for each compressor: status, pressure, power, load, flow, efficiency, priority and mode. These data points give a general overview of how the system is working, the plant air setpoint, which compressors are producing air, and the efficiency of each compressor. Also on this screen is a drop down list which allows the user to switch from one facility to another. Using FCMS, the entire North American Vehicle Operations compressor system is accessible from this screen, with each facility listed in the drop down list, dependent upon user rights.

The current operation plan for the selected facility is also listed at the bottom of the screen. This table shows the different operating plans for non-production versus production time periods, including changes in system pressure set points and compressor operation.

Compressor View

Selecting the name of an individual compressor displays the individual compressor unit monitoring points screen. Here, every piece of monitored data routed through the ProTech controller is shown. This view gives a very comprehensive overview of the compressor's state, showing the current readings and alarm status for all monitoring points.

Additional data for the individual compressor are accessed from this page, including a list of control points, ProTech Register values, an operation schedule, and an export screen. The export screen allows a history of selected monitor, control or register points to be saved to a file for use in a spreadsheet.

Operator and Protection Events

Finally, histories of all operator events and protection events for this compressor are also available. Operator events consist of compressor control actions initiated by manual intervention, such as start/stop or setpoint changes. A protection event occurs when an instrument reading exceeds a predetermined level, indicating potential trouble with the compressor. Access to these historical records allows a precise examination of what events and actions occurred prior to any compressor problems; this information can often help resolve the issue more quickly.

BaWatch

| | Ford Chicago Assembly | Unit 1 | Monitor Points | -History- | Admin | Profile |

Monitor Points - Unit 1   Thu Jun 1 09:48:59 2006

	Description	Tag	Start Low	Trip Low	Alarm Low	Data	Alarm High	Trip High	Start High	Units
1:	System Pressure	PT-101				93.6				psi
2:	Inlet Filter Pres Drop	PDT-107				1.9	10			in H2O
3:	Discharge Pressure	PT-102				99.1				psi
4:	2nd Stage Inlet Temp	TE-103				84	130	140		Deg F
5:	3rd Stage Inlet Temp	TE-104				93	130	140		Deg F
6:	Discharge Temperature	TE-105				191				Deg F
7:	1st Stage Vibration	VT-110				0.27	1.5	2	2.4	mils
8:	2nd Stage Vibration	VT-111				0.36	1.5	2	2.4	mils
9:	3rd Stage Vibration	VT-112				0.41	1.5	2	2.4	mils
10:	Oil Pressure	PT-105	80	80	100	141.3	200			psi
11:	Oil Press Before Filter	PT-108				142.7				psi
12:	Oil Filter Pressure Drop					1.4	4	5		psi
13:	Bearing Pressure	PT-6		300	351	533.2	699			psi
14:	Oil Temperature	TE-101	60	60	70	115	140	150	150	Deg F
15:	Stator Temp A	TE-106				157	330	340		Deg F
16:	Stator Temp B	TE-107				157	330	340		Deg F
17:	Stator Temp C	TE-108				155	330	340		Deg F
18:	IB Motor Bearing Temp	TE-109				164	203	212		Deg F
19:	OB Motor Bearing Temp	TE-110				164	203	212		Deg F
20:	Drive Motor Current	IT-114		10		106.8				Amps
21:	Power Consumption	JT-9				736				kW
22:	Flow Element DP	PDT-113				2.32				in H2O
23:	System Flow	FT-13				3,580				CFM

Figure 41-3. Compressor view.

Ba Watch

| Ford Chicago Assembly | ⌄ | | Unit 1 | ⌄ | | -View- | ⌄ | | Combined | ⌄ | | Admin | Profile |

Start: 5 / 25 / 2006
Stop: 6 / 1 / 2006
Search: [   ]
[Go]

Date	Time	Type	Source	Description	Data
06/01/2006	06:02:11	Sensor Fault	Vantage	SurgeWatch Motor	
06/01/2006	06:02:11	Sensor Fault	Vantage	Power Consumption	
06/01/2006	06:02:00	Start	Network		
06/01/2006	01:38:18	Stop	Vantage		
05/31/2006	13:55:40	Set Points	Keypad		
05/31/2006	13:55:23	Set Points	Keypad		
05/30/2006	00:24:42	Set Points	Keypad	Operator	
05/30/2006	00:24:21	Set Points	Keypad	Operator	
05/30/2006	00:23:50	Set Points	Keypad	Operator	
05/30/2006	00:23:27	Set Points	Keypad	Operator	
05/30/2006	00:22:56	Set Points	Keypad	Operator	
05/30/2006	00:22:44	Set Points	Keypad	Operator	
05/30/2006	00:22:18	Set Points	Keypad	Operator	
05/30/2006	00:22:06	Set Points	Keypad	Operator	
05/30/2006	00:21:01	Set Points	Keypad	Operator	

Figure 41-4. Events log.

Energy Management Report

Ford Chicago Assembly

Wednesday May 31st, 2006

Operational savings of 7,070 kWh (13.3%), $417 over a 1 day period. The energy usage for the period was 49,951 kWh, at a cost of $2,947. The baseline energy usage adjusted for the period was 57,021 kWh, a cost of $3,364.

Independent of operational savings, changes in ambient temperature resulted in an increase in energy usage of 4,064 kWh (7.7%), a cost of $240 for the period.

Changes in system pressure resulted in a decrease in energy usage of 1,591 kWh (3.0%), a savings of $94 for the period. Changes in system pressure caused a decrease in the net amount of leak flow, resulting in a decrease in energy usage of 1,545 kWh (2.9%), a savings of $91 for the period. Changes in the net amount of compressor blowoff resulted in a decrease in energy usage of 4,955 kWh (9.3%), a savings of $292 for the period.

Compressor Operation (CFM vs Time)

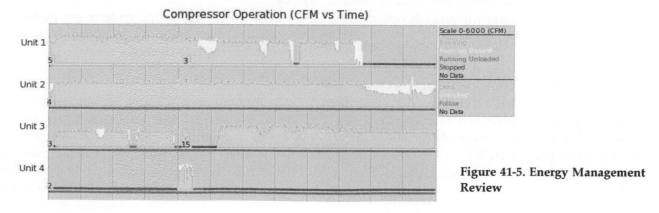

Figure 41-5. Energy Management Review

ENERGY MANAGEMENT REPORT

From the system overview screen, a drop down list provides links to several other reports, including daily, monthly, and custom time durations. After selection, a small calendar appears, which allows the user to choose the day or month they would like to review. Once the date is selected, the report screen is shown.

The report screen is a comprehensive performance and operating report for summary of the facility's entire compressed air system. A brief summary describes any changes in system energy costs that occurred during the report period. A series of graphs shows the details of compressor operation, plant pressure, power and flow, system efficiency and isothermal efficiency for the given time frame. An additional series of tables shows performance and operational data in numerical form for all the compressors in the facility's system. These reports are comprehensive and give an excellent overview of compressed air system efficiency and performance.

Administration Screens

Also available as a link from the system overview screen are several administrative screens. Each facility has a separate list of authorized users; those for the current facility are shown in the access list section. The management report screen documents new issues discovered during the daily performance review. Any exceptions or notes about the facility operations, such as monitoring issues or abnormal energy savings, are shown in this section. Finally, the site logs screen allows the user to select a range of dates to display a list of all history and protection events that occurred for the entire facility.

FCMS Performance Enhancement Service

A description of FCMS would not be complete without examining how the system is used on a daily basis. An integral component of FCMS is the performance enhancement service performed by Bay Controls. Monitored data are reviewed by noon each working day, which consists of the following steps:

System/Log Overview
- Review of alarms, faults, trips, abnormal or unexpected cfm/kW on all compressors, and any manual operator changes at the compressor's panel (i.e., manual start/stop, manual set point changes, unlink a compressor, etc.).
- Verify communication network is working properly.

- Generate the daily report and review
- This report is generated based on most recent operational and historical data.
- The report is reviewed for operational changes and the reasons for these changes.
- The graphs and tables on pressure, flow, blow off (if applicable), power, efficiency, cfm/kW and cost are reviewed for abnormal operation and opportunities for improvement.

Operational Plan Review

The operational plan created is a collaborative effort between Bay Controls and each site.

This plan specifies which units to operate, the run priority of those units and the set points for production, non-production and weekends.

Event Analysis

Items discovered in the daily reviews are discussed, prioritized and assigned for follow-up:

i. Technical Support—Direct immediate contact with the site.
ii. Engineering—Changes to operational plan and other design related changes.

A day end review is performed of important open items to make sure critical events or actions have been resolved or an adequate course of action is being pursued.

Documentation

Management reports are updated for each BayWatch site to include/document new issues discovered during the daily review. This information is accessible online for each BayWatch site under the same pull down menu as the daily and monthly reports.

Additionally, the following weekly and monthly services are also completed:

Weekly

Each week engineering reviews are performed of the operation plan based on identified improvements and/or plant schedule changes.

Monthly

Monthly Reports for each site are generated summarizing the energy savings obtained and documenting the key operating parameters. Bay engineers use a variety of analytical, quantitative tools and methods to model each compressed air system's performance under different operating scenarios, and determine optimum operating plans.

These services could only be performed with the use of an enterprise-wide system, which makes it possible for remote engineers to see a comprehensive view of each compressor and air generation system. The main purposes behind the daily reviews are to quickly identify issues for immediate response, identify incremental energy saving opportunities and monitor progress.

IMPLEMENTATION PROBLEMS

With a project this extensive, encompassing a large number of facilities in locations across North America, implementation difficulties were expected. For this project, the primary issues were technical conflicts with preexisting control systems, connectivity problems, and management issues with system operation.

A number of plants had preexisting system integration solutions that connected an existing compressor control system to a customized head-end computer. Most of these installations were fairly old and in various stages of disrepair. Although the installations were dated, plant personnel were accustomed to and comfortable with the antiquated front-end application. Fear of change and loss of functionality were key concerns from operations personnel. These challenges were slowly overcome through education, product demonstration, and other change management techniques. At plants that truly used the existing system integration solution for local monitoring and control, project underrun funds were used to install the Bay Controls BayView system integration software. This application restored local monitoring and control in a manner that married perfectly with the Bay hardware and web-based system. At locations that used the integration solution only for monitoring and not control, the plant was migrated to the web-based solution.

Internet connectivity issues occurred due to the geographic locations of the plants and central data collector. A few plants located in Mexico experienced difficulty obtaining reliable internet connectivity in a timely fashion. Other plants which had compressors located at a centralized powerhouse were frequently located far from the centralized main phone board. These locations either required a new broadband line installed from the nearest main drop or had to settle for using a dial-up connection and modem.

Management issues included gaining approval from corporate and plant IT departments to install a remote monitoring system. Concerns centered around these primary categories: preventing connection to the Ford local area network, business value of the data being transmitted, fail-safe capabilities of the system, and network protection strategy. Formal review of each issue with corporate, divisional, and plant IT contacts ensured that all concerns were addressed and documented.

ECONOMIC, TECHNICAL AND OPERATIONAL BENEFITS OF USING FCMS

Industrial manufacturers, such as Ford and other automotive companies, prefer to make data-driven business decisions. However, having access to accurate, consistently reliable data is often challenging or not possible. Thus, the ability to use actual data from field instrumentation has been extremely useful to Ford and has added credibility to compressor system analyses. Easily accessible performance reports and analysis from FCMS provide Ford with the tools necessary to maintain and operate an efficient compressed air system. The following sections illustrate how Ford Motor Company uses FCMS to manage compressed air and further drive energy and operational savings.

Efficiency Improvements

The FCMS system has enabled BayWatch engineers to incrementally improve compressor system efficiency by identifying beneficial operational modifications. The BayWatch service is able to recognize the most energy efficient compressors and what is the optimal mix of compressors for a given facility during production or non production periods.

Figure 41-6 shows an example of a compressor system that was operating with different system settings for each shift; the facility had not settled on a most effective, standardized operating plan. The FCMS engineers were able to identify the missed energy saving opportunities and developed a more efficient approach to running the air system.

Verify Current System Efficiency and Efficiency Gains

The ability of FCMS to accurately and consistently calculate the cost of compressed air at each facility has enabled the ability to benchmark compressed air efficiency and production between plants and against other manufacturers. As Ford decides to make future investments in compressed air systems, FCMS will be used to justify capital expenditures using hard data.

An interesting note is that the Vehicle Operations project used field verification and paper records to de-

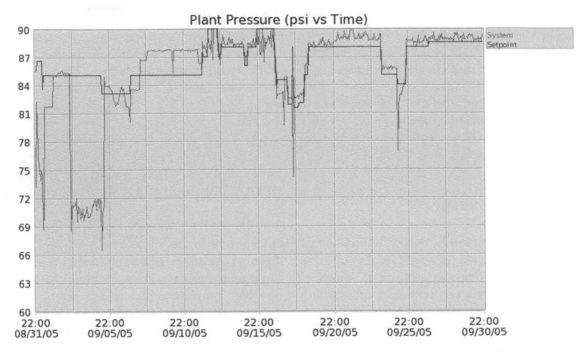

Figure 41-6. This chart shows numerous pressure setpoint changes, identifying a possible area for efficiency improvements.

termine pre-project energy use baselines. This was due to a lack of both historical performance data and time to meter existing conditions. A similar compressor project currently underway at Ford Powertrain Operations is installing BayWatch Virtual Gateways prior to the controller change in order to measure the baseline using the same instruments that the final controller will use. Any resulting efficiency gains will be highly documented thanks to the BayWatch system.

TVM Energy Program

One of the ways that Ford management uses the features of FCMS is in their Team Value Management (TVM) energy program. The TVM program is an ongoing initiative that seeks to maximize energy savings throughout Ford's manufacturing facilities by implementing low-cost and no-cost efficiency actions. One TVM goal is to reduce the non-production consumption of compressed air to 25% of the normal production consumption. Non-production compressed air shutdown performance is measured weekly at Bay using collected data from FCMS, which is then packaged and sent to Ford for internal publication and incorporation into the TVM database. Progress is tracked with the database to determine how close facilities are to meeting targets. What once was a tedious in-plant process of tallying weekly energy expenses and compressor system operating data, is now a simple matter of generating an appropriate report using the web based features of FCMS.

Measurement of Air Leakage

In one example of an efficiency and performance issue, FCMS reporting was used to identify the minimum cfm flow rate during Christmas day. Since it was assumed that there were no activities underway in the Ford plants, the measured flow rate was a close approximation to the leakage rate of the compressed air distribution system. Leakage is a notoriously hard quantity to measure in normal circumstances; the use of FCMS made this measurement relatively easy.

Preventive Monitoring Service

The constant monitoring service performed by Bay has been helpful in identifying and resolving problems with compressors before they become serious. In another example, there was a sensor calibration issue at a plant. BayWatch engineers identified the problem during a normal review and notified the plant before any serious issues with compressor performance or safety could occur. Figures 41-7 and 41-8 show how the power, flow and system efficiency readings were out of their normal ranges, alerting the BayWatch engineers to the problem.

Real Time, Data Driven Troubleshooting

When a problem with a compressor does occur, FCMS data recording and real time monitoring allow Bay engineers to aid with troubleshooting, resulting in a faster resolution to the problem.

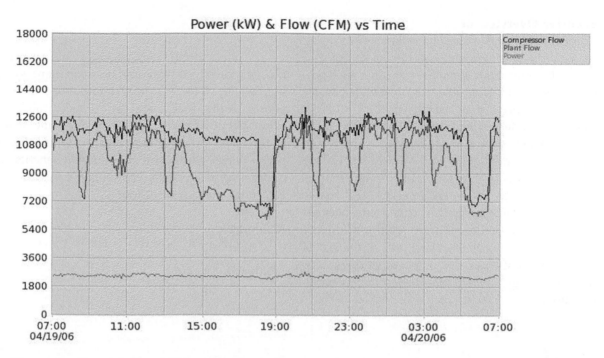

Figure 41-7. Power & Flow vs Time. Starting at about 15:00 hours, the compressor and plant flow readings diverted. The issue was resolved by 19:00 hours.

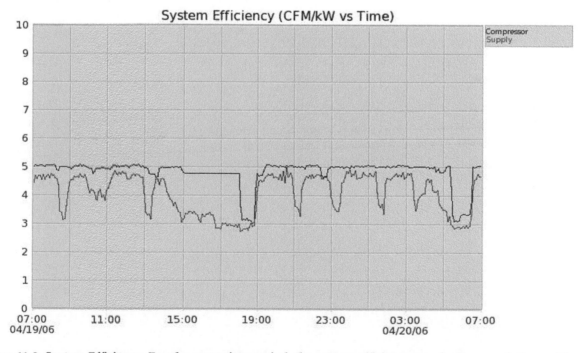

Figure 41-8. System Efficiency. For the same time period, the system efficiency graph also revealed a problem.

In one instance, after a compressor mechanical failure, a plant worked in real time with the BayWatch Center to troubleshoot the machine during restart attempts. This dynamic interaction with off site personnel would be impossible without FCMS capabilities.

Another troubleshooting problem occurred at another plant when a compressor experiencing an increas-

ing number of start/stops. A compressor motor failed and initially the compressor control system was thought to be the root cause. Analysis of the motor windings indicated it was a preexisting condition in the motor. In this case, BayWatch was used to identify the number of start/stops and other performance issues that were taking place before the failure occurred.

Comprehensive Overview of
Entire Compressed Air System Operation

Ford has found it invaluable to have a centralized overview of all compressor systems. No other tool provides such an easy and accessible way to see the current status of every compressor on the network.

A soon to be released feature of FCMS is the Tactical Overview, which shows an at-a-glance synopsis of compressor system status at all Ford Vehicle Operations facilities. The columns on the Tactical Overview list the state of the remote network connection (WAN), the state of the local compressor intranet (LAN), the current event status of the system (faults, alarms, shut-offs for any compressors in the system), a running tally of the last 24 hours of system events, positive or negative deviations from projected energy savings, and tasks that currently need resolution.

CONCLUSION

From a broad perspective, this project proved that advanced control and management systems can significantly reduce facility compressed air energy expenses. Additionally, and more specifically, the FCMS component of the project proves the value of web based compressed air management systems. The ongoing performance gains and operational benefits make the underlying BayWatch technology an extremely useful enterprise management tool.

References:

[1] Capehart, B.; Kennedy, W.; Turner, W. Process energy management. In *Guide to Energy Management*; 5th Ed.; The Fairmont Press: Georgia, 2006; 418.

About the Authors

Bill Allemon is an Energy Program Manager at Ford Land, a wholly owned subsidiary of the Ford Motor Company. His responsibilities include managing capital projects, administering energy performance contracts, and driving energy awareness activities for North American Vehicle Operations. Bill represents Ford on the ENERGY STAR Motor Vehicle Manufacturing Focus Group. Bill's team submitted a winning application for a 2006 ENERGY STAR Partner of the Year Award in Energy Efficiency. During his 16 years with Ford, Bill has held various positions in the design, construction, operation and maintenance of industrial, institutional, and commercial facilities. Bill holds a Bachelors degree in Electrical Engineering from Lawrence Technological University and a Master of Science in Administration from Central Michigan University.

Paul J. Allen P.E. is the Chief Energy Management Engineer at Reedy Creek Energy Services (a division of the Walt Disney World Co.) and is responsible for the development and implementation of energy conservation projects throughout the Walt Disney World Resort. Paul is a graduate of the University of Miami (BS degrees in Physics and Civil Engineering) and the University of Florida (MS degrees in Civil Engineering and Industrial Engineering). Paul is also a registered Professional Engineer in the State of Florida. The Association of Energy Engineers (AEE) inducted Paul into the Energy Managers Hall of Fame in 2003. (paul.allen@disney.com)

Rick Avery is the Director of Marketing at Bay, LLC. In this role, Rick is responsible for bringing to market Bay's control and management products, while promoting the energy savings message to industry. Prior to joining Bay in 2004, Rick held sales and marketing management positions in the consumer products sector. Rick holds a Bachelors degree from the University of Rochester and a MBA from the Weatherhead School of Management at Case Western Reserve University.

John Avina, Director of Abraxas Energy Consulting, has worked in energy analysis and utility bill tracking for over a decade. Mr. Avina performed M&V for Performance Contracting at Johnson Controls. In later positions with SRC Systems, Silicon Energy and Abraxas Energy Consulting, he has taught well over 200 software classes, handled technical support for nearly a decade, assisted with prod-uct development, and written manuals for Metrix Utility Accounting System™ and Market-Manager™, a building modeling program. Mr. Avina managed the development of new analytical software that employed the weather regression algorithms found in Metrix™ to automatically calibrate building models. In October 2001, Mr. Avina, and others from the defunct SRC Systems founded Abraxas Energy Consulting. Mr. Avina has a MS in Mechanical Engineering from the University of Wisconsin-Madison, where he was a research assistant at the Solar Energy Lab. He is a Member of the American Society of Heating Refrigeration and Air-Conditioning Engineers (ASHRAE), the Association of Energy Engineers (AEE, and a Certified Energy Manager (CEM).

Michael R. Brambley, Ph.D., manages the building systems program at Pacific Northwest National Laboratory (PNNL), where his work focuses on developing and deploying technology to increase the energy efficiency of buildings and other energy using systems. His primary research thrusts in recent years have been in development and application of automated fault detection and diagnostics and wireless sensing and control. He has been with PNNL for over 20 years before which he was an assistant professor in the Engineering School at Washington University in St. Louis. Michael is the author of more than 60 peer-reviewed technical publications and numerous research project reports. He holds M.S. (1978) and Ph.D. (1981) degrees from the University of California, San Diego, and the B.S. (1976) from the University of Pennsylvania. He is an active member of the American Society of Heating, Refrigerating, and Air-Conditioning Engineers (ASHRAE) for which he has served on technical committees for computer applications and smart building systems. He has been the organizer of numerous seminars and symposia at ASHRAE's semi-annual meetings and is a member of ASHRAE's Program Committee. In addition to several other professional organizations, Michael is also a member of the Instrumentation, Systems, and Automation Society (ISA) and Sigma Xi, The Scientific Research Society.

Barney L. Capehart, Ph.D., CEM is a Professor Emeritus of Industrial and Systems Engineering at the University of Florida in Gainesville, FL. He has broad experience in the commercial/industrial sector having served as the founding director of the University of Florida Energy Analysis and Diagnostic Center/Industrial Assessment Center from 1990 to 1999. He personally conducted over 100 audits of industrial and manufacturing facilities, and has helped students conduct audits of hundreds of office buildings, small businesses, government facilities, and apartment complexes. He regularly taught a University of Florida course on energy management, and currently teaches energy management seminars around the country for the Association of Energy Engineers (AEE). He is a Fellow of IEEE, IIE and AAAS, and a member of the Hall of Fame of AEE. He is editor of the recently published *Encyclopedia of Energy Engineering and Technology*, Taylor and Francis/CRC Publishing Company, 3 volumes, 190 articles, July 2007. He is also the editor of *Information Technology for Energy Managers—Understanding Web Based Energy Information and Control Systems*, Fairmont Press, 2004; senior co-editor of *Web Based Energy Information and Control Systems—Case Studies and Applications*, Fairmont Press, 2005; senior co-editor of *Web Based Enterprise Energy Management and BAS Systems*, Fairmont Press, 2007; senior co-author of the Seventh Edition of the **Guide to Energy Management**, Fairmont Press, 2011; and author of the chapter on *Energy Management in the Handbook of Industrial Engineering, Second Edition*, by Salvendy. He also wrote the chapter on **Energy Auditing** for the *Energy Management Handbook, Sixth Edition* by Wayne C. Turner and Steve Doty. He can be reached at Capehart@ise.ufl.edu.

Lynne C. Capehart, BS, JD, is a consultant in energy policy and energy efficiency, and resides in Gainesville, FL. She received a B.S. with High Honors in mathematics from the University of Oklahoma, and a JD with Honors from the University of Florida College of Law. She is co-author of *Florida's Electric Future: Building Plentiful Supplies on Conservation*; the co-author of numerous papers on PURPA and cogeneration policies; and the co-author of numerous papers on commercial and industrial energy efficiency. She is the co-editor of *Web Based Energy Information and Control Systems—Case Studies and Applications*, Fairmont Press, 2005; and co-editor of *Web Based Enterprise Energy Management and BAS Systems*, Fairmont Press, 2007; she was project coordinator for the University of Florida Industrial Assessment Center from 1992 to 1999. She is a member of Phi Beta Kappa, Alpha Pi Mu, and Sigma Pi Sigma. She is past president of the Quilters of Alachua County Day Guild, and has two beautiful grandchildren. Her email address is Lynneinfla@aol.com.

Alex Chervet is the product marketing manager responsible for connectivity and internet products at Echelon

Corporation. Mr. Chervet joined Echelon from BASF, a manufacturer of chemical and pharmaceutical products. Prior to joining BASF he was a development engineer at Schindler Elevator Corporation.

David Clute joined Cisco's Advisory Services group in June 2005. Mr. Clute has served in several capacities during his tenure at Cisco Systems including Manager, eSolutions and Manager, WPR Global Operations. In his current role, he provides consulting expertise for Cisco-Connected Real Estate and "Next Generation" building design for converged real estate and information technology solutions. In addition to his primary role within Advisory Services, Mr. Clute also serves on the Executive Board for OSCRE Americas, the Open Standards Consortium for Real Estate, promoting data exchange standards for the real estate industry. Mr. Clute has over 25 years of experience in architecture, engineering, systems development and implementation of applications for the infrastructure management and corporate real estate industry. He is recognized in the industry as a leading authority involving the integration of Computer Aided Design (CAD) Computer-Integrated Facilities Management (CIFM) and Geographic Information Systems (GIS) for large-scale corporate, government and military clients. Clute received his B.S. Architectural Engineering from the University of Colorado-Boulder.

Gregory Cmar is cofounder and CTO of Interval Data Systems, Inc. Greg is one of the most knowledgeable people on the planet when it comes to how interval data can be used to manage energy systems. He brings 35 years of experience in facility operations, energy conservation, energy analytics, energy auditing, monitoring and control systems, and utility billing, as well as database and software technologies to IDS. Greg leads the product definition and development effort as well as the energy management services team. Greg was a cofounder and director of engineering at ForPower, an energy conservation consulting firm; engineering manager at Coneco, an energy services company and subsidiary of Boston Edison; vice president of Enertech Systems, an energy monitoring and control systems contractor; and various roles at Johnson Controls, the Massachusetts Energy Office, and Honeywell. Greg holds patent #5,566,084 for the process for identifying patterns of electric energy, effects of proposed changes, and implementing such changes in the facility to conserve energy.

Toby Considine has been playing with computers since the New England Time Share in the 60s and first worked professionally with computers when microcomputers required user-written device drivers in the late 70s. He has developed systems in manufacturing, distribution, decision support, and quality assurance for clients who ranged from Digital Equipment Corporation to Reebok. Mr. Considine helped develop and support what grew into Boston Citinet, the largest free public access system of its day, in the mid 1980s. For the last 20 years, Mr. Considine has worked as an internal consultant to the facilities services division of UNC-Chapel Hill. The difficulty of supporting current control systems in a wide area environment and in bringing information from those systems to the enterprise have been a constant challenge. For the last four years, he has been working to build interfaces to make control systems transparent to the enterprise based on internet standards-based protocols.

Carla Fair-Wright is an award winning author and business consultant with over 20 years of experience implementing and supporting projects for clients such as the US Air Force, Shell Oil, Pitney Bowes, and Cameron.

She is a Certified Software Quality Engineer (CSQE) and Microsoft Certified Professional (MCP) with a BS in Computer Science, and Associate Degrees in Electronic Technology and Technical Management. A Project Management Professional (PMP), Carla has also carried out graduate work in Personnel Management and served as a Technical Reviewer for ReviewNet Corporation, an internet-based provider of pre-employment testing for IT personnel.

Carla has been featured in the National Society of Black Engineers (NSBE) magazine, Maintenance Technology, and CODE Magazine. She is the owner of Optimal Consulting LLC, a small company that provides project management services.

A Gulf War veteran, Carla has been honored for her contributions to the Armed Services as seen by her many military decorations which include Joint Service Commendation, Air Force Commendation Medal, Joint Service Achievement Medal, three Air Force Achievement Medals, and the National Defense Service Medal.

Carla is a long-time advocate for bringing young woman into the sciences. She appeared on Fox News 26 as a guest expert on the subject of Gender Bias. Her most recently conducted lectures on leadership were at Rice University, DeVry University, Lamar University and the High School For Engineering Professions. Carla is the Past President of Society of Woman Engineers (Houston).

Carla Fair-Wright can be reached at Optimal Consulting LLC, 12520 Westheimer Road, Suite 142, Houston, TX, 77077, email: fair@optimalconsulting.biz, (800) 723-6120.

Kevin Fuller is Executive Vice President and General Manager for Interval Data Systems, IDS. He is responsible for marketing and product development for IDS. He brings over 20 years of technical and marketing experience in database, data warehouse, OLAP, and enterprise applications to his role as executive vice president. Kevin has a strong appreciation of how businesses use data to their advantage, and focuses on how to apply technology to solve real business problems. He can be reached at kevin@intdatsys.

Dr. Clifford Federspiel is the president of Federspiel Controls, a consulting firm that provides energy services and energy management control products to the commercial buildings industry. Previously he held an academic staff appointment at UC Berkeley, where he was affiliated with the Center for the Built Environment (CBE) and the Center for Information Technology Research in the Interest of Society (CITRIS). At Berkeley, Dr. Federspiel managed several projects on the application of wireless sensor networks (motes) to building automation. Prior to his appointment at UC Berkeley, Dr. Federspiel was a senior member of the technical staff at Johnson Controls. Dr. Federspiel received his Ph.D. And SMME from the Massachusetts Institute of Technology, and his BSME from Cal Poly, San Luis Obispo. Cf@federspielcontrols.com.

Girish (Rish) Ghatikar is a Research Associate in the Energy Efficiency Standards (EES) Group of the Energy Analysis Department (EAD) at the Ernest Orlando Lawrence Berkeley National Laboratory. He works with Jim McMahon, et al. To design, develop and maintain systems to store archival records, lower operational costs, and facilitate transparent and robust scientific research. His other tasks are to develop new technologies for energy efficiency and environmental activities and to evaluate cost benefit analysis. He can be reached at Gghatikar@lbl.gov.

Bill Gnerre is the cofounder and CEO of Interval Data Systems, Inc. With an engineering background and 25-plus years of enterprise sales, marketing, and entrepreneur experience, Bill leads the overall company management and growth activities. Bill has an exemplary record of bringing enterprise software applications to market and helping customers the value and accomplishments possible through the use of data and the adoption of technology. His previous roles include being a partner at Monadnock Associates, a consulting organization specializing in assisting startup software companies; cofounder of ChannelWave Software; director of sales & marketing at Wright Strategies; and product marketing roles at Formtech and Computervision, both vendors of CAD technologies. Earlier in his career Bill worked in various mechanical engineering positions.

Dr. Jessica Granderson is a Post-doctoral Research Fellow in the Environmental Energy Technologies Division at the Lawrence Berkeley National Laboratory, and is a member of the Commercial Buildings and Lighting research groups. Dr. Granderson holds an AB in Mechanical Engineering from Harvard University, and a PhD in Mechanical Engineering from UC Berkeley. She has a background in intelligent lighting controls, and whole-building energy performance monitoring and diagnostics. She can be reached at Jgranderson@lbl.gov.

David C. Green has combined experience in Intranet/Internet technology and database queries and has developed programming for Energy Information Systems. David has been the president of his own consulting company, Green Management Services, Inc., since 1994. He has a Bachelor of Science degree in Chemistry and a Master of Arts degree in Computer Science. David is also a Lieutenant Colonel in the Illinois Army National Guard and has 18 years of military service. David has successfully completed major projects for The ABB Group, Cummins Engine Company, ECI Telematics, M.A.R.C. Of the Professionals, Walt Disney World and The Illinois Army National Guard. (dcgreen@dcgreen.com).Paul Green is the Marketing Partner of ThornProducts LLC. He has over 25 years experience in international marketing and sales of wireless communications products for Harris, Skydata Inc (a Harris spin-out in joint partnership with Matra Marconi /Samsung), and a co-inventor of Cognitive Radio technology for Adapt4. (pgreenis@cfl.rr.com)

Rusty T. Hodapp, P.E., CEM, CEP, GBE, LEEDTM AP, has over 25 years of experience in energy, facility and infrastructure asset engineering and management with two Fortune 100 companies and one of the world's premier commercial airports. He is the Vice President of Energy & Transportation Management at the Dallas/ Fort Worth International Airport where he is responsible for the operation, maintenance, repair and renewal of the airport's energy, utility and transit systems. Under his leadership, DFW's energy efficiency and air quality initiatives have been widely acclaimed winning a U.S. Department of Energy Clean Cities Excellence award in 2004 and the prestigious Star of Energy Efficiency award from the Alliance to Save Energy in 2005. Hodapp holds a Bachelor of Science in Chemical Engineering from Colorado State University and a Master of Business Administration from

the University of Texas at Arlington. He is a Registered Professional Engineer in the State of Texas, holds professional certifications in Energy Management, Energy Procurement, Green Building Engineering and is a LEED Accredited Professional. In 2003, the Association of Energy Engineers named him "International Corporate Energy Manager of the Year." For more information, contact Rusty Hodapp at: rhodapp@dfwairport.com.

Terry Hoffmann is Director of Marketing, Building Automation Systems, for Johnson Controls building efficiency business. His responsibilities include defining and developing materials for new product deployment, strategic brand management and the identification of leading edge technology. Terry has worked extensively in sales with experience in the Building Automation, Fire and Security markets. He served as Marketing Manager of Johnson Controls' International Division and as Manager for International Performance Contracting. E has written numerous articles in various industry trade publications including most recently: the *ASHRAE Journal, Today's Facility Manager,* the *Refrigeration Systems Engineering and Service (RSES) Journal,* and the *HVAC Systems Maintenance and Operations Handbook.* In his 33 year tenure with Johnson Controls, Terry has spoken extensively at numerous conferences and industry forums including the American Society of Heating and Air Conditioning Engineers (ASHRAE) and the Association of Energy Engineers (AEE). He represents Johnson Controls as a board member of LonMark International. A native of Milwaukee, Wisconsin, Terry holds a Bachelor of Science degree in Electrical Engineering from Marquette University and a Master's degree in Engineering Management from the Milwaukee School of Engineering where he serves as an adjunct professor. Contact information: Terry.Hoffmann@jci.com.

H.A. (Skip) Ingley, Ph.D., P.E., is an Associate Professor in the Mechanical Engineering department at the University of Florida in Gainesville FL. He has functioned as project manager and lead mechanical engineer for over 700 engineering projects for the time period 1983-present. As lead mechanical engineer and project manager, Ingley conducted over 500 comprehensive technical assistance energy studies for several institutions in the State of Florida. As a researcher during the 1973-2007-time period, conducted extensive experimental-based research at the University of Florida Solar Energy and Energy Conversion Laboratory. The topics of this research included solar powered heating and air conditioning systems, absorption air conditioning, energy conserving building technology, IAQ and energy considerations in the design and construction of modular housing, solar fenestration studies, fuel cell air contaminant studies, wind turbine life cycle energy/carbon analyses, combined heat and power system integration with solar PV and thermal systems, hydrogen generation using salt water algae, hydrogen production using a combined ammonia/water Rankine cycle, solar distillation, and the study of thermo-physical properties of vegetable oils in their use as heat transfer media. He was Co-director of the UF SEECL for the time period 1980—1983. He is a member of American Society of Heating, Refrigerating & Air Conditioning Engineers, Inc; the American Society of Mechanical Engineers; the Mechanical Contractors Association of America; and the American Solar Energy Society. He can be reached at ingley@ufl.edu.

Safvat Kalaghchy is the program director for the computing and information technology group at the Florida Solar Energy Center (FSEC). He is responsible for the design, development, and implementation of energy related information technology and scientific computing projects at FSEC. He is the architect for the www.infomonitors.com and the backend engine, the experimental management database system (EDBMS) that enables automated field-monitoring project. He co-developed the first version FlaCom, state of Florida's commercial energy code compliance software, now known as EnergyGauge/FlaCom. He has also developed a number of other complex scientific software to analyze the behavior of thermal systems. Safvat has a BS and MS in mechanical engineering from the Florida Institute of Technology.

Srinivas Katipamula, Ph.D., got his M.S. And Ph.D. In mechanical engineering in 1985 and 1989, respectively, from Texas A&M University. He has been working as a senior research scientist at Pacific Northwest National Laboratory, in Richland, WA, since January 2002. He managed the analytics group at the Enron Energy Services for 2 years (2000 through 2001). Before joining EES, he worked at PNNL for 6 years and prior to that he worked for the energy systems lab at the Texas A&M University from 1989 to 1994. He has authored or co-authored over 60 technical publications, over 25 research reports, and made several presentations at national and international conferences. He has recently written a chapter, "Building Systems Diagnostics and Predictive Maintenance," for *CRC Handbook on HVAC*. He is an active member of both ASHRAE and the American Society of Mechanical Engineers (ASME).

Sila Kiliccote is a scientific engineering associate at Lawrence Berkeley National Laboratory in Building Technologies Department with the Lighting Group. She has an electrical engineering degree fro University of New Hampshire, with a minor in illumination engineering and a master's in building science degree from Carnegie Mellon University.

Michael Kintner-Meyer, Ph.D., has been a staff scientist at the Pacific Northwest National Laboratory since 1998. His research focus is on building automation technology for optimal control strategies of HVAC equipment for improving the energy-efficiency of buildings and to enhance the reliability during emergency conditions on the electric power grid. At PNNL, he leads the "Load-As-A-Reliability Resource" research activity that focuses on technology development and analyses of Grid-friendly Appliances™ and load management strategies. Michael holds a M.S. (1985) from the Technical University of Aachen, Germany and a Ph.D. (1993) from the University of Washington in Seattle, WA. He is an active member of ASHRAE for which he serves on technical committees as well as in the local chapter. He is member of the American Society of Mechanical Engineers (ASME) and the German Engineering Society, Verein Deutscher Ingenieure (VDI). He has authored and co-authored numerous papers and reports in U.S. And international technical journals.

Bill Kivler, director of Global Engineering for Walt Disney World, is a 30-year veteran of facilities construction and operation. Bill has been with the Walt Disney World Company since 1993. He has held positions of increasing responsibility through the present-day role of technical director of Global Engineering. In his current role Bill supports the WDW property in several ways. Global contract administration, I.T. Administration of the computerized maintenance management system, technology initiatives, metrics reporting, critical communications and support, hurricane coordination support, communications strategies, productivity initiatives support, energy and utility conservation strategies. Operationally Bill has 7 departments reporting to him supporting the maintenance of technologies such as, Office Machine Systems, I.T. Hardware, Video systems, Radio systems, Support Systems, Key Control Systems, Access Control Systems, Alarm and Monitoring Systems, Energy Management Systems. Bill is also responsible for leading the Resorts Engineering and Downtown Disney Engineering Divisions. Prior to coming to WDW, Bill spent 13 years in the U.S. Virgin Islands as executive director of engineering and program manager for the largest resort community on the island. His responsibilities included facilities maintenance, power plant design, construction, and maintenance, new construction, and capital renewal. He was also responsible for all governmental regulations local, state, and federal regarding air, water, fuel, and building code permits. The resort operated an autonomous power plant which produced power, chilled water, steam, drinking water from sea water, and sewage treatment. Preceding the Virgin Islands, Bill managed the facilities for over 30 hotels/resorts along the U.S. East Coast. Bill attended Franklin and Marshall College, The Center for Degree Studies, and RCA Institute.

James M. Lee—Chief Executive Officer, is the founder of Cimetrics and has acted as its CEO since its formation. Mr. Lee has been a leader in the embedded control networking and building automation community for 20 years. As founder and former President of the BACnet Manufacturers Association, the leading open systems networking consortium in the building automation industry, Mr. Lee's aggressive promotion of the BACnet open protocol standard has helped make Cimetrics a high-profile player in the arena. Mr. Lee has a B.A. In Physics from Cornell University.

Mr. Jim Lewis is the CEO and co-founder of Obvius, LLC, in Portland, OR. He was the founder and president of Veris Industries, a supplier of current and power sensing products to BAS manufacturers and building owners. Prior to founding Veris, Mr. Lewis held several positions at Honeywell including Branch Manager. He has extensive experience in knowing the needs of building owners, integrating existing metering and sensing technologies and developing innovative products for dynamic markets. For more information or a demonstration, contact Obvius Corporation at (503) 601-2099, (866) 204-8134 (toll free), or visit the website at: http://www.obvius.com

Fangxing Li is presently a senior consulting R&D engineer at ABB Inc. He received his B.S. And M.S. Degrees in electric power engineering from Southeast University, China, in 1994 and 1997 respectively. He received his Ph.D. Degree in computer engineering from Virginia Tech in 2001. His areas of interests include Web applications in power systems, power distribution analysis, and energy market simulation. Dr. Li is a member of IEEE and Sigma Xi. He can be reached at fangxing.li@us.abb.com or fangxing.li@ieee.org.

Eric Linkugel works for Pacific Gas & Electric Company as a Business Customer Specialist for Demand Response Programs. He graduated from California State Polytechnic University—San Luis Obispo, with a B.S. In Indus-

trial Technology and an M.S. In Industrial and Technical Studies.

Gerald Mimno has seven years experience developing wireless internet applications for energy measurement, information management, and controls. Mr. Mimno has a BA and MCP from Harvard University followed by experience in economic development, real estate development, and business development. He has 20 years of practical experience in building systems. He is a licensed Construction Supervisor in the Commonwealth of MA. Presently he is Principal of Victoria Properties and can be reached at gerald@mimno.net

Timothy Middelkoop, Ph.D., C.E.M., is an Adjunct Professor in the Industrial and Systems Engineering Department and the Assistant Director of the Industrial Assessment Center at the University of Florida in Gainesville FL. He has industry experience in designing web-based applications and embedded systems and regularly teaches a course on web based decision support systems (web, databases, and optimization). His research focuses on clean energy optimization and control, cyberinfrastructure, large-scale (HPC) and multi-core scientific computing, distributed sensor networks, integrated design systems, multi-agent systems, and computational optimization. His current focus is to engineer distributed systems by developing and understanding the fundamental design patterns required to build robust scalable systems that are beneficial to society. In the area of clean energy optimization control and integration, he is leading a multi-disciplinary research team that is working closely with industry to develop models and systems to use energy more efficiently and intelligently. He is a member of Industrial Engineering honor society Alpha Pi Mu, the Institute for Operations Research and the Management Sciences (INFORMS), and the Institute of Industrial Engineers (IIE). He can be reached at t.middelkoop@ufl.edu.

Gerald Mimno has seven years experience developing wireless internet applications for energy measurement, information management, and controls. Mr. Mimno has a BA and MCP from Harvard University followed by experience in economic development, real estate development, and business development. He has 20 years of practical experience in building systems. He is a Licensed Construction Supervisor in the Commonwealth of MA. He is responsible for developing new markets and relationships based on wireless and internet energy technologies and has written extensively on the value of interval data. Gerald Mimno, General Manager Advanced AMR

Technologies, LLC 285 Newbury Street Peabody, MA 01960, TEL (978)826-7660, FAX (978)826-7663, gmimno@ AdvancedAMR.com

Naoya Motegi is a graduate student research assistant in the Commercial Buildings Systems Group in the Building Technologies Department at LBNL. He is currently a graduate student in the Department of Architecture, University of California, Berkeley. He has a Bachelor of Architecture and Master of Engineering in Architecture and Civil Engineering from Waseda University in Tokyo, Japan.

Patrick J. O'Neill, Ph.D., co-founded NorthWrite, and leads corporate operations. Before joining NorthWrite, Patrick spent 10 years at Honeywell International, where he most recently served as vice president of Technology and Development for e-Business. Patrick defined technology strategy, prioritized developments, allocated resources, and operated the infrastructure for Honeywell's stand-alone e-ventures. Patrick also co-founded and acted as chief technology officer for Honeywell's *myFacilities.com,* an application service provider targeting the facility management and service contracting industries. Previously, Patrick was director of development for Honeywell's Solutions and Service business, managing global product research and development worldwide, with development teams in the U.S., Australia, India, and Germany. Before joining Honeywell, Patrick worked at the Department of Energy's Pacific Northwest National Laboratory and the University of Illinois at Urbana- Champaign. He holds Bachelor's, Master's and Doctoral degrees in mechanical and industrial engineering from the University of Illinois, Urbana-Champaign. Patrick is a member of numerous professional organizations including ASHRAE, where he has held leadership positions in the computer applications, controls, and smart building systems technical committees. He has written and published many articles on software, systems and controls, and building operations and management.

Mary Ann Piette is the research director of the California Energy Commission's PIER Demand Response Research Center and the deputy group leader of the Commercial Building Systems Group. She has been at Berkeley Laboratory for more than 20 years, with research interests covering commercial building energy analysis, commissioning, diagnostics, controls, and energy information system. Her recent work has shifted toward developing and evaluating techniques and methods to improve demand responsiveness in buildings and industry. She has

a Masters in mechanical engineering from UC Berkeley, and a Licentiate in building services engineering at the Chalmers University of Technology in Sweden.

Phillip Price is a Staff Scientist at Lawrence Berkeley National Laboratory. Phil has bachelors degrees in physics and math, and a Ph.D. in theoretical atomic physics. Phil has worked in LBNL's Indoor Environment Department since 1992, as a post-doctoral fellow and as a scientist. He has varied interests and has worked on Bayesian hierarchical modeling; decision analysis; computed tomography; model-measurement comparison for complicated models; and model optimization. He can be reached at pnprice@lbl.gov.

Sam Prud'homme is a freelance technical writer and computer programmer. He has been affiliated with Bay, LLC, since 1993, where he produced operation manuals and product literature for the company's line of air compressor controls. While at Bay, he also created software utilities used to estimate compressed air system energy expenses and the potential savings from updated system controls. In 1994, he authored the EPRI Compressed Air Handbook, a joint project between the Electric Power Research Institute and Bay. Sam has a computer science degree from Yale University.

Partha Raghunathan has been the VP of Specialty Solutions for the last five years at Net.Orange, a Dallas TX-based healthcare information technology service vendor that helps hospitals and clinics improve productivity and quality of care. Prior to Net.Orange, Partha spent 2 years as VP of Business Development at a bio-pharmaceutical company using an informatics platform to discover new drugs for the treatment of neurological disorders; and 8 years at i2 Technologies, the leading provider of supply chain optimization software, where he managed helped Fortune 500 companies improve operational efficiencies. Partha has a Bachelors degree in Mechanical Engineering from the Indian Institute of Technology and an MS in Industrial & Systems Engineering, with a specialization in Operations Research, from the University of Florida. While at UF, he worked under the guidance of Dr. Barney Capehart in the Industrial Assessment Center (IAC) to help small to medium businesses implement measures to identify and implement energy conservation measures. He can be reached at praghunathan@gmail.com or at (214) 507-7385.

Rich Remke is commercial controls product manager for Carrier Corporation in Syracuse, NY. Rich has been the product manager for Carrier for the past four years and is responsible for controls product marketing and new product development. He holds a B.S. In information system management from the University of Phoenix. Rich has been in the HVAC and controls industry for over 20 years. Rich started his control work as a SCADA technician for Reedy Creek Energy Services at Walt Disney World, FL. He then moved into controls system engineering, project management, sales, and technical support for United Technologies/Carrier Corporation. Rich also spent several years supporting Carrier's Marine Systems group, providing controls technical support and system integration engineering. Rich has created several custom user applications, including a facility time schedule program, a DDE alarm interface, integration of Georgia Power real time pricing data to Carrier CCN, and a custom tenant billing application. (richard.remke@carrier.utc.com)

Chris Sandberg is Principal Engineer, Energy Management Systems at Reedy Creek Energy Services (a division of the Walt Disney World Co.) and is responsible for the design, installation and commissioning of building automation systems for Disney theme park projects throughout the world. Chris is a graduate of Purdue University (BS in Construction Engineering and Management) and the University of Florida (Master of Building Construction). (chris.d.sandberg@disney.com)

Osman Sezgen is a Staff Research Associate in the Indoor Environment Department at Lawrence Berkeley National Laboratory. He has been with LBNL since 1990. He worked on projects characterizing and forecasting end-use demand in the commercial and residential buildings sectors. He also worked with the Electricity Markets Group before he went to Enron in 1999. At Enron, as a member of the Corporate Research Group, he supported Enron Energy Services managing a team of researchers developing quantitative models that are used for product development, pricing and risk management. Aosezgen@lbl.gov.

Blanche Sheinkopf has been the national coordinator of the United States Department of Energy's EnergySmart Schools program since 2001. An educator and curriculum writer for more than 25 years at levels ranging from pre-kindergarten through university, she has been a college of education faculty member at the University of Central Florida, the George Washington University, and American University, and was the coordinator of education and training programs at the Florida Solar Energy Center. She was founder and CEO of Central Florida Research Services, a full-service marketing research company for 11 years. She currently serves on the boards of several

organizations including the American Solar Energy Society and the Educational Energy Managers Association of Florida.

Ken Sinclair has been in the building automation industry for over 35 years as a service manager, building owner's representative, energy analyst, sub-consultant and consultant. Ken has been directly involved in more than 100 conversions to computerized control. Ken is a founding member and a past president of both the local chapter of AEE and the Vancouver Island chapter of ASHRAE. The last five years his focus has been on *AutomatedBuildings. com*, his online magazine. Ken also writes a monthly building automation column for *Engineered Systems* and has authored three industry automation supplements: *Web-Based Facilities Operations Guide, Controlling Convergence* and *Marketing Convergence.*

Michael R. Tennefoss is the vice president of product marketing & customer services at Echelon Corporation. Mr. Tennefoss joined Echelon from Stellar Systems, a manufacturer of intrusion detection sensors and alarm monitoring systems, where he served as director of monitor & display products. Prior to joining Stellar Systems he was the director of marketing at ETP and vice president of marketing at Vindicator Corporation.

Greg Thompson is the Chief Architect for the EcoStruxure program within Schneider Electric. EcoStruxure provides "active energy management" systems to datacenter, buildings, and industry customers globally. These systems are simplified, save money, and most importantly, reduce waste by enabling a guaranteed compatibility between the management of power, white space, process and machines, building control, and security. Greg resides in Nashville, TN.

Terrence Tobin is a senior communications project manager for Schneider Electric. He has worked in the high technology sector for over 25 years in a variety of marcom, branding and media relations roles, and has written extensively on energy management topics.

Steve Tom, PE, Ph.D., is the director of technical information at Automated Logic Corporation, Kennesaw, Georgia, and has more than 30 years experience working with HVAC systems. At ALC Steve has coordinated the training, documentation, and technical support programs, and frequently works with the R&D engineers on product requirements and usability. Currently Steve is directing the development of *www.CtrlSpecBuilder.com*, a free web-based tool for preparing HVAC control system specifications. Prior to joining Automated Logic, Steve was an officer in the U.S. Air Force where he worked on the design, construction, and operation of facilities (including HVAC systems) around the world. He also taught graduate level courses in HVAC design and HVAC controls at the Air Force Institute of Technology. (Stom@automatedlogic.com)

Jason Toy graduated from Northeastern University in 2005 with a dual degree in computer science and mathematics. He has worked at many different startups with the passion of trying to move technology forward. He is expert in the data mining and analysis field.

His latest work is a startup called socmetrics, a platform to help companies build word of mouth by identifying relevant bloggers/influencers. This is done by analyzing enormous amounts of public data available on the internet to learn insights about people.

Contact info: jtoy@jtoy.net; 6176064373

Wolfgang Wagener, Ph.D., Architect AIA, RIBA is Head of Real Estate and Construction solutions within Cisco's global Real Estate and Workplace Resources organization. Wolfgang's primary area of expertise is in working with occupiers, developers, and property owners to deliver innovative real estate, design and technology solutions that enhance business performance. An architect by profession, he lectures regularly across Europe, North America, and Asia. Prior to joining Cisco Systems, Wolfgang was a practicing architect, urban planner, educator and author. He worked with Murphy/Jahn in Chicago, Richard Rogers Partnership in London, and he had a private practice in Los Angeles, where he was also a Visiting Professor at the University of California in Los Angeles (UCLA) and the University of Southern California. He managed research, planning, design and construction of mixed use urban and residential developments, transportation buildings, and corporate headquarters throughout Europe, Asia Pacific and North America. His research and education areas are 19th and 20th century architecture and urban development, workplace design, environmental sustainability and the impact of technology innovations in the real estate and construction industry. Wolfgang Wagener received a Ph.D. in Architecture from the RWTH Aachen, one of Europe's leading technology institutions, and an Advanced Management Degree in Real Estate Development from Harvard University.

David Watson has 20 years experience designing, programming, and managing the installation of control and communications systems for commercial buildings, industrial processes and remote connectivity solutions. At LBNL, he is working with innovative building technologies such as demand response systems, energy information systems and wireless control networks. Prior to joining LBNL, David held engineering, project management and product development positions at Coactive Networks, Echelon, York International and Honeywell. He designed and managed the installation of hundreds of projects including: internet based control and monitoring of thousands of homes and businesses, communication systems for micro turbine based distributed power generation systems and industrial process controls for NASA wind tunnels and biotech manufacturing. Mr. Watson graduated from California Polytechnic University, San Luis Obispo with a degree in mechanical engineering.

Joel Weber is the Chief Information Officer of Weber & Associates, a financial services firm servicing postsecondary education institutions. Before joining Weber & Associates, Joel was a software engineer and consultant to numerous Fortune 100 firms largely in the energy exploration, production and trading sectors. He has an MBA from the University of Texas at Austin, where he concurrently studied digital forensics with Larry Leibrock, PhD and taught a course on digital forensic investigations. He also holds an MS in Industrial & Systems Engineering, where he served as a project team leader in the Industrial Assessment Center under the tutelage of Barney Capehart. Joel holds numerous technology related certifications including several in ITIL (Information Technology Infrastructure Library), an internationally recognized framework for IT Services Management best practices.

John Weber is president and founder of Software Toolbox Inc. Prior to founding Software Toolbox in 1996, John spent 6 years with GE Fanuc Automation and their distribution channel in a variety of technical and commercial field positions. He has been working with communications systems and developing software for over 15 years. He has spoken at numerous ISA and other shows domestically and internationally on subjects including communications, OPC, HMI configuration, and others. John holds a Bachelor of Science in industrial and systems engineering (1989) from the University of Florida and a Masters Degree in business from Clemson University (1995).

Tom Webster, PE, is a Project Scientist at the Center for the Built Environment (CBE) at UC Berkeley. He has been engaged in building research and development for over thirty years and currently co- leads several advanced integrated systems research projects covering laboratory and field testing, energy simulations, and monitoring systems development. His experience includes research on solar energy systems and building energy analysis and simulation, HVAC control system product development, air handler diagnostics, and monitoring system development. Tom has spent the last 10 years conducting research on underfloor air distribution (UFAD) systems including room air stratification laboratory testing, simulation model development, UFAD system cost analysis, and field testing.

Jeff Yeo, P.Eng., is a Senior Software Developer with the PowerStruxure Software group at Schneider Electric. He has twenty years of power-related experience in distribution, process control, system administration, power quality, product testing, field service engineering, and software development.

Index

Printed and bound by CPI Group (UK) Ltd, Croydon, CR0 4YY

23/10/2024

01777682-0016